Laser-Based Measurements for Time and Frequency Domain Applications

A Handbook

SERIES IN OPTICS AND OPTOELECTRONICS

Series Editors: **E. Roy Pike**, Kings College, London, UK
Robert G. W. Brown, University of California, Irvine, USA

Laser-Based Measurements for Time and Frequency Domain Applications

A Handbook

Pasquale Maddaloni • Marco Bellini
Paolo De Natale

CRC Press
Taylor & Francis Group
Boca Raton London New York

CRC Press is an imprint of the
Taylor & Francis Group, an **informa** business

A TAYLOR & FRANCIS BOOK

over by Rita Cuciniello: "Generation of a mid-infrared optical frequency comb, P. Maddaloni et al., *New Journal of Phys s* 8 (2006) 262," in the artist's concept.

riginal: Oil on canvas.

CRC Press
Taylor & Francis Group
000 Broken Sound Parkway NW, Suite 300
Boca Raton, FL 33487-2742

First issued in paperback 2020

© 2013 by Taylor & Francis Group, LLC
CRC Press is an imprint of Taylor & Francis Group, an Informa business

No claim to original U.S. Government works

ISBN 13: 978-0-367-57647-9 (pbk)
ISBN 13: 978-1-4398-4151-8 (hbk)

Visit the Taylor & Francis Web site at
http://www.taylorandfrancis.com

and the CRC Press Web site at
http://www.crcpress.com

To the young people I wish the same luck that led me to disinterest myself from my person, but to always pay attention to everything that surrounds me, to everything in the scientific world, without neglecting the values of society.

Rita Levi Montalcini

to our children:
Maria and Bernadette,
Alessia and Andrea,
Rufael and Tsion.

Contents

Foreword

Since its first realization in 1960, the laser has quickly parted from its initial definition of "a solution looking for a problem" to become the solution to many and incredibly different problems, both in our everyday lives and in the most advanced fields of science and technology.

Thanks to its unique properties, i.e., the capability to generate intense, highly directional, and highly monochromatic radiation in different regions of the electromagnetic spectrum, the laser has found applications in fields as diverse as communications, medicine, fundamental physics, as well as supermarket counters....

In particular, thanks to the possibility of producing either ultrashort laser pulses, lasting just a few femtoseconds, or continuous beams with an extremely well-controlled frequency, the two fields of accurate time and frequency measurements have literally boomed in the last decades. Measuring times and delays with femtosecond accuracy has allowed us, for example, to observe the real-time movement of atoms in molecules, and to follow and steer chemical reactions. On the other hand, measuring frequencies with a very high precision has given us unprecedented access to the most intimate structure of matter, is revolutionizing precision metrology of time and space, and is providing new tools for many important applied fields, like environmental monitoring.

These two intertwined subjects, time and frequency, have continued to evolve independently for many years, with the impressive parallel evolution of ultrafast and ultra-stable laser sources, until it was recently realized that they can be seen as two different faces of the same medal. In fact, the frequency spectrum of the train of ultrashort pulses emitted by a (properly phase-stabilized) mode-locked laser is remarkably simple. It is made of millions of extremely narrow spectral lines whose frequencies are exactly spaced by the laser repetition rate. Such a special laser source can thus combine the best of the two worlds: on one side, by giving access to a huge number of ultra-stable laser lines, all with precisely controlled frequencies, it serves as a perfect ruler in the spectral domain; on the other side, by making it possible to control the absolute optical phase of ultrashort light pulses, it discloses new, highly nonlinear phenomena to experimental investigation.

This experimentalist's dream came true with the development of frequency combs around the beginning of this century, and it was a strike of serendipity on the hill of Arcetri in Florence, where the LENS and INO laboratories used to be, that started it. There, on a lucky afternoon in 1997, Marco Bellini and I could surprisingly observe stable interference fringes from the white-light supercontinua independently produced by two identical ultrashort pulses. Since that moment, the evolution of comb-based measurements has seen no sign of slowing down, and has made possible some of the most accurate measurements ever performed by mankind, allowing to measure frequencies in the simplest atom in the universe, hydrogen, as well as to calibrate spectra coming from the borders of the universe, itself.

Indeed, if a stable and precisely determined frequency reference is available, it can be used as a clock to measure time intervals with high accuracy by just counting the number of cycles in the interval. Furthermore, from the definition of the speed of light, any distance measurement can be referred to a time or frequency measurement. Since the measure of many physical properties, beyond time and distance, can be often converted to a phase or frequency measurement, the incredible precision made possible by frequency combs can be readily exported to a number of different fields.

Laser-based precision measurements are now facing a new era, where the ever-growing accuracy in the determination of times and frequencies will not only allow us to unveil some of the best hidden secrets of nature, but also impact our everyday lives.

In this context, the publication of this book is particularly timely and welcome. The authors have made a considerable effort to make this book useful and interesting to different kinds of readers: they provide a detailed treatment of the basic concepts of time and frequency measurements, carefully describe different kinds of lasers and some of the most advanced laser-based measurement techniques, and, finally, present the latest developments in the field, with a hint to the possible future trends in applications and fundamental science.

Being among the many important actors in this long story, the authors of this book are privileged witnesses of the evolution of time and frequency measurements, and can provide an informed and wide vision of this developing field from many different viewpoints.

Theodor W. Hänsch
Co-Recipient of the 2005 Nobel Prize in Physics

Preface

Time and light or, in other words, frequency and photons: two key ingredients that are making possible the most accurate and sensitive measurements ever performed by mankind. This was the flywheel to writing this book and the "leitmotiv" throughout it. However, dealing with this subject is an impervious task both in a historical perspective and in the viewpoint of contemporary developments. Indeed, on the one hand, the art of timekeeping has deep roots, dating back to the origins of civilization, and light from the sky was soon recognized as a vital element for measuring time. On the other hand, trying to describe the furious activity and progress that have been characterizing this field for the last decades is a bit like taking a picture of a very fast object that keeps moving at increasingly high speed. In such a fascinating adventure, the laser-era undoubtedly represented a turning point, marking the birth of optical frequency metrology. In this wake, a new metrological tool emerged around 1999, when it was realized that the pattern of equally spaced pulses generated by a mode-locked laser in the time domain is equivalent to a precisely spaced comb of frequencies in the frequency domain, and that the phase of light is the same throughout the broad-covered spectrum. This eventually merged decades of independent technologies, namely those of ultrastable and ultrafast laser sources, emitting, respectively, continuous-wave radiation and trains of very short pulses. In parallel, another real breakthrough was represented by the advent of laser-based optical clocks that, by progressively reducing the fluctuations in the emitted frequency, have now reached impressive stability and accuracy levels. But many other milestones have been achieved in the field of laser-based measurements, since the introduction of the laser itself. Among the most remarkable, we should mention guiding and delivering laser-light with optical fibers as well as accessing additional portions of the electromagnetic spectrum with frequency-tunable coherent sources, based on novel materials and operation principles, or nonlinear optical phenomena. Apart from these spectacular technological achievements, the quantum nature of light and matter has opened other new scenarios, like that of measurements based on entanglement of photons and macroscopic objects. As a whole, such a scientific fervor has revolutionized the branch of atomic, molecular, and optical physics allowing, as the first immediate consequence, to devise increasingly ambitious experiments of fundamental character, but also engendering tremendous progress in terms of high-tech, everyday-life applications, perhaps going so far as to change even the way we think. This book is based on first-hand, laser-based measurements and direct experimental work performed by the authors during the last 25 years. Such activities are strictly related to the rise and development of the European Laboratory for Nonlinear Spectroscopy-LENS in Florence and the Istituto Nazionale di Ottica-INO (now part of the Italian National Research Council-CNR) in Florence and Naples, Italy. These labs and activities have flourished on the hill of Arcetri (Florence), where Galileo Galilei spent the last part of his life and Enrico Fermi conceived quantum statistics, but also in front of the breathtaking gulf of Naples, where, in his quick passage, Ettore Majorana left an indelible legacy. In fact, also thanks to their irresistible charm, these two cities have always attracted scientists from all over the world, so years of collaborations, discussions, and joint work are somehow reflected in the text of this book. The purpose is to offer a detailed account of the most recent results obtained for time- and frequency-domain applications of lasers, while providing all the background information on the main kinds of sources and techniques developed thus far. Moreover, the theoretical framework necessary to understand the experimental applications is fully developed throughout the book. Therefore, most of the matter is intended to be accessible to final-year undergraduates, but also post-docs and scientists actively working in the field can find a wide, fresh, and balanced overview of conquests in the field of laser-based measurements together with the main related references. A detailed outline is given at the end of Chapter 1, as a natural outcome of the historical introduction. Also preliminary in character, Chapter 2 provides the basic concepts and the mathematical tools that are necessary to address the physics of oscillators, at the heart of the whole treatment. Likewise relevant to the self-sufficiency of the book, microwave, and particularly, optical resonators are extensively discussed in Chapter 3. Crucial aspects of operation and fundamental properties of lasers are presented in Chapter 4, while precision spectroscopy and absolute frequency metrology are dealt with in Chapters 5 and 6, respectively. Then, Chapter 7 is devoted to microwave and optical frequency standards and their dissemination. Finally, Chapter 8 dwells upon the variegated speculative landscape opened by the field of laser-based frequency measurements, outlining the most exciting, current, and forthcoming research directions. Due to the large amount of unfolded work, we apologize in advance for any mistakes, inaccuracies, and inevitable limitations, hoping that the reader may appreciate our approach and share our enthusiasm.

The authors wish to thank Luca Lorini for careful reading of Chapter 2; Simone Borri and Gianluca Gagliardi for their contribution to two sections in Chapter 4 and Chapter 8, respectively; Maurizio De Rosa for stimulating discussions; Gianluca Notariale for preparing many of the figures; Elisabetta Baldanzi for editing part of the text and caring about permission requests; and Rita Cuciniello for creating the cover. Also, the authors are immensely grateful to Prof. Theodor W. Hänsch for writing the Foreword.

Pasquale Maddaloni
Marco Bellini
Paolo De Natale

Authors

Pasquale Maddaloni was born in Castellammare di Stabia, Italy, on February 16, 1975. He owes much of his professional accomplishment to his parents, Catello and Maria, for their constant and amorous support. As a completion of his studies, Pasquale received the diploma (*cum laude*) and the Ph.D. degree in Physics from the University of Naples (1999) and the University of Padua (2003), respectively, for experimental research in the field of ultracold atoms and Bose-Einstein Condensation. In 2003 he joined the National Institute for Optics-INO (now part of the National Research Council-CNR) where he worked, in a postdoctorate position, on the development of innovative mid-infrared coherent radiation sources. Since 2009 he has been a research scientist at INO in Naples, much of his interest relating to nonlinear optics and precision spectroscopy assisted by optical frequency comb synthesizers (OFCSs). In this framework, his main achievement is represented by the first fiber-based mid-IR generation of an OFCS and its application to absolute frequency metrology. More recently, he extended his experimental activities to the world of cold stable molecules. When not working, Pasquale shares his passion for gardening, traveling, and photography with his beloved wife, Rosa, and daughters, Maria and Bernadette.

Marco Bellini graduated (*cum laude*) and got his Ph.D. degree in Physics at the University of Florence, Italy. He later joined the European Laboratory for Nonlinear Spectroscopy (LENS) and, since 1999, he has been a researcher at the Istituto Nazionale di Ottica (INO) CNR. Since the early times of his scientific career, he has been dealing with light possessing extreme features. His research tools range from single photons to ultraintense laser pulses; from ultrabroadband supercontinuum to highly monochromatic THz sources, and to radiation in the extreme ultraviolet end of the electromagnetic spectrum. On one side, he used ultrashort and ultraintense laser pulses to produce highly nonlinear interactions with matter, leading to fundamental studies on supercontinuum sources (thus contributing to the development of the femtosecond frequency comb), and to the production and applications of high-order laser harmonics. On the other side, the ability to arbitrarily manipulate and characterize light at the single-photon level has allowed him to perform experiments on the foundations of quantum mechanics, and to develop new tools for the emerging quantum technologies. On all these topics he maintains active collaborations with several leading scientists around the world. He is also involved in several national and international research projects and collaborates, as a reviewer and as an editor, with top international scientific journals. When not working, he likes traveling and exploring Nature in all its forms. This book (and most of the rest) would not have been possible without the constant love and support of his parents, Silvana and Antonio, and his wife, Francesca.

Paolo De Natale graduated in Physics (*cum laude*) at Federico II University, Naples, Italy. He joined the European Laboratory for Nonlinear Spectroscopy (LENS) in 1988 and, since 1996, he has been staff scientist at Istituto Nazionale di Ottica (INO)-CNR (formerly INOA) which, at the time of writing, he directs. His research activity has always been focused on atomic, molecular, and optical physics. Pioneering results included tests of theories, measurements of fundamental constants, using state-of-the-art spectroscopic techniques and originally developed coherent sources, often harnessing nonlinear optical phenomena. He has often carried on research in quite unexplored spectral regions, like infrared and far-infrared (THz), for the lack of suitable sources. Another part of PDN research has been devoted to the study and design of novel optoelectronic devices, especially based on ferroelectric crystals. To trigger such multidisciplinary research, PDN summoned a group combining unusual skills in spectroscopy, interferometric diagnostics, mastering of nonlinear sources and techniques, that enabled quick achievement of unique results. Such activity is still ongoing in the Napoli section of the Istituto Nazionale di Ottica-CNR (formerly INOA), producing internationally recognized breakthrough results. The most recent best results to which he has given significant contributions include the first fiber-based mid-IR generation of a frequency comb as well as demonstration of its suitability for frequency metrology; pioneering work on sub-Doppler molecular spectroscopy with an absolute frequency scale in the IR, using coherent sources based on difference frequency generation with smart phase-locks, using optical frequency-comb synthesizers for building highly coherent mid-IR radiation sources; development of a novel spectroscopic technique (SCAR), based on saturated absorption cavity ring-down, overcoming background-related sensitivity limits of standard cavity ring-down; pioneering work on the intrinsic noise properties of Quantum Cascade Lasers; discovery of the intrinsic sensitivity limits of fiber-based optical sensors; molecular gas sensing at parts per quadrillion, achieving the highest sensitivity (i.e., minimum detectable gas pressure) ever observed in a spectroscopic experiment on a gas of simple molecules, taking to detection of radiocarbon-dioxide ($^{14}CO_2$) below its natural abundance. Paolo De Natale has authored about 200 papers, has edited 10 books and special journal issues, and holds 5 patents. Finally, most of the time away from the family due to his scientific activity was compensated by the strong and loving committment of his wife Roberta.

1

Shedding light on the art of timekeeping

The Present contains nothing more than
the Past, and what is found in the effect
was already in the cause.

Henri L. Bergson - Creative Evolution

1.1 The great show of Time and Light, the curtain rises!

We have always lived in a world illuminated by Light and marked by the relentless flow of
Time. In spite of the difficulty of finding a universal definition for them, Light and Time
are essential elements of our existence. They govern countless aspects of our practical life
and accompany us in various cultural and sentimental experiences. The earliest people on
the planet naturally entrusted the organization of their activities in the light coming from
celestial bodies. The periodic character of the most basic astronomical motions was immediately recognized: the Sun rise and set (Earth's rotation around its axis), the appearance
of the highlighted portion of the Moon (Moon's revolution around the Earth), and the
weather periodical behavior that seemed to be related to the movement of the Sun with
respect to the stars (Earth's revolution around the Sun). The units of days, months, and
years accordingly followed. The main disadvantage of Nature's clocks resided in that the
scale unit was too large for many practical purposes. Consequently, *natural oscillators* soon
began to be supplemented by those constructed by mankind. Around 3500 BC Egyptians
already divided the time of the day into shorter sections by observing the direction of the
shadow cast from obelisks or sundials by the Sun, depending on its position in the sky
[1]. It is amazing to note that the ancient and honored Earth-Sun clock met many of the
most demanding requirements that the scientific community today exacts from an acceptable standard: first, it is universally available and recognized; second, it involves neither
responsibility nor operation expenses for anyone; third, it is pretty reliable and we cannot
foresee any possibility that it may stop or *lose* the time. In spite of all these nice features,
however, this clock does not represent an extremely stable timepiece. According to our current knowledge in astronomy, first, Earth's orbit around the Sun is elliptical rather than a
perfect circle, which means that Earth travels faster when it is closer to the Sun than when
it is farther away. In addition, Earth's axis is tilted with respect to the plane containing its
orbit around the Sun. Finally, Earth spins at an irregular rate around its axis of rotation

and even wobbles on it. The latter effect is due to the circumstance that, as Earth is neither perfectly symmetrical in shape, nor homogeneous, nor ideally rigid (its mass distribution constantly changes over time), its rotation axis does not coincide exactly with the figure axis. For the same reason, even natural disasters of exceptional importance may perturb the clock mechanism. For example, it has been calculated that the recent earthquake in Japan (March 2011) has moved Earth's figure axis by a few milliarcseconds or, in other words, rearranged Earth's mass bringing more of it a bit closer to the rotation axis. This should have slightly increased Earth's rate of spin, thus shortening the length of the day by less than 2 microseconds [2]. Such a small variation has no practical effect in daily life, but it is of interest for precision measurements of space and time.

Although the interaction between sunlight and Earth's swinging is far from being considered a resonant phenomenon, Earth-Sun clocks constitute the first example of a timepiece in which a light source interrogates a frequency reference. Quite surprisingly, however, in the following five millennia, Light and Time travelled rather distinct roads in the advancement of human thought and abilities.

1.2   Brief history of timekeeping: time-frequency equivalence

Clepsydrae based on controlled flows of water (either into or out of a vessel) were available in Egypt, India, China, and Babylonia from about 1500 BC and represented the first non-astronomical means of measuring time. Sand clepsydrae were introduced only in the late fourteenth century AD. By using the integrated quantity of moved substance to provide a measurement of the elapsed time, this type of timekeeper did not rely on counting the number of cycles of an oscillatory event. The resort to light was abandoned too. In the last part of the thirteenth century mechanical clocks began to appear in Europe [3]. The first prototypes, representing the natural progression of wheel clocks driven by water (already introduced in China after the 8th century), were just geared machines based on the fall of a weight regulated by a verge-and-foliot escapement. Variations of this design reigned for more than 300 years, but all had the same basic problem: the period of oscillation of the escapement was heavily affected both by the amount of force and the extent of friction in the drive. Like water flow, the rate was difficult to adjust.

A significant advance occurred in the 17th century when Galileo Galilei discovered that the period T of a pendulum swing virtually does not depend on the excursion, provided that the latter is not too large:

$$T \simeq 2\pi \sqrt{\frac{l}{g}} \qquad (1.1)$$

Here l is the pendulum length, and g is the acceleration due to gravity. Galilei, in fact, recognized the value of the pendulum as a time-keeping device and even sketched out a design for a clock. However, it was Christiaan Huygens in 1656 to realize the first successful operational pendulum clock. Reaching an error of less than 1 minute a day, such device recovered and definitively consecrated the idea that the most accurate way of keeping the time was to employ an oscillatory system operating at a specific resonance frequency ν_0. Hence, any time interval could be measured by counting the number N of elapsed cycles and then multiplying N by the period $T = 1/\nu_0$. Light, however, was still excluded from the time-keeping saga.

From then onwards, time and frequency became the quantities that humanity could measure with the highest precision. Indeed, during the next three centuries, continuous refinements improved considerably the accuracy of pendulum clocks. In 1671, William Clement began building clocks with the new *anchor* escapement, a substantial improvement over the verge because it interfered less with the motion of the pendulum. In 1721, George Graham improved the pendulum clock accuracy to 1 second per day by compensating for changes in the pendulum length due to temperature variations. John Harrison, a carpenter and self-taught clockmaker, developed new methods for reducing friction. By 1761, he had built a marine chronometer with a spring-and-balance-wheel escapement that kept time on board a rolling ship to about one fifth of a second a day, nearly as well as a pendulum clock could do on land. Over the next century, refinements led in 1889 to Siegmund Riefler's clock with a nearly free pendulum, which attained an accuracy of a hundredth of a second a day and became the standard in many astronomical observatories. A true free-pendulum principle was demonstrated by R.J. Rudd around 1898. This gave birth to a generation of superior timepieces that culminated in 1920 with the realization by William H. Shortt of a clock consisting of two synchronized pendulums. One pendulum, the master, swung as unperturbed as possible in an evacuated housing. The slave pendulum driving the clockwork device was synchronized via an electric linkage and in turn, every half a minute, initialized a gentle push to the master pendulum to compensate for the dissipated energy. Keeping time better than 2 milliseconds a day, Shortt clocks almost immediately replaced Riefler ones for time distribution on local and eventually national scale.

The performance of such clocks was overtaken as soon as the technology of quartz crystal oscillators became mature for the construction of the first timekeeper (W. Marrison and J.W. Horton, 1927). Quartz clock operation hinges on piezoelectricity that is the capability of some materials to generate electric potential when mechanically stressed or, conversely, to strain when an electric potential is applied. Due to this interaction between mechanical stress and electric field, when placed in a suitable oscillating electronic circuit, the quartz will vibrate at a specific resonance frequency (basically depending on its size and shape) and the frequency of the circuit will become the same as that of the crystal. Such a signal is eventually used to operate an electronic clock display. As they had no escapements to disturb their regular frequency, quartz crystal clocks soon proved their superiority with respect to pendulum-based ones. A serious source of systematic error, namely the dependence of the period on the strength of the local gravity vector (and hence on the pendulum location), was overcome too. Although quartz oscillators had provided a major advance in timekeeping, so as to become, in the late 1930s, the new timekeeping standards, it was apparent that there were limitations to that technology. These devices could provide frequency with a precision of about 10^{-10}, but going beyond proved to be a real challenge. Operationally, fundamental mode crystals could be made to provide frequencies up to 50 MHz. Higher frequencies capable of providing more precise timekeeping were possible using overtones but were not commonly used. Moreover, aging and changes in the environment, including temperature, humidity, pressure, and vibration, affected the crystal frequency. In order to compensate for these problems, different systems were designed, including temperature-compensated and oven-controlled crystal oscillators.

To make a significant advance in precision timekeeping of laboratory standards, however, a fundamental change was required [4, 5, 6]. Scientists had long realized that atoms (and molecules) have resonances; each chemical element and compound absorbs and emits electromagnetic radiation at its own characteristic frequencies. An unperturbed atomic transition is identical from atom to atom, so that, unlike a group of quartz oscillators, an ensemble of atomic oscillators should all generate the same frequency. Also, unlike all electrical or mechanical resonators, atoms do not wear out. Additionally, all experimental observations in spectroscopy have proved compatible with the hypothesis that atomic properties are the same at all times and in all places, when they are assessed by an observer situated close to the atom and accompanying it in the same motion. It is therefore possible to build instruments, which, using a specified atomic transition, are all able to deliver a signal in real time with the same frequency, anywhere and at any time, provided that relativistic effects due to non-coincidence of atom and observer have been properly taken into account [7]. These features were appreciated by Lord Kelvin who suggested using transitions in hydrogen as a time-keeping oscillator. However, it wasn't until the mid 20th century that technology made these ideas possible. The first atomic clocks owe their genesis to the explosion of advances in quantum mechanics and microwave electronics before and during the Second World War. Indeed, the sudden development of the radar and very high frequency radio communications made possible the generation of the kind of electromagnetic waves (microwaves) needed to interact with atoms. Atomic oscillators use the quantized energy levels in atoms and molecules as the source of their resonance frequency. The laws of quantum mechanics dictate that the energies of a bound system, such as an atom, have certain discrete values. An electromagnetic field at a particular frequency can excite an atom from one energy level to a higher one. Or, an atom at a high energy level can drop to a lower level by emitting energy. The resonance frequency of an atomic oscillator is the difference between the two energy levels, E_1 and E_2, divided by Planck's constant, h:

$$\nu_0 = \frac{E_1 - E_2}{h} \qquad (1.2)$$

The basic idea of atomic clocks is the following. First, a suitable energy transition is identified in some atomic species (microwave atomic frequency standards are commonly based on hyperfine transitions of hydrogen-like atoms, such as rubidium, cesium, and hydrogen). These provide transition frequencies that can be used conveniently in electronic circuitry (1.4 GHz for hydrogen, 6.8 GHz for rubidium, and 9.2 GHz for cesium). Then, an ensemble of these atoms is created (either in an atomic beam, or in a storage device, or in a *fountain*). Next, the atoms are illuminated with radiation from a tunable source that operates near the transition frequency ν_0. The frequency where the atoms maximally absorb is sensed and controlled. When the absorption peak is achieved, the cycles of the oscillator are counted: a certain number of elapsed cycles generates a standard interval of time. Most of the basic concepts of atomic oscillators were developed by Isidor Rabi and his colleagues at Columbia University in the 1930's and 40's. Although he may have suggested using cesium as the reference for an atomic clock as early as 1945, research aimed at developing an atomic clock focused first on microwave resonances in the ammonia molecule. In 1949, the National Bureau of Standards (NBS) built the first atomic clock, which was based on ammonia (at 23.8 GHz). However, its performance wasn't much better than the existing standards, and attention shifted almost immediately to more promising atomic-beam devices based on cesium. The first practical cesium atomic frequency standard was built at the National Physical Laboratory (NPL) in England in 1955 by Dr. Louis Essen. In collaboration with the U.S. Naval Observatory (USNO), it was immediately noted that observations of the Moon over a period of several years would be required to determine Ephemeris Time with the same precision as was achieved in a matter of minutes by the first cesium clock. For the benefit of the reader, we recall here that the ephemeris second is based on the period of revolution of the Earth around the Sun which is more predictable than the rotation of Earth itself (for more details, refer to Chapter 7).

While NBS was the first to start working on a cesium standard, it wasn't until several years later that NBS completed its first cesium atomic beam device. By 1960, cesium standards had been refined enough to be incorporated into the official timekeeping system of NBS. Standards of this sort were also developed at a number of other national standards laboratories, leading to wide acceptance of this new timekeeping technology. Then, pressure mounted for an atom-based definition of time. This change occurred in 1967 when, by international agreement,

---

*the **second** was defined as the duration of 9,192,631,770 periods of the radiation corresponding to the transition between two hyperfine levels $|F = 4, m_F = 0\rangle \leftrightarrow |F = 3, m_F = 0\rangle$ in the ground state $^2S_{1/2}$ of the ^{133}Cs atom.*

This definition made atomic time agree with the second based on Ephemeris Time, to the extent that measurement allowed. As of 2011, the definition of the SI second remains the same, except for a slight amendment made in 1997. Calculations made by Wayne Itano of NBS in the early 1980s revealed that blackbody radiation can cause noticeable frequency shifts in cesium standards [8], and his work eventually resulted in an addendum to the definition: the Comité International des Poids et Mesures (CIPM) affirmed in 1997 that the definition refers to *a cesium atom at rest at a thermodynamic temperature of 0 K*. Thus, a perfect realization of the SI second would require the cesium atom to be in a zero magnetic field in an environment where the temperature is absolute zero and where the atom has no residual velocity.

Excluding the pendulum, quartz and microwave atomic clocks are far from being relegated to history and their study will be resumed in Chapter 7. We close this section by observing that the development of increasingly more accurate frequency standards was paralleled by an augmented frequency of the employed oscillator: from Earth's rotation ($\sim$ 10 µHz), via pendulum clocks ($\sim$ 1 Hz) and quartz oscillators ($\sim$ 1 MHz), to microwave atomic standards ($\sim$ 1 GHz). More strictly, the accuracy performance of a frequency standard is characterized by the so-called quality factor (Q) which is defined, in general, as the oscillator resistance to disturbances to its oscillation period. This notion can be grasped in the case of a pendulum clock, where, in order to replace the energy lost by friction, pushes must be applied by the escapement. These pushes are the main source of disturbance to the pendulum motion. The smaller the fraction of the pendulum energy that is lost to friction, the less energy needs to be added, the less the disturbance from the escapement, the more the pendulum is *independent* of the clock mechanism, and the more constant its period is. In other words, the Q factor is related to the ratio of the total energy in the system to the energy lost per cycle or, equivalently, to how long it takes for the swings of the oscillator to die out:

$$Q \equiv \frac{\tau}{T} = \frac{\nu_0}{\Gamma} \qquad (1.3)$$

where $\tau \equiv 1/\Gamma$ is the time constant describing the (exponential) decay of the swing amplitude. Hence, the Q of pendulum clocks is increased by maximizing τ or, equivalently, minimizing the overall frictional losses (Γ). As it will be shown in Chapter 2, for a damped harmonic oscillator, Γ equals the full width at half maximum $\Delta\nu$ of the system response function (resonance curve) in the frequency domain. With this in mind, the above formula can be generalized to all types of oscillators as

$$Q \equiv \frac{\nu_0}{\Delta\nu} \qquad (1.4)$$

Concerning quartz oscillators, here we just mention that, starting from the electric equivalent of the crystal, ν_0 and $\Delta\nu$ are respectively calculated as the resonance frequency and width of an oscillatory circuit. In the case of atoms, finally, $\Delta\nu$ is calculated in the frame of quantum mechanics. Actually, as we will see in Chapter 4, atoms absorb or emit energy over a small frequency range surrounding ν_0: this spread of frequencies is referred to as the linewidth and its ultimate limit is related to Heisenberg's uncertainty principle. Since the response of a high-Q system decays much more rapidly as the driving frequency moves away from ν_0, the oscillator with the highest Q would be desirable as a frequency standard, Q^{-1} being roughly proportional to its limiting accuracy. This can be achieved either by using an atomic transition where ν_0 is as high as possible, or by making $\Delta\nu$ as narrow as possible.

1.3   The parallel story of the speed of light

In order to fully appreciate the significance of the modern physical measurements of the speed of light, just think that even today we are usually not aware of any delay between the occurrence of an event and its visual appearance in the eye of a distant observer. In fact, a single visual *snapshot* is probably, for most people, the basis for the intuitive notion of an *instant*. Therefore, it is of great interest to shortly trace the history of ideas concerning the finiteness of the velocity of light [9, 10]. Among the ancient Greeks, there was a general belief that this speed was infinite. An exception is represented by Empedocles from Acragas (490-435 BC) who, according to Aristotle (384-322 BC), "was wrong in speaking of light as travelling or being at a given moment between the earth and its envelope, its movement being unobservable to us". So powerful is Aristotele's cosmology that it compels him to declare that "...light is due to the presence of something, but it is not a movement".

An interesting *proof* that the velocity of light must be infinite is given by Heron of Alexandria (I century BC). According to him, you turn your head to the heaven at night, keeping the eyes closed; then suddenly open them, at which time you see the stars. Since no sensible time elapses between the instant of opening the eyes and the instant of sight of the stars, light must travel instantaneously. Since the causal direction of an instantaneous interaction is inherently ambiguous, it's not surprising that ancient scholars considered two competing models of vision, one based on the idea that every object is the source of images of itself, emanating outwards to the eye of the observer, and the other claiming that the observer's eye is the source of visual rays emanating outwards to *feel* distant objects. Indeed, at that time, the problem of the speed of light was secondary, whereas there was much more interest in catoptrics and vision matter.

Amidst the Islamic scientists, Avicenna (980-1073) was perhaps the most famous: his thought represents the climax of medieval philosophy. Avicenna observed that, if the perception of light is due to the emission of some sort of particles by the luminous source (as he believed), then the speed of light must be finite. Alhazen (965-1039), another Muslim physicist and one of the greatest scholars of optics of all time, came to the same conclusion. In his treatise on optics he states that light is a movement and, as such, is at one instant in one place and at another instant in another place. Since light is not in both these places at the same time, there must be a lapse of time between the two: hence the transmission cannot be instantaneous.

Nevertheless, Aristotle's point of view was echoed by many thinkers in western history: John Peckam (1230-1292), Thomas Aquinas (1225-1274), and Witelo (1230-1275) to name a few. It is curious to note that Roger Bacon (1214-1292), although in perfect agreement with Alhazen's conclusions on this subject, felt the need to show in "Opus Majus" that the sort of reasoning used by Alhazen was identical to that of the scientists who attempted to prove the opposite view. Bacon's remarks afford a striking example of the confusion exhibited by a first rate mind attempting to be reasonable with no genuine scientific or experimental basis as guide. The debate continued into the beginning of the *scientific revolution* of the 17th century. Such giants as Francis Bacon (1561-1626), Johannes Kepler (1571-1630), and René Descartes (1596-1650) adhered to the idea of instantaneous propagation. Descartes considered an eclipse of the moon, caused by the moon, earth, and sun being in a straight line, with Earth interposed between the other two: "Now suppose that it requires an hour for light to travel from the earth to the moon. Then the moon will not become dark until exactly one hour after the instant of collinearity of the three bodies. Similarly, here on the earth, we will not observe Moon's darkening until the passage of another hour, or until two hours after the moment of collinearity. But during this time, the moon will have moved in its orbit and the three bodies will no longer be collinear. But clearly, this is contrary to experience, for one always observes the eclipsed moon at the point of the ecliptic opposite to the sun. Hence light does not travel in time, but in an instant".

In the face of all this, the remarks by Galileo (1564-1642) seem like a breath of fresh air in a stale room. In his great treatise on mechanics there is a conversation about the velocity of light during which Salviati claims that the general inconclusiveness of observations on this subject had led him to conceive an experiment. He says: "Let each of two persons take a light contained in a lantern such that, by the interposition of the hand, the one can shut off or admit the light to the vision of the other. Next, let them stand opposite each other at a distance of a few cubits and practice until they acquire such skill (in uncovering and occulting their lights) that the instant one sees the light of his companion, he will uncover his own. After acquiring this skill, the two experimenters were to perform the same operations at greater distances, ten miles if necessary (using telescopes). If the exposures and occultations occur in the same manner as at short distances, we may safely conclude that the propagation is instantaneous; but if time is required at a distance of three miles, which, considering the going of one light and the coming of the other, really amounts to six, then the delay ought to be easily observable". This experiment was executed by the Florentine Academy and their account of it is as follows: "We tried it at a mile's distance and could not observe any. Whether in a greater distance it is possible to perceive any sensible delay, we have not yet had an opportunity to try".

The first experimental evidence of the finite speed of light was due to Ole Christensen Roemer in 1676 by observing the eclipses of the inner-most moon of Jupiter (Io) [11]. Discovered by Galileo in 1610, detailed tables of the movements of these moons had been developed by Borelli (1665) and Cassini (1668). Io has a period of about 42.5 hours and, if Earth were stationary, it would show an eclipse at regular intervals of 42.5 hours. But Earth revolves about the sun and, in so doing, assumes positions 1 and 2 in Figure 1.1. Roemer noticed that when the Earth was close to Jupiter (position 1), the eclipses occurred 8.5 minutes ahead of the time predicted on the basis of yearly averages. The eclipses were late by the same amount when the Earth was opposite (position 2). Roemer concluded that twice that difference was the time it took the light to traverse the diameter of Earth's orbit ($\sim 3 \cdot 10^8$ km), which gave a figure of ~ 227000 km/s.

Despite the force of Roemer's analysis, and the early support of both Huygens and Newton, most scientists remained skeptical of the idea of a finite speed of light. Alternative explanations were provided by Cassini and later by his nephew Giacomo Filippo Maraldi. They suggested that Jupiter's orbit and the motion of its satellites might explain the observed inequalities. It was not until 50 years later, when the speed of light was evaluated in a completely different way, arriving at nearly the same value, that the idea became widely accepted. Such measurement was performed in 1728 by James Bradley by observing stellar aberration, that is the apparent displacement of stars due to the motion of the Earth around the Sun. A useful analogy to help understand aberration is to imagine the effect of motion on the angle at which rain falls. If you stand still in the rain (when there is no wind), it comes down vertically on your head. If you run through the rain it appears to come to you

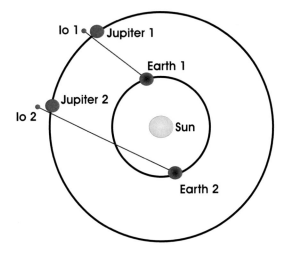

FIGURE 1.1
Roemer's evaluation of the speed of light.

from an angle and hit you on the front. It is worth pointing out that all stellar positions are affected equally in the aberration phenomenon, which distinguishes this effect from parallax where nearby stars are influenced more noticeably. By observing a star in Draco, and recording its apparent position during the year, Bradley argued that stellar aberration is approximately the ratio of the orbital speed of the Earth (around the Sun) to the speed of light (see Figure 1.2). Based on the best measurement of the limiting starlight aberration (20.5 arcseconds $\simeq 0.0001$ rad) by Otto Struve, and taking the speed of Earth to be about 30 km/s from Encke's estimate, this implied a light speed of about 301000 km/s [9].

Unfortunately, measurements of the speed made in this way depended on the astronomical theory and observations used. Better determinations of the speed might be made if both source and observer were terrestrial. The first measurement of c on Earth was by Armand Fizeau in 1849 [12]. His method measured the time needed for light to travel to a flat mirror

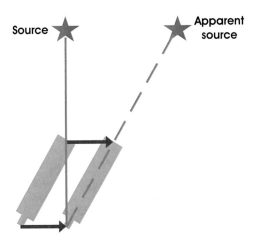

FIGURE 1.2
Bradley's determination of the light speed.

at a known distance and return. For that purpose he designed a set-up where a collimated beam emitted by a limelight passed through a half-mirror and a rotating cogwheel was then reflected back by a mirror situated some 8.633 kilometers away, passed (or not) through the cogwheel again, and was eventually reflected by the half-mirror into a monocular (see Figure 1.3). At a low rotation rate, the light passes through the same blank of the wheel on the way out and on the way back. But with increasing rotation rate, a higher and higher percentage of the transmitted light is cut on its way back by the incoming tooth of the wheel, resulting in a decreasing light intensity collected in the monocular. Total extinction of the returning light is reached when the time duration of the open gate corresponds exactly to the duration of the round-trip, such that the light that has gone through finds the gate closed when it returns. Knowing the precise distance d between the wheel and the mirror, the number of teeth N_t of the wheel, and its rotation rate ω (expressed in radians per second), the speed of light in air can be deduced to be

$$c = (2d)(2N_t)f_c \qquad (1.5)$$

where $f_c = N_t\omega/2\pi$ is the frequency at which the beam is effectively stopped. Obviously, if one increases further the rotating speed of the wheel, light will appear again as the returning light will start passing through the gap situated right after the one it has passed on its way out. Using this method with the cogwheel placed in Montmartre and the reflector in Suresnes, Fizeau obtained a value of $c = 315300$ km/s, limited by the precision of his measurement of ω, but yet better than any measurement realized before. Such a method was subsequently taken up first by Marie Alfred Cornu in 1874 and then by Joseph Perrotin in 1902. Some experimental tricks allowed them to provide the following more accurate results for the speed of light *in vacuum*: 299990 ± 200 km/s and 299901 ± 84 km/s, respectively (their results already included correction for the refractive index of air) [13].

In 1855, Kirchhoff realized that $1/\sqrt{\varepsilon_0\mu_0}$ has the dimension of a speed, where μ_0 (ε_0) is the magnetic permeability (electric permittivity) of free space entering the laws of magnetism (electricity). In 1856 Weber and Kohlrausch measured this constant using only electrostatic and magnetostatic experiments [14]. Incidentally, they were the first to adopt the symbol c (from Latin *celeritas*) for the speed of light. Within experimental accuracy, the value found by them agreed with the speed of light. This remained a coincidence until Maxwell formulated his theory of electromagnetism in 1865 and concluded that "...light is an electromagnetic disturbance propagated through the field according to electromagnetic laws" [15]. Maxwell's equations established that the velocity of any electromagnetic wave (and thus of light) in a *vacuum* is c, where

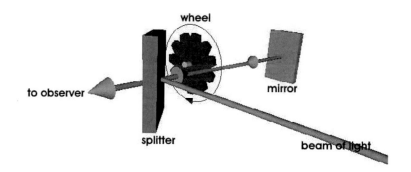

FIGURE 1.3
First terrestrial measurement of c, performed by Fizeau.

$$c^2 = \frac{1}{\varepsilon_0 \mu_0} \tag{1.6}$$

At this stage, the status of c increased tremendously because it became a characteristic of all electromagnetic phenomena.

The method demonstrated by Leon Foucault (1862) relies on the same principle adopted by Fizeau (time of flight technique), but replaces the cogwheel by a revolving mirror (Figure 1.4). Light rays from the source S that strike the revolving mirror R and proceed through the lens will strike the stationary mirror M and return to the source. If, after the light beam first strikes R outbound from S, R can be rotated before it is struck again by the beam returning from M, then the returning beam will no longer return exactly to the source S but will instead be deflected away from S in the direction of the rotation. By rotating the mirror at a constant speed, the amount of deflection will be the same for all light beams which go through the lens, strike M and return. Then, for a continuous beam of light from S and a constant high speed of rotation of R, an image of the source will appear beside S instead of coincident upon it. The faster R rotates or the longer is $\overline{RS}$, the farther the returned image I will be displaced from the source S and the easier it will be to measure the deflection. By carefully measuring the amount of displacement from S to I, and the distance from S to R, the angle of deflection can be determined. Together with the known, fixed speed of rotation, this angle can be used to determine the time it took light to travel the distance from R to M and back. Let $\theta = \arctan(\overline{IS}/\overline{IR})$ denote the angle of deflection (this means that the angle through which the mirror has rotated is $\theta/2$). If the speed of rotation is measured in number of cycles (n_c) per second, then the speed of light is given by

$$c = \frac{2 \cdot \overline{RM}}{\dfrac{\theta}{2} \dfrac{1}{2\pi} \dfrac{1}{n_c}} \tag{1.7}$$

In this arrangement, the distances $\overline{IS}$ and $\overline{SR}$ should be as large as possible to reduce the error in measuring θ. The distance $\overline{IS}$ is maximized by maximizing the speed of rotation of R and the distance $\overline{RM}$. In Foucault's setup, M was spherical with center at R. The greatest distance $\overline{RM}$ achieved by Foucault was 20 m, which produced a displacement $\overline{IS}$ of only about 1 mm. The result was 298000 ± 500 km/s [13].

Going back to the approach by Weber and Kohlrausch, in 1907 Rosa and Dorsey obtained a much more accurate determination of c. As the value of μ_0 is fixed at exactly $4\pi \cdot 10^{-7}$ N·A^{-2} through the definition of the ampere, only ε_0 had to be measured in their experiments. This can be accomplished by determining the ratio of the capacitance of a condenser as measured in electrostatic and electromagnetic units. Rosa and Dorsey used the Maxwell bridge method (employing carefully standardized resistances) to determine the electromagnetic capacitance and standards of length and mass to determine the electrostatic capacitance [16]. They used a variety of shapes (spherical, cylindrical, and plane) and sizes of condensers [17]. Both the calculations and experiments were beset with difficulties, but their result was probably the most reliable up to that time. The final value, $c = 299710$ km/s, was the mean of about 900 individual determinations with an estimated maximum error of 30 km/s, apart from uncertainties in the value taken for the international ohm. In 1941 a more accurate knowledge of the latter standard allowed Birge to apply a correction to their result yielding the value $c = 299784$ km/s [18].

Foucalt's apparatus was perfected by Michelson in several versions till the famous experiment in 1927 [19]. As shown in Figure 1.5, the apparatus involved a rotating octagonal glass prism. When the prism is stationary the light follows the path shown and an image

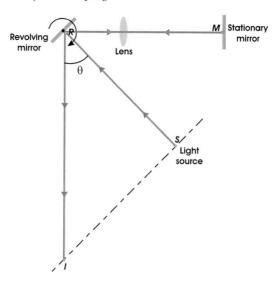

FIGURE 1.4
Focault's apparatus for measuring c.

of the source can be seen through the telescope. If the prism is rotated slowly, the image disappears because either face X is not in a suitable position to direct the outgoing beam to the concave reflector C, or face Y is unable to send the incoming beam to the telescope. However, if the rotation speed of the prism is increased so that it turns exactly one-eighth of a revolution in the same time that it takes light to travel from X to Y, then an image of the source is seen through the telescope. Michelson adjusted the speed of rotation until he was able to observe a stationary image of the source. This occurred when the prism was rotating at $f_{rot} \simeq 530$ Hz (this rate was measured by comparison with a free pendulum furnished by the United States Coast and Geodetic Survey). The experiment was carried out on Mt. Wilson (USA) and the concave reflector C was on Mt. San Antonio $d = 35$ km away. The result was $c = (2d)/[(1/8)(1/f_{rot})] = 299796 \pm 4$ km/s.

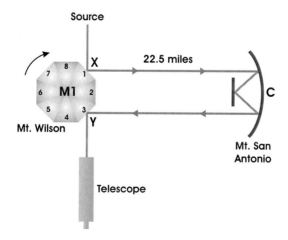

FIGURE 1.5
Michelson's famous experiment for the measurement of c.

An ingenious modification of the toothed-wheel method was used by Karolus and Mittelstaedt in 1928. In their apparatus, a Kerr cell, at the terminals of which an alternating difference of potential was applied, was used to periodically interrupt the passage of a luminous beam. The main advantage of this method is that the frequency of such periodic interruption can be accurately determined, which was not the case when the toothed wheel was used; moreover, a much higher frequency can be used with this method (around 10 MHz), so that a correspondingly short base (in this case 41.4 meters) can be utilized [13]. This approach gave the value 299778 ± 20 km/s.

The completion of this excursus on the measurements of c necessitates now the discussion of experiments whose understanding requires knowledge beyond basic physics, many of the involved concepts being precisely the subject of this book. For the moment, may the reader be satisfied with an intuitive comprehension; a full appreciation will result from revisiting each experiment as the pertinent notions are gradually acquired.

As we will learn in Chapter 3, another valuable option to determine c is to measure the resonance frequencies of a cavity resonator whose dimensions are precisely known. In 1946, Louis Essen and A.C. Gordon-Smith pursued this approach, by establishing the frequency for a variety of normal modes of an evacuated microwave cavity [20, 18]. The latter consisted of a copper cylinder constructed with great uniformity. The resonant frequency ν_n of an evacuated right circular hollow cylinder closed at both ends is given by

$$c = \frac{\nu_n}{\sqrt{\left(\frac{r}{\pi D}\right)^2 + \left(\frac{n}{2L}\right)^2}} \left(1 + \frac{1}{2Q_{cav}}\right) \qquad (1.8)$$

where r is a constant for a particular mode of resonance, n is the number of half-wavelengths in the guide, D is the diameter and L the length of the cylinder, and Q_{cav} is the quality factor of the resonator accounting for the finite conductivity of the cavity walls. The quantities ν_n, D, L, and Q_{cav} could all be measured with a precision of a few parts in 10^6. In particular, the dimensions of the resonator were measured in the Metrology Division of the National Physical Laboratory using gauges calibrated by interferometry. The final result (using the E_{010} and E_{011} modes) was $c = 299792 \pm 9$ km/s, where the estimated maximum error was the sum of different contributions including setting of the frequency to resonance and measurement of the frequency by the spectrum analyzer, uncertainty of the resonator temperature, dimensional measurements, residual effects of coupling holes and probes, non-uniformity of the resonator, and uncertainty of Q_{cav}. Almost simultaneously, a very similar value (299789.3 ± 0.4 km/s) obtained by the same measurement scheme was published by Bol in a short note [21].

In those years, radar systems also began to be used to measure the speed of light. Again, the time-of-flight principle was exploited: twice the known distance to a target was divided by the time it took a radio-wave pulse to return to the radar antenna after being reflected by the target. This was done by Aslakson in 1949 with the result 299792.4 ± 2.4 km/s [22]. Incidentally we mention here (see Chapter 7 for further details) that, today, a Global Positioning System (GPS) receiver measures its distance to GPS satellites based on how long it takes for a radio signal to arrive from each satellite: from these distances the receiver position is calculated.

Then came the geodimeters. Originally intended for use in geodesic surveying, Bergstrand demonstrated their use in accurate measurement of the light speed [23]. With reference to Figure 1.6, the principle can be described as follows: a light beam is emitted through a Kerr cell to a distant mirror and reflected back to a receiving photocell close to the emitter. The two cells are supplied by the same crystal-controlled high frequency voltage (about 10 MHz in the original work). The difference in phase of the emitted and

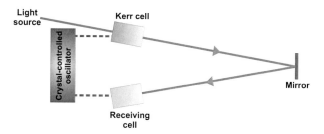

FIGURE 1.6
Bergstrand's method for evaluating c.

received light is compared. If the mirror distance is known, the speed of light can be measured [24]. Bergstrand repeated this kind of measurement several times with the final value 299793.1 ± 0.2 km/s [25]. Such phase-shift method was pushed to the limit of its accuracy by Grosse in 1967: 299792.50 ± 0.05 [26].

Another way to get the speed of light is to independently measure the frequency ν and wavelength λ of an electromagnetic wave in *vacuum*. The value of c can then be found by using the relation $c = \lambda\nu$ (see Figure 1.7).

This approach was followed first in the microwave domain at NPL by Froome in 1958 [27]. The basis of the determination consisted of the simultaneous measurement of the free-space wavelength and frequency of an electromagnetic wave generated by a microwave source. The latter was a frequency-stabilized klystron oscillator operating at 36 GHz (in short, the klystron is a specialized linear-beam electron vacuum tube which converts, via velocity modulation, the kinetic energy of the electron beam into a radio-frequency/microwave signal). The greater part of the output from this oscillator was fed by means of a waveguide switch into one of the two silicon crystal distorter units tuned for maximum harmonic output at 72 GHz (about 0.4 cm wavelength). One harmonic generator was used to supply the interferometer itself, the other for operating the cavity resonator refractometer by means of which the refractive index of the air in the neighborhood of the equipment could be measured. The measurement of the microwave frequency was accomplished by comparing a portion of the klystron output against a high harmonic of a 5-MHz quartz crystal standard. The 5-MHz was multiplied in stages of two to five times up to 600 MHz and then fed into a silicon crystal harmonic generator mounted in waveguide, so that the harmonic at exactly 36 GHz could be mixed with a small fraction of the klystron output. The beat frequency between the two was detected by means of a calibrated communications receiver. The accuracy of frequency determination was at least as good as 1 part in 10^8. The estimated accuracy of the refractive index measurement was 1.1 parts in 10^7. The value of

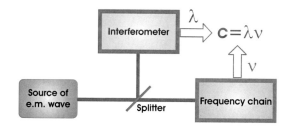

FIGURE 1.7
Determination of $c = \lambda \cdot \nu$ by independent measurements of the free-space wavelength λ and frequency ν of an electromagnetic wave.

microwave wavelength obtained by means of the interferometer, when multiplied by the air refractive index and the microwave frequency, gave a *vacuum* phase velocity which had still to be corrected for the effect of diffraction before the true free-space value could be derived. Basic concepts of interferometry for wavelength measurements will be given in Chapter 5. Here we just recall that the diffraction limit is proportional to λ. It is interesting to note that the greatest single uncertainty in the whole measurement arose from the use of the length standards. At that time, interferometry-based length measurements were ultimately referenced to the cadmium red line (falling at $\lambda \simeq 644$ nm) as emitted by the international specified form of the Michelson lamp. The result was 299792.50 ± 0.10 km/s.

Later, it was discovered that the cadmium line was actually a cluster of closely separated lines, and that this was due to the presence of different isotopes in natural cadmium. Thus, in order to get the most precisely defined line, it was necessary to use a mono-isotopic source. Allowing for easier isotopic enrichment and lower operating temperatures for the lamp (which reduces broadening of the line due to the Doppler effect), the pretty bright orange line of krypton-86 (at $\lambda \simeq 606$ nm) was then selected as the new standard. Krypton-86 offered the additional advantages of having zero nuclear spin.

In the same years, a quite different, spectroscopy-based approach was also pursued to measure c [28]. That was the so-called band spectrum method, involving the simultaneous measurement of the rotational constant B'' of the ground state of a diatomic (or linear) molecule in pure frequency units by means of microwave spectroscopy and in cm^{-1} units by means of infrared (rotation-vibration) spectroscopy. Then, the ratio

$$\frac{B'' \text{ microwave}}{B'' \text{ infrared}} = c \tag{1.9}$$

yields the speed of light. The most precise result obtained by such method was $c = 299792.8 \pm 0.4$ km/s, the main limitation being dictated by the accuracy of the theory of band spectra of molecules.

1.3.1 The laser arrives: length-frequency equivalence and the birth of optical frequency metrology

Up to this point in both stories, that of timekeeping and c, only microwave radiation made its entrance. The future of metrology was changed fundamentally on 12 December 1960 when a small team at Bell Labs, led by Ali Javan, eventually found the right conditions for their Optical Maser to generate the self-sustained optical oscillation that was anticipated by Charles Townes and Arthur Schawlow in a classic paper of 1958 [29]. The emergence of the laser promised to open new scenarios in the field of metrology. Indeed, very soon the wavelength of visible radiation could be measured fairly well by Michelson or Fabry-Perot interferometers. This possibility enabled the development of laser frequency measurement programs at various national standards laboratories such as NBS at Boulder, NPL at Teddington, and National Research Council at Ottawa. In spite of the fact that lasers provided coherent frequency sources in the infrared and visible, optical frequencies could not immediately be measured with the required degree of accuracy. Specifically, two fundamental drawbacks had to be overcome. First, laser frequency stability had to be greatly improved. Indeed, in the case of the gas laser, although its short term linewidth was a few hundred Hz, over a long period, its frequency could vary within the Doppler and pressure broadened gain curve of the laser. By the late 1960s, lasers stabilized in frequency to atomic and molecular resonances were becoming reliable research tools and the development of the technique of saturated absorption had produced lasers with one-second fractional frequency instabilities as small as $5 \cdot 10^{-13}$ [26]. Second, the laser optical frequency was much too high for conventional frequency measurement methods. To remove this limitation, the approach taken was to synthesize signals at progressively higher and higher frequency using harmonic-generation-and-mixing (heterodyne) methods and to lock the frequency of a nearby oscillator or laser to the frequency of this synthesized signal. Photodiodes, as well as metal-insulator-metal (MIM) diodes, fabricated by adjusting a finely tipped tungsten wire against a naturally oxidized nickel plate, were used for harmonic generation and mixing [30]. With this approach, a frequency synthesis chain was constructed linking the microwave output of the cesium frequency standard to the optical region [31], so that the Boulder group could directly measure the frequency of a helium-neon laser stabilized against the 3.39-μm (88 THz) transition of methane [32] (note that the frequency of the methane stabilized helium-neon laser is over 1000 times higher in frequency than that of the oscillator used in Froome's measurement). At the same time, the wavelength of the 3.39-μm line of methane was measured with respect to the Kr-86 6057-Angstrom standard by using a frequency-controlled Fabry-Perot interferometer [33]. In this way, the laser eventually permitted to preserve the small interferometric errors associated with the short optical wavelength, while utilizing microwave frequencies which were still readily manipulated and measured. The extension of frequency measurements into the infrared portion of the electromagnetic spectrum had in a sense solved the dilemma raised raised by Froome's experiment: to measure the frequency, it is best to do the experiment not too far removed from the primary Cs frequency where extremely stable oscillators can be made and frequencies are easily measured with great accuracy. However, to measure the wavelength it is best to do the experiment close to the visible 86-krypton wavelength standard where wavelengths can be more easily compared and where diffraction problems are not severe. Table 1.1 summarizes the most significant milestones in the story of *c* measurements.

When the measurements were completed, the uncertainty limitation was found to be the asymmetry of the krypton line on which the definition of the meter was then based [34]. The experiment thus showed that the realization of the meter could be substantially improved through redefinition. This careful measurement resulted in a reduction of the uncertainty of the speed of light by a factor of nearly 100. The methods developed at NIST were replicated in a number of other laboratories [35, 36, 37], and the experiments were repeated and improved to the point where it was generally agreed that this technology could form the basis for a new definition of the meter. An important remaining task was the accurate measurement of still higher (visible) frequencies which could then serve as more practical realizations of the proposed new definition. The Boulder group again took the lead and provided the first direct measurement of the frequency of the 633 nm line of the iodine-stabilized helium-neon laser [38], as well as a measurement of the frequency of the 576 nm line in iodine [39].

These and similar measurements around the world (frequency and wavelength measurements were refined to the accuracy of few parts in 10^{10} and in 10^9, respectively) were the last ingredients needed to take up the redefinition of the meter. The product of the measured frequency and the wavelength yields a new, definitive value for the speed of light. The new (and current) definition of the meter, accepted by the 17th Conference Generale des Poids et Mesures in 1983, was quite simple and elegant:

---

the **meter** *is the length of the path traveled by light in vacuum during a time interval of 1/299,792,458 of a second*

A consequence of this definition is that the speed of light is now a defined constant, not to be measured again. In subsequent years, measurement of other stabilized-laser systems added to the ways in which the meter could be realized. Furthermore, these experiments definitely demonstrated that, in order to obtain highly precise results, it is necessary to measure the frequency of light rather than its wavelength, marking the birth of optical frequency metrology.

Time-frequency and length-frequency equivalence principles are only two aspects of a more general trend in contemporary metrology, that of establishing measurement units based, rather than on artifacts, on atomic (quantum) standards or on fundamental constants [40]. These are invariant both on a practical scale and as far as can be measured in the laboratory. Let's deepen the significance of this process starting with the second and the meter.

Although the cesium frequency cannot presently be explicitly written in terms of fundamental constants because of the complexity of the atomic theory required, it is a quantum system that will have the stability associated with fundamental constants. The uncertainty in calculating this frequency is many orders of magnitude away from its measurement uncertainty. The Rydberg constant could be considered the natural fundamental constant-based unit of frequency. It is determined with a relative uncertainty of $6.6 \cdot 10^{-12}$, which is currently the limit at which an atomic frequency can be calculated from fundamental constants. The choice of the cesium definition was a good one in the sense that the technology, although superior to the alternative clocks of the day, still had much room for improvement, and the definition has endured to this day, during which time its practical realization has improved by five orders of magnitude. A clear example of the link between fundamental constants and the units is the adoption of the speed-of-light definition of the meter. The meter was originally defined as the length of a prototype meter bar intended to be $1/10,000,000$ of the length of a quadrant of Earth. By 1960, the development of interferometry allowed an atomic redefinition of the meter in terms of the wavelength of light from a specific source, the krypton lamp (the meter was defined as 1650763.73 wavelengths of the orange-red emission line in the spectrum of krypton-86 atom in vacuum). With the invention of the laser, length measurement by interferometry was radically improved and the krypton standard was not accurate enough. The meter definition could then have been revised using the wavelength of a specified stabilized laser. However, the progress in understanding the metrological importance of the speed of light, along with the progress in its accurate measurements, led to the change from defining the meter in terms of the wavelength of light from a specific source, to a fundamental constant-based definition in which the speed of light is a defined quantity. The choice of the speed-of-light definition over the use of a particular stabilized laser should ensure that this definition will endure, whereas the krypton definition lasted only 23 years. In practice, a number of *recommended radiations*, that is, frequencies of particular stabilized lasers, are published accompanying the definition. This means that to realize the meter there is no need to measure the distance that light travels in $1/299,792,458$ of a second by literally timing a light beam. One can, for example, continue to use a laser interferometer and measure the frequency of the laser used, or use a recommended stabilized laser and then use the relationship $c = \lambda \nu$ (as well as corrections for refractive index, if the measurement is not done in a vacuum). In other words, the realization is a method that implements the definition by using the known laws of physics; it allows the experimental production of a known quantity of the same kind as the one defined, but the method used may be dissimilar to the one in the definition. We close this discussion by mentioning another clarifying example, namely the definition of the volt. Electrical quantum metrology started in 1962 when Josephson predicted that in the presence of an applied microwave field, a direct superconducting tunnelling current could pass between superconductors separated by an insulating barrier. This current can only pass when the voltage V across the barrier satisfies the relationship

TABLE 1.1

List of the most significant terrestrial measurements of c.

Year	Investigator	Method	Value (km/s)
1849	Fizeau	Toothed wheel	$315,300$ [12]
1862	Foucault	Revolving mirror	$298,000 \pm 500$ [13]
1906	Rosa and Dorsey	EM constants	$299,710 \pm 30$[18]
1927	Michelson	Revolving mirror	$299,796 \pm 4$[13]
1928	Karolus and Mittelstaedt	Kerr cell	$299,778 \pm 20$[13]
1948	Essen and Gordon-Smith	Cavity resonator	$299,792 \pm 9$[18]
1949	Aslakson	Radar	$299,792.3 \pm 2.4$ [22]
1952	Bergstrand	Geodimeter	$299,793.1 \pm 0.2$ [25]
1958	Froome	Millimiter-wave interferometry	$299,792.50 \pm 0.10$ [27]
1965	Rank	Spectroscopy	$299,792.8 \pm 0.4$ [28]
1972	Evenson	Direct frequency and wavelength measurement of a laser	$299,792.4562(11)$ $299,792.4587(11)$[34]

Note: The two values in Evenson's measurement were due to the asymmetry in the krypton 6057-Angstrom line defining the meter. Except for the first two (in air), all the listed results refer to the value in vacuum.

$$2eV = nh\nu \qquad (1.10)$$

where e is the electron charge, h the Planck constant, ν the applied frequency, and n an integer. It was recognized that voltage standards could be based on this effect. A number of experiments found no corrections to expression 1.10 or dependence on material or experimental conditions at a level of up to parts in 10^{16}. In 1972, a number of countries used the Josephson effect to maintain the volt and agreed on an assigned value for $2e/h$ so that their voltages were in agreement. They are not necessarily the correct SI value; hence, the agreed-upon value is referred to as a *representation* of the volt. Again a frequency measurement played a crucial role.

1.3.2 Role of c in fundamental physics

Besides representing an essential pillar of frequency metrology, the parameter c is ubiquitous in contemporary physics, entering many contexts that are apparently disconnected from the notion of light itself. Our thoughts soon turn to the very famous second postulate of Special Relativity: "The velocity c of light in vacuum is the same in all inertial frames of reference in all directions and depend neither on the velocity of the source nor on the velocity of the observer". The theory of Special Relativity explores the consequences of this invariance of c with the assumption that the laws of physics can be written in the same form in all inertial frames (first postulate). Declared by Einstein in 1905, after being motivated by Maxwell's theory of electromagnetism and the lack of evidence for the luminiferous ether, the invariance of the speed of light and its isotropy has been consistently confirmed by many experiments over the years. Other experimentally verified implications of Special Relativity include length contraction (moving objects shorten), and time dilation (moving clocks run slower). The factor γ by which lengths contract and times dilate is known as the Lorentz factor and is given by $\gamma = 1/\sqrt{1 - \left(\frac{v}{c}\right)^2}$, where v is the speed of the object. Special Relativity also establishes that the energy of an object with rest mass m and speed v is given by $E = \gamma m c^2$. Since the γ factor approaches infinity as v approaches c, it would take an infinite amount of energy to accelerate an object with mass to the speed of light. The speed of light is therefore the upper limit for the speeds of objects with positive rest mass. Experimental Tests of General Relativity for the most part also verify Special Relativity, since the laws of the latter are included as part of the former via the principle of consistency. For a tutorial introduction to Special and General Relativity the reader may refer to [41], while an updated list of experimental verifications inferred from advanced frequency measurements will be given in Chapter 8.

As a clarification of what was just discussed, it is worth adding that, in principle, we should distinguish between the electromagnetism constant $c_{EM} = 1/\sqrt{\varepsilon_0 \mu_0}$, and the space-time constant c_{ST} appearing in the Lorentz transformation that is at the basis of the formulation of Special Relativity [42]. In general, for example, the celebrated equation unifying the concepts of energy and mass should be written in the form $E = mc_{ST}^2$. c_{EM} agrees with c_{ST} insofar as the mass of the photon is zero. If we were to show experimentally that the photon has non-zero mass, then the standard derivation of relativity from electromagnetism would have to be abandoned. Incidentally, extensions of quantum electrodynamics (QED) in which the photon has a mass have been considered [43]. In such a theory, the photon speed would depend on its frequency. No variation of the speed of light with frequency has been observed in rigorous testing, putting stringent limits on the mass of the photon [44]. The same is true for gravitational waves: their speed in vacuum c_{GW}, is equal to c_{ST} as long as we assume that general relativity is valid. Again, if we were able to formulate a theory with light massive gravitons, then the speed of propagation of gravity might be different from c_{ST}. Finally, the space-time-matter constant c_E, introduced by Einstein to describe coupling of gravity to matter, coincides by definition with c_{ST} only in the context of general relativity.

As mentioned, Einstein's relativity treats space and time as a unified structure known as space-time (with c relating the units of space and time) and requires that physical theories satisfy a special symmetry called Lorentz invariance, whose mathematical formulation contains precisely the parameter c. Lorentz invariance is an almost universal assumption for modern physical theories, such as quantum electrodynamics, quantum chromodynamics, and the Standard Model of particle physics. One consequence is that c is the speed at which all massless particles and waves, not only light, must travel. This result is constantly put to the test in different areas of experimental physics. In this respect, great emphasis was given to a high-energy physics experiment according to which beams of neutrinos, fired through the ground from Cern near Geneva to the Gran Sasso lab in Italy 450 miles (720 km) away, seemed to arrive sixty billionths of a second earlier than they should if travelling at the speed of light in a vacuum [45]. Subsequently, however, a discrepancy between the clocks at Cern and Gran Sasso was discovered to be at the root of the observed faster-than-light results. In the future, use of an optical fiber, as opposed to the GPS system used at the moment, should ensure a more accurate synchronization of the two clocks. This gives even more prominence, if any were needed, to the scope of time and frequency measurements.

The most striking feature of Einstein's relativity is undoubtedly the upper limit to velocity of any physical object set by c, albeit there are situations in which it may seem that matter, energy, or information travels at speeds greater than c. A first amazing example is the following. Think about how fast a shadow can move. If you project the shadow of your finger using a nearby lamp onto a distant wall and then wag your finger, the shadow will move much faster than your finger. If your finger moves parallel to the wall, the shadow's speed will be multiplied by a factor D/d where d is the distance from the lamp to your finger, and D is the distance from the lamp to the wall. If the wall is very far away, the movement of the shadow will be delayed because of the time it takes light to get there, but the shadow's speed is still increased by the same ratio. The speed of a shadow is therefore not restricted to be less than the speed of light. Unfortunately, the shadow is not a physical object and it is not possible to send information on a shadow. Also, certain quantum effects appear to be transmitted instantaneously and therefore faster than c. Among these we mention the celebrated Einstein-Podolsky-Rosen (EPR) paradox [46], the Hartman effect [47], and the Casimir effect [48]. It has been pointed out, however, that none of these effects can be used to send information.

Other examples come from the astrophysical/cosmological context. According to Hubble's law, two galaxies that are a distance D apart are moving away from each other at a speed HD, where H is Hubble's constant. This interpretation implies that two galaxies separated by a distance greater than c/H must be moving away from each other faster than the speed of light. Actually, the modern viewpoint describes this situation differently: general relativity considers the galaxies as being at rest relative to one another, while the space between them is expanding. In that sense, the galaxies are not moving away from each other faster than the speed of light; they are not moving away from each other at all! This change of viewpoint is not arbitrary; rather, it agrees with the different but very fruitful view of the universe that general relativity provides. So the distance between two objects can be increasing faster than light because of the expansion of the universe, but this does not mean, in fact, that their relative speed is faster than light.

It is worth stressing that all the experiments performed to date have confirmed that it is impossible for information or energy to travel faster than c. One simple general argument for this follows from the counter-intuitive implication of special relativity known as the relativity of simultaneity. If the spatial distance between two events A and B is larger than the time interval between them multiplied by c, then there are frames of reference in which A precedes B, others in which B precedes A, and others in which they are simultaneous. As a result, if something were travelling faster than c relative to an inertial frame of reference, it would be travelling backwards in time relative to another frame, and causality would be violated. In such a frame of reference, an *effect* could be observed before its *cause*. Such a violation of causality has never been recorded.

In a medium, light usually does not propagate at a speed equal to c; furthermore, different types of light wave will travel at different speeds. The speed at which the individual crests and troughs of a plane wave (a wave filling the whole space, with only one frequency) propagate is called the phase velocity v_p. So, while in *vacuum* we have $c = v_p = \lambda\nu$, in a medium we have $c/n(\nu) = v_p = \lambda\nu$ where $n(\nu)$ is the refractive index of the medium (in general it also depends on the intensity, polarization direction of propagation,...). In actual circumstances such idealized solutions do not arise. Even in the most monochromatic light source or the most sharply tuned radio transmitter or receiver, one deals with a finite spread of frequencies or wavelengths. Since the basic equations are linear, it is in principle an elementary matter to make the appropriate linear superposition of solutions with different frequencies where each monochromatic component has its own phase velocity. Consequently, there is a tendency for the original coherence to be lost and for the pulse to become distorted in shape. At the very least, we might expect it to propagate with a rather different velocity from, say, the average phase velocity of its component waves. The general case of a highly dispersive medium or a very sharp pulse with a wide spread of wave numbers is difficult to treat. But the propagation of a pulse which is not too broad in its wave-number spectrum, or a pulse in a medium for which the frequency depends weakly on wave number, can be handled in an approximate way. In this case, it can be shown that the transport of energy occurs with the group velocity

$$v_g = \left.\frac{d\omega}{dk}\right|_{k=k_0} \tag{1.11}$$

where $\omega(k) = ck/n(k)$ describes the dispersion of the material and k_0 is the center wavenumber of the packet. In general, however, the behavior of the wave packet is much more complicated and the group velocity above defined does not identify with the information velocity. In transparent materials, the refractive index generally is greater than 1, meaning that the phase velocity is less than c. In other materials, it is possible for the refractive index to become smaller than 1 for some frequencies; in some exotic materials it is even possible for the index of refraction to become negative. This reflects in turn on the value of the group velocity. Even in those cases where group velocities exceeding c are observed, it is still valid, according to causality, that it is impossible to transmit information faster than the speed of light in vacuum. Indeed, when the notion of front velocity is introduced and the principle of causality is accounted for, it can be rigorously shown in the frame of classical electrodynamics that information travels at the front velocity that is actually limited to c [49].

The finiteness of the speed of light has implications for the whole realm of sciences and technologies. In some cases, it is considered as a hindrance. For instance, being the upper limit of the speed with which signals can be sent, c provides a theoretical upper limit for the operating speed of microprocessors. In supercomputers, the speed of light imposes a limit on how quickly data can be distributed among processors. If a processor operates at 1 GHz, a signal can only travel a maximum of about 30 centimeters in a single cycle. Processors must therefore be placed close to each other to minimize communication latencies; this represents a trade-off with cooling needs. If clock frequencies continue to increase, the speed of light will eventually become a limiting factor for the internal design of single chips. In other cases, the finiteness of c turns out to be useful. For instance, the finite speed of light is important in astronomy. Due to the vast distances involved, it can take a very long time for light to travel from its source to Earth. For example, photographs taken today in the Hubble Ultra Deep Field capture images of the galaxies as they appeared 13 billion years ago, when the universe was less than a billion years old. The fact that more distant objects appear to be younger, due to the finite speed of light, allows astronomer to infer the evolution of stars, galaxies, and of the universe itself. Moreover, position measurements by GPS systems rely on the finiteness of c.

We close this section by observing that it is generally assumed that fundamental constants such as c have the same value throughout space-time, meaning that they do not depend on location and do not vary with time. However, it has been suggested in various theories that the speed of light may have changed over time. No conclusive evidence for such changes has been found, but this remains a crucial subject of ongoing metrological research [50].

As we will see during this book, and in particular in Chapter 8, advanced laser-based measurements in the frequency and time domains promise to give a new insight into many of the aforementioned issues.

1.4 In the end, time and light met up again: optical atomic clocks and outline of the book

As we have seen, the advent of the laser played a central role in the statement of the meter definition marking, in fact, the beginning of optical frequency metrology. In the following years, the laser became an invaluable source in many research fields. Today, it is the *true light* which enlightens every advanced frequency metrology experiment. Additionally, from the sixties to the present, three major developments were triggered by the laser in the field of fundamental research: ultra-high-resolution spectroscopy, the field of trapping/cooling of atoms, and the realization of optical frequency comb synthesizers based on femtosecond (fs) mode-locked lasers. In turn, these three discoveries have played a crucial role in the timekeeping story leading to the realization of the current optical atomic clocks. Although present-day cesium microwave frequency standards perform at an already remarkable level (fractional uncertainty below 1 part in 10^{15}), a new approach to timekeeping based on optical atomic transitions promises still greater improvements. According to the given definition of Q, by using optical ($\nu_0 \sim 10^{15}$ Hz) rather than microwave ($\nu_0 \sim 10^{10}$ Hz) frequencies, optical standards should be considerably more accurate. Also several key frequency shifts are fractionally much smaller in the optical domain and their investigation will be greatly accelerated by the much smaller instability of the optical standards. A projection of the fractional uncertainty achievable in the new era of optical atomic clocks is made in Figure 1.8 which displays some of the major milestones in the improvement of clocks over the past 400 years.

The potential advantages of optical atomic clocks were recognized in the early days of frequency standards. However, optical standards did not truly begin to experience these potential gains until the past decade, when the above three fields enjoyed an extraordinary growth. First, huge advances in laser cooling techniques made it possible to cool a variety of atoms and ions (including those with narrow clock transitions) to millikelvin temperatures and below. The use of laser-cooled atomic samples enabled, in turn, the extended interaction times required to observe a narrow transition linewidth. To resolve such narrow linewidths, probe lasers need to be spectrally pure. Recent improvements in laser stabilization based on environmentally isolated optical reference cavities have enabled laser linewidths at the subhertz level to be achieved [51]. Finally, and perhaps most critically, compact and reliable optical frequency comb synthesizers (OFCSs) for counting optical frequencies (linking them,

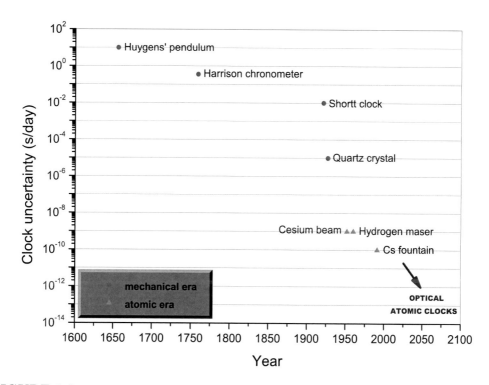

FIGURE 1.8
Major milestones in the improvement of clocks over the past 400 years, including the projected fractional uncertainty of next-generation optical atomic clocks. (Adapted from [5].)

phase coherently and in a single step, to other frequencies in the microwave and optical domains) became available replacing the old and cumbersome frequency chains [52, 53].

Being the most recent and sophisticated objects presented in this book, optical atomic clocks will be our guiding star along the whole treatment. Thus, just to point out all the relevant ingredients, a paradigmatic scheme is given in Figure 1.9 [5].

Schematically, the probe laser, whose frequency is pre-stabilized against an optical cavity, is used to excite transitions in a laser-cooled, trapped sample of atoms (either neutral or ions). A servo system uses the signal from the quantum absorbers to keep the probe laser frequency centered on resonance. Light is sent to the OFCS, which enables counting of the clock cycles. In this picture, as 5000 years ago, a light source (the laser plays here the role of Sun) interrogates an oscillatory phenomenon (the atomic resonance is now the equivalent of Earth's rotation). The abysmal difference lies in the much higher operation frequency (and in the much greater stability compared to Earth's motion), as well as in the resonant character of light-matter interaction. So, optical atomic clocks represent the happy end of the history of Light and Time.

Now, the time is approaching when optical frequency standards will have accuracies and stabilities superior to the best microwave cesium standards. Then it will be necessary to revisit the definition of the second. There are a number of candidate optical frequency standards, but at present no particular standard is clearly superior to the others. The time lag in adopting a new atomic standard mainly reflects the work that is necessary to ensure that one specifically selected system is indeed superior.

As illustrated in these first sections, the history of physics shows that, when the accuracy of measurements is improved, new physics may be discovered and explored. Throughout history, at several moments, the discovery or development of a new type of oscillator with

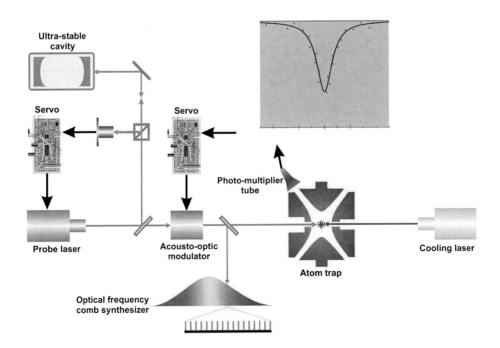

FIGURE 1.9
Schematic layout of an optical atomic clock.

improved performance has meant a huge step forward in our knowledge of physics or even coincided with a scientific revolution. For example, the advent of pendulum clocks provided experimental verification of Galilean laws of mechanics, while, more recently, observation of astronomical oscillators like binary pulsars confirmed many of the predictions of the General Relativity. In the same way, optical atomic clocks are expected to improve our knowledge in fundamental physics.

The emergence of the laser, moreover, made it essential to distinguish its *special* light from the more *incoherent* radiation emitted by hot bodies. This induced Roy Glauber to utilize the quantum theory to describe the properties of light and how these can be observed. His work laid the foundations for the field of research today called Quantum Optics, and earned him the Nobel Prize awarded in 2005 [54]. Quantum Optics is a very wide field and only those aspects that will intersect the main path of our book will be dealt with. In technical applications, the quantum effects are often very small. The field state is chosen so that it can be assigned well-defined phase and amplitude properties. In laboratory measurements, too, the uncertainty of quantum physics seldom sets the limit. But the uncertainty that nevertheless exists appears as a random variation in the observations. This *quantum noise* sets the ultimate limit for the precision of optical observations. In high-resolution frequency measurements, quantum amplifiers, and frequency standards, it is in the end only the quantum nature of light that sets a limit for how precise our apparatuses can be. Such ultimate limits have been explored in recent years by two main protagonists, John L. Hall and Theodor W. Hänsch. For their unceasing and illuminating research within the field of laser-based precision spectroscopy culminating with the realization of OFCSs, they received the other half of the Nobel Prize in Physics in 2005. It now seems possible, with the frequency comb technique, to make frequency measurements in the future with a precision approaching one part in 10^{18}. This will soon lead to actualize the introduction of a new, optical standard clock. What phenomena and measuring problems can take advantage of this extreme precision? Just to mention a few examples, more exact satellite-based navigation systems will become available and novel applications in telecommunication may emerge. Enormous benefits will also come out for navigation on long space journeys and for space-based telescope arrays that are looking for gravitational waves or making precision tests of the theory of relativity. Besides technological applications, this improved measurement precision may also be used in fundamental physical studies like those related to the antimatter-ordinary matter connection (spectroscopic studies of anti-hydrogen), to parity violation in chiral molecules, as well as to the search for possible changes in the constants of Nature over time. These and other fascinating issues will be discussed in Chapter 8.

At the end of this introductory chapter, we hope that, by grasping the concepts here proposed, the reader is in tune with the authors to better appreciate the logical organization of the book. In a sense, we are going to give, chapter by chapter, a detailed description of all the key elements on which the operation of optical atomic clock hinges. Here is the outline:

In **Chapter 2** we shall discuss the general basic features of harmonic oscillators and introduce the mathematical background for their characterization. An introductory overview of the most commonly used techniques for measuring and suppressing the phase noise in oscillators will be also given, together with a few elementary notions on feedback systems. The issue of accurate optical frequency synthesis will be also addressed in the last sections.

Chapter 3 is entirely devoted to passive resonators working both in the microwave and optical domain. Greater emphasis is given to the latter: besides traditional bulk resonators and their most updated developments, guided cavities based on optical fibers as well as micro-resonators relying on whispering gallery modes will be treated into a certain detail.

In **Chapter 4** we shall deal with continuous-wave (cw) coherent radiation sources.

After a short introduction on masers, we will focus on lasers by illustrating, firstly, some key aspects of their operation and a number of fundamental properties in their output. Then, a wide range of laser-based systems will be presented. A few clarifying examples of intensity and frequency stabilized laser sources will close the chapter.

In **Chapter 5** we shall provide a comprehensive treatment of high-resolution and high-sensitivity spectroscopic techniques for ultraprecise frequency measurements. Then, optical frequency standards utilizing either absorption cells or atomic/molecular beams will be described.

In **Chapter 6** the issue of time and frequency measurements with pulsed laser systems will be addressed. Starting with general mode-locking theory and mechanisms, advanced schemes for optical frequency comb synthesis (from mode-locked lasers) and relative stabilization will be presented. The extension of OFCSs into novel spectral regions, from the extreme ultraviolet (XUV) to the far-infrared (FIR), will be also discussed.

Chapter 7 is mainly devoted to microwave frequency standards. This category comprises high-quality crystal-based oscillators, high-performance hydrogen masers, and cutting-edge fountains based on cold alkali atoms. Being fundamental to the understanding of atomic standards, a propaedeutic review on trapping/cooling techniques for atoms, ions, and molecules is also provided. In the last part, a brief account on time and frequency dissemination (including optical frequency transfer) is given.

Chapter 8 starts with optical atomic clocks, ranging from the more established ones, based on single laser-cooled trapped ions, to the newest systems relying on neutral atoms trapped in an optical lattice. Then, based on the wide phenomenology explored thus far, possible research prospects in the field of time and frequency measurements are drawn for the next future.

2

Characterization and control of harmonic oscillators

> Human time does not rotate in a circle,
> but moves fast in a straight line. That is
> why man can not be happy, because
> happiness is the desire for repetition.
> *Milan Kundera - The Unbearable*
> *Lightness of Being*

The fish in the water is silent, the animals
on the earth are noisy, the bird in the air
is singing. But man has in himself the
silence of the sea, the noise of the earth
and the music of the air.
Rabindranath Tagore

2.1 The ideal harmonic oscillator

The purpose of this chapter is to acquaint the reader with the basic concepts and the mathematical tools that are necessary to address and better understand the contents which are at the heart of this book. From the previous chapter we learnt that oscillators are ubiquitous in the field of frequency metrology. We start by reviewing the most relevant features of harmonic (sinusoidal) oscillators which, although known from General Physics, deserve here a discussion devoted to our specific context. We focus on two archetypes: the pendulum and the RLC-series circuit. The former is the paradigm of mechanical oscillators, while the latter embodies the electrical ones. According to the specific property we are interested in, from time to time we will resort to one or the other system, but the conclusions will always be general.

For a point pendulum supported by a massless and inextensible cord of length l, the equation of motion is given by

$$ml^2\ddot{\theta} = -mgl\sin\theta - \beta\dot{\theta} \tag{2.1}$$

where m is the bob mass, g the local acceleration of gravity, θ the angle between the cord and the vertical, and β accounts for the overall friction (basically the resistance by the air and the escapement). For infinitesimal displacements, we replace $\sin\theta$ by θ and get the following second-order linear differential equation

$$\ddot{\theta} + \frac{\beta}{ml^2}\dot{\theta} + \frac{g}{l}\theta = 0 \tag{2.2}$$

In the case of the RLC series circuit, it is the mesh current I obey an equation of the same form

$$\ddot{I} + \frac{R}{L}\dot{I} + \frac{1}{LC}I = 0 \tag{2.3}$$

with the obvious meaning of symbols. Here, the resistor R provides dissipation and is thus responsible for damping, whereas the LC tank sets the oscillation frequency.

Another celebrated example, which actually falls into the category of mechanical oscillators, is offered by the Lorenz model of the atom. Predating the emergence of quantum mechanics, such a classical picture was the first attempt to explain atomic spectra. It rests on the idea that an electron of mass m and charge $-e$ is bound to the nucleus (charge $+e$) by a restoring force that is proportional to the displacement (Hooke's law). To account for the fact that an excited atom loses its energy by emitting electromagnetic radiation, a damping mechanism for the oscillation is also considered by including a viscous term (proportional to velocity) into the equation of motion. Therefore, the electron position turns out to be governed by the law

$$\ddot{x} + \frac{k}{m}x + \frac{\alpha}{m}\dot{x} = 0 \tag{2.4}$$

It is easily recognized (see Figure 2.1) that, with the appropriate identifications, Equations 2.2, 2.3, 2.4 are all of the form

$$\ddot{y} + 2\Gamma\dot{y} + \omega_0^2 y = 0 \tag{2.5}$$

A little more general equation of motion is obtained when a driving term is added to compensate for the slowing down of the oscillation

$$\ddot{y} + 2\Gamma\dot{y} + \omega_0^2 y = D\cos(\omega_D t) \tag{2.6}$$

The driving term may arise from the interaction with an electromagnetic monochromatic plane wave in the case of the Lorentz oscillator, or simply be the periodic push in a pendulum as well as the AC generator in the RLC circuit. Apart from the examples just mentioned, Equation 2.6 appears in a number of different systems ranging from solid-state turbulence to soliton dynamics, from Josephson junctions to phase-locked loops [55]. In order to find a

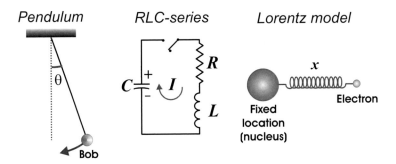

FIGURE 2.1

Pendulum, RLC-series circuit, and Lorentz oscillator as paradigmatic examples of harmonic oscillators. The following identifications return Equation 2.5 for each of the three cases. Pendulum case: $y \equiv \theta$, $\omega_0^2 \equiv g/l$, $2\Gamma \equiv \beta/(ml^2)$; RLC circuit: $y \equiv I$, $\omega_0^2 \equiv 1/(LC)$, $2\Gamma \equiv R/L$; Lorentz oscillator: $y \equiv x$, $\omega_0^2 \equiv k/m$, $2\Gamma \equiv \alpha/m$.

solution for it, first consider the homogeneous Equation 2.5. In the case $\omega_0^2 > \Gamma^2$, from the associated characteristic equation one finds the solution

$$y(t) = e^{-\Gamma t} \left(E e^{i\omega_0' t} + F e^{-i\omega_0' t} \right) \tag{2.7}$$

where $\omega_0' = \sqrt{\omega_0^2 - \Gamma^2}$ and the constants E and F are found by imposing the initial conditions $y(t = 0) = \xi$ and $\dot{y}(t = 0) = \eta$. By defining the quantities $A_1 = A \sin \varphi$ and $A_2 = A \cos \varphi$, Equation 2.7 can be re-written in the more convenient form

$$y(t) = A e^{-\Gamma t} \sin (\omega_0' t + \varphi) \equiv A e^{-\omega_0 g \cdot t} \sin \left(\omega_0 \sqrt{1 - g^2} \cdot t + \varphi \right) \tag{2.8}$$

where $y(t = 0) = A \sin \varphi = \xi$, $\dot{y}(t = 0) = A[-\Gamma \sin \varphi + \omega_0' \cos \varphi] = \eta$, and $g \equiv \Gamma/\omega_0$. Equation 2.8 describes the well-known case of a damped harmonic oscillator (see upper of Figure 2.2) which, in the limit $\Gamma \to 0$, reduces to

$$y(t) = A \sin (\omega_0 t + \varphi) \tag{2.9}$$

To illustrate some interesting properties in the frequency domain, let us take the Fourier transform of Equation 2.8 with $\xi = 0$

$$\hat{y}(\omega) = \int_0^\infty A e^{-\Gamma t} \sin \omega_0' t \; e^{-i\omega t} dt =$$

$$\frac{A}{2i} \int_0^\infty \left\{ e^{[i(\omega_0' - \omega) - \Gamma]t} - e^{[-i(\omega_0' + \omega) - \Gamma]t} \right\} dt =$$

$$\frac{A}{2i} \left[\frac{1}{\Gamma - i(\omega_0' - \omega)} - \frac{1}{\Gamma + i(\omega_0' + \omega)} \right] \simeq$$

$$\frac{\frac{A}{2i}}{\Gamma - i(\omega_0' - \omega)} \tag{2.10}$$

where the lower integration limit has been changed from $-\infty$ to 0 since $y(t) = 0$ for $t \le 0$, and the last equality is valid close to the resonance, that is for $\omega - \omega_0' \ll \omega_0'$. The response function of the oscillator is thus a Lorentzian profile

$$|\hat{y}(\omega)|^2 = \frac{A^2/4}{\Gamma^2 + (\omega_0' - \omega)^2} \tag{2.11}$$

with a full width at half maximum $FWHM = 2\Gamma$. If y is interpreted as the electron position in an atom emitting or absorbing radiation, it can be convenient to find the constant A by normalizing the spectrum 2.11 such that $\int_{-\infty}^{+\infty} |\hat{y}(\omega)|^2 d\omega = 1$. This returns

$$S(\omega) \equiv |\hat{y}(\omega)|^2 = \frac{1}{\pi} \frac{\Gamma}{\Gamma^2 + (\omega_0' - \omega)^2} \equiv \frac{1/(\pi \omega_0)}{g + (1/g)(\sqrt{1 - g^2} - \zeta)^2} \tag{2.12}$$

where $\zeta \equiv \omega/\omega_0$. Equation 2.12 is plotted in the lower frame of Figure 2.2 for two different values of g.

Note that, in the limit $\Gamma \to 0^+$, corresponding to the response curve of the ideal undamped harmonic oscillator (Equation 2.9), Equation 2.12 is one of the representations of the Dirac delta function $\delta(\omega_0)$. Now, the general solution of Equation 2.6 is found by simply adding Equation 2.8 to a particular solution which we will seek in the form of

$$y(t) = a \cdot \cos (\omega_D t + \psi) \tag{2.13}$$

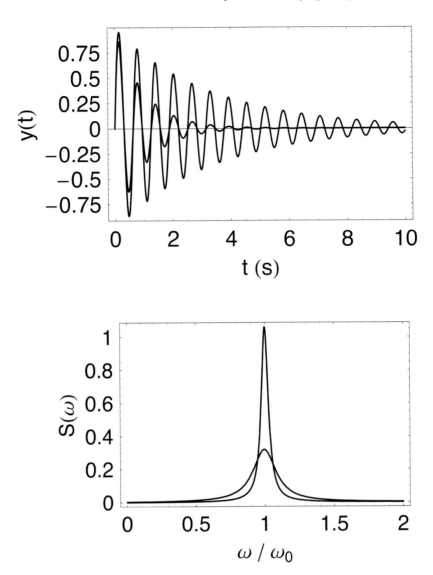

FIGURE 2.2
Representation of a damped harmonic oscillator (with $\omega_0 = 10$ Hz, $A = 1$, and $\varphi = 0$) in time and frequency domain for two different values of $g \equiv \Gamma/\omega_0$ (0.1 and 0.03), according to Equation 2.8 and Equation 2.12.

By substitution of Equation 2.13 into Equation 2.6 we obtain the following system of equations

$$\begin{cases} -a\omega_D^2\cos\psi - 2a\Gamma\omega_D\sin\psi + a\omega_0^2\cos\psi - D = 0 \\ a\omega_D^2\sin\psi - 2a\Gamma\omega_D\cos\psi - a\omega_0^2\sin\psi = 0 \end{cases} \tag{2.14}$$

whose solution yields

$$\psi = \arctan\left(\frac{2\Gamma\omega_D}{\omega_D^2 - \omega_0^2}\right) \tag{2.15}$$

$$a = \frac{-D}{\sqrt{4\Gamma^2\omega_D^2 + \left(\omega_D^2 - \omega_0^2\right)^2}} \tag{2.16}$$

Therefore, by introducing the phase $\psi = \vartheta + \frac{\pi}{2}$, the general solution of Equation 2.6 is finally obtained as

$$y(t) = Ae^{-\Gamma t}\sin\left(\sqrt{\omega_0^2 - \Gamma^2} \cdot t + \varphi\right)$$
$$+ \frac{D}{\sqrt{4\Gamma^2\omega_D^2 + \left(\omega_D^2 - \omega_0^2\right)^2}}\sin\left(\omega_D t + \vartheta\right) \tag{2.17}$$

Under steady-state conditions ($t \to \infty$), the oscillator output, no more damped, is described by

$$\begin{cases} y(\omega_D) = Y_0(\omega_D)\sin\left[\omega_D t + \Phi(\omega_D)\right] \\[2mm] Y_0(\omega_D) = \dfrac{D}{\sqrt{4\Gamma^2\omega_D^2 + \left(\omega_D^2 - \omega_0^2\right)^2}} \equiv \dfrac{D_0}{\sqrt{4g^2\iota^2 + (\iota^2 - 1)^2}} \\[4mm] \Phi(\omega_D) = \arctan\left(\dfrac{2\Gamma\omega_D}{\omega_D^2 - \omega_0^2}\right) - \dfrac{\pi}{2} \equiv \arctan\left(\dfrac{2g\iota}{\iota^2 - 1}\right) - \dfrac{\pi}{2} \end{cases} \tag{2.18}$$

where $\iota \equiv \omega_D/\omega_0$ and $D_0 \equiv D/\omega_0^2$. The oscillation amplitude $Y_0(\omega_D)$ is maximum for $\omega_D = \omega_0$; the corresponding phase is $\Phi(\omega_D = \omega_0) = 0$ (see Figure 2.3). This example makes clear the character of the so-called resonance phenomenon between an external driving source and an oscillator with its own characteristic frequency. An expression for the *FWHM* of such resonance curve can be given by finding an approximate solution for the equation $\left[Y_0^2(\omega_D = \omega_0)\right]/2 = Y_0^2(\omega_D)$. This provides

$$4\Gamma^2\omega_D^2 + \left(\omega_D^2 - \omega_0^2\right)^2 = 8\Gamma^2\omega_0^2 \tag{2.19}$$

which, putting $s = \omega_D^2 - \omega_0^2$, is equivalent to a quadratic algebraic equation in s

$$s^2 + 4\Gamma^2 s - 4\Gamma^2\omega_0^2 = 0 \tag{2.20}$$

which returns

$$\omega_{D,\pm}^2 = \omega_0^2 - 2\Gamma^2 \pm 2\Gamma\sqrt{\omega_0^2 + \Gamma^2} \simeq \omega_0^2 \pm 2\Gamma\omega_0 \tag{2.21}$$

where the last approximate equality holds for high-quality oscillators ($\omega_0^2 \gg \Gamma^2$). Then, we can write

$$\omega_{D,\pm} = \omega_0\sqrt{1 \pm \frac{2\Gamma}{\omega_0}} \simeq \omega_0\left(1 \pm \frac{\Gamma}{\omega_0} \pm \mathcal{O}\left[\frac{\Gamma^2}{\omega_0^2}\right]\right) \tag{2.22}$$

from which the *FWHM* is calculated as

$$FWHM \equiv \omega_{D,+} - \omega_{D,-} = 2\Gamma \tag{2.23}$$

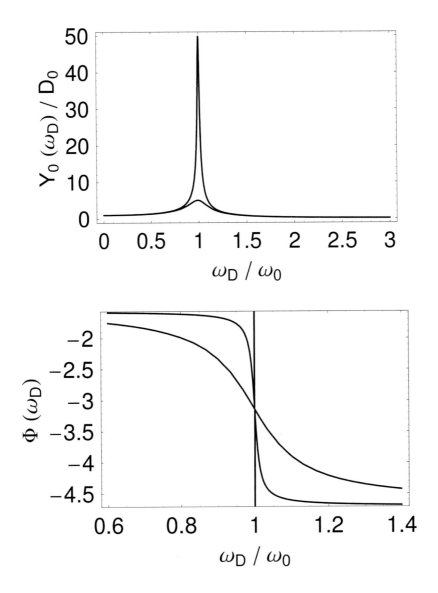

FIGURE 2.3
Amplitude and phase of a driven oscillator for two different values of g (0.1 and 0.01)
according to Equation 2.18.

In order to gain more physical insight, in agreement with the intuitive vision introduced
for the pendulum in the previous chapter, we now define the Q factor of the resonance as

$$Q = 2\pi \frac{\text{max. energy stored in the oscillator at } \omega_0}{\text{energy lost per cycle at } \omega_0} \equiv 2\pi \frac{E(\omega_0)}{W(\omega_0)} \qquad (2.24)$$

In order to evaluate this formula, let us first observe that, at resonance, Equation 2.18 gives

$$y^2(t, \omega_D = \omega_0) = \left(\frac{D}{2\Gamma\omega_0}\right)^2 \sin^2(\omega_0 t) \qquad (2.25)$$

Next, let us exploit the analogy with the RLC circuit. By identification of y with I, we get

$$E(\omega_0) = \frac{1}{2}LI_{max}^2 = \frac{1}{2}L\left(\frac{D}{2\Gamma\omega_0}\right)^2 \qquad (2.26)$$

and

$$W(\omega_0) = \int_0^{2\pi/\omega_0} RI^2(t)dt = R\left(\frac{D}{2\Gamma\omega_0}\right)^2 \int_0^{2\pi/\omega_0} \sin^2(\omega_0 t)dt$$

$$= \frac{\pi R}{\omega_0}\left(\frac{D}{2\Gamma\omega_0}\right)^2 \qquad (2.27)$$

Combining Equations 2.24, 2.26, and 2.27 we finally obtain

$$Q = \frac{\omega_0}{R/L} = \frac{\omega_0}{2\Gamma} \simeq \frac{\omega_0}{FWHM} \qquad (2.28)$$

This derivation relates the two directly observable quantities ω_0 and $FWHM$ to the inner physical meaning of the Q factor. Moreover, from the third of Equations 2.18 and the formula $\arctan(x) = \frac{\pi}{2} - \arctan\left(\frac{1}{x}\right)$ (valid for $x > 0$) we get

$$\Phi(\omega_D) = -\arctan\left(\frac{\omega_D^2 - \omega_0^2}{2\Gamma\omega_D}\right) \qquad (2.29)$$

that, for very high Q (which is equivalent to say close to the resonance), can be expanded as follows

$$\Phi(\omega_D) \simeq \frac{\omega_0^2 - \omega_D^2}{2\Gamma\omega_D} + \frac{1}{3}\left(\frac{\omega_D^2 - \omega_0^2}{2\Gamma\omega_D}\right)^3 + \dots \qquad (2.30)$$

Retaining only the first term in the Taylor expansion, we have

$$\left.\frac{d\Phi(\omega_D)}{d\omega_D}\right|_{\omega_D=\omega_0} \simeq -\frac{1}{\Gamma} = -\frac{2Q}{\omega_0} \qquad (2.31)$$

which suggests that a very rapid phase change is attainable in the vicinity of the resonance frequency of high-Q oscillators. It is left as an exercise to show that, for very high Q, the inflection points of $[Y_0(\omega_D)/D]^2$ are given by $\omega_D^{ip} = \omega_0 \pm \frac{\Gamma}{\sqrt{3}}$ and that, in the vicinity of them, we can write

$$[Y_0(\omega_D)]^2 \simeq D^2\left\{S\left(\omega_0 \pm \frac{\Gamma}{\sqrt{3}}\right) \mp \frac{3\sqrt{3}}{4}\frac{Q^3}{\omega_0^5}\left[\omega_D - \left(\omega_0 \pm \frac{\Gamma}{\sqrt{3}}\right)\right]\right\} \qquad (2.32)$$

This means that, close to the inflection points, the response function of the oscillator acts as an extremely sensitive frequency-to-amplitude converter (see also Section 2.7.2).

2.1.1 Synchronization in coupled oscillators

Many physical situations can create coupling between two or more oscillatory systems. A classical example in electronics is represented by a pair of LC resonant circuits coupled by a mutual inductance, while a paradigmatic mechanical system is that consisting of two spring-and-mass oscillators coupled by a third spring. In all such situations, the frequency of one or both oscillators will be shifted and energy can be transferred from one to the other. In order to introduce the notion of synchronization between two coupled oscillators, we exploit here the analogy with the pendulum. As a matter of fact, the earliest accounts on synchronization are by the Dutch researcher Christiaan Huygens [56]. He studied the motion of two identical clocks (two pendulums of almost same time period) suspended from the same wooden beam. He observed that the motion of the two pendulums in opposite directions was very much in agreement and that the rhythm was maintained without getting spoilt. Even when this rhythmic motion was disturbed by some external means, the pendulums readjusted in a short time. This is credited to the phenomenon of synchronization. He attributed this synchronous motion to the interaction of the two pendulums through the wooden beam supporting them. For a long time, synchronization has also been known to occur in living systems. Examples of such systems abound. Synchronous flashing of fireflies, singing crickets, cardiac pacemakers, and firing neurons are some of them. In recent years, the idea of synchronization has also been extended to systems which are not oscillatory. Synchronization of systems showing aperiodic behavior, such as chaotic systems, is one of the new fields of study. In order to derive some general basic properties of the synchronized behavior of two oscillators, in the following we discuss precisely the phenomenon discovered by Huygens. Let us start with two identical pendulums which interact mutually. Physically, the interaction is introduced by suspending them from a common support. Mathematically, this corresponds to the two coupled equations

$$\begin{cases} \ddot{x} + \omega_0^2 x = \Gamma\left(\dot{y} - \dot{x}\right) \\ \ddot{y} + \omega_0^2 y = \Gamma\left(\dot{x} - \dot{y}\right) \end{cases} \tag{2.33}$$

By defining the error variable $e\left(t\right) = x\left(t\right) - y\left(t\right)$ and subtracting the two equations from each other, we get

$$\ddot{e} + \omega_0^2 e + 2\Gamma\dot{e} = 0 \tag{2.34}$$

Being identical to Equation 2.5, we already have a solution for the above equation. It is given by

$$e\left(t\right) = e_0 e^{-\Gamma t}\sin\left(\sqrt{\omega_0^2 - \Gamma^2} \cdot t + \phi_e\right) \tag{2.35}$$

where e_0 and ϕ_e are determined by the initial conditions on $e\left(t\right)$ (and hence on $x\left(t\right)$ and $y\left(t\right)$). Therefore, for positive Γ, the error must go to zero asymptotically regardless of the initial conditions. This means that, for sufficiently long times, Equation 2.33 reduces to

$$\begin{cases} \ddot{x} + \omega_0^2 x = 0 \\ x\left(t\right) = y\left(t\right) \end{cases} \tag{2.36}$$

which yields $x\left(t\right) = y\left(t\right) = B\sin\left(\omega_0 t + \psi\right)$. The outputs from the two oscillators are coincident in amplitude, frequency and phase. A similar behavior may also arise for two non-identical oscillators

$$\begin{cases} \ddot{x} + \omega_x^2 x = \Gamma\left(\dot{y} - \dot{x}\right) \\ \ddot{y} + \omega_y^2 y = \Gamma\left(\dot{x} - \dot{y}\right) \end{cases} \tag{2.37}$$

provided that the detuning $\delta\omega = \omega_x - \omega_y$ is small in comparison to the coupling Γ. Defining two new variables $u = \dot{x}$ and $w = \dot{y}$, from Equation 2.37 we obtain the first-order system

$$
\begin{cases}
\dot{x} = u \\
\dot{y} = w \\
\dot{u} = -\omega_x^2 x - \Gamma u + \Gamma w \\
\dot{w} = -\omega_y^2 x + \Gamma u - \Gamma w
\end{cases}
\tag{2.38}
$$

which can be expressed in the following matrix form

$$
\begin{pmatrix} \dot{x} \\ \dot{y} \\ \dot{u} \\ \dot{w} \end{pmatrix} = \begin{pmatrix} 0 & 0 & 1 & 0 \\ 0 & 0 & 0 & 1 \\ \omega_x^2 & 0 & -\Gamma & \Gamma \\ 0 & -\omega_y^2 & \Gamma & -\Gamma \end{pmatrix} \begin{pmatrix} x \\ y \\ u \\ w \end{pmatrix} \equiv \mathbb{A} \begin{pmatrix} x \\ y \\ u \\ w \end{pmatrix}
\tag{2.39}
$$

The secular equation $\det(\mathbb{A} - \lambda\mathbb{I}) = 0$ provides the eigenvalues $(\lambda_1, \lambda_2, \lambda_3, \lambda_4)$ and the corresponding eigenvectors (z_1, z_2, z_3, z_4), so that the general solution is given by

$$
\begin{pmatrix} x \\ y \\ u \\ w \end{pmatrix} = c_1 z_1 e^{\lambda_1 t} + c_2 z_2 e^{\lambda_2 t} + c_3 z_3 e^{\lambda_3 t} + c_4 z_4 e^{\lambda_4 t}
\tag{2.40}
$$

where the constants c_i are determined by the initial conditions $x(t=0)$, $y(0)$, $u(0)$, and $w(0)$. The general analytical expression of Equation 2.40 is rather involved, but a numerical solution can be found for any given choice of the initial conditions and of the system parameters. Just as an example, for $\omega_x = 10$ Hz, $\omega_y = 10.1$ Hz, and $\Gamma = 0.5$ Hz, we obtain solutions of the form

$$
\begin{cases}
x(t) = 2\,|A_1|\,e^{-\alpha_1 t} \cos\left[\omega_1 t + \arg(A_1)\right] \\
\quad + 2\,|A_2|\,e^{-\alpha_2 t} \cos\left[\omega_2 t + \arg(A_2)\right] \\
\\
y(t) = 2\,|B_1|\,e^{-\alpha_1 t} \cos\left[\omega_1 t + \arg(B_1)\right] \\
\quad + 2\,|B_2|\,e^{-\alpha_2 t} \cos\left[\omega_2 t + \arg(B_2)\right]
\end{cases}
\tag{2.41}
$$

with $\omega_1 = 10.037$ Hz, $\omega_2 = 10.050$ Hz, $\alpha_1 = 0.495$ Hz, and $\alpha_2 = 0.005$ Hz. As usual, A_1, A_2, B_1, and B_2 are determined by the initial conditions. For instance, the set of conditions $x(0) = -0.5$, $y(0) = 1$, $u(0) = 0.5$, and $w(0) = 1$ yields $|A_1| = 0.383$, $|A_2| = 0.136$, $|B_1| = 0.380$, $|B_2| = 0.136$, $\arg(A_1) = 0.115$, $\arg(A_2) = 0.282$, $\arg(B_1) = -0.084$, and $\arg(B_2) = 0.480$. Such numerical results are summarized in Figure 2.4, where Equations 2.41 are plotted over two consecutive timescales.

Since $\alpha_1 \gg \alpha_2$, after a short transient, Equations 2.41 reduce to

$$
\begin{cases}
x(t) = 2\,|A_2|\,e^{-\alpha_2 t} \cos\left[\omega_2 t - \arg(A_2)\right] \\
y(t) = 2\,|B_2|\,e^{-\alpha_2 t} \cos\left[\omega_2 t - \arg(B_2)\right]
\end{cases}
\tag{2.42}
$$

The phenomenon of two coupled oscillators with different natural frequencies beginning to oscillate at a common frequency owing to coupling is called frequency locking, and this common frequency of oscillation is called locking frequency. In our example, this frequency turns out to be $\omega_2 = 10.050$ Hz, that is the average of ω_x and ω_y, correct to three decimal places. Secondly, the phase difference of the two oscillators settles to a constant value different from zero. This phenomenon is termed as phase locking. The described effect can be understood on the basis of the following argument. Coupling between the two oscillators tries to make their phases equal while detuning tries to drag the phases apart. Hence, the

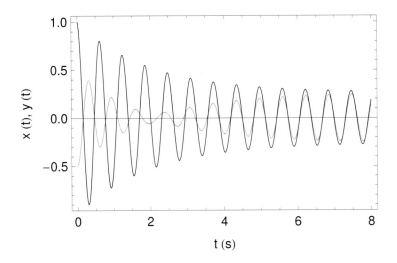

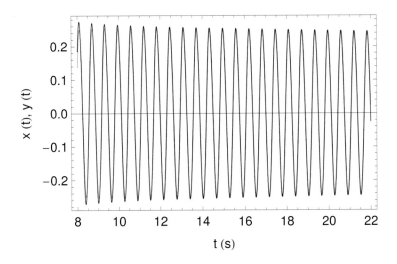

FIGURE 2.4
Oscillation amplitudes for the two coupled oscillators discussed in the text according to Equation 2.41. Two consecutive timescales are displayed.

effects of coupling and detuning are counteractive. So, we get two qualitatively different situations based on the relative strengths of coupling and detuning. When the detuning is small in comparison to the coupling strength, the oscillators settle into a common frequency and a stable relationship between the phases of the two oscillators is established. We then call the two oscillators synchronized. For relatively larger values of detuning, the effect of the coupling is not good enough to force a relation between the phases of the two oscillators. This leads to loss of synchrony.

2.1.2 Beating two oscillators

Suppose now that the outputs (here written for convenience in complex notation) from two independent oscillators $E_1 = E_{01}e^{i(\omega_1 t + \varphi_1)}$ and $E_2 = E_{02}e^{i(\omega_2 t + \varphi_2)}$ superimpose at some point P in the space. Then we have

$$E(P) = E_1 + E_2 = e^{i(\Omega t + \Phi)}\left[Ae^{i(\Delta\Omega \cdot t + \Delta\Phi)} + Be^{-i(\Delta\Omega \cdot t + \Delta\Phi)}\right] \quad (2.43)$$

where $\Omega = (\omega_1 + \omega_2)/2$, $\Phi = (\varphi_1 + \varphi_2)/2$, $\Delta\Omega = (\omega_1 - \omega_2)/2$, and $\Delta\Phi = (\varphi_1 - \varphi_2)/2$. An interesting situation arises when $\Delta\Omega \ll \Omega$. In that case the total amplitude is characterized by a fast oscillation at Ω whose amplitude is modulated at the slow frequency $\Delta\Omega$. Such effect is well known in acoustics, where the term beat is used to describe an interference between two sounds of slightly different frequencies, perceived as periodic variations in volume whose rate is the difference between the two frequencies (see Figure 2.5). This can be analytically seen in the particular case $A = B$, when Equation 2.43 simplifies to

$$\Re[E] = 2A\cos(\Delta\Omega \cdot t + \Delta\Phi)\cos(\Omega t + \Phi) \quad (2.44)$$

If we are measuring the intensity rather than the amplitude, from Equation 2.43 we obtain

$$I(P) = |E|^2 = A^2 + B^2 + 2AB\cos[(\omega_1 - \omega_2)t + (\varphi_1 - \varphi_2)] \quad (2.45)$$

Dropping the DC term, as already shown, the square modulus of the Fourier transform of the signal 2.45 is the Dirac delta function $\delta(\omega_1 - \omega_2)$. However, for real oscillators perturbed by noise processes, the difference $\varphi_1 - \varphi_2$ fluctuates in time causing a spread over a frequency range around $\omega_1 - \omega_2$. This aspect will be taken up later.

Later, we will discover that an important application of such beat-note phenomenon is in frequency metrology, where, for example, one can measure the frequency of some laser by recording its beat note with a close-by optical signal of known frequency. In this scheme, the two light beams with different optical frequencies are superimposed on a photodetector

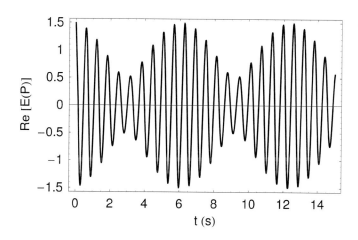

FIGURE 2.5
Beat-note phenomenon between two oscillators obtained by taking the real part of Equation 2.43. The following values are used in the simulation: $\omega_1 = 10$ Hz, $\omega_2 = 11$ Hz, $\varphi_1 = 0$, $\varphi_2 = 0.2$, $A = 1$, and $B = 0.5$.

measuring the optical intensity. As a fast photodetector can have a bandwidth of tens of gigahertz (or even higher), optical frequency differences of this order of magnitude can be measured, e.g., by analyzing the photodetector output with an electronic frequency counter or a spectrum analyzer.

2.2 Self-sustained oscillators

In this section we adopt the language of electronics to show that an effective way of introducing a forcing term in a damped oscillator is to derive the driving source from the oscillator output itself in a positive (regenerative) feedback scheme. We anticipate that this is exactly the working principle of the laser. The basic LC oscillator tank circuit is shown in the left frame of Figure 2.6. The capacitor is charged up to the DC supply voltage V by putting the switch in position A. When the capacitor is fully charged, the switch changes to position B. The charged capacitor is now connected in parallel across the inductive coil L through which it begins to discharge itself. The voltage across C starts falling as the current through the coil begins to rise. This rising current sets up an electromagnetic field around the coil which resists this flow of current. When the capacitor is completely discharged, the energy that was originally stored in it as an electrostatic field is now stored in the inductive coil as an electromagnetic field around the windings. As there is no external voltage in the circuit to maintain the current within the coil, it starts to fall as the electromagnetic field begins to collapse. A back electromotive force is induced in the coil keeping the current flowing in the original direction. This current now charges up the capacitor with the opposite polarity to its original charge. C continues to charge up until the current reduces to zero and the electromagnetic field of the coil has collapsed completely. The energy originally introduced into the circuit through the switch has been returned to the capacitor which again has an electrostatic voltage potential across it, although it is now of the opposite polarity. The capacitor now starts to discharge again back through the coil and the whole process is repeated. The polarity of the voltage changes as the energy is passed back and forth between the capacitor and inductor producing an AC type sinusoidal voltage and current waveform. This forms the basis of an LC oscillator tank circuit and, theoretically, the oscillatory action (at frequency $\omega_0 = 1/\sqrt{LC}$) would continue indefinitely. However, in a practical LC circuit, every time energy is transferred from C to L or from L to C, losses occur basically due to the resistance of the inductor coils and in the dielectric of the capacitor. All the loss sources can be lumped into a resistor R, which brings us back to the RLC-series circuit studied above. As a consequence, the oscillation in the circuit steadily decreases until it dies away completely and the process stops. To keep the oscillations going, we have to replace exactly the amount of energy lost during each cycle. The simplest way of doing this is to take part of the output from the LC tank circuit, amplify it, and then feed it back into the LC circuit again. This process can be achieved using a voltage amplifier like an operational amplifier, FET, or bipolar transistor as its active device. To produce a constant-amplitude oscillation, the level of the energy fed back to the LC network must be accurately controlled. In other words, there must be some form of automatic amplitude or gain control when the amplitude tries to vary from a reference voltage either up or down. Intuitively, a stable oscillation is maintained if the overall gain of the circuit is equal to one. Any less, the oscillations will not start or die away to zero; any more, the oscillations will occur but the amplitude will become clipped by the supply rails causing distortion. Consider the circuit in the right frame of Figure 2.6, where a bipolar transistor is used as the amplifier with the tuned LC tank circuit acting as the collector load. Another coil L_2, whose electromagnetic field is mutually

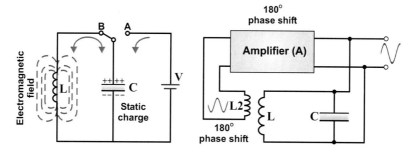

FIGURE 2.6
LC-tank circuit without and with positive (regenerative) feedback.

coupled with that of coil L, is connected between the base and the emitter of the transistor. The changing current flowing in one coil circuit generates, by electromagnetic induction, a potential voltage in the other. In this way, as the oscillations occur in the tuned circuit, electromagnetic energy is transferred from coil L to coil L_2 and a voltage of the same frequency as that in the tuned circuit is applied between the base and emitter of the transistor. This provides the necessary automatic feedback voltage to the amplifying transistor. Also, the amount of feedback can be increased or decreased by altering the coupling between the two coils L and L_2. It is worth pointing out that, when the circuit is oscillating at $\omega_0 = 1/\sqrt{LC}$, its impedance is resistive and the collector and base voltages are 180° out of phase. On the other hand, as dictated by Equation 2.18, at resonance, the voltage applied to the tuned circuit must be in-phase with the oscillations occurring in it. Therefore, we must introduce an additional 180° phase shift into the feedback path between the collector and the base. This is achieved by winding the coil of L_2 in the correct direction relative to coil L or by connecting a phase shift network between the output and input of the amplifier.

The instructive LC example has been used to introduce the general theory of oscillators with positive feedback [57]. The basic scheme (Figure 2.7) is a loop in which the gain A of the sustaining amplifier compensates for the loss $\beta(\omega)$ of the resonator at a given angular frequency ω_0. The condition for the oscillation to be stationary is calculated considering first the open loop (i.e., in the absence of feedback). In this case we have

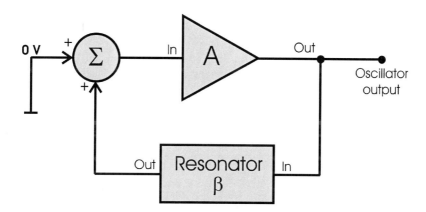

FIGURE 2.7
General scheme of oscillator with positive feedback: if $\omega = \omega_0$, a period is reproduced after a round trip, when $\omega \neq \omega_0$ each round trip attenuates the signal.

$$V_{out} = A \cdot V_{in} \tag{2.46}$$

Then, feedback is allowed and the output voltage re-calculated as

$$V_{out} = A \cdot (V_{in} - \beta V_{out}) \tag{2.47}$$

The closed-loop gain is thus given by

$$G = \frac{V_{out}}{V_{in}} = \frac{A}{1 + A\beta} \tag{2.48}$$

which tells us that $G = \infty$ if $-A\beta = 1$. This means that we have a finite output voltage with zero input, that is a sinusoidal oscillator. The condition $-A\beta = 1$, known as Barkhausen condition, is equivalent to

$$\begin{cases} |A\beta(\omega)| = 1 \\ \arg[-A\beta(\omega)] = 0 \end{cases} \tag{2.49}$$

The unused input (0 V) in Figure 2.7 serves to set the initial condition that triggers the oscillation, and to introduce noise in the loop. It is often convenient to use a constant-gain amplifier (A is independent of frequency), and a bandpass filter as $\beta = \beta(\omega)$ in the feedback path. Some small frequency dependence of the amplifier gain, which is always present in real-world amplifier, can be moved from A to $\beta = \beta(\omega)$. The model of Figure 2.7 is quite general and applies to a variety of systems (electrical, mechanical, lasers...), albeit a little effort may be necessary to identify A and β. Oscillation starts from noise or from the switch-on transient. In the spectrum of such random signal, only a small energy is initially contained at ω_0. For the oscillation to grow up to a desired amplitude, it is necessary that $|A\beta(\omega)| > 1$ at $\omega = \omega_0$ for small signals. In such a condition, oscillation at the frequency ω_0 that derives from $\arg[-A\beta(\omega)] = 0$ rises exponentially. As the oscillation amplitude approaches the desired value, an amplitude control (not shown in the figure) reduces the loop gain, so that it reaches the stationary condition $A\beta(\omega) = 1$. The amplitude can be stabilized by an external automatic gain control, or by the large signal saturation of the amplifier. The latter effect is shown in Figure 2.8: when the input amplitude exceeds the saturation level, the output signal is clipped. In summary, we stress that in real-world oscillators

1. it is necessary that $|A\beta(\omega)| > 1$ for small signals,

2. the condition $|A\beta(\omega)| = 1$ results from large-signal gain saturation,

3. the oscillation frequency is determined only by the phase condition $\arg[-A\beta(\omega)] = 0$.

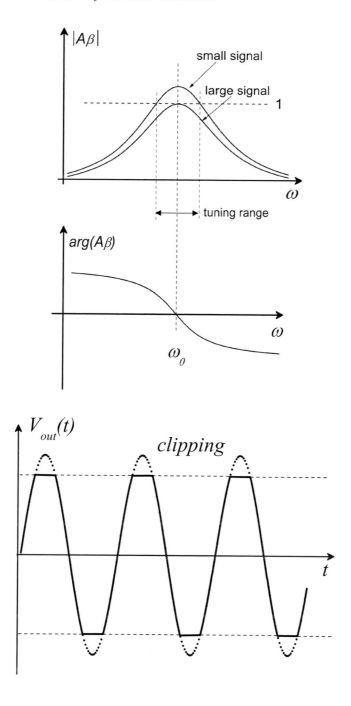

FIGURE 2.8
Some features in the onset of oscillation are illustrated for real-world oscillators with positive feedback. (Adapted from [57].)

We have now all the ingredients needed to analyze into more detail one of the most effective realizations of the LC oscillator, namely the Colpitts oscillator [58]. Such a scheme is shown in Figure 2.9 together with the linear model of the circuit making use of an inverting amplifier with large-signal voltage gain A_v and output impedance R_0. The open loop gain is

$$A \equiv \frac{V_{out}}{V_{in}} = \frac{V_{23}}{V_{13}} = A_v \frac{Z}{Z + R_0} \tag{2.50}$$

where

$$\frac{1}{Z} = \frac{1}{Z_2} + \frac{1}{Z_1 + Z_3} \tag{2.51}$$

while the feedback fraction is given by

$$\beta \equiv \frac{V_f}{V_{out}} = \frac{V_f}{V_{23}} = \frac{Z_1}{Z_1 + Z_3} \tag{2.52}$$

Combining Equations 2.50, 2.51, 2.52, we get

$$A\beta = \frac{A_v Z_1 Z_2}{Z_2 (Z_1 + Z_3) + R_0 (Z_1 + Z_2 + Z_3)} \tag{2.53}$$

Then, the resonance frequency is found by imposing the condition $Z_1 + Z_2 + Z_3 = 0$, which provides

$$\omega_0 = \sqrt{\frac{1}{L} \frac{C_1 + C_2}{C_1 C_2}} \tag{2.54}$$

that, in turn, implies

$$A\beta = \frac{A_v Z_1 Z_2}{Z_2 (Z_1 + Z_3)} = \frac{-A_v Z_1}{Z_2} \tag{2.55}$$

Finally, we express the Barkhausen criterion

$$-A\beta = A_v \frac{C_2}{C_1} = 1 \tag{2.56}$$

which, for a given value of A_v, suggests the sizing of C_1 and C_2.

Other useful schemes of electrical oscillators will be mentioned in Chapter 7 when the quartz frequency reference will be extensively studied.

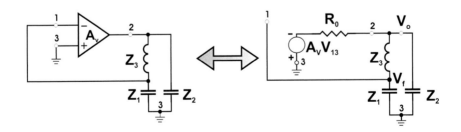

FIGURE 2.9
Scheme of Colpitts oscillator.

2.3 The noisy oscillator

Now let us come to non-ideal oscillators. The frequency and amplitude of even the most advanced oscillators are not really constant in time, but fluctuate due to several factors. These unwanted fluctuations are often referred to as noise or jitter. The noise term, originated in acoustics, is used to name any physical variable that fluctuates over time in an irregular and unpredictable way, as opposed to periodic oscillatory behaviors (sounds in acoustics) for which the initial conditions can be utilized to predict in deterministic manner the future state and which are generically referred to as signals. In most cases, noise is generated by spontaneous fluctuations of microscopic quantities, often related to thermal agitation in the system. For frequency standards one deals in general with the best available oscillators where the quasi-perfect sinusoidal signal is modelled as

$$V(t) = V_0 \left[1 + \alpha(t)\right] \cos\left[2\pi\nu_0 t + \varphi(t)\right] \qquad (2.57)$$

where $\nu_0 = \omega_0/(2\pi)$ is the carrier frequency; the random variables $\alpha(t)$ (dimensionless) and $\varphi(t)$ (having the units of radians) are the fractional amplitude noise and phase noise, respectively. Obviously, we assume $\alpha \ll 1$ and $\varphi \ll 1$, and that the expectation value of the amplitude and frequency are V_0 and ν_0, respectively. In the above equation it is assumed that phase and amplitude fluctuations are orthogonal meaning that no amplitude fluctuations are transferred to phase fluctuations and vice versa. Since, of necessity, all practical oscillators inherently possess an amplitude-limiting mechanism of some kind, amplitude fluctuations are greatly attenuated and phase noise generally dominates. In addition, affecting timing, phase noise is far more important and is first analyzed. Then, a brief treatment of amplitude noise is also given.

2.4 Phase noise

The output-phase performance of real-world oscillators can be characterized by three main gauges: accuracy, stability, and reproducibility (see also Figure 2.10):

- In general, accuracy is the extent to which a given measurement, or the average of a set of measurements for one sample, agrees with the definition of the quantity being measured: it is the degree of *correctness* of a quantity. Thus, in the specific context of frequency standards, the **accuracy** is the capability of an oscillator to provide a frequency that is known in terms of the accepted definition of the second. In short, the frequency accuracy of an oscillator is the offset from the specified target frequency.

- **Stability** is a measure of how much the frequency of the oscillator fluctuates over some period of time and, as we shall see in a short while, is usually characterized in terms of the two-sample Allan variance. In practice, stability is the property of an oscillator to resist, over time, changes in its rate as a function of parameters such as temperature, vibration, and the like. A high-stability oscillator may not necessarily be an accurate one. In a sense, stability can be considered a particular case of precision that is the extent to which a given set of measurements of one sample agrees with the mean of the set.

- **Reproducibility** is the ability of a single frequency standard to produce the same frequency, without adjustment, each time it is put into operation.

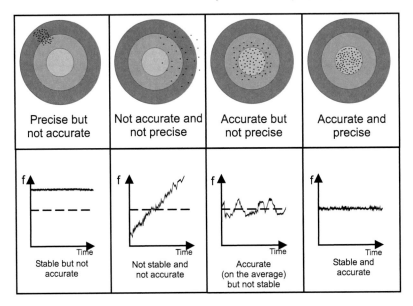

FIGURE 2.10
Accuracy and stability in the classical analogy between the shot of the marksman and the performance of an oscillator. (Adapted from [59].)

The discussion that follows relates to frequency stability [60, 61, 62, 63, 64]. Before getting involved in the analysis of phase and amplitude noise in oscillators, a small digression is needed to introduce some mathematical tools that are useful for statistical analysis. Then, in accordance with these tools, fundamental types of noise will be described.

2.4.1   Review of mathematical tools

Any fluctuating signal $B(t)$ can be decomposed into a purely fluctuating contribution $b(t)$ and a mean (or expectation) value defined as the time average

$$\overline{B(t)} = \lim_{T \to \infty} \frac{1}{2T} \int_{-T}^{T} B(t) \, dt \qquad (2.58)$$

Since in the following we are going to consider only stationary ergodic processes, time averages coincide with ensemble averages (denoted by $\langle \rangle$), so that we can write

$$B(t) = b(t) + \overline{B(t)} = b(t) + \langle B \rangle \qquad (2.59)$$

Now consider the autocorrelation function defined as

$$R_b(\tau) \equiv \overline{b(t + \tau) b^*(t)} = \lim_{T \to \infty} \frac{1}{2T} \int_{-T}^{T} b(t + \tau) b^*(t) dt$$
$$= \langle b(t + \tau) b^*(t) \rangle \qquad (2.60)$$

Since $b(t)$ represents a physical signal, the complex conjugate in the above definition seems unnecessary at first sight. Nevertheless, complex notation is very often adopted to simplify calculations and Definition 2.60 must be used. Note that, if the fluctuations were uncorrelated, then $R_b(\tau)$ would cancel for any τ. Also, by definition, $R_b(\tau) = R_b(-\tau)$.

Another important property is that for a zero-mean ($\overline{B\left(t\right)} = 0$) signal, $R_b\left(0\right)$ coincides with the classical variance of the signal

$$R_b\left(0\right) = \lim_{T \to \infty} \frac{1}{2T} \int_{-T}^{T} |b\left(t\right)|^2 dt = \overline{|b\left(t\right)|^2} \equiv \sigma_b^2 = VAR\left(b\right) \qquad (2.61)$$

Starting from the autocorrelation function, an extremely important observable quantity can be defined. For this purpose, we introduce the quantity $\hat{b}\left(\omega\right)$ defined as

$$\hat{b}\left(\omega\right) = \int_{-\infty}^{\infty} b_T\left(t\right) e^{-i\omega t} dt \qquad (2.62)$$

where the signal $b_T\left(t\right)$ given by

$$b_T\left(t\right) = \begin{cases} b\left(t\right) & |t| < T \\ 0 & \text{otherwise} \end{cases} \qquad (2.63)$$

has finite energy and is thus Fourier integrable. Then we can write

$$\left|\hat{b}\left(\omega\right)\right|^2 = \left|\int_{-T}^{T} b\left(t\right) e^{-i\omega t} dt\right|^2 = \int_{-T}^{T} \int_{-T}^{T} b\left(t\right) b^*\left(\tau\right) e^{-i\omega\left(t-\tau\right)} dt d\tau \qquad (2.64)$$

and, taking the ensemble average,

$$\left\langle \left|\hat{b}\left(\omega\right)\right|^2 \right\rangle = \int_{-T}^{T} \int_{-T}^{T} \langle b\left(t\right) b^*\left(\tau\right)\rangle e^{-i\omega\left(t-\tau\right)} dt d\tau =$$

$$\int_{-T}^{T} \int_{-T}^{T} R_b\left(t-\tau\right) e^{-i\omega\left(t-\tau\right)} dt d\tau = \int_{-2T}^{2T} \left(2T - |\tau|\right) R_b\left(\tau\right) e^{-i\omega\tau} d\tau \qquad (2.65)$$

where the integral property

$$\int_{-T}^{T} \int_{-T}^{T} g\left(t-\tau\right) dt d\tau = \int_{-2T}^{2T} \left(2T - |\tau|\right) g\left(\tau\right) d\tau \qquad (2.66)$$

has been exploited for the last step. Next we consider the quantity

$$\frac{\left\langle \left|\hat{b}\left(\omega\right)\right|^2 \right\rangle}{2T} = \int_{-2T}^{2T} \left(1 - \frac{|\tau|}{2T}\right) R_b\left(\tau\right) e^{-i\omega\tau} d\tau$$

$$= \int_{-\infty}^{\infty} R_{b,T}\left(\tau\right) e^{-i\omega\tau} d\tau \qquad (2.67)$$

where

$$R_{b,T}\left(\tau\right) = \begin{cases} \left(1 - \frac{|\tau|}{2T}\right) R_b\left(\tau\right) & |\tau| < T \\ 0 & |\tau| \geq T \end{cases} \qquad (2.68)$$

has been defined. Finally, we define the 2-sided spectral density as

$$S_b^{2-sided}\left(f\right) \stackrel{\text{def}}{=} \lim_{T \to \infty} \frac{\left\langle \left|\hat{b}\left(\omega\right)\right|^2 \right\rangle}{2T}$$

$$= \int_{-\infty}^{\infty} \lim_{T \to \infty} R_{b,T}\left(\tau\right) e^{-i\omega\tau} d\tau = \int_{-\infty}^{\infty} R_b\left(\tau\right) e^{-i2\pi f\tau} d\tau \qquad (2.69)$$

from which the inverse relationship also holds

$$R_b\left(\tau\right) = \int_{-\infty}^{\infty} S_b^{2\text{-}sided}\left(f\right) \cdot e^{i2\pi f \tau} df \tag{2.70}$$

Therefore, the autocorrelation function $R_b\left(\tau\right)$ and the spectral density function $S_b^{2\text{-}sided}\left(f\right)$ form a Fourier transform pair. This is the content of the Wiener-Khintchine (WK) theorem [65, 66]. From $R_b\left(\tau\right) = R_b\left(-\tau\right)$ it follows that $S_b^{2\text{-}sided}\left(f\right)$ is a real, non-negative, and even function. In experimental work, however, only positive frequencies are of interest. Hence, a one-sided power spectral density is often introduced for non-negative Fourier frequencies

$$S_b^{1\text{-}sided}\left(f\right) = 2S_b^{2\text{-}sided}\left(f\right) \tag{2.71}$$

Finally, it is immediately recognized that

$$VAR\left(b\right) = R_b\left(0\right) = \int_{-\infty}^{+\infty} S_b^{2\text{-}sided}\left(f\right) df \tag{2.72}$$

We anticipate here that, in spite of its rigorous formulation, the power spectral density defined by the first equality of Equation 2.69 is a quantity directly observable by a spectrum analyzer.

2.4.2 Fundamental noise mechanisms

Now we are in a position to give a quantitative description of fundamental types of noise. Though the fundamental *noisiness* of electrical conductors had been known for some time, it was not until 1918 that German physicist Walter Schottky identified and formulated a theory of *tube noise* - a fluctuation in the current caused by the granularity of the discrete charges composing it [67]. Ten years later, Johnson and Nyquist similarly analyzed a different type of noise - one caused by the thermal fluctuations of stationary charge carriers. These are now known as *shot noise* and *Johnson noise*, respectively, and it is a startling fact that neither of them depends on the material or the configuration of the electrical circuit in which they are observed. However, Johnson also measured an unexpected flicker noise at low frequency and shortly thereafter W. Schottky tried to provide a theoretical explanation. In the following we present summaries of the theories of shot [68, 69], Johnson-Nyquist [70, 71], flicker [72], and thermodynamic [73, 74] noise.

Shot noise is due to the corpuscular nature of transport (quantization of the charge carried by electrons). It is always associated with direct current flow. Indeed, the latter is not continuous, but results from the motion of charged particles (i.e., electrons and/or holes) which are discrete and independent. At some (supposedly small, presumed microscopic) level, currents vary in an unpredictable way. If you could *observe* carriers passing a point in a conductor for some time interval, you would find that a few more or less carriers would pass in one time interval versus the next. It is impossible to predict the motion of individual electrons, but it is possible to calculate the average net velocity of an ensemble of electrons, or the average number of electrons drifting past a particular point per time interval. The variation around the mean value (or average) of these quantities is the noise. In order to *see* shot noise, the carriers must be constrained to flow past in one direction only. The carrier entering the *observation* point must do so as a purely random event and independent of any other carrier crossing this point. If the carriers are not constrained in this manner, then the resultant thermal noise will dominate and the shot noise will not be *seen*. A physical system where this constraint holds is a pn junction. The passage of each carrier across the depletion region of the junction is a random event, and, due to the energy barrier, the carrier may travel in only one direction. Since the events are random and independent, Poisson statistics describe this process. Consider the quantized contribution to the current by any given electron. Its current pulse can be approximated as a delta function centered at some time t_k: $I(t) = \sum_k q\delta(t - t_k)$, where q is the elementary charge. The corresponding Fourier transform is given by

$$I(\omega) = \sum_k qe^{-i\omega t_k} \tag{2.73}$$

which provides

$$|I(\omega)|^2 = q^2 \sum_k \sum_h e^{i\omega(t_h - t_k)} = q^2 \left[N_T + \sum_{k \neq h} \sum_h e^{i\omega(t_h - t_k)} \right] \tag{2.74}$$

where N_T is the total number of events occurring in the interval $2T$. Since the times t_k are random, if we take the average of the above equation over an ensemble of a very large number of physically identical systems, the second term on the right side can be neglected in comparison to N_T

$$\left\langle |I(\omega)|^2 \right\rangle \simeq q^2 \langle N_T \rangle = q^2 \langle N \rangle 2T = q\bar{I}2T \tag{2.75}$$

where $\langle N \rangle = \langle N_T \rangle /2T$ is the average rate at which the events occur and $\bar{I} = q \langle N \rangle$ is the corresponding average current. Finally, by definition of power spectral density, we obtain

$$S_I^{1-sided} = 2S_I^{2-sided} = 2 \lim_{T \to \infty} \frac{\left\langle |I(\omega)|^2 \right\rangle}{2T} = 2q\bar{I} \tag{2.76}$$

The counterpart of shot noise in radiation sources, namely intensity noise due to granular character of light (photons), is often referred to as **photonic** noise. As for electrons in a conductor, Poisson statistics also applies in this case. Here, the random arrival times of the photons (at a detector) cause fluctuations in the average number of the detected (per unit time) photons and hence in the detected power. The spectral density associated with such quantum power noise is obtained as

$$S_P^{1-sided} = 2(h\nu)\bar{P} \tag{2.77}$$

where the elementary charge and the average current in Equation 2.76 have been replaced by the energy $h\nu$ carried by a single photon and the average detected optical power $\bar{P}$.

Johnson-Nyquist noise (thermal noise) is the electronic noise generated by the thermal agitation of the charge carriers (usually the electrons) inside an electrical conductor at equilibrium, which happens regardless of any applied voltage. Thermal noise is approximately white, meaning that the power spectral density is nearly equal throughout the frequency spectrum. Additionally, the amplitude of the signal has very nearly a Gaussian probability density function. Here, we give a microscopic derivation for it. Consider a conductor of resistance R, length l, and cross-sectional area A. The voltage across it is

$$V = IR = RAj = RANe \langle u \rangle \tag{2.78}$$

where I is the current, j the current density, e the charge on an electron, N the charge carrier density, and $\langle u \rangle$ the drift speed along the conductor. Noting that NAl is the total number of electrons in the conductor, the following relationships hold

$$\begin{cases} \langle u \rangle = \dfrac{1}{NAl} \sum_i u_i \\[2ex] \langle u^2 \rangle = \dfrac{1}{NAl} \sum_i u_i^2 \end{cases} \tag{2.79}$$

By substitution of Equation 2.79 into Equation 2.78, one gets

$$V = RANe \frac{1}{NAl} \sum_i u_i = \frac{Re}{l} \sum_i u_i = \sum_i V_i \tag{2.80}$$

which allows to define the random variables V_i as

$$V_i = \frac{Re}{l} u_i \tag{2.81}$$

The power spectral density associated with V_i is given by

$$\begin{aligned} S_{i,V}^{1-sided} &= 4 \int_0^\infty \langle V_i(t) V_i(t+\tau) \rangle \cos(2\pi f \tau) d\tau \\ &= 4 \int_0^\infty \langle V_i^2(t) \rangle e^{-\frac{\tau}{\tau_c}} \cos(2\pi f \tau) d\tau \\ &= 4 \left(\frac{Re}{l} \right)^2 \langle u_i^2 \rangle \int_0^\infty e^{-\frac{\tau}{\tau_c}} \cos(2\pi f \tau) d\tau \\ &= 4 \left(\frac{Re}{l} \right)^2 \langle u_i^2 \rangle \frac{\tau_c}{1 + (2\pi f \tau_c)^2} \simeq 4 \left(\frac{Re}{l} \right)^2 \tau_c \langle u_i^2 \rangle \end{aligned} \tag{2.82}$$

from which the total power spectral density is calculated as

$$\begin{aligned} S_V^{1-sided} &= \sum_i S_{i,V}^{1-sided} = 4 \left(\frac{Re}{l} \right)^2 \tau_c \sum_i \langle u_i^2 \rangle \\ &= 4NAl \left(\frac{Re}{l} \right)^2 \tau_c \langle u^2 \rangle = 4NAl \left(\frac{Re}{l} \right)^2 \tau_c \frac{k_B \mathbb{T}}{m} \\ &= 4k_B \mathbb{T} R \frac{Ne^2 \tau_c}{m} \frac{RA}{l} = 4k_B \mathbb{T} R \end{aligned} \tag{2.83}$$

where the equipartition theorem $\langle u^2 \rangle = k_B \mathbb{T}/m$, and the identities $\sigma = Ne^2 \tau_c/m$ and $RA/l = 1/\sigma$, known from solid state physics, have been exploited. Also note that for metals at room temperature we have $\tau_c < 10^{-13}$, hence from the DC through the microwave range $2\pi f \tau_c \ll 1$ is satisfied.

Flicker noise is a type of electronic noise with a $1/f$ spectrum. Its origins are somewhat less understood compared to thermal (Johnson) noise and shot noise. It occurs in almost all electronic devices, and results from a variety of effects, such as impurities in a conductive channel, generation and recombination noise in a transistor due to base current, and so on. In electronic devices, it is a low-frequency phenomenon, as the higher frequencies are overshadowed by white noise from other sources. In oscillators, however, the low-frequency noise is mixed up to frequencies close to the carrier which results in oscillator phase noise.

Since flicker noise is related to the level of DC, if the current is kept low, thermal noise will be the predominant effect.

A simple explanation of the appearance of $1/f$ noise can be stated by considering a single exponential relaxation process

$$N(t, t_k) = \begin{cases} N_0 e^{-\lambda(t-t_k)} & t \geq t_k \\ 0 & t < t_k \end{cases} \tag{2.84}$$

In that case we have

$$F(\omega) = \int_{-\infty}^{+\infty} \sum_k N(t, t_k) e^{-i\omega t} dt = \frac{N_0}{\lambda + i\omega} \sum_k e^{i\omega t_k} \tag{2.85}$$

so that

$$S_N^{2-sided}(\omega) = \lim_{T \to \infty} \frac{\left\langle |F(\omega)|^2 \right\rangle}{2T}$$

$$= \frac{N_0^2}{\lambda^2 + \omega^2} \lim_{T \to \infty} \frac{1}{2T} \left\langle \sum_k e^{i\omega t_k} \sum_h e^{-i\omega t_h} \right\rangle = \frac{N_0^2}{\lambda^2 + \omega^2} \langle N \rangle \tag{2.86}$$

If the relaxation rates are instead distributed according to

$$dP(\lambda) = \frac{A}{\lambda^\beta} d\lambda \tag{2.87}$$

one obtains

$$S_N^{2-sided}(\omega) \propto \int_{\lambda_1}^{\lambda_2} \frac{1}{\lambda^2 + \omega^2} \frac{d\lambda}{\lambda^\beta} = \frac{1}{\omega^{1+\beta}} \int_{\lambda_1}^{\lambda_2} \frac{1}{1 + \frac{\lambda^2}{\omega^2}} \frac{d(\lambda/\omega)}{(\lambda/\omega)^\beta}$$

$$= \frac{1}{\omega^{1+\beta}} \int_{\lambda_1/\omega}^{\lambda_2/\omega} \frac{1}{1 + x^2} \frac{dx}{x^\beta} \simeq \frac{1}{\omega^{1+\beta}} \int_0^\infty \frac{1}{1 + x^2} \frac{dx}{x^\beta} \simeq \frac{1}{\omega^{1+\beta}} \tag{2.88}$$

where the approximate equality holds in the limit $\lambda_1 \ll \lambda \ll \lambda_2$. Thus we obtain a whole class of flicker noise with different exponents.

Finally, it is interesting to note that flicker noise frequently appears in physical nature. For example, a $1/f$ spectral density is found for the fluctuations in the earth's rate of rotation and undersea currents. A study of a common hourglass demonstrated that the flow of sand fluctuates as $1/f$ [75].

Thermodynamic noise The vast majority of electronic components have temperature-dependent parameters. This means that electronic circuits are strongly affected by unavoidable temperature instabilities. As we have just seen, long-term temperature variations (relaxation processes) generate $1/f$ noise. Likewise prominent are relatively fast variations, due to quantization of thermal energy in phonons. The higher the temperature and the lower the heat capacity of the system, the more important these fluctuations are. It is well known from thermodynamics that the total variance of fluctuations of temperature for heat capacity C is described by the following formula

$$\langle \Delta \mathbb{T}^2 \rangle = \frac{k_B \mathbb{T}^2}{C} \tag{2.89}$$

In order to derive the power spectral density, we consider the equation describing the system temperature

$$C\frac{d\mathbb{T}(t)}{dt} + G[\mathbb{T}(t) - \mathbb{T}_0] = W(t) \tag{2.90}$$

where G denotes the thermal conductivity and $W(t)$ is the (possible) power supplied to the system. Note that, if we identify $\mathbb{T}(t)$ with the velocity $v(t)$ and C with the mass m, the above equation also describes the Brownian motion of a particle subject to a frictional force $-Gv(t)$ plus the random (white spectral density) force $W(t)$. This means that the notion of thermodynamic fluctuations in volume coincides with that of Brownian noise (or random walk). Since we are dealing with spontaneous temperature fluctuations, we allow for a fluctuating $W(t)$, positive and negative (added to and subtracted from the system) and completely random (hence with a white noise spectrum). We shall call $H(t)$ this power (Langevin method) and write the following equation for the consequent temperature fluctuations $\Delta\mathbb{T}(t) = \mathbb{T}(t) - \mathbb{T}_0$

$$C\frac{d\Delta\mathbb{T}(t)}{dt} + G \cdot \Delta\mathbb{T}(t) = H(t) \tag{2.91}$$

Taking the Fourier transform of the above equation, we get

$$\Delta\mathbb{T}(f) = \frac{H(f)}{G + i2\pi Cf} \tag{2.92}$$

which translates into the following relationship

$$S_{\Delta T}^{1-sided}(f) = \frac{S_H^{1-sided}}{G^2 + (2\pi Cf)^2} \tag{2.93}$$

To find the value of $S_H^{1-sided}$, we use the fact

$$\frac{k_B \mathbb{T}^2}{C} = \langle \Delta\mathbb{T}^2 \rangle = \int_0^{+\infty} S_{\Delta T}^{1-sided}(f)\, df = \frac{S_H^{1-sided}}{4CG} \tag{2.94}$$

so that

$$S_{\Delta T}^{1-sided}(f) = \frac{4Gk_B \mathbb{T}^2}{G^2 + (2\pi Cf)^2} \tag{2.95}$$

which suggests, in particular, that the only way to reduce this kind of fluctuation is to cool the system.

2.4.2.1 Fluctuation-dissipation theorem (FDT)

We close this digression with another very important consideration concerning fluctuations, that is the so-called fluctuation-dissipation theorem (FDT) [76]. It states that the linear response of a given system to an external perturbation is expressed in terms of fluctuation properties of the system in thermal equilibrium. Onsager proposed a simple derivation of FDT for time-dependent perturbations. This derivation bypasses the more cumbersome analytical developments using linear response theory formalism, the Fokker-Planck equation, or the generalized master equation approach. Onsager derivation is based on the following regression principle: if a system initially in an equilibrium state 1 is driven by an external perturbation to a different equilibrium state 2, then the evolution of the system from state 1 towards state 2 in the presence of the perturbation can be treated as a spontaneous equilibrium fluctuation (in the presence of the perturbation) from the (now) non-equilibrium state 1 to the (now) equilibrium state 2. Suppose that the system is initially in equilibrium with a thermal bath at temperature $\mathbb{T}$, then the probability distribution of system configuration $\mathcal{C}$ in state 1 is given by the canonical ensemble:

$$\mathcal{P}_0\left(\mathcal{C}\right) = \frac{e^{-\beta E(\mathcal{C})}}{\sum_{\mathcal{C}} e^{-\beta E(\mathcal{C})}} \tag{2.96}$$

where $\beta = k_B \mathbb{T}$ and the subscript 0 indicates that the system is unperturbed. At time $t = 0$ a constant perturbation coupled to the observable $B\left(\mathcal{C}\right)$ is applied to the system changing its energy into

$$E_\epsilon\left(\mathcal{C}\right) = E\left(\mathcal{C}\right) - \epsilon\left(t\right) B\left(\mathcal{C}\right) \tag{2.97}$$

where $\epsilon\left(t\right) = \epsilon$ if $t > 0$, and zero otherwise. The effect of the perturbation can be monitored by looking at the evolution of the expectation value $\langle A\left(t\right)\rangle_\epsilon$ of an observable $A(\mathcal{C})$, not necessarily equal to $B\left(\mathcal{C}\right)$, from the equilibrium value in state 1 $\langle A\left(t = 0\right)\rangle_\epsilon = \langle A \rangle_0$ towards the new equilibrium value in state 2. The expectation value of $\langle A\left(t\right)\rangle_\epsilon$ is given by the average over all possible dynamical paths originating from initial configurations weighted with the probability distribution Equation 2.96

$$\langle A\left(t\right)\rangle_\epsilon = \sum_{\mathcal{C},\mathcal{C}_0} A(\mathcal{C})\mathcal{P}_\epsilon\left(\mathcal{C},t\,|\mathcal{C}_0,0\right)\mathcal{P}_0\left(\mathcal{C}_0\right) \tag{2.98}$$

where $\mathcal{P}_\epsilon\left(\mathcal{C},t\,|\mathcal{C}_0,0\right)$ is the conditional probability for the evolution from the configuration $\mathcal{C}_0$ at time $t = 0$ to the configuration $\mathcal{C}$ at time t. The Onsager regression principle asserts that the conditional probabilities after having applied the perturbation are equal to those of spontaneous equilibrium fluctuations in state 2. Hence since the state 2 is still described by the canonical ensemble, but with the energy now including the perturbation term, then

$$\mathcal{P}_\epsilon\left(\mathcal{C},t\,|\mathcal{C}_0,0\right) = \mathcal{P}_0\left(\mathcal{C},t\,|\mathcal{C}_0,0\right) e^{\beta\epsilon[B(\mathcal{C})-B(\mathcal{C}_0)]} \tag{2.99}$$

Inserting Equation 2.99 into Equation 2.98 and expanding the exponential up to linear order we get

$$\langle A\left(t\right)\rangle_\epsilon = \sum_{\mathcal{C},\mathcal{C}_0} A(\mathcal{C})\mathcal{P}_0\left(\mathcal{C},t\,|\mathcal{C}_0,0\right)\mathcal{P}_0\left(\mathcal{C}_0\right) +$$

$$+\beta\epsilon \sum_{\mathcal{C},\mathcal{C}_0} A(\mathcal{C})\mathcal{P}_0\left(\mathcal{C},t\,|\mathcal{C}_0,0\right)[B\left(\mathcal{C}\right)-B\left(\mathcal{C}_0\right)]\mathcal{P}_0\left(\mathcal{C}_0\right) =$$

$$= \langle A \rangle_0 + \beta\epsilon\left[\langle A\left(t\right) B\left(t\right)\rangle_0 - \langle A\left(t\right) B\left(0\right)\rangle_0\right] \tag{2.100}$$

If we define the correlation function, the time-dependent susceptibilty and the response function as

$$C_{A,B}\left(t,s\right) = \langle A\left(t\right) B\left(s\right)\rangle_0 \tag{2.101}$$

$$\chi_{A,B}\left(t\right) = \lim_{\epsilon \to 0} \frac{\langle A\left(t\right)\rangle_\epsilon - \langle A \rangle_0}{\epsilon} \tag{2.102}$$

$$\int_0^t J_{A,B}\left(t,s\right) ds = \chi_{A,B}\left(t\right) \tag{2.103}$$

from Equation 2.100 we get

$$\int_0^t J_{A,B}(t,s)\,ds = \beta\left[C_{A,B}(0) - C_{A,B}(t)\right] = \beta\int_0^t \frac{\partial}{\partial s}C_{A,B}(t,s)\,ds \tag{2.104}$$

which, defining $t - s = y$, becomes

$$
\begin{aligned}
J_{A,B}(y,0) &= -\beta\left[\frac{\partial}{\partial y}C_{A,B}(y,0)\right]\theta(y) = \\
&= -\frac{\beta}{2}\left[\frac{\partial}{\partial y}C_{A,B}(y,0) + \text{sign}(y)\frac{\partial}{\partial y}C_{A,B}(y,0)\right]
\end{aligned}
\tag{2.105}
$$

Taking the Fourier transform of Equation 2.105 one obtains

$$\Im\left[J_{A,B}(\omega)\right] = -\frac{\omega}{2k_B\mathbb{T}}C_{A,B}(\omega) \tag{2.106}$$

For a system described by $X(\omega) = \alpha(\omega)F(\omega)$, X being the *position*, F the external forcing term, and α the response function, Equation 2.106 can be cast in a more familiar form. Indeed, with the identifications $J_{A,B}(\omega) \equiv \alpha(\omega)$ and $C_{A,B}(\omega) \equiv R_X(\tau)$, and using the WK theorem, the power spectral density associated with X can be written as

$$S_X^{1-sided}(f) = 2\cdot\mathbb{F}\left[R_X(\tau)\right] = -\frac{4k_B\mathbb{T}}{\omega}\Im\left[\alpha(\omega)\right] \tag{2.107}$$

where $\mathbb{F}$ denotes here the Fourier transform. Similarly, one has

$$S_F^{1-sided}(f) = -\frac{4k_B\mathbb{T}}{\omega}\frac{\Im\left[\alpha(\omega)\right]}{|\alpha(\omega)|^2} = -\frac{4k_B\mathbb{T}}{\omega}\Im\left[\frac{1}{\alpha^*(\omega)}\right] \tag{2.108}$$

The two above formulas represent the most commonly encountered statements of the fluctuation-dissipation theorem.

Finally, it is worth noting that Johnson noise is a particular case of the FDT. To see this, consider an open circuit consisting of an impedence $Z(\omega)$. Ohm's law is $Q(\omega) = \alpha(\omega)V(\omega)$ where $\alpha(\omega) = 1/[i\omega Z(\omega)]$. Thus we have

$$
\begin{aligned}
S_V^{1-sided}(f) &= -\frac{4k_B\mathbb{T}}{\omega}\Im\left[\frac{1}{\alpha^*(\omega)}\right] = -\frac{4k_B\mathbb{T}}{\omega}\Im\left[-i\omega Z^*(\omega)\right] \\
&= 4k_B\mathbb{T}\cdot\Re\left[Z(\omega)\right] = 4k_B\mathbb{T}R
\end{aligned}
\tag{2.109}
$$

This example also allows one to establish another useful form of FDT. Indeed, one can define the admittance $Y(\omega)$, the conductance $\sigma(\omega)$, and the resistance $R(\omega)$ as

$$
\begin{cases}
Y(\omega) = \dfrac{1}{Z(\omega)} = i\omega\cdot\alpha(\omega) \\
\sigma(\omega) = \Re\left[Y(\omega)\right] \\
R(\omega) = \Re\left[Z(\omega)\right] = \Re\left[\dfrac{1}{Y(\omega)}\right]
\end{cases}
\tag{2.110}
$$

In this way, Equation 2.107 can be thus re-written as

$$S_X^{1-sided}(f) = -\frac{4k_B\mathbb{T}}{\omega}\Im\left[\alpha(\omega)\right] = \frac{k_B\mathbb{T}}{\pi^2 f^2}\Re\left[Y(\omega)\right] = \frac{k_B\mathbb{T}}{\pi^2 f^2}\sigma(\omega) \tag{2.111}$$

Now, when the applied force is periodic $F(\omega) = F_0 \cos(\omega t)$, we can write

$$S_X^{1-sided}(f) = \frac{k_B \mathbb{T}}{\pi^2 f^2} \sigma(\omega) = \frac{2 k_B \mathbb{T}}{\pi^2 f^2} \frac{W_{diss}}{F_0^2} = \frac{8 k_B \mathbb{T}}{\omega^2} \frac{W_{diss}}{F_0^2} \qquad (2.112)$$

where the real part of the admittance (i.e., the conductance) has been related to the average power dissipated by the system, W_{diss}, through the relationship

$$\sigma(\omega) = 2 \frac{W_{diss}}{F_0^2} \qquad (2.113)$$

In order to justify the above relationship, again the analogy with the circuit is useful. Indeed, in this case we have $W_{diss} = \overline{V^2} \cdot \sigma \Rightarrow W_{diss} = \overline{F^2} \cdot \sigma = F_0^2/2$. Equation 2.112 will prove very useful in the application of FDT for the treatment of thermal noise in optical cavities (see next Chapter).

2.5 Phase noise modelling

In previous section we have learnt that fundamental types of noise exhibit power spectral densities with a power-law behavior. The next step is to model an oscillator as a system with n inputs (each associated with one noise source) and two outputs represented by $\alpha(t)$ and $\varphi(t)$ of Equation 2.57 [77]. In the electrical equivalent of the oscillator, noise inputs are in the form of current sources injecting into circuit nodes and voltage sources in series with circuit branches (frame a of Figure 2.11). In this way, circuit noise evolves into amplitude and phase noise of the oscillator output voltage. To better understand this, consider the specific example of an ideal parallel LC oscillator shown in frame b of Figure 2.11. If we inject a current impulse as shown, the amplitude and phase of the oscillator will have responses similar to that shown in the lower frame of Figure 2.11. The instantaneous voltage change ΔV is given by

$$\Delta V = \frac{\Delta q}{C_{tot}} \qquad (2.114)$$

where Δq is the total injected charge due to the current impulse and is the total capacitance at that node. Note that the current impulse will change only the voltage across the capacitor and will not affect the current through the inductor. It can be seen (frame c of Figure 2.11) that the resultant change in $\alpha(t)$ and $\varphi(t)$ is time dependent. In particular, if the impulse is applied at the peak of the voltage across the capacitor, there will be no phase shift and only an amplitude change will result. On the other hand, if this impulse is applied at the zero crossing, it has the maximum effect on the excess phase and the minimum effect on the amplitude. An impulse applied sometime between these two extremes will result in both amplitude and phase changes.

Focusing our attention on the phase, in the light of the above considerations, one can assume the unit impulse response for excess phase as

$$h_\varphi(t, \tau) = \frac{\Gamma(\omega_0 \tau)}{q_{max}} u(t - \tau) \qquad (2.115)$$

where $u(t)$ is the unit step function and Γ is the impulse sensitivity function (ISF) (dividing by q_{max}, the maximum charge displacement across the capacitor, makes Γ independent of signal amplitude). It is a dimensionless, frequency- and amplitude-independent periodic

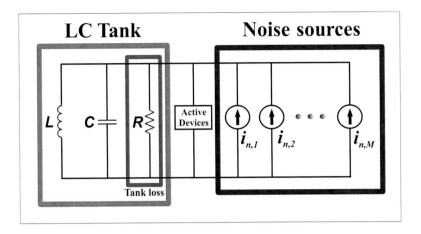

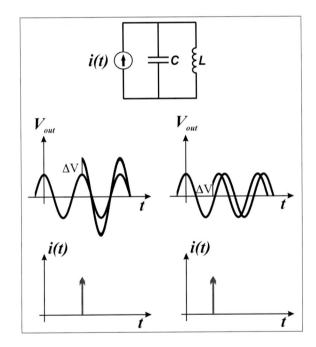

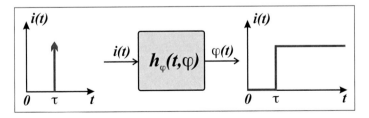

FIGURE 2.11
Generic model for a self-sustained noisy LC oscillator. (Adapted from [77].)

function with period 2π which describes how much phase shift results from applying a unit impulse at time $t = \tau$. Given the ISF, the output excess phase $\varphi(t)$ can be calculated using the superposition integral

$$\varphi(t) = \int_{-\infty}^{+\infty} h_{\varphi}(t, \tau) i(\tau) \, d\tau = \frac{1}{q_{max}} \int_{-\infty}^{t} \Gamma(\omega_0 \tau) i(\tau) \, d\tau \tag{2.116}$$

where $i(\tau)$ represents the input noise current injected into the node of interest. This equation, to be solved numerically, allows computation of $\varphi(t)$ for an arbitrary input current, provided that the ISF has been determined. Therefore, once all the noise sources $i_1(\tau), \ldots, i_N(\tau)$ are properly accounted for, $\varphi(t)$ and hence the power spectral density $S_{\varphi}(f)$ can be determined. Rather than computing $S_{\varphi}(f)$ from Equation 2.116 for each of the forementioned fundamental types of noise, we take a heuristic approach where, starting from the dominant noise mechanisms known (from experience) for the single components (amplifier and resonator), a model is developed that is able to explain the general behavior observed for $S_{\varphi}(f)$ in a variety of practical self-sustained oscillators [57]. Before embarking on this study, starting from the phase $\varphi(t)$, let us define other useful quantities. First, from Equation 2.57, the instantaneous frequency is given by

$$\nu(t) \equiv \frac{1}{2\pi} \frac{d}{dt} [2\pi\nu_0 t + \varphi(t)] = \nu_0 + \frac{1}{2\pi} \frac{d\varphi(t)}{dt} \tag{2.117}$$

In order to compare frequency standards operating at different frequencies ν_0, it is also helpful to define the normalized phase noise (also referred to as phase time)

$$x(t) = \frac{\varphi(t)}{2\pi\nu_0} \tag{2.118}$$

and the instantaneous normalized frequency deviation (also referred to as fractional frequency fluctuation)

$$y(t) = \frac{\nu(t) - \nu_0}{\nu_0} = \frac{\dot{\varphi}(t)}{2\pi\nu_0} = \dot{x}(t) \tag{2.119}$$

From this definition, it follows that the expectation value of $y(t)$ is zero. Now, if $\varphi(t)$ identifies with $b(t)$ of Equation 2.60, we have

$$
\begin{aligned}
R_{\varphi}(\tau) &= \lim_{T \to \infty} \frac{1}{2T} \int_{-T}^{T} \varphi(t + \tau) \varphi^*(t) \, dt \\
&= \int_{-\infty}^{+\infty} S_{\varphi}^{2-sided}(f) \cdot e^{i2\pi f \tau} df
\end{aligned}
\tag{2.120}
$$

Similarly, for the variable $x(t)$ we have

$$
\begin{aligned}
R_x(\tau) &= \lim_{T \to \infty} \frac{1}{2T} \int_{-T}^{T} x(t + \tau) x^*(t) \, dt \\
&= \frac{1}{(2\pi\nu_0)^2} \lim_{T \to \infty} \frac{1}{2T} \int_{-T}^{T} \varphi(t + \tau) \varphi^*(t) \, dt \\
&= \frac{1}{(2\pi\nu_0)^2} \int_{-\infty}^{+\infty} S_{\varphi}^{2-sided}(f) \cdot e^{i2\pi f \tau} df \\
&\equiv \int_{-\infty}^{+\infty} S_x^{2-sided}(f) \cdot e^{i2\pi f \tau} df
\end{aligned}
\tag{2.121}
$$

where

$$S_x^{2-sided}(f) = \frac{1}{(2\pi\nu_0)^2} S_\varphi^{2-sided}(f) \tag{2.122}$$

has been introduced. Finally, we have for the variable $y(t)$

$$
\begin{aligned}
R_y(\tau) &= \lim_{T\to\infty} \frac{1}{2T} \int_{-T}^{T} y(t+\tau) y^*(t) dt \\
&= \frac{1}{(2\pi\nu_0)^2} \lim_{T\to\infty} \frac{1}{2T} \int_{-T}^{T} \dot{\varphi}(t+\tau) [\dot{\varphi}(t)]^* dt \\
&= \frac{(2\pi f)^2}{(2\pi\nu_0)^2} \int_{-\infty}^{+\infty} S_\varphi^{2-sided}(f) \cdot e^{i2\pi f\tau} df \\
&\equiv \int_{-\infty}^{+\infty} S_y^{2-sided}(f) \cdot e^{i2\pi f\tau} df
\end{aligned}
\tag{2.123}
$$

where we have used the property that the time-domain derivative maps into a multiplication by $i\omega = i2\pi f$ in the Fourier transform domain and the function

$$S_y^{2-sided}(f) = \left(\frac{f}{\nu_0}\right)^2 S_\varphi^{2-sided}(f) \tag{2.124}$$

has been introduced. Equation 2.124 defines the well-known relationship between the PSD of phase fluctuations and the PSD of fractional frequency fluctuations. The latter quantity is also referred to as the frequency modulation (PM) noise. In the laser literature, one often sees the frequency noise expressed as the PSD of frequency fluctuations

$$S_{\delta\nu}^{1-sided}(f) = \nu_0^2 S_y^{1-sided}(f) = f^2 S_\varphi^{1-sided}(f) \tag{2.125}$$

2.5.1 The Leeson effect

Now, coming back to our main concern, let us start by studying the case of an oscillator in which the feedback circuit β is an ideal resonator, free from frequency fluctuations and with a large merit factor Q. In order to find an effective model for the function $S_\varphi(f)$, let us make the following considerations. If a static phase ψ is inserted in the loop (Figure 2.12), the Barkhausen phase condition (Equation 2.49 with $A = 1$) becomes

$$\arg \beta(\omega) + \psi = 0 \tag{2.126}$$

and the loop oscillates at the frequency

$$\omega_0 + \Delta\omega \qquad \text{at which} \qquad \arg \ \beta(\omega) = -\psi \tag{2.127}$$

Within the accuracy of linearization, the effect of ψ on the oscillation frequency is obtained as

$$\Delta\omega = -\frac{\psi}{\frac{d}{d\omega} \arg \beta(\omega)} \tag{2.128}$$

which, together with Equation 2.31, yields

$$\Delta\omega = \frac{\omega_0}{2Q} \psi \tag{2.129}$$

Next, let us replace the static phase ψ with a random phase fluctuation $\psi(t)$ that accounts for all the phase noise sources in the loop. In this picture, the phase fluctuation $\varphi(t)$ in Equation 2.57 is just the effect of $\psi(t)$. We analyze now the mechanism by which the power spectrum density of ψ is transferred to φ. For fluctuations of ψ faster than the inverse of the relaxation time $\tau_{rel} = 2Q/\omega_0$, the resonator is an open circuit for the phase fluctuation. The fluctuation $\psi(t)$ crosses the amplifier and shows up at the output, without being fed back at the amplifier input. No noise regeneration takes place in this conditions, thus $\varphi(t) = \psi(t)$, and

$$S_\varphi(f) = S_\psi(f) \tag{2.130}$$

For the slow components of $\psi(t)$, ψ can be treated as quasi-static perturbation according to Equation 2.129

$$\Delta\nu = \frac{\nu_0}{2Q}\psi(t) \tag{2.131}$$

such that

$$S_{\Delta\nu}(f) = \left(\frac{\nu_0}{2Q}\right)^2 S_\psi(f) \tag{2.132}$$

On the other hand, the instantaneous output phase is

$$\varphi(t) = 2\pi \int (\Delta\nu)dt \tag{2.133}$$

which, as the time-domain integration maps into a multiplication by $1/(2\pi f)^2$ in the spectrum, provides

$$S_\varphi(f) = \frac{1}{f^2}S_{\Delta\nu}(f) = \frac{1}{f^2}\left(\frac{\nu_0}{2Q}\right)^2 S_\psi(f) \tag{2.134}$$

Under the assumption that there is no correlation between fast and slow fluctuations, we can add the effects stated by Equation 2.130 and Equation 2.134 to obtain

$$S_\varphi(f) = \left[1 + \left(\frac{f_L}{f}\right)^2\right] S_\psi(f) \tag{2.135}$$

where the Leeson frequency has been introduced

$$f_L = \frac{\nu_0}{2Q} \tag{2.136}$$

By inspection of Equation 2.135, the oscillator behavior is that of a first-order filter with a perfect integrator (a pole in the origin in the Laplace transform domain) and a cut-off frequency f_L (a zero on the real left-axis), as shown in Figure 2.12. It is worth stressing that Equation 2.135 has still a general form since the amplifier noise has not been specified. Moreover, it accounts only for the phase-to-frequency conversion inherent in the loop: the resonator noise is still to be added for the noise spectrum to be correct. Therefore, in order to accomplish this treatment, next we have to specify the amplifier noise $S_\psi(f)$ and finally introduce the effect of the resonator noise.

Concerning the amplifier noise, it has been experimentally observed that, for a given amplifier, the total phase noise spectrum results from adding the white and the flicker noise spectra, as in Figure 2.13. This relies on the assumption that white and flicker phenomena are independent, which is true for actual amplifiers. It is important to understand that b_0

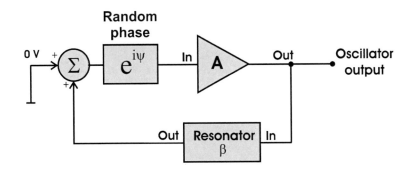

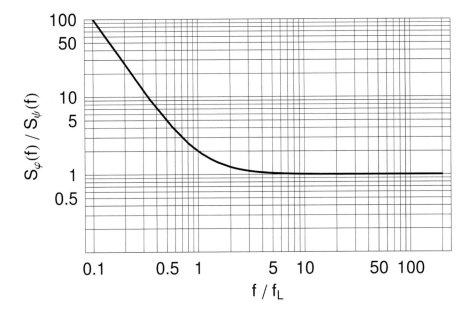

FIGURE 2.12
Model for a noisy self-sustained oscillator and Leeson-effect transfer function. (Adapted from [57].)

(white) is proportional to the inverse of the carrier power P_0: $S_\varphi(f) = b_0 = Fk_BT_0/P_0$ where F is the amplifier noise figure, while b_{-1} (flicker) is about independent of P_0. The factor F is an empirical fitting parameter and therefore must be determined by measurements. The position of the corner frequency f_c depends on the input power. The belief that f_c is a noise parameter of the amplifier is a common mistake. When such an amplifier is inserted into an oscillator, it interacts with the resonator according to Leeson's formula (Equation 2.135)

$$S_\varphi(f) = \left[1 + \left(\frac{f_L}{f}\right)^2\right] S_\psi(f) = \left[1 + \left(\frac{f_L}{f}\right)^2\right] \left[b_0 + \frac{b_{-1}}{f}\right] \qquad (2.137)$$

Concerning the resonator noise, the dissipative loss of the resonator, inherently, originates white noise. Yet, the noise phenomena most relevant to the oscillator stability are

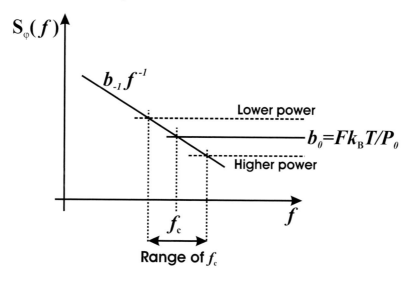

FIGURE 2.13
Typical phase noise of an amplifier on a Log-Log plot (b_{-1} about constant). (Adapted from [57].)

the flicker and the random walk of the resonant frequency ν_0. Thus, the spectrum $S_y(f)$ of the fractional frequency fluctuation $y = \Delta\nu/\nu_0$ shows a term $h_{-1}f^{-1}$ for the frequency flicker, and $h_{-2}f^{-2}$ for the frequency random walk. According to the relationship between $S_\varphi(f)$ and $S_y(f)$ established by Equation 2.124, the term $h_{-1}f^{-1}$ of the resonator fluctuation yields a term proportional to f^{-3} in the phase noise, and the term $h_{-2}f^{-2}$ yields a term f^{-4}. The resonator fluctuation is independent of the amplifier noise, for it adds to the oscillator noise.

In light of the above treatment, a model that is found useful in describing the noise spectra in an arbitrary oscillator is the power law

$$S_\varphi(f) = \sum_{i=0}^{-4} b_i f^i \tag{2.138}$$

in the range $0 \le f \le f_h$, where f_h is the high-frequency cutoff of an ideal (sharp-cutoff) low-pass filter. According to Equation 2.124, the corresponding expansion of $S_y(f)$ is given by

$$S_y(f) = \sum_{i=2}^{-2} h_i f^i \tag{2.139}$$

with the h_i coefficients given in Table 2.1.

Experimental determinations of spectral densities for different frequency sources reaching from quartz oscillators to atomic frequency standards have indeed confirmed the behavior described by Equation 2.138 (or equivalently by Equation 2.139).

2.5.2   The Allan variance

Another tool often used in the characterization of oscillators is the two-sample or Allan variance $\sigma_y^2(\tau)$, as a function of the measurement time τ [78, 65]. The Allan variance is

TABLE 2.1

Power spectral density and Allan variance for the different noise processes ($\gamma = 0.577216$ is Euler's constant).

noise type	$S_\varphi(f)$	$S_y(f)$	$S_\varphi(f) \leftrightarrow S_y(f)$	$AVAR(y)$	$MAVAR(y)$
white φ	b_0	$h_2 f^2$	$h_2 = \dfrac{b_0}{\nu_0^2}$	$\dfrac{3 f_h}{4\pi^2 \tau^2} h_{+2}$	$\propto \tau^{-3}$
flicker φ	$b_{-1} f^{-1}$	$h_1 f$	$h_1 = \dfrac{b_{-1}}{\nu_0^2}$	$\dfrac{3[\gamma + \ln(2\pi f_h \tau)]}{4\pi^2 \tau^2} h_{+1}$	$\propto \tau^{-2}$
white f	$b_{-2} f^{-2}$	h_0	$h_0 = \dfrac{b_{-2}}{\nu_0^2}$	$\dfrac{1}{2\tau} h_0$	$\propto \tau^{-1}$
flicker f	$b_{-3} f^{-3}$	$h_{-1} f^{-1}$	$h_{-1} = \dfrac{b_{-3}}{\nu_0^2}$	$2 \ln 2 h_{-1}$	$\propto \tau^0$
rand. walk f	$b_{-4} f^{-4}$	$h_{-2} f^{-2}$	$h_{-2} = \dfrac{b_{-4}}{\nu_0^2}$	$\dfrac{2\pi^2 \tau}{3} h_{-2}$	$\propto \tau$

always estimated by averaging. Given a stream of M data $\bar{y}$, each representing a measure of the quantity $y(t)$ averaged over a duration τ, ending at the time $t_k = k\tau$, the estimated two-sample Allan variance is

$$AVAR(y) \equiv \sigma_y^2(\tau)$$

$$= \frac{1}{2(M-1)} \sum_{k=1}^{M-1} \left(\bar{y}_{k+1} - \bar{y}_k\right)^2 \equiv \frac{1}{2} \left\langle \left(\bar{y}_{k+1} - \bar{y}_k\right)^2 \right\rangle \tag{2.140}$$

where

$$\bar{y}_k = \frac{1}{\tau} \int_{t_k}^{t_k + \tau} y(t)\, dt \tag{2.141}$$

Practically, to calculate the Allan variance, a set of M independent frequency offset measurements must be acquired, by M successive averages of the signal over a time τ. The process repeats for growing values of τ, to finally obtain the Allan variance trend. Clearly, the τ value corresponding to the minimum of this function corresponds to the averaging time that provides the best stability for the system under investigation. It is worth noting that the Allan variance is based on differences of adjacent frequency values rather than on frequency differences from the mean value, as is the classical standard deviation. Then, the Allan deviation is defined as the square root of the Allan variance.

By definition, it also follows

$$AVAR(y) = \left\langle \frac{1}{2} \left(\frac{1}{\tau} \int_{t_{k+1}}^{t_{k+2}} y(t')\, dt' - \frac{1}{\tau} \int_{t_k}^{t_{k+1}} y(t')\, dt' \right)^2 \right\rangle \tag{2.142}$$

This means that a single sample is obtained by one-half of the squared difference of the mean values of the function $y(t)$ derived from two adjacent intervals of duration τ, and the Allan variance is then the expectation value of this quantity (that is the ensemble average on many samples). To obtain many samples it is not necessary to divide the function $y(t)$ into discrete time intervals but rather we derive a sample for each instant t as follows

$$AVAR\left(y\right) = \left\langle \frac{1}{2}\left(\frac{1}{\tau}\int_{t}^{t+\tau} y\left(t'\right)dt' - \frac{1}{\tau}\int_{t-\tau}^{t} y\left(t'\right)dt'\right)^{2}\right\rangle =$$

$$= \left\langle \left(\int_{-\infty}^{+\infty} y\left(t'\right)h_{\tau}\left(t-t'\right)dt'\right)^{2}\right\rangle \tag{2.143}$$

where the function

$$h_{\tau}\left(t\right) = \begin{cases} -\dfrac{1}{\sqrt{2}\tau} & -\tau < t < 0 \\ \dfrac{1}{\sqrt{2}\tau} & 0 \leq t < \tau \\ 0 & \text{otherwise} \end{cases} \tag{2.144}$$

has been introduced. The square modulus of its Fourier transform $H_{\tau}\left(f\right)$ is given by

$$\left|H_{\tau}\left(f\right)\right|^{2} = 2\frac{\sin^{4}\left(\pi\tau f\right)}{\left(\pi\tau f\right)^{2}} \tag{2.145}$$

Consequently, the Allan variance is the mean square of the following convolution integral

$$\vec{y}\left(t\right) = \int_{-\infty}^{+\infty} y\left(t'\right)h_{\tau}\left(t-t'\right)dt' \equiv y * h_{\tau} \tag{2.146}$$

that can be interpreted as the temporal response of a hypothetical linear filter with impulse response $h_{\tau}\left(t\right)$ to an input signal $y\left(t\right)$. Thus, in formula, we can write

$$AVAR\left(y\right) = \left\langle \left(\vec{y}\left(t\right)\right)^{2}\right\rangle \tag{2.147}$$

Note that, from the convolution theorem, we have

$$\mathbb{F}\left[\vec{y}\left(t\right)\right] = \mathbb{F}\left[y\left(t\right)\right]\cdot\mathbb{F}\left[h_{\tau}\left(t\right)\right] \equiv \hat{y}\left(f\right)H_{\tau}\left(f\right) \tag{2.148}$$

Now, by virtue of the definition of $y\left(t\right)$ (see Equation 2.119), we have

$$\overline{\vec{y}\left(t\right)} = 0 \tag{2.149}$$

so that

$$\left\langle \left(\vec{y}\left(t\right)\right)^{2}\right\rangle = VAR\left(\vec{y}\right) \tag{2.150}$$

Then, Equation 2.147 becomes

$$AVAR\left(y\right) = VAR\left(\vec{y}\right) = \int_{-\infty}^{+\infty} S_{\vec{y}}^{2-sided}\left(f\right)df \tag{2.151}$$

where the expression 2.72 for the classical variance has been exploited. Now, by definition of power spectral density and by virtue of Equation 2.148 and Equation 2.145, we obtain

$$S_{\vec{y}}^{2-sided}(f) = \lim_{T \to \infty} \frac{\left\langle \left| \mathbb{F}\left[\vec{y}(t)\right] \right|^2 \right\rangle}{2T} = \lim_{T \to \infty} \frac{\left\langle |H_\tau(f)|^2 |\hat{y}(f)|^2 \right\rangle}{2T}$$

$$= 2 \frac{\sin^4(\pi\tau f)}{(\pi f)^2} \lim_{T \to \infty} \frac{\left\langle |\hat{y}(f)|^2 \right\rangle}{2T} = 2 \frac{\sin^4(\pi\tau f)}{(\pi f)^2} S_y^{2-sided}(f) \tag{2.152}$$

that, inserted into Equation 2.151, finally yields

$$AVAR(y) = 2 \int_{-\infty}^{+\infty} S_y^{2-sided}(f) \frac{\sin^4(\pi\tau f)}{(\pi f)^4} df =$$

$$= 2 \int_0^{+\infty} S_y^{1-sided}(f) \frac{\sin^4(\pi\tau f)}{(\pi f)^4} df \tag{2.153}$$

This relationship allows to compute the Allan variance directly from the (one-sided) power spectral density. The time-domain description of the instability of oscillators by the Allan variance is often chosen as it is easily calculated from the time series measured with simple electronic counters. The description of fluctuations by power spectral densities in the Fourier frequency domain, however, contains the full information about the noise process if properly determined. Furthermore, $AVAR(y)$ can always be calculated from $S_y(f)$; in contrast, the calculation of the power spectral density from the measured Allan variance requires the solution of an integral equation which is possible only in simple cases. Substitution of each of the five terms of the expansion of $S_y^{1-sided}(f)$ (Equation 2.139) separately into Equation 2.153 gives a very clear picture of the effect of the bandwidth (see Table 2.1). We see that terms in h_1 and h_2 tend to diverge when $f \to \infty$. Practically, however, an infinite $AVAR$ will not be observed as the case $f \to \infty$ would require an infinite bandwidth of the measurement equipment. More rigorously, for h_1 and h_2, we have to write

$$AVAR(y) = 2 \int_0^{f_h} S_y^{1-sided}(f) \frac{\sin^4(\pi\tau f)}{(\pi f)^4} df \tag{2.154}$$

As one can see from the above table, $AVAR$ is not very sensitive to distinguish between flicker phase noise and white phase noise. To overcome this deficiency the so-called modified $AVAR$ ($MAVAR$) has been introduced [79, 80]

$$MAVAR(y) \equiv \text{mod } \sigma_y^2(\tau)$$

$$= \frac{1}{2m^4(M-3m+2)} \sum_{j=1}^{M-3m+2} \left\{ \sum_{i=j}^{j+m-1} \left[\sum_{k=i}^{i+m-1} (\overline{y}_{k+m} - \overline{y}_k) \right] \right\}^2 \tag{2.155}$$

It is estimated from a set of M frequency measurements for averaging time $\tau = m\tau_0$, where m is the averaging factor and τ_0 is the basic measurement interval. With the use of $MAVAR$ the first four types of power spectral densities have the same dependence on τ as $AVAR$, while, for $h_2 f^2$, $MAVAR$ is proportional to τ^{-3}. Analytical expressions for $MAVAR$ can be found in [80]. In a doubly logarithmic plot the particular contributions to the above equation can be identified readily by their slope, thereby allowing identification of the causes of the noise mechanisms in the oscillators (see Figure 2.14).

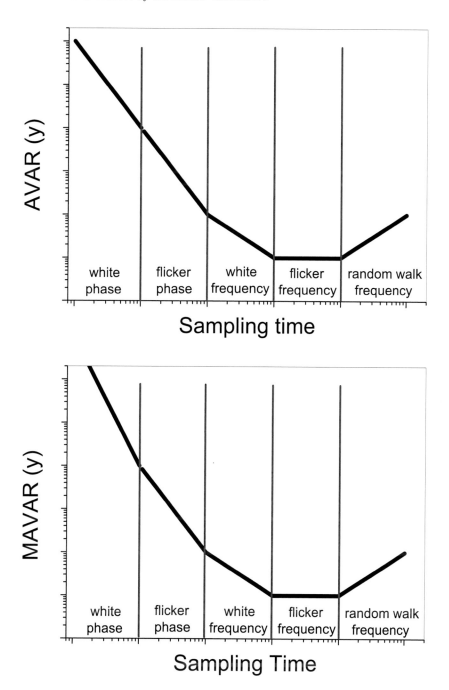

FIGURE 2.14

Representation of different noise processes in a Log-Log plot of $AVAR$ and $MAVAR$ as a function of sampling time.

It is worth pointing out that, besides stochastic fluctuations, deterministic variations of the frequency of a given oscillator have a strong impact on the measured Allan variance. Just as an example, consider an oscillator whose normalized frequency deviation exhibits a linear drift $y(t) = at$. In this case one can easily calculate the Allan variance as $\sigma_y(\tau) = (a/\sqrt{2})\,\tau$. Hence, a linear frequency drift leads to an Allan deviation that linearly increases with measuring time τ.

Throughout this book, we will have occasion to study different types of oscillators, each displaying a characteristic power spectral density or Allan variance behavior, according to the specific dominant noise mechanisms (i.e. to the specific values of the b_i coefficients in Table 2.1).

2.5.3   Oscillator power spectrum in the carrier frequency domain

Often, when dealing with laser or microwave frequency standards, one is interested in the power spectrum of the oscillator in the carrier frequency domain. From previous general considerations, we already know that an ideal oscillator operating at the frequency ν_0 would consist of a delta function at ν_0 in the carrier frequency domain. Actually, we also learnt that for a real oscillator (perturbed by noise processes) the power is spread over a frequency range around the center frequency ν_0.

Now, we establish a link between carrier frequency domain and Fourier frequency domain [81]. For this purpose, the output signal of the oscillator is expressed here in the complex form

$$V(t) = A_0 e^{i[\omega_0 + \varphi(t)]} \tag{2.156}$$

Under the usual assumption of stationary ergodic processes, we calculate the autocorrelation function as

$$R_V(\tau) = \frac{1}{2}\Re\langle V(t)V^*(t+\tau)\rangle = \frac{|A_0|^2}{2}\cos(\omega_0\tau)\,\Re\left\langle e^{i[\varphi(t)-\varphi(t+\tau)]}\right\rangle$$

$$= \frac{|A_0|^2}{2}\cos(\omega_0\tau)\,e^{-\frac{\left\langle[\varphi(t)-\varphi(t+\tau)]^2\right\rangle}{2}} \tag{2.157}$$

where the last step follows from the Gaussian moment theorem. Now recalling that

$$\left\langle[\varphi(t)]^2\right\rangle = \left\langle[\varphi(t+\tau)]^2\right\rangle \overset{\text{def}}{=} R_\varphi(\tau=0)$$

$$= \int_{-\infty}^{+\infty} S_\varphi^{2-sided}(f)\,df = \int_0^{+\infty} S_\varphi^{1-sided}(f)\,df \tag{2.158}$$

and that

$$\langle\varphi(t)\varphi^*(t+\tau)\rangle \overset{\text{def}}{=} R_\varphi(\tau) = \int_0^{+\infty} S_\varphi^{1-sided}(f)\cos(2\pi f\tau)\,df \tag{2.159}$$

we have

$$\frac{\left\langle[\varphi(t)-\varphi(t+\tau)]^2\right\rangle}{2} = 2\int_0^{+\infty} S_\varphi^{1-sided}(f)\sin^2(\pi f\tau)\,df \tag{2.160}$$

By substitution of Equation 2.160 into Equation 2.157, one obtains

$$R_V(\tau) = \frac{|A_0|^2}{2}\cos(2\pi\nu_0\tau)\,e^{-2\int_0^{+\infty} S_\phi^{1-sided}(f)\sin^2(\pi f\tau)\,df} \tag{2.161}$$

Finally, from the WK theorem we have

$$
S_V^{1-sided}(f) = 2S_V^{2-sided}(f) = 2 \int_{-\infty}^{+\infty} R_V(\tau) e^{-i2\pi f \tau} d\tau =
$$

$$
2 \int_{-\infty}^{+\infty} R_V(\tau) \cos(2\pi f \tau) \, d\tau = 4 \int_0^{+\infty} R_V(\tau) \cos(2\pi f \tau) \, d\tau
$$

$$
= |A_0|^2 \int_0^{+\infty} \{ \cos[2\pi(\nu + f_0)\tau]
$$

$$
+ \cos[2\pi(\nu - f_0)\tau] \} e^{-2 \int_0^{+\infty} S_\varphi^{1-sided}(f) \sin^2(\pi f \tau) df} d\tau \tag{2.162}
$$

that, neglecting the rapidly varying term $\cos[2\pi(\nu + f_0)\tau]$, becomes

$$
S_V^{1-sided}(f)
$$

$$
= |A_0|^2 \int_0^{+\infty} \cos[2\pi(f - \nu_0)\tau] \cdot e^{-2 \int_0^{+\infty} S_\varphi^{1-sided}(f) \sin^2(\pi f \tau) df} d\tau \tag{2.163}
$$

This equation allows to calculate the power spectrum if the frequency-noise power spectral density is known. This formula will prove particularly useful in Chapter 4 in the study of the laser linewidth. Note that the real part of the signal defined by Equation 2.156 may also result from the beating of two oscillators (Equation 2.45) provided that we make the identifications $\omega_0 = \omega_1 - \omega_2$ and $\varphi(t) = \varphi_1(t) - \varphi_2(t)$. In that case, the power spectrum of the observed beat note $S_V^{1-sided}(f)$ gives information about the relative phase noise between the two beating oscillators. Of course, if one of the two oscillators is a reference ($\varphi_1(t) \ll \varphi_2(t)$), this procedure returns the power spectral density of the oscillator under test $S_\varphi^{1-sided}(f)$.

It is also useful (see below) to develop an approximate expression of Equation 2.163 for very low phase fluctuations ($\int_0^{+\infty} S_\varphi^{1-sided} df \ll 1$). To this aim, by virtue of Equations 2.158 and 2.159, let us rewrite Equation 2.157 as

$$
R_V(\tau) = \frac{|A_0|^2}{2} \cos(\omega_0 \tau) e^{-R_\varphi(0) + R_\varphi(\tau)}
$$

$$
\simeq \frac{|A_0|^2}{2} \cos(\omega_0 \tau)[1 - R_\varphi(0) + R_\varphi(\tau)] \tag{2.164}
$$

such that $S_V^{1-sided}$ now takes the form

$$
S_V^{1-sided}(f) \simeq |A_0|^2 \int_0^{+\infty} \cos[2\pi(f - \nu_0)\tau][1 - R_\varphi(0) + R_\varphi(\tau)] d\tau
$$

$$
= \frac{|A_0|^2}{2} \int_{-\infty}^{+\infty} e^{i2\pi(f - \nu_0)\tau}[1 - R_\varphi(0) + R_\varphi(\tau)] d\tau
$$

$$
= \frac{|A_0|^2}{2} \left\{ [1 - R_\varphi(0)] \int_{-\infty}^{+\infty} e^{i2\pi(f - \nu_0)\tau} d\tau + S_\varphi^{2-sided}(f - \nu_0) \right\}
$$

$$
= \frac{|A_0|^2}{2} \left\{ [1 - R_\varphi(0)] \delta(f - \nu_0) + S_\varphi^{2-sided}(f - \nu_0) \right\}
$$

$$
= \frac{|A_0|^2}{2} \left\{ [1 - \int_0^{+\infty} S_\varphi^{1-sided}(f)] \delta(f - \nu_0) + S_\varphi^{2-sided}(f - \nu_0) \right\}
$$

$$
\simeq |A_0|^2 \left\{ e^{-\int_0^{+\infty} S_\varphi^{1-sided}(f) df} \cdot \delta(f - \nu_0) + S_\varphi^{2-sided}(f - \nu_0) \right\} \tag{2.165}
$$

Hence, the spectrum in the carrier frequency domain consists of a delta function at $f = \nu_0$ (the carrier) plus two symmetric sidebands with the level of the phase-noise spectral density at $|f - \nu_0|$.

As a first example of calculation of Equation 2.163, we consider a source whose power spectral density in the Fourier frequency domain can be represented as white (frequency independent) frequency noise. By virtue of Table 2.1 one can write

$$S_\varphi^{1-sided}(f) = \frac{\nu_0^2 h_0}{f^2} \tag{2.166}$$

In this case, using $\int_0^{+\infty} [\sin^2(Af)]/f^2 df = \pi |A|/2$, one obtains from Equation 2.163

$$S_V^{1-sided}(f) = |A_0|^2 \int_0^{+\infty} \cos\left[2\pi(f - \nu_0)\tau\right] e^{-\pi^2 \nu_0^2 h_0 \tau} d\tau =$$

$$= |A_0|^2 \frac{\nu_0^2 h_0 \pi^2}{\nu_0^4 h_0^2 \pi^4 + 4\pi^2 (f - \nu_0)^2} \tag{2.167}$$

Hence, the power spectrum (of frequency fluctuations) in the carrier frequency domain of an oscillator with white frequency noise in the Fourier frequency domain is a Lorentzian whose FWHM is given by

$$FWHM = \pi \nu_0^2 h_0 \equiv \pi S_{\delta\nu}^0 \tag{2.168}$$

This suggests that the intrinsic (quantum-limited) linewidth of real oscillators, including lasers, can be simply estimated by measuring the floor $S_{\delta\nu}^0$ in the PSD of $S_{\delta\nu}(f)$ (see also Chapter 4 for a more detailed discussion in the framework of lasers).

In general, Equation 2.163 can be computed numerically for each of the other noise terms in Table 2.1, provided that $S_\varphi(f)$ is multiplied by the square modulus of a suitable low-pass transfer function $\mathcal{L}(if)$ (always present in practical situations): $S'_\varphi(f) \equiv S_\varphi(f) \cdot |\mathcal{L}(if)|^2$. Some notable cases, like white and flicker phase noise, have also been analytically solved [82]. In the former, particularly meaningful case, a single-pole RC filter is used such that

$$S'_\varphi(f) = \frac{b_0}{1 + (f/f_n)^2} \tag{2.169}$$

where f_n is the -3dB filter bandwidth. Then, by substitution of Equation 2.169 into Equation 2.163, we obtain

$$S_V^{1-sided}(f) = \frac{|A_0|^2}{2\pi f_n} e^{-\sigma_\varphi^2} \left\{ \pi \delta(\Delta) \right.$$

$$\left. + \sigma_\varphi^2 \Re \left[\frac{{}_2F_2(1, 1 - i\Delta; 2, 2 - i\Delta; \sigma_\varphi^2)}{1 - i\Delta} \right] \right\} \tag{2.170}$$

where $\Delta = (f - \nu_0)/f_n$, ${}_2F_2(a_1, a_2; b_1, b_2; \sigma_\varphi^2)$ is a generalized hypergeometric function and

$$\sigma_\varphi^2 = \int_0^{+\infty} S'_\varphi(f) df = \frac{\pi b_0 f_n}{2} \tag{2.171}$$

is the total amount of phase noise. Then, if the latter quantity is well below 1 rad ($\sigma_\varphi^2 \to 0$), the spectrum consists of an infinitely narrow carrier (Dirac delta function) sitting on a Lorentzian pedestal. If the phase noise increases, the pedestal expands in height in relation to the carrier, preserving the linewidth as determined by the filter. Further, as the integral of phase noise approaches 1 rad, an interesting phenomenon occurs: the pedestal starts to spread out and the carrier suddenly falls into the pedestal and disappears. If the noise level increases further, the pedestal spreads out proportionally to n^2 [83]. This phenomenon, called carrier collapse, was experimentally demonstrated more than two decades ago [84].

2.5.3.1 Effect of frequency multiplication on the power spectrum

When the frequency of a signal is multiplied (divided) by a factor n, the resulting $S_\varphi(f)$ is multiplied (divided) by a factor n^2. This can be readily seen as follows [85]. Since the only measurable quantity is the accumulated phase, independently of the implemented multiplication technique, the resulting signal takes the form

$$V_n(t) = V_0 \cos\left[n2\pi\nu_0 t + n\varphi(t)\right] \tag{2.172}$$

in the case of multiplication, and

$$V_{1/n}(t) = V_0 \cos\left[(1/n)2\pi\nu_0 t + (1/n)\varphi(t)\right] \tag{2.173}$$

in the case of division. In deriving the above equations, Equation 2.57 with $\alpha \ll 1$ was used. Now, the remainder of the proof is straightforward using the definition of $S_\varphi(f)$ as the Fourier transform of the autocorrelation function of $\varphi(t)$.

In the following, we focus on the multiplication process. In practice, the noise added by the specific multiplication process must also be accounted for, whereupon

$$S_\varphi(n\nu_0, f) = n^2 S_\varphi(\nu_0, f) + \text{Multiplication PM} \tag{2.174}$$

Firstly, consider white frequency noise. In this case, it is trivial to show that the multiplication process simply increases the linewidth of the Lorentzian shape by a factor n^2. In the case of white phase noise, instead, as discussed above, the increase in noise causes the carrier collapse. More in general, under the assumption of small PM noise, equation 2.165 can be used: according to Equation 2.174, it predicts that the power in the carrier decreases exponentially as e^{-n^2} such that, after a sufficiently large multiplication factor n, the carrier power density is less than the PM noise power. The calculation of the collapse frequency of a signal is particularly important because it represents the maximum frequency that can be coherently achieved when multiplying a given signal.

Finally, consider frequency translation. This latter has the effect of adding the PM noise of the input signal at ν_{input} and the reference signal at ν_{ref} to that of the PM noise in the non-linear device (or any other mechanism) providing the translation

$$S_\varphi(\nu_2, f) = S_\varphi(\nu_{input}, f) + S_\varphi(\nu_{ref}, f) + \text{Translation PM} \tag{2.175}$$

Incidentally, this suggests that dividing a high-frequency signal, rather than mixing two high-frequency signals, generally produces a low frequency reference signal with less residual noise.

In the field of modern absolute frequency metrology, frequency multiplication represented a crucial issue in the design of femto-second (fs) dividers which are, in essence, very high order frequency multipliers. Indeed, as we will see in Chapter 6, a mode-locked fs laser provides ultra-short pulses (~ 50 fs) at a repetition rate f_r (typically 100 MHz to a few GHz). In the frequency domain this corresponds to a comb

$$\nu_n = n\nu_r + \nu_{ceo} \qquad (2.176)$$

where n is the harmonic number and ν_{ceo} is the offset frequency from zero. Thus, in the light of the above discussion, we have

$$S_\varphi(\nu_n, f) = n^2 S_\varphi(\nu_r, f) + S_\varphi(\nu_{ceo}, f) + (\text{Mode-Locking PM}) \qquad (2.177)$$

where all the noise contributions arising from the mode-locking process itself have been lumped into the last term. The relative weight of the various noise terms in the above equation may change for different configurations. In particular, for ν_n to fall in the optical domain, an extremely high harmonic number is required ($n \sim 500000$, $\nu_r = 1$ GHz). This implies that the phase noise of the RF reference is increased by 115 dB when it is effectively multiplied up to the optical domain using the fs laser comb. For example, a high-quality quartz-based RF reference at 1 GHz might have a typical noise floor of -110 dBc/Hz, but, when it serves as reference for ν_r, there will be nothing remaining of the phase-coherent carrier in the optical mode ν_n (carrier collapse). There are, however, a few ways to minimize this phase-noise multiplication problem [86]. First of all one can rely on the relatively good short-term stability of the fs laser itself. On time scales of less than ~ 1 ms, the noise of a typical Ti:Sa laser has been measured to fall below that of the most high-quality microwave sources. This means that if one controls ν_r relative to a microwave reference, a control bandwidth of ≤ 1 kHz is all that is required to remove the low-frequency thermal and acoustic noise of the mode-locked laser. Although the observed optical linewidth may still be on the order of $0.1 - 1$ MHz, one generally finds that the fractional stability of the comb elements can be equal to the fractional stability of the microwave reference. Of course, if a microwave reference with lower phase noise is available (such as a cryogenic sapphire microwave oscillator), then one can potentially narrow the optical linewidth further.

2.6   Noise reduction in oscillators

In this section we take up the issue of frequency and phase locking between two oscillators, but from a different point of view with respect to that of Section 2.1.1. Indeed, here we focus on noisy self-sustained oscillators and discuss two relevant schemes that, hinging on phase locking between a reference (master) oscillator and a noisier one (slave), can be used for reduction of noise in the latter. We are speaking of the phase-locked loop and the injection locking scheme.

2.6.1   Phase-locked loops

A phase-locked loop (PLL) is a circuit that synchronizes the signal from an oscillator with a reference signal, so that they operate at the same frequency [87]. The synchronized signal is usually a voltage-controlled oscillator (VCO), that is an oscillator whose frequency changes in response to a control voltage (typically an astable multivibrator). The loop synchronizes the VCO to the reference by comparing their phases and controlling the VCO in a manner that tends to maintain a constant phase relationship between the two. In some types of PLLs this phase relationship is held constant, while in other types it is allowed to vary somewhat (in any case the frequency is always synchronized). A block diagram of the basic PLL is shown in Figure 2.15.

The phase detector (a flip-flop phase detector or a doubly balanced mixer) compares

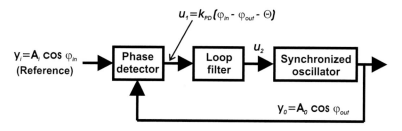

FIGURE 2.15
Block diagram of the basic PLL. (Adapted from [87].)

the reference to the VCO output and produces a signal u_1 that is proportional to the difference in their phases. This is processed by the loop filter to provide the oscillator control signal u_2. The loop filter can be as simple as a flat amplifier ($u_2 = K_{LF}u_1$), but it is usually designed to provide some advantageous response characteristic. If the output frequency, $\omega_{out} = d\varphi_{out}/dt$, should be greater than the reference frequency, $\omega_{in} = d\varphi_{in}/dt$, then $u_1 \sim (\varphi_{in} - \varphi_{out} - \Theta)$ would necessarily decrease with time, causing u_2 to decrease, which, in turn, would cause ω_{out} to decrease, bringing ω_{out} down toward ω_{in}. Thus the PLL provides negative feedback to keep the output frequency ω_{out} equal to the reference frequency ω_{in}. The output amplitude A_0 is constant and independent of the input amplitude A_i. Figure 2.16 shows the type of response that we would like to get from a phase detector (PD).

It produces a voltage proportional to the difference in phases of the reference φ_{in} and the VCO output, which is also the loop output, φ_{out}. The constant of proportionality, K_{PD}, is the gain of the phase detector

$$K_{PD} = \frac{du_1}{d(\Delta\varphi)} \tag{2.178}$$

where $\Delta\varphi = \varphi_{in} - \varphi_{out}$. In its simplest form, the loop filter consists merely of an amplifier, so that

$$\frac{du_2}{du_1} = K_{LF} \tag{2.179}$$

Analogously, the VCO can be characterized by

$$\frac{d\omega_{out}}{du_2} = K_{VCO} \tag{2.180}$$

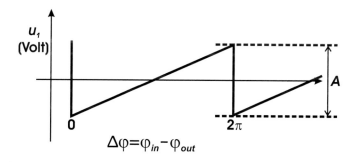

FIGURE 2.16
Response of an ideal phase detector. (Adapted from [87].)

The open-loop gain is thus given by

$$\frac{d\omega_{out}}{d\left(\Delta\varphi\right)} = \frac{du_1}{d\left(\Delta\varphi\right)}\frac{du_2}{du_1}\frac{d\omega_{out}}{du_2} = K_{PD}K_{LF}K_{VCO} \equiv K \tag{2.181}$$

In order to calculate the range of ω_{in} over which lock can be maintained, we observe that if the PD output has a sawtooth characteristic which can vary over a total range of A, this equals $2\pi K_{PD}$ rad or, equivalently, $\pm\pi K_{PD}$ rad about the midpoint. We obtain the corresponding change in ω_{out} by multiplying the change in PD output by $K_{LF}K_{VCO}$, giving a total hold-in (or synchronization) range of

$$\pm\Omega_{H,sawtooth} = \pm\pi K_{PD}K_{LF}K_{VCO} \text{ rad} = \pm\pi K \text{ rad} \tag{2.182}$$

Now, we re-write Equation 2.181 as

$$\frac{d\omega_{out}}{dt} = K\frac{d\left(\Delta\varphi\right)}{dt} = K\left(\omega_{in} - \omega_{out}\right) \tag{2.183}$$

If the reference ω_{in} is constant, it may be subtracted on the left side, so that we can write

$$\frac{d\left(\Delta\omega\right)}{\Delta\omega} = -Kdt \tag{2.184}$$

where $\Delta\omega \equiv \omega_{out} - \omega_{in}$. The solution of this equation is

$$\Delta\omega\left(t\right) = \Delta\omega\left(0\right)e^{-Kt} \tag{2.185}$$

Finally, the output frequency is given by

$$\omega_{out} = \omega_{in} + \Delta\omega\left(t\right) = \omega_{in} + \Delta\omega\left(0\right)e^{-Kt} \tag{2.186}$$

Thus we see that this simple loop responds to an initial frequency error $\Delta\omega\left(0\right)$ by exponentially decreasing the error with a time constant equal to $1/K$. The mathematical relationships in the Laplace domain for the loop can be shown by means of the block diagram in Figure 2.17.

Here the input variable is the reference, or input, instantaneous phase $\varphi_{in}\left(t\right)$. A block is shown preceding φ_{in} in order to establish the integral relationship $\left(1/s\right)$ between $\omega_{in}\left(s\right)$ and $\varphi_{in}\left(s\right)$. The output from the adder (a subtracter in this case) is the phase error $\varphi_e\left(s\right)$, the difference between the input phase and the VCO phase, φ_{out}. The phase error is converted to voltage in the phase detector, which is represented by the gain $K_{PD} = u_1\left(s\right)/\varphi_e\left(s\right)$. For this simple case the loop filter is merely an amplifier with gain K_{LF}; K_{VCO} represents the tuning sensitivity $\omega_{out}\left(s\right)/u_2\left(s\right)$ of the VCO. The output from this block is the output frequency ω_{out}. However, φ_{out} is needed to complete the loop so $\omega_{out}\left(s\right)$ is integrated (multiplied by $1/s$) to produce it. Then the loop is completed by subtracting $\varphi_{out}\left(s\right)$ from $\varphi_{in}\left(s\right)$ in the summer. Note that the minus sign at the summer represents $-180°$ of phase shift around

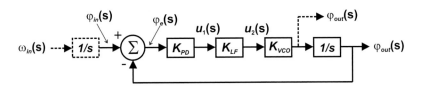

FIGURE 2.17
PLL mathematical relationships in the Laplace domain. (Adapted from [87].)

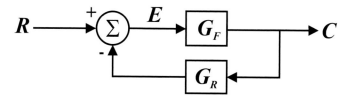

FIGURE 2.18
Generic control system block diagram. (Adapted from [87].)

the loop that does not appear in the transfer functions of the individual blocks. The generic control system block diagram is shown in Figure 2.18.

The well-known equations describing its transfer function are the response of the controlled variable C to the reference R,

$$\frac{C}{R} = \frac{G_F}{1 + G_R G_F} \tag{2.187}$$

where G_F and G_R are forward and reverse transfer functions, respectively, and the response of the error E to the reference

$$\frac{E}{R} = \frac{1}{1 + G_R G_F} \tag{2.188}$$

Note that $-G_R G_F$ is the open-loop gain and C/R and E/R are closed-loop gains. So we identify C as $\varphi_{out}(s)$ and R as $\varphi_{in}(s)$. Since the feedback path has unity gain, we have also $G_R = 1$ and $-G_F$ is the entire open-loop transfer function $-G(s)$

$$G(s) = \frac{K_{PD} K_{LF} K_{VCO}}{s} \equiv \frac{K}{s} \tag{2.189}$$

Thus from Equation 2.187 we obtain

$$\frac{\varphi_{out}(s)}{\varphi_{in}(s)} = \frac{G(s)}{1 + G(s)} = \frac{1}{1 + s/K} \tag{2.190}$$

and from Equation 2.188

$$\varphi_e(s) = \varphi_{in}(s) - \varphi_{out}(s) \Rightarrow \frac{\varphi_e(s)}{\varphi_{in}(s)} = \frac{1}{1 + G(s)} = \frac{s}{s + K} \tag{2.191}$$

Equation 2.190 represents a low-pass characteristic with a cutoff (-3 dB) frequency of $\omega = K$. This is analogous to the low-pass filter of Figure 2.19 in which the voltages have been given names corresponding to the phases. Equation 2.191 says that the error phase has a high-pass characteristic, analogous to Figure 2.19.

It is generally true, even in more complex loops, that the output has a low-pass relationship to the input while the error has a high-pass relationship. This reflects the fact that the error responds immediately to a change in the input while, and because, the output response is delayed. At frequencies well below $\omega = K$, Equation 2.190 can be approximated as

$$\frac{\varphi_{out}(s)}{\varphi_{in}(s)} = 1 \tag{2.192}$$

whereas at frequencies well above $\omega = K$ the equation is approximately

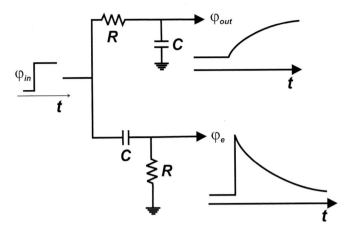

FIGURE 2.19
Low-pass (high-pass) behavior of the output-to-input ratio (error) in a PLL. (Adapted from [87].)

$$\frac{\varphi_{out}(s)}{\varphi_{in}(s)} = \frac{K}{s} = -i\frac{K}{\omega} \tag{2.193}$$

This equals the open-loop transfer function, Equation 2.189. Thus, at low frequencies, where the gain is high, the output follows the input faithfully, whereas, at high frequencies, where the gain becomes low, the loop is essentially open and the response is as if there were no loop (except that the low-frequency gain keeps it locked - otherwise none of these equations would be valid). At the loop corner frequency, $\omega = K$, it is easy to show from Equation 2.190 that the closed-loop gain is $1/\sqrt{2}$, or -3 dB, and the phase shift is $-45°$, just as in the case of the low-pass filter.

For educational purposes we have started our study with first-order PLLs (for which the loop filter transfer function is a constant, i.e., a wide-bandwith amplifier) that are never used in practice. Effective locking circuits make use of second-order PLLs (i.e., characterized by a transfer function with two poles) which can be obtained by using, for example, a low-pass filter of the type

$$H_{LPF} = \frac{1}{1 + \dfrac{s}{\omega_{LPF}}} \tag{2.194}$$

where ω_{LPF} is the cutoff frequency. This can be provided, for instance, by a RC network ($\omega_{LPF} = RC$). In this case we have for the open-loop transfer function

$$G(s) = \frac{K_{PD}K_{VCO}}{s}\frac{1}{1 + \dfrac{s}{\omega_{LPF}}} = \frac{K_{PD}K_{VCO}\omega_{LPF}}{\omega_{LPF}s + s^2} \tag{2.195}$$

and hence for the closed-loop transfer function

$$H(s) \equiv \frac{\varphi_{out}(s)}{\varphi_{in}(s)} = \frac{G(s)}{1 + G(s)} = \frac{K_{PD}K_{VCO}\omega_{LPF}}{\omega_{LPF}s + s^2 + K_{PD}K_{VCO}\omega_{LPF}} \tag{2.196}$$

By comparison with a second-order damped oscillator with unitary DC gain ($\lim_{s\to 0} H_{2or}(s) = 1$)

$$H_{2or}(s) = \frac{\omega_0^2}{2\zeta\omega_0 s + s^2 + \omega_0^2} \tag{2.197}$$

one recognizes the PLL natural angular frequency (roughly corresponding to the circuit bandwidth) as

$$\omega_n = \sqrt{K_{PD}K_{VCO}\omega_{LPF}} \tag{2.198}$$

and the damping constant

$$\zeta = \frac{1}{2}\sqrt{\frac{\omega_{LPF}}{K_{PD}K_{VCO}}} \tag{2.199}$$

Now, if we want a fast system, but without over-elongations, it is necessary to choose the critical damping condition $\zeta = 1/\sqrt{2}$. This yields a well-defined relationship between the low-pass filter bandwidth and the gain

$$\frac{\omega_{LPF}}{2} = K_{PD}K_{VCO} \tag{2.200}$$

Next, we calculate the steady-state phase error for a ramp input $\Delta\omega\cdot t$. The corresponding Laplace transform is $\varphi_{in}(s) = \Delta\omega/s^2$. Therefore, by virtue of Equation 2.196, we can write

$$\begin{aligned}
\varphi_{err}(s) &= \varphi_{in}(s)\left[1 - \frac{\varphi_{out}(s)}{\varphi_{in}(s)}\right] = \frac{\Delta\omega}{s^2}\left[1 - H(s)\right] \\
&= \frac{\Delta\omega}{s^2}\frac{\omega_{LPF}s + s^2}{\omega_{LPF}s + s^2 + K_{PD}K_{VCO}\omega_{LPF}}
\end{aligned} \tag{2.201}$$

By applying the final value theorem we finally get

$$\varphi_{err}(t \to \infty) = \lim_{s\to 0} s\varphi_{err}(s) = \frac{\Delta\omega}{K_{PD}K_{VCO}} \tag{2.202}$$

This means that the PLL mantains a non-zero error, inversely proportional to the gain. The latter however cannot be increased ad libitum due to the proportionality with ω_{LPF} which should be not too high for efficient filtering of the output of the phase comparator.

Another kind of loop filter, denoted as lag-lead filter, is represented in Figure 2.20.

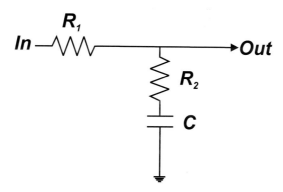

FIGURE 2.20
Passive lag-lead loop filter.

Its transfer function is

$$K_{LF}(s) = \frac{v_2(s)}{v_1(s)} = \frac{1 + R_2 Cs}{1 + (R_1 + R_2) Cs} = \frac{1 + s/\omega_1}{1 + s/\omega_2} = \frac{1 + \tau_1 s}{1 + \tau_2 s} \tag{2.203}$$

In this case, the open-loop transfer function is

$$G(s) = \frac{K_{PD} K_{VCO}}{s} K_{LF}(s) = \frac{K_{PD} K_{VCO}}{s} \frac{1 + s/\omega_1}{1 + s/\omega_2} \tag{2.204}$$

while the closed-loop transfer function is

$$H(s) = \frac{\varphi_{out}(s)}{\varphi_{in}(s)} = \frac{1}{1 + \dfrac{1}{G(s)}}$$

$$= \frac{\dfrac{K_{PD} K_{VCO}}{\tau_2} + \dfrac{\tau_1}{\tau_2} K_{PD} K_{VCO} s}{s^2 + \dfrac{K_{PD} K_{VCO}}{\tau_2} + \left(\dfrac{1 + K_{PD} K_{VCO} \tau_1}{\tau_2} \right) s} \tag{2.205}$$

which yields

$$\begin{cases} \omega_n = \sqrt{\dfrac{K_{PD} K_{VCO}}{\tau_2}} \\[4mm] \zeta = \dfrac{1}{2\omega_0 \tau_2} + \dfrac{\omega_0 \tau_1}{2} \end{cases} \tag{2.206}$$

Therefore, in this case it is possible to choose ζ and ω_0 independently by properly choosing R_1, R_2, and C.

As a final example, consider the charge-pump PLL. It is one of the most popular PLL structures since the 1980s. Instead of a standard phase comparator, it employs a more sophisticated digital circuit called a phase and frequency detector (PFD). Unlike classical phase detectors, the latter gives a non-zero output signal even if the two input frequencies are different. The PFD has two outputs, UP and DOWN, that can alternatively have a high signal. Such outputs are connected to a circuit referred to as a charge pump whose output drives the VCO. Charge-pump PLLs have several advantages over the traditional ones. As already mentioned, the PFD output is non-zero even in the presence of input signals with different frequencies. In this way, it is not necessary to keep the loop filter bandwidth high to allow the PLL to lock signals with sudden frequency changes. In addition, the charge pump acts as a further integrator (besides the VCO), thus giving rise to a zero steady-state phase error (for a ramp input). The block diagram for a charge-pump PLL is shown in Figure 2.21. The charge pump output is, in fact, a train of current pulses I_p which charge the capacity of the low-pass filter.

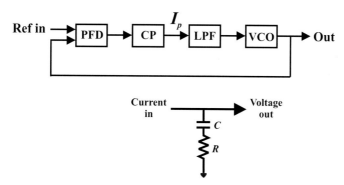

FIGURE 2.21
Block diagram of a charge-pump PLL and scheme of the corresponding low-pass filter.

In this configuration, the open-loop transfer function is calculated as

$$G\left(s\right) = \frac{I_p}{2\pi C} K_{VCO} \frac{sRC + 1}{s^2} \tag{2.207}$$

from which the corresponding closed-loop function is calculated as

$$H(s) = \frac{\dfrac{I_p}{2\pi C} K_{VCO}(sRC + 1)}{s^2 + \dfrac{I_p}{2\pi C} K_{VCO} + \dfrac{I_p}{2\pi} K_{VCO} R s}$$

$$\equiv \frac{\omega_n^2 + 2\zeta\omega_n s}{s^2 + \omega_n^2 + 2\zeta\omega_n s} \tag{2.208}$$

that is again a second-order with

$$\begin{cases} \omega_n = \sqrt{\dfrac{I_p}{2\pi C} K_{VCO}} \\[4mm] \zeta = \dfrac{R}{2}\sqrt{\dfrac{I_p C}{2\pi} K_{VCO}} \end{cases} \tag{2.209}$$

with a zero at $\omega_{zero} = -1/(RC)$. Now we can address the issue of phase noise reduction. For this purpose, consider a noiseless reference and allow the VCO output to have a spurious component φ_{VCO} which can be modeled as a noise input added to the output of a noiseless VCO (see Figure 2.22).

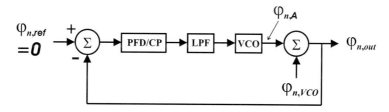

FIGURE 2.22
Block diagram for calculating the noise transfer function in a PLL.

Considering only the noisy components of phases we have

$$\begin{cases} \varphi_{n,out} = \varphi_{n,A} + \varphi_{n,VCO} \\ \\ \varphi_{n,A} = -G(s)\,\varphi_{n,out} \end{cases} \tag{2.210}$$

from which one gets

$$N(s) = \frac{\varphi_{n,out}}{\varphi_{n,VCO}} = \frac{1}{1+G(s)} \tag{2.211}$$

By substitution of Equation 2.207 into Equation 2.211 one obtains the noise transfer function as

$$N(s) = \frac{s^2}{s^2 + 2\omega_n \zeta s + \omega_n^2} \tag{2.212}$$

Note that $N(s) \equiv 1 - H(s)$. As shown in Figure 2.23, Equation 2.212 tells us that the phase noise of the VCO is reduced within a bandwith defined by ω_n, while above this frequency, it is reported to the output without corrections.

Thus, if we assume for the VCO pure white frequency noise

$$S_\varphi^{1-sided}(f) = (\nu_0^2 h_0)/f^2 \equiv \Delta\nu/\pi f^2 \tag{2.213}$$

then the residual phase variance can be calculated as

$$\sigma^2 \equiv \int_0^{+\infty} S_\varphi^{1-sided}(f)|N(f)|^2 df \tag{2.214}$$

By plugging Equation 2.213 and Equation 2.212 into Equation 2.214, we obtain

$$\sigma^2 = \frac{\Delta\nu}{\pi}\int_0^{+\infty} \frac{f^2/f_n^4}{1+(f/f_n)^4}df = \frac{\Delta\nu}{\pi} = \frac{\Delta\nu}{\pi f_n}\int_0^{+\infty} \frac{z^2}{1+z^4}dz$$

$$= \frac{\Delta\nu}{\sqrt{2}f_n} \tag{2.215}$$

where $s = i2\pi f$, $\omega_n = 2\pi f_n$ have been used, $\zeta = 1/\sqrt{2}$ has been chosen (critically damped loop), and $z = f/f_n$ has been introduced.

2.6.1.1 Optical phase-locked loops

When studying the beat-note phenomenon, we have already mentioned that, if the output of two lasers is combined upon the surface of a photodiode, then the inherent non-linearity of the detector will give rise to an internal oscillating field at a frequency equal to the difference frequency of the two lasers. Then, the photodiode output signal can be used as the input to a phase or frequency locking control system which can maintain the frequency separation between the lasers as a constant. In this case, the beat-note signal plays the role of the synchronized oscillator (VCO), its frequency being locked against the rf/microwave reference. Here, the width of the beat-note signal plays the role of $\Delta\nu$ in Equation 2.215.

One major difficulty of such electronic phase locking is the high level of phase fluctuations of typical laser sources with respect to conventional signal sources in the rf and microwave domains [88]. This causes two potential problems: first, the phase-locked loop needs a wide bandwidth to suppress the phase fluctuations to an acceptable level; and, second, there is a

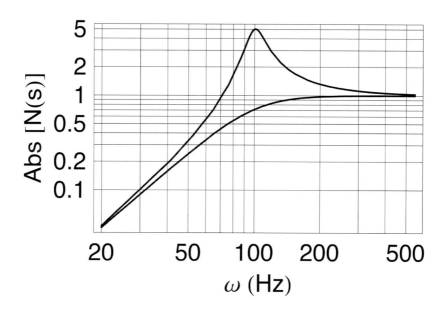

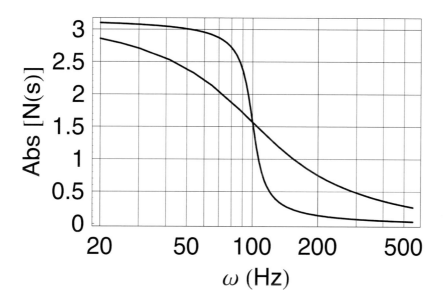

FIGURE 2.23
Log-Log plot and Log-Linear plot of the amplitude and the argument of Equation 2.212 with $\omega = 100$ Hz and for two different values of ζ: 0.1 (gray curve) and 0.7 (black curve.)

probability that the instantaneous residual phase error at the phase detector of the PLL may exceed the detector's dynamic range. If this occurs, then there may be a permanent loss or gain of a whole number of phase cycles between the slave laser diode and the reference laser. This event termed as *cycle slip* arises because the output of the phase detector is a periodic function of the phase. It is clear that when the residual phase variance is significantly smaller than the range of the phase detector, the probability of a cycle slip is predicted to become negligible.

Clearly, if the in-loop phase fluctuations are to be below 1 rad, then it is necessary for the loop bandwidth to exceed the linewidth of the lasers by a significant fraction. For example, the linewidth of typical free-running laser diodes ranges from a few megahertz to several hundreds of megahertz, thus placing severe technical requirements on the control systems. Several approaches have been used to overcome these difficulties [88]:

- **High-Bandwidth and High-Gain Electronics.** Using state-of-the-art electronics, free-running diode lasers with linewidths on the order of 10 MHz have been phase-locked. The widest bandwidth loop achieved was 130 MHz [89].

- **Large Dynamic Range Phase Detectors.** Instead of using extremely high bandwidths to suppress the large free-running fluctuations within the dynamic range of a simple mixer, one can, alternatively, design more complex phase detectors that have broader dynamic ranges.

  Prevedelli et al. designed a digital phase detector with a range of $\pm 32\pi$ rad to lock two diode lasers with a linewidth in the range 25-50 kHz [89]. Using a maximum bandwidth of 1.7 MHz, the measured mean square phase fluctuations, σ^2, were reduced to around 0.2 rad^2. More importantly, though, they demonstrated that even using PLL bandwidths of around 20 kHz, the residual Allan frequency deviation was still reduced to the order of $7 \cdot 10^{-15}/\tau$. Under similar conditions a normal multiplying mixer-based PLL would have been cycle slipping so frequently that it would not have been useful. These *rubber-band* phase locks allow the experimenter to trade off the phase-lock circuitry speed against short-term phase fluctuations without needing to compromise long-term accuracy. Such a system was later perfected by use of an analog+digital phase detector. The analog and digital detectors are mutually exclusive so that only one of them is active at a given time, resulting in a phase detector with both the broad capture range of digital circuits and the high speed and low noise of analog mixers [91]. The rms phase error of the phase lock was about 100 mrad in a 5 Hz–10 MHz bandwidth. More recently, using an ultralow-noise, only-digital phase detector and appropriate loop filters, a phase-locked diode laser system with a residual phase-noise variance less than 0.02 rad^2 was also reported [92].

- **Pre-stabilization** By pre-stabilizing the laser to a reference optical cavity, one can reduce the free-running linewidth (see Chapter 3 and 4) by $3-5$ orders of magnitude, thus relaxing the requirements of the phase-locking circuit bandwidth.

2.6.2   Injection locking

Injection locking is a non-linear phenomenon that can be observed in many natural oscillators when an oscillator is perturbed by a weak signal whose frequency is close to the oscillator free-running frequency. In some sense, injection locking is a special type of forced oscillation in non-linear oscillators. Experimentally one observes the following effects: suppose that a signal of frequency ω_{inj} is injected into an oscillator, which has a self-oscillation (free-running) frequency ω_0. When ω_{inj} is quite different from ω_0, beats of the two frequencies are observed. When ω_{inj} enters some frequency range very close to ω_0 (and if coupling is strong enough), the beats suddenly disappear, and the oscillator starts to oscillate at ω_{inj} instead of ω_0. The frequency range in which injection locking happens is called the locking range. Injection locking also happens when ω_{inj} is close to a harmonic or sub-harmonic of ω_0, i.e., $n\omega_0$ or $1/n\omega_0$. They are called harmonic (or super-harmonic) and sub-harmonic injection locking, respectively. On the other hand, when the second oscillator merely disturbs the first but does not capture it, the effect is called injection pulling. Injection locking and pulling effects are observed in numerous types of physical systems; however, the terms are most often associated with electronic oscillators or laser resonators. In optics, injection locking has been used in lasers to improve the frequency stability and reduce the frequency noise of laser diodes. In electronic systems, injection locking is a well known and practical technique for phase-locked loops (PLLs) to increase pull-in range and reduce output phase jitter. As a result, fast and accurate prediction of injection locking is very important.

Here we follow the approach given in [93], by modelling the oscillator as a one-port circuit consisting of a parallel tank, whose loss is represented by G_1, and a mildly nonlinear negative conductance $-G_m$. The injecting (master) oscillator is represented by a current driver (see Figure 2.24).

In this circuit

$$C_1 \frac{d^2 V_{osc}}{dt^2} + (G_1 - G_m)\frac{dV_{osc}}{dt} + \frac{1}{L_1}V_{osc} = \frac{dI_{inj}}{dt} \tag{2.216}$$

Now let us assume

$$I_{inj}(t) = I_{inj,p}\cos\omega_{inj}t = \Re\left\{I_{inj,p}e^{i\omega_{inj}t}\right\} \tag{2.217}$$

and

$$V_{osc}(t) = V_{env}(t)\cos(\omega_{inj}t + \theta) = \Re\left\{V_{env}(t)e^{i\omega_{inj}t + i\theta(t)}\right\} \tag{2.218}$$

where $V_{env}(t)$ denotes the envelope of the output. In other words, the output is assumed to track the input except for a (posssibly time-varying) phase difference. Substituting the exponential terms in Equation 2.216 and separating the real and imaginary parts, we have

$$\begin{cases} C_1 \dfrac{d^2 V_{env}}{dt^2} - C_1\left(\omega_{inj} + \dfrac{d\theta}{dt}\right)^2 V_{env} \\ \quad + (G_1 - G_m)\dfrac{dV_{env}}{dt} + \dfrac{1}{L_1}V_{env} = \omega_{inj}I_{inj,p}\sin\theta \\[2ex] 2C_1\left(\omega_{inj} + \dfrac{d\theta}{dt}\right)\dfrac{dV_{env}}{dt} + C_1\dfrac{d^2\theta}{dt^2}V_{env} \\ \quad + (G_1 - G_m)\left(\omega_{inj} + \dfrac{d\theta}{dt}\right)V_{env} = \omega_{inj}I_{inj,p}\cos\theta \end{cases} \tag{2.219}$$

To develop more insight, let us study these results within the lock range, i.e., when $d\theta/dt = dV_{env}/dt = 0$. Moreover, we assume that the magnitude of the envelope can be

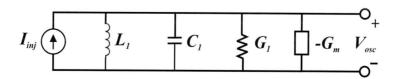

FIGURE 2.24
One-port representation of an oscillator under injection. (Adapted from [93].)

approximated as the tank peak current produced by the $-G_m$ circuit, $I_{osc,p}$, multiplied by the tank resistance $G_1^{-1} = QL_1\omega_0$ ($V_{env} = QL_1\omega_0 I_{osc,p}$), Q being the quality factor of the tank (note that in a RLC-parallel circuit the quality factor is the reciprocal of that in a RLC-series circuit (Equation 2.28), while $\omega_0 = 1/\sqrt{L_1 C_1}$ in both cases). Equations 2.219 thus reduce to

$$\begin{cases} \omega_0 - \omega_{inj} = \dfrac{\omega_{inj} I_{inj,p}}{2Q I_{osc,p}} \sin\theta \\[2mm] (G_1 - G_m) V_{env} = I_{inj,p} \cos\theta \end{cases} \tag{2.220}$$

where $\omega_{inj}^2 - \omega_0^2 \approx 2\omega_0 (\omega_0 - \omega_{inj})$ has also been exploited. The first is Adler's equation [94], from which the injection locking range is easily retrieved by using $\omega_{inj} \approx \omega_0$

$$\omega_0 - \omega_{inj} = \frac{\omega_0 I_{inj,p}}{2Q I_{osc,p}} \sin\theta \tag{2.221}$$

Since the values of $\sin\theta$ can only be between -1 and $+1$, the maximum locking range is

$$\omega_L = \pm \frac{\omega_0 I_{inj,p}}{2Q I_{osc,p}} \tag{2.222}$$

and the threshold condition is expressed as

$$\frac{I_{inj,p}}{I_{osc,p}} > 2Q \frac{|\omega_0 - \omega_{inj}|}{\omega_0} \tag{2.223}$$

Now, by squaring Equations 2.220 and adding the results to each other, we get

$$\left(\frac{\omega_0 - \omega_{inj}}{\omega_L}\right)^2 + \left(\frac{G_1 - G_m}{I_{inj,p}} V_{env,p}\right)^2 = 1 \tag{2.224}$$

For $\omega_{inj} = \omega_0$

$$G_m = G_1 - \frac{I_{inj,p}}{V_{env,p}} \tag{2.225}$$

that is, the circuit responds by weakening the $-G_m$ circuit (i.e., allowing more saturation) because the injection adds in-phase energy to the oscillator. On the other hand, for $|\omega_0 - \omega_{inj}| = \omega_L$, we have $G_m = G_1$, as if there is no injection. Figure 2.25 illustrates the behavior of G_m across the lock range. While derived for a mildly non-linear oscillator, the above result does suggest a general effect: the oscillator must spend less time in the linear regime as ω_{inj} moves closer to ω_0. A *linear* oscillator therefore does not injection lock (a linear oscillator can be defined here as one in which the loop gain is exactly unity for all signal levels).

The phase noise of oscillators can be reduced by injection locking to a low-noise source. Using the one-port model and the identity expressed by Equation 2.225, we can estimate the phase noise reduction in a mildly non-linear oscillator that is injection-locked to a noiseless source. As depicted in Figure 2.26, the noise of the tank and the $-G_m$ cell can be represented as a current source I_n. In the absence of injection, the (average) value of $-G_m$ cancels G_1, and I_n experiences the following transimpedance

$$\left|\frac{V_{out}}{I_n}(i\omega_n)\right| \approx \frac{1}{|2(\omega_n - \omega_0) C_1|} \tag{2.226}$$

Thus, I_n is amplified by an increasingly higher gain as the noise frequency approaches

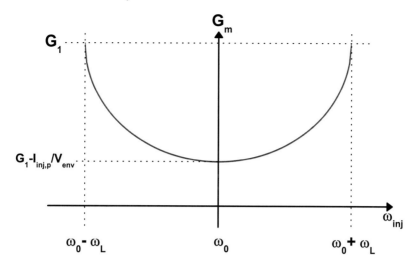

FIGURE 2.25
Behavior of G_m across the lock range according to Equation 2.224.

ω_0 (for very small frequency offsets, the noise shaping function assumes a Lorentzian shape and hence a finite value). Now suppose a finite injection is applied at the center of the lock range, $\omega_{inj} = \omega_0$. Then, Equation 2.225 predicts that the overall tank admittance rises to $G_1 - G_m = I_{inj,p}/V_{env,p}$. In other words, the tank impedance seen by I_n at ω_0 falls from infinity (with no injection) to $V_{env,p}/I_{inj,p}$ under injection locking. As the frequency of I_n deviates from ω_0, $V_{env,p}/I_{inj,p}$ continues to dominate the tank impedance up to the frequency offset at which the phase noise approaches that of the free-running oscillator (Figure 2.27). To determine this point, we equate the free-running noise shaping function of Equation 2.226 to $V_{env,p}/I_{inj,p}$ and note that $C_1/G_1 = Q/\omega_0$ and $V_{env,p} \approx I_{osc,p}/G_1$

$$|\omega_n - \omega_0| = \frac{\omega_0}{2Q} \frac{I_{inj,p}}{I_{osc,p}} \tag{2.227}$$

Thus, the free-running and locked phase noise profiles meet at the edges of the lock range. If the input frequency deviates from ω_0, the resulting phase noise reduction becomes less pronounced. In fact, as ω_{inj} approaches either edge of the lock range, $G_1 - G_m$ drops to zero, raising the impedance seen by the noise current.

In Chapter 4, with the proper identifications, these results will be straightforwardly extended to the case of a master laser injecting a noisier, but more powerful slave laser.

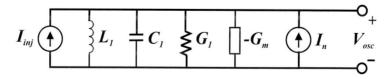

FIGURE 2.26
Model for studying phase noise in an injection-locked oscillator. (Adapted from [93].)

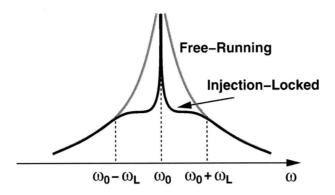

FIGURE 2.27
Reduction of phase noise due to injection locking. (Courtesy of [93].)

2.7 Phase noise measurements

Finally, let us discuss the measurement techniques which are usually employed to character-ize the phase and the amplitude in the output of a real (noisy) oscillator. Different methods are presented which apply to the general oscillator. In the next chapters, we will see how the described schemes specialize to the case of lasers.

2.7.1 Frequency counting

The frequency of electric signals up to a few GHz can be measured directly with an electronic counter. In this case, it is the Allan deviation, calculated on the collected data, to provide information about phase noise. From the previous chapter we learnt that since Galilei and Huygens invented the pendulum clock, time and frequency have been the quantities that we can measure with the highest precision. Measuring a frequency, i.e., counting the number of cycles during a given time interval, is intrinsically a digital procedure that is immune to many sources of noise. Moreover, the precision of time measurements can be increased essentially without limit, by increasing the measurement duration and simply counting the increased number of cycles of some regularly spaced events. However a stronger information growth with measurement duration is possible if we have a nice source that has coherence from the beginning of the measurement until the end. For the present purpose we may take this *coherence* to mean that, if we know the oscillation cycle phase early in the measurement, the coherent source is so steady that the oscillation phase could be predicted at later times near the measurement end to a precision of 1 radian of phase. In this case we can have a measurement precision which will grow with the measurement interval τ according to $\tau^{3/2}$ [52]. A simple way to explain this assertion is to suppose we divided the measurement duration into 3 equal sections, each with $N/3$ measurements. In the starting zone we compare the reference clock and the unknown clock, with a relative phase precision which scales as $(N/3)^{1/2}$. Next, in the middle section, we merely note the number of events, $N/3$. In the last section we again estimate the analog phase relationship between test and reference waves, with a relative imprecision which is again $(N/3)^{1/2}$. Subtracting the two analog phases increases the uncertainty of one measurement by a factor $2^{1/2}$ so, altogether, the relative precision increases as $(1/2^{1/2}) \cdot (N/3)^{3/2}$.

The conventional counter is a digital electronic device which measures the frequency of an input signal [95]. The latter is initially conditioned to a form that is compatible with the internal circuitry of the counter. The conditioned signal appearing at the door of the main gate is a pulse train where each pulse corresponds to one cycle or event of the input signal. With the main gate open, pulses are allowed to pass through and get totalized by the counting register. The time t_{gate} between the opening to the closing of the main gate is controlled by the time base (usually a 10-MHz quartz oscillator). The frequency, f, of a repetitive signal is measured by the conventional counter by counting the number of cycles, n_{cycle}, and dividing it by the time interval, t_{gate}

$$f = n_{cycle}/t_{gate} \tag{2.228}$$

The basic block diagram of the counter in its frequency mode of measurement is shown in Figure 2.28. The time base divider takes the time base oscillator signal as its input and provides as an output a pulse train whose frequency is variable in decade steps made selectable by the gate time switch. The time, t_{gate}, of Equation 2.228 or gate time is determined by the period of the selected pulse train emanating from the time base dividers. The number of pulses totalized by the counting register for the selected gate time yields the frequency of the input signal. The frequency counted is displayed on a visual numerical readout.

By summarizing, a frequency counter measures the input frequency averaged over a suitable time, versus the reference clock. High resolution is achieved by interpolating the clock signal. Further increased resolution is obtained by averaging multiple frequency measurements highly overlapped. In the presence of white phase noise, the square uncertainty improves from $\sigma_\nu^2 \propto 1/\tau^2$ to $\sigma_\nu^2 \propto 1/\tau^3$ [96]. In this case, however, care must be taken if the two-sample (Allan) variance is estimated through measurements by such a counter. Indeed, if the algorithm to retrieve the Allan variance is applied so naive, without properly accounting for the specific counter internal process, then a distorted representation of the Allan variance is obtained [97].

2.7.2 Homodyne techniques

For higher frequencies, electronic counters fail and alternative techniques must be used.

One possibility is to use a phase-detector, that is a device capable of converting phase to amplitude: such a discriminator can be the slope of a Fabry-Perot interferometer, of an

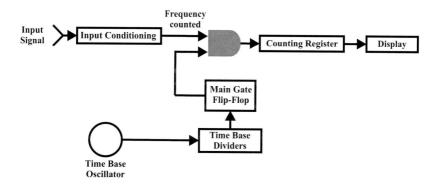

FIGURE 2.28
Basic block diagram of a frequency counter.

absorption lineshape, or of an electronic filter (Figure 2.29). As already shown (see Equation 2.32), if the oscillator or the filter is tuned such that the carrier frequency of the oscillator is at the slope, preferably near the inflection point ν_{ip}, the power transmitted by the filter varies, to first order, linearly with the frequency of the signal as

$$V(\nu - \nu_{ip}) = V(\nu_{ip}) + (\nu - \nu_{ip})K_\phi \tag{2.229}$$

The detector behind the filter converts the power fluctuations into the fluctuations of a voltage that can be analyzed by means of a spectrum analyzer. The application of this method requires knowledge of K_ϕ and that the contributions of other noise sources do not affect the measurement. For instance, fluctuations of the center frequency of the filter or fluctuations of the signal amplitude may mimic a higher spectral density of frequency fluctuations.

2.7.3 Heterodyne techniques

Alternatively, heterodyne techniques can be used. In this case the oscillator under test is compared against a reference one. The basic scheme is shown in Figure 2.30 [98]. The double-balanced mixer, saturated at both inputs, works as a phase-to-voltage converter. For example, an electronic mixer multiplies two input signals, often termed the RF and LO, leading to an output signal called the IF. When needed, an optional synthesizer makes the nominal frequencies equal.

For two input harmonic signals, the output of a doubly balanced mixer contains the sum and the difference of the two input signals but not the input signals or their harmonics

$$\begin{aligned} S_{out} &\propto \cos(\omega_{RF}t + \varphi_{RF})\cos(\omega_{LO}t + \varphi_{LO}) = \\ &= (1/2)\cos[(\omega_{RF} + \omega_{LO})t + (\varphi_{RF} + \varphi_{LO})] + \\ &\quad + (1/2)\cos[(\omega_{RF} - \omega_{LO})t + (\varphi_{RF} - \varphi_{LO})] \end{aligned} \tag{2.230}$$

When $\omega_{RF} = \omega_{LO} = \omega$, the output signal consists of an ac component of twice the input frequency (which can be removed by a low-pass filter) on a dc signal that depends on the phase difference between the reference and under test signals. Consequently, the mixer can be used as a discriminator to detect phase fluctuations, provided that the proportionality

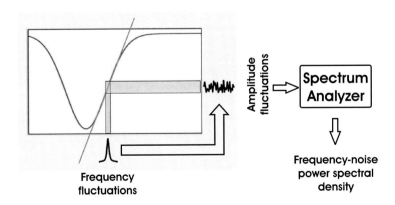

FIGURE 2.29
Homodyne measurement of frequency-noise power spectral density by means of a phase detector.

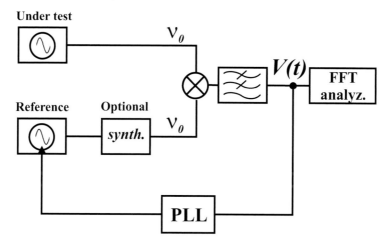

FIGURE 2.30
Heterodyne approach for characterizing phase noise in oscillators. (Adapted from [98].)

constant contained in Equation 2.230 has to be determined a priori. To have a constant slope for the phase discriminator, it has to be operated close to 90° during the measurement, where the cosine near zero can be approximated by a linear discriminant curve. This is the case only if a PLL is used to realize the condition $\omega_{RF} = \omega_{LO}$. The PLL uses the signal from the mixer as the error signal. If the voltage fluctuations from the mixer are measured by a spectrum analyzer, they are a measure of the fluctuations of the phase difference provided that amplitude modulation is negligible. These in turn can be attributed to the phase fluctuations of the oscillator under test only if the phase of the reference oscillator can be regarded as more stable. Modern spectrum analyzers show a quantity that is directly related to the spectral density of the fluctuations of the signal. Care has to be taken if a continuous signal is sampled digitally to obtain spectral densities. It is well known that a harmonic signal can be sampled digitally unambiguously only if at least two samples are taken per period T. The corresponding minimal sampling frequency $\nu_N = 2/T$ is referred to as the Nyquist frequency. Corruption of the power spectral density resulting from insufficient sampling is referred to as aliasing.

2.7.4 Self-heterodyning

Another useful approach is based on a delay-line measurement (Figure 2.31). The phase fluctuations of a single oscillator are determined by comparing a portion of the signal under study with a second part of the the same signal at a previous epoch. To this aim, after splitting the oscillator output and delaying the signal in one path with respect to the other one, the two portions are mixed again and amplified.

By inspection of Figure 2.31, in the Laplace domain we have

$$\Phi_0\left(s\right) = H_\Phi\left(s\right)\Phi_i\left(s\right) \tag{2.231}$$

where $H_\Phi\left(s\right) = 1 - e^{-s\tau}$. Turning the Laplace transforms into power spectra, the above equation becomes

$$S_{\Phi_0}\left(f\right) = \left|H_\Phi\left(f\right)\right|^2 S_{\Phi_i}\left(f\right) \tag{2.232}$$

where

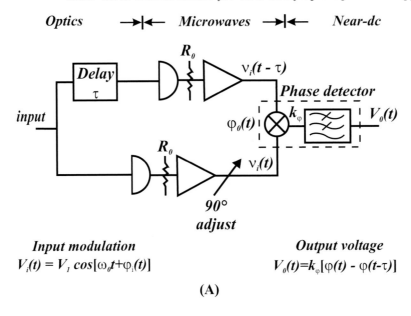

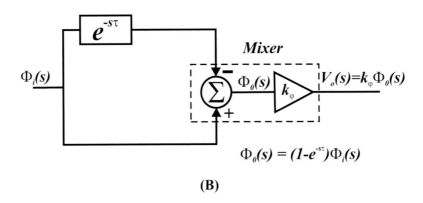

FIGURE 2.31
Self-heterodyne, or delay-line method to investigate phase noise in oscillators. (Adapted from [98].)

$$|H_\Phi\,(f)|^2 = 4\sin^2\,(\pi f\tau) \tag{2.233}$$

Equation 2.232 is used to derive the phase noise $S_{\Phi_i}\,(f)$ of the oscillator under test.

2.8   Amplitude noise measurements

In the case of amplitude noise, generally the spectrum contains only the white noise $h_0 f^0$, the flicker noise $h_{-1}f^{-1}$, and the random walk $h_{-2}f^{-2}$ [99]

$$S_\alpha\,(f) = h_0 + h_{-1}f^{-1} + h_{-2}f^{-2} \tag{2.234}$$

From Equation 2.57, the signal power is obtained as

$$P = \frac{V_0^2}{2R}(1+\alpha)^2 \cong \frac{V_0^2}{2R}(1+2\alpha) \qquad (2.235)$$

The resistance R is needed in the definition if one identifies $V(t)$ in Equation 2.57 with a voltage. It is convenient to write P as $P = P_0 + \delta P$ with $P_0 = V_0^2/2R$ and $\delta P = 2P_0\alpha$. The amplitude fluctuations are measured through the measurement of the power fluctuation δP

$$\alpha(t) = \frac{1}{2}\frac{\delta P}{P_0} \qquad (2.236)$$

and of its power spectrum density

$$S_\alpha(f) = \frac{1}{4}S_{P/P_0}(f) = \frac{1}{4P_0^2}S_P(f) \qquad (2.237)$$

Figure 2.32 shows the basic scheme for the measurement of AM noise.

The detector characteristic is $v_d = k_d P$; hence, the ac component of the detected signal is $\tilde{v}_d = k_d \delta P$ so that $\tilde{v}_d(t) = 2k_d P_0 \alpha(t)$. Turning voltages into spectra, the latter relationship becomes

$$S_\alpha(f) = \frac{1}{4k_d^2 P_0^2}S_v(f) \qquad (2.238)$$

Due to linearity of the network that precedes the detector (directional couplers, cables, etc.), the fractional power fluctuation $\delta P/P_0$ is the same in all the circuits, thus α is the same. As a consequence, the separate measurement of the oscillator power and of the attenuation from the oscillator to the detector is not necessary. The straightforward way to use Equation 2.238 is to refer P_0 at detector input, and $\tilde{v}_d$ at the detector output. Interestingly, phase noise has virtually no effect on the measurement. This happens when the bandwidth of the detector is much larger than the maximum frequency of the Fourier analysis; hence, no memory effect takes place. Of course, this measurement gives the total noise of the source and of the instrument, which can not be separated. However, if a reference source is available whose AM noise is lower than the detector noise, the latter can be previously estimated. If this is not the case, it is useful to compare two detectors, as in Figure 2.32B. The trick is to measure a differential signal $g(P_b - P_a)$, which is not affected by the power fluctuation of the source (this, of course, relies upon the assumption that the two detectors are about equal). The lock-in helps in making the output independent of the power fluctuations. Some residual PM noise has no effect on the detected voltage.

2.8.1 AM noise in optical systems

As we shall see in Chapter 4, Equation 2.57 also describes a quasi-perfect optical signal, when the voltage is replaced by the electric field

$$E(t) = E_0[1 + \alpha(t)]\cos[2\pi\nu_0 t + \varphi(t)] \qquad (2.239)$$

Yet, the preferred physical quantity used to describe the AM noise is the Relative Intensity Noise (RIN), defined as

$$RIN = S_{\delta I/I_0}(f) \qquad (2.240)$$

that is, the power spectrum density of the normalized intensity fluctuation $[I(t) - I_0]/I_0$. The RIN includes both fluctuation of power and the fluctuation of the power cross-section

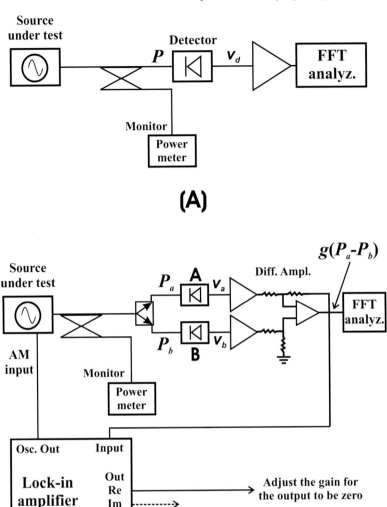

FIGURE 2.32
Basic and differential scheme for the measurement of AM noise. (Adapted from [99].)

distribution. If the latter is constant in time, the optical intensity is proportional to power $\delta I/I_0 = \delta P/P_0$. Then, in low-noise conditions ($|\delta I/I_0| \ll 1$), the power fluctuations are related to the fractional amplitude noise α by $\delta I/I_0 = \delta P/P_0 = 2\alpha$. Therefore we have

$$RIN\,(f) = 4S_\alpha\,(f) \tag{2.241}$$

Generally laser sources show a noise spectrum of the form

$$RIN\,(f) = h_0 + h_{-1}f^{-1} + h_{-2}f^{-2} \tag{2.242}$$

in which the flicker noise can be hidden by the random walk. Additional fluctuations induced by the environment may also be present.

3

Passive resonators

A tympanic resonance, so rich and
overpowering that it could give an air of
verse to a recipe for stewed hare.

John McPhee

Some people will believe anything if you
whisper it to them.

Miguel de Unamuno

For many contemporary physics experiments, the use of microwave and optical resonators has become a powerful tool for enhancement in detection sensitivities, nonlinear interactions, and quantum dynamics. Most often, the term cavity is used to describe such electromagnetic resonators. This term has been taken over from microwave technology, where resonators really look like closed cavities, whereas optical resonators traditionally have an *open* kind of setup. That difference in geometry is related to the fact that optical resonators are usually very large compared with the optical wavelength, whereas microwave cavities are often not much longer than a wavelength. As discussed in the course of this book, microwave and optical cavities allow one to extend the interaction length between matter and field, to build up the optical power, to impose a well-defined mode structure on the electromagnetic field, to implement extreme nonlinear optics, and to study manifestly quantum mechanical behavior associated with the modified vacuum structure and/or the large field associated with a single photon confined to a small volume [100].

Here we shall examine the basic properties of microwave and optical resonators from which we will draw at the appropriate time in the next chapters. We start by considering microwave cavities which represent, in our context, the basis of operation for masers, cesium fountains, and cryogenic sapphire dielectric resonators. Then, we will deal with optical cavities. In a quite general way, these can be defined as arrangements of optical components allowing a beam of light to circulate in a closed path. Such resonators can be made in very different forms. In a traditional approach, an optical resonator can be made from bulk optical components, like dielectric mirrors or prisms. Another common realization is based on fiber components, where the light is guided rather than sent through free space. Often, waveguide resonators are also implemented in the form of integrated optics. Mixed types of resonators, containing both waveguides and parts with free-space optical propagation, also exist. This happens, for instance, in some fiber lasers, where bulk optical components need to be inserted into the laser resonator for various purposes. In the past two decades, whispering gallery mode (WGM) microcavities, featuring the highest Q-factors, have attracted a lot of interest. They will be discussed in certain detail in the last part of this chapter. As one can easily guess, countless experimental activities have benefited from the use of optical cavities in their diversified realizations. Of course, the most important utilization is in laser physics itself, where the resonator losses are compensated by a gain medium to maintain or build up optical power. Other significant applications will be mentioned in a short while as soon as properties of each specific kind of resonator are derived.

3.1   Microwave cavities

The microwave range of the e.m. spectrum covers frequencies between about 1 GHz and 100 GHz. Such high-frequency waves cannot be confined in electric wires; they are usually transmitted through hollow conducting tubes known as waveguides. Accordingly, microwave cavities are completely enclosed metal boxes in which a standing wave can be produced. In general, the dimensions of the cavity are of the order of a wavelength and thus for 30 GHz the dimensions are of the order of 1 cm.

The electric and magnetic fields within a waveguide are given by solutions of Maxwell's equations with the boundary conditions set by the conducting walls of the waveguide [65, 101]. Basically, if the waveguide walls are assumed to be perfect conductors, the electric (magnetic) field cannot have a component along (perpendicular to) the boundary surface. A microwave cavity is usually constructed either by closing off the ends of a short section of a rectangular (or cylindrical) waveguide or from high-permittivity dielectric material (dielectric resonator). Radiation can be coupled from a waveguide into a microwave cavity by means of a suitable aperture or coupling device (a small wire probe or a loop). For specific frequencies, reflections within the cavity can set up standing waves. At these resonant frequencies, the cavity acts like a termination nearly matched to the characteristic impedance of the waveguide and the power reflected back into the waveguide is reduced. The cavity resonance can be precisely studied by measuring the ratio of reflected to incident power as a function of frequency near the resonance. A practical situation of great importance is the propagation or excitation of electromagnetic waves in hollow metallic right cylinders (see Figure 3.1). As already mentioned, such geometry is widespread in frequency standards, e.g., in the hydrogen maser or the cesium fountain clock.

For the moment, the boundary surfaces are presumed to be perfect conductors. In this configuration, assuming for the fields inside the cavity a sinusoidal time dependence $e^{i\omega t}$, Maxwell's equations (in a volume of space which is free of currents and charge and is filled with a uniform, non-dissipative medium) take the form

$$\begin{cases} \boldsymbol{\nabla} \times \boldsymbol{E} = -i\omega \boldsymbol{B} \\ \boldsymbol{\nabla} \times \boldsymbol{B} = i\omega\varepsilon\mu \boldsymbol{E} \\ \boldsymbol{\nabla} \cdot \boldsymbol{B} = 0 \\ \boldsymbol{\nabla} \cdot \boldsymbol{E} = 0 \end{cases} \tag{3.1}$$

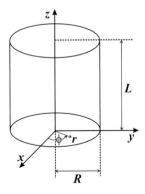

FIGURE 3.1
Hollow metallic right cylinder for the study of microwave cavities.

where $\varepsilon = \varepsilon_0 \varepsilon_r$ and $\mu = \mu_0 \mu_r \simeq \mu_0$ are the permittivity and permeability of the medium, respectively. From Equation 3.1 and by a little algebra one obtains the well-known Helmholtz equation

$$\left(\nabla^2 + k_0^2 \right) \left\{ \begin{array}{c} E \\ B \end{array} \right\} = 0 \tag{3.2}$$

where $k_0^2 = \varepsilon \mu \omega^2$. Because of the cylindrical geometry, it is useful to single out the spatial variation of the fields in the z direction and to assume

$$\left\{ \begin{array}{l} E\left(r \right) = E\left(r, \phi \right) e^{\pm ikz} \\ B\left(r \right) = B\left(r, \phi \right) e^{\pm ikz} \end{array} \right. \tag{3.3}$$

with k the wavenumber in the waveguide. Recalling the explicit expression for the rotor in cylindrical coordinates

$$\nabla \times G = \left(\frac{1}{r} \frac{\partial G_z}{\partial \phi} - \frac{\partial G_\phi}{\partial z} \right) \hat{e}_r + \left(\frac{\partial G_r}{\partial z} - \frac{\partial G_z}{\partial r} \right) \hat{e}_\phi$$
$$+ \left[\frac{1}{r} \frac{\partial \left(r G_\phi \right)}{\partial r} - \frac{1}{r} \frac{\partial G_r}{\partial \phi} \right] \hat{e}_z \tag{3.4}$$

and using the first two Maxwell's equations together with Equation 3.3, we are able to express the radial and azimuthal components of the fields in terms of the axial ones

$$E_r \left(r, \phi, z \right) = \frac{ik}{k^2 - k_0^2} \left(\mp \frac{\partial E_z \left(r, \phi, z \right)}{\partial r} + \frac{\omega}{k} \frac{1}{r} \frac{\partial B_z \left(r, \phi, z \right)}{\partial \phi} \right) \tag{3.5}$$

$$E_\phi \left(r, \phi, z \right) = \frac{ik}{k^2 - k_0^2} \left(\mp \frac{1}{r} \frac{\partial E_z \left(r, \phi, z \right)}{\partial \phi} - \frac{\omega}{k} \frac{\partial B_z \left(r, \phi, z \right)}{\partial r} \right) \tag{3.6}$$

$$B_r \left(r, \phi, z \right) = \frac{ik}{k^2 - k_0^2} \left(\mp \frac{\partial B_z \left(r, \phi, z \right)}{\partial r} - \mu \varepsilon \frac{\omega}{k} \frac{1}{r} \frac{\partial E_z \left(r, \phi, z \right)}{\partial \phi} \right) \tag{3.7}$$

$$B_\phi \left(r, \phi, z \right) = \frac{ik}{k^2 - k_0^2} \left(\mp \frac{1}{r} \frac{\partial B_z \left(r, \phi, z \right)}{\partial \phi} + \mu \varepsilon \frac{\omega}{k} \frac{\partial E_z \left(r, \phi, z \right)}{\partial r} \right) \tag{3.8}$$

Therefore, the wave equations have to be solved only for the z components. Then, Equations 3.5, 3.6, 3.7, 3.8 allow retrieval of the remaining components. On the other hand, by substitution of Equation 3.3 into Equation 3.2, we have for the z components

$$\nabla^2 \left[E_z \left(r, \phi \right) \right] + \left(k_0^2 - k^2 \right) E_z \left(r, \phi \right) = 0 \tag{3.9}$$

$$\nabla^2 \left[B_z \left(r, \phi \right) \right] + \left(k_0^2 - k^2 \right) B_z \left(r, \phi \right) = 0 \tag{3.10}$$

Since the boundary conditions on the electric and magnetic fields cannot be satisfied simultaneously, solutions to the above equations divide themselves into two distinct categories:

- **Transverse magnetic (TM)** waves:
  $B_z = 0$ everywhere; *boundary condition: E_z perpendicular to $\mathcal{S}$*

- **Transverse electric (TE)** waves:
  $E_z = 0$ everywhere; *boundary condition: B_z parallel to $\mathcal{S}$*

where $\mathcal{S}$ is the surface of the cylinder. By using the explicit form for the Laplacean operator in cylindrical coordinates

$$\nabla^2 = \frac{\partial^2}{\partial r^2} + \frac{1}{r}\frac{\partial}{\partial r} + \frac{1}{r^2}\frac{\partial^2}{\partial \phi^2} + \frac{\partial^2}{\partial z^2} \tag{3.11}$$

and introducing the quantity $\gamma^2 = k_0^2 - k^2$, Equations 3.9, 3.10 provide

$$\left[\frac{\partial^2}{\partial r^2} + \frac{1}{r}\frac{\partial}{\partial r} + \frac{1}{r^2}\frac{\partial^2}{\partial \phi^2} + \gamma^2\right] E_z(r,\phi) = 0 \tag{3.12}$$

and the same equation for $B_z(r,\phi)$. Now, let us look for solutions of the form

$$E_z(r,\phi) = A(r)\,\Phi(\phi) \tag{3.13}$$

By insertion of Equation 3.13 into Equation 3.12, one derives

$$\frac{r^2}{A(r)}\left[\frac{\partial^2}{\partial r^2}A(r) + \frac{1}{r}\frac{\partial}{\partial r}A(r) + \gamma^2 A(r)\right] = -\frac{\dfrac{\partial^2\Phi(\phi)}{\partial\phi^2}}{\Phi(\phi)} \tag{3.14}$$

whose left (right) side depends only on r (ϕ). This implies that both sides are equal to the same real constant m^2, providing the two following differential equations

$$\frac{\partial^2}{\partial r^2}A(r) + \frac{1}{r}\frac{\partial}{\partial r}A(r) + \left(\gamma^2 - \frac{m^2}{r^2}\right)A(r) = 0 \tag{3.15}$$

$$\frac{\partial^2\Phi(\phi)}{\partial\phi^2} + m^2\Phi(\phi) = 0 \tag{3.16}$$

The latter equation has solution $e^{\pm im\phi}$; moreover, for the potential to be single-valued when the full azimuth is allowed, m must be an integer ($m = 0, 1, 2, \ldots$). Concerning Equation 3.15, it can be put in a standard form by the change of variable $\xi = \gamma r$:

$$\frac{\partial^2}{\partial \xi^2}A(\xi) + \frac{1}{\xi}\frac{\partial}{\partial \xi}A(\xi) + \left(1 - \frac{m^2}{\xi^2}\right)A(\xi) = 0 \tag{3.17}$$

that is Bessel's differential equation. Its general solution is formed by means of two sets of functions, the Bessel functions of first kind $J_m(\xi)$, and the Bessel functions of second kind (also known as Weber functions) $Y_m(\xi)$:

$$A(\xi) = C_1 J_m(\xi) + C_2 Y_m(\xi) \tag{3.18}$$

However, since $Y_m(\xi)$ is divergent at $\xi = 0$, in order to obtain a physically meaningful result, the associated coefficient C_2 is forced to be zero. As a consequence, for each m, the solutions of Equation 3.17 are given by the Bessel functions of the first kind $J_m(\gamma r)$. Let us consider first TM solutions. By virtue of the boundary condition, among these we have to choose only those for which $E_z(r = R, \phi) = 0$. Thus, if we denote with $\xi_{mn} = \gamma_{mn}R$ the n-th root of the equation $J_m(\gamma R) = 0$, the solution for the z component of the electric field is

$$E_z(r,\phi,z) = E_z(r,\phi)\,e^{-ikz} = J_m\left(\frac{\xi_{mn}}{R}r\right)\cdot e^{\pm im\phi}e^{\pm ikz} \equiv \Psi e^{\pm ikz} \tag{3.19}$$

Then, let us search for TE solutions. These have the same form ($B_z(r,\phi) \propto J_m e^{\pm im\phi}$) but the argument of J_m is found by imposing that $E_\phi(r = R) = 0$ and $B_r(r = R) = 0$. By

TABLE 3.1
Roots of the Bessel function of the first kind $(J_m(x) = 0)$.

m	x_{m1}	x_{m2}	x_{m3}	x_{m4}
0	2.405	5.520	8.654	11.792
1	3.832	7.016	10.173	13.324
2	5.136	8.417	11.620	14.796
3	6.380	9.761	13.015	16.223

TABLE 3.2
Maxima or minima of the Bessel function of the first kind.

m	x_{m1}	x_{m2}	x_{m3}	x_{m4}
0	3.832	7.016	10.173	13.324
1	1.841	5.331	8.536	11.706
2	3.054	6.706	9.936	13.170
3	4.201	8.015	11.346	14.586

virtue of Equation 3.6 (or Equation 3.7) and the fact that $E_z = 0$, this is equivalent to the condition $\frac{\partial J_m}{\partial \xi}\Big|_{\xi = \gamma R} = 0$. Therefore, if we denote with ξ'_{mn} the n-th root of the equation $\frac{\partial J_m}{\partial \xi}\Big|_{\xi = \gamma R} = 0$, the solution for the z component of the magnetic field is given by

$$B_z(r, \phi, z) = B_z(r, \phi) e^{-ikz} = J_m\left(\frac{\xi'_{mn}}{R} r\right) \cdot e^{\pm im\phi} e^{\pm ikz} \equiv \Upsilon e^{\pm ikz} \tag{3.20}$$

The first few values of the roots ξ_{mn} and ξ'_{mn} are given in Tables 3.1 and 3.2.

Now, we have to impose the boundary conditions at the plane surfaces. To this aim, we must superimpose waves moving in both the positive and negative z direction. Starting with TM fields, from Equation 3.19 we get

$$E_z(r, \phi, z) = \frac{\Psi}{2}\left(e^{+ikz} + e^{-ikz}\right) = \Psi \cos kz \tag{3.21}$$

On the other hand, from Equations 3.5, 3.6, we obtain

$$E_\perp^+(r, \phi, z) = \frac{-ik}{k^2 - k_0^2} \nabla_\perp [E_z(r, \phi, z)]$$

$$= \frac{-ik}{k^2 - k_0^2} e^{+ikz} \nabla_\perp \Psi \qquad \text{for} \quad e^{+ikz} \tag{3.22}$$

and

$$E_\perp^-(r, \phi, z) = \frac{ik}{k^2 - k_0^2} e^{-ikz} \nabla_\perp \Psi \quad \text{for} \quad e^{-ikz} \tag{3.23}$$

where the subscript $\perp$ denotes the transverse components. For clarity sake, the vectors $\nabla_\perp \equiv \left(\frac{\partial}{\partial r}, \frac{1}{r}\frac{\partial}{\partial \phi}\right)$ and $E_\perp(r, \phi, z) \equiv (E_r(r, \phi, z), E_\phi(r, \phi, z))$ have been introduced. From the above equations it follows

$$E_\perp (r, \phi, z) = \frac{1}{2} \left[E_\perp^+ (r, \phi, z) + E_\perp^- (r, \phi, z) \right] = \frac{k (\nabla_\perp \Psi)}{k^2 - k_0^2} \sin kz \qquad (3.24)$$

If the plane boundary surfaces are at $z = 0$ and $z = L$, Equation 3.24 together with the boundary condition $E_\perp (r, \phi, z = 0) = E_\perp (r, \phi, z = L) = 0$ implies that $k = q\pi/L$ ($q = 0, 1, 2, \ldots$). By summarizing these results, eigenoscillations being TM with respect to z are given by

$$\begin{cases} E_z (r, \phi, z) = E_0 J_m \left(\frac{\xi_{mn}}{R} r \right) e^{\pm im\phi} \cos \left(\frac{q\pi z}{L} \right) & \begin{array}{l} m = 0, 1, 2, \ldots \\ n = 1, 2, 3, \ldots \\ q = 0, 1, 2, \ldots \end{array} \\ \\ B_z = 0 \quad \text{everywhere} \end{cases} \qquad (3.25)$$

Similarly, for TE fields, the condition $B_z (r, \phi, z = 0) = B_z (r, \phi, z = L) = 0$ requires that eigenoscillations being TE with respect to z are given by

$$\begin{cases} B_z (r, \phi, z) = B_0 J_m \left(\frac{\xi'_{mn}}{R} r \right) e^{\pm im\phi} \sin \left(\frac{q\pi z}{L} \right) & \begin{array}{l} m = 0, 1, 2, \ldots \\ n = 1, 2, 3, \ldots \\ q = 1, 2, 3, \ldots \end{array} \\ \\ E_z = 0 \quad \text{everywhere} \end{cases} \qquad (3.26)$$

In Equations 3.25, 3.26 E_0 and B_0 are the field amplitudes. The field configurations determined by the integers m, n, and q are called the modes of the cavity. The corresponding eigenvalues are found by expressing the relationship $\gamma^2 = k_0^2 - k^2$ as

$$\gamma_{mn}^2 = \mu \varepsilon \omega^2 - \left(\frac{q\pi}{L} \right)^2 \qquad (3.27)$$

which returns

$$\nu_{mnq}^{(TM)} = \frac{1}{2\pi \sqrt{\mu\varepsilon}} \sqrt{\frac{\xi_{mn}^2}{R^2} + \frac{q^2 \pi^2}{L^2}} \qquad (3.28)$$

and

$$\nu_{mnq}^{(TE)} = \frac{1}{2\pi \sqrt{\mu\varepsilon}} \sqrt{\frac{(\xi'_{mn})^2}{R^2} + \frac{q^2 \pi^2}{L^2}} \qquad (3.29)$$

These formulas show that resonant cavities have discrete frequencies of oscillation with a definite field configuration for each resonance frequency. This implies that, if one were attempting to excite a particular mode of oscillation in a cavity by some means, no fields of the right sort could be built up unless the exciting frequency were exactly equal to the chosen resonance frequency. In actual fact, there will not be a delta function singularity, but rather a narrow band of frequencies around the eigenfrequency over which appreciable excitation can occur. An important source of this smearing out of the sharp frequency of oscillation is the dissipation of energy in the cavity walls and perhaps in the dielectric filling the cavity. For finite conductivity of the walls of the resonator, the high-frequency electromagnetic field penetrates into the metallic walls of the cavity. At the same time the electric currents in the walls suffer from ohmic losses and the eigenoscillations are damped. The field distribution inside cylindrical cavities, however, does not differ considerably from that of an ideal resonator with infinite conductivity. A fraction of the energy of the electromagnetic wave is dissipated continuously to heat by the ohmic losses of the wall currents and hence the energy flux in the wall decreases exponentially with a characteristic length. In a regular conducting material such as, e.g., copper, this characteristic length is given by the skin depth δ_S of a few micrometers. In a superconducting material the penetration depth is given by the much smaller London depth of, e.g., $\lambda_L \approx 30$ nm in niobium. As extensively discussed in Chapter 2, a measure of the sharpness of the cavity response to an external excitation is given by the Q factor, defined as the ratio between a particular resonance frequency and the corresponding $FWHM$ ($Q = \omega_0/FWHM = \omega_0/2\Gamma$). The cavity losses also have the effect to shift the resonance frequency from the ideal (in the absence of losses) value ω_0 by the amount (see Equation 2.8)

$$\Delta\omega = \sqrt{\omega_0^2 - \Gamma^2} - \omega_0 = \omega_0\sqrt{1 - \left(\frac{1}{2Q}\right)^2} - \omega_0$$

$$\simeq \omega_0\sqrt{\left(1 - \frac{1}{2Q}\right)^2} = -\frac{\omega_0}{2Q} \tag{3.30}$$

By combination of Equation 3.30 and Equation 3.28, we get

$$\nu_{mnq,Q}^{(TM)} = \frac{1}{2\pi\sqrt{\mu\varepsilon}}\sqrt{\frac{\xi_{mn}^2}{R^2} + \frac{q^2\pi^2}{L^2}}\left(1 - \frac{1}{2Q}\right) \tag{3.31}$$

that, for an evacuated cavity, provides

$$c = \frac{\nu_{mnq,Q}^{(TM)}}{\frac{1}{2\pi}\sqrt{\frac{\xi_{mn}^2}{R^2} + \frac{q^2\pi^2}{L^2}}\left(1 - \frac{1}{2Q}\right)} \simeq \frac{\nu_{mnq,Q}^{(TM)}}{\frac{1}{2\pi}\sqrt{\frac{\xi_{mn}^2}{R^2} + \frac{q^2\pi^2}{L^2}}}\left(1 + \frac{1}{2Q}\right) \tag{3.32}$$

which proves Equation 1.8 introduced in Chapter 1 when discussing the experiment by Louis Essen.

It is worth anticipating here that in Cs fountains and hydrogen masers the TE_{011} resonance is most often used to interrogate or to excite magnetic hyperfine transitions: atoms supposed to interact with the magnetic field enter and leave the resonator through small holes in the center of the bottom and top of the right circular cylinder (see Chapter 7).

Let us close this section with a few considerations about the typical performance of microwave cavities. Dielectric loss of air is extremely low for high frequency electric or magnetic fields. Generally, losses originating from wall currents are small too. Just to give an idea, copper cavities exhibit a typical Q on the order of 10^5. Nevertheless, they tend to oxidize, which increases their loss. For this reason, copper cavities are usually plated with silver. Even though gold is not a good conductor compared to copper, it still prevents oxidation and the resulting deterioration of Q factor over time. However, because of its high cost, it is used only in the most demanding applications. For example, some satellite resonators are silver plated and covered with a gold flash layer. Then, the current mostly flows in the high-conductivity silver layer, while the gold flash layer protects the silver one from oxidizing. When higher Q factors are required, like in advanced fundamental research, superconducting materials are employed; for example, superconducting niobium at 1.8 K leads to a quality factor as high as 10^{10}. This application will be detailed in Chapter 7, where, based on the above treatment, sapphire dielectric microwave resonators will be presented. In such a frame, two particular configurations may serve as further illustrative examples for the current and remaining discussion. The first type is a low-order TE-mode resonance of a cylindrical sapphire puck having its surface coated with a superconducting material. This arrangement can be thought of as a vacuum cavity filled with the dielectric sapphire material. The second type is the whispering gallery mode configuration. It consists of a sapphire ring which confines the electromagnetic energy to the dielectric region by a physical mechanism not unlike total internal reflection in optical systems (see Section 3.6).

3.2   Basic properties of bulk optical cavities

As we try to increase the frequency from microwave to light, we realize that, using the closed cavity argument, we must have a cavity whose dimensions should be of the order of 1 micron. At the beginning of laser research, this was an important hurdle; however, scientists realized that there is no necessity for a closed cavity. We can use an open cavity which, for the case of lasers, is usually nothing other than two facing mirrors. As mentioned, we will divide the class of resonators operating in the optical frequency domain into three main subcategories: bulk resonators, optical-fiber cavities, and WGM microcavities.

Since in traditional optical cavities the wavelength of about one micrometer is typically very small compared to the dimensions of the resonator, diffraction effects are often not very relevant and the resonator structures need not be confined in all three dimensions but can be set up using discrete mirrors. The most simple arrangement consists of two reflecting mirrors facing each other (separated by the distance L), but more than two mirrors can be arranged in a ring or bow-tie configuration (Figure 3.2).

Two approaches are useful for describing the operation of an optical resonator:

- The simplest approach is based on ray optics. Optical rays are traced as they reflect within the resonator; the geometrical conditions under which they remain confined are determined.

- Wave optics is used to determine the resonator modes, i.e., the resonance frequencies and wavefunctions of the optical waves that exist self-consistently within the resonator.

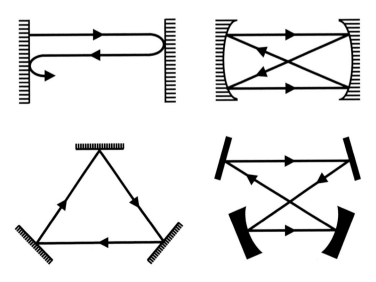

FIGURE 3.2
Some configurations of bulk, mirror-based optical cavities.

3.2.1 Fabry-Perot etalon

Before attacking with either the ray optics or the wave optics treatment, let us first consider a paradigmatic one-dimensional configuration consisting of two flat mirrors at a distance L. This simple resonator, known as Fabry-Perot (FP) etalon, indeed allows one to understand much of the physics of traditional optical cavities. Note that, as we will see in a short while, such a cavity is not stable and the light walks off perpendicular to the cavity axis. Nevertheless, this case can be considered *physically meaningful* if the flat mirrors have effectively infinite extent and the input light can be approximated by a perfect plane wave. Any given mirror may be characterized by its field amplitude reflection coefficient r and its transmission coefficient t. In general these are complex quantities. Typically we are interested in the intensity reflectivity $\mathcal{R} = |r|^2$ and transmission $\mathcal{T} = |t|^2$. Let us neglect mirror losses for the moment and let the input (output) mirror coefficients be r_1 and t_1 (r_2 and t_2). It is convenient to write these in polar form, such as $r_1 = |r_1| e^{i\phi_{r1}} = \sqrt{\mathcal{R}_1} e^{i\phi_{r1}}$. If the input field is E_0, with reference to Figure 3.3, the amplitude of the electric field internal to the cavity is given by

$$E_i = t_1^* E_0 \sum_{n=0}^{\infty} \left(r_1 r_2 e^{2ikL}\right)^n = \frac{t_1^* E_0}{1 - r_1 r_2 e^{2ikL}} \qquad (3.33)$$

where $k = 2\pi/\lambda$ (we assume unitary refractive index inside the cavity). The corresponding intensity is

$$I_i = |E_i|^2 = \frac{\mathcal{T}_1}{\left|1 - \sqrt{\mathcal{R}_1 \mathcal{R}_2} e^{2ikL + i\phi_{r1} + i\phi_{r2}}\right|^2} I_0 = \frac{\mathcal{T}_1}{\left|1 - \sqrt{\mathcal{R}_1 \mathcal{R}_2} e^{i\Omega}\right|^2} I_0 \qquad (3.34)$$

where $\Omega = 2kL + \phi_{r1} + \phi_{r2}$. The transmitted field is easily calculated from the internal field as

$$E_t = E_i e^{ikL} t_2 = \frac{t_1^* t_2 e^{ikL}}{1 - r_1 r_2 e^{2ikL}} E_0 \qquad (3.35)$$

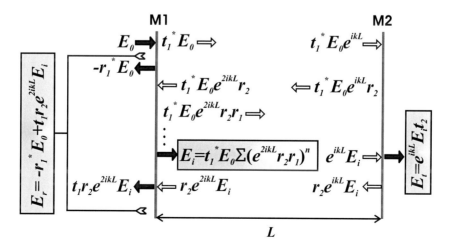

FIGURE 3.3
Fabry-Perot etalon.

such that the transmitted intensity is

$$I_t = |E_t|^2 = \frac{\mathcal{T}_1 \mathcal{T}_2}{\left|1 - \sqrt{\mathcal{R}_1 \mathcal{R}_2} e^{i\Omega}\right|^2} I_0 \tag{3.36}$$

The field reflected from the resonator is actually the coherent sum of two different fields: the promptly reflected beam, which bounces off the first mirror and never enters the cavity; and a leakage beam, which is the small part of the standing wave inside the cavity that leaks back through the input mirror:

$$E_r = -r_1^* E_0 + e^{ikL} r_2 e^{ikL} t_1 E_i = \left[-r_1^* + \frac{\mathcal{T}_1 r_2 e^{2ikL}}{1 - \sqrt{\mathcal{R}_1 \mathcal{R}_2} e^{i\Omega}} \right] E_0 \tag{3.37}$$

The fact that the directly reflected light is described by $-r_1^* E_0$ rather than simply $r_1 E_0$ arises from imposing energy conservation constraints on a single mirror (Stokes relations) [102]. By a little algebra, the reflected intensity is obtained as

$$I_r = |E_r|^2 = \frac{\left| -1 + \left(1 + \dfrac{\mathcal{T}_1}{\mathcal{R}_1} \right) \sqrt{\mathcal{R}_1 \mathcal{R}_2} e^{i\Omega} \right|^2}{\left| 1 - \sqrt{\mathcal{R}_1 \mathcal{R}_2} e^{i\Omega} \right|^2} \mathcal{R}_1 I_0 \tag{3.38}$$

- Let us first consider the transmitted intensity. For normal incidence onto an ideal dielectric mirror we can assume $\phi_{r1} = \phi_{r2} = \pi$, whereupon

$$\begin{aligned}
I_t &= \frac{\mathcal{T}_1 \mathcal{T}_2 I_0}{\left| 1 - \sqrt{\mathcal{R}_1 \mathcal{R}_2} e^{i\Omega} \right|^2} = \frac{\mathcal{T}_1 \mathcal{T}_2 I_0}{1 + \mathcal{R}_1 \mathcal{R}_2 - 2\sqrt{\mathcal{R}_1 \mathcal{R}_2} \cos(2kL)} \\
&= \frac{\mathcal{T}_1 \mathcal{T}_2 I_0}{\left(1 - \sqrt{\mathcal{R}_1 \mathcal{R}_2}\right)^2 + 4\sqrt{\mathcal{R}_1 \mathcal{R}_2} \sin^2\left(\dfrac{2\pi L}{c}\nu\right)}
\end{aligned} \tag{3.39}$$

From the above expression, one can easily recognize that I_t peaks at wavenumbers $k_m L = m\pi$ or equivalently at frequencies $\nu_m = mc/2L$, where m is an integer and c the speed of light. The separation between adjacent peaks, called the free spectral range, is given by

$$FSR = \nu_{m+1} - \nu_m = \frac{c}{2L} \tag{3.40}$$

that is the inverse of the round-trip travel time τ_r (equivalently, the path traveled by light in a single round-trip is an integer number of wavelengths of light). This condition can be used to instantly derive the expression for the free spectral range of any geometric arrangement of plane mirrors (provided that each mirror reflection introduces a phase shift of π). Just as an example, for the equilater ring cavity shown in Figure 3.2 we have $FSR = 1/\tau_r = 1/3d/c = c/3d$. Alternatively, the longitudinal modes can be derived by imposing the condition that a mode is a self-reproducing wave, i.e., a wave that reproduces itself after a single round-trip. In the case of the ring cavity, for instance, the phase shift imparted by a single round-trip of propagation is $3kd$ while the phase shift imparted by reflection is 3π (π at each mirror). Thus, the self-reproducing condition on the phase translates into: $3k_m L + 3\pi = m2\pi$ which yields $3L(k_{m+1} - k_m) = 2\pi$ that is $\nu_{m+1} - \nu_m = c/(3d)$. Note the strict analogy of this condition with the Barkhausen criterion in a positive feedback oscillator requiring that the output of the system be fed back in phase with the input.

The linewidth of the peaks in the FP transmission can be readily derived by observing that in the proximity of the m-th peak we have

$$\sin^2\left(\frac{2\pi L}{c}\nu\right) \simeq \frac{4\pi^2 L^2}{c^2}\left(\nu - \nu_m\right)^2 \tag{3.41}$$

In this approximation, the resonance curve becomes a Lorentzian whose FWHM is

$$\delta\nu = \frac{1 - \sqrt{\mathcal{R}_1\mathcal{R}_2}}{\pi\left(\mathcal{R}_1\mathcal{R}_2\right)^{1/4}}\frac{c}{2L} = \frac{FSR}{\mathcal{F}} \tag{3.42}$$

where the cavity finesse

$$\mathcal{F} = \frac{\pi\left(\mathcal{R}_1\mathcal{R}_2\right)^{1/4}}{1 - \sqrt{\mathcal{R}_1\mathcal{R}_2}} \tag{3.43}$$

has been defined. For a symmetrical ($\mathcal{R}_1 = \mathcal{R}_2 = \mathcal{R}$), high-finesse cavity, the above formula simplifies to

$$\mathcal{F} = \frac{\pi\sqrt{\mathcal{R}}}{1 - \mathcal{R}} \simeq \frac{\pi\mathcal{R}}{1 - \mathcal{R}} \simeq \frac{\pi}{1 - \mathcal{R}} \tag{3.44}$$

Now one can express the cavity transmission in terms of the cavity finesse and free spectral range

$$I_t = \frac{I_{max}}{1 + \left(\frac{2}{\pi}\mathcal{F}\right)^2 \cdot \sin^2\left(\frac{\pi \cdot \nu}{FSR}\right)} \tag{3.45}$$

where $I_{max} = \mathcal{T}_1\mathcal{T}_2 I_0/\left(1 - \sqrt{\mathcal{R}_1\mathcal{R}_2}\right)^2$. Then, for high finesse values, the transmission peaks will be narrow compared with FSR (see Figure 3.4).

This means that a Fabry-Perot etalon may be used as a sharply tuned optical filter or a spectrum analyzer. Indeed, if we scan the cavity length (for example by attaching one mirror to a piezoelectric transducer element), the resonant frequencies are scanned too. As a consequence, if the injected light contains frequencies in a range around some resonant frequency, then one can record the source spectrum. Because of the periodic nature of the spectral response, however, the spectral width of the measured light must be narrower than the free spectral range in order to avoid ambiguity. Otherwise, it is necessary to use the FP in series with some other wavelength selective device such as, for instance, a grating or prism monochromator. Moreover, it is worth observing that, if the spectral width Γ (let us say the full width at half maximum) of the light source under test is larger than the ratio $FSR/\mathcal{F}$, the signal at the FP output is a convolution between the FP transfer function and the input spectrum. To see this, consider a Lorentzian profile for the input light source

$$L\left(\nu\right) = \frac{L_0}{\left(\nu - \nu_0\right)^2 + \Gamma^2/4} \tag{3.46}$$

where ν_0 is the emission center frequency. On the other hand, close to a resonance, Equation 3.45 can be written in the form

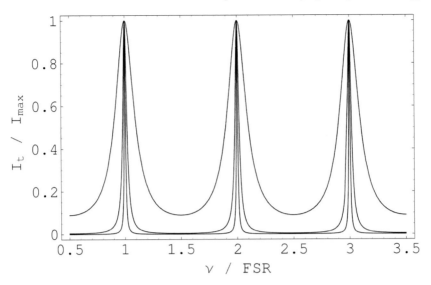

FIGURE 3.4
Fabry-Perot transmission according to Equation 3.45 for three different finesse $\mathcal{F}$ values: 5, 25, 80. Higher finesse values correspond to narrower transmission peaks.

$$I_t = \frac{I_{max}}{1 + \dfrac{4\mathcal{F}^2}{FSR^2} \cdot (\nu - \nu_m)^2} \tag{3.47}$$

Then, the effect of scanning the FP resonance around ν_0 is simulated by taking $\nu_m = \nu_0 + y$ with y varying in the interval $(-\Delta, +\Delta)$ with $\Delta \gg \Gamma$. Without loss of generality we can take $\nu_0 = 0$, such that the FP output is determined as

$$Conv(y) \equiv \frac{V(y)}{I_{max} \cdot L_0} = \int_{-\Delta}^{+\Delta} \frac{1}{1 + \dfrac{4\mathcal{F}^2}{FSR^2} \cdot (\nu - y)^2} \cdot \frac{1}{\nu^2 + \Gamma^2/4} d\nu \tag{3.48}$$

A plot of Equation 3.48 is shown in Figure 3.5. As a further example, consider a two-mode input light source modelled as

$$L(\nu) = \frac{L_1}{(\nu - \nu_1)^2 + \Gamma^2/4} + \frac{L_2}{(\nu - \nu_2)^2 + \Gamma^2/4} \tag{3.49}$$

with $L_2 < L_1$ and $|\nu_2 - \nu_1| > 2\Gamma$. In this case, Equation 3.48 becomes

$$Conv'(y) \equiv \frac{V'(y)}{I_{max} L_1} = \int_{-\Delta'}^{+\Delta'} \frac{1}{1 + \dfrac{4\mathcal{F}^2}{FSR^2} (\nu - y)^2}$$
$$\cdot \left[\frac{1}{(\nu - \nu_1)^2 + \Gamma^2/4} + \frac{L_2/L_1}{(\nu - \nu_2)^2 + \Gamma^2/4} \right] d\nu \tag{3.50}$$

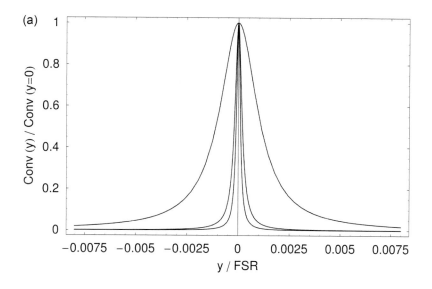

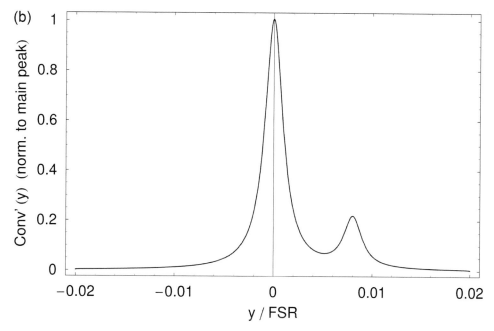

FIGURE 3.5
(a) Plot of $Conv(y)$ for $\mathcal{F} = 5000$ and three different values of Γ/FSR ($2 \cdot 10^{-3}$, $2 \cdot 10^{-4}$ and $2 \cdot 10^{-5}$); integration was performed with $\Delta/\text{FSR} = 0.01$ MHz. (b) Plot of $Conv'(y)$ for $L_2/L_1 = 0.2$, $\mathcal{F} = 5000$, $\nu_1 = 0$, $\nu_2/FSR = 8 \cdot 10^{-3}$, and $\Gamma/FSR = 2 \cdot 10^{-3}$; integration was performed with $\Delta'/FSR = 0.02$.

The result of this integration is shown in frame b of Figure 3.5.

In light of the above discussion, it is obvious that another application of a FP etalon is to filter the frequency content of the incident optical radiation. As we will see in the next chapter, etalons are used, for example, to construct single-mode lasers. Without an etalon, a laser will generally produce light over a wavelength range corresponding to a number of cavity modes. Inserting an etalon into the laser cavity, with well-chosen finesse and free spectral range, can suppress all cavity modes except for one, thus changing the operation of the laser from multi-mode to single-mode.

Note that, besides shifting the resonance position, a change ΔL in the cavity length also affects the value of the free spectral range, albeit to a much lower extent

$$\Delta \nu_m = -\nu_m \frac{\Delta L}{L} \qquad \Delta FSR = -FSR \frac{\Delta L}{L} \Rightarrow \frac{\Delta \nu_m}{\Delta FSR} = \frac{\nu_m}{FSR} \qquad (3.51)$$

Hence, for a typical optical frequency (100 THz) and $FSR = 1$ GHz, we have $\Delta \nu_m = 10^5 \Delta FSR$.

- In order to study the behavior of the intensity internal to the cavity, Equation 3.34 must be used. On resonance and for a symmetrical cavity, we have

$$I_i = \frac{\mathcal{T}_1 I_0}{\left(1 - \sqrt{\mathcal{R}_1 \mathcal{R}_2}\right)^2} = \frac{I_0}{1 - \mathcal{R}} \simeq \frac{\mathcal{F}}{\pi} I_0 \qquad (3.52)$$

For example, a finesse $F \simeq 30000$ gives an enhancement in the intra-cavity intensity of about 10000. This can be exploited in several applications for which concentration of a huge optical power around a particular frequency may be required. Cavity-enhanced sub-Doppler saturation spectroscopy (Chapter 6) or non-linear optics processes (Chapter 4) just represent two examples.

- A detailed discussion on the behavior of the reflected intensity (Equation 3.38) will be given in Chapter 4 when the Pound-Drever-Hall method will be presented.

The results so far describe an ideal cavity, in which there is no absorption or other losses of light inside the cavity. In practice, no mirror is perfect: the mirror material absorbs some amount of the incident light; furthermore the mirror surface is also always imperfect and scatters some amount of light. Moreover, the finite size of the mirrors causes a fraction of the light to leak around the mirror and thereby to be lost. All these loss sources can be lumped into an effective reduced mirror reflectance ($\mathcal{R}'_1$, $\mathcal{R}'_2$). Finally, losses due to absorption and scattering in the medium between the mirrors must also be included. If α_s denotes the absorption coefficient of the medium and Lambert-Beer law is used, the overall intensity attenuation factor in a single round-trip is

$$a = \mathcal{R}'_1 \mathcal{R}'_2 e^{-2\alpha_s L} \equiv e^{-2\alpha_r L} \qquad (3.53)$$

where the last equality holds by definition of the following effective overall distributed-loss coefficient

$$\alpha_r = \alpha_s + \frac{1}{2L} \ln \frac{1}{\mathcal{R}'_1} + \frac{1}{2L} \ln \frac{1}{\mathcal{R}'_2} \equiv \alpha_s + \alpha_{m1} + \alpha_{m2} \qquad (3.54)$$

In this case the cavity finesse is calculated from Equation 3.43 by replacing the product $\mathcal{R}_1 \mathcal{R}_2$ by the factor a

$$\mathcal{F} = \frac{\pi e^{-\frac{\alpha_r L}{2}}}{1 - e^{-\alpha_r L}} \simeq \frac{\pi}{\alpha_r L} \tag{3.55}$$

where $\alpha_r L \ll 1$ has been assumed. By substitution of Equation 3.55 into Equation 3.42, we obtain

$$\delta\nu = \frac{c\alpha_r}{2\pi} = \frac{1}{2\pi\tau_p} \tag{3.56}$$

where the photon lifetime $\tau_p = 1/(c\alpha_r)$ has been defined, $c\alpha_r$ representing the loss per unit time. Being a manifestation of the time-frequency uncertainty relation, Equation 3.56 states that the resonance line broadening is governed by the decay of optical energy arising from resonator losses. In summary, three parameters are convenient for characterizing the losses in an optical resonator of length L: the finesse $\mathcal{F}$, the loss coefficient α_r (expressed in cm^{-1}), and the photon lifetime τ_p (in seconds). In addition, we introduce the quality factor Q defined, as usual, as

$$Q = \frac{\nu_0}{\delta\nu} = \frac{\nu_0}{FSR}\mathcal{F} \tag{3.57}$$

Since optical resonator frequencies ν_0 are typically much greater than the mode spacing FSR, we have $Q \gg \mathcal{F}$. The quality factor of an optical resonator is usually far greater than that of a typical resonator at microwave frequencies. Just as an example, for $\mathcal{F} = 10000$, $\nu_0 = 100$ THz, and $FSR = 1$ GHz, we have $Q=10^9$.

3.2.2 Paraxial ray analysis

As already mentioned, the planar-mirror resonator configuration discussed above is highly sensitive to misalignment. If the mirrors are not perfectly parallel, or the rays are not perfectly normal to the mirror surfaces, they undergo a sequence of lateral displacements that eventually causes them to wander out of the resonator. Spherical-mirror resonators, in contrast, provide a more stable configuration for the confinement of light that renders them less sensitive to misalignment under certain geometrical conditions. Such cavities consist of two spherical mirrors of radii R, separated by a distance L. The centers of the mirrors define the optical axis, about which the system exhibits circular symmetry. Each of the mirrors can be concave ($R < 0$) or convex ($R > 0$). Obviously, the planar-mirror resonator is a special case for which $R_1 = R_2 = \infty$. We first examine the conditions for the confinement of optical rays [103]. Then, we shall determine the resonator modes. The propagation of paraxial rays through various optical structures can be described within the ABCD formalism [102], where the ray is characterized by its distance x from the optical axis and by its slope x' (assumed to be small) with respect to that axis, and a single optical element is characterized by its own ABCD matrix (generally with determinant 1).

The ray path through a given optical element is given by

$$\begin{pmatrix} x_2 \\ x_2' \end{pmatrix} = \begin{pmatrix} A & B \\ C & D \end{pmatrix} \begin{pmatrix} x_1 \\ x_1' \end{pmatrix} \tag{3.58}$$

where x_1, x_1' are the input quantities and x_2, x_2' are the output quantities. Now, the most simple arrangement for an optical resonator consists of two reflecting mirrors facing each other separated by the distance L. The effect on the light rays that bounce back and forth between these mirrors is the same as in a periodic sequence of identical optical systems. The ray transfer through n consecutive elements of the sequence is described by the n-th power of the matrix. This can be evaluated by means of Sylvester's theorem

$$\begin{pmatrix} A & B \\ C & D \end{pmatrix}^n = \frac{1}{\sin \Theta} \begin{pmatrix} A\sin[n\Theta] - \sin[(n-1)\Theta] & B\sin[n\Theta] \\ C\sin[n\Theta] & D\sin[n\Theta] - \sin[(n-1)\Theta] \end{pmatrix} \quad (3.59)$$

where $\cos \Theta = \frac{1}{2}(A + D)$. Sequences are stable when the trace obeys the inequality

$$-1 < \frac{1}{2}(A + D) < 1 \quad (3.60)$$

Inspection of Equation 3.59 shows that rays passing through a stable sequence are periodically refocused. For unstable systems, the trigonometric functions in that equation become hyperbolic functions, which indicates that the rays become more and more dispersed the further they pass through the sequence. In the case of laser resonator with spherical mirrors, one can choose, as an element of the periodic sequence, a spacing followed by one mirror plus another spacing followed by the second mirror. From this one can obtain the trace, and write the stability condition in the form

$$0 < \left(1 - \frac{L}{R_1}\right)\left(1 - \frac{L}{R_2}\right) < 1 \quad (3.61)$$

For a cavity consisting of two identical mirrors ($R_1 = R_2 = R$), that is the most common arrangement, this condition simplifies to $0 < L < 2R$.

Among the different configurations (summarized in Figure 3.6), an interesting degeneracy occurs if we choose the cavity length to be equal to the radius of curvature of the FP cavity. In this case, each resonant mode in the cavity can be thought of as a *bow-tie* mode, which transverses the cavity twice before retracing its path, hence $\nu_{FSR}^{confocal} = c/(4L)$.

3.2.3   Wave analysis

In a previous section, we have learnt that in a FP etalon the maxima of the Airy function (Equation 3.39) occur at the eigenfrequencies $\nu_m = mc/(2L)$. The corresponding electromagnetic field components, called the longitudinal (or axial) modes of the resonator, are plane waves along the optical axis of the resonator. As a plane wave would extend to infinity in the transverse direction, the energy contained in such a wave would be unlimited. A more realistic treatment of the electromagnetic field in the resonator has to include an amplitude dependence on the transverse coordinates and a transverse confinement of the wave inside the resonator. On the other hand, any transverse confinement will result in diffraction effects. Now, let us take another step by taking into account the wave nature of the laser beams. Yet, we neglect for the moment diffraction effects due to the finite size of apertures. The results derived here are applicable to optical systems with apertures that intercept only a negligible portion of the beam power. An *ab initio* treatment would require the solution of the Helmholtz equation for each field component under proper assumptions. Here, instead, we choose a heuristic approach, based on the following considerations [104].

A Gaussian beam (refer to Figure 3.7) reflected from a spherical mirror will retrace the incident beam if the radius of curvature of its wavefront is the same as the mirror radius.

Thus, if the radii of curvature of the wavefronts of a Gaussian beam at planes separated by a distance L match the radii of two mirrors separated by the same distance, a beam incident on the first mirror will reflect and retrace itself to the second mirror, where it once again will reflect and retrace itself back to the first mirror, and so on. The beam can then exist self-consistently within the resonator, satisfying the Helmholtz equation and the boundary conditions imposed by the mirrors. The Gaussian beam is then said to be a mode of the spherical-mirror resonator, provided that the phase also retraces itself. Now we

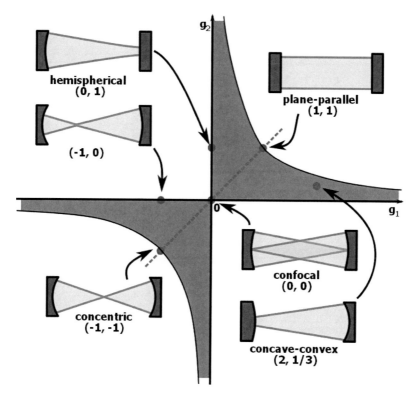

FIGURE 3.6
Resonator stability diagram with $g_1 = (1 - L/R_1)$ and $g_2 = (1 - L/R_2)$. A spherical-mirror resonator is stable if the parameters g_1 and g_2 lie in the regions bounded by the lines $g_1 = 0$ and $g_2 = 0$, and the hyperbola $g_2 = 1/g_1$. All symmetrical resonators lie along the line $g_1 = g_2$.

proceed to determine the Gaussian beam that matches a spherical-mirror resonator. The center of the beam, which is yet to be determined, is assumed to be located at the origin $z = 0$; mirrors R_1 and R_2 are located at z_1 and $z_2 = z_1 + L$. The values of z_1 and z_2 are determined by matching the radius of curvature of the beam, $R(z) = z + z_0^2/z$, to the radii R_1 at z_1 and R_2 at z_2

$$\begin{cases} R_1 = z_1 + \dfrac{z_0^2}{z_1} \\ -R_2 = z_2 + \dfrac{z_0^2}{z_2} \end{cases} \tag{3.62}$$

Solving for z_1, z_2, and z_0, we obtain

$$z_1 = \frac{-L(R_2 + L)}{R_1 + R_2 + 2L} \tag{3.63}$$

$$z_2 = z_1 + L \tag{3.64}$$

$$z_0^2 = \frac{-L(R_1 + L)(R_2 + L)(R_1 + R_2 + L)}{(R_1 + R_2 + 2L)^2} \tag{3.65}$$

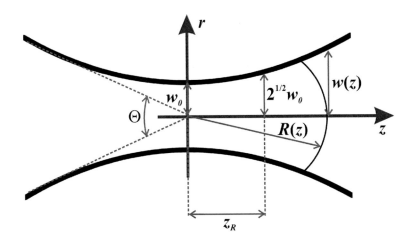

FIGURE 3.7
Some parameters of a Gaussian beam are summarized: Wavelength: λ; Waist radius: w_0;
Spot size: $w^2(z) = w_0^2[1 + (\lambda z/\pi w_0^2)^2]$; Radius of curvature: $R(z) = z[1 + (\pi w_0^2/\lambda z)^2]$;
Rayleigh range: $z_R = \pi w_0^2/\lambda$; Confocal parameter: $b = 2z_R$; Beam divergence: $\Theta \simeq \lambda/\pi w_0$;
Gouy phase: $\psi(z) = \arctan(z/z_R)$.

Having determined the location of the beam center and the depth of focus $2z_0$, everything
about the beam is known. The waist radius is $w_0 = \sqrt{\lambda z_0/\pi}$ and the beam radii at the
mirrors are

$$w_{1,2} = w_0\sqrt{1 + \left(\frac{z_{1,2}}{z_0}\right)^2} \tag{3.66}$$

For the solution to indeed represent a Gaussian beam, z_0 must be real. Using Equation
3.65, it is easy to show that, by imposing $z_0^2 > 0$, Equation 3.61 is retrieved, that is exactly
the confinement condition required by ray optics. As already mentioned, a Gaussian beam
is a mode of the spherical-mirror resonator provided that the wavefront normals reflect onto
themselves, always retracing the same path, and that the phase retraces itself as well. Since
the phase of a Gaussian beam is given by [104]

$$\varphi(r, z) = kz - \zeta(z) + \frac{kr^2}{2R(z)} \tag{3.67}$$

with $\zeta(z) = \arctan(z/z_0)$ and $r^2 = x^2 + y^2$, on the optical axis we have $\varphi(0, z) = kz - \zeta(z)$ (the phase retardation relative to a plane wave is $\zeta(z)$). As the beam propagates
from mirror 1 to mirror 2, its phase changes by

$$\varphi(0, z_2) - \varphi(0, z_1) = k(z_2 - z_1) - [\zeta(z_2) - \zeta(z_1)] = kL - \Delta\zeta \tag{3.68}$$

Note that, since the mirror surface coincides with the wavefront, all points on each mirror
share the same phase. For the beam to truly retrace itself, the round-trip phase change must
be a multiple of 2π, i.e., $2kL - 2\Delta\zeta = m2\pi$. This yields

$$\nu_m = \frac{c}{2L}\left(m + \frac{\Delta\zeta}{\pi}\right) \tag{3.69}$$

The frequency spacing of adjacent modes is the same as that obtained for the planar-mirror resonator and is independent of the curvatures of the mirrors. The second term, which does depend on the mirror curvatures, simply represents a displacement of all the resonances. The Gaussian beam is not the only beam-like solution of the paraxial Helmholtz equation. An entire family of solutions, the Hermite-Gaussian family, exists. Although a Hermite-Gaussian beam of order (p, q) has the same wavefronts as a Gaussian beam, its amplitude distribution differs. The design of a resonator that *matches* a given beam is therefore the same as in the Gaussian-beam case, regardless of (p, q). It follows that the entire family of Hermite-Gaussian beams represents modes of the spherical-mirror resonator (Figure 3.8). The resonance frequencies of the (p, q) mode do, however, depend on the indices (p, q). This is because of the dependence of the axial phase delay on p and q. It can be shown that the phase of the (p, q) mode on the beam axis is [104]

$$\varphi(0, z) = kz - (p + q + 1)\zeta(z) \tag{3.70}$$

which provides

$$\nu_{p,q,m} = m\frac{c}{2L} + (p + q + 1)\frac{\Delta\zeta}{\pi}\frac{c}{2L} \tag{3.71}$$

As already mentioned, modes of different m, but the same (p, q) have identical intensity distributions and are known as longitudinal or axial modes. The indexes (p, q) label different spatial dependences on the transverse coordinates (x, y); these represent different transverse modes.

Next, let us consider a symmetrical resonator with concave mirrors. By substitution of $R_1 = R_2 = -|R|$ into Equation 3.63, we get $z_1 = -L/2$ and $z_2 = L/2$, that is the beam center lies at the center of the resonator. Moreover we obtain

$$\begin{cases} z_0 = \dfrac{L}{2}\sqrt{\dfrac{2|R|}{L} - 1} \\[2ex] w_0 = \sqrt{\dfrac{\lambda L}{2\pi}}\left(\dfrac{2|R|}{L} - 1\right)^{\frac{1}{4}} \\[2ex] w_{1,2} = \dfrac{\sqrt{\dfrac{\lambda L}{\pi}}}{\left[\dfrac{L}{|R|}\left(2 - \dfrac{L}{|R|}\right)\right]^{1/4}} \end{cases} \tag{3.72}$$

and the confinement condition

$$0 \le \frac{L}{|R|} \le 2 \tag{3.73}$$

is again found. From the above expressions, it is easily recognized that the radius of the beam at the mirrors has its minimum value when $|R|/L = 1$, i.e., for the symmetrical confocal resonator. In this case Equations 3.72 further simplify to

$$\begin{cases} z_0 = \dfrac{L}{2} \\[2ex] w_0 = \sqrt{\dfrac{\lambda L}{2\pi}} \\[2ex] w_{1,2} = \sqrt{\dfrac{\lambda L}{\pi}} = \sqrt{2}w_0 \end{cases} \tag{3.74}$$

which imply $\Delta\zeta = \pi/2$. In this case from Equation 3.71 we have

$$\nu_{p,q,m} = m\frac{c}{2L} + (p+q+1)\frac{c}{4L} = \frac{c}{2L}\left[m + \frac{p+q+1}{2}\right] =$$
$$= \frac{c}{4L}(2m+p+q+1) = \frac{c}{4L}\cdot \text{integ.} \qquad (3.75)$$

which means that in a symmetrical confocal resonator, the resonance frequencies of the transverse modes either coincide or fall halfway between those that result from a change of the longitudinal mode index m. Consequently, in the confocal FP resonator, modes with frequencies can be excited that differ by $c/4L$. As already discussed, an electromagnetic wave impinging onto one of the mirrors of the optical resonator can only excite those modes whose frequencies coincide with that of the wave. Now, in a confocal resonator an infinite number of axial modes have the same eigenfrequencies but have different distribution of the transverse field distribution on the surface of the mirror. Consequently, a wave of a given field distribution will predominantly excite that particular mode inside the resonator whose field distribution coincides with the one of the impinging wave. Mathematically speaking, the incident wave will be decomposed into a linear combination of the modes, i.e., of the eigenfunctions representing the field inside the resonator. The coupling of the incident wave to the particular modes is determined by the coupling coefficients which are determined by the overlap integrals between the modes in the resonator and the incident wave. If only one mode is to be excited, the field distributions of the incident wave and of the resonator mode have to coincide exactly at the surface of the resonator mirror.

To conclude the wave-analysis treatment, we briefly comment on diffraction losses by observing that, since Gaussian and Hermite-Gaussian beams have finite transverse extent and since the resonator mirrors are of finite extent, a portion of the optical power escapes from the resonator on each pass. An estimate of the power loss may be determined by calculating the fractional power of the beam that is not intercepted by the mirror. If the beam is Gaussian with radius w and the mirror is circular with radius $R_a = 2w$, for example, a small fraction, $\exp(-2R_a^2/w^2) = 3.35\cdot 10^{-4}$, of the beam power escapes on each pass, the remainder being reflected (or absorbed in the mirror). Higher-order transverse modes suffer greater losses since they have greater spatial extent in the transverse plane. This fact is used to prevent the oscillation of higher-order modes by inserting apertures into the laser resonator whose opening is large enough to allow most of the fundamental $(m, 0, 0)$ mode energy through, but small energy to increase substantially the losses of the higher-order modes.

3.3 Cavity-design considerations

As one can guess, the finesse is a parameter of utmost importance in an optical cavity. For example, it determines the enhancement factor for the intra-cavity absorption signal that can be recovered from the cavity transmission; as the finesse increases, the cavity resonances become more sensitive to optical loss: as a result, small absorption signals become amplified by the cavity response. Precise knowledge of the cavity finesse is crucial, for instance, in spectroscopy experiments for making accurate measurements of the intra-cavity absorption. As we will learn in Chapter 5, for high-finesse cavities ($\mathcal{F} >1000$), the finesse is usually

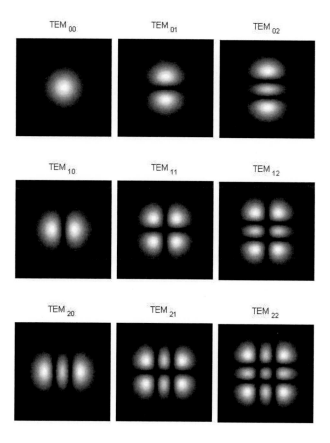

FIGURE 3.8

Intensity profiles of the lowest-order Hermite-Gaussian modes.

measured via cavity ring-down spectroscopy (CRDS). CRDS measurements are performed by injecting laser light into a resonant mode of the cavity and then rapidly switching off the incident light. It is straightforward to use a single-frequency, continuous-wave (cw) laser for CRDS measurements. In this case, for fast switching of the light source, the laser frequency can be swept quickly across the cavity resonance, such that the time the laser spends on it is shorter than the cavity lifetime. Another way of switching the incident light is to use an acousto-optic or electro-optic modulator that can provide switching times of less than a few hundred nanoseconds, sufficiently shorter than a typical high-finesse cavity lifetime.

A second relevant property of an optical cavity is represented by the spectral bandwidth. The latter refers to a spectral window for which the cavity has a high finesse and nearly uniform cavity FSR.

3.3.1 Quarter wave stack reflectors

Both the cavity finesse and the spectral bandwidth are determined by the reflectors used to construct the cavity. In the following, we analyze the properties for a type of highly reflective mirror used in most cavity-enhanced spectroscopic measurements, including the brand new cavity-enhanced direct-frequency-comb spectroscopy (CE-DFCS). These mirrors, commonly referred to as quarter wave stack (QWS) reflectors, are constructed from alternating layers of high (n_h) and low (n_l) index of refraction dielectric material (see Figure 3.9).

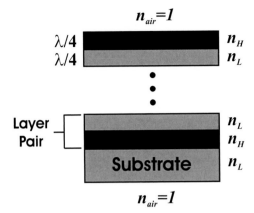

FIGURE 3.9
Structure of quarter wave stack high-reflectivity mirrors.

Indeed, as we will see in a short while, a condition of high reflectivity is produced from alternating dielectric layers with an optical thickness of $\lambda/4$, where λ is the wavelength that the mirror is designed to reflect. The reflectivity of a QWS mirror, based on the transfer matrix formalism, can be found in [105]. At normal incidence, the transfer matrix for a single film of refractive index n and thickness l is given by

$$M = \begin{pmatrix} A & B \\ C & D \end{pmatrix} = \begin{pmatrix} \cos\dfrac{2\pi nl}{\lambda} & -\dfrac{i}{n}\sin\dfrac{2\pi nl}{\lambda} \\ -in\sin\dfrac{2\pi nl}{\lambda} & \cos\dfrac{2\pi nl}{\lambda} \end{pmatrix} \tag{3.76}$$

and the reflection coefficient is determined as

$$r = \frac{An_0 + Bn_Tn_0 - C - Dn_T}{An_0 + Bn_Tn_0 + C + Dn_T} \tag{3.77}$$

where n_0 and n_T are the refractive indices of the two infinite media located at the left and the right of the film, respectively. The film reflectivity is eventually calculated as $\mathcal{R} = |r|^2$. In the case of a QWS mirror, the transfer matrix of a stack of $2N$ layers (or N layer pairs) is calculated as

$$M_{QWS} = \left[\begin{pmatrix} \cos\dfrac{2\pi n_l l_l}{\lambda} & -\dfrac{i}{n_l}\sin\dfrac{2\pi n_l l_l}{\lambda} \\ -in_l\sin\dfrac{2\pi n_l l_l}{\lambda} & \cos\dfrac{2\pi n_l l_l}{\lambda} \end{pmatrix} \right.$$
$$\left. \cdot \begin{pmatrix} \cos\dfrac{2\pi n_h l_h}{\lambda} & -\dfrac{i}{n_h}\sin\dfrac{2\pi n_h l_h}{\lambda} \\ -in_h\sin\dfrac{2\pi n_h l_h}{\lambda} & \cos\dfrac{2\pi n_h l_h}{\lambda} \end{pmatrix} \right]^N \tag{3.78}$$

from which the A, B, C, D elements and thus r are determined. This procedure allows one to calculate $\mathcal{R}_{QWS}(\lambda)$ for given values of $n_l l_l$ and $n_h l_h$. If the QWS mirror is in air ($n_0 = n_T = 1$) and $n_l l_l = n_h l_h = \lambda/4$, Equation 3.78 and use of Equation 3.77 yields

$$\mathcal{R}_{QWS} = |r|^2 = \left[\frac{1 - \left(\dfrac{n_h}{n_l}\right)^{2N}}{1 + \left(\dfrac{n_h}{n_l}\right)^{2N}}\right]^2 \tag{3.79}$$

As can be seen from Equation 3.79, the reflectivity of a quarter wave stack depends on the number of layer pairs in the stack and the ratio n_h to n_l. For example, in the case of 20 layer pairs of fused silica ($n_l = 1.45$, $l_h = 276$ nm) and niobium oxide ($n_h = 1.9$, $l_h = 210$ nm), the peak reflectivity at a center wavelength of 1.6 micron is $\mathcal{R}_{QWS} = 0.99992$. The wavelength dependence of the reflectivity, shown in Figure 3.10, is found by explicitly calculating $\mathcal{R}_{QWS}(\lambda)$: first, the products $n_l l_l$ and $n_h l_h$ are fixed by the condition $n_l l_l = n_h l_h = \lambda_c/4$, where λ_c is the desired center wavelength; then, the matrix elements (Equation 3.78) are computed as a function of λ and inserted into Equation 3.77; finally, 3.79 is used to obtain $\mathcal{R}_{QWS}(\lambda)$. From this calculation, we find that the spectral bandwidth of the mirror reflectivity is determined by the ratio n_h to n_l. Larger n_h/n_l lead to broader spectral bandwidths of the mirror reflectivity. In Figure 3.10, we can see that the combination of fused silica and niobium oxide provide high reflectivity over a spectral range covering 1.5 to 1.7 micron. Outside of this region, the mirror reflectivity drops very rapidly. Therefore, the useful spectral bandwidth of these mirrors is about 200 nm or about 13% of the center optical frequency of the coating. However, the spectral coverage of modern broadband coherent radiation sources (optical frequency comb synthesizers) can be very large (around 90% of the central frequency).

Another important issue related to the design of a high-finesse optical cavity is its dispersion which determines, in essence, the uniformity of the FSR. Actually, due to a variety of frequency-dependent phase shifts inside an optical cavity, the mode spacing is also frequency-dependent. The cavity schematic in Figure 3.11 shows several sources of frequency-dependent phase shifts including the mirror phase shift $\phi_r(\omega)$, the diffraction phase shift $\phi_D(\omega)$, and the phase shift due to intra-cavity media $\phi_m(\omega)$ [106].

To derive an expression for the frequency dependence of the cavity mode spacing, we

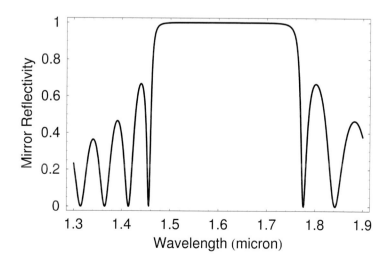

FIGURE 3.10
Reflectivity against wavelength for a QWS mirror ($N = 20$, $n_L = 1.45$, and $n_H = 1.9$).

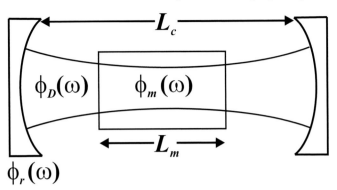

FIGURE 3.11
Several sources of frequency-dependent phase shifts in a QWS mirror-based high-finesse cavity. (Adapted from [106].)

begin by writing an expression for the phase shift accumulated during one round-trip inside the cavity

$$\phi_{rt}\left(\omega\right) = \frac{\omega}{c}\left(L_c - L_m\right) + \frac{\omega n_m}{c}L_m + \phi_D\left(\omega\right) + \phi_{m0}\left(\omega\right) + \phi_r\left(\omega\right) \qquad (3.80)$$

Here, the phase accumulation inside the intra-cavity medium has been separated into a frequency-independent component, $\omega n_m c/L_m$, and a frequency dependent component, $\phi_{m0}\left(\omega\right)$. Cavity modes exist at frequencies where the round-trip phase is equal to a multiple of 2π such that

$$2\pi q = \phi_{rt}\left(\omega\right) \qquad (3.81)$$

where q is the mode number. The change in mode number as a function of frequency is found by taking the derivative of the mode number with respect to ω

$$\frac{\partial q}{\partial \omega} = \frac{1}{2\pi}\left[\frac{L_c + L_m\left(n_m - 1\right)}{c} + \frac{\partial}{\partial \omega}\left(\phi_D + \phi_{m0} + \phi_r\right)\right] \qquad (3.82)$$

The FSR is defined as the frequency interval between adjacent modes; therefore, setting $\Delta q = 1$ yields an expression for the frequency dependent mode spacing

$$FSR\left(\omega\right) = \frac{c}{2L + c\dfrac{\partial \phi\left(\omega\right)}{\partial \omega}} \qquad (3.83)$$

where $L = L_c + L_m\left(n_m - 1\right)$ and $\phi\left(\omega\right) = \phi_D\left(\omega\right) + \phi_{m0}\left(\omega\right) + \phi_r\left(\omega\right)$. The degree to which the cavity FSR is wavelength (or optical frequency)-dependent is determined by the intra-cavity dispersion term. Low intracavity dispersion is crucial, for example, when coupling a broad-bandwidth comb spectrum to respective cavity modes. To better understand this aspect, we shall anticipate here that the frequency spectrum of an optical frequency comb synthesizer consists of a series of discrete, sharp lines typically covering a full optical octave. Such spectrum is described by the formula $f_{n_t} = f_0 + n_t f_r$, where f_0 and f_r are two fixed, well-stabilized frequencies falling in the rf domain and n_t is a huge integer number (10^5-10^6) labeling the different comb teeth. As a consequence, the comb modes f_{n_t} are regularly spaced in the optical frequency domain whereas, as just shown, the cavity FSR is frequency-dependent. It is now clear that intra-cavity dispersion ultimately determines the spectral bandwidth over which the cavity modes and the comb components can be overlapped simultaneously (see also Chapter 6).

3.3.2 Prism-based cavities

In response to this, efforts toward improved designs for low-loss and low-dispersion mirrors have been intense over the past few years including the use of different dielectric materials and specialized coating designs. Recently, new designs for high-finesse cavities that use prism retro-reflectors instead of mirrors have been under development [107, 108]. Since prism retro-reflectors are based on broad bandwidth effects, such as Brewster's angle and total internal reflection, these cavities could provide spectral bandwidths of >80% of the center optical frequency with a finesse of up to 50000. The major advantage of using such prisms is that the bandwidth of the high finesse cavity is practically limited only by the spectral regions of low internal transmission loss of the material used to construct the prisms. Using a variety of materials, it is potentially possible to construct prism cavities capable of covering a region stretching from the UV into the mid-IR. For example, calcium fluoride could be used in the UV/Vis region, fused silica in the visible to the near-IR, and barium fluoride in the near-IR to the mid-IR. In the following we describe the working principle and the main features of this novel cavity design. A thorough analysis including a discussion of the expected loss as well as of the effects of both prism misalignment and errors in construction (that is suitable for specification of prism manufacture and design of the mounting and alignment system) is given in [107]. Figure 3.12 shows a top view of the retro-reflective prisms.

Let the short side (AB) have a length a and the prisms be made of isotropic material, such as fused silica, with refractive index n, relative to the medium surrounding the prisms. A light ray enters the long face (AD) at R_0 with an external angle of incidence nearly at Brewster's angle ($\theta_B = \arctan(n)$) (we will neglect this deviation below). The internal ray strikes face AB at R_1 with an angle of incidence equal to $45°$. This requires that the angle $\angle DAB = 135° - \theta_B$. As long as n is greater than $\sqrt{2}$, this ray will undergo total internal reflection when it strikes side AB. It will then propagate to the next side (BC) and experience a second total internal reflection off this face at R_2 with an angle of incidence equal to $45°$ as long as $\angle ABC = 90°$. This ensures that this reflected ray, $R_2 R_3$, is exactly parallel in the horizontal plane to the ray incident on side AB, $R_0 R_1$, and this ray will leave the prism at Brewster's angle for the prism to air interface at R_3, which is just the complement of Brewster's angle for the air to prism interface. The prism exiting ray is parallel to the incident one. By the law of sines, the ray will be centered on the AB face if the length AR_0 is given by

$$AR_0 = \frac{a \sin 45°}{2 \sin \theta_B} = \frac{a\sqrt{1+n^2}}{2\sqrt{2}n} \qquad (3.84)$$

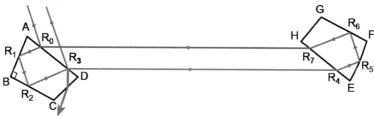

FIGURE 3.12

A schematic of the Brewster's-angle retroreflector-based ring cavity showing the optical beam path. Light is coupled into the cavity at R_0 and decoupled at R_5. All surfaces are flat except EF which has a 6 m convex curve. The effective reflectivity of the cavity is controlled by tuning the input prism around Brewster's angle. (Courtesy of [108].)

The second equality results from $\tan\theta_B = n$. Likewise, the propagation distance R_0R_1 is given by

$$R_0R_1 = \frac{a}{2}\frac{\sin\left(135^\circ - \theta_B\right)}{\sin\theta_B} = \frac{a}{2\sqrt{2}}\left(1 + \frac{1}{n}\right) \tag{3.85}$$

The propagation distance R_1R_2 is $R_1R_2 = a/\sqrt{2}$, and the propagation distance R_2R_3 is

$$R_2R_3 = 3\left(R_0R_1\right) - R_1R_2 = \frac{a}{2\sqrt{2}}\left(1 + \frac{3}{n}\right) \tag{3.86}$$

Thus the total path inside each prism is

$$L_p = R_0R_1 + R_1R_2 + R_2R_3 = \sqrt{2}a\left(1 + \frac{1}{n}\right) \tag{3.87}$$

Therefore, the distance R_3A is three times R_0A, and in order to maintain the maximum aperture, the distance AD should be

$$AD \geq 4R_0A = \frac{2a}{\sqrt{2}\sin\theta_B} = \frac{2a\sqrt{1+n^2}}{\sqrt{2}n} \tag{3.88}$$

We will take this to be an equality. The aperture will be maintained if BC$\geq$AB.

1. For $n \geq \sqrt{3}$, this is realized if we take BC=AB, $\angle$BCD=135°, and $\angle$CDA=θ_B. In this case $CD = \sqrt{2}a/n$.

2. If $n < \sqrt{3}$, it is preferred to take $\angle$BCD=$3\theta_B - 45°$. This will allow the input ray to the cavity, which will strike surface AD at R_3 near Brewster's angle, to leave the prism by striking surface CD near Brewster's angle and minimize reflections and thus scattered light inside the prism. In this case, the prism will have $\angle$CDA = $180° - 2\theta_B$, $\angle$BCD = $2\theta_B - 45°$ and some tedious geometry allows one to calculate that

$$BC = a\frac{5 + 3n + n^2 - n^3}{(1 + n)(4n - n^2 - 1)} \tag{3.89}$$

$$CD = a\frac{\sqrt{2}(1 + n^2)^{3/2}}{n(n - 1)(4n - n^2 - 1)} \tag{3.90}$$

In this second case, the condition

$$BC = a\frac{5 + 3n + n^2 - n^3}{(1 + n)(4n - n^2 - 1)} > AB = a \tag{3.91}$$

returns $n < \sqrt{3}$.

An optical cavity is formed by facing two such retroreflectors towards each other such that corresponding faces are nearly parallel as shown in Figure 3.12. Vertices E, F, G, and H in the second prism correspond to A, B, C, and D, respectively, in the first prism. Likewise, the intersection points of the optic axis with the surfaces in the second prism, R_4, R_5, R_6, and R_7, correspond to R_0, R_1, R_2, and R_3 in the first prism. Let L_g be the distance along the optical axis from Brewster face (AD) to Brewster face (EH); this will be the distance of propagation in the sample per pass of the cavity. Further, the prisms must be placed such that a ray leaving R_3 at Brewster's angle will strike R_4, which is the point on the second prism that corresponds with R_0. This will allow for a ring optical cavity with the optical path passing through the points $R_0 \ldots R_7$, R_0. As described such a cavity would be on the borderline of optical stability. To form a stable optical cavity, some focusing element is needed. A convenient approach is to construct a convex curved surface on face EF and center at R_5. Because the optic axis strikes this surface at $45°$, there will be considerable astigmatism and it would be ideal to construct an astigmatic surface, with different horizontal and vertical radii of curvature, to compensate. Since this typically proves too difficult for fabricators, one is often obliged to select a convex spherical surface with radius R_c. Using the ABCD-matrix formalism we can derive the condition for a stable cavity in the tangential direction (in the plane of incidence). For this purpose, we use the expressions for matrices for a tilted flat interface given in [109]. With reference to Figure 3.13, the tangential matrices describing refraction at interface AD from air to glass and from glass to air are respectively given by

$$M_t^{a-g} = \begin{pmatrix} \dfrac{\cos\theta_2}{\cos\theta_1} & 0 \\ 0 & \dfrac{\cos\theta_1}{n \cdot \cos\theta_2} \end{pmatrix} \tag{3.92}$$

$$M_t^{g-a} = \begin{pmatrix} \dfrac{\cos\theta_4}{\cos\theta_3} & 0 \\ 0 & \dfrac{n \cdot \cos\theta_3}{\cos\theta_4} \end{pmatrix} \tag{3.93}$$

In our case, $\theta_1 = \theta_B$ and $\sin\theta_2 = \sin\theta_3 = \sin\theta_1/n$, such that the matrix describing propagation through the first prism is (note that total internal reflection is described by the identity matrix)

$$M_{p1} = M_t^{g-a} \begin{pmatrix} 1 & L_p \\ 0 & 1 \end{pmatrix} M_t^{a-g} = \begin{pmatrix} 1 & \dfrac{L_p}{n^3} \\ 0 & 1 \end{pmatrix} \tag{3.94}$$

For the second prism, the effect of focusing due to reflection (at an angle of $45°$) at the curved interface EF must be included via the matrix

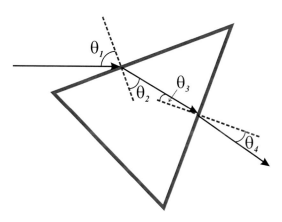

FIGURE 3.13
Propagation of a ray through a prism.

$$\begin{pmatrix} \dfrac{1}{2} & 0 \\ -\dfrac{1}{R_c \cos 45^\circ} & 1 \end{pmatrix} \tag{3.95}$$

whereupon

$$M_{p2} = M_t^{g-a} \begin{pmatrix} 1 & L_p - R_4 R_5 \\ 0 & 1 \end{pmatrix} \begin{pmatrix} \dfrac{1}{2} & 0 \\ -\dfrac{1}{R_c \cos 45^\circ} & 1 \end{pmatrix} \begin{pmatrix} 1 & R_4 R_5 \\ 0 & 1 \end{pmatrix} M_t^{a-g}$$

$$= \begin{pmatrix} \dfrac{-2\sqrt{2}L_p + 2\sqrt{2}R_4 R_5 + R_c}{R_c} & \dfrac{L_p \left(-2\sqrt{2}R_4 R_5 + R_c\right) + 2\sqrt{2}(R_4 R_5)^2}{n^3 R_c} \\ -\dfrac{2\sqrt{2}n^3}{R_c} & 1 - \dfrac{2\sqrt{2}R_4 R_5}{R_c} \end{pmatrix} \tag{3.96}$$

Finally, the matrix describing one round-trip in the cavity is calculated as

$$M_{round\text{-}trip} = M_{air} M_{p2} M_{air} M_{p1} \tag{3.97}$$

with

$$M_{air} = \begin{pmatrix} 1 & L_g \\ 0 & 1 \end{pmatrix} \tag{3.98}$$

Then, the trace of $M_{round\text{-}trip}$ is

$$Tr\left(M_{round\text{-}trip}\right) = \frac{2\left(-2\sqrt{2}L_p - 2\sqrt{2}L_g n^3 + R_c\right)}{R_c} \tag{3.99}$$

Recalling that the cavity is stable if $-1 \leq Tr\left(M_{round\text{-}trip}\right)/2 \leq 1$, we obtain the condition

$$L_g < \frac{R_c - \sqrt{2}L_p}{\sqrt{2}n^3} \tag{3.100}$$

We will now briefly discuss the various sources of optical loss in an ideally constructed cavity, which determines the cavity finesse. The first loss to consider, and the dominant term as one goes to the shorter wavelengths, is the bulk loss of the prism material, due to absorption and scatter. In the near-IR, it is essential to use low OH^- fused silica in order that bulk absorption loss does not dominate, especially near 1.4 μm. For such fused silica, in much of the visible and near-IR, the intrinsic material loss is dominated by Rayleigh scattering, which scales as λ^{-4}. Loss of 1.362 ppm/cm has been reported for fused silica at $\lambda_0 = 1064$ nm. Given a fused silica path of $L_p = 38.2$ mm in each prism, this amounts to a scattering loss of about 10 ppm per both prisms (at 1 μm wavelength). Therefore, we can model the Rayleigh scattering loss as

$$L_R\left(\lambda\right) = \frac{1.362 \cdot 1.064^4}{\lambda^4} \text{ ppm/cm} \tag{3.101}$$

with λ expressed in micron. Due to the wavelength dependence, this loss becomes unacceptably large (compared to what can be obtained with dielectric mirrors) for wavelengths shorter than 400–500 nm, which correspond to scattering loss of 350–150 ppm/pass, respectively. Another source of loss is scatter at each of the interfaces where one has either transmission or total internal reflection. Using super polished substrates, with surface roughness $\sigma = 0.1$ nm root mean squared, the surface scattering can be estimated as 0.15 and 0.79 ppm at $\lambda_0 = 1$ μm for transmission at Brewster's angle and total internal reflection, respectively, for an estimated loss of 1.88 ppm per prism. In both cases, the surface scattering loss scales as λ_0^{-2} and thus becomes less important, compared to Rayleigh scattering, at shorter wavelength. The dominant loss source is due to the fact that P polarized light is coupled into the cavity by deviating the input prism slightly $(\delta\theta)$ from Brewster's angle (θ_B). Then, the fractional Fresnel input and loss per surface is [102]

$$
\begin{aligned}
\mathcal{R}_P\left(\lambda\right) &= \left[\frac{\sqrt{1-\left[1/n\sin\left(\delta\theta+\theta_B\right)\right]^2}-n\cos\left(\delta\theta+\theta_B\right)}{\sqrt{1-\left[1/n\sin\left(\delta\theta+\theta_B\right)\right]^2}+n\cos\left(\delta\theta+\theta_B\right)}\right]^2 \\
&\simeq \frac{\left[n^4\left(\lambda\right)-1\right]^2}{4n^6\left(\lambda\right)}\left[\delta\theta\left(\lambda\right)\right]^2=\frac{\left[n^4\left(\lambda\right)-1\right]^2}{4n^6\left(\lambda\right)}\left[\theta_i-\theta_B\left(\lambda\right)\right]^2 \\
&= \frac{\left[n^4\left(\lambda\right)-1\right]^2}{4n^6\left(\lambda\right)}\left[\theta_i-\arctan\left[n\left(\lambda\right)\right]\right]^2
\end{aligned}
\tag{3.102}
$$

Here θ_i is a fixed incidence angle, very close to Brewster's angle for a given wavelength value, let us say $\theta_i = -0.7^\circ + \arctan\left[n\left(\lambda_0\right)\right]$. Figure 3.14 shows a plot of $2\mathcal{R}_P\left(\lambda\right)$ (when both prisms are considered) vs. wavelength obtained by using Sellmeier equation for fused silica and taking $\lambda_0 = 1$ micron, such that $\theta_i = -0.7^\circ + \arctan\left(1.45\right)$. The function $2L_R\left(\lambda\right)\cdot 3.82$ and the total round-trip loss $2\mathcal{R}_P\left(\lambda\right)+2L_R\left(\lambda\right)\cdot 3.82$ (neglecting surface scattering loss) are also plotted. A loss ranging from 240 to 160 ppm is found between 600 and 750 nm, corresponding to a finesse between 13100 and 19600. This behavior has been experimentally confirmed [108]. In aligning the cavity, one has a trade-off of reducing cavity loss (by operating close to Brewster's angle) and cavity transmission, which is linearly proportional to the Fresnel reflectivity in the common limit that the excitation laser linewidth is much larger than the width of the cavity resonances. An alternative way to couple the optical beam into the cavity is to use frustrated total internal reflection at one of the flat surfaces. This method has the advantage that both prisms can be tuned precisely to Brewster's angle, which will minimize loss. Also, the coupling into the cavity can be adjusted to achieve impedance matching (input loss matching the sum of all other losses), which maximizes the amplitude of the wave coupled into the cavity. The disadvantage of this approach is that alignment is now much more difficult. Finally, we observe that the change in Brewster's angle with temperature is given by

$$
\frac{d\theta_B}{dT} = \frac{d}{dT}\left[\arctan\left(n\right)\right] = \frac{1}{1+n^2}\frac{dn}{dT} \simeq 40 \ \mu\text{rad/K}
\tag{3.103}
$$

for fused silica with $\lambda > 500$ nm, which is negligible for typical ambient temperature changes.

The dispersion of the prism cavity, which will limit the effective spectral range that can be coupled into the cavity as fixed prism spacing, can be written as

$$
\begin{aligned}
FSR\left(\lambda\right) &= \frac{c}{2\left\{L_g[n\left(\lambda\right)]+n\left(\lambda\right)L_p[n\left(\lambda\right)]\right\}} \\
&\simeq \frac{c}{\left[L_0-2\lambda\left(\dfrac{dn}{d\lambda}\right)\left(L_p+n\dfrac{dL_p}{dn}+\dfrac{dL_g}{dn}\right)\right]}
\end{aligned}
\tag{3.104}
$$

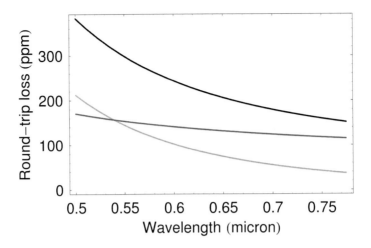

FIGURE 3.14
Total round-trip loss against wavelength (black curve). Individual loss contributions from the dominant loss mechanisms are also shown: Rayleigh scattering (light gray) and Fresnel loss (gray).

where $L_0 = 2\left(L_g + nL_p\right)$ is the round-trip optical path length, dL_p/dn and dL_g/dn are the changes in the intraprism and extraprism path lengths when the prism index, n, changes. Experimentally, one finds that $L_p \gg ndL_p/dn + dL_g/dn$, i.e., the direct material dispersion dominates

$$FSR\left(\lambda\right) \simeq \frac{c}{\left[L_0 - 2\lambda\left(\dfrac{dn}{d\lambda}\right)L_p\right]} \tag{3.105}$$

A plot of Equation 3.105 is given in Figure 3.15, showing a fractional change in the FSR of about 0.2% over a 1.5-micron wavelength span. This result is considerably better than what can be obtained with QWS mirrors.

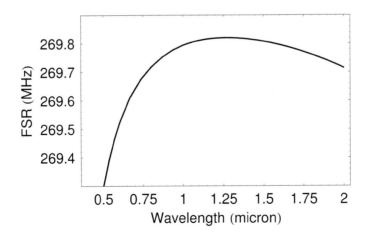

FIGURE 3.15
Prism cavity FSR as a function of wavelength.

3.4 Ultrastable cavities

As discussed before and in particular stated by Equation 3.51, the resonance frequencies of cavities are determined by their length. On the other hand, as we will see in the next chapters, many advanced experiments require that the level of frequency stability of lasers is improved by some external means. Just to mention a few applications, highly stable narrow-linewidth lasers are essential in high-resolution spectroscopy, optical frequency metrology, and fundamental physical tests as well as in interferometric measurements including gravitational wave detection and future space missions. Most often, frequency stability in lasers is achieved exactly by locking the laser to a narrow resonance of a high-finesse cavity. This transfers the length stability of the cavity to frequency stability of the laser. In this task, feedback control is employed to guide the laser frequency to be within a small fraction of the linewidth of one of the cavity resonances, eliminating the intrinsic noises of the laser and replacing them with the measurement noise associated with the FP cavity resonance. However, changes in cavity temperature, in optical power dissipated within the cavity, and mechanical forces deform the cavity, affecting its resonance frequencies. Thus, the prime challenge in the use of cavities for frequency stabilization lies in ensuring that their length be minimally affected by external perturbations. This goal is tackled from two sides: by reducing the level of external disturbances and by making the cavity itself less sensitive to such disturbances. In the following, we will first focus on the effects of thermal instabilities; then, we shall deal with undesired *fast* mechanical vibrations.

3.4.1 Thermal stability

One of the most significant environmental parameters that affect the stability of the mechanical dimensions of macroscopic resonators is the temperature. Temperature fluctuations ΔT around the working temperature T_0 of the resonator result in a variation of the length L_0 to $L(T)$ which can be described as a Taylor series with linear α, quadratic β... coefficients of thermal expansion

$$L(T) = L(T_0) + L(T_0)\alpha\,\Delta T + L(T_0)\beta\,(\Delta T)^2 + \dots \tag{3.106}$$

Since $\Delta\nu_m/\nu_m = -\Delta L/L$, we have to first order

$$\frac{\Delta\nu_m}{\nu_m} = -\alpha\,\Delta T \tag{3.107}$$

High frequency stability thus asks for minimization of the temperature fluctuations and the employment of materials with a low coefficient of thermal expansion (CTE). At room temperature, the linear coefficient of thermal expansion of copper is $\alpha_{Cu} = 1.65 \cdot 10^{-5}$ K^{-1} and that of the temperature compensated nickel iron steel Invar is about an order of magnitude lower and comparable to that of fused silica. Much lower values of the thermal expansion are provided by a mixture of glass and ceramic materials such as Zerodur or temperature compensated glasses called Ultralow Expansion glass (Corning ULE 7971, comprising about 80% SiO_2 and 20% TiO_2) [65].

However, at non-zero temperatures, the rigid reference cavity has an inevitable mechanical thermal fluctuation which fundamentally limits the frequency stability, even if the coefficient of thermal expansion (CTE) is zero.

In the following, we discuss the calculation of such thermal fluctuation in an FP cavity where two mirrors are optically contacted to the ends of a rigid spacer [110, 111]. To get a rough order estimate, we calculate thermal fluctuation in the spacer and mirror with coating separately and then add the results, assuming that every noise component is uncorrelated. Although this is not a rigorous treatment, the process may help our understanding. In principle, the thermal noise in each cavity component is the sum of two contributions: the thermoelastic noise and the Brownian noise.

The former (associated with thermoelastic dissipation) is caused by heat flow along the temperature gradients around the laser beam spot [112]. Thermoelastic noise is interpreted as fluctuations due to thermoelastic damping. The mechanism of the damping in solids was found in the 1930s by Zener. When a non-uniform temperature distribution is applied onto an elastic body, it is deformed through its finite thermal expansion coefficient. On the other hand, when the deformation of the volume is applied by an external force, a non-uniform temperature distribution arises. The temperature distribution relaxes through thermal conductivity, accompanying a reactive deformation of the elastic body. The relaxation is done within a finite time determined by the thermal capacitance and the thermal conductivity of the material. If the applied force is cyclic, the phase of the deformation by the temperature gradient presents a delay against the phase of the applying deformation. The phase delay causes mechanical loss — this is called thermoelastic damping. Since coupling between temperature distribution and elastic deformation is provided by the thermal expansion coefficient, thermoelastic noise is negligible when materials with CTEs close to zero are used.

So we are left with the Brownian noise. The latter is associated with all forms of background dissipation that are homogeneously distributed impurities and dislocations within a material. In fact, the dissipation mechanism is not theoretically calculated, but rather handled as an intrinsic constant dissipation through the FDT. Since the calculation of the dissipated energy is simpler than that of the imaginary part of the transfer function, here we use the FDT in the form stated by Equation 2.112

$$G_x\left(f\right) = \frac{8k_BT}{\omega^2}\frac{W_{diss}}{F_0^2} \tag{3.108}$$

where W_{diss} is the average dissipated power when the generalized oscillatory pressure

$$\boldsymbol{p}\left(\boldsymbol{r}\right) = F_0\cos\left(2\pi ft\right)\boldsymbol{P}\left(\boldsymbol{r}\right) \tag{3.109}$$

is applied on the system. Here, $\boldsymbol{P}\left(\boldsymbol{r}\right)$ is a weighting function (having the units of the inverse of a surface) whose meaning will become clear in a short while. For homogeneous materials, the dissipation is accounted for using the complex Young's modulus [113]

$$E = E_0\left[1 + i\phi(f)\right] \tag{3.110}$$

such that the dissipated power can be written in the form

$$W_{diss} = \omega\mathcal{E}_{max}\phi(f) \equiv 2\pi f\mathcal{E}_{max}\phi(f) \tag{3.111}$$

where $\mathcal{E}_{max}$ is the elastic energy when the strain is at its maximum. Let us start by estimating the contribution from the spacer, regarding the mirrors as small accompaniments of it. In other words, the system is assumed to be a free cylindrical elastic bar of length l, radius R (cross-section area $A = \pi R^2$), and $\boldsymbol{P}\left(\boldsymbol{r}\right) = \hat{x}$ (see Figure 3.16).

The coordinate (x) at the left-hand and right-hand sides are zero and l, respectively. The thermal longitudinal vibration at the left-hand side $(x = 0)$ is estimated. The frequency range of interest is lower than the first longitudinal resonant frequency. The equation of the motion of this elastic bar without the dissipation is described as

$$\rho\frac{\partial^2 u}{\partial t^2} = E_0\frac{\partial^2 u}{\partial x^2} \tag{3.112}$$

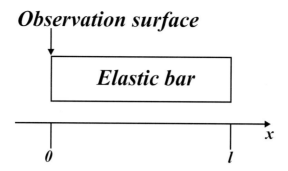

FIGURE 3.16
Schematic of elastic bar for the calculation of the thermal noise contribution from the cavity spacer.

where u is the longitudinal displacement and ρ is the density. This equation is solved by the ansatz $u(x,t) = u(x)e^{i\omega t}$ which leads to

$$\frac{\partial^2 u(x)}{\partial x^2} - k^2 u(x) = 0 \tag{3.113}$$

with the wavenumber $k = \omega\sqrt{\rho/E_0}$. The above equation is solved by a linear combination of a sine and a cosine function

$$u(x) = B\cos(kx) + C\sin(kx) \tag{3.114}$$

where B and C are determined by the boundary conditions. Since the thermal noise at the left-hand side ($x = 0$) is calculated, the generalized force, F, is applied on the left-hand end. The first boundary condition is expressed as

$$F_0 = -E_0 A \frac{du}{dx}\bigg|_{x=0} \tag{3.115}$$

The other end is free. The second boundary condition is written as

$$0 = -E_0 A \frac{du}{dx}\bigg|_{x=l} \tag{3.116}$$

It is worth stressing that the effect of $\cos(2\pi f t)$ in the force has already been accounted for in the formulation of the FDT. Therefore, we get $C = -F/kE_0 A$ and $B = C\cos(kl)/\sin(kl)$, whereupon

$$u(x) = \frac{C}{\sin(kl)}\left[\cos(kl)\cdot\cos(kx) + \sin(kl)\cdot\sin(kx)\right] =$$

$$\frac{-F_0}{kE_0 A\cdot\sin(kl)}\cos[(x-l)k] \tag{3.117}$$

Therefore, the maximum of the elastic energy is given by

$$\mathcal{E}_{max} \equiv \frac{1}{2}E_0 \int_0^l \left(\frac{\partial u(x)}{\partial x}\right)^2 A \cdot dx$$

$$= \frac{A}{2}E_0 \int_0^l \frac{F_0^2}{E_0^2 A^2} \frac{\sin^2\left[(x-l)\,k\right]}{\sin^2(kl)} dx \simeq$$

$$\simeq \frac{F_0^2}{2E_0 A} \int_0^l \left(\frac{x-l}{l}\right)^2 dx \simeq \frac{lF_0^2}{6E_0 A} \tag{3.118}$$

such that

$$W_{diss} = \omega\phi\mathcal{E}_{max} = \frac{\phi\omega F_0^2 l}{6E_0 A} \tag{3.119}$$

which, inserted into Equation 3.108, gives

$$G_{spacer}(\omega) = \frac{4k_B T}{\omega} \frac{l}{3\pi R^2 E_0} \phi_{spacer} \tag{3.120}$$

Next, we evaluate the contribution from the mirror [114]. A simple analytic expression is obtained only when we regard the mirror body as an infinite-half volume (let us say $z > 0$). This is a good approximation when the mirror substrate is sufficiently larger than the Gaussian beam. In the case, when the beam profile is Gaussian and the center of the light spot coincides with the center of the transverse coordinates, we have for the weighting function

$$\boldsymbol{P}(\boldsymbol{r}) = \frac{1}{\pi r_0^2} e^{-\frac{r^2}{r_0^2}} \hat{z} \tag{3.121}$$

where r_0 is the radius of the laser beam. Let $y_z(\boldsymbol{r})$ be the normal displacement of the surface $(z = 0)$ at location $\boldsymbol{r}$ under the action of the applied pressure. In the linear approximation of small strains, from basic elasticity theory we have [115]

$$y_z(\boldsymbol{r}) = \int G(\boldsymbol{r}, \boldsymbol{r}')\, p(\boldsymbol{r}')\, d^2 r' \tag{3.122}$$

where $G(\boldsymbol{r}, \boldsymbol{r}')$ is a Green's function given by

$$G(\boldsymbol{r}, \boldsymbol{r}') = \frac{1 - \sigma^2}{\pi E_0} \frac{1}{|\boldsymbol{r} - \boldsymbol{r}'|} \tag{3.123}$$

where σ is the Poisson ratio of the material. In this case, we have for the maximum elastic energy stored in the material

$$\mathcal{E}_{max} \equiv \frac{1}{2} \int p\left(\boldsymbol{r}\right) y_z\left(\boldsymbol{r}\right) d^2 r$$

$$= \frac{F_0^2}{2} \frac{1-\sigma^2}{E_0 \pi^3 r_0^4} \int_{r=0}^{\infty} \int_{r'=0}^{\infty} \int_{\varphi=0}^{2\pi} \int_{\varphi'=0}^{2\pi} \frac{rr' e^{-\frac{r^2+r'^2}{r_0^2}}}{\sqrt{r^2 + r'^2 - 2rr' \cos\left(\varphi' - \varphi\right)}} dr dr' d\varphi d\varphi'$$

$$= \frac{F_0^2}{4} \frac{1-\sigma^2}{E_0 \pi^3 r_0} \int_{t=0}^{\infty} \int_{\alpha=0}^{\pi/2} \int_{\varphi=0}^{2\pi} \int_{\varphi'=0}^{2\pi} \frac{t^2 e^{-t^2} \sin\left(2\alpha\right)}{\sqrt{1 - \sin\left(2\alpha\right)\cos\left(\varphi' - \varphi\right)}} dt d\alpha d\varphi d\varphi'$$

$$= \frac{F_0^2}{4} \frac{1-\sigma^2}{E_0 \pi^3 r_0} \frac{\sqrt{\pi}}{4} \int_{\alpha=0}^{\pi/2} \int_{\varphi=0}^{2\pi} \int_{\varphi'=0}^{2\pi} \frac{\sin\left(2\alpha\right)}{\sqrt{1 - \sin\left(2\alpha\right) \cos\left(\varphi' - \varphi\right)}} d\alpha d\varphi d\varphi'$$

$$= \frac{F_0^2}{8} \frac{1-\sigma^2}{E_0 \pi^{3/2} r_0} I \tag{3.124}$$

where $\varphi' - \varphi = \theta$ is the angle between $\boldsymbol{r}$ and $\boldsymbol{r}'$ and *polar* coordinates $r = tr_0 \cos\alpha$, $r' = tr_0 \sin\alpha$ have been introduced. In principle, $y_x\left(\boldsymbol{r}\right)$ and $y_y\left(\boldsymbol{r}\right)$ should also be considered in the derivation of $\mathcal{E}_{max}$, but by virtue of the geometry of the problem, the dominant contribution comes from $y_z\left(\boldsymbol{r}\right)$. The integral I in Equation 3.124 is evaluated as follows

$$
\begin{aligned}
I &= \int_{\alpha=0}^{\pi/2} \int_{\theta=0}^{2\pi} \frac{\sin\left(2\alpha\right)}{\sqrt{1 - \sin\left(2\alpha\right)\cos\left(\theta\right)}} d\alpha d\theta \\
&= \int_{\alpha=0}^{\pi/2} \int_{\theta=0}^{2\pi} \sin\left(2\alpha\right) \left[1 + \sum_{n=0}^{\infty} \frac{(2n+1)!!}{(2n+2)!!} \cdot \left[\sin\left(2\alpha\right)\cos\left(\theta\right)\right]^{n+1}\right] d\alpha d\theta \\
&\simeq 2.7\pi \tag{3.125}
\end{aligned}
$$

where the series expansion of $1/\sqrt{1-x}$ has been used. Finally, we obtain $\mathcal{E}_{max} \simeq [2.7 F_0^2 (1 - \sigma^2)]/(8 E_0 \sqrt{\pi} r_0)$ which inserted into Equation 3.111 and then into Equation 3.108 yields

$$G_{mirror}\left(f\right) = 2.7 \frac{k_B T}{\omega} \frac{1-\sigma^2}{E_0 \sqrt{\pi} r_0} \phi_{sub} \tag{3.126}$$

When the full calculation (including contributions from $y_x\left(\boldsymbol{r}\right)$ and $y_y\left(\boldsymbol{r}\right)$) is carried out, the only difference is that the factor 2.7 is replaced by a factor 4 [116]. Equations 3.120, 3.126 suggest that, when selecting a suitable material for a resonator, attention has to be given also to Young's modulus of elasticity E_0 of the material. A higher E_0 value corresponds to a lower displacement fluctuation. A larger spot size may also improve the stability. In addition, the higher E_0 the smaller the deformation and, hence, the smaller the variation of the eigenfrequencies due to tilt or acceleration. Also, when the beam radius decreases, the thermal noise increases because canceling happens only in fluctuations on a smaller scale than the beam radius. In practice, the mirror reflective coating introduces additional losses. Thus, for small coating thickness ($d \ll r_0$), the total loss angle ϕ can be approximated as [117]

$$\phi \simeq \phi_{sub} + \frac{d}{r_0} \cdot \phi_{coat} = \phi_{sub} \left(1 + \frac{d}{r_0} \frac{\phi_{coat}}{\phi_{sub}}\right) \tag{3.127}$$

such that

TABLE 3.3
Relevant properties of materials suitable for ultrastable bulk resonators.

Symbol	Units	Copper	Invar	Fused silica	ULE	Zerodur M	Sapphire (4.2 K)
α	10^{-8} /K	1650	150	55	0.3	< 1	$5 \cdot 10^{-4}$
E_0	10^9 N/m^2	130	145	73	68	89	435
ρ	10^3 Kg/m^3	8.92	8.13	2.2	2.21	2.52	4.0
c_p	J/(Kg K)	385	500	703	0.77	0.81	$5.9 \cdot 10^{-6}$
λ	W/(m K)	400	10.5	1.38	1.31	1.63	280
σ		0.33	0.30	0.17	0.18	0.24	0.25-0.30

Note: The following legend holds: α=linear coefficient of thermal expansion, E_0= Young's modulus of elasticity, ρ=density, c_p=specific heat, λ=heat conductivity, σ=Poisson's ratio.

$$G_{coated\ mirror}\left(f\right) \simeq G_{mirror}\left(f\right)\left(1 + \frac{d}{r_0}\frac{\phi_{coat}}{\phi_{sub}}\right) \tag{3.128}$$

Finally, if we add contributions from the two mirrors and the two ends of the spacer, assuming everything is uncorrelated, we have for the total length fluctuation

$$G_{cavity}\left(f\right) = 2 \cdot G_{coated\ mirror}\left(f\right) + 2 \cdot G_{spacer}\left(f\right) \tag{3.129}$$

Assuming $T = 300$ K, $R = 4$ cm, $l = 24$ cm, $\phi_{spacer} = \phi_{sub} = 1/(6 \cdot 10^4)$, $E_0 = 6.8 \cdot 10^{10}$ Pa, $\sigma = 0.18$ (ULE values reported in table 3.3), $d = 2$ μm, $\phi_{coat} = 4 \cdot 10^{-4}$, and $r_0 = 240$ micron, we obtain $\sqrt{G_{cavity}} \equiv \sqrt{G_L} = 5 \cdot 10^{-17}$ m/$\sqrt{\text{Hz}}$ for $f = 1$ Hz, the dominant contribution coming from the mirror: $G_{coated\ mirror}/G_{spacer} \simeq 110$. This displacement fluctuation results in a frequency noise $\sqrt{G_\nu}/\nu = \sqrt{G_L}/L$ or

$$\sqrt{G_\nu} = \frac{\sqrt{G_L}}{L}\frac{c}{\lambda} \tag{3.130}$$

For 563 nm light, this yields a frequency noise $\sqrt{G_\nu} = 0.1$ Hz/$\sqrt{\text{Hz}}$. This estimate is a good approximation of the world-highest level stabilization result using a cavity.

However, to calculate an accurate thermal-noise level, we must take everything into account simultaneously and precisely: the two mirrors are finite sized and optically contacted onto the spacer, which is a three-dimensional cylinder. Again we have to calculate W_{diss} (to be inserted in Equation 3.108) when a generalized oscillatory pressure with a Gaussian profile for the weighting function is applied on the system. This problem is adequately solved numerically by the help of commercial finite element software. Table 3.4 summarizes eight cases of calculation results (corresponding to the cavity shapes in Figure 3.17), showing the displacement noise level at 1 Hz along with contributions from each component. Corresponding frequency stability limits are also shown, assuming 563 nm light. These results suggest that the thermal-noise limitation is about $0.1 \cdot (1\text{Hz}/f)^{1/2}\text{Hz}/\sqrt{\text{Hz}}$, although the selection of materials, dimensions, and/or beam radius may nominally change this result. In every case, the mirror is the dominant thermal-noise source. This is because, at a frequency region well below the mechanical resonance, only the losses around the beam spot contribute to thermal noise. Therefore, the cavity's overall shape and/or the loss of the spacer do not greatly affect thermal noise when other conditions around the mirror, such as the beam radius, are kept identical (case No. 2/6/7 and case No. 2/3). Because the mirror contribution is dominant, noise decreases with a larger beam radius or with a lower-loss mirror substrate (e.g., case No. 2/4 and case No. 1/2). In the case of the fused silica mirror substrate, the contribution from the coating becomes the most dominant source (e.g., case

TABLE 3.4
Calculation results for the displacement and frequency noise noise level at 1 Hz [110].

Case	Sh	Sp	Sub	r_0 (μm)	$\sqrt{G_L}$ (m/$\sqrt{\text{Hz}}$)	Contr. in G (%) Sp	Sub	Coat	$\sqrt{G_\nu}$ (Hz/$\sqrt{\text{Hz}}$)
1	A	ULE	ULE	240	6.0	3	84	13	0.13
2	A	ULE	Silica	240	2.6	14	24	62	0.059
3	A	Silica	Silica	240	2.5	1	28	71	0.055
4	A	ULE	Silica	370	2.0	25	28	47	0.044
5	A	Zerodur	Zerodur	240	21	3	96	1	0.47
6	B	ULE	Silica	240	2.6	14	24	62	0.058
7	C	ULE	Silica	240	2.8	25	21	54	0.063
8	D	ULE	Silica	200	3.1	16	21	63	0.11

Note: Frequency noise is calculated for 563-nm light. Assumed losses, ϕ, for ULE, fused silica, Zerodur, and coating are $1/(6 \cdot 10^4)$, $1/(1 \cdot 10^6)$, $1/(3 \cdot 10^3)$, and $4 \cdot 10^{-4}$, respectively. The following legend holds: Sh=shape, Sp=Spacer material, Sub=Substrate material, Coat=Coating. Shapes refer to Figure 3.17.

No. 8). Because one cannot greatly alter the coating thickness, the coating loss, or the beam radius, thermal noise from the coating becomes the practical limitation. Cooling the cavity will also help, according to the FDT temperature dependence and assuming material losses which decrease with temperature (not true for fused silica).

3.4.2 Vibration insensitive optical cavities

By summarizing, the ultimate stability limit for a cavity is set by thermomechanical noise, but this limit can only be reached when care is taken to suppress the coupling of temperature fluctuations and mechanical vibrations to the cavity length. Thus, the dimensional stability of optical cavities is paramount too. Depending on the frequency f of the vibrations compared to the mechanical eigenfrequencies f_i of the system (in the 10 kHz range), two regimes can be distinguished. In the high-frequency regime $f \geq f_i$ the external vibrations excite the eigenmodes: the response can be obtained by a decomposition of the individual modes and the reaction of the mode to the applied forces. In the low-frequency limit $f \ll f_i$, the forces that are coupled to the solid accelerate the solid as a whole and lead

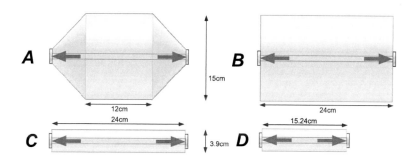

FIGURE 3.17
Assumed cavity shapes. Common parameters are axial hole diameter: 10 mm, mirror diameter: 25.4 mm, mirror thickness: 5 mm, and coating thickness: 2 micron. The thick arrows show positions of Gaussian forces in the calculation. (Courtesy of [110].)

to quasi-static deformation of the solid under the external force (typically the sensitivity is around 100 kHz/ms^{-2}). Vibration isolation systems can minimize the noise level, but compact commercial systems are generally not sufficient to reach a sub-hertz laser linewidth (in particular, they are not effective at reducing seismic noise below ~ 1 Hz). One way to improve the spectral performance of stabilized lasers is to reduce vibration sensitivity by carefully designing the cavity geometry and its mounting. Several groups have proposed and implemented low vibration sensitivity cavities [118, 119, 120, 121].

Here we describe in some detail the most recent design in this context. As already explained, a high-finesse ultrastable optical cavity typically has axial symmetry, comprising two highly reflective concave mirrors bonded to a spacer with a central bore. The optical mode coincides with the symmetry axis and has a small diameter at the mirror surface relative to the cavity dimensions. To achieve vibration insensitivity, one has to ensure that the distance between the two points at the centers of the mirror surfaces remains invariant when a force is applied to the support. This may simply be achieved through symmetrical mounting: a force applied through the mount will cause one half of the cavity to contract while the other half expands with the result that there is no net change in dimension. An obvious implementation of this is to hold the cavity about its plane of symmetry with its axis aligned with gravity and, using this approach, a reduced acceleration sensitivity of 10 kHz/ms^{-2} has been demonstrated [120]. A cavity mounted with its axis horizontal sags asymmetrically under its own weight. However, through optimization of the support positions, the distance between the centers of the mirrors can be made invariant, even though the cavity still deforms on application of a vertical force and, by this method, a vertical acceleration sensitivity of 1.5 kHz/ms^{-2} has been demonstrated [119].

A similar result is achieved by removing material from the underside of the cavity and using this approach a vertical acceleration sensitivity < 0.1 kHz/ms^{-2} has been reported [118]. The geometry is shown in Figure 3.18. Square *cutouts* are made to the underside of a cylindrical spacer and the cavity is supported at four points. The *cutouts* compensate for vertical forces. Vibration insensitivity in the horizontal plane is achieved through symmetrical mounting. Generally, finite element analysis of cavity deformation can be carried out using commercially available software (COMSOL Multiphysics). In such a model, spacer and mirror substrates are considered to be a single rigid body. Also, a static stress-strain model can be used, because the frequency of accelerations that make a significant contribution to the frequency noise are less than 10 Hz and, therefore, to a good approximation, can be considered to be at dc relative to the first structural resonance (~ 10 kHz). For a detailed description of finite element-based analysis of cavity deformation under the influence of vibration noise, for a number of different cavity and support configurations, the reader is referred to [122]. In the specific case of the design presented in Figure 3.18, the solution for total displacement is shown in Figure 3.19. Although the cavity deforms when loaded, the axial displacement at the center of the mirror surface, where the laser beam is reflected, remains unchanged. There is, in fact, no unique null condition: the locus of the displacement zero is a plane in xzc space. The existence of zero crossings in the displacement as a function of support position is an important aspect of the design: it offers an experimental parameter that can be adjusted to obtain the null condition.

A second important type of mirror deformation is, in principle, represented by the tilt, where the mirrors are shifted through an angle θ. For an ideal cavity (where optical and mechanical axes coincide), however, the tilt-induced length variation is a second-order effect and can thus be safely neglected. By contrast, in real cavities, mechanical and optical axes are not coincident due to imperfections in the construction (e.g., mirror polishing, spacer machining, and contacting of the mirrors onto the spacer). In this case, the effects of tilt-induced length variations must be carefully accounted for. This task was accomplished

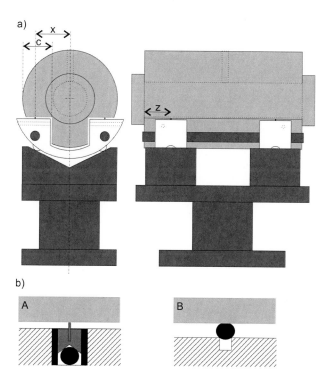

FIGURE 3.18

(a) Cutout cavity on mount. The cavity is represented by the light gray shaded area; the mirror substrates and spacer are made from ULE glass (the mirrors have a concave surface with a radius of curvature of 350 mm and a reflectivity of 0.99998). The support *yokes* are shown in white. Spacer length, 99.8 mm; diameter, 60 mm; axial bore diameter, 21.5 mm; vent-hole diameter, 4 mm. The mirrors are 31 mm in diameter and 8 mm thick. The parameters relevant to the design, the cut depth c, and the support coordinates x and z are indicated. (b) Details of the support point in cross section. The black shaded areas represent rubber tubes and spheres. Case A: Diamond stylus set into a cylindrical ceramic mount, cushioned on all sides by a rubber tube, and from below by a rubber sphere. The stylus digs into the underside of the cavity, and can be considered to be in rigid contact with it. Case B: 3-mm diam rubber sphere recessed into a cylindrical hole in the yoke. (Courtesy of [118].)

in [121], where the above design was further improved. Incidentally, a progression in the thermal noise level was also achieved by using fused silica for the mirror substrates, which provides a higher mechanical Q factor in comparison to ULE. Based on the results of extensive simulations using finite element software, two ultrastable optical cavities were designed and constructed: one horizontal, and the other vertical. The spacers for both configurations were machined from ULE glass rods. The wavelength range of the high-reflection coating mirrors allowed operation at both 1064 and 1062.5 nm (Nd:YAG and Yb-doped fiber laser). Each cavity was optically contacted with a flat mirror and a concave mirror with radius of curvature of 500 mm. Both cavities had a finesse of 800000. As already mentioned, the mirror substrates were made from fused silica, giving rise to a thermal noise floor of $\simeq 4 \cdot 10^{-16}$ (for a 100-mm-long cavity), dominated by the thermal noise of

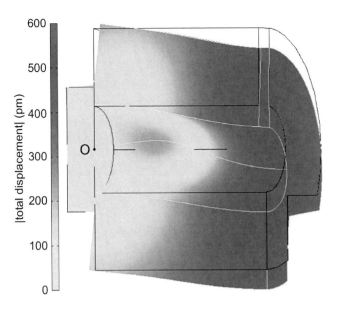

FIGURE 3.19
Plot showing the deformation of a quarter section of the cavity. The point O marks the center of the mirror surface. $c = 18.45$ mm, $x = 22$ mm, and $z = 17$ mm. The scale of the distortion is amplified by a factor of 10^7 and a uniform downward displacement of 10 nm is subtracted. The white (black) lines indicate the outline of the loaded (unloaded) geometry. (Courtesy of [118].)

the high-reflection coatings. Compared to an all-ULE cavity, this gave an improvement greater than a factor of 2. However, since fused silica shows a larger CTE than ULE, the overall effective CTE was much larger than that of an all-ULE cavity, and the zero thermal-expansion coefficient was shifted to well below 0 °C, instead of 10-20 °C. Due to this increased temperature sensitivity, a high thermal shielding factor coupled with a tight temperature control was necessary to minimize the impact of environmental temperature fluctuations. Reported vibration sensitivities were equivalent to the above horizontal cavity design: better than $\simeq 1 \cdot 10^{-11}$ $(\text{ms}^{-2})^{-1}$, but with strongly reduced dependence on support points' position. The vertical cavity, instead, showed a much lower sensitivity than previous vertical cavity designs: $\simeq 3.5 \cdot 10^{-12}$ $(\text{ms}^{-2})^{-1}$ in the vertical direction and $\simeq 1.4 \cdot 10^{-11}$ $(\text{ms}^{-2})^{-1}$ in the horizontal ones.

More recently, an optical cavity design that is insensitive to both vibrations and orientation has also been reported [123]. The design is based on a spherical cavity spacer that is held rigidly at two points on a diameter of the sphere (see Figure 3.20). Coupling of the support forces to the cavity length is reduced by holding the sphere at a *squeeze insensitive angle* with respect to the optical axis. Finite element analysis was used to calculate the acceleration sensitivity of the spherical cavity for the ideal geometry as well as for several varieties of fabrication errors. The measured acceleration sensitivity for an initial, sub-ideal version of the mounted cavity was $4.0(5) \cdot 10^{-11}/\text{g}$, $1.6(3) \cdot 10^{-10}/\text{g}$, and $3.1(1) \cdot 10^{-10}/\text{g}$ (with $g = 9.81$ m/s^2) for accelerations along the vertical and two horizontal directions. This low acceleration sensitivity, combined with the orientation insensitivity that comes with a rigid mount, indicates that this cavity design could allow frequency stable lasers to operate in non-laboratory environments.

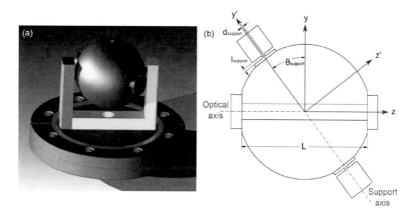

FIGURE 3.20
CAD rendering of a spherical cavity mounted at the squeeze insensitive angle with Viton Oring contacts. Important coordinate systems and dimensions are labelled. The sphere is 50.8 mm in diameter and has a 6 mm diameter bore drilled through it along the optical axis. The mirrors are optically contacted to the sphere on flats separated by $L = 48.5$ mm and are 12.7 mm in diameter and 4.2 mm thick. The two support contacts are attached to the sphere along the support axis which is oriented at $\theta_{support} = 37.31°$ with respect to the y axis and have dimensions $d_{support} = l_{support} = 1$ mm. (Courtesy of [123].)

Let us conclude this section by describing in more detail the procedure by which the thermal noise-driven and vibration-induced cavity length fluctuations are systematically investigated from an experimental point of view (see Figure 3.21).

First, a free-running laser is servo-locked to the resonance of a Fabry-Perot cavity according to the Pound-Drever-Hall technique (see next chapter): the measured remaining instability of the locked laser is a combination of the instability in the length of the frequency reference itself (environmental noise sources that modulate the cavity length are mainly of acoustic origin in the range above 50 Hz and seismic origin from 1 to 50 Hz; for Fourier frequencies less than 1 Hz, temperature changes affect the cavity length as well as other dimensions in the experiment that may inadvertently contribute to the instability) and the defects of the servo-locking system. Therefore, if the lock is sufficiently tight (high electrical bandwidth), measuring the frequency noise power spectral density (FNPSD) of the locked laser in the proper frequency range provides information on the instability of the length cavity. Practically, in order to measure the laser FNPSD, the beat note with a second identical (or better) system is detected and sent to a FFT analyzer after passing through a frequency-to-voltage converter. Alternatively, the beat note signal can be processed by a counter to provide the Allan standard deviation. In particular, to measure the three vibration sensitivity components of the cavity under investigation, the latter is shaken (by means of active vibration isolation platforms) with sinusoidal signals in the frequency range of 1-10 Hz, the amplitude of the imparted acceleration being measured by means of a three-axis seismometer. Then, modulation at the drive frequency is resolved in the power spectrum (Figure 3.22). Finally, a response is derived from the modulation amplitude and the measured acceleration.

Reference System

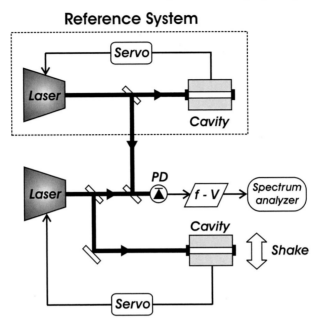

FIGURE 3.21
Schematic of experimental setup for measuring the cavity vibration response. PD: photodiode; f-V: frequency-to-voltage converter. The thicker lines indicate beam paths; the thinner lines indicate electronic signal paths. (Adapted from [118].)

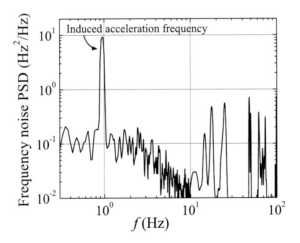

FIGURE 3.22
Frequency noise power spectral density for induced acceleration in a vibration-insensitive cavity. Measurement is between two independent identical systems. (Courtesy of [121].)

3.5 Fiber cavities

Single-mode optical fibers represent another attractive approach to realize high-finesse optical resonators. Basically there are three ways to implement a fiber-based optical cavity:

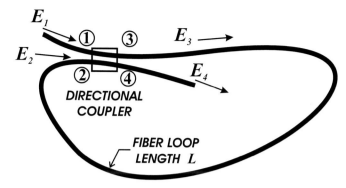

FIGURE 3.23
Schematic of an all single-mode fiber resonator. (Adapted from [124].)

1. One scheme makes use of a single fiber strand and a directional coupler;
2. In a second configuration the two end facets of a single strand of fiber, properly worked out and coated, are faced to each other;
3. The third method relies on a couple of facing Bragg gratings imprinted into the fiber.

3.5.1 Directional coupler-based fiber cavities

With the recent advances in single-mode fiber directional couplers, a fiber ring can be closed in a low-loss manner [124]. A schematic of such a resonator is shown in Figure 3.23.

If the directional coupler has large coupling, light trapped in the fiber ring will couple from port 2 to port 3 and will continue to circulate. Similarly, light introduced to the input port 1 will couple mostly to the output port 4. Consider the case in which the fiber-loop length is adjusted for constructive interference (addition) between coherent components entering port 3 from ports 1 and 2. The small fraction of light from port 2 to port 4 will destructively interfere with the light coupling from port 1 to port 4. The circulating field will grow until an equilibrium is reached. With an optimum value of coupling that depends on losses, the two destructively interfering components emerging from port 4 are equal in amplitude and completely cancel each other. From an energy-conservation point of view, the circulating power grows until the power dissipated by losses in the loop equals the input power at port 1. If the light frequency is now varied continuously, the power emerging from port 4 will show a series of sharp minima whenever the input optical frequency matches the resonant condition. The behavior is similar to a Fabry-Perot-type resonator whose reflected power has sharp minima at resonance. For a resonator of this type to function properly, the directional coupler must have a low insertion loss. As the optimum value of coupling depends on losses, a variable coupling coefficient is desirable. For instance, an evanescent field coupler can be used (see later in this chapter). A single strand of optical fiber is bonded into two slotted quartz blocks a distance L apart. Each fiber-block unit is ground and polished to within a few micrometers of the fiber core. Placing the two units in contact, oriented as in Figure 3.23, produces a ring resonator of perimeter L. Coupler tuning is accomplished by sliding one block over the other to vary the core-to-core separation and optimize the coupling coefficient. The directional coupler is modelled as a perfect (lossless) device with an added lumped loss that is independent of the coupling coefficient. Referring to Figure 3.23, the fractional coupler intensity loss γ_0 is given by

$$|E_3|^2 + |E_4|^2 = (1 - \gamma_0)\left(|E_1|^2 + |E_2|^2\right) \tag{3.131}$$

where E_i is the complex filed amplitude at the i-th port. The complex amplitudes in the fibers after the coupled-mode interaction are related to the incident-field amplitudes by

$$E_3 = \sqrt{1 - \gamma_0}\left[\sqrt{1 - \chi}E_1 + i\sqrt{\chi}E_2\right] \tag{3.132}$$

$$E_4 = \sqrt{1 - \gamma_0}\left[i\sqrt{\chi}E_1 + \sqrt{1 - \chi}E_2\right] \tag{3.133}$$

where χ is the intensity coupling coefficient (no coupling corresponds to $\chi = 0$ whereas $\chi = 1$ gives complete cross coupling). E_2 and E_3 are further related by

$$E_2 = E_3 e^{-\alpha L} e^{i\beta L} \tag{3.134}$$

with

$$\beta = \frac{n\omega}{c} \tag{3.135}$$

where α is the fiber amplitude attenuation coefficient and n the refractive index. From Equations 3.132, 3.133, 3.134 we obtain

$$\begin{cases} \left|\dfrac{E_3}{E_1}\right|^2 = \dfrac{(1 - \gamma_0)(1 - \chi)}{1 + \chi\kappa_r + 2\sqrt{\chi\kappa_r}\sin\beta L} \\[4mm] \left|\dfrac{E_4}{E_1}\right|^2 = (1 - \gamma_0)\dfrac{\chi + \kappa_r + 2\sqrt{\chi\kappa_r}\sin\beta L}{1 + \chi\kappa_r + 2\sqrt{\chi\kappa_r}\sin\beta L} \end{cases} \tag{3.136}$$

where $\kappa_r = (1 - \gamma_0)e^{-2\alpha_0 L}$. For a resonant situation we require $|E_4/E_1|^2$ to vanish, which can be obtained if $\chi = \kappa_r$. Indeed, in that case we get

$$\begin{cases} \left|\dfrac{E_3}{E_1}\right|^2 = \dfrac{(1 - \gamma_0)(1 - \kappa_r)}{(1 + \kappa_r)^2 - 4\kappa_r\sin^2\left(\dfrac{\beta L}{2} - \dfrac{\pi}{4}\right)} \\[5mm] \left|\dfrac{E_4}{E_1}\right|^2 = (1 - \gamma_0)\left[1 - \dfrac{(1 - \kappa_r)^2}{(1 + \kappa_r)^2 - 4\kappa_r\sin^2\left(\dfrac{\beta L}{2} - \dfrac{\pi}{4}\right)}\right] \end{cases} \tag{3.137}$$

such that $|E_4/E_1|^2 = 0$ when $\sin^2\left(\beta L/2 - \pi/4\right) = 1$. This is obtained when

$$\frac{2\pi n L}{c}\nu_q = q2\pi - \frac{\pi}{2} \tag{3.138}$$

with $q = 0, 1, 2, \ldots$. Thus, the free spectral range is

$$FSR = \frac{c}{nL} \tag{3.139}$$

The circulating intensity is given by

$$\left|\frac{E_3}{E_1}\right|^2_{max} = \frac{(1 - \gamma_0)}{(1 - \kappa_r)} \tag{3.140}$$

By equating the first of Equations 3.137 to $1/2\left|E_3/E_1\right|^2_{max}$, the $FWHM$ is found to be

$$FWHM = \frac{c}{nL}\left\{\frac{1}{2} - \frac{1}{\pi}\arcsin\left[1 - \frac{(1-\kappa_r)^2}{2\kappa_r}\right]\right\} \qquad (3.141)$$

that for κ_r near unity, simplifies to

$$FWHM \simeq \frac{c}{nL}\frac{1-\kappa_r}{\pi\sqrt{\kappa_r}} \qquad (3.142)$$

where $\arcsin(1-x) \simeq \pi/2 - \sqrt{2x}$ has been exploited. The cavity finesse is therefore

$$\mathcal{F} = \frac{FSR}{FWHM} = \frac{\pi\sqrt{\kappa_r}}{1-\kappa_r} \qquad (3.143)$$

A finesse of 80, corresponding to $\kappa_r = 0.962$, was achieved in the first experimental realization [124], limited by the coupler insertion loss $\gamma_0 = 3.2\%$. With today's couplers, finesse values of several hundreds are possible.

3.5.2 Resonators based on closely faced fiber tips

In a simple analogy to the free-space optical cavity made of two mirrors, use of gold or multilayer dielectric coatings on the facet, either flat or concave, of a fiber (applied by vacuum evaporation or by pull-off films) has been reported in literature; a stable fiber cavity is then formed from two closely spaced fiber tips placed face to face. Limited by the methods used to fabricate the concave surface on the fiber, so far such cavities have been built with moderate finesse (up to a few 1000) [125]. In this frame, a fiber-based Fabry-Perot cavity combining very small size, high finesse (> 30000), small waist and mode volume, and good mode matching between the fiber and cavity modes was recently realized, mainly for cavity quantum electrodynamics (CQED) studies [126]. In that experiment, a CO_2 laser pulse was used to shape the concave mirror surface, which was then coated with a high-performance dielectric coating. In particular, a parameter regime was found where a single pulse of CO_2 laser radiation focused on the cleaved end face of an optical fiber produced a concave surface with extremely low roughness. In this regime, thermal evaporation occurs, while melting is restricted to a thin surface region, avoiding global contraction into a convex shape. Incidentally, this sets it apart from the regimes used in CO_2 laser-based fabrication of microspheres and transformation of microdisks into high-Q microtoroid resonators (see last section in this chapter). The section through the center in Figure 3.24 shows a profile that is reasonably well approximated by a Gaussian over a wide parameter range.

Close to the center of the profile, the shape is well approximated by a circle, the central radius of curvature (ROC) defining the mirror curvature R. The full width at $1/e$ of the Gaussian profile gives an estimate of the useful mirror diameter D. Because of the approximately Gaussian shape of the depression, R, D, and the total structure depth z_t are related by $z_t \approx D^2/8R$. For example, a mirror structure with $R = 50$ micron and useful diameter $D = 10$ micron is only $z_t = 0.25$ micron deep. With typical laser parameters, the resulting structures are 0.01-4 micron deep and have diameters D between 10 and 45 μm. ROCs measured at the bottom of the depression are between 40 and 2000 μm. Surface roughness was extracted by atomic force microscopy (AFM) measurements $\sigma_{sc} = 0.2\pm0.01$ nm and used to obtain an estimate of scattering losses according to $\mathcal{S} = (4\pi\sigma_{sc}/\lambda)^2$. Thus, $\mathcal{S} = 10$ ppm is obtained for $\lambda = 780$ nm, assuming a high-quality mirror coating that does not significantly increase this roughness. A realistic value for the absorption loss of a supermirror coating is $\mathcal{A} = 2$ ppm [127]. Therefore, with a transmission equaling the losses ($\mathcal{T} = \mathcal{S} + \mathcal{A}$) and using Equation 3.44 together with the relationship $\mathcal{R} + \mathcal{A} + \mathcal{S} + \mathcal{T} = 1$, a maximum finesse $\mathcal{F} = \pi/(\mathcal{T} + \mathcal{S} + \mathcal{A}) = 130000$ is expected for a cavity made from two

(a)

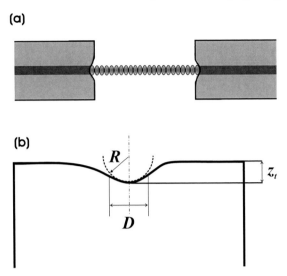

(b)

FIGURE 3.24
(a) A Fabry-Perot cavity made from two closely faced single-mode fiber strands. (b) Mirror geometry and parameters. The profile is not spherical: R designates the ROC in the center and D the structure diameter as defined in the text; z_t is the depth of the structure. (Adapted from [126].)

fiber mirrors with identical coatings. In the pioneering experiment, however, a standard dielectric coating, consisting of 14 layers of SiO_2 and 15 layers of Ta_2O_5 (2.102 refractive index), was employed. As a consequence, a maximum finesse of only 37000 was measured. One key advantage of this configuration is that coupling to and from the cavity is robust and stable over time, as the cavity mirrors are part of the in-coupling and out-coupling fibers. Also, there are no mode-matching lenses, so coupling efficiency is given directly by the mode matching between the mode leaving the fiber and the cavity mode. If single-mode operation is not required, a multi-mode fiber can be used on the output side. This virtually eliminates coupling loss and also makes the cavity more robust against various types of misalignment. But even for single-mode fibers, coupling efficiencies as high as 85% were measured. Finally, it should be noted that, since the cavity length can be made very small (a few microns down to hundreds nm), the part of field penetrating into the multilayer stack contributes significantly to the effective cavity length which enters the formula for the cavity free spectral range (see below).

3.5.3 FBG-based fiber resonators

Instead of coatings applied to the fiber surface, one may also use fiber Bragg gratings (FBGs) as internal mirrors. FBGs are periodic modulations (of period Λ) of the refractive index written into the core of a sensitized optical fiber [128] (Figure 3.25). More in detail, the fabrication of FBGs typically involves the illumination of the core material with ultraviolet laser light, which induces some structural changes and thus a permanent modification of the refractive index. The photosensitivity of the core glass is actually strongly dependent on the chemical composition and the UV wavelength: silica glass has a very weak photosensitivity, whereas germanosilicate glass exhibits a much stronger effect, making possible a refractive index contrast up to 10^{-3}. A significant further increase in photosensitivity is possible by loading the fiber with hydrogen (*hydrogenated fibers*). The first fiber Bragg gratings were

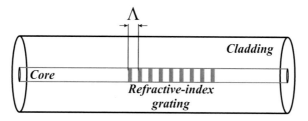

FIGURE 3.25

Structure of the basic fiber bragg grating. The core refractive index is spatially modulated as $n(z) = n + n_1 \cos(2\beta_B z)$.

fabricated with a visible laser beam propagating along the fiber core, but in 1989 a more versatile technique was demonstrated using the interferometric superposition of ultraviolet beams which come from the side of the fiber (*transverse holographic technique*). The angle between the ultraviolet beams determines the period of the light pattern in the fiber core and thus the Bragg wavelength. The two ultraviolet beams are often generated by exposing a periodic phase mask (photomask) with a single UV beam (*phase mask technique*), using the two first-order diffracted beams. Another approach is the *point-by-point* technique, where the regions with increased refractive index are written point by point with a small focused laser beam. Ultimately, the level of background losses depends on the fiber type, the photo-sensitization process, and writing conditions. Advantages of fiber gratings over competing technologies include all-fiber geometry, low insertion loss, high return loss or extinction, and potentially low cost. But the most distinguishing feature of fiber gratings is the flexibility they offer for achieving desired spectral characteristics. Numerous physical parameters can be varied, including: induced index change, length, apodization, period chirp, fringe tilt, and whether the grating supports counterpropagating or copropagating coupling at a desired wavelength. By varying these parameters, gratings can be made with normalized bandwidths $\Delta\lambda/\lambda$ between 0.1 and 10^{-4}, extremely sharp spectral features, and tailorable dispersive characteristics.

In essence, an FBG is simply the realization of a QWS mirror inside a fiber and, in principle, the same formalism could be adopted. In this case, however, the jump in the refractive index $\delta n = n_h - n_l$ is only between 10^{-5} and 10^{-3}. As a consequence, a fairly large number of periods is necessary to reach high reflectivity, which exaggeratedly increases the computation time. Moreover, in dielectric quarter-wave stacks, the larger ratio of high index to low index leads to bandwidths several orders of magnitude higher than Bragg gratings. Typically, the period Λ is of the order of hundreds of nanometers (or much longer for *long-period fiber gratings* not considered here) and the modulation extends over a few millimeters or centimeters. Most fiber Bragg gratings are used in single-mode fibers, and in that case the physical modelling is often relatively handy [129]. It is based on mode coupling, leading to a pair of differential equations with a coupling term, the magnitude of which is related to the local strength of the index modulation. The coupling is then effectively assumed to be smoothly distributed, and the numerical integration is done with a step size which can be much larger than the grating period. Such methods can be used for calculating the frequency-dependent complex amplitudes for transmission and reflection of light as well as, via numerical differentiation, the chromatic dispersion. Numerical models become substantially more complicated if many propagation modes are involved, which happens either for multimode fibers or in single mode fibers if birefringence is relevant. The simplest case is presented here. Let us express the refractive index and the loss within the grating as

$$\begin{cases} n(z) = n + n_1 \cos(2\beta_B z) & \beta_B = \dfrac{\pi}{\Lambda} & n_1 \ll n \\ \alpha(z) = \alpha \end{cases} \tag{3.144}$$

The Bragg wavelength is thus defined as $\lambda_B = 2n\Lambda$ and the spatially varying propagation constant is

$$k(z) = k_0 n(z) + i\alpha = k_0 n + k_0 n_1 \cos(2\beta_B z) + i\alpha \tag{3.145}$$

with $k_0 = \omega_0/c = 2\pi/\lambda_0$ and $\alpha \ll k_0 n \equiv \beta_0$. Neglecting terms in α^2, αn_1, and n_1^2 we can write

$$k^2(z) \simeq \beta_0^2 + 4\beta_0 k_c \cos(2\beta_B z) + 2\beta_0 i\alpha \tag{3.146}$$

where the coupling constant $k_c = \pi/\lambda_0 n_1$ has been defined. When Bragg scattering dominates ($\beta_0 \simeq \beta_B$), the scalar wave equation

$$\nabla^2 E(z) + k^2(z) \cdot E(z) = 0 \tag{3.147}$$

can be analytically solved because only two of the infinite set of diffraction orders are in phase and have significant amplitude. So the total electric field can be written as the sum of two counterpropagating waves of complex amplitudes $A(z)$ and $B(z)$, namely

$$E(z) = E_f(z) + E_b(z) = A(z)e^{i\beta z} + B(z)e^{-i\beta z} \tag{3.148}$$

where β is the propagation constant of the uncoupled waves given by

$$\beta \equiv k(z)|_{n_1=0} = k_0 n + i\alpha = \beta_0 + i\alpha \tag{3.149}$$

whereupon

$$\beta^2 \simeq \beta_0^2 + 2\beta_0 i\alpha \tag{3.150}$$

which allows one to re-write Equation 3.146 as

$$k^2(z) = \beta^2 + 2\beta_0 k_c \left(e^{i2\beta_B z} + e^{-i2\beta_B z}\right) \tag{3.151}$$

By insertion of Equations 3.148, 3.151 into Equation 3.147 and using the slowly-varying envelope approximation (amounting to neglecting the second derivatives) we obtain the following pair of coupled-wave equations

$$\begin{cases} \dfrac{dA}{dz} \simeq ik_c e^{i2\Delta z} B \\ \dfrac{dB}{dz} \simeq -ik_c e^{-i2\Delta z} A \end{cases} \tag{3.152}$$

where $\Delta = \beta_B - \beta = \beta_B - \beta_0 - i\alpha$ has been defined. The integration of this system with boundary conditions $E_b(L) = 0$ and $E_f(0) = E_{inc}$ leads to the following expressions of the forward and backward components of the electric field

$$E_f(z) = E_{inc} \frac{p\cosh[p(z-L)] - i\Delta \sinh[p(z-L)]}{p\cosh[pL] + i\Delta \sinh[pL]} e^{i\beta_B z} \tag{3.153}$$

$$E_b(z) = -iE_{inc}k_c \frac{\sinh[p(z-L)]}{p\cosh[pL] + i\Delta \sinh[pL]} e^{-i\beta_B z} \tag{3.154}$$

with $p = \sqrt{k_c^2 - \Delta^2}$. The reflection and transmission coefficients are then given by

$$r = \frac{E_b(0)}{E_f(0)} = ik_c \frac{\sinh[pL]}{p\cosh[pL] + i\Delta\sinh[pL]} \tag{3.155}$$

$$t = \frac{E_f(L)}{E_f(0)} \tag{3.156}$$

whereupon the reflectivity and the transmittivity are calculated as

$$\mathcal{R} = |r|^2 \quad \text{and} \quad \mathcal{T} = |t|^2 \tag{3.157}$$

Figure 3.26 shows a plot of the first of Equations 3.157 *vs.* λ_0 for $n = 1.45$, $n_1 = 10^{-4}$, $L = 2$ cm, $\alpha = 0.023$ cm^{-1}, and $\lambda_B = 1550$ nm. The ideal reflectivity (corresponding to $\alpha = 0$) is also plotted, clearly showing that, though FBGs may have reflectivity of up to 99.99%, scattering and absorption losses (here lumped into α) ultimately limit the attainable finesse (a few hundreds) in FBG-based optical cavities.

A schematic of an FBG-based fiber cavity is shown in Figure 3.27. For resonators constructed with distributed reflectors, like FBGs, in order to account for the penetration depth, the free spectral range is defined as [130]

$$FSR = \frac{c}{2n(L_0 + L_{eff1} + L_{eff2})} = \frac{c}{2n(L_0 + 2L_{eff})} \tag{3.158}$$

where two identical FBGs have been considered. To simplify the calculation of L_{eff}, first let us define the detuning $\delta = \beta_B - \beta_0$ such that $\Delta = \delta - i\alpha$. Then, let's stay in a neighborhood of the Bragg wavelength (small detuning) where

$$p = \sqrt{k_c^2 - \Delta^2} = \sqrt{k_c^2 - (\delta - i\alpha)^2} = \sqrt{k_c^2 - \delta^2 + \alpha^2 - 2i\delta\alpha} \simeq k_c \tag{3.159}$$

Then, from Equation 3.155 one obtains

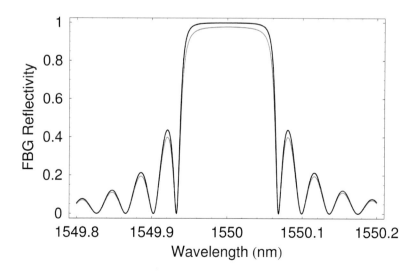

FIGURE 3.26
FBG reflectivity acording to Equation 3.157 for some typical parameters. The black curve corresponds to the ideal lossless case, while the gray one describes the lower real reflectivity.

$$\varphi_r = \arctan\left[\frac{\Im(r)}{\Re(r)}\right] = \arctan\left[\frac{k_c\cosh(k_cL) + \alpha\sinh(k_cL)}{\delta\sinh(k_cL)}\right]$$

$$= \arctan\left[\frac{k_c\coth(k_cL) + \alpha}{\delta}\right] \simeq \arctan\left[\frac{k_c}{\delta}\coth(k_cL)\right]$$

$$= \frac{\pi}{2} - \arctan\left[\frac{\delta}{k_c}\tanh(k_cL)\right] \simeq \frac{\pi}{2} - \frac{\delta}{k_c}\tanh(k_cL) \tag{3.160}$$

whereupon the reflection delay time is calculated as

$$\tau_r \equiv \left.\frac{d\varphi_r}{d\omega_0}\right|_{\delta=0} = \left.\frac{1}{2\pi}\frac{d\varphi_r}{d\nu_0}\right|_{\delta=0} = \left.\frac{\lambda_0^2}{2\pi c}\cdot\frac{d\varphi_r}{d\lambda_0}\right|_{\delta=0}$$

$$= \frac{\lambda_0^2}{2\pi c}\frac{2n}{n_1\lambda_0}\tanh(k_cL) = \frac{n}{ck_c}\tanh(k_cL) \tag{3.161}$$

Finally, denoting with v_g the group velocity and noting that for conventional fibers n_g equals the refractive index n no less than 1%, we can write

$$L_{eff} = \frac{v_g\tau_r}{2} = \frac{c\tau_r}{2n_g} = \frac{c\tau_r}{2n} = \frac{1}{2k_c}\tanh(k_cL) \tag{3.162}$$

Just as an example, for $n_1 = 10^{-4}$, $L = 2$ cm, and $\lambda_0 = 1550$ nm, we obtain $L_{eff} \simeq 2.5$ mm. This effect may be not negligible if the distance between L_0 between the two FBGs is small.

While we have focused here solely on the basic aspects of fiber Bragg gratings, nowadays the realm of FBG-based devices and applications is, to say the least, vast. Just to give an idea, telecom applications of FBGs often involve wavelength filtering, e.g., for combining or separating multiple wavelength channels in wavelength division multiplexing systems. Extremely narrow-band filters can be realized, e.g., with long FBGs (having a length of tens of centimeters) or with combinations of such gratings. Also, FBGs are commonly used as end mirrors of fiber lasers (distributed Bragg reflector fiber lasers), then typically restricting the emission to a very narrow spectral range. Even single-frequency operation can be achieved, e.g., by having the whole laser resonator formed by an FBG with a phase shift in the middle (distributed feedback lasers). Outside a laser resonator, an FBG can serve as a wavelength reference, e.g., for stabilization of the laser wavelength (refer also to next chapter).

The reader who is interested in further exploring this topic may refer to the review papers [131], [132] and references therein.

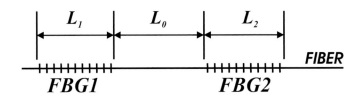

FIGURE 3.27
Fabry-Perot fiber cavity formed by two FBGs. L_1 and L_2 are the physical lengths of the FBGs and L_0 is the distance between them. (Adapted from [130].)

3.6 Whispering gallery mode resonators

Despite their versatility, traditional Fabry-Perot resonators and their folded or ring varieties have remained fairly complex and expensive devices, large in size, difficult in assembly, and prone to vibration instabilities because of relatively low-frequency mechanical resonances. For certain applications, stability and small modal volume are of great importance; however, miniaturization of conventional FP resonators is either very complicated (when high-finesse mirrors are utilized), or yields rather low quality Q-factors [133]. We focus here on the rapidly growing field of monolithic resonators in which the closed trajectories of light are supported by any variety of total internal reflection in curved and polygonal transparent dielectric structures. Following the traditional term of microwave electronics, we call these resonators open dielectric resonators. Extremely high values of Q-factor can be achieved in optical modes of very small volume with appropriately designed high precision dielectric interface, and with use of highly transparent materials. The simplest geometry of such resonators is either a ring, or a cylinder, or a sphere.

Among all resonant geometries, surface tension-induced microcavities, such as silica microspheres, exhibit the highest Q-factor to date of nearly 9 billion [134]. The modes of a spherical dielectric particle were first investigated by Mie at the beginning of the 19th century, in the context of light scattering from spherical particles. The scattering spectrum exhibited sharp features, which can be attributed to resonant circulation of optical energy within the sphere. These optical modes are confined by continuous total internal reflection at the dielectric air interface and are often referred to as *whispering gallery modes*. This description originated from the *problem of the whispering gallery* which Lord Rayleigh published in 1912, describing the phenomenon of acoustical waves he had observed propagating around the interior gallery of the Saint Paul's Cathedral [135]. Before Rayleigh, this effect was assigned to the reflection of acoustic *rays* from a surface near the dome apex. It was assumed that the rays propagated along different large arcs of the dome in the form of a hemisphere should concentrate only at the point that is located diametrically opposite to a sound source. However, Rayleigh found that, along with this effect, another effect exists: sound *clutches* to the wall surface and *creeps* along it. The concave surface of the dome does not allow the beam cross section to expand as fast as during propagation in free space. While in the latter case the beam cross section increases and the radiation intensity decreases proportionally to the square of distance from a source, the radiation in the whispering gallery propagates within a narrow layer adjacent to the wall surface. As a result, the sound intensity inside this layer decreases only directly proportionally to the distance, i.e., much slower than in free space. Rayleigh confirmed his explanation by direct experiments using a whistle as a sound source and a burning candle as a detector. It was found much later, at the beginning of the 20th century, that in dielectric spheres the electromagnetic waves can exist, which have the same spatial structure as whispering gallery acoustic waves. The waves of this type did not attract much attention until the last decade when they suddenly became the objects of wide studies and applications in optics.

3.6.1 Wave theory of whispering gallery modes

First of all, let us try to better understand the intimate character of whispering gallery waves and under what conditions they can appear. This requires the study of the structure of fields in dielectrics. Below, we consider the electrodynamics of a dielectric sphere. The optical modes of a spherical dielectric particle can be calculated by solving the Helmholtz equation in spherical coordinates. While the vector equation in Cartesian coordinates is simply a set of three scalar equations (one in each of the rectangular components of the vector), in other coordinate systems the situation is more complicated. Nevertheless, in Section 3.1 it was shown how the analytic difficulties involved in the treatment of vector differential equations in curvilinear coordinates might be overcome in cylindrical systems by a resolution of the field into two partial fields, each derivable from a purely scalar function satisfying the wave equation. In that case, the two scalar functions were naturally taken as E_z and H_z. In a short while, we will see that in spherical coordinates the two corresponding modes are those in which the electric and magnetic fields, respectively, are transverse to the radius vector from the center of coordinates. So, following [136], [137], let the scalar function $\Psi = \psi(r, \theta, \varphi) e^{-i\omega t}$ be a solution of the Helmholtz equation in a homogeneous and isotropic medium

$$\nabla^2 \psi + k^2 \psi = 0 \tag{3.163}$$

which in spherical polar coordinates is expanded as

$$\frac{1}{r^2} \frac{d}{dr} \left(r^2 \frac{\partial \psi}{\partial r} \right) + \frac{1}{r^2 \sin\theta} \frac{\partial}{\partial\theta} \left(\sin\theta \frac{\partial\psi}{\partial\theta} \right) + \frac{1}{r^2 \sin^2\theta} \frac{\partial^2\psi}{\partial\varphi^2} + k^2\psi = 0 \tag{3.164}$$

The equation is separable, so that upon placing $\psi(r, \theta, \varphi) = \psi_1(r)\,\psi_2(\theta)\,\psi_3(\varphi)$ one finds

$$r^2 \frac{d^2\psi_1}{dr^2} + 2r\frac{d\psi_1}{dr} + \left(k^2 r^2 - p^2 \right) \psi_1 = 0 \tag{3.165a}$$

$$\frac{1}{\sin\theta} \frac{d}{d\theta} \left(\sin\theta \frac{d\psi_2}{d\theta} \right) + \left(p^2 - \frac{q^2}{\sin^2\theta} \right) \psi_2 = 0 \tag{3.165b}$$

$$\frac{d^2\psi_3}{d\varphi^2} + q^2\psi_3 = 0 \tag{3.165c}$$

The parameters p and q are separation constants whose choice is governed by the physical requirement that at any fixed point in space the field must be single-valued. If the properties of the medium are independent of the equatorial angle φ, it is necessary that ψ_3 be a periodic function with period 2π and q is therefore restricted to the integers $m = 0, \pm 1, \pm 2, \ldots$. To determine p, we first identify the solution ψ_2 as an associated Legendre function. Upon substitution of $\eta = \cos\theta$, Equation 3.165b transforms to

$$\left(1 - \eta^2 \right) \frac{d^2\psi_2}{d\eta^2} - 2\eta \frac{d\psi_2}{d\eta} + \left(p^2 - \frac{m^2}{1 - \eta^2} \right) \psi_2 = 0 \tag{3.166}$$

that, if we choose $p^2 = l(l+1)$ where $l = 0, 1, 2, \ldots$, has periodic solutions in θ (which are finite at the poles $\eta = \pm 1$) known as the associated Legendre polynomials. For positive m they are given by

$$\psi_2(\eta) = P_l^m(\eta) = (-1)^m \frac{\left(1 - \eta^2 \right)^{m/2}}{2^l l!} \frac{d^{l+m} \left(\eta^2 - 1 \right)^l}{d\eta^{l+m}} \tag{3.167}$$

while, for negative m, they are given by

$$P_l^{-m}(x) = (-1)^m \frac{(l-m)!}{(l+m)!} P_l^m(x) \tag{3.168}$$

There remains the identification of the radial function $\psi_1(r)$ satisfying Equation 3.165a. If we write $\psi_1(r) = 1/\sqrt{kr}\,\chi(r)$, it is readily shown that $\chi(r)$ obeys the equation

$$\frac{d^2\chi}{d(kr)^2} + \frac{1}{kr}\frac{d\chi}{d(kr)} + \left[1 - \frac{\left(l+\frac{1}{2}\right)^2}{k^2r^2}\right]\chi = 0 \tag{3.169}$$

whose solution, already discussed in Section 3.1, is given by the first-order Bessel function $\chi = J_{l+\frac{1}{2}}(kr)$. Finally, inside the sphere we have

$$\psi_s(r,\theta,\varphi) = \psi_{1s}(r)\,\psi_2(\theta)\,\psi_3(\varphi)$$

$$= C_s \frac{J_{l+\frac{1}{2}}(k_s r)}{\sqrt{k_s r}} \cdot P_l^m(\cos\theta) \cdot \left\{ \begin{array}{c} \cos(m\varphi) \\ \sin(m\varphi) \end{array} \right. \tag{3.170}$$

where C_s is an arbitrary constant and $k_s = 2\pi\nu n_s/c$, n_s being the refractive index of the sphere. Physically, the solution outside the sphere should have the asymptotic form of a runaway wave, because a wave coming from infinity cannot exist. This means that the solutions outside the sphere should be expressed in terms of the Hankel functions of the first kind (which are also solutions to the Bessel equation), which have for large arguments the asymptotic form of a runaway wave with the amplitude decreasing inversely proportional to the distance. Thus, outside the sphere, we have

$$\psi_a(r,\theta,\varphi) = \psi_{1a}(r)\,\psi_2(\theta)\,\psi_3(\varphi)$$

$$= C_a \frac{\mathcal{H}_{l+\frac{1}{2}}(k_a r)}{\sqrt{k_a r}} \cdot P_l^m(\cos\theta) \cdot \left\{ \begin{array}{c} \cos(m\varphi) \\ \sin(m\varphi) \end{array} \right. \tag{3.171}$$

where another arbitrary constant has been introduced (C_a) and $k_a = 2\pi\nu n_a/c$, n_a being the refractive index of the medium containing the sphere (typically air). Now, as in the case of cylindrical coordinates, one can deduce solutions of the vector wave equation directly from the scalar solution $\psi_{s,a}$. In spherical coordinates the following relations

$$\boldsymbol{M}_{s,a} = \mathrm{rot}\,(\boldsymbol{r}\psi_{s,a}) \quad \text{and} \quad k_{s,a}\boldsymbol{N}_{s,a} = \mathrm{rot}\boldsymbol{M}_{s,a} \tag{3.172}$$

must be used, where $\boldsymbol{M}$ and $\boldsymbol{N}$ identify, respectively, with the magnetic and electric field for the TM solution (conversely, the TE solution is found by identifying $\boldsymbol{M}$ and $\boldsymbol{N}$ with the electric and magnetic field, respectively). As an example, we consider the TM modes. In this case, the electric field is allowed to have a radial component proportional to $\psi_{s,a}$ (by contrast, the magnetic field will have only transverse components) such that

$$\boldsymbol{H}_{s,a} = \mathrm{rot}\,(\boldsymbol{r}\psi_{s,a}) = \mathrm{rot}\,(r\psi_{s,a},0,0) = \left(0, \frac{1}{\sin\theta}\frac{\partial\psi_{s,a}}{\partial\varphi}, -\frac{\partial\psi_{s,a}}{\partial\theta}\right)$$

$$\equiv \left(H_r^{s,a}, H_\theta^{s,a}, H_\varphi^{s,a}\right) \tag{3.173}$$

and

$$\boldsymbol{E}_{s,a} = \frac{1}{k_{s,a}}\,\mathrm{rot}\,(\boldsymbol{H}_{s,a})$$

$$= \left(\frac{l(l+1)}{k_{s,a}r}\psi_{s,a}, \frac{1}{k_{s,a}r}\frac{\partial^2(r\psi_{s,a})}{\partial r\partial\theta}, \frac{1}{k_{s,a}r\sin\partial}\frac{\partial^2(r\psi_{s,a})}{\partial r\partial\varphi}\right)$$

$$\equiv \left(E_r^{s,a}, E_\theta^{s,a}, E_\varphi^{s,a}\right) \tag{3.174}$$

Note that the above expressions for the field components were derived just by use of the explicit expression for the rotor in spherical polar coordinates, except for $E_r^{s,a}$ for which Equations 3.163, 3.165a were also exploited. To obtain explicit expressions for the vector wave functions we only need to carry out the differentiations

$$
\left\{
\begin{aligned}
H_r^s &= 0 \\
H_\theta^s &= \frac{mC_s}{\sin\theta}\frac{J_{l+\frac{1}{2}}(k_s r)}{\sqrt{k_s r}} \cdot P_l^m(\cos\theta) \cdot \left\{ \begin{aligned} -\sin(m\varphi) \\ \cos(m\varphi) \end{aligned} \right. \\
H_\varphi^s &= -C_s \frac{J_{l+\frac{1}{2}}(k_s r)}{\sqrt{k_s r}} \cdot \frac{\partial P_l^m(\cos\theta)}{\partial\theta} \cdot \left\{ \begin{aligned} \cos(m\varphi) \\ \sin(m\varphi) \end{aligned} \right. \\
E_r^s &= C_s l(l+1) \frac{J_{l+\frac{1}{2}}(k_s r)}{(k_s r)^{3/2}} \cdot P_l^m(\cos\theta) \cdot \left\{ \begin{aligned} \cos(m\varphi) \\ \sin(m\varphi) \end{aligned} \right. \\
E_\theta^s &= \frac{C_s}{k_s r}\frac{\partial}{\partial r}\left[\sqrt{\frac{r}{k_s}} J_{l+\frac{1}{2}}(k_s r) \right] \cdot \frac{\partial P_l^m(\cos\theta)}{\partial\theta} \cdot \left\{ \begin{aligned} \cos(m\varphi) \\ \sin(m\varphi) \end{aligned} \right. \\
E_\varphi^s &= \frac{mC_s}{k_s r\sin\theta}\frac{\partial}{\partial r}\left[\sqrt{\frac{r}{k_s}} J_{l+\frac{1}{2}}(k_s r) \right] \cdot P_l^m(\cos\theta) \cdot \left\{ \begin{aligned} -\sin(m\varphi) \\ \cos(m\varphi) \end{aligned} \right.
\end{aligned}
\right. \tag{3.175}
$$

The corresponding expressions for the field components outside the sphere are found by replacing k_s with k_a, C_s with C_a, and $J_{l+\frac{1}{2}}(k_s r)$ with $\mathcal{H}_{l+\frac{1}{2}}(k_a r)$

$$
\left\{
\begin{aligned}
H_r^a &= 0 \\
H_\theta^a &= \frac{mC_a}{\sin\theta}\frac{\mathcal{H}_{l+\frac{1}{2}}(k_a r)}{\sqrt{k_a r}} \cdot P_l^m(\cos\theta) \cdot \left\{ \begin{aligned} -\sin(m\varphi) \\ \cos(m\varphi) \end{aligned} \right. \\
H_\varphi^a &= -C_a \frac{\mathcal{H}_{l+\frac{1}{2}}(k_a r)}{\sqrt{k_a r}} \cdot \frac{\partial P_l^m(\cos\theta)}{\partial\theta} \cdot \left\{ \begin{aligned} \cos(m\varphi) \\ \sin(m\varphi) \end{aligned} \right. \\
E_r^a &= C_a l(l+1) \frac{\mathcal{H}_{l+\frac{1}{2}}(k_a r)}{(k_a r)^{3/2}} \cdot P_l^m(\cos\theta) \cdot \left\{ \begin{aligned} \cos(m\varphi) \\ \sin(m\varphi) \end{aligned} \right. \\
E_\theta^a &= \frac{C_a}{k_a r}\frac{\partial}{\partial r}\left[\sqrt{\frac{r}{k_a}} \mathcal{H}_{l+\frac{1}{2}}(k_a r) \right] \frac{\partial P_l^m(\cos\theta)}{\partial\theta} \cdot \left\{ \begin{aligned} \cos(m\varphi) \\ \sin(m\varphi) \end{aligned} \right. \\
E_\varphi^a &= \frac{mC_a}{k_a r\sin\theta}\frac{\partial}{\partial r}\left[\sqrt{\frac{r}{k_a}} \mathcal{H}_{l+\frac{1}{2}}(k_a r) \right] \cdot P_l^m(\cos\theta) \cdot \left\{ \begin{aligned} -\sin(m\varphi) \\ \cos(m\varphi) \end{aligned} \right.
\end{aligned}
\right. \tag{3.176}
$$

The above expressions should satisfy the conditions on the sphere boundary. The continuity condition for the tangential components provides

$$
H_\theta^a(R) = H_\theta^s(R) \quad \Rightarrow \quad C_a = C_s \frac{J_{l+\frac{1}{2}}(k_s R)}{\mathcal{H}_{l+\frac{1}{2}}(k_a R)} \sqrt{\frac{n_a}{n_s}} \tag{3.177}
$$

and

$$
\begin{aligned}
E_\varphi^a(R) = E_\varphi^s(R) \Rightarrow & \frac{\left[\sqrt{k_s r}J_{l+\frac{1}{2}}(k_s r) \right]'_{r=R}}{\sqrt{k_s R}J_{l+\frac{1}{2}}(k_s R)} \\
& = \frac{n_s}{n_a}\frac{\left[\sqrt{k_a r}\mathcal{H}_{l+\frac{1}{2}}(k_a r) \right]'_{r=R}}{\sqrt{k_a R}\mathcal{H}_{l+\frac{1}{2}}(k_a R)}
\end{aligned} \tag{3.178}
$$

where R is the radius of the sphere and the prime indicates the total derivative over r. The former condition allows us to find only the ratio between C_s and C_a, so that one of them remains free. It is determined by the power of the optical source exciting the waves. Solution of the equation stated by the latter condition yields, instead, the eigenvalues of the problem, namely the resonance frequencies of the WGMs. Before solving this equation, let us first identify which of the above modes actually correspond to the WGMs. According to the description of Rayleigh, a whispering gallery wave should be *pressed down* to the sphere surface. Let us look attentively at the radial dependence of the field described by the fourth of Equations 3.175. For a small index l, the function fills almost the entire volume of the sphere. Such modes cannot be the WGMs. However, for a large index l, the Bessel function is very small up to $r \simeq l/k_s$. For $r > l/k_s$, the Bessel function begins to oscillate with a decreasing amplitude. Therefore, for given k_s and R, the first l index corresponding to a WGM is given by $l \simeq Rk_s$. In other words, if we choose the value of $k_s R$ that is closest to the first root of the Bessel function, the field near the sphere surface will have the structure without oscillations (Figure 3.28).

Next we focus on the polar structure of the wave described by $P_l^m (\cos\theta)$. For $m = l$, we have $P_l^l (\cos\theta) \propto \sin^l\theta$ such that the polar structure is peaked at $\pi/2$ that is on the equatorial plane (by contrast, if $m \neq l$ the mode acquires an oscillating transverse structure and the oscillations increase with increasing the difference $l - m$); in addition, the higher is the index l, the narrower is the peak width around $\pi/2$. For very large l, $P_l^l (\cos\theta) \propto \sin^l\theta \simeq \sin^l \pi/2 \simeq 1$. These results are illustrated in Figure 3.29.

In light of the above considerations, it is reasonable to call a WGM a wave for which the Bessel function has no roots inside a sphere and which has identical and large indices l and m. Formally, to construct a WGM, the following instructions must be followed. Given the optical wavelength λ, the sphere radius R, and the refractive indices (n_s and n_a), the index l must be found for which the first maximum of $J_{l+\frac{1}{2}} (k_s r)/(k_s r)^{3/2}$ falls at $r = R$. Then, E_r^s, C_a, and hence E_r^a can be explicitly computed (one can assume, for example, $\cos(m\varphi)$ for the equatorial structure and $P_l^l (\cos\theta) \simeq \text{const}$ as already discussed) such that

$$\frac{E_r}{C_s} = \begin{cases} l(l+1) \dfrac{J_{l+\frac{1}{2}} (k_s r)}{(k_s r)^{3/2}} \cdot \cos(m\varphi) & r < R \\[4mm] \dfrac{J_{l+\frac{1}{2}} (k_s R)}{\mathcal{H}_{l+\frac{1}{2}} (k_a R)} \sqrt{\dfrac{n_a}{n_s}} l(l+1) \dfrac{\mathcal{H}_{l+\frac{1}{2}} (k_a r)}{(k_a r)^{3/2}} \cdot \cos(m\varphi) & r > R \end{cases} \tag{3.179}$$

The associated intensity is given by

$$I \propto \begin{cases} n_s^2 \left[l(l+1) \dfrac{J_{l+\frac{1}{2}} (k_s r)}{(k_s r)^{3/2}} \cdot \cos(m\varphi) \right]^2 & r < R \tag{3.180a} \\[5mm] n_a^2 \left| \dfrac{J_{l+\frac{1}{2}} (k_s R)}{\mathcal{H}_{l+\frac{1}{2}} (k_a R)} \sqrt{\dfrac{n_a}{n_s}} l(l+1) \dfrac{\mathcal{H}_{l+\frac{1}{2}} (k_a r)}{(k_a r)^{3/2}} \cdot \cos(m\varphi) \right|^2 & r > R \quad (3.180b) \end{cases}$$

In Figure 3.30 the above expression for the intensity is plotted for the case $R = 11.5$ micron, $\lambda = 1$ micron, $n_a = 1$, and $n_s = 1.453$. By looking at the behavior of Equation 3.6.1a along r, it is easily recognized that for $l = 100$ the intensity has its maximum just at $r = R$. Then, for this l value, the intensity is calculated over all r by computing Equation 3.6.1a inside the sphere and Equation 3.6.1b outside. One can see that the optical intensity falls down very rapidly (within one optical wavelength) outside the sphere (evanescent field). By taking into account also the azimuthal dependence, we can have a nice 3-D picture of the intensity corresponding to this mode (Figure 3.31). Finally, it is worth pointing out that, by plotting the last three of Equations 3.175, it is readily seen that the dominant electric-field

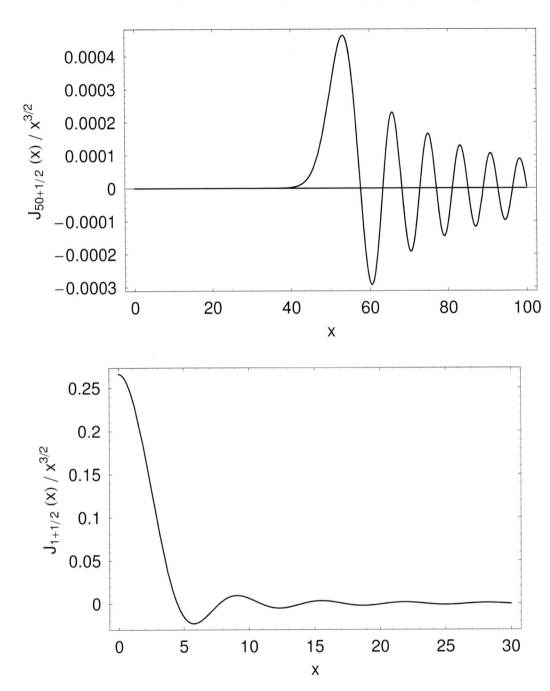

FIGURE 3.28
Radial dependence in the radial component of the electric field inside the sphere for $l = 50$ (upper frame) and $l = 1$ (lower frame), according to the fourth of Equations 3.175.

component in a WGM (i.e., for sufficiently $l = m$ values) is the radial one. Conversely, for TE modes, the electric field is prevalently polarized azimuthally.

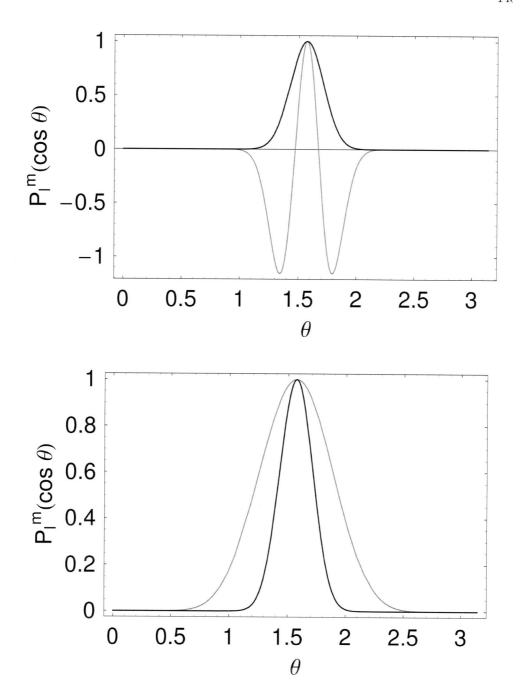

FIGURE 3.29
Polar dependence in the radial component of the electric field inside the sphere according to the fourth of Equations 3.175. Upper frame: $l = m = 50$ corresponds to the black curve, while ($l = 50$, $m = 48$) to the gray one. Lower frame: $l = m = 50$ corresponds to the black curve, while $l = m = 10$ to the gray one.

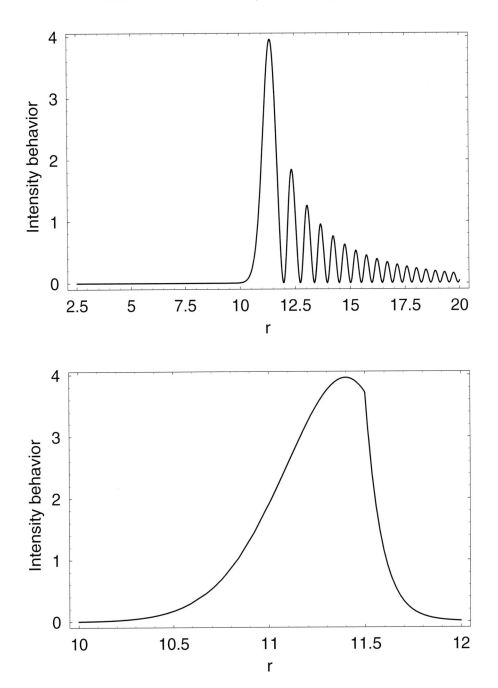

FIGURE 3.30
Upper frame: plot of Equation 3.6.1a for all r values. Lower frame represents the actual
intensity behavior: plot of Equation 3.6.1a inside the sphere and of Equation 3.6.1b outside.

Now, we have to calculate the eigenvalues (resonance frequencies). For this purpose,

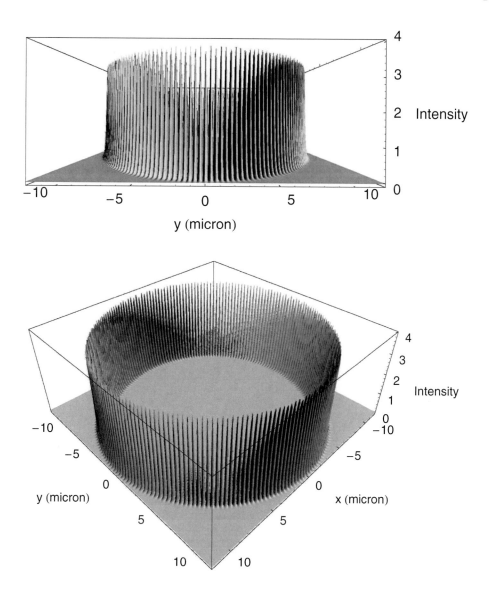

FIGURE 3.31
3-D intensity profile, from two different angles, corresponding to the same mode considered in Figure 3.30.

rather than solving numerically Equation 3.178, we will find the eigenvalues from the solution of an appropriate approximate equation. Thus we come back to Equation 3.165a

$$\frac{d^2\psi_1}{dr^2} + \frac{2}{r}\frac{d\psi_1}{dr} + \left[k^2 - \frac{l(l+1)}{r^2}\right]\psi_1 = 0 \tag{3.181}$$

First, we pose $\psi_1 = u/r$ to get

$$\frac{d^2u}{dr^2} + \left[k^2 - \frac{l(l+1)}{r^2}\right]u = 0 \tag{3.182}$$

which can be approximated near the sphere surface $(r = R)$ as

$$\frac{d^2 u\left(r'\right)}{dr'^2} + \left[k^2 - \frac{l\left(l+1\right)}{R^2} - \frac{2l\left(l+1\right)}{R^3}r'\right] u\left(r'\right) = 0 \tag{3.183}$$

where $r' = R - r$. Upon placing $\alpha = k^2 - l\left(l+1\right)/R^2$, $\beta = 2l\left(l+1\right)/R^3$ $(\sigma = 1/\beta)$, and $t = -\alpha + \beta r'$, we obtain

$$\frac{d^2 u\left(t\right)}{dt^2} - \sigma^2 t \cdot u\left(t\right) = 0 \tag{3.184}$$

that is the Airy equation which has two linearly independent solutions Ai and Bi, such that

$$u\left(t\right) = A' \cdot Ai\left(\sigma^{2/3}t\right) + B' \cdot Bi\left(\sigma^{2/3}t\right) \tag{3.185}$$

Since the Bi function is divergent for $t \to \infty$, we set $B' = 0$ and get

$$u\left(r'\right) = A' \cdot Ai\left(\beta^{-2/3}t\right) = A' \cdot Ai\left(-\alpha\beta^{-2/3} + \beta^{1/3}r'\right) \tag{3.186}$$

The boundary condition $u\left(r' = 0\right)$ implies $Ai\left(-\alpha\beta^{-2/3}\right) = 0$ that yields

$$-\alpha_q\beta^{-2/3} = z_q = -|z_q| \tag{3.187}$$

where z_q are the roots of the Airy function $Ai\left(z\right) = 0$ (for example $z_1 = -2.338$, $z_2 = -4.088$, $z_3 = -5.521$). Finally we have

$$u\left(r'\right) = A' \cdot Ai\left(-|z_q| + \frac{\left[2l\left(l+1\right)\right]^{1/3}}{R}r'\right) \tag{3.188}$$

The second boundary condition $u\left(r' = R\right) \simeq 0$ is automatically satisfied. Indeed, for $l \gg 1$, the above equation provides

$$u\left(r' = R\right) = A' \cdot Ai\left(-|z_q| + \left[2l\left(l+1\right)\right]^{1/3}\right) \simeq A' \cdot Ai\left[\left(\sqrt{2l}\right)^{2/3}\right]$$

$$= A' \cdot \frac{e^{-\frac{2}{3}\sqrt{2l}}}{2\sqrt{\pi}\left(\sqrt{2l}\right)^{1/6}} \simeq 0 \tag{3.189}$$

where the asymptotic expansion of the Ai function has been exploited. Coming back to the secular equation, we get

$$k_{l,q}^2 - \frac{l\left(l+1\right)}{R^2} = \left[\frac{2l\left(l+1\right)}{R^3}\right]^{2/3}|z_q| \tag{3.190}$$

that provides

$$k_{l,q}^2 \simeq \frac{l^2}{R^2} + \frac{|z_q|}{R^2}\left(2l^2\right)^{2/3} = \frac{l^2}{R^2}\left(1 + \frac{|z_q| 2^{2/3}}{l^{2/3}}\right) \tag{3.191}$$

which yields

$$k_{l,q} \simeq \frac{l}{R}\left(1 + \frac{|z_q| 2^{2/3}}{2l^{2/3}}\right) = \frac{1}{R}\left[l + |z_q|\left(\frac{l}{2}\right)^{1/3}\right] \tag{3.192}$$

such that

$$\nu_{l,q} = \frac{c}{2\pi n_s R} \left[l + |z_q| \left(\frac{l}{2} \right)^{1/3} \right] \tag{3.193}$$

Therefore, a WGM can be characterized by its polarization (TE or TM), i.e., the direction of the dominant electric field component, and three integer orders (q, l, m) where q denotes the radial order (it counts the number of field maxima in the radial pattern), l the angular mode number (corresponding to the number of wavelengths around the circumference), and m the azimuthal mode number (in the range $-l \leq m \leq l$, it gives $l - m + 1$ maxima in the polar distribution of the field). The above equation is independent on the azimuthal mode number m, which means that the polar modes in a perfect sphere cavity are degenerate. For definiteness we will apply the term WGM to the modes with large indices l, $l = m$ and $q = 1$. Note, however, that modes with indices $l \neq m$, but close to l, and with $q > 1$, but close to unity, have properties that are close to those of WGMs. This means that there is no sharp difference between WGMs and other modes with nearest indices and, for this reason, the modes with a small difference $l - m$ are sometimes also assigned in the literature to WGMs. Tuning of the eigenfrequencies can be achieved by temperature variation or by strain. The resonance frequency is shifted by $\Delta\nu/\nu = -\Delta R/R - \Delta n_s/n_s$: the temperature dependence reduces the mode frequency by a few GHz per degree; larger tuning over several hundred GHz can be achieved by applying a compressive force near the polar regions in a *microvice* or by stretching a microsphere with two attached stems. For very large l, the free spectral range can be approximated as

$$FSR \simeq \frac{c}{2\pi n_s R} \tag{3.194}$$

This is just the free spectral range expected for a wave that propagates along the inner surface of the sphere that is continuously internally reflected. The total internal reflection process also gives rise to the frequency shift between the nearly identical TM and TE spectra. To see this formally, a similar treatment should be carried out to find the TE eigenvalues. However, a simple argument can be used to find this result [138]. Consider a plane wave propagating along a closed equatorial polygonal ring path with N total (internal) reflections. Addition of the individual phase shifts (given by the Fresnel formulas) that occur during one round trip leads to a frequency shift between TE and TM modes given by [105]

$$\Delta_{TM} - \Delta_{TE} = 2N \arctan \left[\frac{\cos\theta \sqrt{\sin^2\theta - (n_a/n_s)^2}}{\sin^2\theta} \right] \tag{3.195}$$

with $\theta = (1 - 2/N)\pi/2$. In the limit of large N, this returns

$$\Delta_{TM} - \Delta_{TE} = 2\pi \sqrt{1 - (n_a/n_s)^2} \tag{3.196}$$

correponding to

$$\Delta FSR = FSR_{TM} - FSR_{TE} = \frac{c}{2\pi n_s R} \sqrt{1 - (n_a/n_s)^2} \tag{3.197}$$

that is exactly the result one would find by calculating the eigenvalues for the TE modes according to the procedure followed to find those for the TM modes. As mentioned above, the eigenfrequencies for modes with the same index l but different indices m are degenerated.

This degeneration is removed when the shape of a dielectric body deviates from a sphere. For example, in the case of a spheroid with a small eccentricity, we can calculate a correction to the frequency using perturbation theory. For this purpose, we will use the nuclear energy levels calculated for a spheroidal model [139] written as

$$\frac{\epsilon\left(x^2+y^2\right)}{R^2}+\frac{z^2}{\epsilon^2 R^2}=1 \tag{3.198}$$

where R is the radius of the undistorted sphere. By comparison with the standard form of the spheroidal equation $(x^2+y^2)/a^2+z^2/b^2=1$, we get this expression for the eccentricity $e \equiv a - b/R = (R/\sqrt{\epsilon}-\epsilon R)/R = -\epsilon + 1/\sqrt{\epsilon} \simeq -(2/3)(\lambda-1)$ (when $|\lambda-1| \ll 1$). With this notation, a potential well of constant depth in a spheroidal region of constant volume is defined as

$$V\left(s\right)=\left\{\begin{array}{cc} -D & s<R \\ 0 & s \geq R \end{array}\right. \tag{3.199}$$

with $s=\sqrt{\epsilon\left(x^2+y^2\right)+z^2/\epsilon^2}$. For small values of e, the eigenvalues can be computed by first-order perturbation theory employing the normalized solution $\psi_0\left(x,y,z\right)$ of the unperturbed problem

$$E\left(e\right)-E\left(0\right)=\iiint|\psi_0(x,y,z)|^2[V(s)-V(r)]dv$$

$$=\iiint V(r)\left[\left|\psi_0\left(\frac{x}{\sqrt{\epsilon}},\frac{y}{\sqrt{\epsilon}},\epsilon z\right)\right|^2-|\psi_0(x,y,z)|^2\right]dv$$

$$=-D\iiint_{sphere}\left[\left|\psi_0\left(\frac{x}{\sqrt{\epsilon}},\frac{y}{\sqrt{\epsilon}},\epsilon z\right)\right|^2-|\psi_0(x,y,z)|^2\right]dv$$

$$\simeq -D\iiint_{sphere}\left\{\frac{\partial\left|\psi_0\left(\frac{x}{\sqrt{\epsilon}},\frac{y}{\sqrt{\epsilon}},\epsilon z\right)\right|^2}{\partial\left(\frac{x}{\sqrt{\epsilon}}\right)}\right|_{\frac{x}{\sqrt{\epsilon}}=x}\cdot\left(\frac{x}{\sqrt{\epsilon}}-x\right)dv$$

$$+\left.\frac{\partial\left|\psi_0\left(\frac{x}{\sqrt{\epsilon}},\frac{y}{\sqrt{\epsilon}},\epsilon z\right)\right|^2}{\partial\left(\frac{y}{\sqrt{\epsilon}}\right)}\right|_{\frac{y}{\sqrt{\epsilon}}=y}\cdot\left(\frac{y}{\sqrt{\epsilon}}-y\right)$$

$$+\left.\frac{\partial\left|\psi_0\left(\frac{x}{\sqrt{\epsilon}},\frac{y}{\sqrt{\epsilon}},\epsilon z\right)\right|^2}{\partial\left(\epsilon z\right)}\right|_{\epsilon z=z}\cdot\left(\epsilon z-z\right)\Bigg\}dv$$

$$\simeq -D\iiint_{sphere}\left\{\frac{\partial\left|\psi_0\left(x,y,z\right)\right|^2}{\partial x}\cdot\left(\frac{1}{\sqrt{\epsilon}}-1\right)x\,dv\right.$$

$$+\frac{\partial\left|\psi_0\left(x,y,z\right)\right|^2}{\partial y}\cdot\left(\frac{1}{\sqrt{\epsilon}}-1\right)y+\frac{\partial\left|\psi_0\left(x,y,z\right)\right|^2}{\partial z}\cdot\left(\epsilon-1\right)z\Bigg\}dv$$

$$= -D\frac{e}{3}\iiint_{sphere}\left\{\frac{\partial\,|\psi_0\,(x,y,z)|^2}{\partial x}x + \frac{\partial\,|\psi_0\,(x,y,z)|^2}{\partial y}y - \frac{\partial\,|\psi_0\,(x,y,z)|^2}{\partial z}2z\right\}dv$$

$$= -D\frac{e}{3}\iiint_{sphere}\left\{\frac{\partial}{\partial x}x + \frac{\partial}{\partial y}y - 2\frac{\partial}{\partial z}z\right\}|\psi_0(x,y,z)|^2 dv$$

$$= -D\frac{e}{3}\iiint_{sphere}\text{div}\,\mathbf{\Psi}\,dv$$

$$= -D\frac{e}{3}\iint_{SS}\mathbf{\Psi}\cdot\mathbf{n}\,d\sigma \tag{3.200}$$

where SS is the surface of the sphere, the vector $\mathbf{\Psi}\equiv|\psi_0\,(x,y,z)|^2\,(x,y,-2z)$ has been introduced, and the Gauss theorem has been exploited ($\mathbf{n}$ is the versor normal to the surface of the sphere). Therefore, by introducing spherical coordinates one gets

$$E(e) - E(0) = -D\frac{e}{3}\iint_{SS}|\psi_0(R,\theta,\varphi)|^2[(x,y,-2z)\cdot(n_x,n_y,n_z)]d\sigma$$

$$= -D\frac{e}{3}R^3|\mathcal{R}_{n,l}(R)|^2 2\pi\int_0^\pi (1 - 3\cos^2\theta)|Y_l^m(\theta,\varphi)|^2\sin\theta d\theta \tag{3.201}$$

where $\psi_0\,(x,y,z) = \psi_0\,(r,\theta,\varphi) = \mathcal{R}_{n,l}\,(r)\,Y_l^m\,(\theta,\varphi)$ has been used. Recalling that

$$Y_l^m\,(\theta,\varphi) = \sqrt{\frac{2l+1}{4\pi}\frac{(l-m)!}{(l+m)!}}P_l^m\,(\cos\theta)\,e^{im\varphi} \tag{3.202}$$

we finally obtain

$$E\,(e) - E\,(0) = DR^3|\mathcal{R}_{n,l}\,(R)|^2\frac{e}{2}\frac{(1/3)l\,(l+1) - m^2}{(l+3/2)\,(l-1/2)}$$

$$\simeq DR^3|\mathcal{R}_{n,l}\,(R)|^2\frac{e}{6}\left(1 - 3\frac{m^2}{l^2}\right) \tag{3.203}$$

where $l\gg 1$ has been exploited in the last step.

Finally, let us consider an even more crude approximation of Equation 3.182 for $r > R$ and near the sphere surface

$$\frac{d^2u}{dr^2} + \left[k_0^2 - \frac{l\,(l+1)}{a^2}\right]u = 0 \tag{3.204}$$

Its solution is

$$u\,(r) = C_1 e^{-\sqrt{\frac{l(l+1)}{a^2} - k_0^2}r} + C_2 e^{+\sqrt{\frac{l(l+1)}{a^2} - k_0^2}r} = C_1 e^{-\sqrt{\frac{l(l+1)}{a^2} - k_0^2}r}$$

$$\simeq C_1 e^{-\sqrt{\frac{l^2}{a^2} - k_0^2}r} = C_1 e^{-\sqrt{\frac{l^2}{k_0^2 a^2} - 1}k_0 r}$$

$$= C_1 e^{-\sqrt{\frac{l^2 n_s^2}{k_s^2 a^2} - 1}\cdot\frac{2\pi}{\lambda}r} \simeq C_1 e^{-2\pi\sqrt{n_s^2 - 1}\cdot\frac{r}{\lambda}} \tag{3.205}$$

where C_2 has been set to zero to avoid divergence, and the relations $k_0 = 2\pi/\lambda = k_s/n_s$ and $k_s R \simeq l$ have been used (see Equation 3.193). Since the field amplitude becomes very small at distance $r > Rn_s$, the intensity of radiation emitted from the sphere is very low (very high Q). As a result, we can imagine the following picture. The WGM field occupies a volume bounded by a spherical surface of radius Rn_s. Radiation is emitted outside from this volume in the form of a runaway wave with very small amplitude. However, the field occupies, in fact, not the entire volume of a sphere with radius Rn_s but it is *pressed down* to the surface of the dielectric sphere, extending outside the sphere by the distance $r = \lambda/2\pi\sqrt{n_s^2 - 1}$. For such materials as glass and quartz, this distance is smaller than the wavelength in free space, not to mention dielectric with even larger refractive index like diamond or certain semiconductor materials.

We close this section by observing that for other interesting geometries (like microtoroids, microdisks, ...), Helmholtz's equation is not separable. Thus, unlike spherical cavities, there is still no analytical theory for the structure and positions of WGMs in these configurations. Recently, several methods have been developed to numerically solve Maxwell's equations based on techniques like Finite Difference Time Domain (FDTD) or Finite Element Method (FEM) [140].

3.6.2 WGMs in a ray-optical picture

After this rigorous, tutorial discussion on WGMs, let us a introduce a simple approach to gain a deeper physical insight into this phenomenon. For a perfect microsphere, as that previously treated, the problem can be explored by employing geometric optics on its cross section because the incident plane is conserved. Considering a sphere with a radius R and refractive index n_s surrounded by air as shown in Figure 3.32, when light is incident at the interface with an angle i larger than i_c, where $i_c = \arcsin(1/n_s)$ is the critical angle, the ray is then totally reflected. Due to the circular symmetric, it therefore keeps the same incident angle for the following reflections. As a result, the light ray is trapped inside the cavity by successive total internal reflections. If the travel distance in one round-trip is an integral multiple of the wavelength, the WGM resonance mode will be formed. For a large circle where $R \gg \lambda$ and incident angle $i \simeq \pi/2$, the ray travels very close to the circle interface. Thus an approximate condition for a WGM resonance can be expressed as $2\pi R = \lambda l/n_s$ where l is a interference order and λ is the wavelength in vacuum. The resonance condition for frequency is $\nu = lc/(2\pi N_s R)$ and thus the free spectral range is $FSR = c/(2\pi n_s R)$. For the cavity with $n_s = 1.45$ (fused silica refractive index) and radius $R = 20$ micron, its FSR is about 1.6 THz in frequency or 3.5 nm at $\lambda = 800$ nm in wavelength. It should be stressed that this approach works for the case where the cavity is much larger than the operation wavelength (l is large). On the other hand, we also introduce the angular momentum $\boldsymbol{L}$ as shown in Figure 3.32, which is defined as follows: $\boldsymbol{L} = \boldsymbol{r} \times \boldsymbol{k}$. In the case of the WGM resonance shown in Figure 3.32, its angular momentum can be easily derived: $L = r_1 k = r_1 N k_0$ where k_0 is the wave number in vacuum and $r_1 = R \cdot \sin(i)$. In the condition where $\sin(i) \simeq 1$, we have $L \simeq N k_0 R = l$. So the interference order l is often called angular order. As already mentioned, in an ideal sphere the optical modes possess a $2l + 1$ degeneracy with respect to the azimuthal mode number m. This can be understood by using classical ray optical interpretation, in which the optical modes with same l, but different m, orbit around the equatorial plane by alternating reflections from the lower to the upper hemisphere (and vice versa), thereby taking different excursions away from the equator. In other words, along the surface of the sphere, a mode can be thought as tracing out a *zig-zag* path around the sphere with the equatorial plane being the mean plane of propagation: the mode is confined to a belt around the equatorial plane by the curvature of the sphere in the polar direction (Figure 3.33). The wavevector associated with this trajectory is $|\beta_l| = l(l+1)/a$ and the projection onto the equatorial plane (i.e., the propagation constant) is given by $|\beta_m| = m/a$. Different values of m imply that the modes travel in zig-zag paths with different inclinations with respect to the equatorial plane. As already rigorously shown, when $m = l$ (fundamental mode), the inclination is the smallest. Although modes with decreasing values of m propagate at larger inclinations, for the same value of l they all have the same resonant wavelength. This is because modes that take larger excursions away from the equator need to propagate over shorter distances to complete a revolution around the sphere (from the fact that higher latitude circles have less circumference than the equator).

By summarizing, in the ray optics picture, the optical modes within a microsphere are confined by continuous total internal reflection at the dielectric cavity-air interface. However,

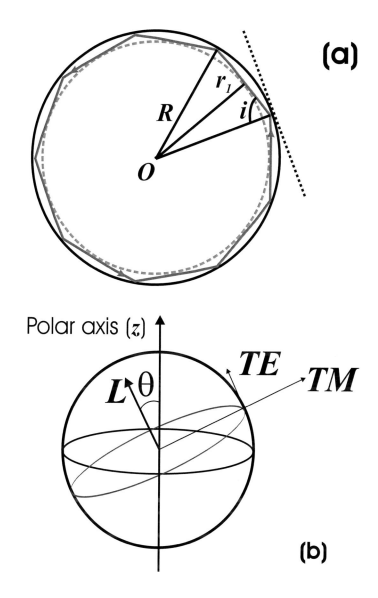

FIGURE 3.32
Sketch of the cross section plane of a sphere where a light ray is travelling by total internal reflections (frame a) and of a WGM in a perfect microsphere (frame b).

it is a general property that total internal reflection at a curved interface is incomplete, and leads to a transmitted wave, which for the case of a resonator causes loss of optical energy. A good intuitive explanation of this effect can be given by considering the phase velocities in a ray optical picture [134]. For total internal reflection at a planar interface, the exponentially decaying field component has a constant phase velocity $u_0 < c$ in the evanescent region. For a curved surface, however, the phase velocity increases with increasing separation from the boundary i.e., $u(r) = u_0 r/a$. At the point where the phase velocity exceeds the phase velocity in air $(u(r) > c)$ the evanescent field becomes propagating, leading to tunnel losses. This loss mechanism is called whispering gallery loss, and is due to tunneling of the photons out of their bound states. This tunneling process can be understood by drawing an analogy to the quantum mechanical treatment of a 1-D particle in a central potential [140]. To this aim, one can re-write the first of Equation 3.165 as an eigenvalue problem using the transformation $\psi_1 = u/r$ and introducing the energy term $E = k_0^2$

$$-\frac{d^2 u}{dr^2} + U_{eff} u = Eu \qquad (3.206)$$

where the effective potential is given by

$$U_{eff} = \left\{ \frac{l(l+1)}{r^2} + k_0^2 [1 - n^2(r)] \right\} \qquad (3.207)$$

with

$$n(r) = \begin{cases} n_s & r \leq R \\ 1 & r > R \end{cases} \qquad (3.208)$$

Thus, the effective potential is discontinuous at the cavity-air interface, giving rise to a potential well. Furthermore the characteristic radii R_a and R_b are given by

$$R_a = \frac{\sqrt{l(l+1)}}{n_s k_0} \qquad (3.209)$$

$$R_b = R_a n_s \qquad (3.210)$$

In the well region $R_a < r < R_b$ discrete bound states exist which correspond to the whispering gallery modes. The region $r < R_a$ as well as $R < r < R_b$ corresponds to a potential barrier, in which the optical modes are exponentially decaying (i.e., evanescent). The region $R_b > r$ supports a continuum of modes, which are unbound. Due to the finite height and finite width of the potential barrier in the region $R < r < R_b$, the optical modes can tunnel from their bound well states into the continuum, giving rise to a tunnel loss. The height and width of the potential barrier decreases as a function of the polar mode number l, causing an increase in tunnel loss.

3.6.3 Mode Q and volume

As already stressed, the extent to which dissipation is present in a resonant system is commonly expressed by the Q-factor of the mode ω which is related to the lifetime τ of light energy in the resonator mode as $Q = \omega\tau$. Just to give an idea, the ring down time corresponding to a mode with $Q = 10^{10}$ and wavelength $\lambda = 1.3$ μm is 7 μs, thus making ultrahigh Q resonators potentially attractive as light storage devices. The total Q-factor is comprised of several loss contributions: intrinsic material absorption, scattering losses (both intrinsic, as well as inherent to the surface of the cavity), surface absorption losses (e.g., due

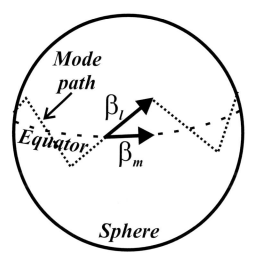

FIGURE 3.33
Schematic of the mode propagation constants along the surface of a sphere.

to the presence of adsorbed water), whispering gallery loss (or tunnel loss), and external coupling losses to a *useful* external mode (such as a prism or a waveguide)

$$Q_{tot}^{-1} = Q_{mat}^{-1} + Q_{scatt}^{-1} + Q_{surf}^{-1} + Q_{ext}^{-1} + Q_{WGM}^{-1} \tag{3.211}$$

The highest Q-factor up to the date, $Q = 2 \cdot 10^{10}$, was achieved in crystalline WGRs [133]. The oblate spheroidal resonators with diameter 0.5-12 mm and thickness 0.03-1 mm were fabricated out of single-crystal blocks by standard diamond cutting and lapping, and optical polishing techniques. It is believed that the current Q-factor values are limited by extrinsic losses in particular specimens of material (originating from uncontrolled residual doping non-stoichiometry). The highest measured Q-factor in amorphous WGM resonators is $Q = 8 \cdot 10^9$ at 633 nm [141]. Approximately the same WGM Q-factors were observed in near infrared (1.55 μm) [142] and at 780 nm [143]. The near-spherical WGR radius varied from 60 μm to 200 μm in [143], and from 600 to 800 μm in [141], [142]. The measured Q-factors were close to the maximum achievable value for the fused silica at 630 nm [144]. Q-factors measured in liquid WGRs (e.g., free-flying or trapped droplets of liquid aerosols) are less than 10^5. The problem is in the difficulty of excitation and detection of WGMs with larger Q using free beam technique [145]. Theoretical implications from the experimental data for the Q factors in liquid WGRs are more optimistic: $Q \geq 10^6$ for $\sim 20\mu$m droplets [146]. On the other hand, it was shown that a pendant 400 μm liquid-hydrogen droplet can achieve high-Q values that exceed 10^9 for WGMs in the ultraviolet [147]. Quality factor $Q = 2 \cdot 10^8$ at λ=2.014 μm was reported earlier for a multiple total-internal reflection resonator, analogous to a WGR, used in optical parametric oscillators pumped at 1064 nm [148]. Quality factors of microring and microdisk WGRs typically do not exceed 10^5. For instance, all epitaxial semiconductor 10 μm diameter microring resonators vertically coupled to buried heterostructure bus waveguides have $Q = 2.5 \cdot 10^3$ [149]. Unloaded Q-factor of the order of 10^5 was demonstrated in 80 μm microdisks [150]. Micron-size microdisk semiconductor resonators have Qs of order of 10^4 [151].

In many applications, like non-linear optics and cavity quantum electrodynamics, not only temporal confinement of light (i.e., the Q-factor), but also the extent to which the light is spatially confined is an important performance parameter. The mode volume of a WGM mode is defined as

$$V_{mode} = \frac{\int w\left(\boldsymbol{r}\right)d^3\boldsymbol{r}}{w_{max}} \tag{3.212}$$

where

$$w\left(\boldsymbol{r}\right) = \frac{1}{2}\left[\varepsilon_0\varepsilon\left(\boldsymbol{r}\right)\boldsymbol{E}\left(\boldsymbol{r}\right)\cdot\boldsymbol{E}^*\left(\boldsymbol{r}\right) + \frac{1}{\mu_0}\boldsymbol{B}\left(\boldsymbol{r}\right)\cdot\boldsymbol{B}^*\left(\boldsymbol{r}\right)\right] \tag{3.213}$$

is the electromagnetic energy density. Numerically calculated mode volumes show that the effective WGM volume occupies only a small fraction of the total volume of the microresonator (a few percent/per miles) [134]. Indeed, WGRs can have mode volumes orders of magnitude less than in Gaussian-mode resonators.

3.6.4   WGM evanescent coupling

Unlike Fabry-Perot cavities, laser beams propagating in free space cannot induce efficient excitation of high-Q WGMs in microcavities. Indeed, in the geometric-optics point of view, the WGMs being confined by successive total internal reflections, no incoming ray can directly excite them. Then, the right approach is that of evanescent-wave coupling, which exploits the WGM evanescent field outside the geometrical boundaries of the microcavity [140]. In last two decades, several evanescent coupling techniques have been developed for this purpose.

The approach used in early studies was based on prism couplers (frame a of Figure 3.34). The incident light strikes the inner prism interface with a specified angle to undergo total internal reflection, and the resulting evanescent field is used to excite the WGMs by placing the microcavity in this field. When it is carefully optimized, this technique can achieve rather good coupling efficiency up to 80 percent, but this requires difficult beam shaping, and it has a bulky size.

At present, in addition to this well-known approach, coupler devices include side-polished fiber couplers (having limited efficiency owing to residual phase mismatch), fiber tapers (almost 100% coupling achieved), hollow fibers, *pigtailing* technique with an angle-polished fiber tip in which the core-guided wave undergoes total internal reflection, and special technique of coupling of the cavities and semiconductor lasers [133].

The most efficient coupling to date was realized with tapered fiber couplers (frame b of Figure 3.34). In general, the evanescent field in a commercial step index single mode fiber locates in its cladding part (as propagation occurs through total internal reflection at the core-cladding interface). To be able to utilize this field, one can either remove the cladding part by chemical etching or by tapering. Tapering is an easy and efficient way to produce an adiabatic coupler with low losses. In this case, the fiber is typically tapered down to micrometer or even nanometer scale. For instance, a tapered fiber can be produced by gently stretching an optical fiber while it is heated over a flame. Under these conditions, the original fiber core becomes so small that it has no significant influence any more, and the light is guided only by the air-glass interface. Provided that the transition regions from the full fiber diameter to the small waist and back again are sufficiently smooth, essentially all the launched light can propagate in the taper region and (more surprisingly) find its way back into the core of the subsequent full-size fiber region. We want now to describe the coupling mechanism occurring when the evanescent field of a fiber taper (or a prism) and a WGM microcavity are brought together. Whatever scheme is used, the coupling efficiency (coupling coefficient α) as a function of the various experimental parameters can be calculated, in principle, by computing the overlap intregral between the wavefunctions describing, respectively, the input field and the particular WGM to be excited. Then, the physics of

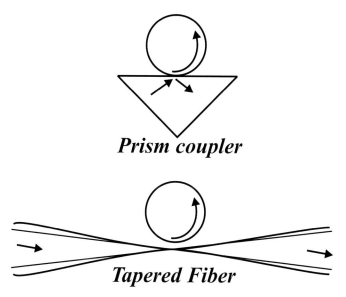

FIGURE 3.34

Two examples of microsphere resonator WGM couplers.

the coupling mechanism is often analyzed in the frame of coupled-mode theory [152]. However, a better physical insight is obtained by another approach, called *Evanescently coupled Fabry-Perot model* [140].

Figure 3.35 shows the schematic of a fiber taper coupled WGM microcavity system, where E_{in} denotes the amplitude of the input optical field, and E_{out} is the amplitude of transmitted or output field. g represents the coupling gap between the taper coupler and a cavity. E_{cav} is the amplitude of internal field just after the input. The schematic of Evanescent F-P model is also presented in Figure 3.35. In this model, an input mirror with transmission $T(g)$ as a function of g represents the evanescent field coupling between a fiber and a WGM cavity. The other mirror can describe the radiation losses or other coupling components, like a second fiber taper. Also shown is the round-trip internal absorption loss coefficient $P/2$. The optical field in the cavity after one round-trip is noted E'_{cav}. In this model, the input mirror is characterized by the reflection coefficient $-r$ (outside), r (inside), and the transmission coefficient t (for both outside and inside). For the second mirror, the corresponding coefficients are given as r' and t'.

First, we consider only the case of perfect mode matching, where the whole incoming field can enter into the cavity and excite the mode under study. In this case, the amplitudes of the optical field can be written as

$$\begin{cases} E_{cav} = tE_{in} + rE'_{cav} \qquad E'_{cav} = r'e^{-P/2}e^{i\phi}E_{cav} \\[2mm] E_{out} = -rE_{in} + tE'_{cav} \qquad E'_{out} = t'E_{cav} \end{cases} \tag{3.214}$$

where $e^{-P/2}$ represents the internal absorption loss in one round-trip, r' contains the radiation losses, and $e^{i\phi}$ represents the round-trip phase. In the following, E'_{out} will be ignored as it is not significant here. From Equation 3.214 we obtain

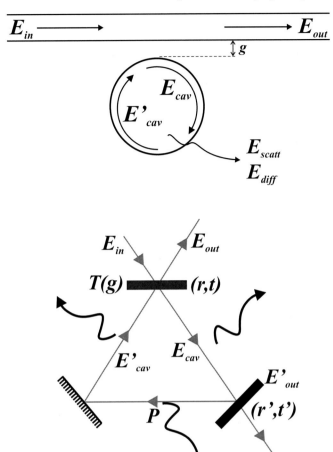

FIGURE 3.35
Schematic of the fiber taper-coupled WGM microcavity system. (Adapted from [140].)

$$\begin{cases} E_{cav} = \dfrac{t}{1 - rr'e^{-P/2}e^{i\phi}} E_{in} \\[4mm] E_{out} = \left(-r + \dfrac{t^2 r' e^{-P/2}e^{i\phi}}{1 - rr'e^{-P/2}e^{i\phi}} \right) E_{in} \end{cases} \tag{3.215}$$

So the amplitude reflection coefficient of the cavity is given by

$$r_{FP} = \frac{E_{out}}{E_{in}} = \frac{-r + r'e^{-P/2}e^{i\phi}}{1 - rr'e^{-P/2}e^{i\phi}} \tag{3.216}$$

where $r^2 + t^2 = 1$ has been exploited. This coefficient has a resonance when the round-trip phase ϕ is an integer multiple of 2π. Therefore, if we consider only one given resonance, we can replace $e^{i\phi}$ by $e^{i\delta\phi}$ where $\delta\phi = \phi - 2m\pi$. Since the WGM microcavities have a high finesse, we can assume that the mirror transmissions $T = |t|^2$, $T' = |t'|^2$ and the internal losses P verify $T, T', P \ll 1$, whereupon one can write

$$\begin{cases} e^{-P/2} \simeq 1 - \dfrac{P}{2} \\ r = \sqrt{1 - T^2} \simeq 1 - \dfrac{T}{2} \\ r' \simeq 1 - \dfrac{T'}{2} \end{cases} \tag{3.217}$$

Moreover, in the neighborhood of a resonance, as long as $\delta\phi \ll 2\pi$, one has $e^{i\delta\phi} \simeq 1 + i\delta\phi$. So, neglecting all the second order terms, the expression of the reflection coefficient simplifies to

$$r_{FP} = \dfrac{\dfrac{T - (T' + P)}{2} + i\delta\phi}{\dfrac{T + (T' + P)}{2} - i\delta\phi} \tag{3.218}$$

Here, the reflected signal is the field that escapes from the coupling region, which should be identified to the field transmitted by the taper. This leads to the normalized (intensity) transmission $\mathcal{T}_{out}$ at the taper output

$$\mathcal{T}_{out} = \frac{\mathcal{P}_{out}}{\mathcal{P}_{in}} = \left| \frac{E_{out}}{E_{in}} \right|^2 = 1 - \frac{T(T' + P)}{\left(\dfrac{T + T' + P}{2} \right)^2 + (\delta\phi)^2} \tag{3.219}$$

One notes that, on resonance, when the total internal losses $T' + P$ match the coupling losses T, the transmission $\mathcal{T}_{out}$ drops to 0. This effect is known as *critical coupling*. Moreover, we can observe that the internal losses P can not be completely distinguished from the radiation losses T', and we will merge them in the so-called *intrinsic* losses, in contrast with the *coupling losses* represented by T. It is more useful to write these expressions in terms of angular frequency by using the fact that $\phi = N\omega/cL$, where L is the round-trip length and N the internal refraction index. This introduces the FSR of the cavity $\Delta\omega_{FSR} = c/NL$, such that $\delta\omega = \Delta\omega_{FSR}\delta\phi$. We will furthermore write $N = N_S$ for silica, and $L = 2\pi a$, so that $\Delta\omega_{FSR} = c/2\pi N_S a$ (here a denotes the microsphere radius). Using the same scaling factor, we introduce the intrinsic and coupling linewidths

$$\begin{cases} \gamma_I = \Delta\omega_{FSR}(T' + P) \\ \gamma_C = \Delta\omega_{FSR}T \end{cases} \tag{3.220}$$

With these notations, the taper transmission is written as

$$\mathcal{T}_{out} = 1 - \frac{\gamma_I \gamma_C}{\left(\dfrac{\gamma_C + \gamma_I}{2} \right)^2 + (\delta\omega)^2} \tag{3.221}$$

However we also need to take account for a small mismatch which is difficultly avoided between the incoming mode and the mode of the cavity. For this purpose we introduce a phenomenological complex overlap parameter α, with $0 < |\alpha| < 1$ which measures the fraction of the incoming field that actually contributes to excite the cavity. Therefore, Equation 3.214 should be rewritten as follows

$$\begin{cases} E_{cav} = \alpha t E_{in} + r E'_{cav} \qquad E'_{cav} = r' e^{-P/2} e i\phi E_{cav} \\ \\ E_{out} = -r E_{in} + \alpha^* t E'_{cav} \qquad E'_{out} = t' E_{cav} \end{cases} \tag{3.222}$$

So the following equations are obtained for the taper transmission and the cavity build-up

$$\begin{cases} \mathcal{T}_{out} = 1 - |\alpha|^2 \dfrac{\gamma_I \gamma_C}{\left(\dfrac{\gamma_C + \gamma_I}{2}\right)^2 + (\delta\omega)^2} \\[3em] \dfrac{\mathcal{P}_{cav}}{\mathcal{P}_{in}} = \left|\dfrac{E_{cav}}{E_{in}}\right|^2 = \Delta\omega_{FSR} \dfrac{|\alpha|^2 \gamma_C}{\left(\dfrac{\gamma_C + \gamma_I}{2}\right)^2 + (\delta\omega)^2} \end{cases} \tag{3.223}$$

In general, we investigate the microcavity by detecting the throughput of the fiber taper on the output photodetector: $I_{PD}(\delta\omega) = \mathcal{T}_{out}(\delta\omega) I_{in}$. In the wavelength range where the WGMs are not excited (out of resonance or weak coupling), one has $\mathcal{T}_{out} = 1$. When a WGM is excited, the transmitted signal decreases. Therefore, to characterize the effect of a signal coupling into a WGM resonance, we can introduce the so-called *dip* parameter, defined by $\mathcal{D}(\delta\omega) = 1 - \mathcal{T}_{out}$ which provides

$$\mathcal{D}(\delta\omega) = |\alpha|^2 \frac{\gamma_I \gamma_C}{\left(\dfrac{\gamma_C + \gamma_I}{2}\right)^2 + (\delta\omega)^2} \tag{3.224}$$

According to Equation 3.224, we recognize that a WGM resonance has a Lorentzian shape of full width at half maximum (FWHM) $\gamma_{tot} = \gamma_C + \gamma_I$. Equation 3.224 also allows one to re-write Equation 3.223 as

$$\mathcal{P}_{cav} = \frac{\Delta\omega_{FSR}}{\gamma_I} \mathcal{D}(\delta\omega) \mathcal{P}_{in} = \mathcal{F}_I \mathcal{D}(\delta\omega) \mathcal{P}_{in} \tag{3.225}$$

where the intrinsic cavity finesse $\mathcal{F}_I = \Delta\omega_{FSR}/\gamma_I$ has been introduced. On resonance, the dip is

$$\mathcal{D}(\delta\omega = 0) = |\alpha|^2 \frac{4\gamma_I \gamma_C}{(\gamma_C + \gamma_I)^2} \tag{3.226}$$

From Equation 3.226 we deduce that two conditions should be fulfilled at the same time to achieve critical coupling condition ($\mathcal{D}(\delta\omega) = 1$). The first one is the condition $\gamma_I = \gamma_C$, which will be discussed here, while the second one is the mode matching condition (mostly gap independent). For more simplicity we will assume that it is properly achieved, ensuring the condition $|\alpha| = 1$. Note that value of the intrinsic Q-factor of a given microcavity is a fixed parameter, leading to a fixed γ_I, and the coupling condition will thus be analyzed through the tuning of γ_C, which is related to the evanescent gap g. Therefore, we introduce the exponential dependence of the transmission T as a function of the coupling gap g. This can be written as follows

$$T = T_0 e^{-2\kappa g} \quad \Rightarrow \quad \gamma_C = \gamma_{C0} e^{-2\kappa g} \tag{3.227}$$

where $\kappa^{-1} = \lambda/2\pi \left(N_S^2 - 1\right)^{-1/2}$ represents the evanescent wave characteristic depth (see Equation 3.205). Here γ_{C0} denotes the coupling losses when the coupler is in contact with the cavity. By insertion of Equation 3.227 into Equation 3.226 (with $|\alpha| = 1$), one obtains

$$\mathcal{D}(\delta\omega = 0) = \frac{1}{\cosh^2 \kappa (g - g_c)} \tag{3.228}$$

where the critical coupling gap g_c ensuring $\mathcal{D}(\delta\omega = 0) = 1$ is defined as

$$g_c = \frac{1}{2\kappa} \log \frac{\gamma_{C0}}{\gamma_I} \qquad (3.229)$$

The critical coupling condition $g = g_c$ (or $\gamma_C = \gamma_I$) is of great importance for both active and passive devices. The corresponding dip on resonance reaches its maximum value 1, where the output signal drops down to zero. Also, the loaded linewidth at this position is twice the intrinsic linewidth. The circulating power in the cavity is expressed as $\mathcal{P}_{cav} = \mathcal{F}_I \mathcal{P}_{in}$. Thus, for a finesse on the order of 10^6 and a modest input power 1 μW, the resulting circulating power in the cavity can be larger than 100 mW.

Another important feature, though often ignored, is that, due to phase change on reflection experienced by evanescent waves, coupling introduces a resonance frequency shift. The evanescent coupled FP model can account for this effect too. In this case, the reflection coefficient involved in Equations 3.222 has a given non-zero phase. When calculations are carried out considering this phase change, a resonance frequency shift $\Delta\omega_C = \Delta\omega_{C0} e^{-2\kappa g}$ is found, where $\Delta\omega_{C0}$ is, as for γ_{C0}, the value achieved in contact.

We close this section by observing that in 1995 it was discovered that high-Q (above 10^8) WGMs typically split into doublet mode structures [153]. It was suggested that such a splitting is due to the coupling between clockwise and counter-clockwise WGMs, which results from the internal backscattering caused by surface roughness or density fluctuations in silica which behave as Rayleigh scatterers. These two components, corresponding to standing waves, are called symmetric and asymmetric modes. Several papers have been devoted to refine this interpretation [154]. Such phenomenon can be accommodated in the *Evanescent F-P model* too [140].

3.6.5 Fabrication and applications of whispering gallery resonators

Today, WGM resonators exist in several geometrical structures like cylindrical optical fibers, microspheres, microfiber coils, microdisks, microtoroids, photonic crystal cavities, etc. up to the most exotic structures, such as bottle and bubble microresonators (Figure 3.36). In the remainder of this chapter we give a brief overview on both the fabrication techniques and applications of such novel resonators.

Concerning the former issue, several techniques have been developed over the past decades to fabricate ultrahigh-Q dielectric microresonators [156]. In the case of microspheres, melting is the favorite technique, as it can easily produce, thanks to surface tension, dielectric microspheres of both good sphericity and surface smoothness. To achieve the melting point for glass materials, several heating methods have been successfully applied:

- Gas flame - Using a microtorch with propane or hydrogen is the most ancient and still rather common technique to melt glass.

- Carbon dioxide laser - It has become the most common technique. Indeed, the CO_2 laser has a working wavelength in mid-infrared region (typically 10.6 μm), which is efficiently absorbed and transformed to heat glass.

- Electric arc - Electric arc is generally used with fiber-splicing equipment.

- Wafer-based fabrication - A variety of other geometries (microdisks, microtoroids, ...) can be realized thanks to the standard wafer-processing technology used for the fabrication of integrated circuits and optical waveguides.

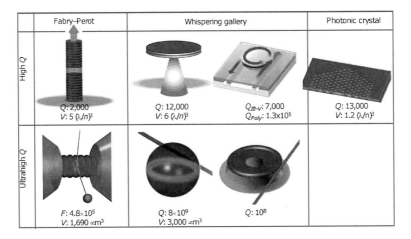

FIGURE 3.36
Several microcavity designs. The microcavities are organized by column according to the confinement method used and by row according to high-Q and ultrahigh-Q. Upper row: micropost, microdisk, add/drop filter (here, one Q value is for a polymer design and the second for a III-V semiconductor design), photonic crystal cavity. Lower row: Fabry-Perot bulk optical cavity, microsphere, microtoroid. n denotes the material refractive index, V the resonator volume, F the finesse, λ the optical wavelength. (Courtesy of [155].)

Next, we focus on applications. Unique spectral properties of WGMs, including narrow linewidth, tunability, and high stability under environmental conditions, make WGRs attractive for numerous practical applications including dynamic filters in optical communication, frequency stabilization, strong-coupling cavity quantum electrodynamics, enhancement and suppression of spontaneous emission, novel sources, spectroscopy, and sensing. Due to space limitations, for each of the following items only the basic principles will be elucidated [157, 155]. However, some of these applications will be resumed in future chapters.

- Passive filters - As already shown, the simplest resonator-based filter includes a WGR and an optical coupler, e.g., a prism coupler. Unfortunately, the Lorentzian lineshape of the filter function associated with a single microresonator represents a limitation for its application in many systems that require large sidemode rejection, in addition to a narrow bandpass and a large tuning range. Thus, cascaded resonators must be used. WGRs offer new possibilities for multipole filtering because of their small size, low losses, and integrability into optical networks. A variety of multiresonator and tunable filters have already been demonstrated in compact and robust packages. Based on this technology, filters with 10-100 GHz bandwidths and optical Qs as high as 10^5 are commercially available.

- Continuous-wave WGM lasers - Miniature lasers are among the most obvious applications of WGRs. The high quality factor of the resonators leads to the reduced threshold of the lasing. The first WGM lasers were realized in solid materials. However, probably because of the lack of input-output techniques for WGMs, the work was discontinued at that point. The next development of WGM-based lasers was in liquid aerosols and individual liquid droplets. Then, during the last decade, the lasers based on sole solid-state WGRs were rediscovered, demonstrated experimentally, and intensively studied.

For example, one effective way to create a WGR laser is the use of solids doped with active elements; e.g., rare earth ions as a WGR host material. Alternatively, a passive WGR can be coated with gain medium: for instance, erbium-doped solgel films can be applied to the surface of high-Q silica microspheres to create low-threshold WGR lasers. Furthermore, lasing in capillaries (cylindrical resonators) has also been demonstrated.

Moreover, WGM-based lasers can be created with semiconductor quantum dots coupled to the WGMs. Such a microlaser made by capturing the light emitted from a single InAs-GaAs quantum dot in the WGM of a glass microsphere was proposed theoretically. One of the most important problems here is fabrication of a single quantum dot microlaser. In addition, WGRs can significantly improve operation of semiconductor quantum well lasers. Just to mention one realization in this frame, an InGaAs-InGaAsP room temperature quantum well disk laser operating at 1.542 micron and using 0.85-micron optical pumping has been demonstrated.

Finally, Raman lasers also represent an interesting option for WGM-based microlasers. On one hand, substantial optical power enhancement within a high-finesse optical cavity has recently yielded bulk CW Raman lasers with low threshold and large tunability. On the other side, an enhancement of stimulated Raman scattering (SRS) is one of the effects demonstrated in spherical microcavities. For this reason, SRS in ultrahigh-Q surface tension-induced spherical and chip-based toroid microcavities is considered both theoretically and experimentally.

- WGRs are also used in optical parametric as well as hyper-parametric wave mixing processes. In the following we only dwell upon the latter issue. Hyperparametric optical oscillation is based on four-wave mixing (FWM) among two pump, signal, and idler photons, and results in the growth of the signal and idler optical sidebands from vacuum fluctuations at the expense of the pumping wave. The hyperparametric oscillations are different from the parametric ones. In fact, the parametric oscillations are based on $\chi^{(2)}$ non-linearity coupling three photons, and have phase matching conditions involving far separated optical frequencies that can only be satisfied in birefringent (or quasi-phase matchable) materials in the forward direction. In contrast, the hyperparametric oscillations are based on $\chi^{(3)}$ non-linearity coupling four photons, and have phase matching conditions involving near-lydegenerate optical frequencies that can be satisfied in most of the materials, both in the forward and backward directions. Recently, the study of hyperparametric oscillations had a new stage connected with the development of WGM, as well as photonic crystal microresonator technology, leading to the realization of microresonator-based optical frequency combs. In Chapter 6 such an application will be further investigated.

- WGRs offer interesting possibilities from both classical as well as quantum points of view. High Q-factors as well as small mode volumes of WGMs result in a multitude of interesting and important phenomena.

    1. Chaos - One of the fundamental problems is related to WGMs in an asymmetric WGR. It was shown that departure from an axial symmetry results in the occurrence of chaotic behavior of light in the resonator. The lifetime of light confined in a WGR can be significantly shortened by a process known as *chaos-assisted tunneling*. Surprisingly, even for large deformations, some resonances were found to have longer lifetimes than predicted by the ray chaos model due to the phenomenon of *dynamical localization*.

    2. Photonic atoms - Another fundamental area of application of WGRs is based on the ability of the resonators to mimic atomic properties. Indeed, it was pointed out that WGM mode numbers correspond to angular, radial, and the azimuthal quantum numbers, respectively, the same as in atomic physics. Thus, WGMs can be thought of as a classical analogy of atomic orbitals.

3. Cavity QED - There is great activity in both theoretical and experimental investigations of cavity quantum electrodynamics effects in WGRs. For instance, spontaneous emission processes may be either enhanced or inhibited in a cavity due to a modification of the density of electromagnetic states compared with the density in a free space. This effect was studied theoretically as well as experimentally in WGRs. In this frame, several methods have been proposed for control of atomic quantum state in atoms coupled to microspheres. These include excitation, decay control, location-dependent control of interference of decay channels, and decoherence control by *conditionally interfering parallel evolutions*.

- Spectroscopy - Starting from liquid WGRs used for resonator-enhanced spectroscopy, solid-state WGRs were utilized to enhance the interaction between light and atoms/molecules. One of the first experiments on the subject was realized in the frame of cavity-QED. The radiative coupling of free atoms to the external evanescent field of a WGM was detected. The coupling manifested itself as a narrow absorption line observed in the resonator transmission spectrum. It was proven that the evanescent field of a high-Q and small mode volume-fused silica microsphere enables velocity-selective interactions between a single photon in the WGM and a single atom in the surrounding atomic vapor. The next stage in the sensor development was related to WGR-based biosensors. The basic detection scheme is that binding of molecules to the microresonator surface induces an optical change proportional to the quantity of bound molecules. The paradigm for this process is a change in the cavity Q as the surface bound molecules affect the photon storage time, either through increased scattering or absorption. In effect, the analyte spoils the Q, and the resulting change can be measured.

- WGM resonators for laser frequency stabilization - Optical feedback from a high-Q microsphere resonator can also be used to narrow the emission linewidth of a laser in a Pound-Drever-Hall(PDH)-type scheme. For further details, see the next chapter.

- In another relevant application, optical WGRs may also serve as sources of coherent microwave radiation, i.e., as optoelectronic oscillators. This will be discussed in Chapter 7.

4

Continuous-wave coherent radiation sources

> I knew, of course, that trees and plants
> had roots, stems, bark, branches and
> foliage that reached up toward the light.
> But I was coming to realize that the real
> magician was light itself.
>
> *Edward Steichen*

> The genuine coherence of our ideas does
> not come from the reasoning that ties
> them together, but from the spiritual
> impulse that gives rise to them.
>
> *Nicolas Gomez Davila*

The invention of the laser, which stands for light amplification by stimulated emission of radiation, can be dated to 1958 with the publication of the scientific paper, *Infrared and Optical Masers*, by Arthur L. Schawlow, then a Bell Labs researcher, and Charles H. Townes, a consultant to Bell Labs [158, 159] . The work by Schawlow and Townes, however, can be traced back to the 1940s and early 50s and their interest in the field of microwave spectroscopy. In 1953, Townes, Gordon, and Zeiger demonstrated a working device, which Townes called the maser, which stands for microwave amplification by stimulated emission of radiation [160]. Then, Townes realized that the shorter wavelengths beyond those of microwaves - the wavelengths of infrared and optical light - probably offered even more powerful tools for spectroscopy than those produced by the maser. Schawlow's idea was to arrange a set of mirrors, one at each end of the cavity, to bounce the light back and forth on-axis, which would eliminate amplifying any beams bouncing in other directions. In the fall of 1957, they began working out the principles of a device that could provide these shorter wavelengths. While Schawlow was working on the device, Townes worked on the theory. After eight months of work, the collaboration was fruitful. In 1958, the two men wrote the aforementioned paper, although they had not yet made an actual laser. Two years later, Schawlow and Townes received a patent for the invention of the laser; the same year a working laser (a pulsed ruby laser at λ=694 nm) was built by Theodore Maiman at Hughes Aircraft Company [161]. Since then, lasers have been developed spanning the spectral range from the far infrared to the vacuum ultraviolet region. They have proved to be invaluable tools not only for the solution of a myriad of scientific problems but also for countless technical applications. In this chapter, we discuss the basic physical principles of lasers as well as the most important classes of both direct laser systems and laser-based coherent radiation sources, with particular regard to the applications in the field of optical frequency measurements.

4.1   Principles of masers

While in many modern textbooks the maser is relegated to history, here we give a fairly detailed description of it, both for educational purposes and because the hydrogen maser is still today a relevant frequency standard. We start by describing the first maser which was an ammonia-beam-based one [46]. To this aim, a model for the ammonia molecule is first needed. The NH_3 molecule is pyramidal in shape: the nitrogen atom is at the apex, while the 3 hydrogen atoms form an equilateral-triangle-shape base (Figure 4.1). Let's denote with $\mathcal{B}$ the plane of the 3 hydrogen atoms, with $\mathcal{R}$ the perpendicular to it passing through the nitrogen and with x the position of the intersection of $\mathcal{B}$ with $\mathcal{R}$. The position of the nitrogen atom is chosen as the origin of the x axis. For low excitation energies, the molecule

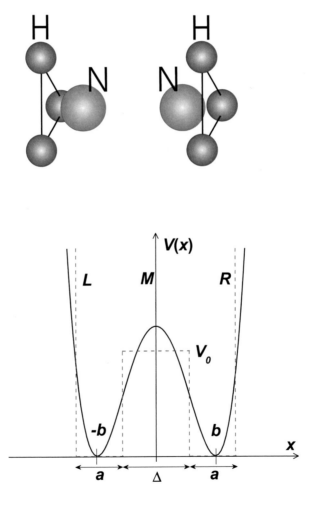

FIGURE 4.1

The ammonia molecule. Upper frame: the two classical configurations. Lower frame: the actual potential (full line) and the simplified (dashed line) describing the inversion [46].

retains its pyramidal shape and the nitrogen atom remains fixed. Qualitatively, the potential energy $\mathcal{V}(x)$ is shown in Figure 4.1. At the equilibrium position $x = b$, $\mathcal{V}(x)$ exhibits a minimum. If x is forced to become smaller, the energy increases; then, it encounters a maximum for $x = 0$, corresponding to an unstable state with the four atoms being in the same plane. When x becomes negative, the molecule is reversed like an umbrella in the wind: for symmetry reasons, $\mathcal{V}(x) = \mathcal{V}(-x)$. In the following, the actual potential $\mathcal{V}(x)$ is replaced by a simplified square-well potential $V(x)$, for which the quantum motion of a particle with mass $3m_H$ is studied, with m_H being the mass of the hydrogen atom. In other words, a collective motion is considered for the 3 H atoms, assuming that they always lay in the same plane. In this frame, it is straigthforward to find the stationary states of the system: for $E < V_0$, the solutions are sinusoids in the regions L and R and exponentials in the middle region M. Since the wave functions have to vanish for $x = \pm(b + a/2)$, the eigenstates of the Hamiltonian can be written as

$$\psi_s(x) = \begin{cases} \lambda \sin k(b + a/2 + x) & \text{L} \\ \mu \cosh Kx & \text{M} \\ \lambda \sin k(b + a/2 - x) & \text{R} \end{cases} \tag{4.1}$$

for the symmetric solutions and

$$\psi_a(x) = \begin{cases} -\lambda \sin k(b + a/2 + x) & \text{L} \\ \mu \sinh Kx & \text{M} \\ \lambda \sin k(b + a/2 - x) & \text{R} \end{cases} \tag{4.2}$$

for the anti-symmetric solutions, with $k = \sqrt{2mE}/\hbar$ and $K = \sqrt{2m(V_0 - E)}/\hbar$. These two types of solutions are sketched in Figure 4.2. The continuity equations for the wave-function and its derivative at the points $x = \pm(b - a/2)$ yield the conditions

$$\begin{cases} \tan ka = -\dfrac{k}{K} \coth K(b - \dfrac{a}{2}) & \text{for} \quad \psi_s \\[4mm] \tan ka = -\dfrac{k}{K} \tanh K(b - \dfrac{a}{2}) & \text{for} \quad \psi_a \end{cases} \tag{4.3}$$

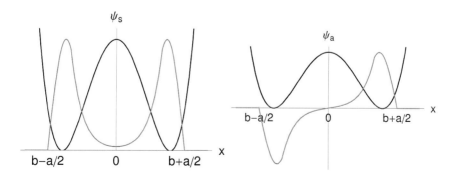

FIGURE 4.2
Qualitative behavior of the symmetric and anti-symmetric solution (gray curve). The original double-well model potential (black curve) is also shown.

Next, in order to gain some physical insight with simple algebra, let's consider the case $E \ll V_0$, which leads to $K \simeq \sqrt{2mV_0}/\hbar \gg k$. We also assume that the width of the central potential barrier, $\Delta = 2b - a$, is such that $K\Delta \gg 1$. Then we obtain

$$\begin{cases} \tan ka \simeq -\dfrac{k}{K}(1 + 2e^{K\Delta}) & \text{for} \quad \psi_s \\[3mm] \tan ka \simeq -\dfrac{k}{K}(1 - 2e^{K\Delta}) & \text{for} \quad \psi_a \end{cases} \tag{4.4}$$

whereupon

$$\begin{cases} \tan ka = -\varepsilon_s ka & \text{for} \quad \psi_s \\ \tan ka = -\varepsilon_a ka & \text{for} \quad \psi_a \end{cases} \tag{4.5}$$

where the two quantities $\varepsilon_s = 1/Ka(1 + 2e^{-K\Delta})$ and $\varepsilon_a = 1/Ka(1 - 2e^{-K\Delta})$ have been introduced. Then, by using $\tan ka \simeq ka - \pi$, Equations 4.5 provide

$$\begin{cases} k_s \simeq \dfrac{\pi}{a(1 + \varepsilon_s)} \\[3mm] k_a \simeq \dfrac{\pi}{a(1 + \varepsilon_a)} \end{cases} \tag{4.6}$$

such that the wavefunctions corresponding to the two lowest energy levels $|\psi_s\rangle$ and $|\psi_a\rangle$ are separated by the energy difference

$$\begin{aligned} 2A \equiv E_a - E_s &= \frac{\hbar^2}{2m}(k_a^2 - k_s^2) \\ &= \frac{\hbar^2 \pi^2}{2ma^2}\left[\frac{1}{(1 + \varepsilon_a)^2} - \frac{1}{(1 + \varepsilon_s)^2}\right] \simeq \frac{\hbar^2 \pi^2}{ma^2}\frac{4e^{-K\Delta}}{Ka} \end{aligned} \tag{4.7}$$

In the basis $\{|\psi_s\rangle, |\psi_a\rangle\}$ the Hamiltonian operator can thus be written in the form

$$\hat{H}_{s,a} = \begin{pmatrix} E_0 - A & 0 \\ 0 & E_0 + A \end{pmatrix} \tag{4.8}$$

Starting from $|\psi_s\rangle$ and $|\psi_a\rangle$, we now define two other acceptable states of the system

$$|\psi_R\rangle = \frac{1}{\sqrt{2}}(|\psi_s\rangle + |\psi_a\rangle) \tag{4.9}$$

$$|\psi_L\rangle = \frac{1}{\sqrt{2}}(|\psi_s\rangle - |\psi_a\rangle) \tag{4.10}$$

or, in matrix form,

$$|\psi_R\rangle = \frac{1}{\sqrt{2}}\begin{pmatrix} 1 \\ 1 \end{pmatrix} \tag{4.11}$$

$$|\psi_L\rangle = \frac{1}{\sqrt{2}}\begin{pmatrix} 1 \\ -1 \end{pmatrix} \tag{4.12}$$

As illustrated in Figure 4.3, these wavefunctions correspond to states for which the probability density is concentrated almost entirely on the left (right) for ψ_L (ψ_R), thus portraying the *classical* configurations for which the molecule is oriented towards either the left- or right-hand side (it is worth stressing that, classically, for $E < V_0$, there are two ground states of equal energy, one in the L configuration and the other in the R configuration, and that no transition $L \leftrightarrow R$ is possible). Now, let's introduce the Bohr

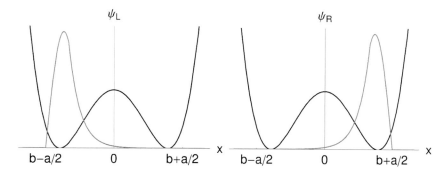

FIGURE 4.3
Classical configurations of the ammonia molecule. Qualitative behavior of ψ_L and ψ_R (gray curve). The original double-well model potential (black curve) is also shown. (Adapted from [46].)

frequency $\hbar\omega_0 = h\nu_0 = E_a - E_s = 2A$ and consider the time evolution of a wave function which is equal to ψ_R at time $t = 0$:

$$\psi(x,t) = \frac{1}{\sqrt{2}}\left[\psi_s(x)e^{-\frac{iE_st}{\hbar}} + \psi_a(x)e^{-\frac{iE_at}{\hbar}}\right]$$

$$= \frac{e^{-\frac{iE_st}{\hbar}}}{\sqrt{2}}\left[\psi_s(x) + \psi_a(x)e^{-i\omega_0 t}\right] \tag{4.13}$$

Then, one recognizes that, after a time $t = \pi/\omega_0$, the wave function $\psi(x,t)$ is proportional to ψ_L: the molecule is in the left configuration; at time $t = 2\pi/\omega_0$, the wave function $\psi(x,t)$ is again proportional to ψ_R: the molecule is back to the right configuration. In other words, due to quantum tunneling, the NH_3 molecule turns over periodically. This inversion phenomenon is precisely at the basis of the ammonia laser operation. In particular, the quantity A determines the frequency at which the transition between the two minima of the potential occurs ($\nu_0 \simeq 24$ GHz).

Next, we define an operator that measures the side of the double well in which the particle (the plane of the 3 H atoms) is located

$$\hat{X}_{s,a} = \begin{pmatrix} 0 & 1 \\ 1 & 0 \end{pmatrix} \tag{4.14}$$

Indeed, it is easily seen that $\hat{X}|\psi_R\rangle = |\psi_R\rangle$ and $\hat{X}|\psi_L\rangle = -|\psi_L\rangle$. If the result of a measurement of $\hat{X}$ is $+1$ (-1), the particle is in the right (left)-hand well. The two vectors $|\psi_R\rangle$ and $|\psi_L\rangle$ also form a basis of the Hilbert space under consideration. Obviously, the above operator is diagonal in this basis

$$\hat{X}_{R,L} = \begin{pmatrix} 1 & 0 \\ 0 & -1 \end{pmatrix} \tag{4.15}$$

On the other hand, the Hamiltonian is no longer diagonal, but has the form

$$\hat{H}_{R,L} = \begin{pmatrix} E_0 & -A \\ A & E_0 \end{pmatrix} \tag{4.16}$$

whose eigenvectors are ψ_s and ψ_a, with eigenvalues $E_0 \mp A$ (notice that the non-diagonal terms are responsible for the inversion phenomenon). From now on, we work with the basis $\{|\psi_s\rangle, |\psi_a\rangle\}$. The electric dipole moment operator is then defined as

$$\hat{D}_{s,a} = d_0 \hat{X}_{s,a} = \begin{pmatrix} 0 & d_0 \\ d_0 & 0 \end{pmatrix} \tag{4.17}$$

where $d_0 \sim 3 \cdot 10^{-11} \text{eV·m/V}$ is a characteristic measurable parameter of the molecule. Next, we apply a static electric field $\mathcal{E}$ to the molecule (parallel to the x axis). Then, the natural choice for the potential energy observable in the molecule in an electric field is then

$$\hat{W}_{s,a} = -\mathcal{E}\hat{D}_{s,a} = \begin{pmatrix} 0 & -\eta \\ -\eta & 0 \end{pmatrix} \tag{4.18}$$

where $\eta = \mathcal{E}d_0$. The total Hamiltonian of the molecule is therefore

$$\hat{H}_{s,a}^{tot} = \hat{H}_{s,a} + \hat{W}_{s,a} = \begin{pmatrix} E_0 - A & -\eta \\ -\eta & E_0 + A \end{pmatrix} \tag{4.19}$$

whose eigenvalues are easily found as

$$E_- = E_0 - \sqrt{A^2 + \eta^2} \tag{4.20}$$

$$E_+ = E_0 + \sqrt{A^2 + \eta^2} \tag{4.21}$$

that, in the weak field limit ($\eta/A \ll 1$), reduce to

$$E_{\mp} = E_0 \mp \left(A + \frac{\mathcal{E}^2 d_0^2}{2A} \right) \tag{4.22}$$

In the same approximation the corresponding eigenfunctions are

$$|\psi_-\rangle \simeq |\psi_s\rangle + \frac{\eta}{2A}|\psi_a\rangle \sim |\psi_s\rangle \tag{4.23}$$

$$|\psi_+\rangle \simeq |\psi_a\rangle - \frac{\eta}{2A}|\psi_s\rangle \sim |\psi_a\rangle \tag{4.24}$$

Now, consider a molecular beam which is made to travel along some direction x and then to cross a region where an inhomogeneous electric field ($\mathcal{E}^2 = y^2 + z^2$) is applied. Since the molecules are "large" objects, their motion in space can be treated as classical, so that, to a good approximation, they are subjected to a force given by

$$\boldsymbol{F}_{\mp} = \pm\nabla\left(\frac{\mathcal{E}^2 d_0^2}{2A} \right) \tag{4.25}$$

As a result, a beam in the pure state $|\psi_-\rangle$ will be defocused and a beam in the pure state $|\psi_+\rangle$ will be channelled; thus, the field inhomogeneity represents a means of selecting molecules whose (internal) state is $|\psi_+\rangle \sim |\psi_a\rangle$ (Figure 4.4). Then, the maser effect is achieved by forcing these molecules to release their energy $2A$ by falling down to the ground state $|\psi_s\rangle$. Since the lifetime for spontaneous transitions from the state $|\psi_a\rangle$ to $|\psi_s\rangle$ is very long (1 month), one valuable option is to stimulate this emission by application of an oscillating field $\mathcal{E}_0 \cos \omega t$, provided that ω is tuned to the Bohr frequency ω_0 of the system. The mechanism of stimulated emission allows this transition to occur very rapidly ($T \approx 70$ ns for a field $\mathcal{E}_0 \approx 10^{13}$ V/m). In this case, the Hamiltonian 4.19 becomes

$$\hat{H}_{tot-s,a} = \begin{pmatrix} E_0 - A & -\eta_0 \cos \omega t \\ -\eta_0 \cos \omega t & E_0 + A \end{pmatrix} \tag{4.26}$$

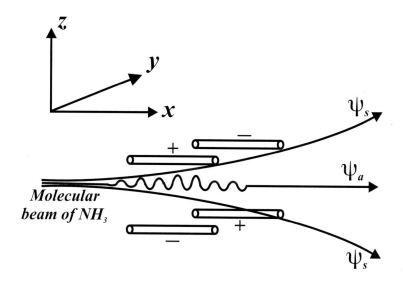

FIGURE 4.4
Channelling (removing) of the beam $|\psi_+\rangle \sim |\psi_a\rangle$ ($|\psi_-\rangle \sim |\psi_s\rangle$) in an electrostatic quadrupole field. (Adapted from [46].)

where $\eta_0 = d_0 \mathcal{E}_0$. Now, if we write the state vector of a single molecule as

$$|\psi(t)\rangle = \begin{pmatrix} a(t) \\ b(t) \end{pmatrix} \tag{4.27}$$

the time-dependent Schrödinger equation leads to the following first-order coupled linear differential system

$$\begin{cases} i\hbar\dot{a} = (E_0 - A)a - \eta_0 b \cos \omega t \\ i\hbar\dot{b} = (E_0 + A)b - \eta_0 a \cos \omega t \end{cases} \tag{4.28}$$

which, by setting $a(t) = e^{-i(E_0-A)t/\hbar}\alpha(t)$ and $b(t) = e^{-i(E_0+A)t/\hbar}\beta(t)$, becomes

$$\begin{cases} 2i\dot{\alpha} = -\omega_1\beta \left[e^{i(\omega-\omega_0)t} + e^{-i(\omega+\omega_0)t} \right] \\ 2i\dot{\beta} = -\omega_1\alpha \left[e^{-i(\omega-\omega_0)t} + e^{i(\omega+\omega_0)t} \right] \end{cases} \tag{4.29}$$

where $\omega_1 = \eta_0/\hbar$ and the initial conditions are $\alpha(t = 0) = 0$ and $\beta(0) = 1$. This system does not admit analytical solution, but a good approximation can be obtained near the resonance ($\omega \approx \omega_0$) by neglecting the rapidly oscillating terms $e^{\pm i(\omega+\omega_0)t}$ whose effect averages to zero after a time $\approx 2\pi/\omega$. In this case, the transition probability $P_{A\to S}(t)$ to find (at time t) a molecule in the state $|\psi_s\rangle$, having thus released the energy $2A = E_a - E_s$, is given by

$$p_{A\to S}(t) = |\alpha|^2 = \frac{\omega_1^2}{(\omega - \omega_0)^2 + \omega_1^2} \sin^2 \sqrt{(\omega - \omega_0)^2 + \omega_1^2} \frac{t}{2} \tag{4.30}$$

This probability oscillates in time between 0 and the maximum value

$$p_{max} = \frac{\omega_1^2}{(\omega - \omega_0)^2 + \omega_1^2} \tag{4.31}$$

As the frequency ω is varied, P_{max} exhibits a characteristic resonant behavior (with a maximum equal to 1 for $\omega = \omega_0$), with the HWHM of the resonance being ω_1. Thus, if an ω value is chosen which satisfies $|\omega - \omega_0| \ll \omega_1$, practically all the molecules have released their energy $2A$ at a time $T = \pi/\omega_1$.

A schematic description of a maser is given in Figure 4.5. Summarizing, an electrostatic quadrupole field is first used to separate the molecules in the state $|\psi_a\rangle$ within a molecular beam of mean velocity v. The resulting beam then enters a microwave cavity, where the oscillating field $\mathcal{E}_0 \cos \omega t$ is applied (in the original work a cylindrical copper cavity was used, operating in the TE_{011} mode). The cavity length, L, is adjusted so that $L/v = T = (2n+1)\pi/\omega_1$ (in practice, a servo mechanism constantly tunes the cavity so that the signal is maximum). In this way, when leaving the cavity, the molecules are in the state $|\psi_s\rangle$ and have released their energy $2A$ in the form of electromagnetic radiation. The energy emitted by such a maser oscillator is usually in an extremely monochromatic wave. The total width at half-power of the spectral distribution of the oscillation is given by [162]

$$\Delta\nu_{ST} = 2\pi k_B T \frac{(\Delta\nu_0)^2}{P_0} \tag{4.32}$$

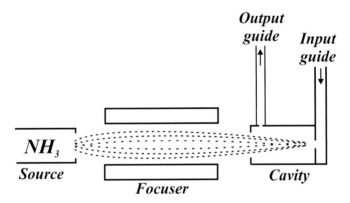

FIGURE 4.5
Sketch of an NH_3 maser device. A heater gives energy to the ammonia molecules in the source. At this stage about half of the molecules are in the excited state. Then, the ammonia molecules stream into the focuser (also called the separator) which is evacuated. The latter removes molecules in the lower quantum state, which would absorb rather than emit photons at the desired frequency, while channelling those in the upper state. Thus, the ammonia molecules that enter the resonant cavity (tuned to 24 GHz) are almost all in the excited state (inverted population). Since the cavity has a very high Q, there is sufficient noise power to initiate transitions from the upper to the lower state. Photons from these transitions can then stimulate emission from other molecules. This radiation reflects back and forth inside the cavity, whose size is specially chosen and regulated to reinforce waves of just this frequency. If the power emitted from the beam is enough to maintain the field strength in the cavity at a sufficiently high level to induce transitions in the following beam, then self-sustained oscillations will result (in the first realization such oscillations were produced with an output power around 10^{-8} W and a linewidth of 8 kHz). By contrast, under conditions such that oscillations are not maintained, the device may serve as an amplifier for an external microwave signal (close to the molecular resonance) injected into the cavity via an input waveguide. Recall that the criterion for self-sustained oscillation is that the power delivered to the cavity by the atomic beam must equal that lost by the cavity.

where $\Delta\nu_0$ is the full width of the molecular emission line and P_0 is the power emitted from the beam. Inserting values corresponding to typical experimental conditions into Equation 4.32 ($T = 300$ K, $\Delta\nu_0 = 3 \cdot 10^3$ Hz, $P_0 = 10^{-10}$ W) we find $FWHM \simeq 0.01$ Hz. Another interesting effect, known as frequency pulling, was also pointed out in [162]: the center frequency ν of the oscillation is given to a good approximation by the equation

$$\nu = \nu_0 + \frac{\Delta\nu_0}{\Delta\nu_C}(\nu_C - \nu_0) \tag{4.33}$$

where $\Delta\nu_C$ is the FWHM of the cavity mode, and $\nu_C - \nu_0$ is the difference between the cavity resonant frequency and the line frequency. Reversing the original treatment by Schawlow and Townes, we will prove Equations 4.32 and 4.33 first for lasers and then extend them to the case of masers. Sometimes, it is useful to express Equation 4.33 in the form

$$\nu = \nu_0 + \frac{Q_C}{Q_0}(\nu_C - \nu_0) \tag{4.34}$$

where the quality factors of the cavity (Q_C) and the atomic resonance line (Q_0) have been introduced, and $\nu_0/\nu_C \simeq 1$ has been used.

4.1.1   The hydrogen maser

The hydrogen maser was invented on the basis of research carried out by Ramsey's team at Harvard, USA, at the end of the 1950s. Its operation relies on the same principle of the ammonia maser. In this case, the ground-state ($1^2S_{1/2}$) hyperfine levels (labelled by the quantum number F) of atomic hydrogen are exploited. In particular, the frequency of the Zeeman sub-levels (identified by their F, m_F numbers), as a function of the applied magnetic field, is given by the well-known Breit-Rabi formula [163]

$$\nu\big|_{F=I\pm\frac{1}{2},m_F\rangle} = \frac{-\nu_0}{2(2I+1)} + \frac{\mu_B g_I m_F B}{h} \pm \frac{\nu_0}{2}\sqrt{1 + \frac{4m_F x}{(2I+1)} + x^2} \tag{4.35}$$

with $x = (g_J - g_I)\mu_B B/(h\nu_0)$. It is worth noting that the Breit-Rabi formula applies for intermediate values (i.e., between the so-called anomalous-Zeeman and Paschen-Back regimes) of the magnetic field. Here ν_0 ($\simeq 1.42$ GHz) is the hyperfine separation in zero field between the states $F = I + 1/2$ and $F = I - 1/2$, while g_J and g_I are the appropriate fine structure Landé factor and the nuclear g-factor, respectively. In the case of the hydrogen atom, the angular momentum of the nucleus is $I = 1/2$ and the two hyperfine levels $F = 1$ (for which m_F can take the values -1, 0 and 1) and $F = 0$ (for which only $m_F = 0$ is allowed) are obtained. Also, since $g_J \gg g_I$ and $g_J = 2$ for $S = J = 1/2$, we can approximate $x \simeq 2\mu_B B/(h\nu_0)$. On such basis, Figure 4.6 shows the level structure of the $1^2S_{1/2}$ ground state under the influence of an applied magnetic field ($g_I = 0.0030$ for the proton). For use as a frequency standard, the hydrogen maser oscillates on the transition $F = 1, m_F = 0\rangle \rightarrow F = 0, m_F = 0\rangle$ at a very low magnetic field value B_H (on the order of 10^{-7} T).

Today, a hydrogen maser does not differ much from the first realization in Norman Ramsey's laboratory [164, 65]. Figure 4.7 provides a schematic diagram. Molecular hydrogen is dissociated in the source and is formed into an atomic beam which passes through a state-selecting magnet. The emergent beam contains only atoms in the states ($F = 1, m = 1$) and ($F = 1, m = 0$). Then, the beam passes into a storage bulb which is located in the center of a cylindrical rf cavity, operating in the TE_{011} mode, tuned to the hyperfine transition frequency. Stimulated emission occurs if the beam flux is sufficiently high and a signal is

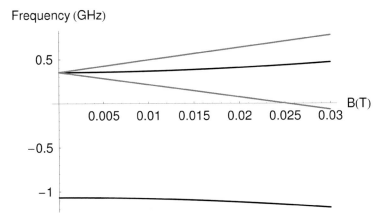

FIGURE 4.6

The active hydrogen maser is an oscillator which obtains its energy from transitions in the ground state $1^2S_{1/2}$ of the hydrogen atom (for such state, the total angular momentum $J = L+S$ of the electronic shell results entirely from the spin $S = 1/2$ of the electron since its orbital angular momentum is $L = 0$). The latter consists of a manifold of 4 sub-levels. Under the influence of the electron-nucleus interaction, the sub-levels split into a triplet and a singlet, the latter level lying lowest. These two groups are identified by the quantum numbers $F = 1$ (highest) and $F = 0$ (lowest). Their unperturbed energy difference corresponds to the well-known 21-cm hydrogen wavelength or to the frequency $\nu_H \simeq 1.42$ GHz. In the presence of a magnetic field the degeneracy of the sub-levels is removed, giving rise to the Zeeman structure. The hydrogen maser operates between the levels $|F = 1, m_F = 0\rangle$ and $|F = 0, m_F = 0\rangle$, represented by the black lines.

produced in the cavity. This signal is detected by means of a small coupling loop. The cavity is surrounded by magnetic shields to reduce the ambient field and the B_H field is produced at the storage bulb by a solenoid. Until the first realization, the main difficulty to produce a self-sustained maser oscillation with gaseous atoms resided in the weakness of

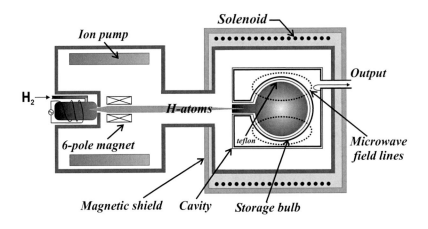

FIGURE 4.7

Schematic of an active hydrogen maser.

the magnetic dipole radiation matrix elements. However, Ramsey and co-workers guessed that with sufficiently long interaction times with the radiation field, such oscillation could be achieved. This increased interaction time was obtained by retaining the atoms in a specially designed storage box with paraffin-covered walls, where the atoms remained for approximately 0.3 seconds before escaping. Also, due to the small interaction with the paraffin surface, atoms were not seriously disturbed for at least 10^4 collisions with the walls. In this section we only illustrate the general theoretical aspects, while a more detailed discussion concerning both the practical implementations and the achievable performance as a frequency standard will be given in Chapter 7.

Following the treatment given in [7, 65, 165, 163], let ΔI denote the difference in the flux of atoms entering the storage bulb in the $(F = 1, m = 1)$ and $(F = 1, m = 0)$ states. Then the average power radiated by this beam is $P = \Delta I h \nu p$, where ν represents the maser oscillation frequency and p the averaged probability of finding the atoms in the excited state. For typical quality factor values ($Q_C \simeq 5 \cdot 10^4$ and $Q_0 \simeq 1.5 \cdot 10^9$) and a small detuning ($\nu_C - \nu_0 \simeq 10$ Hz), Equation 4.34 assures a fractional shift $(\nu - \nu_0)/\nu_0$ no larger than $2.5 \cdot 10^{-13}$. Thus, in the following we can consider $\nu \simeq \nu_C \simeq \nu_0$. Neglecting for the moment any losses, we can use Equations 4.30 and 4.31 to derive $p = \langle p_{A \to S}(t) \rangle = (1/2) p_{max}$ ($\langle \rangle$ denotes time averaging) and write

$$P_{no-losses} = \frac{1}{2} \Delta I h \nu \frac{b^2}{b^2 + [2\pi(\nu - \nu_0)]^2} \tag{4.36}$$

where the Rabi frequency b is defined here as

$$b = \mu_B \frac{\langle B_z \rangle_b}{\hbar} \tag{4.37}$$

with $\langle B_z \rangle_b$ being the B_z component of the rf field averaged over the bulb. The above arguments lay their foundation on the formalism of the state vector for a system whose evolution is described by the Schrödinger equation. Actually, such an approach is not well suited to the case in which the coupling between an atom and its environment (such as, for instance, through collisions with other atoms) is not negligible [166]. If we are interested exclusively in the evolution of the atom (rather than in the correlations induced by the atom/environment interactions), then the formalism of the density matrix can be adopted. In this frame, the effect of the environment on the atom is described by the introduction of suitable relaxation terms in the equation of evolution of the density matrix. A paradigmatic application of the density matrix is just to the case we are considering here, i.e. a two-level atomic system in which the relaxation terms determine its de-excitation. Incidentally, in this simple case, the density matrix can be represented by the so-called Bloch vector, which provides handy geometrical pictures of the system evolution. Summarizing, to correctly describe the hydrogen maser, the relaxation of the atoms in the upper state has to be taken into account. In particular, we shall distinguish between processes which relax the population difference between the two states of interest and those which relax the oscillating moment. In analogy with nuclear magnetic resonance (NMR) terminology, the decay times for such processes are designated, respectively, by T_1 (longitudinal relaxation time) and T_2 (transverse relaxation time). These can be expressed in terms of the following dominant relaxation processes [167]

$$\frac{1}{T_1} \equiv \Gamma_b + \Gamma_w + \Gamma_{se} = \Gamma_1 \tag{4.38}$$

$$\frac{1}{T_2} \equiv \Gamma_b + \Gamma_w + \frac{\Gamma_{se}}{2} = \Gamma_2 \tag{4.39}$$

where Γ_b represents the relaxation rate due to atoms escaping through the entrance hole of the bulb (essentially depending on the bulb geometry), Γ_w is related to the losses occurring when atoms hit the wall (basically due to absorption on the surface), and Γ_{se} is the relaxation rate describing spin-exchange collisions. Typical values are $T_1 = 0.3$ s and $T_2 = 0.5$ s. Then, it can be shown that, including such decay times, Equation 4.36 becomes [165, 166]

$$P = \frac{1}{2}\Delta I h\nu \frac{b^2}{\Gamma_1\Gamma_2 + b^2 + \frac{\Gamma_1}{\Gamma_2}[2\pi(\nu - \nu_0)]^2} \tag{4.40}$$

which, at resonance, reduces to

$$P_0 = \frac{1}{2}\Delta I h\nu \frac{T_1 T_2 b^2}{1 + T_1 T_2 b^2} = \frac{\Delta I h\nu}{2} - \frac{1}{T_1 T_2 b^2}P_0 =$$

$$\frac{\Delta I h\nu}{2} - \frac{P_0}{b^2}(\Gamma_b + \Gamma_w)^2 \left[1 + \frac{3\Gamma_{se}}{2(\Gamma_b + \Gamma_w)} + \frac{\Gamma_{se}^2}{2(\Gamma_b + \Gamma_w)^2}\right] \tag{4.41}$$

Now, a necessary condition for self-sustained oscillation of the maser is that the power delivered by the beam P_0 is equal to the power dissipated in the resonator dW/dt, where W is the energy stored in the magnetic field of the cavity. Thus we can write

$$P_0 = \frac{\Delta I h\nu}{2} - \frac{dW}{dt}\frac{1}{b^2}(\Gamma_b + \Gamma_w)^2 \left[1 + \frac{3\Gamma_{se}}{2(\Gamma_b + \Gamma_w)} + \frac{\Gamma_{se}^2}{2(\Gamma_b + \Gamma_w)^2}\right] \tag{4.42}$$

Next, let's find an explicit expression for dW/dt. To this aim, we first observe that

$$W = \frac{1}{2\mu_0}\int_{V_C} B^2 dV = \frac{V_C}{2\mu_0}\langle B^2\rangle_C \tag{4.43}$$

where $\langle B^2\rangle_C$ is the mean squared amplitude over the cavity volume V_C. Also, according to Chapter 2, the average stored energy W and the average dissipated energy $-dW/dt$ are related by $-dW/dt = (\omega_C/Q_C)W$. To contextualize, starting from Equation 2.8 and using $\omega_0 \gg \Gamma$ we have

$$W \propto \langle e^{-2\Gamma t}\sin^2(\omega_0' t)\rangle \simeq \frac{e^{-2\Gamma t}}{2} \tag{4.44}$$

and

$$-\frac{dW}{dt} = \Gamma e^{-2\Gamma t} \tag{4.45}$$

such that

$$\frac{W}{-dW/dt} = \frac{1}{2\Gamma} \equiv \frac{Q}{\omega_0} \tag{4.46}$$

which, with the equalities $Q \equiv Q_C$ and $\omega_0 \equiv \omega_C$, proves the above assertion. Thus, we obtain

$$\frac{dW}{dt} = \frac{\omega_C V_C \hbar^2}{2\mu_0 Q_C \eta \mu_B^2}b^2 \tag{4.47}$$

where the filling factor $\eta \equiv \langle B_z\rangle_b^2 / \langle B^2\rangle_C$ has been introduced and Equation 4.37 has been exploited. By substitution of Equation 4.47 into Equation 4.42 we get

$$P_0 = \frac{\Delta I h\nu}{2} - \frac{\omega_C V_C \hbar^2}{2\mu_0 Q_C \eta \mu_B^2}(\Gamma_b + \Gamma_w)^2 \left[1 + \frac{3\Gamma_{se}}{2(\Gamma_b + \Gamma_w)} + \frac{\Gamma_{se}^2}{2(\Gamma_b + \Gamma_w)^2}\right] \tag{4.48}$$

which can be put in the form

$$\frac{P_0}{P_c} = \frac{\Delta I h\nu}{2P_c} - \left[1 + \frac{3\Gamma_{se}}{2(\Gamma_b + \Gamma_w)} + \frac{\Gamma_{se}^2}{2(\Gamma_b + \Gamma_w)^2}\right] \tag{4.49}$$

if the quantity

$$P_c \equiv (\Gamma_b + \Gamma_w)^2 \frac{\omega_C V_C \hbar^2}{2\mu_0 Q_C \eta \mu_B^2} \tag{4.50}$$

is introduced. Next, we observe that $\Gamma_{se} = n\sigma_{se}\overline{v}_r$ where σ_{se} is the hydrogen "spin-flip" cross section $(2.85 \cdot 10^{-15}\mathrm{cm}^2)$, $\overline{v}_r = 4(k_B T/\pi m)^{(1/2)}$ the average relative velocity of the hydrogen atoms $(3.58 \cdot 10^5 \mathrm{cm/s}$ at $T = 308$ K), and $n = N/V_b$ the hydrogen density (N is the total number of atoms in the volume V_b of the bulb). Therefore, we have

$$\Gamma_{se} = n\sigma_{se}\overline{v}_r = \frac{I_{tot}}{V_b \Gamma_b}\sigma_{se}\overline{v}_r \tag{4.51}$$

where I_{tot} denotes the total flux of atoms entering the storage bulb. Note that, in order to explicit n in Equation 4.51, the condition stating that the incident beam flux must equate the escaping one $(I_{tot} = N\Gamma_b)$ has also been used. Finally, we define a threshold flux

$$I_{th} = 2P_c/(h\nu) \simeq 2P_c/(\hbar\omega_C) \tag{4.52}$$

i.e., the minimum flux that is required to sustain oscillation in the cavity. In conclusion, by substitution of Equation 4.51 and Equation 4.52 into Equation 4.49 we obtain

$$\frac{P_0}{P_c} = \frac{\Delta I}{I_{th}} - \left[1 + 3q\frac{\Delta I}{I_{th}} + 2q^2\left(\frac{\Delta I}{I_{th}}\right)^2\right] \tag{4.53}$$

with

$$q = \frac{\sigma\overline{v}_r\hbar}{2\mu_B^2\mu_0}\frac{V_C}{V_b}\frac{1}{Q_C\eta}\frac{I_{tot}}{\Delta I}\frac{\Gamma_b + \Gamma_w}{\Gamma_b} \tag{4.54}$$

Therefore, self-sustained oscillation $(P/P_c > 0)$ of the H maser occurs only if the flux of atoms falls in the range between ΔI_{min} and ΔI_{max}. These limits are derived by solving Equation 4.53 for $P/P_c = 0$

$$\frac{\Delta I_{min,max}}{I_{th}} = \frac{1 - 3q \mp \sqrt{1 - 6q + q^2}}{4q^2} \tag{4.55}$$

The maser quality parameter q is usually less than 0.1 and the power dissipated in the cavity is close to 1 pW, while the power coupled to external circuits is usually equal to one tenth of this value. At the end of this section it is worth pointing out some advantages of the hydrogen maser:

- Due to the long transition time (exceeding one second), the resonance line is extremely narrow: its width is about 1 Hz without maser operation.

- Since the diameter of the storage bulb is smaller than the transition wavelength, this guarantees confinement to the Lamb-Dicke regime which suppresses the first-order Doppler effect of the atoms interacting with the standing-wave field in the microwave resonator. In other words, the velocity of the hydrogen atom in the bulb, when suitably averaged, is close to zero;

- The hydrogen atom spends most of its time in free space where it has a simple unperturbed hyperfine spectrum and the effects of wall collisions are small due to the low electric polarizability;

- The ability of the device to operate as a self-excited maser oscillator provides the advantages of low noise amplification which characterize masers.

A deeper understanding and appreciation of these features will arise in the next sections and future chapters.

4.2   Compendium of laser theory

As stated by the acronym, by definition, the laser is an amplifier, but, as we shall soon see, it is really an oscillator. Lasers need an active medium which can amplify light; this amplifier, using a suitable cavity for feedback, becomes an oscillator. An amplifier is nothing but an energy converter from one form of energy to another. Just as an example, consider an integrated circuit amplifier and say it can amplify input power from 10 mW to 1000 mW at radio frequency. As well known, extra power comes from the dc power supply. Thus, in a sense, the amplifier is converting dc power to rf power. The same happens in a laser amplifier, that is, every laser must have what we call a pump (most commonly electrical or optical); through the laser, this pump power is converted into light energy. Let us go further into the analogy with electrical oscillators. Recall that for the basic feedback oscillator, the Barkhausen condition (Equation 2.49) for oscillation is $-A\beta = 1$, meaning that we have a finite output with no input signal. A small noise signal (spontaneous emission) will start and grow until a steady-state situation arises. All the frequencies that satisfy this condition can oscillate. However, since both A and β are functions of frequency, we can select the frequency of oscillation by proper choice of their dependence. In a nutshell, A is determined by the active medium while the feedback fraction β is decided by the optical cavity. Summarizing, a laser basically consists of three components (Figure 4.8):

1. the **active medium** where an inverted population, strongly deviating from a thermal Boltzmann distribution, is created by selective energy transfer. This acts as an amplifier for the incident light.

2. The **energy pump** (flashlamp, gas discharge, electric current, or another laser) that generates the mentioned population inversion.

3. The **optical resonator** that stores the fluorescence (spontaneous emission) emitted by the active medium in a few modes of the radiation field (cavity resonances). The fluorescent emission is isotropic, but some of these photons will travel along the cavity optical axis. Then, the optical resonator allows for them to go back and forth through the active medium, thus realizing a long amplification path. In these modes the photon number becomes $q \gg 1$ and, therefore, the induced emission becomes much larger than the spontaneous one. At the same time, the resonator makes the ouput laser beam highly directional.

One of the challenges in understanding the behavior of atoms in cavities arises from the strong feedback deliberately imposed by the cavity designer. This feedback means that a small input can be amplified in a straightforward way by the atoms, but not indefinitely. Simple amplification occurs only until the light field in the cavity becomes strong enough to affect the behavior of the atoms. Then, the strength of the light, as it acts on the amplifying atoms, must be taken into account in determining the strength of the light itself. This sounds like circular reasoning and in a sense it is. The responses of the light and the atoms to each other can become so strongly interconnected that they cannot be determined independently but only self-consistently. Strong feedback also means that small perturbations can be rapidly magnified. Thus, lasers are potentially highly erratic and unstable devices, sometimes exhibiting a truly chaotic behavior that has long been studied. Anyway, for our purposes, we want to describe lasers when they operate in a stable regime, with well-determined output intensity and frequency as well as spatial mode structure. In this frame, a short compendium of laser theory will be given, inspired by a few *classic* textbooks [168, 169, 68, 170, 171, 172, 104].

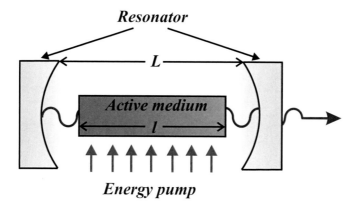

FIGURE 4.8
Schematic setup of a laser.

The main phenomena taking place in a laser are the propagation process of the light generated in the resonator and the light-matter interaction in the active medium. The former can be understood in the wave picture of light and is, in its simplest form, described by the theory of the Fabry-Perot interferometer, summarized in Chapter 3. In the following, the light-matter interaction processes which are relevant to laser operation are described.

4.2.1 The active medium

For a two-level system, when light of frequency ν corresponding to the energy difference $\Delta E = E_2 - E_1 = h\nu_0$ passes through the medium, three forms of photon-atom interaction are possible (Figure 4.9). If the atom is in the lower energy level, the photon may be absorbed, whereas if it is in the upper energy level, a clone photon may be emitted by stimulated emission. These two processes lead to attenuation and amplification, respectively. In particular, the stimulated emission provides a phase-coherent amplification mechanism for an applied signal. The signal extracts from the atoms a response that is directly proportional to, and phase coherent with, the electric field of the stimulating signal. Thus the amplification process is phase-preserving. The stimulated emission is, in fact, completely indistinguishable from the stimulating radiation field. This means that the stimulated emission has the same directional properties, same polarization, same phase, and same spectral characteristics as the stimulating emission. These facts are responsible for the extremely high degree of coherence that characterizes the emission from lasers. The third form of interaction, spontaneous emission, in which an atom in the upper energy level emits a photon independently of the presence of other photons, is, instead, responsible for amplifier noise (in addition to triggering the laser action).

4.2.1.1 Einstein A and B coefficients for absorption and emission

The classical treatment of the radiation field accounts only for the effects of absorption and stimulated emission, whereas a quantum mechanical approach introduces the additional concept of spontaneous emission. Here, we follow Einstein's original formulation and consider these three spectroscopic events to be experimentally observable transitions for which we can write rate equations. To start, it is convenient to consider the radiation in which the sample is bathed to be that of a blackbody absorber-emitter. Although the blackbody is an ideal gas of photons, the use of a convenient expression for the energy density $\rho(\nu)$ will

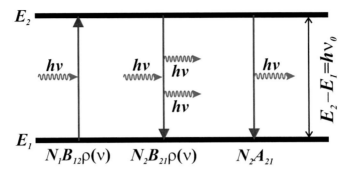

FIGURE 4.9
Absorption, induced emission, and spontaneous emission in a two-level system.

not restrict the applicability of the general rate expressions obtained: the rate constants are molecular properties, so the relationships among them do not depend on the use of the blackbody model. Let us designate the phenomenological rate constants for absorption, stimulated emission, and spontaneous emission with B_{12}, B_{21}, and A_{21}, respectively. Let us designate the population density of the upper (lower) level by N_2 (N_1). Then, we have the following expressions for the rates (per unit volume) of upward (W_{12}) and downward (W_{21}) transitions

$$W_{12} = N_1 B_{12} \rho(\nu) \tag{4.56a}$$
$$W_{21} = N_2 B_{21} \rho(\nu) + N_2 A_{21} \tag{4.56b}$$

Note that in the absence of radiation, Equation 4.56b predicts a first order decay of the excited state population, that is $N_2(t) = N_2(0)e^{-A_{21}t}$. This leads to the conclusion that the radiative lifetime is the inverse of the A coefficient

$$\tau_{spont} = \frac{1}{A_{21}} \tag{4.57}$$

If there is more than one downward transition, the radiative rate is the sum of the rates for all downward transitions. To maintain equilibrium, the rates of upward and downward transitions must be balanced

$$N_1 B_{12} \rho(\nu) = N_2 B_{21} \rho(\nu) + N_2 A_{21} \tag{4.58}$$

Next, since in linear spectroscopy experiments the Boltzmann populations of the two states are practically unperturbed, we write

$$\frac{N_2}{N_1} = \frac{g_2}{g_1} e^{-\frac{E_2 - E_1}{k_B T}} = \frac{g_2}{g_1} e^{-\frac{h\nu}{k_B T}} \tag{4.59}$$

Solving Equation 4.58 for the energy density and using Boltzmann's law gives

$$\rho(\nu) = \frac{A_{21}}{B_{12} \frac{g_1}{g_2} e^{\frac{h\nu}{k_B T}} - B_{21}} \tag{4.60}$$

Then, as already mentioned, we assume that $\rho(\nu)$ is a blackbody

$$\rho(\nu) = \frac{8\pi h\nu^3}{c^3} \frac{1}{e^{\frac{h\nu}{k_B T}} - 1} \qquad (4.61)$$

Now, Equations 4.60 and 4.61 are in agreement if

$$g_1 B_{12} = g_2 B_{21} \qquad (4.62\text{a})$$

$$A_{21} = \frac{8\pi h\nu^3}{c^3} B_{21} \qquad (4.62\text{b})$$

In material media other than gases, Equation 4.62b can be generalized by replacing c with c_0/n where n is the refractive index and c_0 is the vacuum light speed. From here and in the subsequent discussion, we adopt the notation $c = (c_0/n)$ with $c_0 = \lambda\nu$ where λ is the vacuum wavelength (in gases $n \simeq 1$). Equation 4.62b has the important consequence that, as the frequency increases, spontaneous emission competes more and more effectively with stimulated emission. Thus in systems at equilibrium at room temperature, the spontaneous emission of light at optical frequencies is greatly favored over stimulated emission. But as the frequency decreases into the far-IR and microwave regions, the process of stimulated emission becomes more favorable. In this sense, it is no coincidence that *laser action* was first achieved in masers. Indeed, the above advantage goes with a second boon: according to Boltzmann's distribution, the population ratio of the higher energy state to the lower one is about 1:1 in the microwave range ($\Delta E \sim 10^{-5}$ eV); this means that thermal energy ($k_B T_{room} = 0.0258$ eV) is enough to generate a large population of atoms in the upper, inherently long-lived, energy state, which implies no need for pumping.

Before continuing the main discussion, we incidentally mention that, in the frame of quantum mechanics, one can show that the B_{12} coefficient is related to the transition dipole moment $\boldsymbol{\mu}_{12} = \int \psi_2^* \boldsymbol{\mu} \psi_1 dr$, $\boldsymbol{\mu}$ being the electric dipole moment operator, by the relationship

$$B_{12} = \frac{|\boldsymbol{\mu}_{12}|^2}{6\varepsilon_0 \hbar^2} \qquad (4.63)$$

whereupon one obtains

$$\tau_{spont} = \frac{1}{A_{21}} = \frac{3\varepsilon_0 \hbar}{8\pi^2} \frac{g_2}{g_1} \frac{\lambda^3}{|\boldsymbol{\mu}_{12}|^2} \qquad (4.64)$$

So far, we have considered the transition rates due to a field with a uniform (white) spectrum. However, in order to derive the rate equations for a laser system, in the following we focus the attention on the transition rates that are induced by a monochromatic (i.e., single-frequency) field. On the other hand, as we will see later, absorption or emission of radiation on an atomic transition $\Delta E = E_2 - E_1 = h\nu_0$ does not result in a strictly monochromatic spectral line, but rather in a frequency distribution around the central frequency ν_0. This gives a line profile $g(\nu - \nu_0)$, normalized such that $\int g(\nu - \nu_0)d\nu = 1$, whose shape depends on basic physical properties. For example, optical frequency transitions in gases can be broadened, e.g., by lifetime, collisions, or Doppler broadening, whereas transitions in solids can be broadened by lifetime, dipolar or thermal broadening, as well as by random inhomogeneities. Therefore, we can establish here that the rates of absorption and stimulated emission due to the interaction between a monochromatic field of frequency ν with an atomic transition centered at ν_0 are proportional to the field strength times the lineshape function $g(\nu - \nu_0)$. This allows us to re-write $\rho(\nu_0)$ in Equations 4.56a and 4.56b as $\rho(\nu_0) = \rho_\nu g(\nu - \nu_0)$

$$W_{12} = N_1 B_{12} \rho_\nu g(\nu - \nu_0) \tag{4.65a}$$

$$W_{21} = N_2 B_{21} \rho_\nu g(\nu - \nu_0) + N_2 A_{21} \tag{4.65b}$$

Now, let us suppose that the monochromatic field is described by a plane wave traveling in the z-direction through the active medium. Then, since the processes of absorption and stimulated emission, respectively, subtract from and add to the energy of the radiation field, the elemental change $d\rho_\nu$ in the elemental time $dt = dz/c$ is calculated from Equations 4.65b and 4.65a as

$$\frac{d\rho_\nu}{dt} = \left[\left(N_2 - \frac{g_2}{g_1} N_1\right) B_{21} \rho_\nu g(\nu - \nu_0)\right] h\nu \tag{4.66}$$

whereupon

$$\frac{d\rho_\nu}{\rho_\nu} = \left(N_2 - \frac{g_2}{g_1} N_1\right) \sigma_{21}(\nu - \nu_0) dz \tag{4.67}$$

where the stimulated-emission cross section

$$\sigma_{21}(\nu - \nu_0) \equiv B_{21} g(\nu - \nu_0) \frac{h\nu}{c} = \frac{A_{21} \lambda^2}{8\pi n^2} g(\nu - \nu_0) = \frac{\lambda^2}{8\pi n^2 \tau_{spont}} g(\nu - \nu_0) \tag{4.68}$$

has been defined. Similarly, we have

$$\sigma_{12}(\nu - \nu_0) \equiv B_{12} g(\nu - \nu_0) \frac{h\nu}{c} = \frac{g_2}{g_1} B_{21} g(\nu - \nu_0) \frac{h\nu}{c} = \frac{g_2}{g_1} \sigma_{21}(\nu - \nu_0) \tag{4.69}$$

Note that in Equation 4.66 (and hence in Equation 4.67), we have neglected the small contribution of spontaneous emission in the same direction of the beam, as spontaneous emission is emitted in all directions. Since the intensity field I_ν is simply proportional to the energy density, integration of Equation 4.67 between $z = 0$ and z yields

$$I_\nu(z) = I_\nu(z = 0) e^{\left(N_2 - \frac{g_2}{g_1} N_1\right) \sigma_{21}(\nu - \nu_0) z} \tag{4.70}$$

This equation shows that the material behaves as an amplifier if $N_2 > (g_2/g_1)N_1$, while it behaves as an absorber (attenuator) if $N_2 < (g_2/g_1)N_1$. On the other hand, as already mentioned, at thermal equilibrium, populations are described by Boltzmann statistics. As a consequence, for light frequencies (500 THz) and room temperature (T=300 K), we have $N_2 \ll N_1(g_2/g_1)$. However, if a non-equilibrium condition is achieved for which $N_2 > N_1(g_2/g_1)$ (population inversion), then the material acts as an amplifier (active medium).

4.2.1.2   Line-broadening mechanisms

The term broadening is used to denote the finite spectral width of the response of atomic systems to electromagnetic fields. Line broadening may be typically noticed in a plot of the absorption as a function of frequency or in the frequency dependence of the gain of a laser medium. This has a major effect on the laser operation.

For resonances observed on a large group of atoms, it is useful to distinguish between line-broadening mechanisms according to whether all atoms have the same broadened spectrum, or the spectrum of the whole group is broadened because each atom has a slightly different frequency and the global spectrum merely reflects the distribution of frequencies among the particles. The former is called homogeneous broadening, as exemplified by broadening, common to all atoms, due to a finite radiative lifetime, while the latter is inhomogeneous broadening, as exemplified by a group in which each atom has a slightly different frequency because of its differing environment.

Homogeneous broadening

As just mentioned, the essential feature of a homogeneously broadened atomic transition is that a signal applied to it has exactly the same effect on all atoms in the ensemble. Mechanisms which result in a homogeneously broadened line are lifetime broadening, collision broadening, dipolar broadening, and thermal broadening.

- Lifetime broadening - Already treated in Chapter 2, this type of broadening is caused by the decay mechanisms of the atomic system. Spontaneous emission or fluorescence has a radiative lifetime. Broadening of the atomic transition due to this process is related to the fluorescence lifetime τ_{21} by $\Delta\omega_a\tau_{21} = 1$, where $\Delta\omega_a$ is the bandwidth. Actually, since the natural or intrinsic linewidth of an atomic line often gives a minor contribution, physical situations in which the lineshape is determined by the spontaneous emission process itself are quite rare.

- Collision or pressure broadening - Collisions of radiating particles (atoms or molecules) with one another and the consequent interruption of the radiative process in a random manner also lead to broadening. This can be understood with the following simple argument. Since a collision interrupts either the emission or the absorption of radiation, the long wave train, that otherwise would be present, becomes truncated. After the collision, in fact, the process is restarted without memory of the phase of the radiation prior to the collision. Thus, qualitatively, the result of frequent collisions is the presence of many truncated radiative or absorptive processes. Since the spectrum of a wave train is inversely proportional to its length, the radiation linewidth in the presence of collisions is greater than that of an individual uninterrupted process. Collision broadening can be significant in gas lasers.

- Dipolar broadening - Dipolar broadening arises from interactions between the magnetic or electric dipolar fields of neighboring atoms. This interaction leads to results very similar to collision broadening, including a linewidth that increases with increasing density of atoms. Since dipolar broadening represents a kind of coupling between atoms, so that excitation applied to one atom is distributed or shared with other atoms, it is a homogeneous broadening mechanism.

- Thermal broadening - This is brought about by the effect of the thermal lattice vibrations on the atomic transition. Such vibrations modulate the resonance frequency of each ion in the active medium at a very high frequency. As this represents a coupling mechanism between the ions, a homogeneous linewidth is obtained in this case, too. For example, thermal broadening is the mechanism responsible for the linewidth of the Nd:YAG laser.

Whatever the originating mechanism, the normalized ($\int g_L d\nu = 1$) lineshape in the case of homogeneous broadening is given by a Lorentzian

$$g_L(\nu - \nu_0) = \frac{\Gamma_L/2\pi}{(\nu - \nu_0)^2 + \Gamma_L^2/4} \tag{4.71}$$

where Γ_L is the FWHM. By virtue of Equation 4.68 and Equation 4.71, the peak value for the Lorentzian case is obtained as

$$\sigma_L^{peak} = \frac{\lambda^2}{4\pi^2 n^2 \tau_{spont}\Gamma_L} \tag{4.72}$$

Inhomogeneous broadening

Mechanisms which cause inhomogeneous broadening tend to displace the center frequencies of individual atoms, thereby broadening the overall response of a collection (without broadening the response of individual atoms). For example, owing to Doppler shifts, different atoms have slightly different resonance frequencies on the same transition. An applied signal, at a given frequency within the overall linewidth, interacts strongly solely with those atoms whose shifted resonance frequencies lay close to the signal frequency. Examples of inhomogeneous frequency-shifting mechanisms include Doppler broadening and crystal inhomogeneities.

- Doppler broadening - This is due to the Doppler effect caused by a distribution of velocities in the atomic/molecular sample. Different velocities of the emitting (or absorbing) particles result in different (Doppler) shifts, the cumulative effect of which is precisely the line broadening. A particular and perhaps the most important case is the thermal Doppler broadening related to the thermal motion of the particles.

- Crystal inhomogeneities - Solid-state lasers may be inhomogeneously broadened by crystalline defects. This happens only at low temperatures where the lattice vibrations are small. Random variations of dislocations, lattice strains, and so forth may cause small shifts (via the Stark effect) in the exact energy level spacings from ion to ion. Just like Doppler broadening, these variations do not broaden the response of an individual atom, but they do cause the exact resonance frequencies of distinct atoms to be slightly different.

In order to determine the lineshape for an inhomogeneous line, we can imagine the medium as made up of classes of atoms each designated by a center frequency ν_ξ. Furthermore, we define a function $p(\nu_\xi)$ so that the a priori probability that an atom has its center frequency between ν_ξ and $\nu_\xi + d\nu_\xi$ is $p(\nu_\xi)d\nu_\xi$ (with $\int_{\infty}^{+\infty} p(\nu_\xi)d\nu_\xi = 1$). The atoms within a given ν_ξ are considered homogeneously broadened having a lineshape function $g^\xi(\nu)$ that is normalized so that $\int_{-\infty}^{+\infty} g^\xi(\nu)d\nu = 1$. Then we obtain

$$g_{inhom}(\nu)d\nu = \left[\int_{-\infty}^{+\infty} p(\nu_\xi)g^\xi(\nu)d\nu_\xi \right] d\nu \tag{4.73}$$

If we assume that in each class ξ all the atoms are identical (homogeneous broadening) we can use

$$g^\xi(\nu) = \frac{\Delta_\nu/2\pi}{(\nu - \nu_\xi)^2 + (\Delta_\nu/2)^2} \tag{4.74}$$

where Δ_ν is called the homogeneous linewidth of the inhomogeneous line. Atoms with transition frequencies that are clustered within Δ_ν of each other can be considered indistinguishable. The term homogeneous packet is often used to describe them. Now, let us consider the specific case of thermal Doppler broadening in a gaseous system. An atom is in thermal motion, so that the frequency of emission or absorption in its own frame corresponds to a different frequency for an observer. The change in frequency associated with an atom with velocity component v_z along the line of sight (say z axis) is, to lowest order in v/c, given by

$$\nu = \nu_0(1 + \frac{v_z}{c}) \tag{4.75}$$

where ν_0 is the rest-frame frequency. The number of atoms having velocities in the range v_z to $v_z + dv_z$ is proportional to the Maxwellian distribution

$$e^{-\frac{mv_z^2}{2k_BT}} dv_z \tag{4.76}$$

where m is the mass of the atom. From the above two equations we have that the strength of the emission in the frequency range ν to $\nu + d\nu$ is proportional to

$$e^{-\frac{mc^2(\nu-\nu_0)^2}{2\nu_0^2 k_BT}} d\nu \tag{4.77}$$

The normalized ($\int g_D d\nu = 1$) shape function for the Doppler-broadened line can therefore be written as

$$g_D(\nu - \nu_0) = \sqrt{\frac{\ln 2}{\pi}} \frac{2}{\Gamma_D} e^{-\frac{4\ln 2(\nu-\nu_0)^2}{\Gamma_D^2}} \tag{4.78}$$

where the FWHM given by

$$\Gamma_D = \sqrt{8\ln 2} \frac{\nu_0}{c} \sqrt{\frac{k_BT}{m}} \tag{4.79}$$

has been introduced. By virtue of Equation 4.68 and Equation 4.78, the peak value for the Doppler case is obtained as

$$\sigma_D^{peak} = \sqrt{\frac{\ln 2}{\pi}} \frac{\lambda^2}{4\pi n^2 \tau_{spont} \Gamma_D} \tag{4.80}$$

Now, by using Equations 4.77 and 4.74 into Equation 4.73 and putting $\Delta_\nu = \Gamma_L$, we obtain a composite lineshape known as Voigt profile

$$g_V(\nu) = \frac{\Gamma_L}{\Gamma_D} \frac{\sqrt{\ln 2}}{\pi^{3/2}} \int_{-\infty}^{+\infty} \frac{e^{-\frac{4\ln 2(\nu_\xi - \nu_0)^2}{\Gamma_D}}}{(\nu - \nu_\xi)^2 + (\Gamma_L/2)^2} d\nu_\xi \tag{4.81}$$

Obviously, Equation 4.81 reduces to Equation 4.71 (4.78) in the limit $\Gamma_L \gg \Gamma_D$ ($\Gamma_D \gg \Gamma_L$). Finally, it is left for exercise to show that the convolution of a Lorentzian line of width Γ_{L1} with another Lorentzian curve of width Γ_{L2} again gives a Lorentzian whose width is $\Gamma_L = \Gamma_{L1} + \Gamma_{L2}$. Also, the convolution between a Gaussian line of width Γ_{D1} and a second Gaussian of width Γ_{D2} is again a Gaussian curve, this time of width $\Gamma_D = \sqrt{\Gamma_{D1}^2 + \Gamma_{D2}^2}$. For any combination of broadening mechanisms, it is therefore always possible to reduce the problem to a convolution of a single Lorentzian line with a single Gaussian line. Sometimes, however, one mechanism predominates, and it is then possible to speak of a pure Lorentzian or Gaussian line.

4.2.2 The pump

We now consider how to produce a population inversion in a given material. Thus far, we have not specified where levels 1 and 2 appear in the overall energy-level scheme of the lasing atoms. We might imagine that level 1 is the ground level and level 2 the first excited level of an atom. In this case, when we attempt to achieve continuous laser oscillation, we encounter a serious difficulty: the mechanism we use to excite atoms to level 2 can also de-excite them. For example, if we try to pump atoms from level 1 to level 2 by irradiating the medium, the radiation will induce both upward transitions $1 \to 2$ (absorption) and downward transitions $2 \to 1$ (stimulated emission). The best we can do by this optical pumping process is to produce nearly the same number of atoms in level 2 as in level 1; we cannot obtain a positive steady-state population inversion using only two atomic levels in the pumping process. It is not necessary to show this by writing down equations; since $g_1 B_{12} = g_2 B_{21}$, such an argument is quite intuitive.

A possible way to circumvent this problem is to make use of a third level, as shown in the left frame of Figure 4.10. In such a laser, some pumping process acts between level g and level 3. An atom in level 3 cannot stay there forever. As a result of the pumping process, it may return to level g, but for other reasons such as spontaneous emission or a collision with another particle, the atom may drop to a different level of lower energy. In the case of collisional de-excitation, the energy lost by the atom may appear as internal excitation in a collision partner, or as an increase in the kinetic energy of the collision partners, or both. The key to the three-level inversion scheme is to have atoms in the pumping level 3 drop very rapidly to the upper laser level 2. This accomplishes two purposes. First, the pumping from level g is, in effect, directly from level g to the upper laser level 2, because every atom finding itself in level 3 converts quickly to an atom in level 2. Second, the rapid depletion of level 3 does not give the pumping process much chance to act in reverse and repopulate the ground level g.

Another useful model for achieving population inversion is the four-level laser scheme shown in the right frame of Figure 4.10. Pumping proceeds from the ground level g to the level 3, which, as in the three-level laser, decays rapidly into the upper laser level 2. In this model the lower laser level 1 is not the ground level, but an excited level that can itself decay into the ground level. This represents an advantage over the three-level laser, as the depletion of the lower laser level obviously enhances the population inversion on the laser transition. That is, a decrease in N_1 results in an increase in $N_2 - N_1$. In addition, level 1 is generally not populated (if $\Delta E_{1g} \gg k_B T$), which immediately gives a population inversion.

It is obvious from Figure 4.10 that the minimum energy input per output photon is $h\nu_{3g}$; so the power efficiency of the laser cannot exceed

$$\eta_q = \frac{\nu_{21}}{\nu_{3g}} \equiv \frac{\nu_0}{\nu_{3g}} \tag{4.82}$$

which is known as atomic quantum efficiency. According to this formula, in an efficient laser system ν_0 and ν_{3g} must be of the same order of magnitude and, as a consequence, the laser transition should involve low-lying levels.

In this frame, it is customary to define another quantity, referred to as the quantum defect q, which sets an upper limit to the power efficiency:

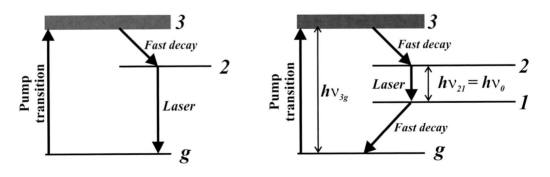

FIGURE 4.10
Scheme of a three-level (left frame) and four-level (right) laser.

$$q = h\nu_{pump} - h\nu_{laser} = h\nu_{pump} \left(1 - \frac{\lambda_{pump}}{\lambda_{laser}} \right) \qquad (4.83)$$

that is different from the quantum efficiency. The latter refers, in fact, to the average number of output photons per pump photon, rather than to the photon energies. In other words, the above q definition tells us that the power efficiency of the laser could not be 100% even if every pump photon could be converted into a laser photon. As we will see later, some laser crystals (e.g. those doped with ytterbium) have a particularly small quantum defect of only a few percent of the pump photon energy, leading to potentially very high power efficiency. However, a small q also leads to quasi-three-level behavior of the gain medium, which makes certain aspects of laser design more sophisticated, and may even make it more difficult to achieve a high wall-plug efficiency.

We close this section by shortly reviewing the most common pump sources employed in lasers, referring the reader to [168] for a more comprehensive discussion.

- Optical pumping, i.e., by cw or pulsed light emitted by a powerful lamp or laser beam. Optical pumping by an incoherent source is particularly suited to solid-state or liquid lasers (i.e., dye lasers). Line-broadening mechanisms in solids and liquids produce, in fact, considerable broadening, so that we usually deal with pump bands rather than sharp lines. These bands can therefore absorb a sizeable fraction of the broadband light emitted by the lamp. As a general rule, an efficient pump system utilizes a radiation source having a spectral output that closely matches the absorption bands of the gain medium, transfers the radiation with minimum losses from the source to the absorbing laser material, and creates an inversion which spatially matches the mode volume of the resonator mode.

- Electrical pumping, i.e., by a cw, radio frequency, or pulsed current flowing in a conductive medium, such as a semiconductor or an ionized gas. In the latter case, indeed, line-broadening mechanisms are weaker than in solids. For gases at low pressure, often used in lasers (a few tens of Torr), in fact, collision-induced broadening is very small, and the linewidth is essentially determined by Doppler broadening (on the order of a few GHz or less). In conclusion, due to the absence of broad absorption bands in the active medium of gas lasers, optical pumping would be inefficient.

A common means of pumping gas lasers is an electric discharge, which may be produced in a gas contained inside a glass tube by applying a high voltage to electrodes on either side of the tube. Electrons are ejected from the negative electrode (the cathode) and drift toward the positive electrode (the anode). When an electron collides with an atom (or molecule), there is some probability that the atom makes a transition to a higher energy state. In an inelastic collision of the first kind, the energy lost by the electron is converted to internal excitation energy of the atom. Such a process is often referred to as electron-impact excitation (in electron-impact excitation of molecules, the internal energy added to the molecule may be in the form of vibrational and rotational energy as well as electronic energy). If an electron collides with an already excited atom or molecule, it can cause the atom or molecule to drop to a lower level of excitation, the energy difference now going into an increase in the kinetic energy of the system. This type of inelastic collision is called a collision of the second kind (or a *superelastic* collision).

A gas laser may be pumped directly by electron-impact excitation, in the sense that collisions of the active atoms with the electrons are the sole source of the population inversion. In this case, the rates for the various excitation (collisions of the first kind) and de-excitation (collisions of the second kind) processes enter into the population rate equations as pumping and decay rates, respectively. Frequently, however, electron-impact excitation produces a population inversion indirectly, in the sense that it sets the stage for another process that acts more directly to produce a positive gain. The most important of these other processes is excitation transfer from one atom (or molecule) to another. One way for an excited atom to transfer energy to another atom is by photon transfer: the photon spontaneously emitted by one atom is absorbed by the other. In this way, the first atom drops to a lower level and the second atom is raised to a higher level, i.e., there is an excitation transfer between the two atoms. This process has a negligible probability of occurrence unless the photon emitted by the donor atom is within the absorption linewidth of the acceptor atom, that is, there must be a resonance (or near-resonance) of the atomic transitions. Actually, the process of excitation transfer via spontaneous emission is quite negligible compared to other transfer processes that result from a direct non-radiative (e.g., collisional) interaction between two atoms. The calculation of excitation transfer rates between atoms (and molecules) is usually very complicated, and experimental determinations of transfer rates are essential; such studies form an entire field of research.

- Chemical pumping, i.e., by an exothermic chemical reaction. Two such reactions can be used, namely: associative reactions, i.e., $A + B = (AB)^*$, which results in the molecule AB being left in an excited vibrational state; dissociative reactions, where dissociation may be induced by a photon, i.e, $AB + h\nu = A + B^*$, which results in species B (atom or molecule) being left in an excited state. Chemical pumping usually applies to materials in the gas phase, and it generally requires highly reactive and often explosive gas mixtures. Energy generated in an exothermic reaction is often quite large, and high powers (for cw operation) or energies (for pulsed operation) can be available for laser action.

4.2.3 The resonator

As already explained, to make an oscillator from an amplifier, it is necessary to introduce suitable positive feedback. This is accomplished by placing the active medium between two highly reflecting mirrors (optical cavity). In this case, a plane e.m. wave travelling in a direction perpendicular to the mirrors bounces back and forth between them, and is amplified on each passage through the active material. Also, if one of the two mirrors is partially transparent, a useful output beam is obtained from it.

Consider a resonator of length L in which an active medium of length l is inserted. Let I represent the intensity of the beam just after mirror 1 at a given cavity position at time $t = 0$. The intensity I' after one cavity round-trip is

$$I' = R_1 R_2 (1 - L_i)^2 I e^{2(N_2 - \frac{g_2}{g_1} N_1)\sigma_{21} l} = R_1 R_2 (1 - L_i)^2 I e^{2N\sigma_{21} l} \qquad (4.84)$$

where $N = N_2 - (g_2/g_1)N_1$, $R_1 = 1 - a_1 - T_1$ and $R_2 = 1 - a_2 - T_2$ are the power reflectivities of the two mirrors, and L_i is the single-pass internal loss of the cavity. Note that, in order to lighten the notation, in Equation 4.84 we have dropped the frequency dependence of the cross section $(\sigma_{21}(\nu - \nu_0) \equiv \sigma_{21})$. The change of intensity for a cavity round-trip is then

$$\Delta I \equiv I' - I = \left[(1 - a_1 - T_1)(1 - a_2 - T_2)(1 - L_i)^2 e^{2N\sigma_{21} l} - 1 \right] I \qquad (4.85)$$

Next we assume that the mirror losses are equal $a_1 = a_2 = a$ and so small that we can set $1 - a_1 - T_1 \simeq (1 - a)(1 - T_1)$ such that

$$\Delta I = \left[(1 - T_1)(1 - T_2)(1 - a)^2 (1 - L_i)^2 e^{2N\sigma_{21} l} - 1 \right] I \qquad (4.86)$$

Now we define $\gamma_1 = -\ln(1 - T_1)$, $\gamma_2 = -\ln(1 - T_2)$, $\gamma_i = -\ln(1 - a) - \ln(1 - L_i)$ and $\gamma = \gamma_i + (\gamma_1 + \gamma_2)/2$ such that

$$\Delta I = \left[e^{2(N\sigma_{21}l-\gamma)} - 1\right]I \simeq 2(N\sigma_{21}l - \gamma)I \qquad (4.87)$$

Next, if we divide both sides by the time $\Delta t = 2L_e/c_0$ where $L_e = L + (n-1)l$ is the optical length of the resonator, and make the approximation $\Delta I/\Delta t \simeq \dfrac{dI}{dt}$ we get

$$\frac{dI}{dt} = \left(\frac{c_0\sigma_{21}lN}{L_e} - \frac{\gamma c_0}{L_e}\right)I \qquad (4.88)$$

Obviously, the above equation is valid in the same form for the number of photons q, such that we can write

$$\frac{dq}{dt} = \left(\frac{c_0\sigma_{21}lN}{L_e} - \frac{1}{\tau_c}\right)q \qquad (4.89)$$

where

$$\frac{1}{\tau_c} = \frac{\gamma c_0}{L_e} = \frac{\gamma_i c_0}{L_e} + \frac{\gamma_1 c_0}{2L_e} + \frac{\gamma_2 c_0}{2L_e} \qquad (4.90)$$

that is the decay time for photons in the optical resonator, has been introduced. Once again, we have neglected the small contribution of spontaneous emission in the same direction of the beam. With reference to Figure 4.11, this could be included in Equation 4.89 by the term $R_{sp} \propto (1/\tau_{2g} + 1/\tau_{21})N_2(Al)$, A being the area of the active medium. However, since, when the laser is oscillating, the number of photons in the cavity can be as high as 10^{16} (and much more than this value for pulsed lasers), we drop the term R_{sp}, and instead assume that an arbitrarily small number of photons is initially present in the cavity just to allow laser action to start.

Finally, from Equation 4.90 we recognize that the last term gives the rate of photon loss due to transmission through mirror 2. The output power through this mirror is thus given by

$$P_{out} = h\nu\frac{\gamma_2 c_0}{2L_e}q \qquad (4.91)$$

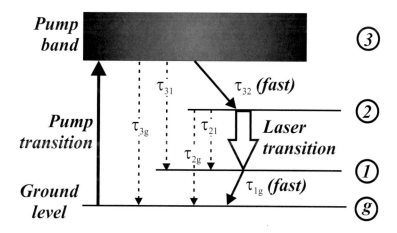

FIGURE 4.11
Pumping-oscillation cycle for a typical 4-level laser.

4.2.4   Rate equations for a four-level system

Now we have all the ingredients to write down the rate equations describing the laser behavior. We will consider only a four-level system (corresponding to the most commonly used configuration) in the space-independent model (both the mode distribution and the pump rate are assumed to be spatially independent). We will also assume that the transition from the pump band into the upper laser level and from the upper to the lower laser level occurs very rapidly. Therefore, referring again to Figure 4.11, we can write

$$\begin{cases} N_3 \simeq 0 \qquad N_1 \simeq 0 \\ \dfrac{dN_2}{dt} = w_p N_g - \dfrac{N_2}{\tau_{21}} - \dfrac{N_2}{\tau_{2g}} - DqN \\ N_{tot} = N_g + N_1 + N_2 + N_3 \simeq N_g + N_2 \end{cases} \qquad (4.92)$$

where the effective pumping rate

$$w_p = \left(1 + \frac{\tau_{32}}{\tau_{31}} + \frac{\tau_{32}}{\tau_{3g}} \right)^{-1} w_{g3} \leq w_{g3} \qquad (4.93)$$

addresses the fact that some of the absorbed pump photons will decay to levels other than the upper laser level. The last term in the second of Equations 4.92 deserves some explanation. As already discussed, in one cavity round-trip, the rate of stimulated-emission transitions (per unit volume) minus that of absorption transitions is given by

$$2 \left(N_2 - \frac{g_2}{g_1} N_1 \right) F_\nu \sigma_{21} = 2 N F_\nu \sigma_{21} \qquad (4.94)$$

where F_ν is the flux of photons (number of photons per unit surface and unit time) at frequency ν interacting with the active medium. Thus we have

$$F_\nu = \frac{q}{A_b(2L_e/c_0)} = \frac{ql}{V_a(2L_e/c_0)} \qquad (4.95)$$

where $V_a = A_b l$ represents the so-called active volume, A_b being the cross-sectional area of the cavity mode. By insertion of Equation 4.95 into Equation 4.94 we get

$$2 N F_\nu \sigma_{21} = \frac{l c_0 \sigma_{21}}{V_a L_e} \equiv Dqn \qquad (4.96)$$

where $D = l c_0 \sigma_{21}/(V_a L_e)$ has been defined. Now, $-Dqn$ is exactly the rate at which N_2 decreases due to the combined effect of absorption and stimulated emission. Next we define the fluorescence decay time of the upper laser level as

$$\frac{1}{\tau_f} = \frac{1}{\tau_{21}} + \frac{1}{\tau_{2g}} \equiv \frac{1}{\tau_{spont}} + \frac{1}{\tau_{2g}} \qquad (4.97)$$

Also, since $N_g \gg N_2$ we can write $N_{tot} = N_g + N_2 \simeq N_g$ such that the quantity $w_p N_g \simeq w_p N_{tot} \equiv R_p$ is constant. Finally we obtain

$$\frac{dN}{dt} = R_p - \frac{N}{\tau_f} - DqN \qquad (4.98)$$

where $N = N_2 - (g_2/g_1)N_1 \simeq N_2$ has also been used. Now, by also considering Equation 4.89, we get the following set of coupled equations

$$\begin{cases} \dfrac{dN}{dt} = R_p - \dfrac{N}{\tau_f} - NqD \\[3mm] \dfrac{dq}{dt} = \left[DV_a N - \dfrac{1}{\tau_c} \right] q \end{cases} \tag{4.99}$$

The above coupled equations describe the static and dynamic behavior of the considered 4-level laser. If pumping is initiated at $t = 0$, they must be solved with the initial conditions $q(t = 0) \simeq 1$ and $N(t = 0) \simeq 0$. Anyway, here we are only interested in the steady-state solution, while the time-dependent solution will be discussed later on. We begin by considering the threshold condition for laser action. Suppose that at time $t = 0$ an arbitrarily small number of photons (e.g., $q(t = 0) \simeq 1$) is present in the cavity due to spontaneous emission. From the second master equation we then see that in order to have $dq/dt > 0$, we must have $DV_a N > 1/\tau_c$. Laser action is therefore initiated when the population inversion N reaches a critical value N_{th} given by

$$N_{th} = \frac{\gamma}{l\sigma_{21}} \equiv \frac{\gamma'}{\sigma_{21}} \tag{4.100}$$

where $\gamma' = \gamma/l$. Being equivalent to the first Barkhausen condition (see Equation 2.49), Equation 4.100 expresses the circumstance that, at the threshold, the gain

$$G = N\sigma_{21} = N\frac{\lambda^2}{8\pi n^2 \tau_{spont}} g(\nu - \nu_0) \tag{4.101}$$

must equal the losses γ': $G_{th} = \gamma'$. Next, we find the steady-state solution ($dN/dt = dq/dt = 0$) as

$$N_0 = \frac{\gamma}{l\sigma_{21}} = N_{th} \tag{4.102}$$

and

$$q_0 = V_a \tau_c \left(R_p - \frac{N_0}{\tau_f} \right) = V_a \tau_c \left(R_p - \frac{N_{th}}{\tau_f} \right) \tag{4.103}$$

The condition $q_0 > 0$ defines the critical pumping rate as

$$R_{p,th} = \frac{N_{th}}{\tau_f} = \frac{\gamma}{l\tau_f \sigma_{21}} \tag{4.104}$$

If we define $x = R_p/R_{p,th} = P_p/P_{p,th}$ where P_p is the pumping power and $P_{p,th}$ is its threshold value, then we can write

$$q_0 = V_a \frac{\tau_c}{\tau_f} N_{th}(x - 1) \tag{4.105}$$

According to Equation 4.91, the laser output power is given by

$$P_{out,0} \equiv P_{4-lev} = A_b \frac{h\nu}{\tau_f \sigma_{21}} \frac{\gamma_2}{2} \left(\frac{P_p}{P_{p,th}} - 1 \right) \tag{4.106}$$

We can define the laser slope efficiency as

$$\eta_s \equiv \frac{dP_{4-lev}}{dP_p} = A_b \frac{h\nu}{\tau_f \sigma_{21}} \frac{\gamma_2}{2} \frac{1}{P_{p,th}} \tag{4.107}$$

which can be re-arranged in a very meaningful form if an explicit expression for $P_{p,th}$ is found. For this purpose, to fix the ideas, we consider the specific case of electrical pumping, and define the pump efficiency η_p as the ratio between the pump power $P_p = R_p Alh\nu_{3g}$ required to produce the pump rate R_p and the actual electrical power feeding the lamp P_{el}:

$$P_p = \eta_p P_{el} \tag{4.108}$$

Therefore, we have

$$P_{el} = \frac{R_p}{\eta_p} Alh\nu_{3g} \tag{4.109}$$

which, at threshold, becomes

$$P_{p,th} \equiv P_{el,th} = \frac{R_{p,th}}{\eta_p} Alh\nu_{3g} = \frac{\gamma}{\eta_p} \frac{h\nu_{3g}}{\tau_f} \frac{A}{\sigma_{21}} \tag{4.110}$$

where Equation 4.104 has been used. By insertion of Equation 4.110 into Equation 4.107 we get

$$\eta_s = \eta_p \frac{\gamma_2}{2\gamma} \frac{\nu}{\nu_{3g}} \frac{A_b}{A} \equiv \eta_p \eta_c \eta_q \eta_t \tag{4.111}$$

where η_c represents the fraction of generated photons coupled out of the cavity (output coupling efficiency); η_t, being the fraction of the active medium cross section used by the beam cross section, is called the transverse efficiency.

For a fixed rate, there is some value for the transmission, T_2, of the output mirror that maximizes the output power. Physically, the reason for this optimum arises from the fact that, as the transmission T_2 is increased, we have the following two contrasting effects: (1) the output power tends to increase due to the increased mirrorr transmission; (2) the output power tends to decrease due to the decreased number of cavity photons arising from the increased cavity loss. The optimum transmission is obtained by imposing the condition (for a given value of pump power P_p)

$$\frac{dP_{4-lev}}{d\gamma_2} = 0 \tag{4.112}$$

Of course, we must take into account that $P_{p,th}$ is also function of γ_2. To this aim, we re-write

$$P_{p,th} = \frac{\gamma}{\eta_p} \frac{h\nu_{3g}}{\tau_f} \frac{A}{\sigma_{21}} = P_{p,th}(\gamma_2 = 0) \frac{\gamma}{\gamma_i + \frac{\gamma_1}{2}} \tag{4.113}$$

where the minimum threshold pump power $P_{p,th,min} \equiv P_{p,th}(\gamma_2 = 0)$ has been introduced and

$$P_{4-lev} = \frac{h\nu}{\tau_f \sigma_{21}} A_b \left(\gamma_i + \frac{\gamma_1}{2}\right) S \left[\frac{x_m}{S+1} - 1\right] \tag{4.114}$$

with $S = (\gamma_2/2)/(\gamma_i + \gamma_1/2)$ and $x_m = P_p/P_{p,th,min}$. Now, the only term in 4.114 that depends on γ_2 is the quantity S which is simply proportional to γ_2. Therefore, the optimum-coupling condition can be obtained by setting $dP_{4-lev}/dS = 0$ which returns

$$S_{opt} = \sqrt{x_m} - 1 \tag{4.115}$$

The corresponding expression for the output power is obtained as

$$P_{4-lev,opt} = A_b \frac{h\nu}{\tau_f \sigma_{21}} \left(\gamma_i + \frac{\gamma_1}{2}\right) (\sqrt{x_m} - 1)^2 \tag{4.116}$$

We close this section by remarking that the above rate-equations-based treatment implicitly assumes a single-mode oscillating laser. In next sections we shall make some clarifications about this point.

4.2.4.1 Transient behavior and relaxation oscillation

In general, the set of coupled rate Equations 4.99 derived for a 4-level laser does not admit an analytic solution and one has to resort to numerical computation. As a representative example, Figure 4.12 shows a computed plot of $N(t)$ and $q(t)$ obtained by using for the parameters R_p, D, τ_f, τ_c, and V_a values which can be considered typical for a Nd:YAG laser (see figure caption).

Qualitatively, the behavior is the following. While the population inversion immediately starts growing due to the pumping process, the photon number remains at its initial low value until the population inversion crosses the threshold value. When the photon number becomes large enough, the stimulated emission process becomes dominant over the pumping process. The population then begins to decrease and, after the photon peak, population inversion is driven below its threshold by the continuing high rate of stimulated emission. Thus the laser goes below threshold, and the photon number decreases. When this photon number decreases to a sufficiently low value, the pumping process again becomes dominant over the stimulated-emission process. The population inversion can now begin growing again and the whole series of events just considered repeats itself. The photon number is then seen to display a regular sequence of peaks of decreasing amplitude; consecutive peaks are approximately equally spaced in time. The output power therefore shows a similar time behavior. This aspect of regular oscillation for the output power is usually referred to as a damped relaxation oscillation. Indeed, for times long enough, the behavior of $q(t)$ is well described by that of a damped sinusoidal oscillation. In the specific case of Figure 4.12, such oscillation occurs with a period $T_{ro} \simeq 10$ μs and an exponential decay time $t_{ro} \simeq 100$ μs. Recalling the basic properties of harmonic oscillators (discussed in Chapter 2), in the relative intensity noise (RIN) spectrum (see Section 2.8) this will correspond to a Lorentzian peak centered at $\nu_{ro} = 1/T_{ro} \simeq 100$ kHz and having a width of $\Gamma_{ro} = 1/(2\pi)t_{ro} \simeq 1.6$ kHz.

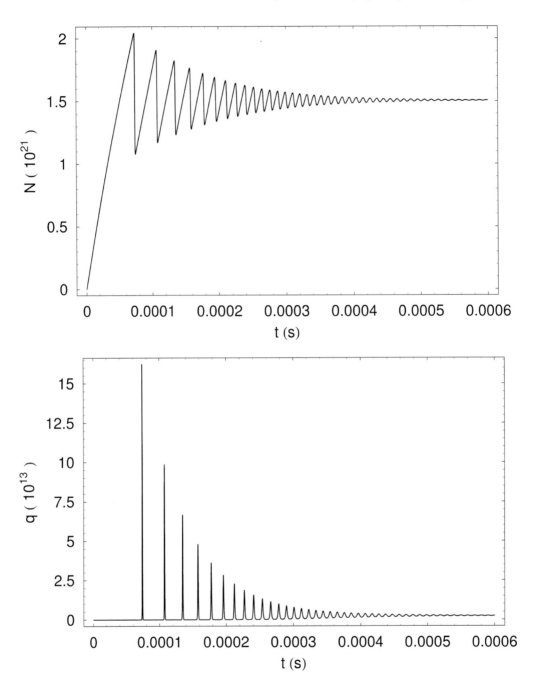

FIGURE 4.12
Computed plot of $N(t)$ and $q(t)$ for the following parameters: $\sigma_{21} = 9.2 \cdot 10^{-23}$ m^2, $\tau_f = 2.3 \cdot 10^{-4}$ s, $L_e = 0.123$ m, $\gamma = 0.009$, and $A_b = 30 \cdot 10^{-6}$ m^2.

Finally, it is interesting to note that the oscillation of $N(t)$ precedes that of $q(t)$ by about half the oscillation period, since we must first produce an increase in population to have a corresponding increase in the photon number.

4.3 Frequency pulling

In order to find the laser oscillation frequency we require that the phase shift imparted to a light wave completing a round-trip within the resonator must be a multiple of 2π. In the absence of population inversion we have

$$2\frac{2\pi\nu}{c_0}L_e + \text{phase due to mirror refl.} = 2\pi m \tag{4.117}$$

Now, if a population inversion is created in the active medium, then this will change the real part of the refractive index by the amount Δn^r, whereupon the frequencies of the cavity modes will become

$$2\frac{2\pi\nu}{c_0}L_e + 2\frac{2\pi\nu}{c_0}\Delta n^r l + \text{phase due to mirror refl.} = 2\pi m \tag{4.118}$$

Incidentally, note that Equation 4.118 expresses the second Barkhausen condition (see Equation 2.49). The m-th resonant frequency, ν_m, of the cold resonator is obtained by putting $\Delta n^r = 0$ into Equation 4.118. Thus we have

$$\nu + \frac{l}{L_e}\Delta n^r \nu = \nu_m \tag{4.119}$$

In order to progress beyond this point, it is necessary to make some assumptions about the gain medium. It is well known that for a homogeneously broadened transition (Lorentzian lineshape) with $FWHM = \Gamma$, the real and imaginary part of the refractive index are related by

$$\frac{\Delta n^i}{\Delta n^r} = \frac{1}{2}\frac{\Gamma}{\nu_0 - \nu} \tag{4.120}$$

Further, the gain $G(\nu)$, as expressed by Equation 4.101, is related to the Δn^i by

$$\frac{G(\nu)}{2} = -\frac{2\pi\nu}{c_0}\Delta n^i \tag{4.121}$$

such that

$$\Delta n^r = \frac{\nu - \nu_0}{\Gamma}\frac{c_0}{2\pi\nu}G(\nu) \tag{4.122}$$

which, inserted into Equation 4.119, yields

$$\nu + \frac{c_0}{2\pi}\frac{l}{L_e}\frac{\nu - \nu_0}{\Gamma}G(\nu) = \nu_m \tag{4.123}$$

This equation can be solved for the oscillation frequency $\nu = \nu'_m$ corresponding to each cold-resonator mode ν_m. Under steady-state conditions, i.e., when the gain equals the loss, $G(\nu_m) = \gamma'(\nu_m) = \gamma(\nu_m)/l = (1/lc_0)(L_e/\tau_c) = (2\pi/c_0)(L_e/l)\delta\nu$ (see Equations 4.90 and 3.56), where $\delta\nu$ is the width of the cold resonator mode. Therefore, an approximate analytic solution can be obtained if $\nu = \nu'_m \simeq \nu_m$

$$\nu'_m \simeq \nu_m - (\nu_m - \nu_0)\frac{\delta\nu}{\Gamma} \qquad (4.124)$$

The cold-resonator frequency ν_m is therefore pulled toward the atomic resonance frequency ν_0 by a fraction $\delta\nu/\Gamma$ of its original distance from the central frequency $(\nu_m - \nu_0)$. The sharper the resonator mode, the less significant the pulling effect. This result is also valid for an inhomogeneous line. In general, the oscillation frequency is well approximated by the weighted average of the two frequencies ν_m and ν_0

$$\nu'_m = \frac{\dfrac{\nu_0}{\Gamma} + \dfrac{\nu_m}{\delta\nu}}{\dfrac{1}{\Gamma} + \dfrac{1}{\delta\nu}} \qquad (4.125)$$

of which, as it can be easily seen, Equation 4.124 represents a particular case (in the limit $\delta\nu/\Gamma \ll 1$).

Frequency pulling is, of course, present in masers too. In this case $(\delta\nu/\Gamma \gg 1)$, Equation 4.123 yields

$$\frac{\nu - \nu_0}{-\nu + \nu_m} = \frac{\Gamma}{\delta\nu} \qquad (4.126)$$

which in the limit $\Gamma/\delta\nu \ll 1$ $(\nu \simeq \nu_0)$ provides

$$\nu = \nu_0 + (\nu_m - \nu_0)\frac{\Gamma}{\delta\nu} \qquad (4.127)$$

that returns Equation 4.33.

4.4   Achieving single-mode oscillation

We have already mentioned that the two coupled rate Equations 4.99 derived for a four-level system assume single mode oscillation. Actually, in lasers, the frequency separation of the cavity modes is usually smaller than the width of the gain profile. Just as an example, if we take $L = 1$ m, the frequency difference between two consecutive longitudinal modes is $\Delta\nu = c/2L = 150$ MHz. The laser linewidth, on the other hand, may range from 1 GHz, for a Doppler-broadened transition of a visible or near-infrared gas laser, to 300 GHz or more for a transition of a crystal ion in a solid-state material. This is at the basis of multi-mode oscillation in lasers. In fact, spontaneous photons, which are responsible for triggering the laser action, are distributed among all the resonator modes $1, ..., M$ falling under the lineshape function $g(\nu - \nu_0)$ associated with the active medium. As a consequence, in principle, we should write M rate equations for the quantities $q_1, ...q_M$. If, however, we are only interested in the steady-state laser operation, the problem greatly simplifies.

Let's start by considering the homogeneously broadened case and assume, for simplicity, that one cavity mode, let us say q_1, is coincident with the peak of the gain curve (we further assume that oscillation occurs on the TEM$_{00}$ mode, so that all mode frequencies are separated by $c/2L$). In this case, since all the atoms in the active medium possess the same lineshape $g_L(\nu - \nu_0)$, the condition expressing that the gain must equal the cavity losses can written as

$$N_{th} \frac{\lambda^2}{8\pi n^2 \tau_{spont}} g_L(\nu - \nu_0) = \gamma' \tag{4.128}$$

which clearly shows that oscillation starts in the central mode q_1 when the inversion $N = N_2 - (g_2/g_1)N_1$ reaches the critical value

$$N_{th} = \frac{\gamma' 4\pi^2 n^2 \tau_{spont} \Gamma_L}{\lambda^2} \tag{4.129}$$

where Equation 4.72 has been used. Moreover, it was demonstrated that the steady-state gain coefficient remains clamped at the threshold value G_{th} even when the pumping rate R_p is increased above its threshold value $R_{p,th}$. Therefore, the peak gain, represented by the length OP in Figure 4.13, remains fixed at the value OP_{th} when $R_p \geq R_{p,th}$. Since the line is homogeneously broadened, its shape cannot change, so the whole gain curve must remain the same for $R_p \geq R_{p,th}$. As a consequence, the gains of other modes, represented by lengths $O'P'$, $O''P''$, etc., always remain smaller than the OP_{th} value for the central mode q_1. The latter is thus the only mode that can oscillate in the steady state.

In practice, however, homogeneously broadened lasers do indeed oscillate on multiple modes because the different modes occupy different spatial portions of the active medium. When oscillation on the most central mode is established, the gain coefficient can still exceed the loss coefficient at those locations where the standing-wave electric field of the most central field vanishes. This phenomenon is called spatial hole burning. It allows another mode, whose peak fields are located near the energy nulls of the central mode, the opportunity to lase as well (Figure 4.14).

By contrast, in an inhomogeneously broadened medium, the lineshape function $g_{inhom}(\nu)$, and hence the gain, represents the composite envelope of several Lorentzian lineshape functions $g^\xi(\nu)$ (see Equation 4.74), and hence of several gains, each corresponding to a cluster of atoms (homogeneous packet) centered at ν_ξ. For each of these packets,

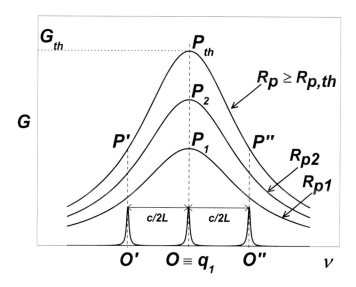

FIGURE 4.13
Frequency dependence of laser gain coefficient versus pump rate R_p for an ideal homogeneously broadened medium.

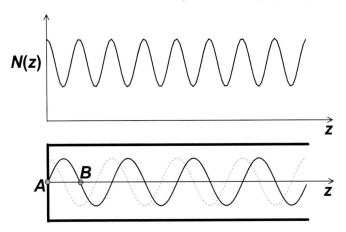

FIGURE 4.14
Spatial hole burning: for simplicity let us consider two modes whose standing wave patterns are shifted by $\lambda/4$ in the active medium. We assume that mode 1 is the first to reach threshold. However, when oscillation on mode 1 starts, inversion around those points where the electric field is zero (points A, B, etc.) is mostly left undepleted, so the inversion can continue growing there even when the laser is above threshold. This situation is clarified in the upper frame, where the spatial distribution of the population inversion in the laser medium is indicated. Accordingly, mode 2, which initially has a lower gain, experiences a growth in gain with increased pump rate, since it uses inversion from regions not depleted by mode 1. Therefore, sufficiently far above threshold, mode 2 can also be set into oscillation; this obviously occurs when its gain becomes equal to its losses.

we can define an its own gain coefficient (independent from that of the other packets) given by

$$G^{\xi}(\nu) = N \frac{\lambda^2}{8\pi n^2 \tau_{spont}} g^{\xi}(\nu) \tag{4.130}$$

Therefore, when R_p (and hence N) increases above $R_{p,th}$, the gain of the central mode remains fixed at OP_{th}; the gains of the other modes $O'P'$, $O''P''$, etc., can, however, keep on increasing to the threshold value (Figure 4.15). In this case, if the laser is operating somewhat above threshold, more than one mode can be expected to oscillate. Accordingly, it is possible to burn holes in the composite gain curve $g_{inhom}(\nu)$ (spectral hole burning).

Then, two additional features are worthy of mention:

- The phenomenon of spatial hole burning does not play a significant role in an inhomogeneous line. In this case, in fact, different modes (with a large enough frequency separation) interact with different sets of atoms, so the hole-burning pattern of one set of atoms is ineffective for the other mode.

- In the case of a homogeneous line, when a few modes are oscillating with frequencies around the center of the gain line, the spatial variation of inversion is essentially smeared out due to the presence of the corresponding, spatially shifted, standing wave patterns of these modes. Therefore, the homogeneous character of the line prevents other modes, further away from the center of the gain line, from oscillating. As a result, compared to an inhomogeneous line, a homogeneous line restricts oscillation to a smaller number of modes centered around the peak of the gain line.

Finally, it should be noted that many types of lasers may reach the oscillation threshold even for several atomic or molecular transitions. In such case, in order to achieve single-mode operation, one has first to select a single transition. Then, several methods for constraining the laser to oscillate in a single transverse and/or longitudinal mode, for either homogeneous or inhomogeneous lines, can be applied.

4.4.1 Line selection

In order to achieve single-line oscillation in laser media that exhibit gain for several transitions, wavelength-selecting elements inside or outside the laser resonator can be used. If the different lines are widely separated in the spectrum, the selective reflectivity of the dielectric mirrors may already be sufficient to select a single transition. In the case of broadband reflectors or closely spaced lines, prisms or gratings are commonly utilized. Left frame in Figure 4.16 illustrates line selection by a prism. The different lines are refracted by the

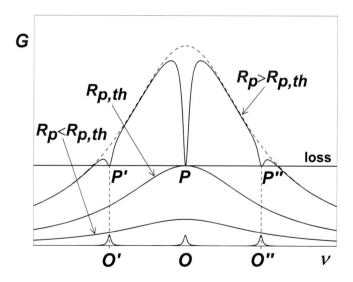

FIGURE 4.15
Frequency dependence of laser gain coefficient versus pump rate R_p for an inhomogeneously broadened medium: phenomenon of spectral hole burning.

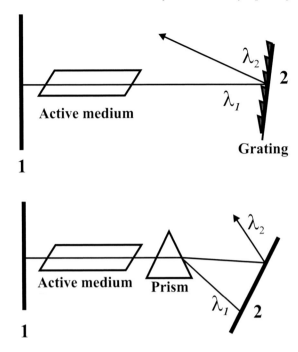

FIGURE 4.16
Line selection in a laser by means of a prism or a grating.

prism, and only the line that is vertically incident upon the end mirror is reflected back into itself and can reach the oscillation threshold, while all other lines are reflected out of the resonator. As most prism materials such as glass or quartz absorb in the infrared region, in this wavelength range it is more convenient to use a Littrow grating (right frame in Figure 4.16). Often the laser beam is expanded by a proper mirror configuration in order to cover a larger number of grating grooves, thus increasing the spectral resolution.

4.4.2   Single-tranverse mode selection

In most applications, oscillation in TEM_{00} mode is desirable and, to achieve this, a diaphragm of suitable aperture size may be inserted at some point on the axis of the resonator. If the radius a of this aperture is sufficiently small, it dictates the value of the Fresnel number of the cavity $a^2/L\lambda$. As a decreases, the difference in loss between the TEM_{00} mode and the next higher order modes (TEM_{01}, or TEM_{10}) increases. Therefore, by choosing an appropriate aperture size, we can enforce oscillation on the TEM_{00} mode. Note that this mode-selection scheme inevitably introduces some loss for the TEM_{00} mode itself.

4.4.3   Single-longitudinal mode selection

A particularly straightforward method consists of employing a laser cavity sufficiently short that the longitudinal mode separation is greater than the width of the gain curve. In this case, if a mode is tuned (for example by mounting one cavity mirror on a PZT) to coincide with the center of the gain curve, the two adjacent longitudinal modes are far enough away from the line center that, for a laser not too far above threshold, they cannot oscillate. The requirement for this mode-selecting scheme can be written as

$$L \leq \frac{c}{\Delta \nu_0} \tag{4.131}$$

where $\Delta \nu_0$ is the total width of the gain curve and L the cavity length. Such a technique works for short helium-neon lasers ($c/L = 1$ GHz for $L = 0.3$ m), where gain line widths are relatively small (a few GHz or smaller), and for some semiconductor lasers ($c/L = 1000$ GHz). For lasers with much larger bandwidths (e.g., dye lasers or tunable solid-state lasers), the cavity length to satisfy the above condition becomes too small to be practical. In this case, and also when longer lengths of the active medium are needed (e.g., for high-power lasers), longitudinal mode selection can be accomplished by means of other techniques.

A common way of achieving single-longitudinal mode oscillation, for both homogeneous or inhomogenous lines, involves inserting one or more Fabry-Perot (FP) etalons within the cavity. As a trivial generalization of what was shown in Chapter 3, the transmission maxima of the etalon occur at frequencies ν_n

$$\nu_n = n \frac{c}{2 n_r d \cos \theta'} \tag{4.132}$$

where θ' is the refraction angle of the beam within the etalon, and n_r and d are the refractive index and the length of the etalon. If d is small enough, the spacing between adjacent resonance frequencies of the etalon will be large compared to the width $\Delta \nu_0$ of the gain profile. By adjusting θ', a resonance frequency can be brought near the center of the gain profile, while the next resonance frequency lies outside the gain profile. In this configuration, it is easy to show that the cavity length must satisfy the condition

$$\frac{c}{\Delta \nu_0} \leq L \leq \frac{c}{\Delta \nu_0} 2F \tag{4.133}$$

where F is the finesse of the etalon. If the cavity length does not satisfy Equation 4.133, then two or more etalons of different thickness are needed. In the case of two etalons, the thicker one is required to discriminate against adjacent longitudinal modes of the cavity; the second, thinner one must then discriminate against adjacent transmission maxima from the first etalon (Figure 4.17).

For a homogeneously broadened transition, single-longitudinal mode operation can automatically be achieved, or at least greatly facilitated, if the laser cavity has the form of a ring and oscillation is constrained to be unidirectional. In this case, in fact, the phenomenon of spatial hole burning within the active medium does not occur, and the laser tends to oscillate on a single mode. Actually, if the transition is only partly homogeneously broadened and the gain profile is very broad, some further bandwidth selecting elements, such as birefringent filters and/or etalons, may also be needed. An additional advantage of this unidirectional ring configuration is that higher output power is available, since the entire active medium, rather than just those regions around the maxima of the standing wave pattern, contributes to laser output. To achieve unidirectional ring operation, a unidirectional device, or optical diode, giving preferential transmission for one direction of beam propagation, must be inserted within the cavity. Figure 4.18 shows a typical example of a folded ring configuration, including a unidirectional device. In this case, pumping is provided by an ion laser and the dye solution is made to flow transversely to the beam in the form of a liquid jet. Single-transverse mode operation is automatically achieved due to the transverse gain distribution from focused pumping. Laser tuning and gain bandwidth reduction are obtained by combining a birefringent filter with two FP etalons (a thin etalon and a scanning etalon of different free spectral ranges). The optical path length of the cavity is

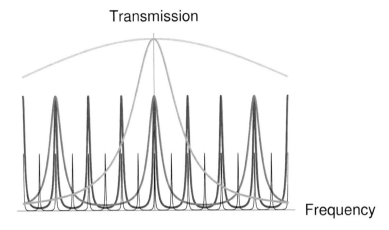

FIGURE 4.17

Transmission curves for different intra-cavity components. For a single-mode operation, the transmission maxima of the different elements have to coincide in one point. In increasing gray-scale order: Mirror&Crystal, Birefringent filter, Thin etalon, Thick etalon, Cavity modes.

conveniently tuned by rotating a tilted, plane-parallel, glass plate inside the resonator (the galvoplate). Single-longitudinal mode operation is then ensured by a unidirectional device consisting of a Faraday rotator and a birefringent plate.

So far, in referring to cavity modes, we have ignored polarization. A very common and convenient way of obtaining a linearly polarized output from a laser is to use various optical elements inclined at Brewster's angle. As an example, Figure 4.19 illustrates a laser in which

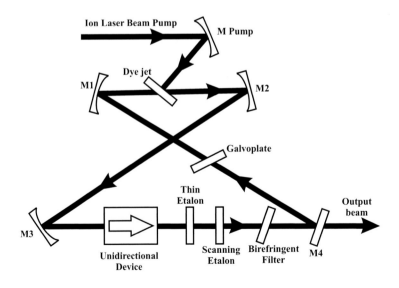

FIGURE 4.18

Typical folded ring configuration: single-longitudinal mode dye laser using a unidirectional ring cavity.

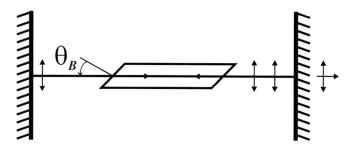

FIGURE 4.19
A laser with the gain-cell windows cut at the Brewster angle. The indicated polarization (parallel to the plane of incidence, which is the plane of the figure) will suffer no reflective loss at the windows, and therefore will lase preferentially. The orthogonal polarization will undergo a greater loss due to reflection at the windows.

the ends of the gain cell are cut at the Brewster angle with respect to the cavity axis. The plane of incidence associated with the cavity field is that of the figure. Laser radiation that is linearly polarized in this plane will not suffer any reflection off the ends of the gain cell. Radiation polarized perpendicular to the plane of the figure, however, will have a greater loss coefficient because it is reflected at the windows. Lasing is therefore more favorable for linear polarization in the plane of incidence.

Another example of unidirectional ring resonator using a non-planar cavity, known as non-planar ring oscillator (NPRO), is shown in Figure 4.20 [173]. In this configuration, the laser light circulates in a single laser crystal, typically made of Nd:YAG or Yb:YAG. The internal surface of the slightly convex, input/output (front) face is provided with a dielectric mirror coating which is highly transmissive for the pump light, while serving as the resonator output-coupler mirror. Thanks to the non-planarity, total internal reflection occurring at the other surfaces induces, in each round-trip, a slight rotation of the light polarization. Moreover, if a small magnet is attached to the laser crystal, an additional polarization rotation can be achieved by exploitation of the Faraday effect. Third, the coating of the output-coupler facet also exhibits a slightly polarization-dependent reflectivity. As a result of the combination of these three effects, one of the two oscillation directions experiences a lower optical loss when the beam hits the output-coupler surface. By contrast, the higher loss experienced by the opposite oscillation direction eventually suppresses the latter. This scheme smartly avoids any standing-wave pattern and hence the spatial hole burning phenomenon. Typically, NPRO cavities exhibit relatively large free spectral range values, allowing for continuous (mode-hope free) frequency tuning over several GHz. Tuning is accomplished either by means of a piezoelectric transducer pressing on the crystal, or by changing the crystal temperature (with a Peltier element), or by adjusting the pump power. By virtue of the high mechanical stability and the low resonator optical losses, in conjunction with the reduced noise levels attainable with pumping diode lasers, the noise of an NPRO-based laser can be very small. Indeed, typical linewidths are in the range of a few kilohertz, while output powers up to several watts can be reached.

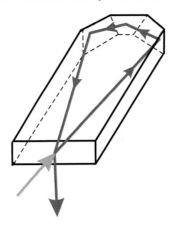

FIGURE 4.20
Setup of a non-planar ring oscillator laser. The front face of the crystal has a dielectric coating, serving as the output coupler and also as a partially polarizing element, facilitating unidirectional oscillation. The light gray beam is the pump beam, normally originating from a laser diode.

4.5 The laser output

The light emitted by a laser has properties that are radically different from those of the light emitted by classical (incoherent) sources. These properties have been the basis for the myriad applications found for lasers since their advent in the 1960s; they have escaped the confines of the research laboratory to become ubiquitous in industrial production and modern society. Lasers now have countless applications in such disparate areas as medicine, metallurgy, and telecommunications and are at the heart of new developments in commercial electronics. Essentially, laser-based applications can be distinguished between those for which the energy delivered by a laser beam is of principal importance and those relying on the unique coherence properties of the light lasers emit. In this book we are interested only in the latter ones. Coherence is one of the most important concepts in optics and is strongly related to the ability of light to exhibit interference effects. A light field is called coherent when there is a fixed phase relationship between the electric field values at different locations or at different times. Partial coherence means that there is some (although not perfect) correlation between phase values.

4.5.1 Spatial coherence

Spatial (or lateral) coherence means a strong correlation (fixed phase relationship) between the electric fields at different locations across the beam profile. Lasers have the potential for generating beams (e.g., Gaussian beams) with very high spatial coherence arising from the existence of resonator modes, which define spatially correlated field patterns. For example, within a cross section of a beam from a laser with diffraction-limited beam quality, the electric fields at different positions oscillate in a totally correlated way, even if the temporal structure is complicated by a superposition of different frequency components. Spatial coherence is the essential prerequisite of the strong directionality of laser beams. In this respect, a laser beam is the closest approximation that we possess to the light ray of geometrical optics. Nevertheless, as already remarked in Chapter 3, resting on the assumption of infinite transverse extent of the cavity (so that the circulating light fields can be represented by plane waves), the simplified description of the laser operation given in previous sections is somewhat unrealistic. Indeed, the various components of a laser cavity are of limited spatial extent and, if the light waves were really plane waves, the diffraction at one of these aperture-limiting components would make impossible reproduction of the form of the wavefront after a complete cavity round trip and would, furthermore, introduce severe losses. We also know that, in practice, diffraction losses are compensated for by the use of focusing elements such as concave mirrors and that the light field inside the cavity (and hence the output beam) is described by a Gaussian transverse profile. Thus, the output laser beam is characterized by the radius w_0 of its waist and the divergence $\Theta = \lambda/\pi w_0$, λ being the laser wavelength. This implies that, in order to reduce the divergence, a shorter wavelength can be employed, or a larger beam diameter can be used by insertion of suitable optics along the path. Conversely, when a laser beam (assumed parallel) with initial waist w_i is focused by a lens of focal length f, the final radius of the spot w_f is given by $w_f = \lambda f/\pi w_i$. By using large numerical aperture lenses for which the induced aberrations have been minimized (for example, microscope objectives) and illuminating their entire aperture by a well-collimated laser beam, it is possible to obtain a focal spot of size of the order of an optical wavelength. It is worth noting, however, that this is a lower limit that can only be achieved for lasers operating on the fundamental TEM_{00} transverse mode. In conclusion, spatial coherence allows the laser light to be focused onto a small area or to be transmitted in a highly parallel beam.

4.5.2 Spectral and temporal coherence

Much more relevant for the subject of this book, the *temporal coherence* of a single-mode laser ensures its *monochromaticity*. Let us start, precisely, by clarifying from the experimental point of view the intimate connection between these two concepts which may seem, at first glance, rather distinct.

The degree of temporal coherence is usually measured by a Michelson interferometer (Figure 4.21). The incident beam is split by a 50:50 beam splitter (BS) into two beams of equal intensity. One of these beams is reflected off mirror M1 and makes its way to BS

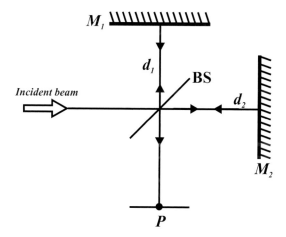

FIGURE 4.21
Basic setup for a Michelson interferometer.

again, where part of it is transmitted. Similarly, the other beam reflects off mirror M2 and propagates back to BS, where part of it is reflected. At a point such as P, there is thus a superposition of two fields. It is well known from general physics that in the case of perfectly monochromatic light and for stationary fields, the total intensity at P is given by

$$\langle I(P) \rangle = \langle I(R) \rangle \cos\left[\frac{2\pi}{\lambda}(d_1 - d_2)\right] \tag{4.134}$$

with constructive interference at P when

$$|d_1 - d_2| = n\lambda \quad n = 0, 1, 2, ... \tag{4.135}$$

and destructive interference when

$$|d_1 - d_2| = (n + \frac{1}{2})\lambda \quad n = 0, 1, 2, ... \tag{4.136}$$

where the angular brackets denote time averaging. Suppose now that the radiation incident upon the Michelson interferometer has a spectral width $\delta\lambda$. Then the total intensity at P is the sum of contributions like Equation 4.134 if we add intensities of different frequency components. Since each wavelength component of the incident radiation is associated with a different pattern of bright and dark spots as $|d_1 - d_2|$ is varied, the pattern will be smeared out if $\delta\lambda$ is large enough. A convenient way to quantitatively characterize the behavior of intensity at point P is to introduce the so-called visibility defined as

$$V = \frac{\langle I(P) \rangle_{max} - \langle I(P) \rangle_{min}}{\langle I(P) \rangle_{max} + \langle I(P) \rangle_{min}} \tag{4.137}$$

where $\langle I(P) \rangle_{max}$ ($\langle I(P) \rangle_{min}$) is the maximum (minimum) intensity recorded at point P as the interferometer path separation is varied. Experimentally, it is found that the visibility decreases with increasing values of the time difference $\tau = 2|d_1 - d_2|/c$ in the arrival at point P (between the wave traveling in the first and second arm). Furthermore, the visibility decreases more rapidly for larger bandwidths of the quasi-monochromatic radiation. In other words, the more nearly monochromatic the radiation, the greater its temporal coherence. To understand this, we will assume for simplicity that the intensity is constant for wavelengths between $\lambda - (1/2)\delta\lambda$ and $\lambda + (1/2)\delta\lambda$, and zero outside this range. Then we expect that the interference pattern is smeared out if $|d_1 - d_2|$ is large enough that the largest wavelength $\lambda + (1/2)\delta\lambda$ corresponds to an intensity maximum, whereas the smallest wavelength $\lambda - (1/2)\delta\lambda$ corresponds to an intensity minimum (or vice versa). From Equations 4.135 and 4.136 we have therefore the two conditions

$$|d_1 - d_2| = n(\lambda + \frac{1}{2}\delta\lambda) \tag{4.138}$$

$$|d_1 - d_2| = (n + \frac{1}{2})(\lambda - \frac{1}{2}\delta\lambda) \tag{4.139}$$

Subtraction of the first of these equations from the second yields

$$|d_1 - d_2| = \left(\frac{1}{\lambda - \frac{1}{2}\delta\lambda} - \frac{1}{\lambda + \frac{1}{2}\delta\lambda}\right) = \frac{1}{2} \tag{4.140}$$

Since $\delta\lambda \ll \lambda$ we then obtain

$$|d_1 - d_2| = \frac{\lambda^2}{2\delta\lambda} \tag{4.141}$$

Since $2|d_1 - d_2|/c = \tau$, where τ is the time difference in the arrival at point P (between the wave traveling in the first and second arm), then we have

$$\tau = \frac{1}{c}\frac{\lambda^2}{\delta\lambda} = \frac{1}{\delta\nu} \tag{4.142}$$

where $c = \lambda\nu$ and $\delta\lambda/\lambda = \delta\nu/\nu$ have also been exploited. Equation 4.142 gives the value of the time difference τ, at which we expect the interference pattern to be smeared out. For separations larger than τ the visibility should be very small or zero. In agreement with experiments, τ decreases with increasing bandwidth $\delta\nu$. It is worth noting that there is some arbitrariness to the boundary between coherence and incoherence. Instead of Equation 4.142 it is conventional to define

$$\tau_{coh} = \frac{1}{2\pi\delta\nu} \tag{4.143}$$

as the coherence time of quasi-monochromatic radiation of bandwidth $\delta\nu$. The distance $c\tau_{coh}$ is called the coherence length. If a beam is divided into two parts, the coherence length is the path difference beyond which there will be very little interference (or fringe visibility) when the two fields are superimposed. Note that the coherence length arises from temporal coherence and is thus unrelated to the coherence area of Section 4.5.1, which is a measure of spatial coherence. A laser operating on a single transverse mode will have excellent spatial coherence, whereas its temporal coherence will be determined by the bandwidth of the output radiation. If it is operated on a single longitudinal mode, $\delta\nu$ is often so small that the coherence length is practically infinite for many purposes. As an example, in the optical region of the spectrum ($\nu \simeq 500$ THz) and for $\delta\nu = 100$ Hz, the fractional spectral purity of a laser beam can be as low as $2 \cdot 10^{-13}$. In this respect, temporal coherence allows the laser energy to be highly concentrated in the frequency domain. A laser operating on more than one longitudinal mode, however, can have a much larger bandwidth, and therefore a much smaller coherence length, than in the single-mode case. Many He-Ne lasers, for instance, operate on two longitudinal modes separated in frequency by $c/2L$ (L being the laser cavity length). In this case $\delta\nu \sim c/2L$. As we will see in Chapter 6, in the case of mode-locked lasers, where many longitudinal modes oscillate in phase, the output is a train of phase-locked pulses and the spectrum is a frequency comb. The coherence length is determined by the duration of the individual pulses, and since they can be extremely short, the coherence length can be very small. For pulses in the femtosecond range, coherence lengths are measured in microns; this makes them useful in optical coherence tomography.

In conclusion, it is worth stressing that the above treatment only refers to the so-called degree of first-order coherence $g^{(1)}$. The degree of second-order coherence $g^{(2)}$, ignored here, is typically used to find the statistical character of intensity fluctuations as well as to differentiate between states of light that require a quantum mechanical description and those for which classical fields are sufficient [174].

4.5.3 The effect of spontaneous emission

Having introduced the concept of spectral coherence from a phenomenological perspective, a fundamental question now arises: what is the spectral width of the output of single-mode lasers? This question was answered by Schawlow and Townes and the fundamental (ultimate) limit to the frequency width of a laser's output is consequently known as the Schawlow-Townes limit. Its cause is spontaneous emission. Whereas stimulated emission adds coherently to the stimulating field, that is, with a definite phase relationship, the spontaneously emitted radiation adds incoherently to the cavity field. Some of the radiation emitted by the spontaneous emission will propagate very nearly along the same direction as that of the stimulated emission and cannot be separated from it. This has two main consequences. First, the laser output will have a finite spectral width. This effect is described in Section 4.5.3.1. Second, the signal-to-noise ratio achievable at the output of laser amplifiers is limited because of the intermingling of spontaneous emission noise power with that of the amplified signal. The issue of amplified spontaneous emission (ASE) and its impact on the signal-to-noise ratio will be addressed in Section 4.5.3.2.

4.5.3.1   Intrinsic laser linewidth

A proper treatment of spontaneous emission requires the quantum theory of radiation. Therefore, the problem of determining the fundamental lower limit to the spectral width of a laser can be solved rigorously only by using quantum electromagnetic theory. However, it is possible to give an argument that leads to the same answer given by the quantum theory of radiation [68]. Let us consider the effect of one spontaneous emission event on the electromagnetic field of a single oscillating laser mode. A field such as

$$E(t) = \Re\left[E_0 e^{i(\omega_0 t + \vartheta)}\right] \tag{4.144}$$

can be represented by a phasor of length E_0 rotating with an angular (radian) rate ω_0. In a frame rotating at ω_0, we would see a constant vector E_0. Since E_0^2 is proportional to the average number of quanta in the mode, $\overline{q}$, we shall represent the laser field phasor before a spontaneous emission event takes place by a phasor of length $\sqrt{\overline{q}}$ as in Figure 4.22.

The spontaneous emission adds one photon to the field, and this is represented, according to our conversion, by an incremental vector of unity length. Since this field increment is not correlated in phase with the original field, the angle ψ is a random variable (i.e., it is distributed uniformly between zero and 2π). The resulting change $\Delta\vartheta$ of the field phase can be approximated for $\overline{q} \gg 1$ by

$$\Delta\vartheta_{one-emission} = \frac{1}{\sqrt{\overline{q}}} \sin\psi \tag{4.145}$$

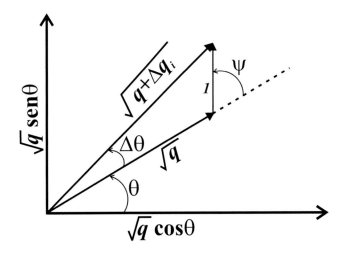

FIGURE 4.22
Effect of the *i*th spontaneous emission event on the electric field of a single oscillating laser mode.

Next, consider the effect of N_{spont} spontaneous emissions on the phase of the laser field. The problem is one of angular random walk, since ψ may assume with equal probability any value between 0 and 2π. We can then write

$$\left\langle [\Delta\vartheta(N)]^2 \right\rangle = \left\langle [\Delta\vartheta_{one-emission}]^2 \right\rangle N = \frac{1}{q} \left\langle \sin^2 \psi \right\rangle N_{spont} \qquad (4.146)$$

where $\langle\rangle$ denotes an ensemble average taken over a very large number of individual emission events. Therefore, to obtain the root mean square (rms) phase deviation in a time t, we need to calculate the average number of spontaneous emission events $N_{spont}(t)$ into a single laser mode in time t. This task is readily accomplished by observing that the total number of spontaneous emissions per second in the active volume V_a within the bandwidth $\Delta\nu$ is $(V_a N_2/\tau_{spont})g(\nu)\Delta\nu$ (recall that N_2 is the total number of atoms per unit volume in the upper laser level, $g(\nu)$ the normalized lineshape function, and τ_{spont} the spontaneous lifetime of an atom in level 2). If N_{mode} is the number of modes within the bandwidth $\Delta\nu$, then the total number of spontaneous emissions per second is

$$N_{spont} = \frac{V_a N_2}{N_{th}} \frac{N_{th}}{N_{mode}\tau_{spont}} g(\nu)\Delta\nu \equiv V_a \mu \frac{N_{th}}{N_{mode}\tau_{spont}} g(\nu)\Delta\nu \qquad (4.147)$$

where $\mu = N_2/N_{th}$ has been introduced and N_{th} is, as usual, the population inversion $N_2 - (g_2/g_1)N_1$ at threshold. According to Equations 4.101 and 4.90, this can be expressed as

$$N_{th} = \frac{8\pi n^2 \tau_{spont}}{g(\nu)} \frac{\gamma'}{\lambda^2} = \frac{8\pi n^2 \tau_{spont}}{g(\nu)\lambda^2} \frac{L_e}{l} \frac{1}{c_0 \tau_c} \qquad (4.148)$$

while the total number of spatial modes (including transverse and longitudinal modes) in the cavity volume V_c is, according to standard theory of blackbody radiation [68], given by

$$N_{mode} = V_c \frac{8\pi n^3}{\lambda^2 c_0} \Delta\nu \qquad (4.149)$$

By substitution of Equations 4.148 and 4.149 into Equation 4.147 we obtain

$$N_{spont} = \frac{V_a}{V_c} \frac{L_e}{nl} \frac{\mu}{\tau_c} = \frac{l[L+(n-1)l]}{Lnl} \frac{\mu}{\tau_c} \simeq \frac{\mu}{\tau_c} \qquad (4.150)$$

It is worth noting that the quantity $l[L+(n-1)l]/(Lnl)$ is exactly equal to 1 either if $n = 1$ (gas laser) or if $L = l$ (the active medium fills the entire cavity). Therefore, the number of spontaneous emissions into a single lasing mode in a time t is given by

$$N_{spont}(t) = \frac{\mu}{\tau_c} t \qquad (4.151)$$

As a consequence, using also $\left\langle \sin^2 \psi \right\rangle = 1/2$, Equation 4.146 becomes

$$\left\langle [\Delta\vartheta(t)]^2 \right\rangle = \frac{1}{2q} \frac{\mu}{\tau_c} t \qquad (4.152)$$

such that, by virtue of Equation 2.157, we can write for the autocorrelation function

$$\begin{aligned}
R_E(\tau) &= \frac{|E_0|^2}{2} \cos(\omega_0\tau)e^{-\frac{\left\langle [\Delta\vartheta(\tau)]^2 \right\rangle}{2}} \\
&= \frac{|E_0|^2}{2} \cos(2\pi\nu_0\tau)e^{-\frac{\mu}{4q\tau_c}\tau}
\end{aligned} \qquad (4.153)$$

from which the frequency-noise power spectral density is calculated as (see Equation 2.162)

$$S_E^{1-sided}(f) = 4 \int_0^\infty R_E(\tau) \cos(2\pi f \tau) d\tau$$

$$= |E_0|^2 \int_0^\infty \left\{ \cos[2\pi(f + \nu_0)\tau] + \cos[2\pi(f - \nu_0)\tau] \right\} e^{-\frac{\mu}{4\bar{q}\tau_c}\tau} d\tau$$

$$\simeq |E_0|^2 \int_0^\infty \cos[2\pi(f - \nu_0)\tau] e^{-\frac{\mu}{4\bar{q}\tau_c}\tau} d\tau \tag{4.154}$$

which, according to Equation 2.167, yields a Lorentzian profile with a FWHM given by

$$\Delta\nu_{ST} = \frac{\mu}{4\pi\bar{q}\tau_c} = \frac{\mu}{2\bar{q}}\delta\nu \tag{4.155}$$

where $\delta\nu = 1/(2\pi\tau_c)$ is, as usual, the width of the cold resonator mode. Finally, according to Equation 4.91, the output power is $P_{out} = h\nu_0\bar{q}(\gamma_2 c_0/2L_e) \simeq h\nu_0\bar{q}/\tau_c = 2\pi h\nu_0\bar{q}\delta\nu$, where the second equality holds if $\gamma_i \simeq \gamma_1 \simeq 0$ (see Equation 4.90). Thus we obtain

$$\Delta\nu_{ST} = \pi\frac{\mu h\nu_0(\delta\nu)^2}{P_{out}} \equiv \pi S_\nu^0 \tag{4.156}$$

which, according to Equations 2.166 and 2.168, corresponds to a white-frequency noise described by the following power spectral density

$$S_\phi^{1-sided}(f) = \frac{S_\nu^0}{f^2} \tag{4.157}$$

Equation 4.156 is the celebrated Schawlow-Townes (ST) formula. Note that in a four-level system $N_1 \simeq 0$ and hence $\mu \simeq 1$. Just as an example, for a typical helium-neon laser ($P_{out} = 1$ mW, $\nu_0 = 4.741 \cdot 10^{14}$ Hz, $\delta\nu \simeq 5 \cdot 10^5$ Hz) one has $\Delta\nu_{ST} \simeq 10^{-3}$ Hz, whilst for a semiconductor laser ($P_{out} = 3$ mW, $\nu_0 = 1.935 \cdot 10^{14}$ Hz, $\delta\nu \simeq 3 \cdot 10^{10}$ Hz) $\Delta\nu_{ST} \simeq 1$ MHz is found. This means that, while in the former case technical broadening is dominant, in the latter case the ST limit is easily accessible to experiments. About semiconductor lasers, the analysis leading to the above ST formula ignores the modulation of the index of refraction of the laser medium, which is due to fluctuations of the electron density caused by spontaneous emission. We will include this effect later.

We close this discussion by observing that, in order to retrieve the ST limit for a maser, we only have to replace the energy in one quantum $h\nu_0$ with the energy due to thermal agitation $k_B T$ and the width of the cold resonator mode with that of the molecular (atomic) emission line. In this way, we exactly find the original result by ST (see Equation 4.32). Indeed, in the case of masers, the atomic response function is narrower than that of the cavity, and $h\nu_0 \ll k_B T$ such that zero-point fluctuations are due to thermal agitation. By contrast, in lasers one has $h\nu_0 \gg k_B T$ and there is essentially no thermal noise. Spontaneous emission is thus responsible for quantum mechanical noise.

Additional considerations on the laser linewidth - It should be pretty clear by now that spectral properties of lasers are conveniently described either in terms of their optical line shape and associated linewidth or in terms of the power spectral density of their frequency noise. A measurement of the laser linewidth (obtained by heterodyning with a reference laser or by self-homodyne/heterodyne interferometry using a long optical delay line) is often sufficient in many applications. Some experiments, though, require more complete knowledge of the Fourier distribution of the laser frequency fluctuations. Knowledge of the frequency-noise power spectral density enables one to retrieve the laser line shape and, thus, the linewidth (while the reverse process, i.e., determining the noise spectral density from the lineshape is not possible). However, this operation, as formalized by Equation 2.163, is most often not straightforward. Here we use the quantity $S_{\delta\nu}^{1\text{-}sided}(f)$ instead of $S_{\phi}^{1\text{-}sided}(f)$. Note however, that, since $\delta\nu \equiv \nu(t) - \nu_0 = (1/2\pi)\dot{\phi}$ (see Equation 2.117) and that the time-domain derivative maps into a multiplication by $i2\pi f$ in the Fourier transform domain, we simply have $S_{\delta\nu}^{1\text{-}sided}(f) = f^2 S_{\phi}^{1\text{-}sided}(f)$. In Chapter 2, few analytically solvable cases were presented. Of course, in the most general case (arbitrary noise spectrum), a numerical computation of Equation 2.163 can be carried out starting from the measured frequency noise power spectral density. Such a numerical analysis reveals that, under some assumptions, the frequency noise spectrum can be separated into two regions that affect the lineshape in a radically different way [175]. In the first region, defined by $S_{\delta\nu}^{1\text{-}sided}(f) > 8\ln(2)f/\pi^2$, the noise contributes to the central part of the lineshape, which is Gaussian, and thus to the laser linewidth. In the second region, defined by $S_{\delta\nu}^{1\text{-}sided} < 8\ln(2)f/\pi^2$, noise contributes mainly to the wings of the lineshape. Therefore, one can obtain a good approximation of the laser linewidth by the following simple expression

$$FWHM = \sqrt{8\ln(2)\,A} \qquad (4.158)$$

where A is the area of the overall surface under the portions of $S_{\delta\nu}(f)$ that exceed the β-separation line

$$A = \int_{1/T_0}^{\infty} H\left(S_{\delta\nu}(f) - 8\ln(2)f/\pi^2\right) S_{\delta\nu}(f)\,\mathrm{d}f \qquad (4.159)$$

where $H(x)$ is the Heaviside unit step function ($H(x) = 1$ if x$\geq$0 and $H(x) = 0$ if x$<$0) and T_0 is the measurement time that prevents the observation of low frequencies below $1/T_0$ (Figure 4.23).

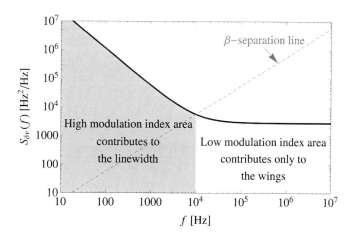

FIGURE 4.23

A typical laser frequency noise spectral density composed of flicker noise at low frequencies and white noise at high frequencies. The dashed line separates the spectrum into two regions whose contributions to the laser line shape is very different: the high modulation index area contributes to the linewidth, whereas the low modulation index area contributes only to the wings of the line shape. (Courtesy of [175].)

As an application of this approach, let us discuss the process of laser linewidth reduction for a simplified laser frequency noise model that still keeps the main features of the problem. In this model, a free-running laser with a constant frequency noise level h_b ($\mathrm{Hz}^2/\mathrm{Hz}$) is considered, and we assume that the frequency noise is reduced to another constant level h_a with a servo loop of bandwidth f_b. The resulting frequency noise power spectral density is given by

$$S_{\delta\nu}(f) = \begin{cases} h_a & \text{if } f < f_b \\ h_b & \text{if } f \geq f_b \end{cases} \tag{4.160}$$

With this model, it is interesting to calculate the evolution of the laser line shape and linewidth with the servo-loop bandwidth. By insertion of Equation 4.160 into Equation 2.163 and subsequent numerical evaluation, the FWHM is determined. Obviously, the result is that the laser linewidth tends toward πh_b when the bandwidth f_b tends toward zero. On the other hand, the linewidth drops down to πh_a when the bandwidth f_b tends toward infinity. In Figure 4.24 we report with a dashed line the linewidth obtained with the approximate formula 4.158, and the agreement with the results of the numerical integration is good, except when the value of the servo bandwidth is between h_a and h_b. In order to understand the origin of this discrepancy, we reported in Figure 4.25 the laser line shape for four particular values of the bandwidth.

We observe that the lineshape changes considerably in this range: the servo loop repels the frequency noise from the center, and, as a consequence, two sidebands appear outside of the servo bandwidth, i.e., at $\delta\nu > f_b$, while the central part strongly narrows and becomes Lorentzian. Because of this radical change of lineshape, the different linewidths at half-maximum are not similar in this range, and comparison with Equation 4.158 loses its significance, which explains the observed discrepancy. Nevertheless, the approximate formula is able to predict the minimum servo-loop bandwidth necessary to efficiently reduce the laser linewidth, which is given by $f_b^{min} = \pi^2 h_b/(8\ln(2))$. It depends on the free-running

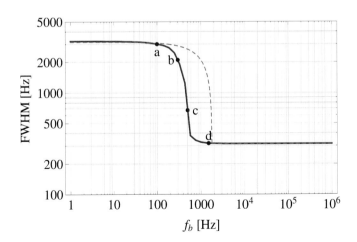

FIGURE 4.24
Evolution of the laser linewidth (FWHM) with the servo-loop bandwidth f_b. Special values are indicated by the following points: a, $f_b = 100$ Hz; b, $f_b = 300$ Hz; c, $f_b = 500$ Hz; and d, $f_b = 1500$ Hz. The continuous line has been obtained by numerical integration of the exact relation (Equation 2.163), while the dashed one results from the approximate formula (Equation 4.159). (Courtesy of [175].)

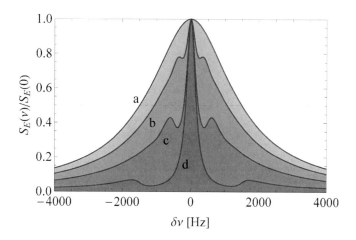

FIGURE 4.25
Evolution of the laser lineshape with the servo-loop bandwidth. (Courtesy of [175].)

laser noise level h_b and corresponds to the situation in which the noise level h_b is entirely be-low the β-separation line for frequencies outside of the servo bandwidth. As a consequence, when $f_b > f_b^{min}$, only the low frequency part with noise level h_a is above the β-separation line and contributes to the laser linewidth, which is given by πh_a. Note that the final laser linewidth depends on the noise level h_a, and thus on the servo-loop gain at low frequency, but is independent of the servo bandwidth, provided that $f_b > f_b^{min}$.

4.5.3.2 Amplified spontaneous emission

As already explained, in addition to the process of stimulated emission, spontaneous emis-sion of photons also occurs in lasers. These spontaneous photons are also amplified by the active medium. Such amplified spontaneous emission (ASE) is a fundamental source of noise. Whereas the amplified signal has a specific frequency, direction, and polarization, the ASE noise is broadband, multidirectional, and unpolarized. As a consequence it is possible to filter out some of this noise by following the amplifier with a narrow bandpass opti-cal filter, a collection aperture, and a polarizer. In this section we will derive the effect of spontaneous emission noise in a traveling-wave optical amplifier in which the gain medium, with no mirrors, is used to amplify a weak input field [68]. The basic problem is to find the degradation of the signal-to-noise power that is caused by the (inevitable) addition of some spontaneous emission (noise) power to the amplified signal.

To evaluate the added noise power from spontaneous emission, we consider an inverted

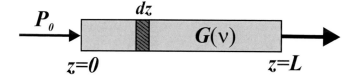

FIGURE 4.26
Schematic drawing of an optical amplifier of length L and cross section A, with a gain coefficient $G(\nu)$.

atomic medium with population densities N_2 and N_1 in the upper and lower transition levels, respectively. Here, of course, we refer to the steady-state values of population densities, but, not to overburden the notation, we drop the subscript th which stands for threshold. The inverted medium occupies the space between $z = 0$ and $z = L$ with a cross sectional area A. An optical beam with power P is propagating through the gain medium. According to Equation 4.101, the coherent amplification due to stimulated emission is given by

$$dP = N\frac{\lambda^2 g(\nu)}{8\pi n^2 \tau_{spont}}Pdz = G(\nu)P \cdot dz \tag{4.161}$$

As photons are emitted via the spontaneous emission process in the medium, they are also being amplified as they propagate. We will next calculate the power of the amplified spontaneous emission at the output of the optical amplifier. Referring to Figure 4.26, we consider a differential volume element Adz at z. The number of excited atoms (in the upper state) inside the volume is $N_2 Adz$. Thus, the total optical power (due to spontaneous emission) from this volume element can be written as

$$dP_N = \frac{N_2 h\nu Adz}{\tau_{spont}} \tag{4.162}$$

Since this power is emitted into many transverse spatial modes, only a small fraction β is emitted into the transverse spatial mode of the input wave. Moreover, only the fraction falling within an optical bandwidth $\Delta\nu$ is of interest to us (physically, $\Delta\nu$ can be the bandwidth of a bandpass filter in front of the detector). Therefore, we have

$$dP_N = \beta\frac{N_2 h\nu A}{\tau_{spont}}g(\nu)\Delta\nu dz \tag{4.163}$$

which, by virtue of Equation 4.161, can be re-arranged as

$$dP_N = \beta\frac{8\pi n^2 A}{\lambda^2}\frac{N_2}{N}G(\nu)h\nu\Delta\nu dz \tag{4.164}$$

Now, within a spectral bandwidth of $\Delta\nu$, the number of longitudinal modes of traveling waves is given by $\Delta\nu/(c_0/nL)$. Thus the fraction of spontaneous emission power going into this group of longitudinal modes all corresponding to a single spatial transverse mode (one direction of propagation, one polarization) is given, according to Equation 4.149, by

$$\beta = \frac{1}{N_{mode}}\frac{\Delta\nu}{(c_0/nL)} = \frac{\lambda^2}{8\pi n^2 A} \tag{4.165}$$

Then, putting $\mu = N_2/N$ we obtain

$$dP_N = \mu G(\nu)h\nu\Delta\nu dz \tag{4.166}$$

Therefore, the total increment of optical power due to stimulated and spontaneous emission into the transverse spatial mode of the input wave is given by (note that, if the two contributions were coherent, we would add their fields)

$$dP = G(\nu)Pdz + \mu G(\nu)h\nu\Delta\nu dz \tag{4.167}$$

which, solved with the boundary condition $P(z = 0) = P_0$, yields

$$P(z = L) = P_0 e^{GL} + \mu h\nu\Delta\nu(e^{GL} - 1) \equiv P_0 G_a + \mu h\nu\Delta\nu(G_a - 1) \tag{4.168}$$

where $G_a = e^{GL}$ is the intensity gain of the amplifier. The ASE power is then recognized as

$$P_{ASE} = \mu h \nu \Delta \nu (G_a - 1) \tag{4.169}$$

The signal-to-noise power ratio at the output of the optical amplifier (in the optical domain) is

$$\left(\frac{S}{N} \right)_{output} = \frac{G_a S_i}{G_a N_i + \mu h \nu \Delta \nu (G_a - 1)} \tag{4.170}$$

where S_i is the input signal power, N_i the noise power of the signal at input, and $N_{amp} \equiv P_{ASE}$ is the noise power added by the amplifier. For a four-level system ($\mu \simeq 1$) with high gain ($G_a \gg 1$) the above equation simplifies to

$$\left(\frac{S}{N} \right)_{output} = \frac{S_i}{N_i + h \nu \Delta \nu} \tag{4.171}$$

which is maximized by reducing the optical bandwidth $\Delta \nu$ as much as possible.

4.6 Laser frequency fluctuations and stabilization techniques

Most often, the Schawlow-Townes limit is masked by the so-called technical noise. In practice, the limit to monochromaticity is set by changes in cavity length induced by vibrations or thermal drifts. We have already seen in Chapter 3 that any alteration in the optical length of the laser cavity causes a corresponding change in the frequency of the selected cavity mode. Thus, for a typical laser cavity length of 1 m, the displacement of a mirror through $\lambda/600$ (1 nm for a laser operating in the yellow region of the visible spectrum) leads to a frequency change of 1 MHz. Such displacements can arise for numerous reasons, for example, thermal expansion of the cavity structure, or changes of pressure (either atmospheric or caused by acoustic waves), which give rise to a change in the refractive index of the air in the laser cavity, as well as pump noise (e.g., injection current noise for diode lasers). These phenomena give rise to a more or less random modulation of the laser frequency, which is known as jitter. Because of this, the short-term frequency stability of a laser, in the absence of active stabilization techniques, is often not better than a few MHz. In the longer term (more than one minute) laser frequency variations arise mainly as a result of slow temperature changes so that the laser frequency is rarely defined to better than the longitudinal mode separation. A good first step, then, would seem to consist in controlling the temperature of the laser cavity. However, this strategy places unrealistic limits on the temperature control for even a modest requirement on the long-term stability of the optical frequency. As an example, if the laser frequency is required to stay within 100 kHz of its desired operating point of 400 THz (750 nm), the requirement on the length L of a laser cavity is

$$\frac{\Delta \nu}{\nu} = \frac{\Delta L}{L} = \alpha T < \frac{10^5 \text{Hz}}{4 \cdot 10^{14} \text{Hz}} = 2.5 \cdot 10^{-10} \tag{4.172}$$

Despite constructing the laser from relatively low expansion materials like Invar and glass, the temperature fluctuations ΔT of the entire cavity would have to be stabilized to less than 0.25 mK, a daunting engineering task, to say the least. While passive stabilization efforts can be used to reduce these harmful effects on the frequency, active feedback control is often necessary to meet the frequency stability required for many applications.

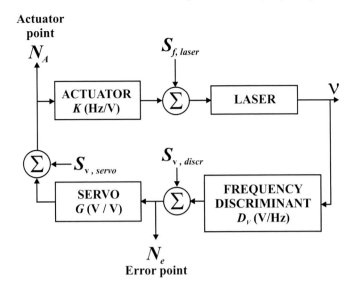

FIGURE 4.27
Laser frequency stabilization cast as a problem in control theory.

A general arrangement for active feedback control of the laser frequency is shown in Figure 4.27 [176].

Schematically: a portion of the light from the laser is sampled by the discriminator, which converts frequency fluctuations into voltage fluctuations with a conversion gain or slope of D_v (V/Hz). A loop filter, having a frequency dependent gain coefficient G (V/V), conditions this error signal for stable, optimal feedback. The control signal from the loop filter is then sent to the transducer (with a conversion gain K (Hz/V)), which makes the correction to the laser frequency. Using active frequency control, the spectral density of laser frequency noise, $S_{f,laser}$, can be suppressed over the bandwidth of the control loop. The closed-loop spectral density of frequency noise $S_{f,cl}$ (Hz/$\sqrt{Hz}$) for the control problem shown in Figure 4.27, where the discriminator noise and servo noise are ignored, is given by

$$S_{f,cl} = \frac{S_{f,laser}}{|1 + KGD_v|} \tag{4.173}$$

This equation represents a somewhat simplified view of active frequency control because it assumes that the closed-loop performance is affected only by the frequency noise of the laser. This is difficult to realize in practice. A more realistic situation is one where each element in the controller is modeled to have a noise contribution that adds to its input. The laser and actuator noise contributions are lumped in Figure 4.27 into one term $S_{f,laser}$ because the noise contribution at the output of the actuator is indistinguishable from the laser noise. $S_{v,discr}$ and $S_{v,servo}$ are the spectral densities of voltage noise associated with the frequency discriminator and servo. The linear spectral density is the square root of the power spectral density and the total noise power is the sum of the individual contributions. Therefore, the total closed-loop linear spectral density of frequency noise is given by

$$S_{f,cl} = \frac{\sqrt{S_{f,laser}^2 + |KS_{v,servo}|^2 + |KGS_{v,discr}|^2}}{|1 + KGD_v|} \tag{4.174}$$

In the limit of very large servo gain G, the discriminator noise contribution dominates all other terms and the minimum closed-loop spectral density of frequency noise is

$$S_{f,cl,min} = \frac{S_{v,discr}}{D_v} \qquad (4.175)$$

This minimum spectral density of frequency noise depends only on the properties of the discriminator D_v, and its noise contribution $S_{v,discr}$. From this result, feedback performance is increased in the following two ways:

- **Increase the discriminator slope**: This is accomplished by using the slope of narrow atomic (molecular) resonances or that provided by the modes of ultrastable (i.e., constructed in such a way as to provide the necessary stability over the time scale of interest) high-finesse optical cavities.

- **Minimize the discriminator noise**: The discriminant noise, however, includes contributions from technical noise associated with the discrimination technique such as fluctuations in the resonant frequency of the Fabry-Perot, $1/f$ noise in the discriminant amplifiers and quantum noise associated with measurement of the laser frequency. For a properly designed frequency controller, the fundamental limit on frequency noise is set by the quantum fluctuations at the discriminant detector (quantum-limited shot noise).

In the remainder of this introduction, we describe the main features of a feedback control system used for laser frequency stabilization. Then, in the following subsections we shall illustrate some specific schemes, each corresponding to a particular way of generating the slope discriminator.

Proportional-Integrative-Derivative control - The reader should note that all aforementioned gains and spectral noise densities are (Fourier) frequency dependent. Each component in the feedback loop has a finite operating bandwidth, and understanding the combined frequency response of these elements is critical to achieving stable feedback. Basically, the requirement for stability restricts the achievable gain in a feedback system. To understand this, imagine that each component in the loop has a frequency noise that is flat until its specified -3 dB bandwidth is reached, after which the response rolls off at -6 dB/octave. This decreasing amplitude response necessarily corresponds to a $-90°$ phase lag (characteristic of an integrator). Since each loop component contributes some phase lag, a certain frequency always exists where the total accumulated phase for the closed loop signal will be $-180°$. At this frequency, the feedback is now positive: the applied correction signal is no longer cancelling the disturbance noise but is instead reinforcing it. If the total loop at this frequency is above unity, the system will become unstable and strongly oscillate. The gain at this frequency needs to be reduced below unity for the feedback to operate in a stable manner. As an alternative to lowering the gain, stable operation can also be achieved by raising the frequency at which the oscillation occurs. A fundamental treatment of control theory is beyond the scope of this book, and only general guidelines will be given here [177, 178].

Consider a system characterized by a single variable S which may vary and drift somewhat over time due to the variation of environmental variables v which we cannot measure or are unaware of. We possess a mechanism for measuring the state of the system as well as a control input u which we can use to modify the state S of the system. Our objective is to set or lock the state of the system to a desired value $S = S_d$ and keep it there without letting it drift or vary over time, regardless of variations in the environmental variables v. In the case of laser frequency stabilization, S is the signal provided by the slope discriminator

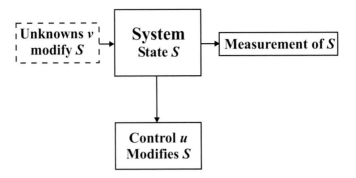

FIGURE 4.28
Generic feedback model.

and the variable v corresponds to the laser frequency that we want to lock to the reference frequency (i.e., the zero of the slope). Thus, in this case we have $S_d = 0$. With reference to Figure 4.28, we will lock the state of the system to $S = S_d$ with the following procedure:

1. Measure the state S of the system.

2. Determine how far the system is from its desired set point by defining an error variable, $e = S - S_d$.

3. Calculate a trial control value $u = u(e)$.

4. Feed the calculated control value, $u(e)$, back into the control input of the system S.

5. The state of the system changes in response to the change in the control value.

6. Return to step.

If we repeat this feedback cycle indefinitely with an appropriately calculated control value $u(e)$, then the system will converge to the state $S = S_d$ and remain there even under the influence of small changes to other variables (i.e., v) which influence the value of the state S. The PID controller is the most common form of feedback. For a PID control, the expression for $u(e)$ depends only on the error signal $e = S - S_d$ and is given by

$$u(e, t) = g_P e(t) + g_I \int_0^t e(t')dt' + g_D \frac{d}{dt}e(t) \tag{4.176}$$

where g_P, g_I, and g_D are, respectively, the proportional, integral, and derivative gains. We now make the simplifying assumption that the control input variable, $u(e)$, controls or modifies the state of the system through the process of addition. If the system has a characteristic response time τ, then the system state variable S evolves according to

$$S(t + \tau) = S(t) + u(e, t) \tag{4.177}$$

For small τ we have

$$S(t + \tau) \simeq S(t) + \tau \frac{d}{dt}S(t) \tag{4.178}$$

whereupon

$$u(e, t) = \tau \frac{d}{dt}S(t) \tag{4.179}$$

By substitution of Equation 4.179 into Equation 4.176, we get (with $S_d = 0$)

$$\tau \frac{d}{dt} S(t) = g_P S + g_I \int_0^t S(t') dt' + g_D \frac{d}{dt} S(t) \tag{4.180}$$

Taking the time derivative of Equation 4.180, we obtain

$$(\tau - g_D) \ddot{S} - g_P \dot{S} - g_I S = 0 \tag{4.181}$$

that is a homogeneous second-order differential equation with constant coefficients. Its solution is of the form

$$S(t) = A_+ e^{\lambda_+ t} + A_- e^{\lambda_+ t} \tag{4.182}$$

where

$$\lambda_\pm = \frac{g_P \pm \sqrt{g_P^2 + 4g_I(\tau - g_D)}}{2(\tau - g_D)} \tag{4.183}$$

and A_+, A_- can be calculated by setting the initial conditions (for instance, $S(t = 0) = S_0$ and $\dot{S}|_{t=0} = 0$). Therefore, one can see that the system will converge to $S_d = 0$ when feedback control is applied, so long as λ_+ and λ_- are negative (negative feedback), otherwise the system will diverge. Since τ is not known accurately a priori, optimizing the gain values for proportional, integral, and derivative values may be done manually or with tuning methods such as the Ziegler-Nichols method [179]. Concerning the practical implementation of PID controllers, operational amplifiers can be used. A complete PID circuit is composed primarily of op-amps which, depending on their placement within a circuit, can subtract, add, invert, amplify, differentiate, integrate, and filter signals. A typical circuit scheme is shown in Figure 4.29. However, as already mentioned, such a realization implies that the three gains g_P, g_I, and g_D are (Fourier) frequency dependent. In this case, it is useful to write the PID regulator in Laplace transform form

$$G(s) = g_P(s) + \frac{g_I(s)}{s} + g_D(s)s \tag{4.184}$$

from which one recognizes that the derivative part mainly takes care of fast peaks in the perturbations, while the primary purpose of the integral part is to provide infinite gain at DC (0 Hz). This ensures that the steady-state value of S is really S_d. Indeed, just due to the actual frequency dependence $g_P(s)$, a pure proportional controller will not settle at its target value, but may retain a steady-state error.

In order to point out additional interesting aspects, we continue the discussion in the Laplace domain (Figure 4.30). The whole system can be described by its complex transfer function $S(s)$ and the servo is represented by $G(s)$. The transfer function, which determines the capability of the system to follow the input signal, is described by

$$H(s) \equiv \frac{Y_{out}(s)}{Y_{in}(s)} = \frac{G(s)S(s)}{1 + G(s)S(s)} \tag{4.185}$$

Another important parameter is the effectiveness of the servo for cancelling perturbations. The transfer function that characterizes this is given by

$$H'(s) \equiv \frac{Y_{out}(s)}{Z(s)} = \frac{S(s)}{1 + G(s)S(s)} \tag{4.186}$$

For laser stabilization this is the interesting term because disturbances have to be completely cancelled in a fast way. We are not interested in optimizing the properties of $H(s)$ as

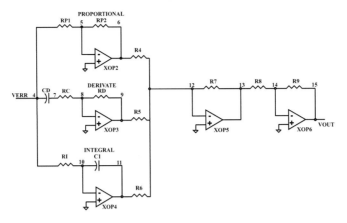

FIGURE 4.29
Simplified scheme of a servo controller for a frequency stabilized laser.

the setpoint is usually kept constant. Obviously, the bigger $G(s)$ is, the better perturbations are cancelled. Now, for the system under control, we use a second-order system with a resonance (this describes quite well the behavior of a piezo attached to a mirror). Its transfer function is given by

$$S(s) = \frac{\omega_0^2}{\omega_0^2 + \dfrac{\omega_0}{Q}s + s^2} \qquad (4.187)$$

where ω_0 is the resonance frequency and Q is the damping factor of the resonance. The points where the phase equals $180°$ and the gain crosses unity gain are of critical importance. Whether a system oscillates can be determined by looking at the open loop gain $G(s)S(s)$. In this respect, the derivative part of the controller adds a leading phase which can compensate the phase lag of the resonance. This controller can be optimized in such a way that it is capable of a fast and smooth settling by increasing the servo bandwidth. The increased bandwidth makes it possible to use higher gain values. Nevertheless, the gain at low frequencies has not increased as much as one would wish. A better approach is to add another PI stage. Figure 4.31 shows the behavior of a PID+PI servo and of a double PI + a

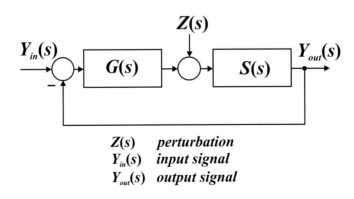

$Z(s)$ *perturbation*
$Y_{in}(s)$ *input signal*
$Y_{out}(s)$ *output signal*

FIGURE 4.30
Basic servo loop with servo $G(s)$ and system $S(s)$.

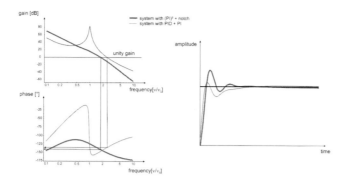

FIGURE 4.31
Left plot: open loop gain of the system with PID + PI servo and system with double PI servo + notch filter. Right plot: reaction of the two systems to a step function at the input. (Courtesy of [177].)

notch filter (band-stop filter). It is apparent that the low frequency gain has been improved by more than 30 dB for the double PI + notch servo. Still, a stable system is maintained as can be seen in the response to a step function. The ringing is only minimal and the response is fast. The PID + PI servo does a comparable job but the gain at low frequencies is roughly 20 dB less. This means that the PI + notch servo can cancel low frequency perturbations better than the PID + PI servo. In practice it turns out that always at least two PI stages and a notch are used to achieve a tight and stable lock.

4.6.1 Side-lock to an atomic/molecular resonance

In this approach, the laser beam is passed through a gas vapor (at relatively low pressure) and the transmission is measured by a photodetector. As the laser frequency is scanned across a given atomic/molecular resonance, the gas absorbs the laser light, and the detected transmission signal exhibits a Doppler-broadened dip. Either side of such transmission feature provides a discriminator slope for locking the laser frequency off resonance. To obtain a steeper discriminator slope, Doppler-free saturation spectroscopy, providing more narrow resonance features, can be performed. A modification of such traditional side-lock technique, exhibiting first-order insensitivity to laser intensity fluctuations, has also been demonstrated [180]. In this approach, the different power dependencies of the Doppler-free saturated-absorption spectrum and the Doppler-broadened background spectrum are exploited. A judicious combination of these components can make the intensity dependences in the error signal cancel and deliver a first-order intensity insensitivity. For further details, refer to the section devoted to Doppler-free saturated-absorption spectroscopy in the next chapter.

Obviously, the side-lock (SL) technique can be applied to a cavity resonance as well. Unlike quantum resonances, an optical cavity does not suffer from saturation effects and its linewidth can be engineered to give much larger discriminator slopes. Although easy to implement, the SL scheme suffers from several drawbacks. First, as it is modulation-free, one necessarily detects the error signal at DC, where there can be significant amplitude noise. Second, amplitude modulation (AM) from the laser directly couples into the error signal; the feedback loop cannot tell the difference between frequency modulation (FM) and AM.

Changes in the laser amplitude will therefore be *written* onto the laser frequency. Also, when a cavity is used, fast frequency fluctuations of the laser will not be detected in transmission through the cavity due to the photon-lifetime of a Fabry-Perot cavity. Moreover, because one locks to the side of the resonance rather than the top, there is reduced build-up of optical power in the cavity, and there may be increased noise on the transmitted intensity. Another limitation is the narrow locking range. A small deviation from the locking point can cause the laser to unlock if the frequency momentarily shifts across the cavity transmission peak.

4.6.2 Pound-Drever-Hall method

All the above complications are circumvented if the Pound-Drever-Hall (PDH) stabilization method is adopted [181, 182]. Named after its inventors and R.V. Pound, who used a corresponding technique in the microwave region, such a scheme hinges on the modulation of the frequency of the laser light. This enables detection of the error signal at a high frequency where the technical noise is close to the shot-noise limit. The resulting demodulated error signal has a high signal-to-noise ratio and a large acquisition range, which can produce robust locks. Furthermore, this error signal has odd symmetry about the line center that enables locking to the top of a cavity fringe.

The idea behind the PDH method is simple. Consider a purely frequency modulated (FM) laser beam impinging on the input mirror of an optical cavity and reflecting back to a detector. For low modulation index, one can view the frequency spectrum of the modulated light as consisting of a carrier with two sidebands: one at higher frequency with a phase relative to the carrier that is in phase with the modulation, and one at lower frequency that is out of phase by $180°$. As long as there is no absorption or phase shift of the laser carrier or modulation sidebands with respect to one another, the detector photocurrent will not have a signal at the modulation frequency. A simple view of this fact is that the beating between the carrier and the upper frequency sidebands creates a photocurrent modulation that is exactly canceled by the out-of-phase modulation from the lower frequency side. If a sideband is attenuated or phase shifted, or the carrier phase is shifted, the photocurrents will not cancel and RF power at the modulation frequency will appear on the detector signal. Near a cavity resonance, the resultant optical reflection of the carrier from the cavity is phase shifted with respect to the sideband components that are further away from the cavity resonance. Consequently, the detector photocurrent will show power at the modulation frequency. The laser frequency noise will then appear as noise sidebands centered around the modulation frequency. When this signal is mixed to base-band (using phase-sensitive detection with the appropriately chosen phase), the result is a frequency discriminator with odd symmetry that may be used to correct the frequency of the laser. Here we note that the light seen by the detector actually consists of two components: the fraction of the input beam that is reflected, plus the fraction of the internal cavity wave that is transmitted back out of the input coupler. The detected photocurrent represents the interference of these two components. For Fourier components of the laser frequency noise below the cavity linewidth, this system acts like a frequency discriminator, as described above. At frequencies above the cavity linewidth, the input field is essentially heterodyned with the cavity wave. Thus, for these Fourier components, the system acts as a phase discriminator, so that the system response to faster frequency fluctuations decreases as $1/f$. In electronic terms, the PDH technique gives us a frequency error signal with a sensitivity that can be measured in volts per hertz of optical frequency. At the Fourier frequency corresponding to the cavity linewidth, the sensitivity starts to decrease, and continues to decrease as $1/f$. At some higher frequency, the error signal will cease to be useful as the magnitude decreases to the

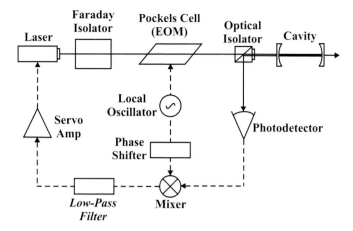

FIGURE 4.32
Basic Pound-Drever-Hall scheme.

level of the background noise, although this limit is usually well above the attainable servo bandwidth.

Figure 4.32 shows a basic setup. The phase of the laser beam is modulated by an electro-optical modulator. The reflected beam is picked off with an optical isolator (a polarizing beamsplitter plus a quarter-wave plate) and sent into a photodetector, whose output is compared with the local oscillator signal via a mixer. A low-pass filter on the output of the mixer isolates the DC (or very low) frequency signal, which then goes through a servo amplifier and into the tuning port on the laser, locking the laser to the cavity. The phase shifter is not essential in an ideal system, but is useful in practice to compensate for unequal delays in the two signal paths.

After the beam has passed through the Pockels cell (electro-optic modulator), its electric field becomes

$$E_{inc} = E_0 e^{i(\omega t + \beta \sin \Omega t)} \tag{4.188}$$

where Ω is the modulation frequency and β is the modulation depth. We can expand this expression, using Bessel functions, to

$$e^{i\beta \sin \Omega t} = \sum_{n=-\infty}^{+\infty} J_n(\beta) e^{in\Omega t}$$

$$= J_0(\beta) + \sum_{n=1}^{+\infty} \left[J_n(\beta) e^{in\Omega t} + (-1)^n J_n(\beta) e^{-in\Omega t} \right]$$

$$\simeq J_0(\beta) + J_1(\beta) e^{i\Omega t} - J_1(\beta) e^{-i\Omega t} \tag{4.189}$$

where the last equality holds in the limit $\beta \ll 1$ and $J_{-n}(\beta) = (-1)^n J_n(\beta)$ has been exploited ($J_0(\beta) \simeq 1$ and $J_1(\beta) \simeq \beta/2$). Therefore, the total reflected beam is given by

$$E_{ref}/E_0 = F(\omega) J_0(\beta) e^{i\omega t} + F(\omega + \Omega) J_1(\beta) e^{i(\omega+\Omega)t}$$

$$- F(\omega - \Omega) J_1(\beta) e^{i(\omega-\Omega)t} \tag{4.190}$$

where $F(\omega)$ is the cavity reflection coefficient at frequency ω. Then, the power in the reflected beam is

$$
\begin{aligned}
P_{ref} = |E_{ref}|^2 = P_c|F(\omega)|^2 + P_s \left\{ |F(\omega+\Omega)|^2 + |F(\omega-\Omega)|^2 \right\} \\
+2\sqrt{P_cP_s}\left\{ \mathrm{Re}[F(\omega)F^*(\omega+\Omega) - F^*(\omega)F(\omega-\Omega)]\cos\Omega t \right. \\
\left. +\mathrm{Im}[F(\omega)F^*(\omega+\Omega) - F^*(\omega)F(\omega-\Omega)]\sin\Omega t \right\} + (2\Omega \quad \text{terms})
\end{aligned} \tag{4.191}
$$

where $P_c = J_0^2(\beta)P_0$, $P_s = J_1^2(\beta)P_0$, $P_c + 2P_s \simeq P_0$, and $P_0 = |E_0|^2$. Now, in order to find a convenient expression for $F(\omega)$, let us recall Equation 3.37. For a symmetric cavity ($r_1 = r_2 = \sqrt{\mathcal{R}}e^{i\phi_r}$, $\mathcal{T}_1 = \mathcal{T}_2 = \mathcal{T}$) with $\phi_r = \phi_{r1} = \phi_{r2} = \pi$ and with no losses ($\mathcal{R}+\mathcal{T}=1$), it simplifies to

$$
\begin{aligned}
F(\omega) = \frac{E_r}{E_0} = \sqrt{\mathcal{R}}\frac{1 - e^{\frac{2\pi i\nu}{FSR}}}{1 - \mathcal{R}e^{\frac{2\pi i\nu}{FSR}}} = \sqrt{\mathcal{R}}\frac{1 - e^{\frac{2\pi i(\nu_m+f)}{FSR}}}{1 - \mathcal{R}e^{\frac{2\pi i(\nu_m+f)}{FSR}}} \\
= \sqrt{\mathcal{R}}\frac{1 - e^{\frac{2\pi if}{FSR}}}{1 - \mathcal{R}e^{\frac{2\pi if}{FSR}}}
\end{aligned} \tag{4.192}
$$

where f is the distance of the laser frequency from the resonance ν_m. So, if $(2\pi f)/FSR \ll 1$ and $\mathcal{F} \simeq \pi\mathcal{R}/(1-\mathcal{R})$ (valid for high finesse) we have

$$
F(\omega) \equiv F(f) = \frac{1}{\sqrt{\mathcal{R}}}\frac{f\left(f - i\frac{\delta\nu}{2}\right)}{\frac{\delta\nu^2}{4} + f^2} \simeq \frac{f\left(f - i\frac{\delta\nu}{2}\right)}{\frac{\delta\nu^2}{4} + f^2} \equiv \frac{x(x - i/2)}{x^2 + 1/4} \tag{4.193}
$$

where $\delta\nu = FSR/\mathcal{F}$ and the variable $x \equiv f/\delta\nu$ has been defined.

Now, we feed one input of the mixer with P_{ref} and the other with a modulation signal at Ω' and recall that the mixer forms the products of its inputs. First, let us consider the effect on the sine term. In this case we have

$$
\sin\Omega t \sin\Omega' t = \frac{1}{2}\left\{ \cos\left[(\Omega-\Omega')t\right] - \cos\left[(\Omega+\Omega')t\right] \right\} \tag{4.194}
$$

that for $\Omega = \Omega'$ simplifies to a dc term plus a 2Ω term. In a similar way, the 2Ω terms give rise to Ω and 3Ω terms, the first term gives rise to a Ω term, and the cosine term to a 2Ω term. Then, if the mixer output is followed by a low-pass filter, we have a PDH error signal of the form

$$
\begin{aligned}
\epsilon_{\text{after mixer}} = K\sqrt{P_cP_s} \cdot \mathrm{Im}[F(\omega)F^*(\omega+\Omega) - F^*(\omega)F(\omega-\Omega)] \\
\equiv K\sqrt{P_cP_s} \cdot \mathrm{Im}[F(x)F^*(x+x_\Omega) - F^*(x)F(x-x_\Omega)]
\end{aligned} \tag{4.195}
$$

where K is a proportionality constant. Figure 4.33 shows a plot of such error signal with $F(f)$ given by Equation 4.193.

According to Equations 4.193, close to the resonance ($f \to 0$) we can write $|F(\omega)|^2 \simeq 0$, $|F(\omega \pm \Omega)|^2 \simeq 1$, and $[F(\omega)F^*(\omega+\Omega) - F^*(\omega)F(\omega-\Omega)] \simeq -2i\mathrm{Im}[F(\omega)] \simeq -4(f/\delta\nu)$, whereupon Equation 4.191 becomes

$$
P_{ref} \simeq 2P_s - 8\sqrt{P_cP_s}\frac{f}{\delta\nu}\sin\Omega t + (2\Omega \quad \text{terms}) \tag{4.196}
$$

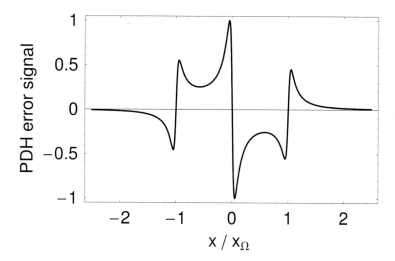

FIGURE 4.33
Plot of Equation 4.195 (normalized to $K\sqrt{P_c P_s}$) against x/x_Ω.

from which one identifies the error signal before the mixer as

$$\epsilon = -8\sqrt{P_c P_s}\frac{f}{\delta\nu} = -8\sqrt{P_c P_s}\frac{\mathcal{F}}{FSR}f = -16\sqrt{P_c P_s}\frac{L\mathcal{F}}{\lambda}\frac{f}{\nu} \tag{4.197}$$

where L is the cavity length and $\lambda = c/\nu$ is the laser wavelength. The slope D of the error signal is thus proportional to

$$D = -8\sqrt{P_c P_s}\frac{\mathcal{F}}{FSR} \tag{4.198}$$

which can be maximized by increasing the ratio $\mathcal{F}/FSR$. It is sometimes useful to maximize D with respect to the power in the sidebands, too. The quantity D has a very simple form when $P_c + 2P_s \simeq P_0$, i.e., when negligible power goes into the higher order sidebands,

$$D \propto \sqrt{P_c P_s} \simeq \sqrt{\frac{P_0}{2}}\sqrt{\left(1 - \frac{P_c}{P_0}\right)\frac{P_c}{P_0}} \tag{4.199}$$

which is maximum at $P_c/P_0 = 1/2$ or at $P_s/P_c = 1/2$. In other words, D is maximized when the power in each sideband is half the power in the carrier.

So far, we have only talked about laser frequency, but it is a straightforward exercise to extend this analysis in terms of both frequency and cavity length

$$\epsilon = -16\sqrt{P_c P_s}\frac{L\mathcal{F}}{\lambda}\left\{\frac{f}{\nu} + \frac{\delta L}{L}\right\} \tag{4.200}$$

where δL is the deviation of the cavity length from resonance, analogous to f. Note that it is not possible to distinguish laser frequency noise from cavity noise just by looking at the error signal. Finally, let us derive an expression for the shot-noise-limited resolution. Any noise in the error signal itself is indistinguishable from noise in the laser's frequency. There is a fundamental limit to how *quiet* the error signal can be, due to the quantum nature of light. On resonance, the reflected carrier will vanish, and only the sidebands will reflect off the cavity and fall on the photodetector. The average power falling on the photodiode is approximately $P_{ref} = 2P_s$. The shot noise in this signal has a flat spectrum with spectral density of

$$S_e = \sqrt{2(h\nu)(2P_s)} = 2\sqrt{\frac{hc}{\lambda}P_s} \tag{4.201}$$

Dividing the error signal spectrum by Equation 4.198 gives us the apparent frequency noise

$$S_f = \frac{S_e}{D} = \frac{1}{8}\frac{1}{\mathcal{F}L}\sqrt{\frac{hc^3}{\lambda P_c}} = \frac{\delta\nu}{4}\sqrt{\frac{h\nu}{P_c}} \tag{4.202}$$

Since you can't resolve the frequency any better than this, you can not get it any more stable than this by using feedback to control the laser. Note that the shot noise limit does not explicitly depend on the power in the sidebands, as you might expect. It only depends on the power in the carrier. The shot noise limit does depend implicitly on the power in the sidebands, since $P_c = P_0 - 2P_s$, but this is a relatively minor effect. It is worth putting in some numbers to get a feel for these limits. For this example we will use a cavity that is 20 cm long and has a finesse of 10^4, and a laser that operates at 500 mW with a wavelength of 1064 nm. If the cavity had no length noise and we locked the laser to it, the best frequency stability we could get would be

$$S_f \simeq 1.2 \cdot 10^{-5} \frac{\text{Hz}}{\sqrt{\text{Hz}}} \tag{4.203}$$

Incidentally we note that, by squaring Equation 4.202 and using Equation 4.156 (with μ=1), we obtain the minimum laser linewidth

$$\Delta\nu_{PDH} = \pi\frac{(\delta\nu)^2}{16}\frac{h\nu}{P_c} = \Delta\nu_{ST}\frac{P_{out}}{16P_c} \tag{4.204}$$

This means that, in principle, the closed-loop spectral density can be suppressed to a level below the ST limit within the bandwidth of the loop. In practice, it is difficult to reach this level of stability due to amplifier noise in the discriminator, non-ideal mode-matching into the interferometer, and residual AM noise in the phase modulators (see below).

The same shot noise would limit your sensitivity to cavity length if you were locking the cavity to the laser. In this case, the apparent length noise would be

$$S_L = \frac{L}{\nu}S_f = \frac{1}{8}\frac{1}{\mathcal{F}}\sqrt{\frac{hc\lambda}{P_c}} \tag{4.205}$$

For the cavity and the laser we used in the example above, this would be

$$S_L = 8.1 \cdot 10^{-21} \frac{\text{m}}{\sqrt{\text{Hz}}} \tag{4.206}$$

In conclusion, it is useful to summarize the advantages inherent to the PDH technique:

- The modulation frequency can be chosen high enough for the technical noise of the laser to be no longer relevant;

- Since we are locking on resonance, where the reflected carrier vanishes, none of the following sources contribute (to first order) to the error signal: variation in the laser power; response of the photodetector used to measure the reflected signal; the modulation depth; the frequency modulation Ω;

- The reference cavity can serve as a very efficient mode-cleaner for the laser beam if the PDH technique is used to lock the cavity resonance frequency to the laser frequency [183].

It is also worth pointing out the following possible deception. As explained, in principle, the error signal of a purely phase modulated laser beam is zero when the laser is in resonance with the cavity mode. The carrier is completely transmitted and the two sidebands are reflected with a 180° phase and thus exactly cancel each other. Every perturbation of this symmetry causes an additional offset so that the zero crossing is off the resonance. This is even worse if the perturbations are time dependent. Possible sources of perturbations creating an offset are the following:

- The polarization of the light does not match the crystal axis of the EOM. In this case, the polarization is rotated with the modulation frequency of the EOM. The subsequent polarizing beam splitter produces an amplitude modulation of the light and thus prevents a perfect cancellation of the sidebands at the photodiode. This produces a DC-offset after the mixer. Thus the polarization of the light has to be carefully aligned with the crystal axes;

- It has been observed that an EOM is capable of producing residual amplitude modulation (RAM). If the crystal is cut at Brewster's angle, internal backscattering is prevented and the residual modulation is minimized;

- Another problem is crosstalk between the signal modulating the EOM and the photodiode detecting the error signal. This again generates a DC signal after the mixer. Shielding of the photodiode and reducing the stray fields of the RF helps to get rid of this problem.

Finally it is worth remarking that, due to the limited bandwidth of electronic servo loops, frequency fluctuations at high Fourier frequency cannot be sufficiently suppressed by the PDH technique. When needed, such high frequency noise can be filtered passively by the narrow transition modes of an optical resonator. To avoid the complexity of an independent second filter cavity, the reference resonator itself (which is already kept in resonance by the servo loop) can be used for filtering. The photon lifetime in the high-finesse cavity and thus the averaging time of the intracavity field is long enough that high frequency laser fluctuations are efficiently filtered in the transmitted light. Thus, the cavity acts as a low-pass filter for the laser noise S_ν^0 which is suppressed in the transmitted light to S_ν^c as

$$S_\nu^c(f) = \frac{S_\nu^0(f)}{1 + \left(\dfrac{f}{\Delta f_{HFHM}}\right)^2} \tag{4.207}$$

where Δf_{HFHM} is the width of the cavity resonance [184]. A typical setup for filtering the laser light is shown in Figure 4.34.

4.6.2.1 10^{-16}-level laser frequency stabilization

As discussed in Chapter 3, state-of-the-art for stabilization techniques usually involves phase-locking a laser source (with electronic feedback) to a single longitudinal and/or transverse mode of a passive, ultrastable Fabry-Perot (FP) cavity. The FP cavity consists of very high-reflectivity mirrors, optically contacted onto a rigid spacer. In the limit of good signal-to-noise ratio and tight phase lock, the length stability of the FP cavity gives the frequency stability of the resulting optical wave. A fundamental limit that many cavity-stabilized lasers have reached is given by Brownian thermal mechanical fluctuations of the FP cavity.

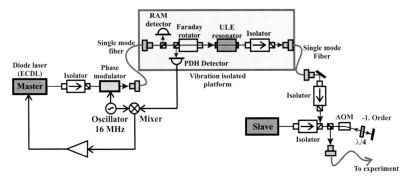

FIGURE 4.34
Setup of the filtered laser system with extended cavity diode laser (ECDL), Pound-Drever-Hall (PDH) lock to the high finesse ultralow expansion glass (ULE) resonator, and additional detector for residual amplitude modulation (RAM). The filtered light is amplified by an injection-locked slave laser diode. The acousto-optic modulator (AOM) in double-pass with quarter wave retardation plate is used to tune the laser frequency. (Adapted from [184].)

As derived in Chapter 3, the fractional frequency instability limit from thermal noise is typically dominated by the two cavity mirrors

$$\sigma_{therm} = \sqrt{\ln 2 \frac{8k_B T}{\pi^{3/2}} \frac{1-\sigma^2}{E w_0 L^2} \left(\Phi_{sub} + \Phi_{coat} \frac{2}{\sqrt{\pi}} \frac{1-2\sigma}{1-\sigma} \frac{d}{w_0} \right)} \qquad (4.208)$$

Here, σ, E, and Φ_{sub} are Poisson's ratio, Young's modulus, and the mechanical loss for the mirror substrate, w_0 is the laser beam ($1/e$ field) radius on the mirror, T is the mirror temperature (K), k_B is Boltzmann's, constant and L is the cavity length. Φ_{coat} and d denote the mechanical loss and thickness of the thin-film reflective coating. The first term in parentheses is the mirror substrate contribution and the second term is the contribution from the coating. High-stability FP cavities are typically made from ultralow expansion (ULE) glass to reduce cavity length changes due to temperature drift around room temperature. Cavity lengths are often 10-20 cm. Under such conditions, the lowest thermal noise instability is typically $3 \cdot 10^{-16}$ to $1 \cdot 10^{-15}$, roughly consistent with the best experimentally observed instability. To reduce thermal noise, the choice of mirror substrate material (E and Φ_{sub}), beam radius (w_0), cavity length (L), and cavity temperature (T) can be modified. Each modification presents different technical challenges. In a very recent experiment, a long cavity featuring a larger beam size and alternative mirror material was fabricated to realize an ultrastable optical atomic clock [185]. Such cavity was composed of a rigid ULE spacer with optically bonded fused-silica mirror substrates. Since the value of Φ_{sub} for fused-silica mirror substrates is more than ten times smaller than that for ULE, the substrate thermal noise term shrinks below that of the thin-film reflective coating. Dominated by the thin-film coating, σ_{therm} could be further reduced by choosing a long cavity ($L = 29$ cm) with a longer radius of curvature ($R = 1$ m), allowing for a larger beam radius w_0. With these cavity parameters, the thermal-noise-limited fractional frequency instability was reduced to $1.4 \cdot 10^{-16}$.

For characterization purposes, two similar cavity systems were constructed (Figure 4.35). Then, several milliwatts of laser light at 578 nm were divided into multiple paths, one to each of the two cavities and one to the ytterbium optical lattice apparatus (see Chapter 7). Each cavity was enclosed in a vacuum chamber, which was single-stage temperature

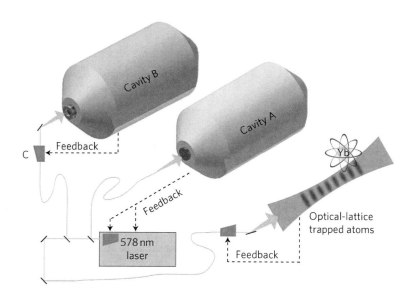

FIGURE 4.35
Laser light at 578 nm is incident on two independent, isolated optical cavities. Each cavity is composed of a rigid ULE spacer with optically bonded fused-silica mirror substrates. Feedback for laser frequency control is usually applied to acousto-optic modulators. The stabilized light probes the narrow clock transition in an ultracold sample of ytterbium, confined in a one-dimensional optical lattice. (Courtesy of [185].)

controlled (fluctuations over 24 h at a few millikelvin). Length changes due to acceleration-induced deformation were strongly suppressed by implementing a cavity mounting design, similar to those described in Chapter 3 [121, 118]. In this way, the measured acceleration sensitivity (along gravity) was as low as $1 \cdot 10^{-11}/\mathrm{ms}^{-2}$. The vacuum chamber and optics coupling light to the cavity sat on vibration isolators (one cavity on an active system, the other passive). Each system was located in different parts of the laboratory in independently closed acoustic-shielded chambers. Optical links between the isolated systems and the laser source were made with optical fibers using active phase stabilization (discussed later in this chapter). Free-space optical paths were generally in closed boxes to reduce air currents. The free-running 578 nm laser light was locked to Cavity A using PDH stabilization, using fast electronic feedback to an AOM common to all optical paths, and slow electronic feedback to a piezoelectric transducer on the laser source. Thus, light incident on both cavities was phase-stabilized to Cavity A. To measure the laser frequency noise spectrum, an additional AOM was used to tune laser light incident on Cavity B into resonance, and the PDH signal of Cavity B served as a frequency discriminator. To measure laser frequency stability, the PDH signal from Cavity B was filtered and fed to an AOM to lock the laser frequency of the second beam to the resonance of this cavity. This AOM frequency thus provided the difference between the two cavities, and was counted to determine the frequency stability. The frequency noise spectrum of one cavity-stabilized laser is shown in Figure 4.36. The noise spectrum approaches the projected thermal noise for Fourier frequencies around 1 Hz; at higher frequencies, the spectrum is approximately white, with several spikes attributed to seismic noise on one of the cavities. In the same figure, the fractional frequency instability, together with σ_{therm}, is also shown. During typical best performance, for averaging times below 10 s an instability as low as $2 \cdot 10^{-16}$ was observed. For averaging times > 10 s,

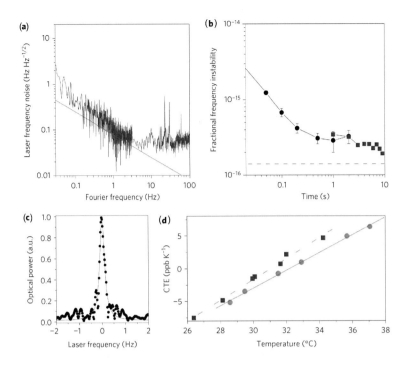

FIGURE 4.36
(a) Frequency noise spectrum (data) and theoretical estimate of the thermal noise (line).
(b) Fractional frequency instability of one cavity. Squares and circles refer to data derived
from two different frequency counters. The dashed line denotes the Brownian thermal-
noise-limited instability. (c) Laser power spectrum (dots) and Lorentzian fit (trace) with
FWHM linewidth of 250 mHz (resolution bandwidth, 85 mHz). (d) Measured CTE_{cav} versus
temperature for Cavity A (dots with solid linear fit) and Cavity B (squares with dashed
linear fit). (Courtesy of [185].)

laser instability typically increased. The measured laser power spectrum with a linewidth
of 250 mHz is displayed too. Moreover, to minimize thermal drift of the cavity resonance,
the cavity coefficient of thermal expansion (CTE) was engineered to cross zero near room
temperature.

   In light of the above discussion, to further improve the laser frequency stability, the
ultimate limit set by thermodynamic fluctuations in the mirror substrates and coatings
must be overcome. In this sense, several proposals are discussed in the literature, including
cryogenic operation of reference cavities [186, 187, 188, 189]. In an alternative method, the
optical reference cavity is replaced by a long ($\sim$ km) all-fiber delay line yielding a simplified
setup at the expense of slightly reduced stability. A different approach consists of using high
Q whispering gallery mode (WGM) resonators. These two latter schemes are discussed in
the following subsections.

4.6.2.2   WGM resonators for laser frequency stabilization

In this approach, the reference for the laser's frequency is provided by fabricating whispering
gallery mode (WGM) resonators from a single crystal of magnesium fluoride [190]. In that
work, the WGM resided in a protrusion of a MgF_2 cylinder, whose compact dimensions (≤ 1

cm^3) and monolithic nature inherently reduced the resonator's sensitivity to vibrations. In principle, it also enables operation in more noisy and/or space-constrained environments, such as a cryostat or a satellite. Furthermore, in contrast to the highly wavelength-selective and complex multilayer coatings required for mirror-based resonators, WGM resonators are intrinsically broadband, limited only by optical absorption in the host material. WGM resonators were fabricated combining a shaping and several polishing steps on a home-built precision lathe. Several MgF$_2$ resonators were produced with a typical radius of 2 mm, the WGMs being located in the rim of the structure. The achieved surface smoothness, together with very low absorption losses in the ultrapure crystalline material (Corning), enabled quality factors in excess of $2 \cdot 10^9$. A high-index prism was used to couple the beam of an external cavity diode laser into the WGM. For the laser stabilization experiments, a MgF$_2$ resonator was mounted into a prism-coupling setup shielded against vibrations and thermal fluctuations (Figure 4.37).

Then, the beam of a commercial Littman-type extended cavity diode laser was focused on the face of the coupling prism through anti-reflection coated windows in the vacuum chamber and small (diameter 1 cm) bore holes in the aluminum shields. The laser was locked to a high-Q WGM using the PDH method, implemented with an external electro-optic phase modulator driven at 11.4 MHz. The obtained error signal was fed back via a two-branch control system actuating both the grating tilt in the laser (via a piezoelectric transducer) and the diode pump current. To assess the frequency fluctuations of the laser locked to the WGM resonator, its frequency was compared to that of an independent diode laser locked to an ultrastable mirror-based resonator in the same laboratory (Figure 4.38).

The extraordinary stability of the latter afforded a direct measurement of the WGM-stabilized laser by analyzing the spectral properties of the radio-frequency (rf) beat generated between the two lasers in a heterodyne detector. Using a rf spectrum analyzer, the width of the beat signal at 586 MHz could be fitted to a Lorentzian of 290 Hz linewidth. For a more systematic characterization, the Allan deviation $\sigma_y(\tau)$ of the stabilized laser's frequency as a function of gate time τ was also determined. The results of these measurements are summarized in Figure 4.39. A minimum Allan deviation of 20 Hz at an optical

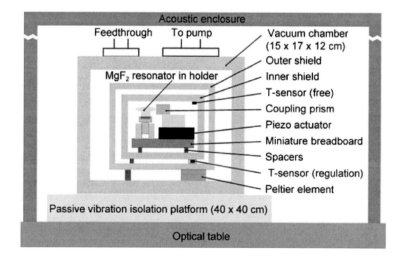

FIGURE 4.37
Thermal and acoustic isolation of MgF$_2$ cavity and coupling setup (not to scale). (Courtesy of [190].)

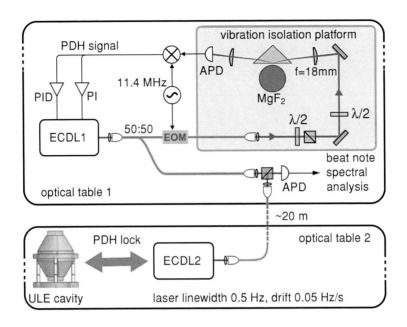

FIGURE 4.38

Laser stabilization to a MgF_2 resonator using the PDH method, and comparison to an ultrastable laser locked to a mirror-based cavity on another optical table. APD: avalanche photodiode; ECDL: external cavity diode laser; EOM: electro-optic modulator; $\lambda/2$: half-wavelength retardation plate; PI(D): proportional-integral-(differential) feedback controller; ULE: ultralow expansion glass. (Courtesy of [190].)

wavelength of 972 nm, corresponding to a relative Allan deviation of $6 \cdot 10^{-14}$, was found at an integration time of 100 ms (this level of instability is compatible with the limits imposed by fundamental fluctuations of the material's refractive index at room temperature).

In conclusion, while the stability does not yet reach the level achieved by optical references based on two-mirror resonators, WGM resonators could be used to significantly reduce the frequency fluctuations of diode lasers or free-running frequency combs in simple, compact setups, or be employed to transfer a first laser's stability to other lasers across a wide wavelength region. Moreover, the performance is expected to improve if the strong temperature sensitivity can be reduced or eliminated. For example, the operation close to temperatures ($\sim 200°C$) at which the thermo-refractive coefficients α_2 of MgF_2 vanish is practically feasible. On the other hand, self-referenced temperature stabilization could dramatically improve temperature stability. Finally, the inherent compatibility of these resonators with cryogenic operation opens a promising approach not only towards an improved temperature stability and reduced sensitivity to temperature fluctuations, but also in view of a strong suppression of thermodynamic fluctuations limiting also today's best optical flywheels.

More recently, with a similar setup, the same group exploited the strong thermal dependence of the difference frequency between two orthogonally polarized TE and TM modes of the optically anisotropic MgF_2 crystal to derive a dual-mode feedback signal. This was used as feedback for self-referenced temperature stabilization to nanokelvin precision, resulting in frequency stability of 0.3 MHz/h at 972 nm, as measured by comparing with an independent ultrastable laser [191].

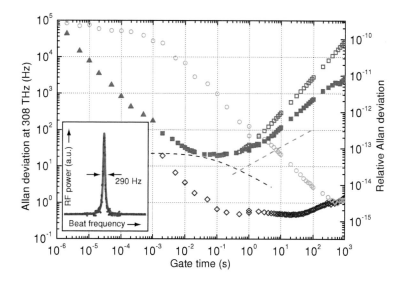

FIGURE 4.39 (SEE COLOR INSERT)
Allan deviation of the optical beat note in Hz, and normalized to the optical carrier at 308 THz. Inset shows a beat note measured at 586 MHz with a spectrum analyzer and a Lorentzian fit. Diamonds are reference measurement of beat-note stability between two lasers locked to two mirror-based resonators. Red symbols are Allan deviation of beat frequency between one laser locked to a mirror-based resonator and the other one to the WGM resonator, as obtained with two different counters (triangles and squares). Full (open) symbols show data after (before) removal of a linear drift of 38 Hz/s. For comparison, circles show the Allan deviation measured between a laser stabilized to a two-mirror cavity and one tooth of an optical frequency comb stabilized by a hydrogen maser. Blue dashed and orange dotted lines indicate the estimated Allan deviation due to thermo-refractive and photo-thermal noise, respectively. (Courtesy of [190].)

4.6.3 Hänsch-Couillaud technique

Polarization spectroscopy of a reflecting anisotropic cavity can also provide dispersion-shaped resonances which are well suited for locking a laser frequency to a fringe center [192]. To explain this method let us consider the setup illustrated in Figure 4.40.

Linearly polarized light from a tunable single mode laser is reflected by a confocal reference cavity used off-axis so that a small angle between incident and reflected beam avoids feedback into the laser cavity. The linear polarizer inside the cavity is rotated so that its transmission axis forms an angle θ with the polarization axis of the incident beam. The incoming light can be decomposed into two orthogonal linearly polarized components with the electric field vector parallel and perpendicular to the transmission axis of the intracavity polarizer. Their field amplitudes are

$$E_\perp^{(i)} = E_0 \sin\theta \qquad (4.209)$$

$$E_\parallel^{(i)} = E_0 \cos\theta \qquad (4.210)$$

where E_0 is the amplitude of the incident beam. The parallel component sees a cavity of low loss and experiences a frequency-dependent phase shift in reflection. The perpendicular component, simply reflected by the mirror M1, serves as a reference. Any relative phase

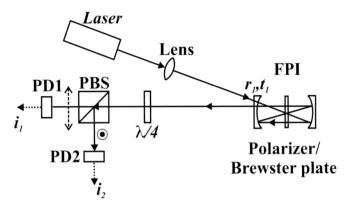

FIGURE 4.40
Scheme for Hänsch-Couillaud laser frequency stabilization.

change between the two reflected components will make the resulting beam elliptically polarized. We first consider the parallel component. The complex amplitude of the reflected wave is obtained by Equation 3.37 with the following identifications $r_1 r_2 = 1$ and $r_1 = \sqrt{R_1}$. Thus we get

$$
E_\parallel^{(r)} = E_\parallel^{(i)} \left\{ \sqrt{R_1} - \frac{T_1}{\sqrt{R_1}} \frac{Re^{i\delta}}{1 - Re^{i\delta}} \right\}
$$

$$
= E_\parallel^{(i)} \left\{ \sqrt{R_1} - \frac{R \cdot T_1}{\sqrt{R_1}} \frac{\cos\delta - R + i\sin\delta}{(1 - R)^2 + 4R\sin^2\frac{\delta}{2}} \right\}
\tag{4.211}
$$

where R_1 and T_1 are the reflectivity and transmittivity of the cavity entrance mirror M1, and $R < 1$ gives the amplitude ratio between successive round-trips, which determines the cavity finesse $\mathcal{F} = \pi\sqrt{R}/(1 - R)$. Such ratio accounts for any attenuation by the internal polarizer and for other losses, including the two extra reflections which are required for one round-trip in a confocal resonator used off-axis. The amplitude of the reflected perpendicular component, on the other hand, is to first approximation simply

$$
E_\perp^{(r)} = E_\perp^{(i)} \sqrt{R_1}
\tag{4.212}
$$

At exact resonance ($\delta = 2m\pi$), both reflection coefficients are real and the reflected wave components remain in phase. Away from resonance, however, the parallel component acquires a phase shift relative to the perpendicular component, owing to the imaginary part of $E_\parallel^{(r)}$, and the reflected beam acquires an elliptical polarization. The handedness of the polarization ellipse depends on the sign of the detuning from resonance.

To detect the ellipticity, the reflected light is sent into an analyzer assembly consisting of a $\lambda/4$ retarder and a linear polarization beam splitter. The fast axis of the retarder is rotated by $45°$ relative to the polarization axis of the beam splitter output a. The light intensities I_a and I_b at the two outputs are monitored by two detectors connected to a differential amplifier. In order to calculate the signal, we assume, without loss of generality, the fast axis of the QWP is parallel to the axis of the intra-cavity polarizer. Then, using the formalism introduced by Jones, we find the field amplitudes of the reflected beam after passing through the retarder and polarization beamsplitter

$$E_{a,b} = \frac{1}{2}\begin{pmatrix} 1 & \pm 1 \\ \pm 1 & 1 \end{pmatrix}\begin{pmatrix} 1 & 0 \\ 0 & i \end{pmatrix}\begin{pmatrix} E_{\parallel}^r \\ E_{\perp}^r \end{pmatrix} \tag{4.213}$$

where the first matrix describes a linear polarizer set at $45°$ and the second one a quarter-wave plate with the fast axis horizontal. Hence the difference signal of the photocurrents is

$$i_1 - i_2 \propto E_0^2 \cos\theta \, \sin\theta \, \frac{T_1 R \sin\delta}{(1-R)^2 + 4R\sin^2\delta/2} \tag{4.214}$$

This function is plotted in Figure 4.41. It combines a steep resonant slope with far reaching wings and provides an ideal error signal for servo locking of a laser frequency. The signal is maximized if $\theta = 45°$, so that $2\cos\theta\sin\theta = 1$. However, the total intensity of the reflected light at resonance is smallest near $\theta = 0$ and operation at smaller angles θ can offer a better signal to noise ratio, if laser intensity fluctuations are the dominant source of noise. Any birefringence due to stress in the dielectric mirror coatings or other optical elements has been ignored in our analysis. Such residual birefringence can produce line asymmetries and should be compensated.

The HC technique is very versatile owing to its simple and inexpensive setup (no frequency modulation of the laser is needed) and is often used for pre-stabilization of a laser. However, as with any dc technique the locking point is sensitive to baseline drifts of the error signal and it is furthermore affected by the technical noise of the laser at low Fourier frequencies.

4.6.4 Laser frequency stabilization by locking to an optical fiber-delay line

Another approach for suppressing frequency noise in a laser is to phase lock its emission to a fiber interferometer [193]. With reference to Figure 4.42, lets denote with E_1 and E_2 the electric fields of the radiation before combining at the second beam splitter. Thus, if $L_1 \gg L_2$, in complex notation, we have

$$E_1 = E_0 e^{i(\Omega' t + \Omega' t_0)} = E_0 e^{i\Omega' t_0} e^{i(\Omega' - \Omega)t} e^{i\Omega t} \tag{4.215}$$

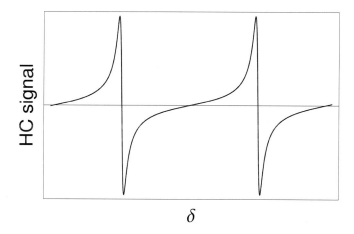

FIGURE 4.41
Dispersive resonances obtained by polarization spectroscopy ($R = 0.9$).

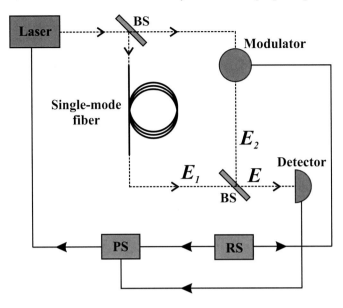

FIGURE 4.42
Scheme of the optical fiber used as optical delay line to stabilize the frequency of the laser. RS=reference signal; PS=phase-sensitive detector; BS=beam splitter. (Adapted from [193].)

$$E_2 = E_0 e^{i(\Omega t + \Gamma \sin \omega t)} \simeq E_0 e^{i\Omega t} \left(1 + \frac{\Gamma}{2} e^{i\omega t} - \frac{\Gamma}{2} e^{-i\omega t} \right) \qquad (4.216)$$

where $\Omega' = 2\pi c/\lambda'$ and Ω are the optical frequencies at the ends of fibers 1 and 2, Γ and ω are the amplitude and frequency modulations, and

$$t_0 = \frac{L_1 - L_2}{c} \qquad (4.217)$$

c being the speed of light. As $\Gamma \ll 1$, the Bessel series approximation (to first order in Γ) has also been used in the last step of Equation 4.216. After these two beams are combined, provided that the polarization of the beams in the fibers can be kept, the electric field is $E = E_1 + E_2$, whereupon the radiation intensity incident on the photodetector is

$$I = |E|^2 \simeq E_0^2 \{ 2 + 2 \cos[(\Omega - \Omega')t - \Omega' t_0]$$
$$+\Gamma \cos[(\Omega - \Omega')t + \omega t - \Omega' t_0] - \Gamma \cos[(\Omega - \Omega')t - \omega t - \Omega' t_0] \} \qquad (4.218)$$

The above signal is first demodulated in the lock-in amplifier (mathematically this corresponds to a multiplication by a term proportional to $\sin \omega t$). Subsequently, the high frequency components in the mixer output is filtered out. Thus, the output S of the lock-in amplifier is proportional to

$$S \propto \sin[(\Omega - \Omega')t - \Omega' t_0] = \sin[(\Omega - \Omega')t - 2\pi \frac{L_1 - L_2}{\lambda'}] \qquad (4.219)$$

After the error signal S is fed back, the phase of the laser is locked and we have

$$\sin \left[(\Omega - \Omega')t - 2\pi \frac{L_1 - L_2}{\lambda'} \right] = 0 \qquad (4.220)$$

whose solution

$$(\Omega - \Omega')t - 2\pi\frac{L_1 - L_2}{\lambda'} = n\pi \tag{4.221}$$

may be further separated into

$$\Omega - \Omega' = 0 \tag{4.222}$$

$$2\frac{L_1 - L_2}{\lambda'} = n\pi \tag{4.223}$$

where n is an integer. Equation 4.223 shows that $L_1 - L_2$ can be used to define λ', while Equation 4.222 shows that, with such a feedback, the light frequency at different points of the optical fiber will follow the defined frequency at any time t. It is obvious that in practice $L_1 - L_2$ is not constant and different kinds of noise may occur. For a low output level laser, we can disregard the noise induced by non-linear scattering processes; so the noise source could be temperature drift, shot noise, mechanical thermal noise, vibration, acoustic noise, strain noise, etc.

The first frequency stabilization experiment of a laser onto a fiber spool used a Mach-Zehnder interferometer (MZI), with phase modulation into one arm, to stabilize a He-Ne laser. Corrections were applied via a Piezoelectric Transducer (PZT)-stretcher. It led to a 5 kHz linewidth on 1 s [193]. Later, a distributed feedback (DFB) erbium-doped fiber laser (EDFL) was stabilized onto a 100 m path imbalance MZI using homodyne electronics and a PZT actuator, reaching about 2 Hz^2/Hz at 1 kHz [194].

In the following, we present a system using an all-fiber 2-km imbalance Michelson interferometer (MI) with heterodyne detection [195]. Referring to Figure 4.43, the input optical wave is split between the two arms by a 50/50 fiber coupler. The first arm is directly connected to a Faraday mirror; the second arm is connected to a 1-km spool of standard SMF-28 fiber followed by an acousto-optic frequency shifter and a Faraday mirror. The Faraday rotator mirror ensures that, in a retracing fiber-optic link, the polarization output state is orthogonal to the entrance state; as a consequence, the two waves in the MI output port have

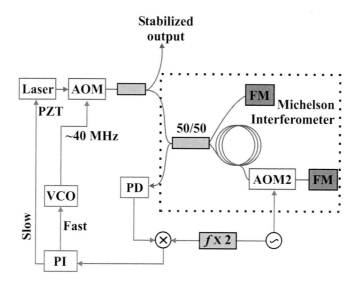

FIGURE 4.43
Scheme of the laser frequency noisereduction system: AOM, acousto-optic modulator; PD, photodiode; VCO, voltage-controlled oscillator; PI, proportional-integrator filter; FM, Faraday mirror. (Adapted from [195].)

always the same state of polarization, which leads to a maximum beat-note signal amplitude without requiring any polarization controller. The fiber spool is situated into a ring-shaped aluminum box, and the interferometer is placed inside a thick aluminum box, recovered by a thermal isolating thermoplastic film, which is set onto a compact seismic vibration isolation platform. The whole experiment is covered by an acoustic isolation box. As discussed in Chapter 2, the frequency-shift fibered MI acts as an optical frequency (ν_{opt}) to rf phase (Φ_{err}) converter with transfer function $\Phi_{err}(f)/\nu_{opt}(f) \equiv H_{Mich}(f) = [1 - \exp(-i2\pi f\tau)]$, where τ is the fiber double-pass delay time and f is the Fourier frequency. For $f \ll 1/\tau$, we have $H_{Mich} \simeq 2\pi\tau$. Then, the optical power photo-detected at the interferometer output port contains an rf carrier at $2f_{AOM}$, phase modulated by Φ_{err}, which is down-converted by an rf mixer driven by the frequency-doubled output of a low-noise reference oscillator at 70 MHz. This provides a low-frequency error signal proportional to $\Phi_{err} + \Delta\theta_{RF}$, where $\Delta\theta_{RF}$ is the local oscillator phase shift. The error signal is amplified, filtered, and converted into optical frequency correction using an AOM operating at 40 MHz, which is driven by a high-modulation-bandwidth voltage-controlled oscillator for fast correction and a piezoelectric element controlling the fiber-laser cavity length for drift compensation. The correction bandwidth (~ 100 kHz) is limited by the round-trip delay in the fiber interferometer. The laser source is a single-longitudinal-mode Er^{3+}-doped fiber Bragg grating laser with an emission wavelength of 1542 nm and a maximum output power of 100 mW. The free-running laser-frequency noise is dominated by a flicker component with 10^4 Hz^2/Hz at 100 Hz. The frequency-noise PSD of the fiber-stabilized laser is measured by comparison with a high-finesse Fabry-Perot cavity-stabilized laser. For this purpose, the rf beat-note signal is down-converted to 700 kHz by a low-phase-noise synthesizer, then frequency-to-voltage converted and analyzed using a fast Fourier transform analyzer (after removing a linear drift of the order of 1 kHz/s). Results are shown in Figure 4.44. With the antivi-

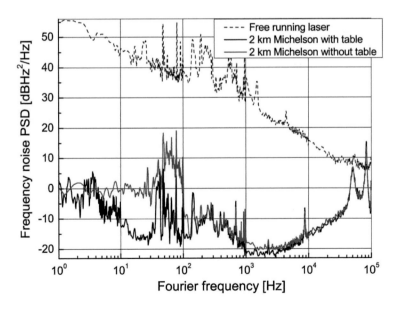

FIGURE 4.44
Frequency-noise power spectral density versus Fourier frequency of the free-running laser (dashed curve) and laser stabilized on a 2 km imbalance Michelson interferometer with (dark curve) and without (gray curve) a passive anti-vibration table. (Courtesy of [195].)

bration platform, the frequency-noise reduction is larger than 40 dB between 1 Hz and 10 kHz, and the frequency noise PSD is (notwithstanding several peaks) below 1 Hz^2/Hz in the same range. The effect of the anti-vibration table is remarkable between 5 Hz and 100 Hz; however, even without this special table, noise reduction is better than 30 dB. For Fourier frequencies between 40 Hz and 30 kHz the measurement is limited by the reference-laser frequency noise and is therefore an upper limit of the fiber-stabilized laser noise. Mainly due to the use of rf heterodyne detection (at a frequency where laser intensity noise and detection noise are negligible), this represents a several-orders-of-magnitude improvement compared to previous results of laser stabilization hinging on fiber-delay lines. In the telecom spectral window, the above method represents a valuable alternative to cavity locking. Indeed, by providing a fibered system without any optical alignment or polarization adjustment, it is inherently more compact and flexible than cavity-based systems. For Fourier frequencies below 40 Hz, the described system is presumably limited by thermal fluctuations and mechanical vibrations, which can be further improved. For example, the interferometer could be installed in a temperature-stabilized vacuum tank with several thermal shields. In addition, special fibers (such as specifically designed photonic-crystal fibers or liquid-crystal polymer coatings) with lower thermal sensitivity could be employed.

4.6.5 Injection locking

The injection locking technique is applied primarily to continuous-wave single-frequency laser sources when a high output power is desired in conjunction with a very low intensity noise and phase noise. Such low-noise performance is hardly achievable in high-power lasers, because these are generally more susceptible to mechanical vibrations and thermal influences; in addition, they cannot utilize very low-noise pump sources. In principle, a valuable option would be to construct a low-noise, low-power laser and then amplify its output. However, this approach is beset with various fundamental and practical problems: in particular, reaching the standard quantum noise level is prevented by the to-some-extent unavoidable amplifier noise. This drawback is precisely overcome in the injection locking scheme. As we learnt in Chapter 2, injecting a weak signal into a more powerful free-running oscillator can produce an interesting and useful set of injection locking effects in any kind of self-sustained periodic oscillator, including the laser. In this specific context, let us consider a high-power laser (called the slave laser) which is initially producing a coherent output intensity I_0 at the free-running frequency ω_0. Suppose now that a very weak signal of intensity I_1 coming from a second laser (called the master laser) is injected into the slave laser via some suitable coupling method, at frequency ω_1 which is sufficiently close but not exactly coincident with ω_0 (Figure 4.45). In these conditions, the injection forces the slave laser to operate exactly on the injected frequency with relatively scant noise (usually, the injection-locked laser operates on a Gaussian resonator mode, but it is also possible to enforce operation on some higher-order mode). As expressed by Equation 2.222, the full locking range is given by

$$\Delta\omega_{lock} = \frac{\omega_0}{Q}\sqrt{\frac{I_1}{I_0}} \qquad (4.224)$$

where Q is the quality factor of the resonance at ω_0. The square root term in Equation 4.224 is due to the fact that currents in Equation 2.222 must now be replaced with intensities, or powers. As already illustrated in Chapter 2, injection locking is a powerful tool to suppress frequency noise in the free-running oscillator (slave laser).

To conclude the analysis of the injection locking phenomenon, lets investigate its physical origin in the specific case of lasers [171]. As already stated, the gain of an oscillating laser is clamped at the threshold value; this implies that the gain for an externally injected signal

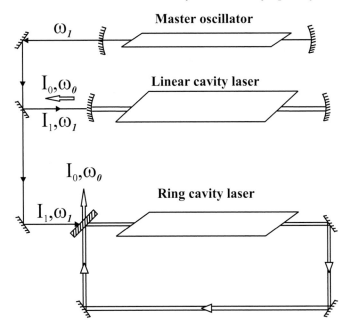

FIGURE 4.45
Schematic of laser injection locking. (Adapted from [171].)

at any other frequency away from the exact resonance frequency remains stable. In other words, in spite of the much stronger oscillation already present at ω_0, the signal at ω_1 can circulate around inside the cavity and be regeneratively amplified by the active medium. In principle, with sufficiently narrow-band filters, one could measure the amplification of the injected signal $\omega_1 \neq \omega_0$, wholly independent of the simultaneous oscillation at ω_0 (this is entirely true only if the signal at ω_1 is weak enough to make interference/beating effects negligible). There will come a time where the amplified intensity at ω_1 approaches the free-running oscillation intensity I_0; at this point, the amplified signal saturates the laser gain down by just enough that the free-running oscillation at ω_0 is turned off, leaving only the injected signal at ω_1. For a comprehensive review on this topic, the reader is referred to [196].

Laser injection locking was first realized with two He-Ne lasers [197] and then extensively applied to diode lasers [198]. More recently, optical injection locking was exploited to combine the unique spectral features of a comb-referenced difference frequency generation (DFG) source with the power scaling capabilities of mid-IR quantum cascade lasers (QCLs) [199]. For an explanation of the working principle of these laser sources, see later in this chapter. A schematic of the apparatus is shown in Figure 4.46.

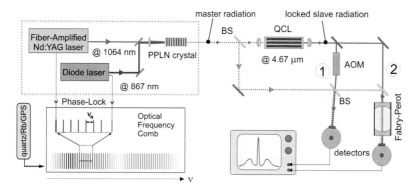

FIGURE 4.46

Injection-locking setup between a comb-referenced DFG source and a QCL.

The slave QCL is a continuous-wave Fabry-Perot type device emitting up to ~ 20 mW near room temperature. The laser facets are not anti-reflection coated. The master radiation (maximum power ~ 100 μW) is produced by nonlinear difference frequency generation in a periodically-poled LiNbO$_3$ (PPLN) crystal using an Yb-fiber-amplified Nd:YAG laser (at 1.064 μm) and an external-cavity diode laser (ECDL) emitting at 867 nm. In order to control the phase/frequency of the generated IR radiation against the fs Ti:sapphire OFCS an electronic scheme based on direct digital synthesis (DDS) was used (see later in this chapter); in particular, this led to a MIR radiation with a kHz-level spectral purity of the monolithic Nd:YAG laser. An accurate mode matching between master and slave, along with precise control of their polarization via quarter-/half-wave plates, was necessary to achieve the injection-locking (IL) regime. The IL condition was checked and optimized by observing the slave radiation transmitted by a Fabry-Perot resonator (free spectral range 650 MHz, finesse $\simeq 20$). The QCL remained injection-locked for several minutes without any active control. In this passive IL regime, the frequency of the single-mode slave laser radiation could be scanned within a locking range as high as 1 GHz (depending on the slave/master power ratio) by tuning the master laser. As expected, the locking range was proportional to the square root of the ratio between master and slave powers. Then, the IL quality was directly studied by analyzing the beat note spectrum between master and slave radiations. For this measurement, the slave radiation was frequency-shifted by 90 MHz via an acousto-optical modulator. Both the master and slave beams were superimposed by means of a beam splitter and finally detected onto a fast HgCdTe detector (nominal bandwidth 200 MHz). The beat note was measured by a real-time FFT spectrum analyzer (see Figure 4.47).

For a deeper understanding of the frequency-noise characteristics of the injection-locked radiation, a measurement of the frequency-noise power spectral density (FNPSD) for both laser sources was also carried out (Figure 4.48). For this purpose, both the master and slave radiations were coupled to the Fabry-Perot cavity, and the slope of a transmission mode was used as a frequency discriminator to convert frequency noise into amplitude fluctuations.

Compared to the free-running case, phase fluctuations were largely reduced (by three to four orders of magnitude in most of the frequency interval). As expected, optical injection was successful in strongly reducing also the noise contribution due to the laser current driver, allowing to overcome one of the main limiting factors to the QCL linewidth and thus loosening the requirements on QCL power supply. As repeatedly emphasized, by integrating the FNPSD autocorrelation function it is possible to reconstruct the laser spectral profile and to measure its linewidth over any desired timescale. To this aim, a numerical integration

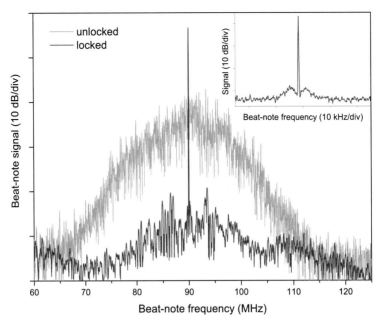

FIGURE 4.47

Recorded beat notes, with the QCL in unlocked and locked conditions. The slave is operated at 618 mA, with an output power of $\sim$ 6 mW, corresponding to a slave/master power amplification $P_S/P_M \simeq 250$. The first trace maps the broad power spectrum of the unlocked QCL, while the second trace shows a narrow peak, whose width is limited by the resolution of the spectrum analyzer, rising about 40 dB above a residual plateau.

of the acquired noise spectra was performed. The whole spectral range of Figure 4.48 (20 Hz to 20 MHz) was considered, obtaining the spectral profiles and linewidths over 50-ms timescale (see the inset of Figure 4.48). By switching from free-running to injection-locking regime, the QCL linewidth was reduced by more than two orders of magnitude, from 2.75 MHz to 23 kHz (HWHM), thus confirming the large improvements in the slave spectral profile previously highlighted by the beat-note analysis.

In conclusion, the above frequency-noise analysis demonstrates that the stability properties of the DFG source are transferred to the slave QCL, which is forced to oscillate at the master frequency, with an effective power amplification up to a factor 1000, within a range of $\sim$ 1 GHz in passive injection-locking conditions.

4.7   Intensity fluctuations

Although significantly less severe than frequency noise, intensity fluctuations in the laser output beam are also present, primarily due to vibrations of the resonator as well as to the other specific sources:

- Gas lasers: fluctuations in the power supply current and instability of the discharge process;

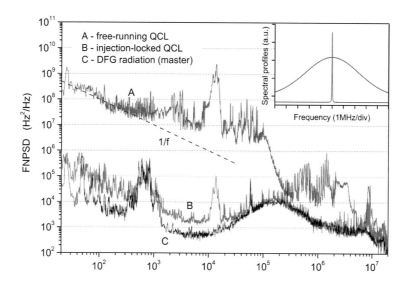

FIGURE 4.48

FNPSD measurements. Above ~ 1 kHz, the free-running QCL exhibits a current-driver dominated frequency noise. Above ~ 10 MHz, the trace corresponding to the injection-locked slave overlaps with the free-running one due to the detector noise floor. For the master radiation, the excess noise around 200 kHz comes from the phase-locking loop of the ECDL to the frequency comb. From this comparison it emerges that, in the injection-locking regime, the QCL well reproduces the noise features of the master source throughout the investigated interval.

- Dye lasers: density fluctuations in the dye jet solution;

- Solid-state lasers: pump fluctuations;

- Semiconductor lasers: fluctuations in the bias current, amplitude fluctuations related to spontaneous emission and electron-hole recombination.

Besides these short-term fluctuations, long-term drifts also exist, generally caused by thermal misalignment in the laser cavity and by degradation (over at least a few thousand hours) of various optical components, including the active medium itself.

Whatever the origin, several schemes have been devised for intensity stabilization. In the following, we shall discuss two that are most often used. The former is schematically depicted in Figure 4.49. A small fraction of the output power is conveyed by a beam splitter (BS) onto a detector, whose output V_D is compared with a reference voltage V_R. Then, the difference $\Delta V = V_D - V_R$ is amplified and fed to the power supply of the laser. Obviously, the upper frequency limit of this stabilization loop is settled by the capacitances and inductances in the power supply as well as by the time interval it takes for the increase of the laser intensity following the current increase. In gas lasers, for instance, this delay is represented by the time required by the gas discharge to reach a new equilibrium after the current change; this means that it is not possible to stabilize the system against fluctuations of the gas discharge. Concerning diode-pumped solid-state lasers, on the other hand, intensity fluctuations can be effectively reduced up to a frequency higher than the

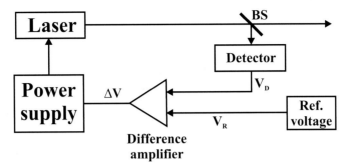

FIGURE 4.49
Intensity stabilization of lasers by controlling the power supply.

relaxation-oscillation one, as the inverse of this latter is much longer than the response time of the diode pump. The above stabilization technique suppresses intensity fluctuations down to less than 0.5%, which can be satisfactory for many purposes. In more challenging experiments, however, the power fluctuations of the free running laser would limit the detection sensitivity. Just as an example, this is the case for the laser interferometers employed in the detection of gravitational waves (GWs): here, power fluctuations in the GW frequency band induce radiation pressure fluctuation which, by shaking the suspended test mass, can mask possible GW signals [200]. For this and other demanding applications, another technique, illustrated in Figure 4.50, is more suitable. The operation principle is that the optical power is reduced by means of an electrically controllable attenuator, and the control signal is derived from the output power as measured with a photodiode: in essence, the attenuation is augmented when the power is measured to be too high, and vice versa. The most common approach is based on a PID-type electronic feedback loop, whose design is crucial for realizing effective noise suppression over a large bandwidth. An EOM or AOM can be employed to control the power throughput with a high servo bandwidth. The ultimate performance of such power stabilization scheme is fixed by the photo-detector sensitivity, with the latter being limited, in principle, by technical noise and quantum noise. Being related to the quantization of the light energy into photons (or the quantization of the photocurrent into electrons after the photodetection process), this latter represents a fundamental limit and can be surpassed only by non-classical states of light, such as squeezed states (however, we

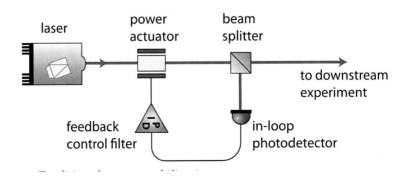

FIGURE 4.50
Traditional noise eater for laser power stabilization. (Courtesy of [200].)

should mention that, due to their complex generation and fragility, these are not used in any power stabilization experiment so far). As already shown in Chapter 2, the single-sided linear spectral density S_q of the relative quantum power noise, often referred to as shot noise, is independent of the Fourier frequency

$$S_q = \sqrt{\frac{2hc}{P\lambda}} \tag{4.225}$$

where λ is the vacuum wavelength and P is the detected laser power. In case a photodetector is used as a power sensor, this quantum-noise level can also be expressed using the photocurrent I

$$S_q = \sqrt{\frac{2e}{I}} \tag{4.226}$$

where e is the elementary charge. Therefore, the quantum-noise-limited sensitivity of a photodetector can be improved by increasing the detected laser power P (and with it the photocurrent I). The drawback is that a high power originates troublesome technical problems, such as saturation and thermal effects in the photodiode, as well as dynamic range limits of the readout electronics. So, in order to successfully increase the photodetector sensitivity, two different, complementary approaches can be undertaken [200].

4.7.1 High-sensitivity photodiode array

One approach consists of scaling the number of photodiodes leading to a high-power high-sensitivity array (Figure 4.51). A Nd:YAG NPRO with an output power of about 2 W is the laser source to be stabilized. The beam emerging from the ring resonator (used as mode cleaner and to reduce beam pointing fluctuations) is split with 50:50 beam splitters into eight partial beams of nearly equal power (57-65 mW) using multiple reflections. The photodiode array comprises eight InGaAs photodiodes with an active diameter of 2 mm (each photodiode is connected to a low-noise transimpedance amplifier with a low current noise 200 Ω resistor; at the operation point, each photodiode detects a photocurrent of about 50 mA). Four signals are added and used to stabilize the laser power (in-loop), while the

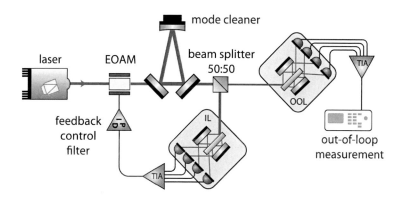

FIGURE 4.51
Setup of the photodiode array power stabilization experiment. The power fluctuations of the laser are measured with the in-loop photodetector IL and are compensated by the feedback control loop using the electro-optic amplitude modulator EOAM. The photodetector OOL is used for an independent measurement of the achieved power stability. (Courtesy of [200].)

remaining four signals are added and utilized to check the power stability (out-of-loop). The in-loop signal is subtracted from a low-pass-filtered voltage reference, amplified in analog servo electronics, and fed back to the EOAM. This dc-coupled feedback control loop has a bandwidth of about 80 kHz with a loop gain of more than 68 dB for frequencies below 1 kHz. With closed feedback loop, the out-of-loop measured power noise (Figure 4.52) is at the expected level defined by the uncorrelated sum of the quantum noise and electronic noise of both the in-loop and out-of-loop detectors in the whole frequency band. For frequencies up to 7 Hz the measured noise is dominated by the electronic noise of the in-loop and out-of-loop detector and for higher frequencies by quantum noise at a level of $1.8 \cdot 10^{-9}$ Hz$^{-1/2}$. At 10 Hz a relative power noise of $2.4 \cdot 10^{-9}$ Hz$^{-1/2}$ is measured at the out-of-loop detector.

4.7.2 Optical ac coupling

The other approach is using optical resonators to enhance the sensitivity of a photodetector without increasing the average photocurrent. Conventionally, electrical ac coupling is exploited to measure small fluctuations on top of large dc signals (or of slowly varying ones). In optical ac coupling, the reflection at an optical resonator is used to create a similar effect, where the signal is attenuated at low Fourier frequencies. The principle is the following: while the carrier and low-frequency sidebands are almost completely transmitted for a nearly impedance-matched resonator, power fluctuation sidebands are mainly reflected by the resonator at high frequencies. In this way, the average power in reflection

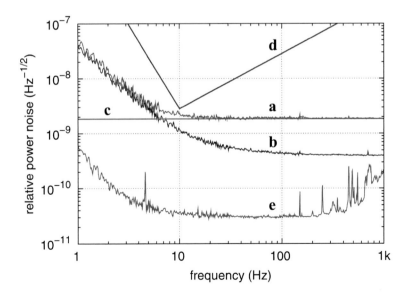

FIGURE 4.52
Power noise measured out-of-loop (a) with the photodiode array. For low frequencies, the measurement is limited by electronic noise (b) and for frequencies above 7 Hz by quantum noise (c) of the in-loop (200 mA photocurrent) and out-of-loop detector (189 mA photocurrent). The in-loop measured power noise (e) is far below these limiting noise sources. The Advanced LIGO power-noise requirement (d), shifted by 3 dB in the plot to account for the noise of the out-of-loop detector, is met in the whole frequency band. (Courtesy of [200].)

is reduced, whereas the high frequency fluctuation sidebands are fully preserved [201]. This effect can be more rigorously described as follows. Let $h(f) = 1/(1 + if/f_0)$ characterize the approximated power fluctuation filter effect (in transmission) of the resonator with the bandwidth f_0. Next, consider the field amplitude U_{in} of a laser beam with the average amplitude U_0 that is modulated at a frequency $f = \omega/2\pi$ with a modulation index $m \ll 1$ in the plane-wave model

$$U_{in} = U_0 \left(1 + me^{i\omega t}\right) = U_0 + U_{fl} \tag{4.227}$$

where $U_{fl} \equiv U_0 m e^{i\omega t}$ represents the fluctuating part of the incident field amplitude. This input beam is coupled into the optical resonator. At resonance, the field amplitude U_{refl} of the reflected beam can be written as the sum of two terms. The first one corresponds to reflection of the average field $U_{refl,1} = U_0 a$, where the parameter a describes the impedance matching of the resonator. The resonator is under-coupled for $a \in (0, 1]$, impedance matched for $a = 0$ (recall that a resonator is said to be *impedance-matched* when the reflected power from the cavity is exactly zero on resonance), and over-coupled for $a \in [-1, 0)$. The second term contributing to U_{refl} corresponds to reflection of the fluctuating field and is thus given, for an *impedance-matched* resonator, by $U_{fl}[1 - h(f)]$. For a non-impedance-matched resonator, the latter expression is readily generalized to $U_{refl,2} = U_{fl}[1 - (1 - a)h(f)]$, whereupon we get

$$U_{refl} = U_{refl,1} + U_{refl,2} = U_0 \left\{a + me^{i\omega t} \left[1 - (1 - a)h(f)\right]\right\} \tag{4.228}$$

Therefore, the transfer function $G(f)$ (at the Fourier frequency f) for relative power fluctuations from the beam upstream of the resonator to the beam in reflection is given by

$$G(f) = \frac{1 - (1 - a)h(f)}{a} = g - (g - 1)h(f) \tag{4.229}$$

$$|G(f)| = \sqrt{\frac{1 + g^2 \frac{f^2}{f_0^2}}{1 + \frac{f^2}{f_0^2}}} \tag{4.230}$$

where $g = 1/a$ represents the maximum gain for very high frequencies: $G(f \gg f_0) \to g$. This transfer function $G(f)$ can be used to realize more sensitive power detectors, provided that a special impedance matching and a very good mode matching are accomplished: with moderate experimental effort, a gain of about $g = 10$ is indeed realizable. Figure 4.53 shows the novel power stabilization setup that can be implemented within this approach. The beam splitter and the in-loop photodetector of the traditional configuration are substituted by a compound detector, comprising the resonator and the photodetector in reflection. Thanks to the mode cleaner, a mode matching exceeding 99% is achieved to the subsequent ACC resonator (used for the optical ac coupling). The latter has a finesse of about 10000, a bandwidth of about 35 kHz, and is slightly under-coupled. The laser frequency is stabilized to a fundamental mode resonance of the ACC. The beam reflected off the ACC is detected with the LPD photodetector. The relative power noise measured with optical ac coupling is compared with that measured in a traditional setup (this latter measurement is accomplished by setting the laser frequency off-resonant to the ACC). Figure 4.54 shows that these two measurements agree to each other (within the measurement and calibration accuracy of $\simeq 1$ dB) up to $\simeq 2$ MHz. At higher frequencies, instead, the traditional (optical ac coupling) measurement is limited by quantum noise at a level of $1 \cdot 10^{-8}$ Hz$^{-1/2}$ ($7 \cdot 10^{-10}$ Hz$^{-1/2}$), 3 mA photocurrent.

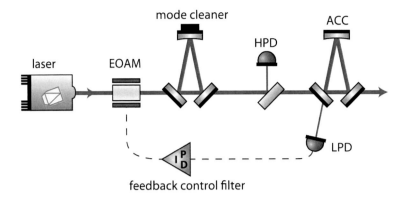

FIGURE 4.53
Simplified experimental setup of the power stabilization experiment with optical ac coupling. About 10% of the beam is sampled and detected with the HPD photodetector; the remaining beam power of up to 900 mW is directed to the ACC. In the power sensing experiment, the free-running power noise of the laser is measured with photodetector LPD. In the power stabilization experiment, this detector is used as an in-loop detector and the power noise is measured with the independent out-of-loop photodetector HPD. (Courtesy of [200].)

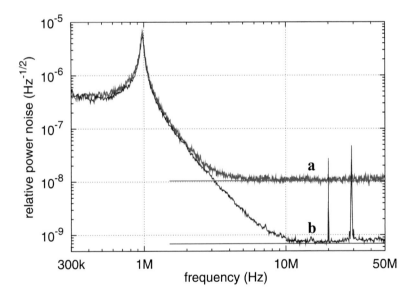

FIGURE 4.54
Measured relative power noise with optical ac coupling. The laser relaxation oscillation at 1 MHz and the steep roll-off towards higher frequencies are apparent. The quantum limits for the different measurements are shown as horizontal lines. The sensitivity using the optical ac coupling (b) is improved by $G(f)$, compared to the same detector with off-resonant ACC (a). The signal at 20 MHz is injected for calibration purposes and that around 29 MHz is due to the modulation for stabilizing the mode cleaner at a resonance. (Courtesy of [200].)

4.7.3 The laser as quasi-ideal oscillator

In light of discussions so far, particularly when frequency/intensity stabilization techniques are also used, we can definitely state that the laser output represents the closest possible approximation to a classical monochromatic, polarized electromagnetic wave. The associated electric field can be described as

$$E(\boldsymbol{r}, t) = E_0 \cos[\omega_0 t + \phi(\boldsymbol{r}) + \varphi(t)] \tag{4.231}$$

where $\phi(\boldsymbol{r})$ characterizes the spatial dependence of the wavefront. In addition, a well-collimated Gaussian laser beam can also be viewed as the closest approximation to the light ray of geometrical optics. In that case, the wavefronts are nearly plane ($\phi(\boldsymbol{r}) = -\boldsymbol{k} \cdot \boldsymbol{r}$). At the same time, if the small random fluctuations (over time) of E_0 and φ are considered, apart from the constant wavefront phase term, Equation 4.231 coincides with that of a quasi-perfect sinusoidal oscillator (see, for instance, Equation 2.57). Therefore, all the formalism developed in Chapter 2 can be exploited. In particular, all the measurement techniques to characterize phase and amplitude noise, described in Section 2.7, apply to lasers as well.

We summarize here the approaches that are most commonly used to ascertain frequency noise of lasers:

- Measurement of the frequency-noise power spectral density (PSD). In this scheme the laser frequency fluctuations are converted into amplitude fluctuations (against an atomic/molecular or cavity resonance having a known slope) which are then processed by a spectrum analyzer. This provides the PSD in the Fourier domain which is related to the power spectrum in the carrier frequency domain, or the laser line shape, by Equation 2.163.

- Beating with a reference laser. This originates a signal of the type expressed by Equation 2.45. Counting the frequency of the resulting beat note gives a time-description of the laser instabilities via the quantity $AVAR$, which is ultimately related to the frequency-noise PSD by Equation 2.153. Alternatively, the width W_{beat} of the Fourier transform of the beat-note signal (as measured by a spectrum analyzer) provides the laser linewidth W_{laser}

$$W_{beat} = \sqrt{W_{laser}^2 + W_{ref}^2} \simeq W_{laser} \tag{4.232}$$

where $W_{ref} \ll W_{laser}$ denotes the width of the reference laser.

- Delayed self-heterodyning (DSH) technique (Figure 4.55) [202, 203, 204]. One portion of the laser beam is transmitted through a long optical fiber which provides some time delay; the other part passes through an AOM which shifts all the optical frequency components by a few tens/hundreds of MHz. The two beams are then superimposed on a second beam splitter, and the resulting beat note (centered at the AOM frequency) is recorded with a sufficiently fast photodetector. If the delay time is much larger than the coherence time of the laser, the electric fields of the two beams are uncorrelated and the frequency-shifted beam is regarded as an independent local oscillator that has the same linewidth as that of the delayed beam. Therefore, the original spectral shape of the laser can be estimated from the observed spectrum. For lasers with a very narrow linewidth (i.e. long coherence

length), let's say below 1 kHz, a more effective scheme to obtain uncorrelated beams is represented by the re-circulating fiber loop (Figure 4.56). Here, a long delay is realized by making the light perform multiple round trips through a moderately long fiber. In order to keep the components corresponding to different numbers of round trips well separated in the frequency domain, an AOM is inserted in the loop, which shifts the optical frequency by the amount ω_s (much larger than the laser linewidth) in each round trip. As a result, the beat signal with a frequency of $n \cdot \omega_s$ is produced by the two beams with a time difference of $n \cdot \tau_d$. Therefore, if we can detect the $n \cdot \omega_s$ beat signal, n-fold improvement

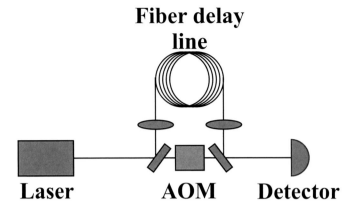

FIGURE 4.55
Delayed self-heterodyning technique.

in the resolution can be expected compared with that of a conventional DSH method. Indeed, if we assume that the laser has a Lorentzian line shape with a FWHM of $\Delta\omega$ and that the delay time is much longer than the coherence time of the laser, one can show that the normalized nth-order beat spectrum $S_n(\omega)$ is given by [202]

$$S_n(\omega) = \frac{(\alpha/2)^n(\Delta\omega/\pi)}{\Delta\omega^2 + (\omega - n\omega_s)^2} \tag{4.233}$$

where α represents the transmission coefficient of the delay fiber and the AOM and $\omega = 2\pi f$, with f being the Fourier frequency. Actually, without any amplifying element in the loop, the losses from both the AOM and the fiber are notable, so that the light intensity quickly decays during the round trips. This heavily restricts the number of round trips that can be used for the linewidth measurement. To overcome this drawback, the effective loss of the loop can be strongly decreased by inserting a fiber amplifier.

However, a major hindrance to the transfer of low-phase-noise signals is represented by the fact that the fiber optical insertion phase is extremely sensitive to environmental

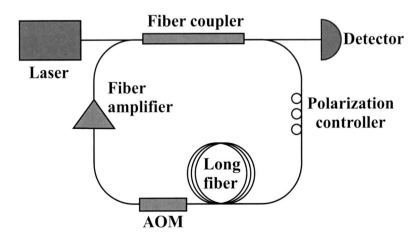

FIGURE 4.56
Recirculating fiber loop for improving the resolution of the delayed self-heterodyne method.

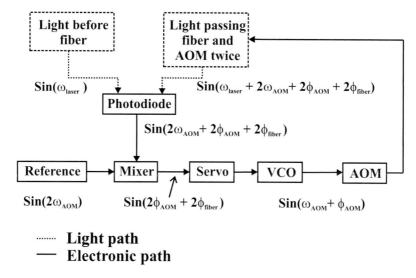

FIGURE 4.57
Scheme of the fiber noise cancellation. (Adapted from [177].)

perturbations, including changes in the refractive index, pressure and temperature. As a consequence of the phase noise written onto the laser beam propagating through the fiber, the original spectrum is broadened (under typical operating conditions, such broadening can be at the kHz level). In the current context, this implies that a beat measurement between two very narrow lasers, where one or both beams travel in fibers, will be dominated by the broadening of the fiber. A simple and effective technique for accurate fiber noise cancellation, which is crucial in demanding applications, such as optical frequency standards and quantum optics, is illustrated in Figure 4.57 [177]. Before entering the fiber, the light is sent through the AOM. A small fraction of the light is back-reflected from the other end fiber and passes again through the AOM. In this way, a double pass setup is realized, where the beat note between the light passing the AOM twice and the un-modulated light can be detected. The beat-signal is then mixed down to DC with a stable reference which also runs at ω_{AOM}. The phase difference between the reference and the beat signal is eventually fed back to the VCO which drives the AOM. The servo tries to set the quantity $\phi_{AOM} + \phi_{fiber}$ to zero, thus compensating the phase fluctuations caused by the fiber. Experimentally, it is found that the cancellation only works, if the fiber length is shorter than the coherence length of the laser.

4.8 Some specific laser systems

In the following we shortly review the most important laser sources, either direct or based on non-linear conversion processes, restricting our attention to those systems that have played, or still play, a crucial role in optical frequency metrology. Moreover, skipping all the aspects related to although important fabrication technologies, we shall deal solely with basic concepts, while those properties that are relevant to the issue of frequency measurements will be discussed in the next chapter in the scope of laser frequency standards. For each type of laser treated here, more details, from both a physical and a technical point of view, can

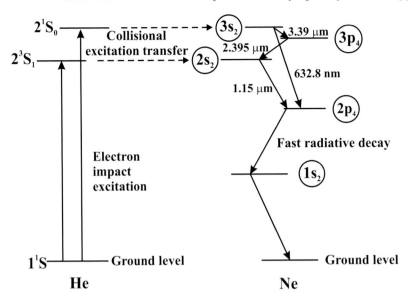

FIGURE 4.58
Energy-level diagrams of He and Ne.

be found in [168, 169, 205, 206]. Such textbooks also offer a good starting point to learn more about some laser categories that are ignored in this context, like chemical, excimer, free-electron, and X-ray lasers.

4.8.1   He-Ne laser

Most early laser developers sought four-level laser materials that could sustain a steady population inversion so that they could produce a cw beam. Years of gas spectroscopy experiments had generated extensive tables of spectral lines, which could be mined for promising transitions. Thus, seven months after Maiman's success with ruby, the helium-neon laser became the first type to emit a continuous beam rather than pulses (at 1.15 μm). The energy-level diagram of He and Ne atoms is shown in Figure 4.58.

Note that actual laser emission takes place through the neon energy levels while helium gas is added to the mixture to greatly facilitate the pumping processes. Ne is excited by the transfer of excitation from He atoms, which in turn are excited by collisions with electrons. The population inversion mechanism in He-Ne lasers thus involves a combination of electron-impact excitation (of He) and excitation transfer (from He to Ne). In Figure 4.58 the 3.39 μm, 1.15 μm, and 632.8 nm lines of neon, which are the strongest lasing transitions, are indicated, the actual oscillating transition basically depending on the wavelength at which the peak reflectivity of the dielectric cavity mirrors is centered. The common upper level of the 3.39-μm and 632.8-nm transitions, designated $3s_2$, is populated by excitation transfer from nearly resonant He atoms excited by electron impact to the 2^1S level. The upper level of the 1.15-μm transition is nearly resonant with the 2^3S level of He, and is populated by excitation transfer from He atoms in that excited state. Actually, Ne is also pumped directly into excited states by electron impacts, but the excitation transfer from He is the dominant pumping mechanism. The excited levels 2^1S_0 and 2^3S_1 of He, in addition to being nearly resonant with levels of Ne (and therefore allowing strong collisional excitation transfer), have the advantage of being forbidden by a selection rule to de-excite by spontaneous emission (S $\rightarrow$ S transitions are electric-dipole-forbidden). This allows these levels to *hold* energy for delivery to Ne during collisions. The partial pressures of Ne and He in typical He-Ne lasers are roughly 0.1 and 1 Torr, respectively. At these low pressures the upper-state lifetimes are determined predominantly by spontaneous emission rather than collisional de-excitation. The $3s_2$ and $2s_2$ levels of Ne have short radiative (i.e., spontaneous emission) lifetimes, roughly 10-20 ns, due to the strong allowed ultraviolet transitions to the ground state. For Ne pressures typical of He-Ne lasers, however, these radiative lifetimes are actually about 10^{-7} s because of radiative trapping. This occurs when the spontaneously emitted photons are re-absorbed by atoms in the ground state, thereby effectively increasing the lifetime of the emitting level. Since the ground state is generally the most highly populated level even when there is population inversion, radiative trapping is significant only from levels connected to the ground level by an allowed transition. Thus, the Ne $3p_4$ and $2p_4$ levels, which are forbidden by a selection rule from decaying spontaneously to the ground level, are not radiatively trapped and have lifetimes of about 10^{-8} s, roughly 10 times shorter than the $3s_2$ and $2s_2$ levels. This means that the s $\rightarrow$ p transitions have favorable lifetime ratios for lasing, that is, their lower (p) levels decay more quickly than their upper (s) levels, making it easier to establish a population inversion. The integrated absorption coefficients of the 632.8-nm and 3.39-μm lines have roughly the same magnitude, but the 3.39-μm line has a Doppler width about 5.4 times smaller than the 632.8-nm line, and consequently a considerably larger line-center gain. Without some mechanism for suppressing oscillation on the 3.39-μm line, therefore, the familiar 632.8-nm line would not lase. The 632.8-nm laser overcomes this problem by introducing into the optical path elements that absorb strongly at 3.39 μm, but not at 632.8 nm, such as glass or quartz Brewster windows. This raises the threshold pumping level for the 3.39 μm oscillation above that of the 632.8-nm oscillation. The 632.8-nm and 1.15-μm transitions have a common lower level, $2p_4$, which decays rapidly into the $1s_2$ level. The latter is forbidden by a selection rule from decaying radiatively into the ground level, and is therefore relatively long-lived. This is bad for laser oscillation on the 632.8-nm and 1.15-μm lines because electron-impact excitation can pump Ne atoms from $1s_2$ to $2p_4$, thereby reducing the population inversion on these lines. However, Ne atoms in the $1s_2$ level can decay to the ground level when they collide with the walls of the gain tube. In fact, it is found that the gain on the 632.8-nm and 1.15-μm lines increases when the tube diameter is decreased; this is attributed to an increase in the atom-wall collision rate with decreasing tube diameter.

A typical gas-laser setup is shown in Figure 4.59. The gas envelope windows are tilted at Brewster's angle such that radiation with the electric field vector in the plane of the paper suffers no reflection losses at the windows. This causes the output radiation to be polarized in the sense shown, because the orthogonal polarization undergoes reflection losses at the windows and, as a consequence, has a higher threshold.

4.8.2   Carbon dioxide laser

The electric-discharge carbon dioxide laser has a population inversion mechanism similar in some respects to the He-Ne laser: the upper CO_2 laser level is pumped by excitation transfer from the nitrogen molecule, with N_2 itself excited by electron impact. The relevant energy levels of the CO_2 and N_2 molecules are vibrational-rotational levels of their electronic ground states. Figure 4.60 shows that the first excited vibrational level ($v = 1$) of the N_2 molecule lies close to the level (001) of CO_2. Because of this near resonance (ΔE is only $\simeq 18$ cm^{-1}), there is a rapid excitation transfer between $N_2(v = 1)$ and $CO_2(001)$, the upper laser level. $N_2(v = 1)$ is itself a long-lived (metastable) level (transition $1 \rightarrow 0$ is in fact electric-dipole-forbidden as a diatomic homonuclear molecule cannot have a net electric dipole moment), so it effectively stores energy for eventual transfer to $CO_2(001)$; it is also efficiently pumped by electron-impact excitation. As in the case of the He-Ne laser,

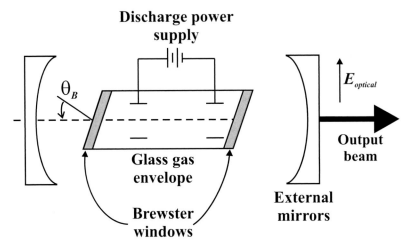

FIGURE 4.59
Schematic setup of a typical gas laser.

therefore, advantage is taken of a fortuitous near resonance between an excited state of the lasing species and an excited, long-lived collision partner. Moreover, the cross section for direct electric collisions $e + CO_2(000) \rightarrow e + CO_2(001)$ is very large and appreciably larger than those for excitation to both the 100 and 020 level. Note also that direct electron impact can also lead to excitation to upper (00n) vibrational levels of CO_2. However, the CO_2 molecule rapidly relaxes from these upper states to the (001) state by near-resonant collisions of the type $CO_2(00n)+CO_2(000) \rightarrow CO_2(00, n-1)+CO_2(001)$. This process tends to degrade all excited molecules to the (001) state. The next point to consider is the decay of both upper and lower laser levels. Note that, although transitions $001 \rightarrow 100$, $001 \rightarrow 020$, $100 \rightarrow 010$, and $020 \rightarrow 010$ are optically allowed, the corresponding decay times τ_{sp}, for spontaneous emission, are very long ($\tau_{spont} \propto 1/\nu^3$). The decay of these various levels is therefore determined essentially by collisions. Taking for example the case of a total pressure of 15 Torr (in a 1:1:8 CO_2:N_2:He partial-pressure ratio), one finds that the upper laser level has a lifetime $\tau_s \simeq 0.4$ ms. Concerning the relaxation of the lower laser level, we note that the 100 and 020 levels are essentially resonant. This accidental degeneracy results in a strong quantum mechanical coupling in which states in effect lose their separate identities (Fermi resonance effect). Furthermore, the 010 and 020 levels undergo a very rapid vibration-to-vibration energy transfer $020+000 \rightarrow 010+010$. For practical purposes, then, the stimulated emission on the $001 \rightarrow 100$ vibrational band takes CO_2 molecules from 001 to 010. The 010 level thus acts in effect like a lower laser level that must be rapidly knocked out in order to avoid a bottleneck in the population inversion. Now we are left with the decay time from 010 to the ground level 000. Since this transition is the least energetic one in any of the molecules in the discharge, relaxation from the 010 level can occur only by transferring this vibrational energy to translational energy of the colliding partners. According to collision theory, energy is most likely to be transferred to lighter atoms, i.e., to helium in this case. For the same partial pressures considered above one obtains a lifetime of about 20 μs. It follows from the preceding discussion that this is also the value of the lifetime of the lower laser level. Therefore, due to the much larger value of the upper state lifetime, population accumulates in the upper laser level, so the condition for cw laser action is also fulfilled. Note that He has another valuable effect: due to its high thermal conductivity, He helps keep the CO_2 cool by conducting heat to the walls of the container. A low translational temperature for CO_2 is necessary to avoid populating the lower laser level by thermal excitation; energy separation between levels is in fact comparable to $k_B T$.

From the preceding considerations we see that laser action in a CO_2 laser may occur either on the $001 \rightarrow 100$ ($\lambda = 10.6$ μm) transition or on the $001 \rightarrow 020$ ($\lambda = 9.6$ μm) transition. The laser transition has a low-pressure (< 5 torr) Doppler linewidth at $T = 300$ K of $\Delta\nu_D \simeq 60$ MHz (pressure broadening sets in at > 5 torr and becomes dominant above 100 torr). The CO_2 laser is one of the most powerful and efficient lasers (several kilowatts and about 30% overall working efficiency). This efficiency results primarily from the fact that the laser levels are all near the ground state and hence the atomic quantum efficiency is about 45%.

So far we have ignored the fact that both upper and lower laser levels consist of many closely spaced rotational levels. Accordingly, laser emission may occur on several equally spaced rotational-vibrational transitions belonging to either P or R branches, with the P branch exhibiting the largest gain. To complete our discussion we must also consider that, as a consequence of the Boltzmann distribution between rotational levels, the $J' = 21$ rotational level of the upper 001 state is the most heavily populated. Laser oscillation then occurs on the rotational-vibrational transition with the largest gain, i.e., originating from the most heavily populated level. This happens because, in a CO_2 laser, the thermalization rate among rotational levels ($\sim 10^7 \text{s}^{-1}$ torr^{-1}) is faster than the rate of decrease in population (due to stimulated emission) of the rotational level from which laser emission occurs. Therefore, the entire population of rotational levels contributes to laser action on the rotational level with the highest gain. In conclusion, laser action in a CO_2 laser normally occurs on the $P(22)$ line. Other lines of the same transition as well as lines belonging to the $001 \rightarrow 020$ transition (the separation between rotational lines in a CO_2 laser is about 2 cm^{-1}) can be selected by a diffraction grating.

4.8.3 Dye lasers

Dye lasers are the original tunable lasers. Discovered in the mid-1960s these tunable sources of coherent radiation span the electromagnetic spectrum from the near-ultraviolet to the near-infrared. In addition to their extraordinary spectral versatility, dye lasers have been shown to oscillate from the femtosecond pulse domain to the continuous wave regime. Dye lasers provide especially interesting examples of optical pumping, accomplished either by lasers or flash-lamps. In particular, recent advances in semiconductor laser technology have made it possible to construct very compact all solid-state excitation sources that, coupled with new solid-state dye laser materials, should bring the opportunity to build compact tunable laser systems for the visible spectrum [206]. In the following, however, by virtue of their relevance in the scope of spectroscopic frequency measurements (particularly for sub-Doppler spectroscopy), we only focus on traditional cw dye lasers utilizing liquid gain media. The active molecules in these lasers are large organic molecules in a solvent such as alcohol or water. Such molecular gain media have a strong absorption generally in the visible and ultraviolet regions, and exhibit large fluorescence bandwidths covering the entire visible spectrum. The general energy level diagram of an organic dye is shown in Figure 4.61.

It consists of electronic singlet and triplet states with each electronic state containing a multitude of overlapping vibrational-rotational levels giving rise to broad continuous energy bands. Indeed, due to the strong interaction of dye molecules with the solvent, the closely spaced rovibronic levels are collision broadened to such an extent that the different fluorescence lines completely overlap: thus, both the absorption and fluorescence spectrum consists of broad continuum which is homogeneously broadened. Absorption of visible or ultraviolet pump light excites the molecules from the ground state S_0 into some rotational-

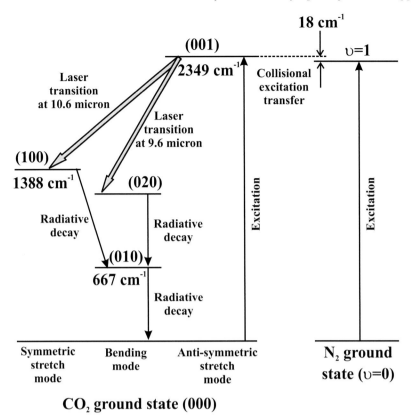

FIGURE 4.60
Vibrational energy levels of CO_2 and N_2.

vibrational level belonging to an upper excited singlet state, from where the molecules decay non-radiatively to the lowest vibrational level of the first excited singlet state S_1 on a picosecond time-scale. From S_1 the molecules can decay radiatively, with a radiative lifetime on the nanosecond time-scale, to a higher lying vibrational-rotational level of S_0. From this level they rapidly thermalize into the lowest vibrational-rotational levels of S_0. Alternatively, from S_1, the molecules can experience non-radiative relaxation either to the triplet state T_1 by an inter-system crossing process or to the ground state by an internal conversion process. If the intensity of the pumping radiation is high enough, a population inversion between S_1 and S_0 may be attained and stimulated emission occurs. Internal conversion and inter-system crossing compete with the fluorescence decay mode of the molecule and therefore reduce the efficiency of the laser emission. The rate for internal conversion to the electronic ground state is usually negligibly small so that the most important loss process is inter-system crossing into T_1 that populates the lower metastable triplet state. Thus, absorption on the triplet-triplet allowed transitions could cause considerable losses if these absorption bands overlap the lasing band, inhibiting or even halting the lasing process. This triplet loss can be reduced by adding small quantities of appropriate chemicals that favor non-radiative transitions that shorten the effective lifetime of the T_1 level. One very important characteristic of laser dye molecules is that their emission spectra are shifted in wavelength from their absorption spectra (the emission wavelength is longer than the absorption wavelength). This fortunate circumstance prevents the laser radiation from being strongly absorbed by the dye itself. This is a consequence of the Franck-Condon principle

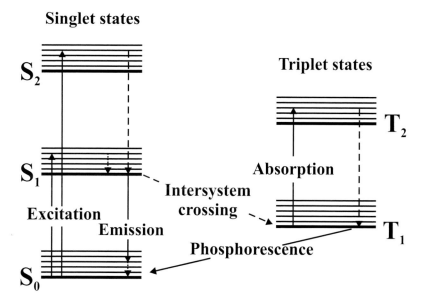

FIGURE 4.61
Schematic energy level diagram for a dye molecule. Full lines: radiative transitions; dashed lines: non-radiative transitions; dotted lines: vibrational relaxation.

and the fact that the vibrational relaxation associated with any electronic state is very rapid. Laser dyes usually belong to one of the following classes: A) Polymethine dyes, which provide laser oscillation in the wavelength range $0.7 - 1.5 \mu m$; B) Xanthene dyes, whose laser operation is in the visible (a celebrated example is represented by rhodamine 6G at 590 nm); C) Coumarin dyes, which oscillate in the blue-green region 400-500 nm (a celebrated example is represented by coumarin 2 at 450 nm). Oxadiazole derivatives, working between 330-450 nm, allow lasing in the UV region.

Cw dye lasers use dye flowing at linear speeds of up to 10 meters per second which are necessary to remove the excess heat and to quench the triplet states. Various resonator configurations, pump geometries, and designs of the dye flow system have been successfully tried to realize optimum performance in terms of output power, tuning range, and frequency stability.

In the original cavity reported by Peterson and colleagues, in 1970, a beam from an Ar^+ laser was focused onto an active region which was contained within the resonator. The resonator comprised dichroic mirrors that transmitted the blue-green radiation of the pump laser and reflected the red emission from the dye molecules. Using a pump power of about 1 W, in a TEM_{00} laser beam, a dye laser output of 30 mW was reported. Subsequent designs replaced the dye cell with a dye jet, an introduced external mirror, and integrated dispersive elements in the cavity. Dispersive elements such as prisms and gratings are used to tune the wavelength output of the laser. Frequency-selective elements, such as etalons and other types of interferometers, are used to induce frequency narrowing of the tunable emission. The most commonly used cw dye laser cavity design is shown in Figure 4.62. It is an eight-shaped ring dye laser cavity comprised of mirrors M1, M2, M3, and M4, which eliminates the spatial hole burning effect.

Two aspects of cw dye lasers are worth emphasizing. One is the availability of relatively high powers in single longitudinal mode emission and the other is the demonstration of very stable laser oscillation [207, 208]. In the area of laser stabilization and ultra-narrow linewidth oscillation it is worth mentioning the work by the group of J.C. Bergquist [209], where a dye laser at 563 nm was locked to a well-isolated, high-finesse (>150000) Fabry-Perot cavity.

Rather than locking the laser directly to the high-finesse cavity, it was first pre-stabilized against a cavity with a finesse of 800 using the PDH technique. To this aim, an intra-cavity EOM in the dye laser provided high-frequency correction of laser frequency noise. A PZT behind one of the dye laser cavity mirrors eliminated long-term frequency drifts between the dye laser and the low-finesse cavity. A loop bandwidth of 2 MHz in this pre-stabilization stage narrowed the dye-laser short-term linewidth to 1 kHz. Then, an optical fiber delivered light from the dye-laser table to a vibrationally isolated table supporting the high-finesse cavity. An AOM mounted on the isolated table shifted the frequency of the incoming light to match a cavity resonance. Again, frequency locking was implemented using the PDH method. The feedback loop performed corrections at frequencies as high as 90 kHz by varying the AOM drive frequency and at low frequencies by adjusting a PZT on the pre-stabilization cavity. The high-finesse cavity, consisting of ULE mirrors optically contacted onto the ends of a ULE spacer, was of the ultrastable type described in Chapter 3. In this way, a linewidth of 0.6 Hz for averaging times up to 32 s was achieved, with a fractional frequency instability of $3 \cdot 10^{-16}$ at 1 s.

From the early days of dye lasers, attempts were made to incorporate the dye molecules into solid hosts and a variety of materials and pumping arrangements were tried for operation of dyes in the solid state, but the lasing efficiencies were low and the dye molecules experienced fast photodegradation, resulting in the laser emission fading rather quickly. In the early 1990s, however, the development of improved host materials with higher laser damage resistance and the synthesis of new high-performance laser dyes spurred a renaissance in the field of solid-state dye lasers. In recent years, approaches involving the use of either new polymeric formulations or silicon-modified organic matrices or organic-inorganic hybrid materials as host materials for the laser dyes, are resulting in novel, improved solid-state dye lasers. These promising results have been obtained with dyes emitting in the green to red spectral region, while much less work has been done with dyes emitting in the blue. Also, the obtained results in solid state are still far from the performance of the same dyes in liquid solution. Additional interesting prospect applications of dye lasers can be envisioned in micro-opto-fluidics systems.

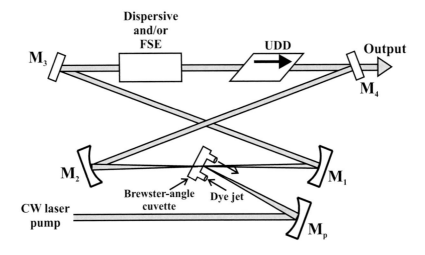

FIGURE 4.62
Cw dye laser ring cavity. FSE=frequency selective element, UDD=uni-directional device.

4.8.4 Ion-doped lasers and optical amplifiers

The first laser, set up by Maiman in 1960, was a ruby laser, using chromium ions (Cr^{3+}) in a solid-state matrix of sapphire (Al_2O_3) [161]. Thereafter, many different combinations of host materials, like $Y_3Al_5O_{12}$ (yttrium aluminum garnet or simply YAG) or glass and metal ions became successful gain media with emission wavelengths from the ultraviolet to the mid-infrared spectral range. In particular, ions of rare earths, like erbium, neodymium, ytterbium, holmium, praseodymium or thulium, represent nowadays worldwide spread dopants of solid-state and fiber gain media to construct efficient lasers and optical amplifiers. Despite the rapid progress in semiconductor laser technology, solid-state lasers still play an important role in many fields in science. An updated review on this topic is represented by [210].

4.8.4.1 Nd:YAG laser

Nd-YAG lasers are probably the most renowned in this family. Besides this oxide medium, other host media include some fluoride (e.g., $YLiF_4$) or vanadate (e.g., YVO_4) materials as well as some phosphate or silicate glasses. Typical doping levels in Nd:YAG, e.g., are ~ 1 atomic %. This laser relies on a classical four-level lasing scheme. A typical feature, shared within the family of rare earth-doped gain media, is that ions' levels energies are quite unperturbed by the host matrix. Energies of ions quantum levels are, in principle, shifted by electric fields within the host material (Stark shift). However, triply ionized rare earth dopants generally have the upper laser levels (4f), that determine the emission wavelength, well shielded from host electric fields by the outer 6s orbitals. Figure 4.63 shows a simplified energy level scheme for Nd:YAG. The two main pump bands of Nd:YAG occur at ~ 730 and 800 nm, respectively. These bands are coupled by a fast non-radiative decay to the $^4F_{3/2}$ level from where decay to the lower I levels occurs (to $^4I_{9/2}$, $^4I_{11/2}$, $^4I_{13/2}$, etc.). The rate of this decay is much slower ($\simeq 230$ μs). This means that level $^4F_{3/2}$ accumulates a large fraction of the pump power, so that it is a good candidate as the upper level for laser action.

From the preceding discussion, one sees that several laser transitions are possible between $^4F_{3/2}$ and several lower lying I levels; among these transitions, $^4F_{3/2} \rightarrow^4 I_{11/2}$ is the strongest one. Level $^4I_{11/2}$ is then coupled by a fast (hundreds of picoseconds) non-radiative decay to the $^4I_{9/2}$ ground level, so that thermal equilibrium between these two levels is very rapidly established. Since the energy difference between $^4I_{11/2}$ and $^4I_{9/2}$ levels is almost an order of magnitude larger than k_BT, then, according to Boltzmann statistics, level $^4I_{11/2}$ may, to a good approximation, be considered empty at all times. Thus laser operation on the $^4F_{3/2} \rightarrow^4 I_{11/2}$ transition corresponds to a four-level scheme. The $^4F_{3/2}$ level is split by the Stark effect into two sub-levels (R1 and R2), while the $^4I_{11/2}$ level is split into six sub-levels. Laser action usually occurs from the upper R2 sub-level to a particular sub-level of the $^4I_{11/2}$ level, since this transition has the highest value for the stimulated emission cross section. The transition occurs at $\lambda = 1.064$ μm, which is the most widely used lasing wavelength for Nd:YAG lasers. Note that laser action can also be obtained on the $^4F_{3/2} \rightarrow^4 I_{13/2}$ transition ($\lambda = 1.319$ μm is the strongest transition wavelength in this case) provided the multilayer dielectric coatings of the cavity mirrors have high reflectivity at $\lambda = 1.319$ μm and sufficiently low reflectivity at $\lambda = 1.064$ μm. In the case of the usual $\lambda = 1.064$ μm transition, the laser transition is homogeneously broadened at room temperature via interaction with lattice phonons. The corresponding width is $\Delta\nu \simeq 4.2$cm^{-1}=126 GHz at T=300 K. Nd:YAG lasers can operate either cw or pulsed, and can be pumped by either a lamp or a semiconductor laser.

With its very narrow free-running linewidth (about 1 kHz) and a power stability of 0.1%

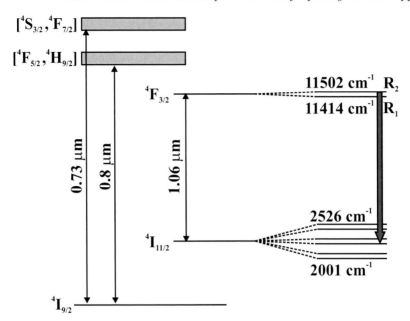

FIGURE 4.63
Simplified energy levels of Nd:YAG.

(when the laser employs an internal amplitude noise eater), the Nd:YAG laser (at 1064 nm), in the non-resonant planar ring oscillator configuration, represents an ideal starting point for developing ultrastable optical oscillators. Indeed, a subhertz-linewidth Nd:YAG laser was recently obtained by locking to a high-finesse, ultrastable reference cavity by means of the PDH technique [211].

4.8.4.2   Ti:Sa laser

In the case of a TiSa laser it is a Ti^{3+} ion which is the dopant into sapphire, Al_2O_3, as host. This small amount of titanium (roughly 0.1%) is then responsible for the lasing emission. It can be operated over a broad tuning range ($\Delta\lambda \simeq 400$ nm), thus providing the largest bandwith of any laser. Figure 4.64 shows the $3d^1$ absorption and emission band of the Ti^{3+} ion. The surrounding crystal field splits this level up into the $^2T_{2g}$ and the ^2E_g levels. These levels get split further by the spin orbit interaction and experience a big broadening caused by the interaction of the ion with lattice vibrations.

Population inversion is achieved by a strong pump laser populating a high vibrational mode of the ^2E_g level (usually an Ar laser or a frequency-doubled Nd:YAG laser). The electrons then decay very quickly by phonon interaction into a lower vibrational mode. A decay by emission of a photon into a vibrational mode of the $^2T_{2g}$ level is followed by a non-radiative decay into the ground state. This effectively corresponds to a four-level laser where two transitions occur by a fast non-radiative decay. Since the gain profile of the active medium is very broad, in order to avoid multi-mode lasing, additional frequency selective elements inside the laser help to select a single mode. A further advantage of the Ti:Sa crystal is the very good mechanical stability and the high heat conductivity which renders it pretty stable in frequency and power output. A comprehensive description of an advanced cw Ti:Sa laser system can be found in [177].

Finally, due to the extremely large gain bandwidth of the Ti^{3+} ion, allowing the gener-

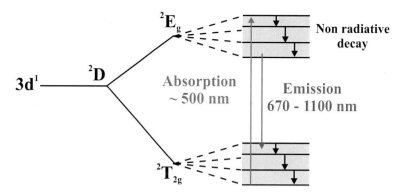

FIGURE 4.64
Level scheme of the Ti^{+3} ion in sapphire.

ation of very short pulses, the Ti:Sa laser has also played a crucial role in the realization of optical frequency comb synthesizers via the Kerr-lens mode-locking effect (see Chapter 6).

4.8.4.3 Erbium-doped fiber amplifiers (EDFAs)

It has been found that the glass optical fibers that are used as waveguides in communications can also be used as optical amplifiers if they are doped with optically active atoms such as erbium. The erbium-doped fiber amplifier(EDFA) is the one that is most widely used because the energy levels of the Er^{3+} ion in a glass host provide a convenient pumping wavelength centered about 0.94 μm and a stimulated emission central wavelength of 1.54 μm. Glass fibers have three wavelength regions of low loss at approximately 0.8 μm, 1.3 μm, and 1.55 μm that have come to be known as the first, second, and third telecommunications windows, respectively. Semiconductor lasers, which are commercially available with emission wavelengths of 810 and 980 μm, are generally used as the pumping source. However, there is also a strong absorption band centered about 1.5 μm, which permits pumping with 1.48 μm semiconductor lasers [212]. The transitions that are involved in producing the stimulated emission of an EDFA are shown in Figure 4.65. Most often a pump wavelength of 980 nm is used to raise ions to the upper level. The lifetime in this state is very short, on the order of 1 μs. Ions then decay to a metastable state with a relatively long lifetime of approximately 10 ms. This long lifetime allows the population of ions in the metastable state to build up so as to produce an inverted population with respect to the ground state.

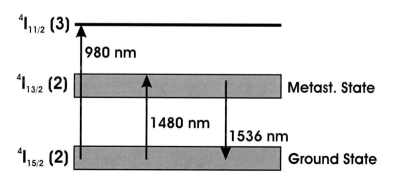

FIGURE 4.65
Energy levels and transitions in EDFA.

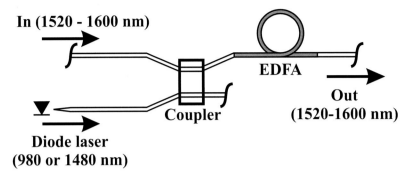

FIGURE 4.66
Basic configuration of an EDFA.

The fact that the metastable and the ground state are not single energy levels, but rather manifolds or bands of energy levels, results in the amplifier having gain over a range of wavelengths from approximately 1520 to 1600 nm. The basic physical implementation of an EDFA is shown in Figure 4.66. A thorough discussion about specific EDFA systems and their performances is given in [213] and [214].

4.8.4.4 Ytterbium-doped fiber amplifiers (YDFAs)

Yb-doped silica fiber gain medium operates efficiently in the 1030-1100 nm wavelength region. As an example, the energy levels of Yb^{3+} ions in Yb:YAG are shown in Figure 4.67. Within the above range, output power levels of several kW have been achieved. Gain media doped with Yb ions have a comparatively higher power efficiency, that is defined as the difference between the energy of each pump photon to that of a laser photon (also called Stokes shift), with respect to Er:Yb-codoped lasers emitting around 1.5 micron wavelength. Indeed, Yb-doped gain media are generally pumped at 976 nm wavelength and laser photons can have an energy differing only a few percent from the pump, providing higher slope efficiency and lower threshold. This also implies that they are quasi-three-level lasers, with the general problem of re-absorption of generated laser photons by unsaturated gain medium. This makes the overall design of Yb-doped lasers and amplifiers a bit more critical, but the impressive scaling of their performance in the last ten years, especially in terms of output powers, justifies their increasing and pervasive use in many applications simultaneously requiring high powers and high coherence. Since the emission cross section of Yb-doped aluminosilicate fiber extends up to 1200 nm, several approaches have been proposed to operate Yb-doped fiber lasers in the long wavelength (up to 1200 nm) range [215, 216, 217, 218, 219].

In addition to a broad-gain bandwidth, Yb-doped fiber amplifiers can offer high output power and power-conversion efficiency. Also, they do not have many of the problems affecting erbium-doped amplifiers. In particular, excited state absorption and concentration quenching by inter-ionic energy transfer do not occur, and high doping levels are possible, leading to high gain in short fiber lengths. Amplification of ultrashort pulses takes advantage from the broad bandwidth and the high saturation fluence allows for high pulse energies.

In principle, pumping sources can be in the 860-1064 nm wavelength range, thus allowing a variety of pumping schemes. To get high-output power from fiber amplifiers, double-clad fibers are used [220, 221]. Such fibers usually have a single-mode core where the signal wave propagates, surrounded by a larger multi-mode undoped inner cladding into which the pump light is launched. The modes of the inner cladding have some overlap with the doped core so that the pump light can be absorbed there. The main advantages of this scheme are the

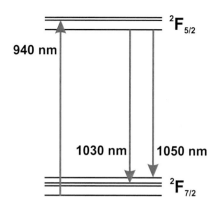

FIGURE 4.67
Energy levels of Yb^{3+} ions in Yb:YAG, and the usual pump and laser transitions.

following: higher powers can be coupled into a multi-mode core (because of the larger spot size); the pump source (e.g., a high-power diode laser) does not need to emit a single spatial mode; the pump launch efficiency can be very high and the alignment tolerances relatively uncritical.

In 1997, two groups recognized the peculiar advantages of Yb-doped fiber amplifiers for precise spectroscopic applications, as compared to semiconductor lasers and amplifiers [222, 223]. Indeed, the combined high-power and narrow-linewidth properties proved to be very appealing for precise atomic and molecular frequency measurements as well as for the development of frequency standards in the visible range by use of frequency-doubled radiation [224, 225, 226]. A very low added phase noise arising in the amplification of highly coherent cw sources was measured for the first time in 2000 [227]. In that work, an upper limit of 300 Hz was measured for the additional phase noise originating in the amplification of the radiation coming from a 5-kHz-linewidth NPRO-based Nd:YAG laser. Moreover, a negligible cross talk for simultaneous amplification of several different laser frequencies was for the first time described, marking a significant difference with respect to semiconductor-based amplifiers. This feature seemed to be in contrast with the dominantly homogeneous transition broadening observed in Yb-doped germanosilicate and silicate glasses [228]. Such effect could be ascribed to the slow gain dynamics of YDFAs. Indeed, transient effects of gain saturation and recovery typically occur on a time scale of approximately 100 ms, depending on signal and pump levels. In the saturation regime, YDFAs are thus virtually free from transient gain modulation effects (referred to as cross talk), that vanish for modulation frequencies above approximately 10 kHz. This is a definite advantage of rare earth-doped fiber amplifiers as compared to semiconductor optical amplifiers, which are characterized by a picosecond gain dynamics and thus show high cross talk effects, which represents a well-known limit for multi-channel transmission in optical communication systems.

To better understand the physics of Yb-doped fiber amplifiers, we will make an example of basic design mainly intended for operation around 1083 nm, that corresponds to a well-known transition of atomic He $(2^3S \to 2^3P)$, thoroughly investigated to determine the fine structure constant and measure QED effects. We make this example by using the original design reported in [222]. As shown in Figure 4.68, the emission cross section of an Yb-doped germanosilicate glass has maxima around 975 and 1027 nm.

In particular, the emission cross section at 1083 nm (where we want to optimize the amplifier emission) is one third lower than that at 1027 nm. As a general rule, in order to get an optimized amplifier configuration, the residual pump power at one end of the fiber

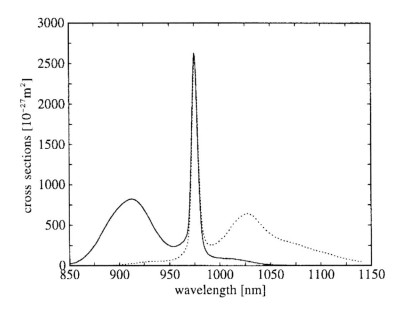

FIGURE 4.68
Absorption and emission cross sections of ytterbium-doped germanosilicate glass, as used in the cores of ytterbium-doped fibers. (Courtesy of [222].)

should be just sufficient to reach transparency (zero gain) for the signal wave at this end [228]. In Figure 4.69 we plot the transparency power versus signal wavelength for various pump wavelengths and typical fiber parameters as given by

$$P_{trans} = \frac{A_C h\nu_p}{\eta_p(\sigma_{21}^{(s)}\sigma_{12}^{(p)}/\sigma_{12}^{(s)} - \sigma_{21}^{(p)})\tau} \tag{4.234}$$

where A_c is the core area, ν_p is the pump frequency, $\sigma_{12}^{(p)}$ and $\sigma_{21}^{(p)}$ are the effective pump absorption and emission cross sections, $\sigma_{12}^{(s)}$ and $\sigma_{21}^{(s)}$ the corresponding values for the signal wavelength, and η_p represents an additional overlap factor accounting for the fact that some fraction of the power propagates in the undoped cladding of the fiber.

By pumping around 910 nm wavelength, where there is a strong absorption peak, one can get a strong population inversion, but most of the amplifier gain would be concentrated around 975 nm, where there is the strongest emission peak. Since the gain at 1083 nm (where the seeding radiation is injected) is much lower, this would imply a strong ASE (Amplified Spontaneous Emission) generation around 975 nm. Of course, one would like to ideally get 100% of the amplified radiation within the linewidth of the seeder and nothing elsewhere. To avoid such a strong ASE generation with a consequent low gain around 1083 nm, a better pumping wavelength is around 975 nm. Around this wavelength the cross sections for absorption and emission are about the same and an inversion approaching 50% can be achieved for powers well above the pump saturation power. In these conditions, most of the ASE comes from the emission peak at 1027 nm, that is anyway much lower than that at 975 nm. Moreover, the expected gain at 1083 nm is approximately one-third of that at 1027 nm. ASE generation can be limited by proper filtering around the emission peak wavelength. It is also worth noting that when the seeding wavelength is far enough from the absorption peaks, as is the case of 1083 nm, some overlength of fiber does not

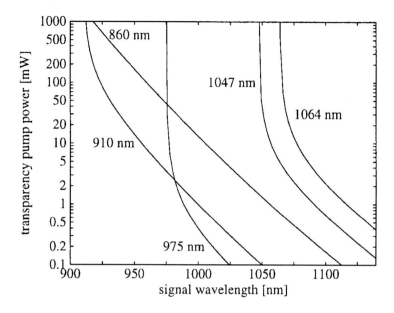

FIGURE 4.69
Transparency pump power versus signal wavelength for various common pump wavelengths, assuming a germanosilicate fiber with 900-nm cutoff and NA=0.2. (Courtesy of [222].)

create significant re-absorption problems and hence does not degrade the amplifier gain performance (indeed, the signal re-absorption at 1083 nm in unpumped fiber is estimated at only 0.4 dB/m, including background loss). Figure 4.70 shows the experimental setup. A pump power of up to 500 mW at 978 nm was generated by a fiber-coupled semiconductor MOPA laser, while the 1083-nm signal was produced by a semiconductor distributed Bragg reflector (DBR) laser with external cavity. In a preliminary experiment 14 dB gain and 60 mW of amplified signal were achieved.

As already mentioned, in a subsequent experiment, such laser source was used for accurate frequency measurement of the fine-structure energy splittings in helium atom [226]. Since obtaining a secondary frequency standard at 1083 nm was a difficult task, a valuable approach was doubling the frequency to 541 nm and then seeking a suitable saturated-absorption dip in the rich I_2 spectrum. Thus, the 1083-nm emission coming from the ytterbium-doped fiber amplifier was applied to a type-I phase matched Mg:LiNbO$_3$ crystal in a high-Q fundamental-resonant cavity for frequency doubling. The doubling cavity Q was $\sim 2 \cdot 10^7$, and about 70% of the incident power was coupled in. The maximum 2nd-harmonic conversion efficiency exceeded 20% and provided an output power of 3.7 mW for making continuous frequency scans of up to 600 MHz in the green. An optical spectrum analyzer at 541.5 nm showed fringes of 4.6 MHz full width half maximum, close to the instrumental width.

4.8.4.5 Narrow-linewidth fiber lasers

The broadband emission of trivalent rare earth ions allows the development of sources emitting either broad continuous-wave (cw) spectra or ultrashort pulses, as well as widely tunable narrow-linewidth radiation, as described in the following [229]. The core diameter of a standard single-mode fiber can vary from 3 to 10 μm, so that a significant intensity can be developed with a modest average power. As a consequence, the intensities necessary to reach

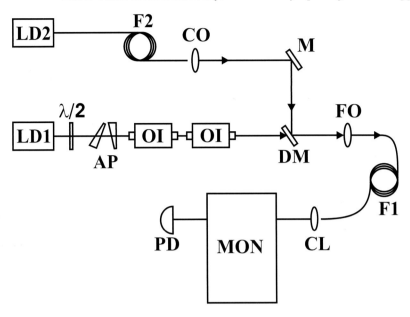

FIGURE 4.70
Set-up of a power amplifier for a single-frequency at 1083 nm. It is based on a Yb-doped silica fiber which is pumped by a MOPA semiconductor laser at 978 nm. LD1=1083-nm laser diode; AP=anamorphic prisms; OI=optical isolator; M=mirror; DM=dichroic mirror; FO=20× microscope objective; LD2=978-nm pump laser (fiber coupled through F2); F1=Yb doped fiber; CO,CL=collimating lenses, MON= monochromator; PD = photodetector.

the cw oscillation threshold for both three-level and four-level systems can be achieved with modest input powers. The inherent waveguiding property associated with the fiber ensures that the intensity is maintained over long distances, thereby providing long interaction lengths between the rare earth dopant and the pump field. Hence, a significant intracavity gain can be developed, and a small-signal gain of 25 dB is not uncommon. A small-signal gain of this amplitude enables elements with a comparatively high insertion loss, such as optical isolators, frequency modulators, and integrated interference filters, to be inserted in the cavity of a fiber laser without significantly increasing the oscillating threshold or reducing the output power. A fiber laser using a trivalent rare earth as the active element has the potential for very narrow linewidth operation compared with other sources that oscillate in the same spectral regions (e.g., semiconductor lasers). The cavity linewidth scales inversely with the cavity length of the laser, and the waveguiding nature of a fiber allows cavity lengths of many meters to be established. In comparison, the cavity length of semiconductor lasers is typically a fraction of a centimeter. The coupling of amplified spontaneous emission to the oscillating mode is determined by the gain cross section of the transition. For most rare earth ions, this cross section is of the order of 10^{-21} cm^2, whereas for a semiconductor laser it is typically 10^{-16} cm^2. This means that the optimum linewidth that can be expected from a fiber laser is significantly smaller than that of a semiconductor laser, making the fiber laser a suitable tool for narrow-linewidth applications. In addition, single-frequency fiber lasers are attractive because they can offer shot-noise-limited operation in the MHz range, whereas for semiconductor lasers this occurs several 100 MHz away from the center frequency. The round-trip optical path length for a typical fiber laser may range from 1 cm to 50 m, giving a longitudinal mode spacing from 30 GHz to 6 MHz. Since rare earth dopants (like Er^{3+} and Nd^{3+}) exhibit fluorescence spectra that extend over hundreds of THz, the gain medium can support many longitudinal modes of the cavity. It is accepted that the spectral broadening for a rare earth dopant in an amorphous host is a result of both homogeneous and inhomogeneous broadening mechanisms. The most effective way to obtain oscillation in only one longitudinal mode is to include bandwidth-limiting elements in the cavity. These can be either wavelength-selective mirrors (such as a Bragg reflector, a diffraction grating,...) or tunable filters (like an interference filter, a Fabry-Perot filter,...). More specifically, several different configurations have proved reliable to induce narrow-linewidth, single-mode operation in fiber lasers for a variety of host and dopant materials [229]. In the simplest realization, a laser cavity is established by two temperature-controlled, spectrally narrow passive FBGs that are fusion spliced to a short piece of active material. In this way, the lasing wavelength is determined by the spectral overlap between the two FBGs. In such a scheme, aside from varying the temperature of the Bragg reflector, the laser wavelength can be tuned by placing the grating under tensile or compressive stress. However, one of the major problems associated with a linear-cavity laser is spatial hole burning, which can preclude single-frequency operation. Various methods to overcome this drawback have been proposed and implemented, ranging from simple short two-mirror cavities to modified Sagnac cavities as well as twisted-mode fiber laser cavities. In the former case, the two outputs from a 50% fused fiber coupler are spliced together to form a fiber Sagnac loop (Figure 4.71). Light launched into one of the coupler's input ports propagates in both directions around the loop and recombines at the coupler. For a lossless 50% coupler and no loop birefringence, all the light is coupled back to the arm into which light was launched. In other words, the loop acts as a mirror with a reflection coefficient of 100%. Then, the Sagnac geometry can be modified by placing a length of gain fiber and an optical isolator inside the fiber loop. The latter suppresses spatial hole burning. If one of the input arms is spliced to a length of fiber containing a reflective device (either a bulk mirror or a bulk diffraction grating or a Bragg reflector) a linear cavity can be configured in which the gain medium is accessed by a travelling wave, the remaining port of the coupler then acting as an output for the cavity.

In the twisted-mode configuration, spatial hole burning is avoided by ensuring that circularly polarized light counter-propagates through the gain medium (Figure 4.72). In general, a standard rare earth doped fiber is weakly birefringent. By applying the correct strain to such a fiber, any desired state of polarization can be generated at a given point. A twisted-mode cavity comprises a polarization-sensitive fused fiber coupler, which defines the polarization state of the cavity and a length of active fiber placed between the two polarization controllers. The cavity reflectors can be provided, for instance, by butt-coupled high reflectors as shown in Figure 4.72. With proper adjustments of the polarization controllers, two counter-propagating waves with the same circular polarization state are generated,

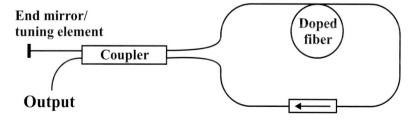

FIGURE 4.71
Diagram of unidirectional Sagnac loop fiber laser. (Adapted from [229].)

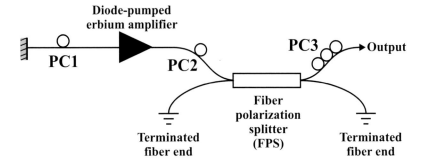

FIGURE 4.72
Schematic of a twisted-mode fiber laser cavity. (Adapted from [229].)

which avoids interference, thus eliminating spatial hole burning. With such lasers, output powers around 1 watt and linewidths of a few kHz can be achieved.

However, one of the major problems associated with both the modified Sagnac loop lasers and the twisted-mode laser is that they are sensitive to environmental perturbations, which can induce random frequency changes and mode hopping. Therefore, the most effective method to eliminate spatial hole burning is to ensure that the laser operates as a unidirectional ring, such that the gain medium is interrogated by a travelling wave. While in a bulk ring laser, a differential loss of $\sim 5\%$ is sufficient to ensure unidirectional operation, in a fiber laser such loss level does not necessarily eliminate bidirectional operation and a more effective method of forcing unidirectional operation is needed. For example, an integrated fiber-optic isolator with a low insertion loss (1 dB) and a high optical isolation (greater than 35 dB) can be used. Because of their importance in optical communications, single-frequency Er-doped fiber lasers operating at 1550 nm have been extensively studied in the unidirectional ring configuration. A paradigmatic scheme is shown in Figure 4.73. In this case, multi-mode operation is inhibited by increasing the spectral selection in the cavity by concatenation of two FP fiber filters. The first of them is broadband (FSR = 4 THz, transmission bandwidth of 26 GHz) and can be used for coarse electrical tuning of the laser. The second filter (FSR = 100 GHz, bandwidth 1.4 GHz) suppresses the multi-mode operation. An optical isolator placed between the two filters eliminated inter-etalon effects. In the first realization, the laser linewidth was measured using a re-circulating, delayed, self-heterodyne interferometer to be 1.4 kHz with a frequency jitter of 2.4 kHz owing to thermal noise [230]. The noise characteristics of the laser system were also quantified: if the amount of ASE power (originating in the gain medium) coupled to the cavity modes is reduced, the dominant noise source will be the detector itself. This can be accomplished by placing the narrow-band filter between the gain medium and the output coupler.

In the remainder we focus on the second main category of fiber lasers, i.e., that of Yb-doped lasing devices. With a gain coefficient of up to 5 dB/cm, optical single-mode fiber made from highly doped phosphate glass is uniquely suited for short, single-frequency fiber lasers based on this design. Indeed, Er/Yb co-doped phosphate glass fiber lasers, emitting around 1.06 or 1.5 μm, have been developed using only a few centimeters of active fiber (which relaxes the requirements for mode selection) with attractive output features: narrow linewidth around 1-3 kHz, fast frequency modulation up to 10 kHz, polarized, single-mode (with a side-mode suppression ratio better than 50 dB) output exceeding 200 mW with very low intensity, and frequency noise [231]. In particular, for an emission wavelength of 1038 nm, coinciding with the maximum of the gain spectrum, a slope efficiency of more than

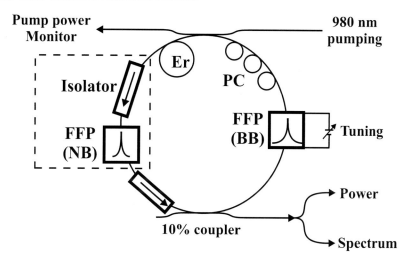

FIGURE 4.73
Erbium-doped unidirectional ring resonator containing two fiber FP filters with different free spectral ranges. FFP=fiber Fabry-Perot, NB=narrow band, BB=broadband, PC=polarization controller, WDM=wavelength-division multiplexer, Er=erbium-doped fiber. (Adapted from [229].)

50 % and a record output power of more than 400 mW was observed from a 1.7-cm long piece fiber.

More recently, an all-fiber, high power, Yb-doped silica fiber laser operating at 1179 nm (when core pumped at 1090 nm and heated at 125°C) was demonstrated [219]. In that case aluminosilicate host fiber, which is preferable for the long wavelength generation, was used. Such fiber had a 10 μm Yb-doped core diameter with an NA of 0.11. The Yb concentration in the core was estimated to be 14000 ppm-wt. The background loss in the fiber was less than 10 dB/km. Figure 4.75 shows the schematic of the Yb-doped fiber laser experimental set up. The fiber laser consisted of a 20-m long Yb-doped fiber that was fusion spliced to a 1090-1179 nm WDM (wavelength division multiplexer) coupler. The pump power was

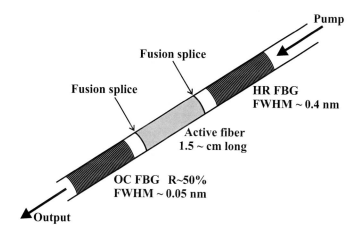

FIGURE 4.74
Cavity geometry for Yb-doped phosphate-glass fiber laser. (Adapted from [231].)

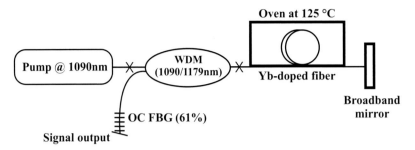

FIGURE 4.75
Experimental setup of the 1179 nm fiber laser. (Adapted from [219].)

coupled into the core of the Yb-doped fiber through the WDM coupler. The pump consisted of a patented Yb-doped fiber laser capable of delivering up to 30 W of output power at 1090 nm. Therefore, the WDM coupler was mounted on a heat sink to remove excess heat and avoid damage during the high power operation. The other end of the doped fiber was flat cleaved and butted to a broadband mirror with high reflectivity both at the pump and signal wavelengths. An FBG, acting as an output coupler, was spliced to the 1179 nm arm of the WDM coupler to select the lasing wavelength. The reflectivity of the FBG was 61% at 1179 nm with 3 dB bandwidth of 0.25 nm. The Yb-doped fiber was placed inside an oven, which was maintained at 125 °C to increase the pump absorption at 1090 nm. With such a system, an output power exceeding 12 W (corresponding to a slope efficiency of 43 %) was obtained, and the 3-dB linewidth of the laser spectrum was less than 0.38 nm.

4.8.5 Semiconductor lasers

The concept of a semiconductor laser was introduced by Basov et al. (1961) who suggested that stimulated emission of radiation could occur in semiconductors by the recombination of carriers injected across a p-n junction [232]. The first semiconductor lasers appeared in 1962, when three laboratories independently achieved lasing. After that, progress was slow for several reasons. One reason was the need to develop a new semiconductor technology. Semiconductor lasers could not be made from silicon where a mature fabrication technology existed. Rather, they required direct bandgap materials which were found in compound semiconductors. There were also problems involving high threshold currents for lasing, which limited laser operation to short pulses at cryogenic temperatures, and low efficiency, which led to a high heat dissipation. A big stride toward solving the above problems was made in 1969, with the introduction of heterostructures. Then, cw operation at room temperature became possible because of better carrier and optical confinement. Two factors are largely responsible for the explosion in the field of semiconductor lasers. One is the exceptional and fortuitous close lattice match between AlAs and GaAs, which allows heterostructures consisting of layers of different compositions of $Al_xGa_{1-x}As$ to be grown. The second is the presence of several important opto-electronic applications where semiconductor lasers are uniquely well suited, as they have the smallest size (several cubic millimeters), highest efficiency (often as much as 50% or even more), and the longest lifetime of all existing lasers.

Today, semiconductor lasers include several types of devices. The most important classes for high-resolution spectroscopy are reviewed in the following.

4.8.5.1 Heterostructure diode lasers

Even if the p-n homojunction is now relegated to history, it is very useful to illustrate the general working principle. Referring to results from quantum mechanical treatments as found in standard textbooks, we first illustrate some basic properties of semiconductor physics [233]. Semiconductors are solid-state materials where the valence band is filled with electrons and the conduction band is empty at zero temperature. In contrast to isolators, the energy width of the gap between these bands is about 1 eV and, hence, at finite temperatures some electrons are thermally activated into the valence band. In a semiconducting material, the Fermi energy (i.e., the energy level separating the filled energy levels from the empty ones) is located in the gap between the conduction and the valence band (close to the middle of the band gap). Also, we can state that it is possible for many purposes to treat electrons in the conduction band and holes in the valence band similar to free particles, but with an effective mass (m_n or m_p) different from elementary electrons not embedded in the lattice. This mass is furthermore dependent on other parameters such as the direction of movement with respect to the crystal axis. The kinetic energy of electrons is measured from the lower edge of the conduction band upwards, that of the holes downward from the upper edge of the valence band (Figure 4.76).

Intrinsic semiconductors are rarely used in semiconductor devices since it is extremely difficult to obtain sufficient purity in the material. Moreover, in most cases one intentionally alters the property of the material by adding small fractions of specific impurities. This procedure, which can be performed either during crystal growth or later in selected regions of the crystal, is called doping. Depending on the type of added material, one obtains n-type semiconductors with an excess of electrons in the conduction band or p-types with additional holes in the valence band. In the former (latter) case, a movement of the Fermi level $E_{F,n}$ ($E_{F,p}$) from the intrinsic level towards the conduction (valence) band occurs. For heavily doped material ($\sim 10^{18}$ atoms/cm^3), the levels $E_{F,n}$ and $E_{F,p}$ are pushed into the conduction and valence bands, respectively (Figure 4.77) [234].

In the GaAs semiconductor, a popular donor impurity is Se from column VI in the periodic table, which has one more valence electron than As, which is in column V, while popular acceptor impurity is Zn from column II, which has one valence electron less than Ga from column III. Note that the doped media by themselves are electrically neutral, that is, they are not charged positively or negatively, even though they have current carriers. To see how an inversion is created at a p-n junction, we plot in Figure 4.77 the energy

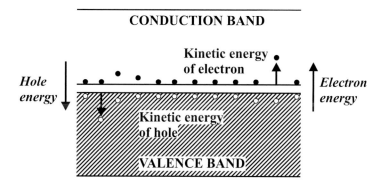

FIGURE 4.76
Potential and kinetic energy for a semiconductor in the band representation.

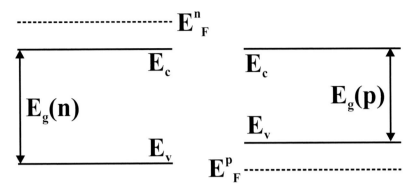

FIGURE 4.77
Band diagram of heavily doped p and n GaAs.

bands and electron occupation as functions of position in the transverse x direction, i.e., perpendicular to the junction plane. This figure shows that, once the two bodies (n-type and p-type) are brought into contact, in the absence of an applied voltage across the electrodes, the chemical potential μ (Fermi energy) is constant throughout the entire structure (the two levels $E_{F,n}$ and $E_{F,p}$ line up through the diffusion of electrons from the n^+ to the p^+ and vice versa to establish thermal equilibrium) resulting in no net flow of carriers. More importantly, there is no region containing both electrons in the conduction band and holes in the valence band, which is necessary to obtain an inverted population. When a voltage V_0 is applied so that the p-doped region is positive relative to the n-doped region, the system is not in thermal equilibrium and the electron energies are altered as shown in the lower frame of Figure 4.78. The shared chemical potential now splits into two levels separated by the amount eV_0 so that a region originates over which population inversion exists (active region).

Inside this region, stimulated emission occurs due to electron-hole recombination. At steady state, the inversion is maintained by the injection of carriers, via the electrodes, by an external power supply. Lasing occurs when the rate of stimulated emission due to electron-hole recombination approximately equals the total rate of optical losses. If it is assumed that the only electrons that recombine are those at the bottom of the conduction band, then electrons injected at the quasi-Fermi level must lose kinetic energy in order to reach the band edge. This is achieved by the emission or absorption of phonons so that both energy and momentum are lost to the lattice. The electrons therefore *cascade* through the states to occupy those made empty by recombination (Figure 4.79). Similar arguments apply to the hole current. Such a *cascade* process is extremely fast, of the order of a picosecond, in the extended states of the semiconductor. Therefore, a semiconductor with an inverted population behaves like a four-level system.

To date, most laser diodes have been made in GaAs, $Ga_{1-x}Al_xAs$, or $Ga_xIn_{1-x}As_{1-y}P_y$, but other materials will no doubt eventually also be used extensively to obtain emission at different wavelengths, once the fabrication technology has been developed.

A p-n junction is usually formed by epitaxial growth of a p-type layer on an n-type substrate. Ohmic contacts are made to each region to permit the flow of electrical current which is the pumping energy source required to produce the inverted population in the active region adjacent to the junction. Two parallel end faces are fabricated to work as mirrors providing the optical feedback necessary for the establishment of a lasing optical mode (Figure 4.80). To provide feedback for laser action, the two end faces are prepared by cleavage along crystal planes. Often, these two surfaces are not provided with reflective

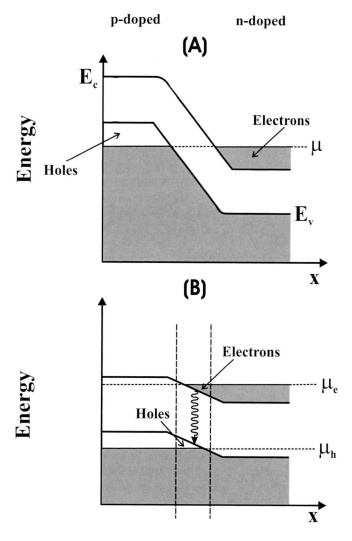

FIGURE 4.78
Electron energy and occupation perpendicular to the p-n junction (a) without an applied voltage and (b) with a forward biased applied voltage.

coatings; in fact, since the refractive index of a semiconductor is very large (e.g., n=3.6 for GaAs), there is already a sufficiently high reflectivity (about 32% for GaAs) from the Fresnel reflection at the semiconductor-air interface [212]. The laser end faces are often referred to as Fabry-Perot (FP) surfaces.

As already explained, when current is passed through the laser diode, light can be generated in the resulting inverted population layer by both spontaneous and stimulated emission of photons. Due to the reflection that occurs at the FP surfaces, some of the photons will pass back and forth many times through the inverted population region and be preferentially multiplied by stimulated emission. Those photons that are travelling exactly in the plane of the layer and exactly perpendicular to the FP surfaces have the highest probability of remaining in the inverted population layer where they can reproduce themselves by stimulated emission. Hence, they become the photons of the optical mode or modes that are established when steady-state operation is achieved at a given current level. The radiation

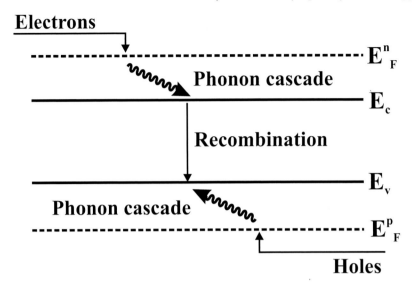

FIGURE 4.79
An equivalent four-level scheme for a diode laser with phonon cascade for the electrons and holes. (Adapted from [234].)

of the lasing mode must also be of uniform frequency and phase to avoid destructive interference. As a result, a standing wave is produced within the laser diode with an integral number of half-wavelengths between the parallel faces

$$m = \frac{2Ln}{\lambda_0} \tag{4.235}$$

where L is the distance between the end faces, n the index of refraction of the laser material, and λ_0 is the vacuum wavelength of the emitted light. The mode spacing is determined by taking $dm/d\lambda_0$, keeping in mind that semiconductor lasers are operated near the bandgap where n is a strong function of wavelength. Thus

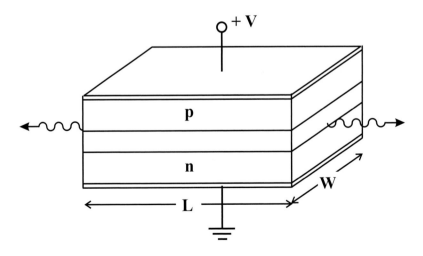

FIGURE 4.80
Basic structure of a p-n junction laser.

$$\frac{dm}{d\lambda_0} = -\frac{2Ln}{\lambda_0^2} + \frac{2L}{\lambda_0}\frac{dn}{d\lambda_0} \tag{4.236}$$

For $dm = -1$, the mode spacing is given by

$$d\lambda_0 = \frac{\lambda_0^2}{2L\left(n - \lambda_0\dfrac{dn}{d\lambda_0}\right)} \tag{4.237}$$

or equivalently

$$d\nu = \frac{c}{2nL\left(1 + \dfrac{\nu}{n}\dfrac{dn}{d\nu}\right)} \tag{4.238}$$

Usually, several longitudinal modes will coexist, having wavelengths near the peak wavelength for spontaneous emission. The mode spacing for a GaAs laser is typically $d\lambda_0 \simeq 0.3$ nm. In order to achieve single mode operation, the laser structure must be modified so as to suppress all but the preferred mode. When a laser diode is forward biased and current begins to flow, the device does not immediately begin lasing. The dynamics can be described as follows [212]. At low current levels, the light that is emitted is mostly due to spontaneous emission and has a characteristic spectral linewidth on the order of hundreds of Angstroms (incoherent light). As the pumping current is increased, a greater population inversion is created in the junction region and more photons are emitted. The spontaneously emitted photons are produced going more or less equally in all directions. Most of these start off in directions that very soon carry them out of the inverted population region where net stimulated emission can occur, and thus are unable to reproduce themselves. However, those few photons that happen to be travelling exactly in the junction plane and perpendicular to the reflecting end faces are able to duplicate themselves many times before they emerge from the laser. In addition, for any given energy bandgap and distribution of holes and electrons, there is one particular energy (wavelength) that is preferred over others. To first order, this wavelength usually corresponds to the peak wavelength at which spontaneous emission takes place in the material. As a result of this preferred energy and direction, when stimulated emission builds up with increasing current, the emitted radiation narrows substantially both in spectral linewidth and spatial divergence. As stimulated emission builds up, the photon density (intensity) of the optical mode increases, thus leading to a further increase in stimulated emission so that more hole-electron pairs are consumed per second. Hence, spontaneous emission is suppressed for any given input hole-electron pair generation rate, because the stimulated emission uses up the generated pairs before they can recombine spontaneously. The transition from non-lasing to lasing emission occurs abruptly as the current level exceeds the threshold value. As the threshold current is exceeded, the onset of lasing can be experimentally observed by noting the sharp break in the slope of the optical power versus pump current curve, which results from the higher quantum efficiency inherent in the lasing process. Also the spectral lineshape of the emitted light abruptly changes from the broad spontaneous emission curve to one consisting of a number of narrow modes. Quantitatively, the lasing threshold corresponds to the point at which the increase in the number of lasing photons (per second) due to stimulated emission just equals the number of photons lost because of scattering, absorption, or emission from the laser. In conventional terms used to describe an oscillator, one would say that the device has a closed loop gain equal to unity. Using this fact, it is possible to develop an expression for the threshold current as a function of various material and geometrical parameters [212]

$$J_{th} = \frac{8\pi e n^2 \Delta\nu D}{\eta_q \lambda_0^2} \left(\alpha + \frac{1}{L} \ln \frac{1}{R} \right) \tag{4.239}$$

where R is the power reflection coefficient of the FP surfaces, D is the thickness of the emitting layer, e the electronic charge, $\Delta\nu$ is the linewidth of spontaneous emission, η_q is the internal quantum efficiency, α is the loss coefficient (including all types of optical loss). Equation 4.239 shows that a homojunction laser has a very high threshold current density at room temperature ($J_{th} \sim 10^5$ A/cm^2). This prevents the laser from operating as cw at room temperature (without suffering destruction in a very short time). The main reason for this high threshold is that the photon distribution extends or spreads into the non-inverted regions on each side of the junction due mainly to diffraction. Thus, there is a light emitting layer of thickness D, which is greater than the thickness d of the inverted population layer. For example, in GaAs diodes, $d \simeq 1$ micron and $D \simeq 10$ micron. This fact suggests that the laser diode should be designed so as to make the ratio $D/d = 1$ for optimum performance. Given these reasons, homojunction lasers only operate cw at cryogenic temperatures (typically at liquid nitrogen temperature T=77 K).

We have just seen that confining the optical field to the region of the laser in which the inverted population exists results in a substantial reduction of the threshold current density and a corresponding increase in efficiency. The method that is now generally adopted for decreasing the active layer thickness involves blocking the carrier flow with a layer of material that has a higher bandgap energy than the active region. The resulting structure is called single heterostructure if only one blocking layer is used, and double heterostructure (DH) if a blocking layer is used on either side of the active region. With a heterostructure laser, the thickness of the active region is determined during growth, and active region thicknesses of 0.1 μm or less can readily be achieved. The physical structure of a typical DH laser in GaAlAs is shown in Figure 4.81, along with a diagram of the index of refraction profile in the direction normal to the p-n junction plane [212].

The double heterostructure also provides an optical waveguide for the laser field, resulting in a higher confinement factor. Equally important, because of their wider bandgap, the blocking layers are transparent to the laser field, thus reducing optical losses. The improvement in laser performance due to the introduction of heterostructures is largely responsible for making semiconductor lasers into practical devices. Indeed, the DH laser has a threshold current density typically from 100 to 400 A/cm^2, and a differential quantum efficiency as high as 91%. More details both on the physical properties and the fabrication techniques of advanced heterojunction laser structures can be found in [212].

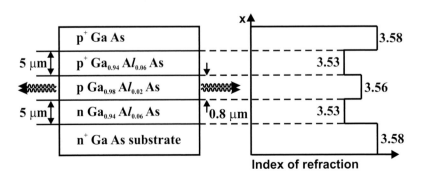

FIGURE 4.81
Double heterostructure laser diode. (Adapted from [212].)

Once lasing is achieved, the next goals are usually to increase output power and lower laser threshold current. For the double heterostructure laser just described, scaling to a higher power is a problem. The reason is that both the carriers and the laser mode are confined to within the same thin region. While we would like the carriers to be in a thin layer to maximize the density, we would also like the radiation field to be in a thick layer to ensure that its intensity is below the material damage threshold. It turns out that we can have both with more complicated heterostructure configurations. One example is the separate confinement heterostructure (SCH), which involves two barrier layers for carrier confinement, and two cladding layers for optical confinement. Present state-of-the-art fabrication techniques allow one to reduce the active layer thickness even to the dimension of the order of or less than an electron de Broglie wavelength, which is about 120 Angstrom in GaAs. We then have a quantum-well laser where the carriers are confined to a square well in the transverse dimension and move freely in the other two dimensions. The change from a three-dimensional to a two-dimensional free-particle density of states causes a quantum-well gain medium to behave differently from a bulk gain medium. A useful property of a quantum-well layer is that it is thin enough to form stable heterostructures with semiconductors of noticeably different lattice constants. The necessary deformation (strain) in the quantum-well lattice structure produces stress in the neighborhood of the interface which significantly alters the band structure. The change in band structure can lead to a reduction of laser threshold current density. Additional improvements of the semiconductor laser performance appear possible if one reduces the dimensionality of the gain medium even further than in the quasi-two-dimensional quantum wells. Instead of having carrier confinement only in one space dimension, one may produce structures where the quantum confinement occurs in two or even all three space dimensions. These quasi-one-dimensional or quasi-zero-dimensional nanostructures are referred to as quantum wires or quantum dots, respectively. Simple density-of-state arguments indicate that the reduced dimensionality leads to a more efficient inversion and, hence, to the possibility of ultralow threshold laser operation. However, more recent studies show that the Coulomb interaction effects among the charge carriers become increasingly more important for a decreased dimensionality of the semiconductor structure. These Coulomb effects seem to, at least partially, remove the advantages gained by the modified density of states. Furthermore, the manufacturing of quantum-wire or quantum-dot laser structures is still in its infancy. Therefore, we do not discuss the potentially very interesting quantum-wire or quantum-dot laser devices in this book. A good starting point to enter this intriguing topic is given, for instance, by [235].

Intensity noise

For a diode laser, which is operated from a low-noise current source and which is isolated against optical feedback, the technical contributions are usually small compared to other tunable sources. Furthermore, as a result of their low cavity Q-factor and due to the short inversion lifetime, the relaxation resonance of diode lasers appears at rather high Fourier frequencies (typically GHz range) and noise enhancement due to this effect can be neglected at low Fourier frequencies [236]. The spectrum of the intensity noise density is usually quite flat in the Fourier frequency range 1 MHz $< f < 1$ GHz. It displays a peak at the relaxation frequency position and a certain increase below 1 MHz where thermal and mode-partition effects become important (see below). The flat portion at intermediate frequencies is about 10-20 dB above the shot-noise level, its actual amplitude critically depending on the number of modes which contribute to the output signal. A typical value of the relative spectral intensity noise density is about 10^{-6} Hz^{-1}.

Mode partition noise

Although the intensity noise of the total light output of a laser diode is small, this does not hold in general for an individual mode. All longitudinal modes make use of essentially the same gain reservoir if we assume a homogeneously broadened gain profile but they are driven independently by spontaneous emission. Hence, if the instantaneous power of an individual mode increases, for instance, the sum of the powers of all other modes decreases by almost the same amount. Due to this cross-saturation effect the power of an individual mode may fluctuate stronger by orders of magnitude than the total power of all modes. This kind of excess intensity noise is usually called mode partition noise [236]. Large mode partition noise levels are predominantly found in diode lasers which oscillate in just a few longitudinal modes of comparable power. The power modulation index of each of these modes may reach values not very different from unity in extreme cases. Such strong fluctuation processes can be understood as some kind of fast switching between these modes, and may have a severe impact on many kinds of measurements.

Spectral linewidth and frequency noise

Like all lasers, diode lasers exhibit unavoidable, random fluctuations in the phase and the intensity of their radiation. We previously derived the (Schawlow-Townes) linewidth arising from spontaneous emission noise, and remarked that for diode lasers, in contrast to most other lasers, this fundamental source of noise typically dominates *technical noise*. Indeed, for diode lasers, correction to the Schawlow-Townes formula, due to a coupling of the laser phase and intensity, can be significant. By following the treatment given in [169], we now discuss in greater detail these two corrections to the ST formula. To simplify the derivation, we assume in the wave equation

$$\nabla^2 E - \frac{1}{c^2}\frac{\partial^2 E}{\partial t^2} = \frac{1}{\varepsilon_0 c^2}\frac{\partial^2 P}{\partial t^2} \tag{4.240}$$

an electric field and a polarization density of the form

$$E(z,t) = \hat{x}\mathcal{E}(t)e^{-i(\omega t - kz)} \qquad P(z,t) = \hat{x}\mathcal{P}(t)e^{-i(\omega t - kz)} \tag{4.241}$$

If the complex amplitudes $\mathcal{E}(t)$ and $\mathcal{P}(t)$ vary slowly compared to $e^{-i\omega t}$, we can drop their second derivatives with respect to time in the wave equation and write

$$\left(\frac{\omega^2}{c^2} - k^2\right)\mathcal{E} + 2i\frac{\omega}{c^2}\dot{\mathcal{E}} = -\frac{\omega}{\varepsilon_0 c^2}(\omega\mathcal{P} + 2i\dot{\mathcal{P}}) \tag{4.242}$$

which, in terms of the refractive index $n(\omega)$ and the group velocity v_g, becomes

$$\frac{d\mathcal{E}}{dt} = \frac{iv_g c}{2n\omega}\left(n^2\frac{\omega^2}{c^2} - k^2\right)\mathcal{E} \tag{4.243}$$

An injection current will change the refractive index of the active medium. We denote by $n_0(\omega)$ the index in the absence of a current and by $\Delta n(\omega) = \Delta n_R(\omega) + i\Delta n_I(\omega)$ the change in the index caused by the injection current, and use $n = n_0 + \Delta n_R + i\Delta n_I$ in 4.243. Taking n_0 to be real and $k = n_0\omega/c$, and assuming that Δn_R and Δn_I are small enough that Δn_R^2, Δn_I^2, $\Delta n_R\Delta n_I$ can be ignored, we replace 4.243 by

$$\dot{\mathcal{E}} = -\frac{v_g\omega}{c}\Delta n_I(1 - i\alpha)\mathcal{E} \tag{4.244}$$

where the enhancement linewidth factor $\alpha = \Delta n_R/\Delta n_I$ has been defined. Now Δn_I

causes changes in the real part of the field amplitude $\mathcal{E}$, i.e., it is associated with gain G and loss γ in the medium

$$-\frac{v_g \omega}{c}\Delta n_I = \frac{G - \gamma}{2} \tag{4.245}$$

such that 4.244 becomes

$$\dot{\mathcal{E}} = \frac{1}{2}(G - \gamma)(1 - i\alpha)\mathcal{E} \tag{4.246}$$

which, in terms of the amplitude $|\mathcal{E}|$ and phase ϕ ($\mathcal{E} = |\mathcal{E}|e^{i\phi}$), provides the two following equations

$$\frac{d}{dt}|\mathcal{E}|^2 = (G - \gamma)|\mathcal{E}|^2 \tag{4.247}$$

$$\dot{\phi} = -\frac{\alpha}{2}(G - \gamma) \tag{4.248}$$

Since $|\mathcal{E}|^2$ is proportional to the field intensity and therefore to the number of photons q, we can write these equations equivalently as

$$\dot{q} = (G - \gamma)q \tag{4.249}$$

$$\dot{\phi} = -\frac{\alpha}{2}(G - \gamma) \tag{4.250}$$

This exhibits the coupling between the field phase and intensity, the strength of which depends on the α parameter. Equations do not include effects of spontaneous emission. To calculate the principal quantity of interest here, i.e., the mean-square phase fluctuation, we will refer again to Figure 4.22. A change $\Delta\phi_i^{(1)}$ in the phase ϕ due to the i-th spontaneous emission event is given by

$$\Delta\phi_i^{(1)} \simeq \frac{1}{\sqrt{q}}\sin\theta_i \tag{4.251}$$

Here, the symbol ψ ($\Delta\vartheta$) is replaced by θ_i ($\Delta\phi_i$). There is another contribution to the change in ϕ in a spontaneous emission event, this one due to the phase-intensity coupling described by Equation 4.250; integrating both sides of that equation from a time before the i-th spontaneous emission event occurs to a time immediately after, we obtain

$$\Delta\phi_i^{(2)} = -\frac{\alpha}{2}\left[\ln(q + \Delta q_i) - \ln(q)\right] \simeq -\frac{\alpha}{2q}\Delta q_i \tag{4.252}$$

if the change Δq_i in the length of the phasor is small compared to the length q of the phasor before the spontaneous emission. Δq_i can be deduced with reference to Figure 4.22 and the law of cosines: $\Delta q_i = 1 + 2\sqrt{q}\cos\theta_i$. Therefore,

$$\Delta\phi_i^{(2)} \simeq -\frac{\alpha}{2q}(1 + 2\sqrt{q}\cos\theta_i) = -\frac{\alpha}{2q} - \frac{\alpha}{\sqrt{q}}\cos\theta_i \tag{4.253}$$

Then for the change in ϕ in the i-th spontaneous emission event we write

$$\Delta\phi_i = \phi_i^{(1)} + \phi_i^{(2)} = -\frac{\alpha}{2q} - \frac{1}{\sqrt{q}}(\sin\theta_i - \alpha\cos\theta_i) \tag{4.254}$$

and the total phase change after N spontaneous emission events is

$$\Delta\phi = \sum_{i=1}^{N} \Delta\phi_i = \frac{1}{\sqrt{q}} \sum_{i=1}^{N} (\sin\theta_i - \alpha\cos\theta_i) \tag{4.255}$$

where we have dropped the constant phase term $-\alpha/(2q)$, as we are only interested in phase fluctuations. Then we have

$$\langle\Delta\phi^2\rangle = \frac{1}{q} \sum_{i=1}^{N} \left(\langle\sin^2\theta_i\rangle + \alpha^2\langle\cos^2\theta_i\rangle\right) = \frac{N}{2q}(1+\alpha^2) \tag{4.256}$$

which is exactly the same as Equation 4.152 multiplied by the factor $(1+\alpha^2)$. This means that Equation 4.156 can be generalized to

$$\Delta\nu_{ST} = \frac{\pi\mu h\nu_0(\delta\nu)^2}{P_{out}}(1+\alpha^2) \tag{4.257}$$

The α factor in the above formula is often referred to as the Henry factor α_H, as C.H. Henry first presented a comprehensive theory of the spectral width of a single-mode semiconductor laser [237]. α_H basically depends on the material of the diode laser. Just to give an idea, for GaAs and $\lambda = 850$ nm, $\alpha \simeq 4$ has been determined. Moreover, the ST derivation is based on the assumption that the laser output coupling is small, that is, that the mirror reflectivities are near unity. More generally the quantum lower limit to the laser linewidth can be written as

$$\Delta\nu_{ST} = K\frac{\pi\mu h\nu_0(\delta\nu)^2}{P_{out}}(1+\alpha^2) \tag{4.258}$$

where

$$K = \left[\frac{(\sqrt{r_1} + \sqrt{r_2})(1 - \sqrt{r_1 r_2})}{\sqrt{r_1 r_2}\ln(r_1 r_2)}\right]^2 \tag{4.259}$$

r_1 and r_2 denoting here the power mirror reflectivities. The K factor, which is ascribed to *excess spontaneous emission noise*, approaches unity as $r_1, r_2 \to 1$, but can be greater than 1 in other configurations.

Substantial deviations of the frequency noise density from the above white noise level have been observed in practice both for the low and high Fourier frequency region [236]. For most diode lasers, one finds a $1/f$ increase at low frequencies resulting from different effects: noise of the current source, thermal fluctuations of the carrier density, mode partition noise, presence of carrier traps in the vicinity of the active layer. Such $1/f$ noise contribution becomes important for Fourier frequencies below a corner which is typically between 100 kHz and 10 MHz. At high Fourier frequencies, the frequency noise is enhanced above the white noise floor due to the carrier density dynamics of the LD (relaxation resonance). The resonance manifests itself in the emission spectrum as satellite peaks, which are separated from the line center by the relaxation resonance frequency (GHz range) and multiples thereof.

Tuning of diode lasers

To find the frequency of the mth longitudinal mode, the phase shifts occurring at the laser facets have to be taken into account [65]. Often, the facet at the rear of the diode laser is coated as a mirror of high reflectivity and the laser field can be thought of as a standing wave with a node at the facet. This is obviously not true for the other facet that serves as an output coupler with a typical coefficient of reflection of $\simeq 35\%$. Consequently, there is a phase shift φ of the wave internally reflected from the output facet that is equivalent to an additional optical path length

$$m\lambda = 2n(\nu)L + \varphi\frac{\lambda}{2\pi} = m\frac{c}{\nu} \qquad (4.260)$$

or

$$\nu_m = m\frac{c}{2n(\nu)L + \frac{\varphi c}{2\pi}} \qquad (4.261)$$

In the case of small variations of these parameters one can assume

$$\frac{\Delta\nu}{\nu} = \frac{\Delta m}{\nu}FSR - \frac{\Delta\varphi}{2\pi\nu}FSR - \frac{\Delta L}{L} - \frac{\Delta n}{n} \qquad (4.262)$$

with the FSR determined according to Equation 4.238. The first contribution can lead to mode jumps of about 100 GHz. The phase φ in the second term can be varied in particular by coupling back a part of the light emitted by the diode laser. This effect can be used for frequency stabilization of diode lasers. On the other hand, spurious radiation reflected back, e.g., by the window of the housing of the diode laser, by the collimating lens or from other optical components, can alter the frequency of the laser. The last two contributions are influenced by the temperature of the laser. For small temperature variations the length $L(T)$ of the laser diode is expected to vary linearly with the temperature. The index of refraction n influences the laser frequency in a complicated way since it varies with the frequency, temperature, the injection current, and the laser power. Temperature fluctuations affect the index of refraction via different effects. In general, raising the temperature of a laser diode increases its wavelength where the monotonic variation is interrupted by discontinuous jumps [238]. The monotonous variation with a typical value of about -30 GHz/K is due to the length variation and the associated shift of the mode frequency. At the same time the temperature-dependent lattice constants of crystal and the associated variation of the band structure result in a shift of the gain profile of the laser. As a result, mode jumps of about 50 GHz - 100 GHz or more occur and the mean wavelength variation with temperature over a wide temperature range amounts to about -100 GHz/K. The wavelength, however, is not an unambiguous function of the temperature but rather shows hysteresis, depending on whether the temperature is raised or lowered. Besides the ambient temperature, the temperature of the diode laser is also affected by the injection current. With a nearly constant voltage drop across the p-n junction the dissipated power, and hence the temperature increase, is proportional to the injection current. Consequently, a smooth increase of the injection current results in a red detuning of the frequency of the laser due to the associated temperature variation. The shift of the frequency of a solitary diode laser as a function of the injection current varies from a few negative to positive GHz/mA, depending on the specific device material. Variation in the current, in general, also affects the index of refraction by the changed number of the free charge carriers. For larger variations of the injection current and higher modulation frequencies this influence of the current prevails, which is of particular importance when the injection current is to be used as a fast input for frequency stabilization.

In light of the above discussion, diode lasers are usually operated under temperature-stabilized conditions and with low-noise power supply. Valuable examples of temperature-control circuits and current drivers for diode lasers can be found in [239].

Modulation characteristics

An important advantage of semiconductor diode lasers (SDL) over other laser systems is that the amplitude and frequency of the emitted radiation can be easily modulated by changing the injection current [240]. The output electric field for a current-modulated SDL can then be written as

$$E(t) = A[1 + m\cos(\omega_m t)]\sin[\omega t + \beta\cos(\omega_m t + \theta)] \qquad (4.263)$$

where $\omega/2\pi$ is the laser carrier frequency, $\omega_m/2\pi$ is the modulation frequency, and m and β are the amplitude and frequency modulation indices, respectively. The FM modulation index is typically more than ten times larger than the AM modulation index. In fact, in those applications where FM modulation is important, the AM modulation can be often ignored. As far as the FM modulation capabilities are concerned, changes in the injection current affect both the diode temperature and the index of refraction which depends on the carrier density. The temperature effect is the dominant one in the low frequency modulation range ($< 10^5 - 10^6$ Hz), where the frequency change is about 3 GHz/mA. For higher modulation frequencies, the change in the index of refraction dominates, leading to a smaller variation of the emission frequency with the injection current. In presence of optical feedback, the frequency modulation capabilities are strongly modified.

Frequency stabilization techniques for diode lasers

The large linewidth of a solitary diode laser of the Fabry-Perot type cannot be reduced by negative electronic feedback alone, as the required large servo bandwidth can hardly be achieved. These difficulties can be overcome by employing an optical feedback loop. The basic idea in this case is to reduce the intrinsic linewidth of the laser by using a cavity with a Q much higher than the one of the solitary laser diode. Depending on the elements and configurations used to accomplish the feedback, several terms have been coined. If an external reflective element is added to the solitary diode laser, the combined arrangement is termed *External Cavity Laser* (ECL), as an external cavity is formed by the reflective element and the output coupler of the solitary laser. If the output coupler has a low reflectivity, e.g., from an anti-reflective coating, the laser cavity is formed by the rear mirror of the laser diode and the external reflector and, hence, acts as an *Extended Cavity Diode Laser* (ECDL). In addition, if a wavelength-selective optical element is used in such a cavity, this system may also allow a control of the emission wavelength.

Resonant optical feedback - The high sensitivity of diode lasers to optical feedback is a well-known phenomenon that generally has a disruptive effect on the lasers' output frequency and amplitude stability. Under certain circumstances this sensitivity to feedback can be put to advantageous use. Frequency stabilization of diode lasers by resonant optical feedback was first realized by [241]. In that work, with the appropriate optical geometry the laser optically self-locked to the resonance of a separate Fabry-Perot reference cavity. The method relies on having optical feedback occur only at the resonance of a high-Q reference cavity. In this case the cavity serves two functions. It provides the optical feedback, which narrows the laser's spectral width, and it provides the center-frequency stabilization to the cavity resonance. Such laser locking system contrasted with the previous systems in that the laser saw optical feedback only when its frequency matched the resonant frequency of the reference cavity. A variety of optical locking geometries were tested; Figure 4.82 diagrams a particularly simple and effective version.

When the confocal reference cavity is operated off axis, it should be viewed as a four-port device. It is important to note that the two ports on the input mirror side (labeled I and II) have different output characteristics. The output beam of type I is a combination of the reflected portion of the input beam with the transmitted portion of the resonant field inside the cavity. This beam has a power minimum when the laser frequency matches a cavity resonance. In contrast, the three outputs of type II contain only the transmitted

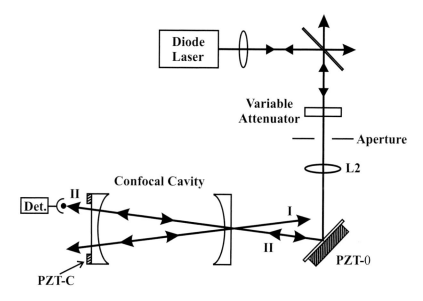

FIGURE 4.82

Schematic of one version of the optical feedback locking system. Lens L2 is used to mode match the laser into the confocal reference cavity. The aperture blocks the unwanted feedback of type I while passing the desired feedback of type II. The variable attenuator is used to study the feedback power dependence of the locking process. The piezoelectric translator PZT-0 is used to optimize the feedback phase relative to the undisturbed laser. PZT-C is used to scan the reference cavity and in turn the optically locked laser frequency. A photodetector (Det.) monitors the transmitted power. (Adapted from [241].)

portion of the cavity resonant field and hence have the desired characteristic of a power maximum on resonance. The geometry of Figure 4.82 is one possible method of arranging to have the resonant optical feedback of type II (maximum feedback on resonance) while avoiding the complications inherent in the directly reflected beam contained in output I. For a wide range of feedback conditions this system tends to self-lock stably, thus forcing the laser frequency to match that of the cavity resonance. The original experiment was carried out using commercial, single-mode, 850-nm GaAlAs lasers without any special preparation of these devices. The free-running, unperturbed laser linewidths were measured to be approximately 20-50 MHz. A variety of reference cavities were used for the optical locking system. These had free spectral ranges varying from 250 MHz to 7.5 GHz and resonance widths ranging from 4 to 75 MHz. With type II feedback if the laser frequency is far from matching a reference cavity resonance there is no optical feedback and the laser frequency scans as usual with changes in the injection current. However, as the laser frequency approaches a cavity resonance, resonant feedback occurs and the laser frequency locks to the cavity resonance, even if the laser current continues to scan. The actual frequency range over which the diode laser locks is a function of the feedback power level and the phase of the feedback light relative to that of the unperturbed laser. The optical locking of the laser frequency to the reference cavity resonance can be observed directly with a photodiode, which monitors the power transmitted through the cavity as a function of the laser current. In the original experiment, such a signal consisted of an unusual Fabry-Perot transmission function, which had a flat top and a width approximately 10 times larger than the actual cavity resonance width of 50 MHz. This shape is a manifestation of the fact that the laser frequency is not scanning with the laser current but is locking to the frequency of peak transmission of the cavity resonance. For low levels of optical feedback the locking range depends on the feedback power ratio. This is the ratio of the external feedback power, coupled into the laser mode, to the power reflected internally by the output facet. Locking ranges of a few hundred megahertz were observed with feedback power ratios typically in the range of 10^{-4} to 10^{-5}. In order to determine the spectral characteristics of the optically locked semiconductor lasers, the beat note between two lasers independently locked to separate reference cavities was also measured. Such analysis revealed a linewidth reduction by a factor of about 1000 (from 20 MHz to 20 kHz) without any electronic control. One additional advantage of the optically self-locked laser system is that it shows some reduction in sensitivity to other sources of optical feedback. Furthermore, the excess intensity noise is reduced by roughly 10 dB when the laser is optically locked to the cavity. Finally, for many applications it is important to be able to scan the laser frequency as well as to have a stable long-term frequency lock. Continuous, narrow linewidth scans of the laser frequency can be made by synchronizing the sweep of the laser current with that of the reference cavity length (PZT-C) and the feedback phase (PZT-0). With this synchronization, continuous scans of several GHz can be made. It is worth noting that it is necessary to control the laser current only accurately enough that the free running laser frequency is within the optical-self-locking range. A detailed theoretical description of optical feedback from a FP interferometer is given in [242]. If we assume a white frequency noise spectral density, the laser linewidth can be expressed as

$$\Delta\nu = \frac{\Delta\nu_0}{\left[1 + \sqrt{1 + \alpha^2}\sqrt{\beta}\dfrac{L_p}{\eta l_d}\dfrac{F_{cfp}}{F_d}\right]^2} \tag{4.264}$$

where $\Delta\nu_0 = (1 + \alpha^2)\Delta_{ST}$ is the linewidth of the free-running diode laser. Here ηl_d is the diode laser optical length, L_p the length of the confocal FPI, α the Henry parameter, β the power mode coupling factor, F_{cfp} and F_d, respectively, the confocal FP finesse and the diode laser cavity finesse. For typical values of $F_d = 2$, $F_{cfp} = 100$, $\eta l_d = 1$ mm, $L_p = 20$ cm, $\alpha = 4$, $\beta = 10^{-3}$, one would expect a linewidth reduction of 10^{-8} and hence a linewidth of few Hz. In practice, however, the minimal achievable linewidth was measured around a few kilohertz which can be attributed to the influence of technical $1/f$ noise at Fourier frequencies below 1 MHz.

Extended Cavity Diode Laser - In the opposite regime, strong feedback from an external diffraction grating allows one to obtain a reduction by about two orders of magnitude of the emission linewidth and to achieve a wavelength tuning range of several nanometers at fixed temperature. The main drawback of this scheme is that it requires the diode facets to be anti-reflection coated in order to increase the amount of light coupled into the diode. However, most of diode lasers available on the market are already provided with a high reflectance coating on the back facet and with a reduced reflectance coating on the output facet. Two particular arrangements are often used [65, 243, 240](Figure 4.83). The first one, called the Littrow configuration, employs the reflection grating as the output coupler of the extended cavity. The grating angle is set such that the first-order reflection coincides with the incident beam from the diode laser. The zero-order reflection is used to couple out the output beam and the wavelength of the diode laser is adjusted by a rotation of the grating. The Littman-Metcalf configuration uses a folded laser cavity. In contrast to the Littrow configuration, the incident beam and the diffracted beam are no longer collinear. The diffracted beam is reflected back from a mirror and is directed into the laser diode after a second diffraction at the grating. Tuning of the wavelength is achieved by rotating the mirror. Again, the zero-order beam is used to couple out a fraction of the power circulating in the cavity. By comparing both configurations, the Littman-Metcalf configuration has the advantage that the tuning does not change the direction of the output beam. The double diffraction, however, leads to increased intra-cavity losses and requires gratings of high reflectivity. The double pass, on the other hand, leads to an increased selectivity. Another advantageous feature of the Littman-Metcalf configuration is the free choice of the angle of incidence independent of the wavelength. Consequently, this configuration allows one to use a large angle of incidence independently thereby illuminating a large number of grooves of the grating with the associated better resolution.

More recently, an enhanced Littrow-configuration ECDL has been developed that can be tuned without changing the direction of the output beam [244]. In this scheme, the output of a conventional Littrow ECDL is reflected from a plane mirror fixed parallel to the tuning diffraction grating. Using a free-space Michelson wavemeter to measure the laser wavelength, the laser can be tuned over a range greater than 10 nm without any alteration of alignment.

For an ECDL with grating, one can show that the minimum achievable linewidth is given by

$$\Delta\nu_{min} = \frac{\Delta\nu_{DL}}{[1 + (L_d/nL_{DL})]^2} \qquad (4.265)$$

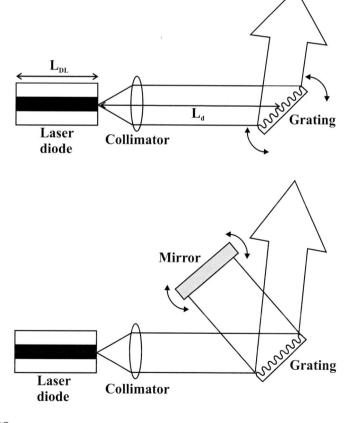

FIGURE 4.83
ECDL using a grating as output coupler in Littrow configuration (top frame) and an intra-cavity grating in Littman-Metcalf configuration.

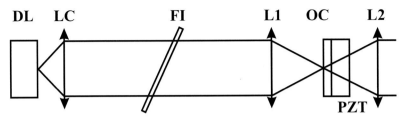

FIGURE 4.84
Schematic of the external cavity laser using an interference filter (FI) for wavelength selection: (DL) laser diode, (LC) collimating lens, (OC) partially reflective out-coupler, (PZT) piezoelectric transducer actuating OC, (L1) lens forming a *cat's eye* with OC, and (L2) lens providing a collimated output beam. (Adapted from [245].)

where $\Delta\nu_{DL}$, n, and L_{DL} are the linewidth, the refractive index, and the length of the original laser, whereas L_d denotes the distance between the grating and the laser output facet.

Being too sensitive to the ambient pressure and optical misalignment (induced by mechanical or thermal deformation), the above grating-based ECDL designs are not well suited for applications where better robustness is required such as, for instance, space-born ones. In that case, ECDLs incorporating narrow-band dielectric filters can be used [245]. A schematic of this setup is given in Figure 4.84.

The light emitted from the diode is collimated by an objective lens with short focal length (3 - 4.5 mm) and high numerical aperture (0.6). The lens is chosen to compensate for aberrations arising from the diode's packaging window. A partially reflecting mirror, here named out-coupler, provides the feedback into the diode. The OC is displaced by a piezoelectric transducer (PZT) in order to vary the cavity length. A narrow-band high-transmission interference filter is introduced into the cavity. The filter (90% transmission and 0.3 nm FWHM) provides the frequency selectivity usually obtained by replacing the out-coupler with a diffraction grating. With this setup, single mode, tunable operation is achieved. In addition much better stability against optical misalignment is achieved by focussing the collimated beam in a *cat's eye* onto the out-coupler. Contrary to the Littrow laser design, reflection and wavelength discrimination are provided by two different elements so that the amount of feedback can easily be optimized (by varying the reflectivity of the out-coupler). The diode back facet is coated for high reflection, while the output facet has no particular high-quality AR coating giving rise to a second cavity formed by the laser chip itself, but making it inexpensive. Finally, it is worth pointing out that the reduction of the wavelength sensitivity against mechanical instabilities is not achieved at the expense of a reduced tunability. Indeed, tuning over 20 nm was demonstrated in the first realization.

Grating enhanced external cavity diode laser

The above approaches lead to complementary results: diode lasers with optical feedback from a cavity provide the smallest linewidth. For these lasers, the frequency noise at large Fourier frequencies is strongly reduced. They are therefore well suited for applications where, e.g. two independent lasers have to be phase locked by means of an optical phase locked loop. However, their continuous tuning ranges are much smaller than those of ECDLs, due to the lack of an *intracavity* element that provides a coarse wavelength pre-selection. Further, ECDLs are much easier to use and are more reliable than external cavity diode lasers. For example, extended cavity diode lasers show much better frequency repeatability than external cavity diode lasers that usually require four dependent actuators to set the wavelength, i.e., the current, an etalon, the length of the feedback cavity, and the position of a mirror used to adjust the path length to the external cavity. For an extended cavity diode laser there are only two actuators (current, grating), which are almost independent, if only modest tuning ranges ($\simeq$ GHz) have to be achieved.

There have been efforts to connect both ideas for combining the advantages of both approaches and overcoming the drawbacks. A pseudo-external cavity was developed to get control on the power used for feedback to the laser and the power available for the experiment. In this setup, a polarizing cube was inserted in the external cavity, together with waveplates, to control power through polarization. Such a laser was used to simultaneously injection lock a more powerful laser diode emitting around 796 nm and lock a bow-tie cavity for frequency doubling. Continuous frequency tuning in excess of 11 GHz in the UV allowed scanning of a Ca II transition at 397 nm, while the fundamental radiation was used to record an Ar I sub-Doppler line [246]. More recently, an approach that truly combines the external cavity and the extended cavity setup has been proposed, referred to as grating-enhanced ECDL (GEECDL) [247, 248]. It basically consists of a grating diode laser in Littman configuration with the retro-mirror replaced by an external cavity (finesse 35). In short, with reference to Figure 4.85, light emitted by the laser diode is diffracted by a transmission grating. The first diffraction order (95% efficiency) is coupled into a folded cavity, which provides optical feedback to the laser diode. Obviously, the optical setup constitutes two coupled cavities, which are defined by, e.g., the two mirrors MP and MF and by MF and the rear facet of the laser diode chip. The coupled cavity configuration necessitates a relative frequency stabilization (*internal* stabilization) to guarantee reliable single-mode operation. Internal stabilization is realized by means of a polarization-sensitive stabilization scheme (Hänsch-Couillaud type) and actuation of mirror MP. With such an arrangement, a continuous tuning range of up to 20 GHz was achieved. Also, a short-term linewidth of about 11 kHz was measured by beating two essentially identical diode lasers. Phase locking between these lasers was achieved with a servo bandwidth as small as 46 kHz, although an analog phase detector was used that required sub-radian residual phase error. Despite small phase error detection range and small servo bandwidth, cycle-slip-free phase locking was accomplished for typically many 10 min, and the optical power was essentially contained in a spectral window of less than 20 mHz relative to the optical reference. Due to the excellent performance this laser concept is well suited for atomic or molecular coherence experiments, which require phase locking of different lasers to each other, and as part of a flywheel for optical clocks.

4.8.5.2 Distributed feedback (DFB) lasers

The lasers that have been described so far depend on optical feedback from a pair of reflecting surfaces, which form a Fabry-Perot etalon. An alternative approach utilizes DFB from a Bragg-type diffraction grating. In DFB lasers, the grating is usually produced by corrugating (by chemically assisted ion beam etching) the interface between two of the semiconductor layers that comprise the laser (Figure 4.86). This corrugation provides 180° reflection at certain specific wavelengths, depending on the grating spacing. The basis for selective reflection of certain wavelengths has already been explained in Chapter 3: the vacuum wavelength of light that will be reflected through 180° by such grating is

$$\lambda_0 = \frac{2\Lambda n_g}{l} \tag{4.266}$$

where Λ is the grating spacing, n_g is the effective index in the waveguide for the mode under consideration, and $l = 1, 2, 3, \ldots$. Although the grating is capable of reflecting many

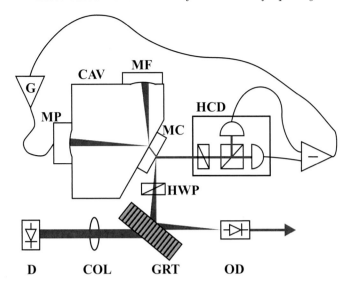

FIGURE 4.85
Schematic of the setup. D, laser diode; COL, collimator; GRT, transmission grating; HWP, half-wave plate; MC (T=8.2%), MP, MF (both R=99.7%), mirrors define the external cavity CAV. HCD, balanced polarization detector is part of the internal stabilization; OD, optical diode. (Adapted from [248].)

different longitudinal modes, corresponding to the various values of l, usually only one mode will lie within the gain bandwidth of the laser. Thus single-longitudinal-mode operation is obtained relatively easily in DFB lasers (in fact, because of the difficulty of fabricating a first order ($l = 1$) grating, usually a third-order grating is used).

The active region can also be isolated from the grating region by using the distributed Bragg reflection (DBR) structure. In such a device, two Bragg gratings are employed, which are located at both ends of the laser and outside of the electrically-pumped active region. In addition to avoiding non-radiative recombination due to lattice damage, placement of two grating mirrors outside of the active region permits them to be individually tailored to produce single-ended output from the laser. In order to achieve efficient, single-longitudinal-

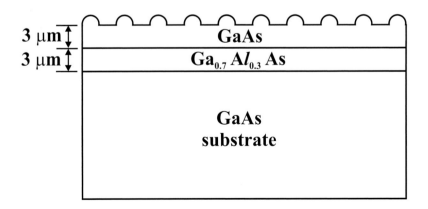

FIGURE 4.86
Cross-sectional view of an optically pumped DFB laser. (Adapted from [212].)

mode operation, one distributed reflector must have narrow bandwidth, high reflectivity at the lasing wavelength, while the other must have relatively low reflectivity for optimal output coupling. DFB and DBR lasers have unique performance characteristics that give them distinct advantages over conventional reflective-end-face lasers in many applications. As already mentioned, first, the spacing between the lth and the $l \pm 1$ modes is generally so large compared to the linewidth of the laser gain curve that only one mode has sufficient gain to lase. Thus single-longitudinal-mode operation is obtained relatively easily in distributed feedback lasers. In addition to providing a means of accurately selecting the peak emission wavelength, grating feedback also results in a narrower linewidth of the optical emission. The spectral width of the emission line is established by a convolution of the laser gain curve with the mode-selective characteristics of the laser cavity. Since the grating is much more wavelength selective than a cleave or polished end-face, the resulting emission linewidth of a DFB or DBR laser is significantly less than that of reflective-end-face laser. While the single-mode linewidth of a conventional cleaved-end-face laser is typically about 50 GHz, values reported for modern DFB and DBR lasers with sophisticated grating structures range from about 50-100 kHz. Finally, DFB lasers offer improved wavelength stability as compared to cleaved-end-face lasers, because the grating tends to lock the laser to a given wavelength. Also, DFB lasers are characterized by an improved temperature stability resulting from the fact that, while the shift of emitted wavelength in the cleaved laser follows the temperature dependence of the energy bandgap, the shift in wavelength of the DFB laser follows only the temperature dependence of the index of refraction.

Fabricated from disparate materials, the market of diode lasers is today very vast. Trying to summarize, the following main categories are available:

- Fabry-Perot diodes: emission between 370 and 1120 nm, output power up to 300 mW from ECDL.

- AR-coated diodes: emission between 650 and 1770 nm, output power up to 150 mW from ECDL, maximum coarse tuning of 100 nm, minimum linewidth (5 μs integration time) of 100 kHz.

With FP diodes, the internal resonator of the diode functions like an etalon, attenuating certain external modes, and therefore participating in the selection of the external mode. The effect of the internal resonator is less pronounced when an AR coating is added on the output facet. AR diodes do not lase without external feedback. The AR coating improves coarse and mode-hope-free tuning of an ECDL and allows for more stable single-mode operation. The internal resonator of both FP and AR diodes can be synchronized with the grating movement by changing the diode current simultaneously. This *feed forward* mechanism moves the internal mode structure of the laser diode along with the external modes, permitting larger mode-hop free tuning.

- DFB/DBR: emission between 640 and 2900 nm, output power up to 150 mW, maximum coarse tuning of 6 nm, minimum linewidth (5 μs integration time) of 500 kHz. Frequency tuning is accomplished by thermally or electrically varying the grating pitch. Thermal tuning offers extremely large mode-hop free scans of hundreds of GHz. Electric modulation, on the other hand, can be employed for fast frequency modulation over a smaller range (several tens of GHz at modulation frequencies from kHz to MHz).

Whether to choose an external cavity diode laser or a DFB/DBR laser depends on the individual application. An ECDL is the preferred choice for applications that require a broad coarse tuning range, or an ultra-narrow linewidth. The main advantage of a DFB laser is its extremely large continuous tuning range. Indeed, mode-hope-free (MHF) scans of several nanometers are routinely attained, while maximum MHF tuning is 50 GHz in ECDLs.

Finally, it is worth spending a few words on the spatial properties of the output beam. In most diode lasers it exhibits non-favorable properties, such as a large divergence, high asymmetry of radius between two perpendicular directions, and astigmatism. It is not always trivial to find the best design for beam shaping optics, that need to be compact, easy to manufacture and align, able to preserve the beam quality and to avoid interference fringes, to remove astigmatism, and with low losses, etc. Typical parts of such diode laser beam shaping optics are collimating lenses (spherical or cylindrical), apertures, and anamorphic prisms. Obviously, when enough power is available, the output spatial mode can be eventually *projected* into the TEM_{00} one by propagation through an optical fiber.

Needless to say, at the same time, incessant search for novel materials and development of more and more effective fabrication technologies is ongoing, aiming at extending the current spectral coverage of semiconductor diode lasers both to the UV [249] and MIR regions [250, 251, 252, 253]. The radiation sources reported in these works (and references therein) still suffer, however, from some limitations, particularly either in terms of cw narrow-linewidth or room-temperature operation. In any case, at the time of writing, accurate characterizations of their spectral features cannot be found in literature and examples of high-resolution spectroscopic measurements certainly cannot either.

Some of these cutting-edge sources fall in two other well-established classes of semiconductor diode lasers, namely vertical-cavity surface-emitting lasers (VCSELs), and multiple-quantum-well (MQW) lasers. After a digression on tapered semiconductor amplifiers, these will be shortly reviewed in the following two subsections, referring the interested reader to [234] for a more extensive treatment.

4.8.5.3   Tapered semiconductor amplifiers

When higher output powers are needed, tapered semiconductor amplifiers (TSA) can be used [254, 255] to amplify the radiation coming from a semiconductor diode laser. In fact, the double heterostructure, p-n junction diode which is used in semiconductor lasers can also function as an optical amplifier. The basic mechanism of amplification is that of stimulated emission, just as in a laser. However, in the amplifier the diode is usually biased somewhat below lasing threshold so that oscillation does not occur. In this frame, offering high power combined with good spatial mode, tapered amplifiers have become the most valuable choice. In addition, when injected with light from a single-mode external-cavity grating-stabilized diode laser, TSA can retain the narrow spectral features of the injected light. The master laser beam is coupled into the small single-mode channel at the AR coated rear facet of the tapered amplifier chip. The single-mode channel acts as a spatial mode filter (like a single-mode fiber). The close-fitting tapered angle is adapted to the diffraction angle of a single-mode laser at a specified wavelength. The laser beam is amplified in a single pass through the tapered region, without losing its high spectral and spatial quality, and leaves the chip through the AR coated large output facet (Figure 4.87). Since the amplifier tapers laterally, this leads to an output aperture that is highly asymmetric and to an astigmatic output beam. The same considerations as for bare diode lasers thus apply.

More recently, a 1-W tapered amplifier requiring only 200 μW of injection power at 780 nm was presented [256]. This was achieved by injecting the seeding light into the amplifier from its tapered side and feeding the amplified light back into the small side (Figure 4.88). The amplified spontaneous emission of the tapered amplifier was suppressed by 75 dB. Such

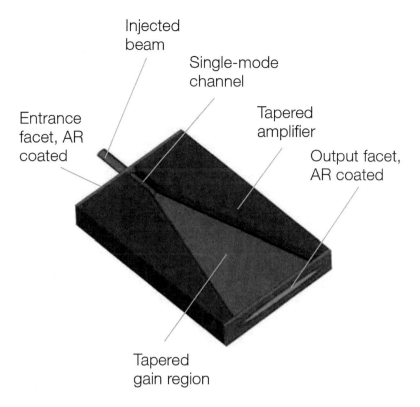

FIGURE 4.87
Geometry of a tapered semiconductor amplifier chip. Such flared geometry maintains a single transverse mode while boosting the power to the watt level (http://www.toptica.com).

a double-passed tapered laser is extremely stable and reliable: the output beam remains well coupled to the optical fiber for a timescale of months, whereas the injection of the seed light did not require realignment for over a year of daily operation.

4.8.5.4 Multiple-quantum-well lasers

In all the lasers previously discussed the dimensions of device structures were large compared to the wavelength of electrons in the device. When the dimensions of the structure are reduced to the point at which they are approaching the same order of magnitude as the electron wavelength some unique properties are observed. This is the case with a class of devices that have come to be known as *quantum well* devices, which feature very thin epitaxial layers of semiconductor material. Improved lasers can be made by employing quantum well structures. A good starting point for a detailed theoretical description is provided by [205], while a more in-depth discussion on specific structures is given in [212]. Here we just explain the main benefit of quantization in the electron and hole states with simple arguments. A quantum well structure consists of one or more very thin layers of a relatively narrow bandgap semiconductor interleaved with layers of a wider bandgap semiconductor, as shown in Figure 4.89.

The special properties of these very thin layers result from the confinement of carriers (electrons and holes) in a manner analogous to the well-known quantum mechanical problem of the *particle in a box*. In this case, the carrier is confined to the narrow bandgap *well* by the larger bandgap *barrier* layers. The magnitude of the wavefunction of the electron (or

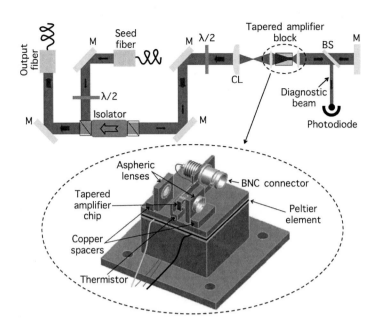

FIGURE 4.88
Schematic diagram of the TA double-pass configuration. The seed light is delivered to the setup via a single-mode optical fiber. It then passes through the auxiliary port of a Faraday isolator, a $\lambda/2$ wave plate, a single cylindrical lens (CL), and a collimation lens until it reaches the tapered region of the TA chip. After its first amplification, it emerges from the small side, where it is collimated and retro-reflected. A microscope cover slip acts as a beam splitter (BS) to monitor the return optical power (P_{back}). After the second amplification, the light is delivered to the experiment via a polarization-maintaining single-mode optical fiber. The TA chip is mounted onto a solid copper support, glued onto a Peltier element, which, in turn, is glued onto a solid base. The aspheric lenses, used for the collimation of the beam, are epoxied to the same support as the TA. (Courtesy of [256].)

hole) must approach zero at the barrier wall because the probability of finding the particle within the wall is very small. Hence, the wavefunction must form a standing wave pattern, sinusoidally varying within the well and damping to near zero at the edge of the barrier. The set of wavefunctions which satisfy these boundary conditions corresponds to only certain allowed states for the carrier. The carrier motion is thus quantized, with discrete allowed energies corresponding to the different wavefunctions. Such quantization reduces the total number of carriers needed to achieve a given level of population inversion. The free carrier absorption coefficient, which is proportional to the number of carriers, is also reduced. As a consequence, the threshold current density is reduced by approximately a factor of 10 as compared to that of a conventional double heterostructure laser diode. Photon generation in a MQW laser occurs through electron transitions from the nth energy level (E_{nc}) in the conduction band to the mth energy level (E_{mv}) in the valence band

$$\hbar\omega = E_g + E_{nc} + E_{mv} \tag{4.267}$$

where E_g is the band-gap energy.
The advent of quantum well structures has changed the direction of semiconductor laser

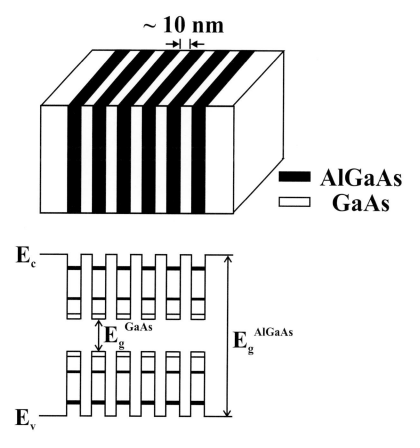

FIGURE 4.89

Multiple-quantum-well (MQW) structure. A typical MQW structure might have about 100 layers. The thickness of the layers is typically 100 Angstrom, or less. The GaAs-AlAs material system is particularly convenient for the growth of MQW devices because GaAs and AlAs have almost identical lattice constants and thus interfacial strain can be avoided. However, GaInAsP is also a suitable material, as long as the concentrations of the constituent elements are properly chosen to produce lattice matching. (Adapted from [212].)

research and development over a period of just a few years. Substantial improvements in threshold current density, linewidth and temperature sensitivity have already been demonstrated, even as compared to the properties of the DFB laser. Combining DFB and MQW structures has produced devices in InGaAsP with linewidth-power products of 1.9-4.0 with minimum linewidths of 1.8-2.2 MHz [257]. In this frame, continuous-wave around-room-temperature operation of type-I quantum well GaInAsSb/ AlGaInAsSb DFB laser diodes emitting from 2900 to 3500 nm has also been reported [258]. Such lasers are characterized by an optical output power > 1 mW and a linewidth < 3 MHz.

It seems certain that continued development of this relatively new device, including the development of quantum wire and quantum dot lasers, will lead to further improvements in the operating characteristics of semiconductor laser diodes.

4.8.5.5　Vertical-cavity surface-emitting lasers

Unlike the commonly known and market dominating edge-emitting structures that emit light perpendicular to the epitaxial growth direction of the semiconductor layers, the so-called vertical-cavity surface-emitting lasers (VCSELs) feature surface emission parallel to the growth direction. This concept, proposed by Iga in 1977 and firstly realized two years later, offers numerous advantages, such as low power consumption, low beam divergence, high fiber coupling efficiency due to a circular output beam, and on-wafer testing. Furthermore, these devices emit inherently longitudinally single-mode due to their small cavity lengths of only several micrometers and, therefore, large longitudinal mode separation. The typical device structure of a VCSEL is shown in Figure 4.90. The key elements are an active layer, two DFB mirrors, and a contact window to allow light to be emitted from the surface. In the particular device shown, a strain-compensated, quantum-well, 7-layer active region is sandwiched between two AlGaAs/GaAs mirrors by wafer fusion [212]. Typically, the active region is a multilayer, double heterostructure diode, as in a conventional Fabry-Perot end-face laser. However, in this case, lightwaves travel in the direction perpendicular to the junction plane. They are reflected by top and bottom *mirrors* which consist of multilayer structures, alternating layers of materials with differing indices of refraction that are approximately one half wavelength thick (at the lasing wavelength in the material). For this spacing, the lightwaves reflected from each interface between layers positively reinforce waves reflected from all of the others in the reverse direction, so an effective *mirror* is formed. The overall reflectivity of the multilayer structure depends on the reflectivity at each interface and on the number of layers. The reflecting semiconductor layers are usually grown by either MOCVD (Metal Organic Chemical Vapor Deposition) or and MBE (Molecular Beam Epitaxy), since sub-micron thickness is required.

The most common wavelength is 850 nm, but devices spanning from blue-green to near-/mid-infrared have also been realized. Both frequency [260] and intensity [261] noise have been investigated into a certain detail for 850-nm selectively oxidized VCSELs. In the former work, the frequency noise power spectral density was measured (against the slope of a Fabry-Perot resonator) revealing a $1/f^n$ part in the low-frequency range, independent of the output power, and a white-noise part in the high-frequency range, inversely dependent of the output power. Considering only white noise, the laser power spectrum, described by a Lorentzian, exhibited a linewidth $\Delta\nu_L = \pi A/P_{out}$ with $A = 180$ MHz·mW. In the latter experiment, the intensity noise characteristics were deeply explored for both a free-running and injection-locked (by a low-noise ECDL) VCSEL. In particular, sub-shot noise operation, resulting from very strong anti-correlations between the transverse modes, was observed. As expected, injection locking reduced the power of the non-injected modes and improved the anti-correlations between the transverse modes.

Finally, a microwave frequency reference based on VCSEL-driven dark line resonances in Cs vapor has also been reported [262]. An external oscillator locked to one of these resonances exhibited a stability of $1.6 \cdot 10^{-12}$ at 100 s. A physics package for a frequency-reference based on this design can be compact, low-power, and simple to implement.

4.8.5.6　Quantum cascade lasers

Quantum cascade lasers (QCLs) are a special kind of semiconductor lasers, emitting in the mid- and far-infrared spectral regions, in which laser action takes place between quantized energy levels in the conduction band [263]. Its innovative conception has been thought to overcome the severe limitations of standard bipolar semiconductor lasers in long wavelength emission. Due to the dependence of the emitted light on the material bandgap, extension of diode-laser operation to the mid-infrared spectral region has proven hard. As the bandgap shrinks, indeed, diode laser operation becomes more critical in terms of maximum operat-

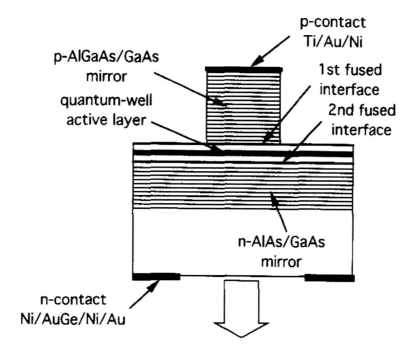

FIGURE 4.90
Cross-sectional view of a VCSEL. Typical layers of GaAs and GaAlAs are shown but other materials such as GaInAsP/InP are often used. In some cases the reflecting layers can be thin films of dielectrics. (Courtesy of [259].)

ing temperature and temperature stabilization, while undesired effects, like thermal runaway and thermal recycling, become more and more significant and chemical bonds become weaker. As an example, semiconductor laser diodes made of lead salts emit in the mid-IR but, due to the problems highlighted above, they must be operated at cryogenic temperatures and can only provide, at most, mW-level output powers, having a small continuous single-mode tuning range. Quantum cascade lasers (QCLs) rely on a completely different process for light emission, independent of the bandgap. They are unipolar devices, as they use only one type of charge carriers (electrons) which allow light emission by means of quantum jumps between conduction band sublevels artificially created by structuring the active region in a series of quantum wells. In Figure 4.91 a schematic of the operation principle for bipolar diode lasers (a) and QCLs (b) is shown for direct comparison.

The schematic energy diagram of an intersubband light emitting device is shown in Figure 4.92. Although this structure shown cannot provide any laser action (for several physical reasons), we refer to it in the schematic description of the working principle of a QCL. Keep in mind that all the following statements are valid for a working structure such as those shown in 4.93.

The device consists of a series of identical stages made of a quantum well and a barrier. Once an electron is injected into a quantum well (the gain region) by an external current driver, it undergoes a radiative transition (yellow arrow) between two sub-levels of a quantum well (level 3 is the upper state of the laser transition and level 2 the lower one). The energy of the emitted photon (equal to the energy difference between these two states) is determined primarily by the thickness of the quantum well. This is why QCLs can be de-

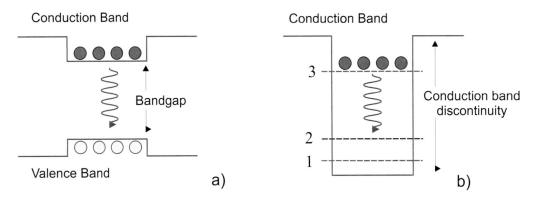

FIGURE 4.91
Schematic of the operation principle for bipolar diode lasers (a) and QCLs (b).

signed, for the same material choice, to emit at any desired wavelength in a wide spectral range. Following the photon emission, a non-radiative transition (electron-optical phonon interaction, brown dashed arrow) quickly depletes level 2 and brings the electron to the ground state (level 1). The population inversion is assured by the longer lifetime of level 3 ($\tau_3 \approx 1$ ps) with respect to level 2 ($\tau_2 \approx 0.1$ ps). The tilting of identical stages provides the peculiar cascade mechanism: thanks to an appropriate bias voltage the energy levels of adjacent stages are aligned according to Figure 4.92. Each electron in the ground state can be recycled by injection into the upper state of the adjacent identical active region through

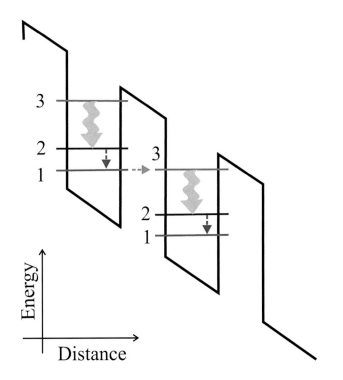

FIGURE 4.92
Schematic energy diagram of an intersubband light emitting device.

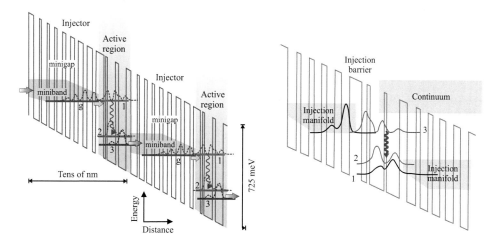

FIGURE 4.93 (SEE COLOR INSERT)
Two different QCL structures: a standard three quantum wells on the left and a bound-to-continuum structure on the right.

quantum tunnelling: here a new photon is emitted, and the process continues along the whole structure. This cascade effect, with one photon emitted per electron at each of these steps, is responsible for the high power which QCLs are able to provide.

In a working device the active region is made of several coupled quantum wells providing the quantum levels involved in laser emission. Active regions are alternated with doped multi-quantum-well regions called injectors which, when an appropriate bias voltage is applied, quickly channel the electrons into the next stage. The number of periods (injector + active region) typically ranges from 20 to 40 for mid-IR devices, but working devices with up to 100 periods have been demonstrated. In Figure 4.93 two different working structures (a standard three quantum wells on the left and a bound-to-continuum structure on the right) are shown.

The first operating QCLs were demonstrated by the group of Federico Capasso, at Bell Labs, in 1994 [264], more than 20 years after the original idea proposed by Kazarinov and Suris in 1971 [265]. These first prototypes, based on AlInAs (barriers)/GaInAs (wells) grown on InP substrate by molecular beam epitaxy (MBE), operated only at mid-IR wavelengths, in pulsed mode and at a maximum temperature of 90 K. In just over 15 years, thanks to advancements in band-engineering and waveguide design, operating temperatures well above 300 K have been achieved, for both pulsed and single-mode cw operation. Output powers up to 3 W [266, 267] cw and 34 W [268] pulsed have been achieved around room temperature for mid-IR Fabry-Perot type devices, corresponding to cw power efficiencies above 15%.

Following the first Fabry-Perot type devices, characterized by multimode operation, the realization of distributed feedback (DFB) devices [269] has to be mentioned as a fundamental step towards the development of single-mode, narrow-linewidth QCLs. The diffraction grating, which helps in selecting a single mode, can be either etched on the upper surface of the laser ridge or in the material just above the injector-active region stack. A great advance of DFB QCLs is their mode-hope-free wide tunability, extending over tens of cm^{-1} by varying operating current and temperature.

Undoubtedly, however, the most important factor in the establishment of QCLs as versatile mid-infrared sources has been played by the development of devices operating near room temperature [270]. As a consequence, they are much more compact, stable and easy to handle than cryogenic ones. One of the advantages of near room temperature operation is the possibility to deposit effective optical coatings on the laser facets, which allows to incorporate QCLs in external cavity configurations. Following this approach, in the last few years, QCLs operating near room temperature with broad gain profiles and mounted in external cavities have been demonstrated, with tuning ranges up to 200 cm^{-1} cw and over 300 cm^{-1} in pulsed operation [271, 272, 273, 274].

Materials

The most used material for mid-IR QCLs is AlInAs/GaInAs (grown on InP substrate), with a conduction band offset (quantum well depth) of 520 meV. These InP-based devices have reached extremely high levels of performance in the mid-infrared spectral range, achieving high power and above room temperature operation, even in continuous wave emission. Moving to longer wavelengths, to the far-IR (or TeraHertz) region, requires to change both the heterostructure composition (typically to AlGaAs/GaAs) and its design (bound-to-continuum structures). AlGaAs/ GaAs QCLs were demonstrated by Sirtori et al. in 1998 [275], proving that the QC concept is not restricted to one material system. Moving to operating wavelengths significantly shorter than ~ 5 μm causes the upper laser state to move up in energy, increasing the rate of electron escape into the quasi-continuum above the barriers. This process is even more critical when the device works in cw mode near room temperature and, as the active region can reach temperatures substantially higher than the laser mount (tens to hundreds of degrees), it represents the main limiting factor for QCLs temperature performance (maximum operating temperature). This limitation can be reduced by increasing the barrier height as happens, for example, in heterostructures with higher Al content in the barriers and lower In content in the quantum wells (strained AlInAs/GaInAs heterostructures). Recently, QCLs have been developed in material systems with very deep quantum wells in order to achieve short wavelength emission. The AlAsSb/InGaAs material system has quantum wells 1.6 eV deep and has been used to fabricate QCLs emitting around 3 μm [276, 277]. The InAs/AlSb material system is attractive for the development of short wavelength QCLs due to the high conduction band offset of 2.1 eV: working devices in the $2.6 - 2.9$ μm interval have been demonstrated [278, 279] and electro-luminescence at wavelengths as short as 2.5 μm has been observed [280]. QCLs may also allow laser operation in materials traditionally considered to have poor optical properties such as Si and Ge. Indeed, Si-based QCLs would be key components for a photonic platform integrated in the complementary metal-oxide semiconductor technology. The absence, in these materials, of the electron-longitudinal optical phonon scattering, presently considered the main source of non-radiative intersubband relaxation for electrons in III-V compounds, should considerably improve carrier lifetimes, leading to larger gain and higher operation temperature. Theoretical proposals for Si/SiGe and Ge/SiGe quantum cascade emitters have been made [281, 282] and intersubband absorption in the conduction band of compressively strained germanium quantum wells bounded by Ge-rich SiGe barriers have been observed in the THz range [283].

THz QCLs

Emission of QCLs at THz frequency was demonstrated only in 2002 [284]. The importance of this result goes in parallel to the interest in the far-infrared spectral region; indeed, the filling of the THz gap has recently raised a wide technological interest, owing to the possibility of realizing novel systems for security, quality control, and medical imaging operating at these frequencies. The first device was based on GaAs/AlGaAs heterostructure and showed single-mode emission at 4.4 THz with output powers of more than 2 mW at relatively low threshold current densities. Since the QCL tunability scales linearly with emission frequency, application to spectroscopy proved to be difficult with the first THz devices. Tuning rates with driving currents in the range of 4-5 GHz/A were measured for 4.6 THz devices [285], about two orders of magnitude less than the typical values for mid-infrared QCLs. Several schemes have been proposed to obtain a wider tunability, in particular by means of extended cavities [286, 287], demonstrating coarse tuning up to 5.5 cm^{-1}, about 4% of the laser center frequency. Operation of extended cavity configurations is quite difficult due to the need for cryogenic cooling (generally close to He-liquid temperatures), that forces use of a vacuum-tight vessel. Priorities in THz QCLs research include an increase of the maximum operating temperatures as well as of the emitted power. At present, the maximum operating temperature of a THz QCL is 195 K [288] and recently approached 225 K in high magnetic fields [289]. Increasing the operating temperatures up to the range of commercially available thermoelectric coolers (about 240 K) will make THz QCLs very attractive for a broad range of potential applications in areas such as biological sensing, pharmaceutical sciences, THz wave imaging, and hazardous materials detection. Nevertheless, QCLs represent the THz solid state radiation sources that actually show the best performance in terms of optical output power which, in the best devices, reaches more than 100 mW (average), still at cryogenic temperatures. A closely correlated goal to the raise of the power is the assessment of the electrical-to-optical power conversion efficiency, the so-called wall-plug efficiency. High power bound-to-continuum QCLs operating at 2.83 THz having a wall-plug efficiency as high as 5.5% in continuous wave at 40 K were demonstrated [290]. These lasers are expected to show excellent spectral characteristics. Previous measurements set an upper limit of few tens of kHz to their fast linewidth [285].

Very recently, following previous studies on mid-IR QCLs [291, 292, 293, 294] that will be discussed below, experimental studies as well as theoretical calculation of the linewidth in a QCL emitting at a frequency of 2.5 THz have been performed [295]. The measured intrinsic linewidth of about 90 Hz represents the narrowest ever observed in any semiconductor laser. A further step towards metrological applications of THz lasers has been the generation of a free-space-propagating THz comb, that has already proven useful for phase-locking onto one of the teeth of a 2.5 THz emitting QCL [296]. An alternative approach for generation of THz frequencies is based on difference frequency generation in dual-wavelength mid-IR QCLs [297, 298]. In a typical scheme, a passive non-linear layer, designed for giant optical non-linearity, is integrated on top of the active region of a dual-wavelength mid-IR QCL. This approach, although strongly power limited, has the advantage to provide THz radiation with room temperature mid-IR devices. In a slightly different scheme, THz radiation is generated inside a QCL starting from IR coherent radiation [299]. A pioneering *hybrid* approach was based on mixing of an IR QCL and radiation from a gas CO_2 laser, at a wavelength around 8 μm, in a MIM diode. Although a power of only a few hundred pW was generated, application to spectroscopy was demonstrated, by recording two lines of hydrogen bromide [300].

Finally, further information concerning both specific structures and applications of THz QCLs are reviewed in [301, 302].

Linewidth in QCLs

Since their first demonstration, QCLs have been expected to exhibit peculiar noise characteristics and high intrinsic spectral purities. Although several theoretical [303, 304, 305, 306] and experimental [304, 307, 308, 309] works have explored the frequency noise of QCLs, only recently a thorough study of their intrinsic noise has been possible [295, 291, 292, 293, 294] due to some stringent requirements: this noise, indeed, is often overwhelmed by the contribution coming from the QCL current source, so that an ultralow-noise current driver (along with a sensitive and fast enough detector) is needed to single out the elusive intrinsic noise.

As previously discussed, the Schawlow-Townes formula describes the linewidth for any laser. However, due to its generality, such formula does not contain the specific parameters (and thus the inner physical mechanisms) that determine linewidth for each specific device. Since the advent of diode lasers in 1962, 20 years of theoretical and experimental activity [310, 311, 312] were needed to formalize the theory of their linewidth [237]. As already detailed, the Schawlow-Townes equation was corrected by the so-called Henry linewidth enhancement factor α_H. The latter, in essence, accounts for the effect of refractive index variations caused by electron density fluctuations, and finally explains the excess line broadening shown by bipolar semiconductor lasers. QCLs, however, are intersubband-based devices and their inner physical mechanisms are very different from bipolar lasers. In 1994, an α_H factor close to zero was predicted for QCLs, because of their negligible refractive index variations at the peak of the gain spectrum [264]. This assumption was confirmed by several experiments [313, 314, 315]. The pioneering theoretical work by Yamanishi et al. [306], based on rate equation analysis of a three-level model of a QCL, reformulated the Schawlow-Townes equation in terms of the characteristic parameters of the QCL medium and the operating conditions. The prediction of an intrinsic linewidth several orders of magnitude smaller than that of bipolar semiconductor lasers operating at the same driving current was subsequently verified by direct experimental observation [291]. A subsequent paper clarified the main differences and analogies, in terms of power noise spectral density, between bipolar and cryogenically cooled intersubband lasers [292] (see Figure 4.94). Three main trends have been highlighted in the measured FNPSD of the QCL: a $1/f$ trend dominating at low frequencies (curve B), a cut-off at about 200 kHz (curve C), and the final flattening onto the white noise level (trace A). Besides fluctuations in the refractive index of the medium (curve A), taken into account by the Henry factor, current fluctuations (and hence temperature fluctuations) also arise in bipolar semiconductor lasers (trace D). This latter contributes much to the total noise at low frequencies, giving rise to curve E in the inset. In QCLs, these two latter mechanisms are ineffective. The former is suppressed by the very small Henry factor, while the latter no longer exists because of the almost instantaneous recovery of electron density fluctuations to the steady-state value (due to the very fast non-radiative relaxation channels), which makes the corresponding current fluctuations negligibly small in the low frequency range. QCLs, instead, show a $1/f$ noise at lower frequencies, for which an explanation has recently been suggested [292]. Imperfections in QCLs' quantum wells would cause current fluctuations internal to the heterostructure: they, in turn, would generate temperature fluctuations, contributing to the frequency noise. In this frame, the slope change observed in the FNPSD (curve C) is due to a thermal cut-off.

For a more intimate understanding of these features, it is important to recall the main physical differences at the basis of lasing in bipolar and QC lasers. The main channel that depopulates the upper laser level in a QCL is non-radiative, with a typical decay time $\tau_{nr} \sim$ ps , i.e., about 3 orders of magnitude faster than the radiative channel $\tau_r \sim$ ns (see Figure 4.95). This implies that the threshold in a QCL is much higher than that of a bipolar laser, where the non-radiative channel is negligible. Now, the intrinsic linewidth is defined as the ratio of the spontaneous emission rate (w_r) to the stimulated emission rate (w_s) effectively coupled into the lasing mode, and

$$w_r \propto \beta \frac{N_3}{\tau_r} \qquad (4.268)$$

$$w_s \propto \beta \frac{(N_3 - N_2)}{\tau_r} N_{ph} \qquad (4.269)$$

where N_2, N_3 are the electronic populations of levels 2, 3 respectively, N_{ph} is the number of photons in the laser cavity at threshold, β is the coupling coefficient of the spontaneous

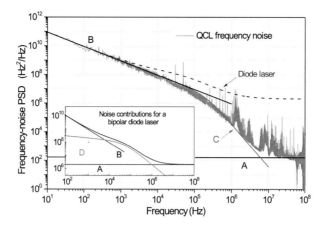

FIGURE 4.94
Expected contributions to the frequency-noise power spectral density (FNPSD) in a QCL. The measured FNPSD is also plotted and compared with the typical noise figure of a diode laser (dashed line and inset).

emission into the lasing mode (see [306] for a complete discussion). Therefore, when a QC starts lasing, the number of photons is several orders of magnitude higher than in a bipolar laser just above threshold and this corresponds to a predominant stimulated emission and thus to a narrower intrinsic linewidth. It is worth noting that, in general, QC lasers can operate at higher currents than bipolar ones, generally becoming unstable when the operating current is much larger than the threshold (typically much lower than that in QCLs): this is another way to understand the narrower intrinsic linewidth in QCLs. Finally, very recent studies [293, 294] show that frequency noise power spectral density and intrinsic linewidth in QCLs are much reduced operating them close to room temperatures rather than cryogenically cooled.

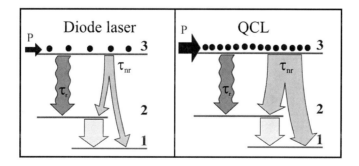

FIGURE 4.95
In a bipolar diode laser a moderate pumping (P) is sufficient to start lasing (left). In a QCL, due to its large non-radiative decay channels, a much larger pump is needed (right). This mechanism is responsible for the high thresholds of QCLs and, consequently, of the much larger number of photons present in the laser cavity with respect to standard diode lasers.

4.8.5.7   Interband cascade lasers

Interband cascade lasers (ICLs) differ from both conventional and cascaded semiconductor radiation sources in that population inversion is essentially created by carriers which are produced internally, at semimetallic interfaces (SMIF) within each stage of the active region (Figure 4.96) [316]. Such internal generation is aided by the alignment of the conduction band (CB) minimum in InAs at an energy ~ 0.2 eV below the valence band (VB) maximum of GaSb. The applied electric field induces a transition from a positive energy gap (procured by quantum confinement) back to the semimetallic alignment with band overlap $E_{SM}(V)$, which allows the electron (hole) states in the InAs (GaSb) QW to be populated in thermal quasi-equilibrium. As the generated carriers are swept away from the junction by the external electric field, with holes (electrons) flowing to the left (right), in order to maintain quasi-equilibrium, equal numbers of carriers must be replenished continuously at the SMIF. Ideally, the single-stage voltage drop, i.e. the separation between the quasi-Fermi levels (QFLs) in adjacent stages, must be such as to produce an optical gain which compensates the photon loss in the cavity, simultaneously generating internally a quasi-equilibrium carrier density. While the former condition is a characteristic of the active QWs, the latter depends on the design and doping of the (multiple) electron and hole QWs on both sides of the SMIF. Ultimately, when joined to any carriers brought about by extrinsic doping, the resulting electron and hole densities determine the desired QFL separation. To go into more detail, with reference to Figure 4.97, let us consider the typical multiple QW. It consists of a chirped InAs/AlSb electron injector, two active InAs electron QWs of the next stage on one side of the SMIF, and a GaSb/AlSb hole injector (whose role, apart from entering the voltage-balancing condition, is to provide a barrier to direct the electron current flow that bypasses the VB) plus an active GaInSb hole QW on the other. In such analysis, carrier thermalization on both sides of the SMIF is assumed to be much faster than the carrier lifetime. Since the injected electrons dwell mostly in the injector, whereas almost all the injected holes move to the active GaInSb QW, the active hole population considerably outnumbers that for electrons. In this scenario, the quasi-equilibrium electron and hole populations in the active QWs are only partially rebalanced, provided that Auger recombination and free-carrier absorption processes are much weaker for holes than for electrons. Then, full rebalancing can be accomplished by n-doping the electron injector QWs much more heavily $(5 \cdot 10^{18}$ cm$^{-3})$. As experimentally confirmed, this substantially lowers the lasing threshold J_{th} and the power density relative to earlier ICL devices. In particular, values of J_{th} as low as 170 A cm^{-2} have been demonstrated in [316], falling within the typical range for near-infrared diodes. By requiring nearly two orders of magnitude less input power to operate in continuous-wave, room-temperature mode than the QCL, the ICL represents a very attractive option for spectroscopic applications asking for low output power and minimum heat dissipation at wavelengths from 3 μm to beyond 6 μm. While referring the interested

reader to [316] (and references therein) and to [317] for a more comprehensive discussion on ICLs, in the remainder we focus on their spectroscopic applications.

In this respect, operation of cw, room-temperature ICLs in external-cavity configuration has also been demonstrated; in particular, tuning across the 3.2-3.5 μm wavelength range was achieved at a maximum power of 4 mW [318]. Formerly, a trace-gas sensor based on a pair of liquid-nitrogen-cooled ICLs was developed for simultaneous detection of formaldehyde and ethane; minimum detection limits of 3.5 ppbv and 150 pptv were obtained, respectively, with a 100-m astigmatic Herriott multi-pass cell and 1-s integration time [319]. More recently, a DFB ICL (operated under liquid-nitrogen cooling at 86 K) with an output power of 4 mW (at a current of 25 mA) was used in an off-axis integrated-cavity-output-spectroscopy experiment for real-time breath ethane measurements; a detection sensitivity of 0.48 ppb/Hz$^{1/2}$ was achieved [320]. In addition, measurements of the

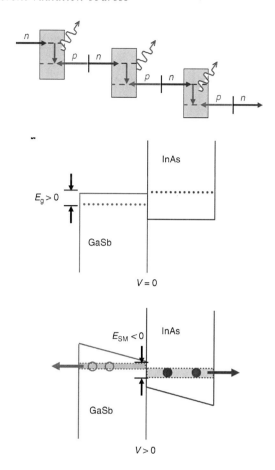

FIGURE 4.96
Schematic diagram of the carrier injection process in an ICL. The black vertical lines point the location of the SMIF within each stage. The energy alignment for adjacent InAs and GaSb QWs in equilibrium and under bias is also shown: the solid blue and red lines represent the conduction and valence-band edges, respectively, in bulk materials. In the absence of a bias, quantum confinement induces an energy gap E_g between the lowest conduction and highest VB states (blue and red dotted lines, respectively), whereas, under bias, a semi-metallic overlap E_{SM} is imposed that procures the generation of equal electron and hole densities, represented by the solid blue and open red circles, respectively. The applied field also causes both carrier types to flow away from the interface (arrows), requiring that they be replenished continuously to preserve quasi-thermal equilibrium populations. (Courtesy of [316].)

relative intensity noise for an ICL were performed as a function of laser current at 30 and 100 K; such characterization revealed that: away from threshold, the laser primarily exhibits a frequency-independent photon noise spectral density in agreement with theory; at threshold, the observed photon noise spectral density exhibits large fluctuations at closely spaced discrete frequencies; thermal effects at 100 K result in a large increase in the photon noise above threshold [321].

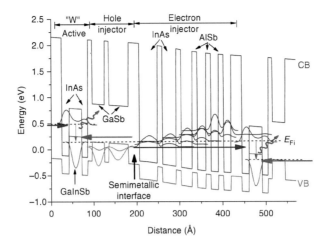

FIGURE 4.97 (SEE COLOR INSERT)
Illustrative band diagrams of the complete ICL active core. Probability densities and zone-center energies (indicated by the wavefunction zero points) for some of the most important subbands are superimposed. The probability densities for the active electron (hole) subbands are indicated with blue (red) lines, while those for the injector-electron (hole) subbands are indicated with wine-coloured (green) lines. The blue (red) arrows indicate the direction of the electron (hole) motion in the structure. The dashed lines indicate the position of the quasi-Fermi levels in each stage. (Courtesy of [316].)

However, neither high-resolution spectroscopic measurements nor accurate characterizations of frequency noise have been reported so far.

4.8.6   Nonlinear laser cw sources

In 1961, by focusing the 694-nm-wavelength pulses emitted from the newly invented ruby laser onto a quartz plate, Franken discovered that ultra violet light of wavelength 347 nm was emerging from the crystal [166]. This observation suggested that, by virtue of some nonlinear phenomenon taking place in the quartz, the intense excitation light of frequency ω had generated the second harmonic, at frequency 2ω. Such a discovery marked the birth of Nonlinear Optics.

A useful approach for a quantitative treatment of non-linear optical phenomena, taking place in a medium that is perfectly transparent to the propagating radiation, is to write down the equations which relate the polarization $\boldsymbol{P}$ to the electric field $\boldsymbol{E}$. Consider that, as a thumb-rule, the equivalent electric field binding an outermost electron in a dielectric is on the order of 1 GV/m. Therefore, for electromagnetic fields much lower than this value, propagating in a dielectric, it is possible to expand $\boldsymbol{P}$ in a power series of the applied electric field, $\boldsymbol{E}$. Thus, neglecting the frequency dependence (small dispersion), the expansion takes the form

$$
\begin{aligned}
P_i = \varepsilon_0 \sum_j \chi_{ij}^{(1)} E_j + \varepsilon_0 \sum_j \sum_k \chi_{ijk}^{(2)} E_j E_k \\
+ \varepsilon_0 \sum_j \sum_k \sum_l \chi_{ijkl}^{(3)} E_j E_k E_l + ...
\end{aligned}
\tag{4.270}
$$

where $\chi_{ijk}^{(2)}$ ($\chi_{ijkl}^{(3)}$) is a real rank 3 (4) tensor, which characterizes the second-order (third-order) non linear susceptibility, and so on. For typical laser intensities, successive terms in the expansion of Equation 4.270 drop off fast, so that it generally suffices to retain up to the second-order term, unless $\chi^{(2)}$ happens to be zero, in which case the third-order term in $\chi^{(3)}$ must be called into question. Next, consider the case in which the point r in the material is subjected to a superposition of two laser fields (i.e., two monochromatic fields of frequency ω_1 and ω_2) propagating along the direction z

$$E_1(t) = \mathcal{E}_1 e^{i(k_1 z - \omega_1 t)} + c.c. \tag{4.271}$$

$$E_2(t) = \mathcal{E}_2 e^{i(k_2 z - \omega_2 t)} + c.c. \tag{4.272}$$

Following a standard procedure contained in all non-linear optics textbooks (see for instance [322, 166]), by substitution of $E(t) = E_1(t) + E_2(t)$ into Maxwell's equations and under suitable approximations, one finally obtains an equation relating the amplitude of the generated field $\mathcal{E}_3$ to the components $\mathcal{P}_{NL}^{\omega_3}$

$$\frac{d\mathcal{E}_3}{dz} = i\frac{\omega_3}{n_3 c}\chi^{(2)}\mathcal{E}_1(z)\mathcal{E}_2(z)e^{i\Delta k z} \tag{4.273}$$

where $n_3^2 \equiv n^2(\omega_3) = 1 + \chi^{(1)}(\omega_3)$ is the refractive index *seen* by the generated field (the medium is supposed to be homogeneous and isotropic at first order) and

- $\omega_3 = \omega_1 + \omega_2$, $\chi^{(2)} \equiv \chi^{(2)}(-\omega_3; \omega_1, \omega_2)$, and $\Delta k = k_1 + k_2 - k_3$ for the sum frequency generation (SFG) process. A particular case is represented by second harmonic generation (SHG) for which $\omega_1 = \omega_2 = \omega$.

- $\omega_3 = \omega_1 - \omega_2$, $\chi^{(2)} \equiv \chi^{(2)}(-\omega_3; \omega_1, -\omega_2)$, and $\Delta k = k_1 - k_2 - k_3$ for the difference frequency generation (DFG) process.

Since the amplitude of the wave at ω_3 is zero at the entry face of the non-linear crystal, we have: $\mathcal{E}_3(z = 0) = 0$. Also, due to the weakness of the non-linear effect, to a first approximation, the intensity of the pump beams does not change over the length of the medium; in other words, we can take $\mathcal{E}_1$ and $\mathcal{E}_2$ to be constant. Then, integration of Equation 4.273 between $z = 0$ and $z = L$ yields

$$|\mathcal{E}_3(L)|^2 = \left(\frac{2\omega_3\chi^{(2)}}{n_3 c}\right)^2 |\mathcal{E}_1|^2|\mathcal{E}_2|^2 \left[\frac{\sin\Delta k \cdot (L/2)}{\Delta k}\right]^2 \tag{4.274}$$

for the intensity of the ω_3 wave leaving the non-linear crystal. According to this equation, for a given mismatch Δk, the non-linear conversion process will exhibit maximal efficiency for a crystal of length $L_{opt} = \pi/|\Delta k|$; the corresponding optimal intensity will be proportional to

$$|\mathcal{E}_3(L_{opt})|^2 = \left(\frac{2\omega_3\chi^{(2)}}{\pi n_3 c}\right)^2 |\mathcal{E}_1|^2|\mathcal{E}_2|^2 L_{opt}^2 \tag{4.275}$$

Obviously, in order to exploit an optimal non-linear crystal that is as long as possible, one has to make Δk as small as possible. The perfect phase matching condition $\Delta k = 0$ can be expressed as

$$k_3 = k_1 \pm k_2 \tag{4.276}$$

for the SFG and DFG process, respectively. Note that in the DFG case, the energy and *linear momentum* conservation relations can be re-written as $\omega_1 = \omega_3 + \omega_2$ and $k_1 = k_3 + k_2$, respectively. These latter are formally identical to the SFG conditions, provided that indexes 1 and 3 are inter-changed. So in the following considerations, we can just refer to one of these

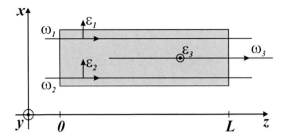

FIGURE 4.98
SFG process in a birefringent non-linear crystal of type I.

two processes, say the SFG one. Since we must have $\omega_3 = \omega_1 + \omega_2$, it would seem impossible to achieve perfect phase matching in a dispersive material, where $k_i = n(\omega_i)(\omega_i/c)$ and the refractive index $n(\omega)$ is a monotonic (increasing) function of ω. However, we can overcome this obstacle by employing a birefringent material, where the refractive index depends on the polarization. For example, in the case of a type-I non-linear crystal, two along-x-polarized waves at ω_1 and ω_2 originate a wave at ω_3 which is polarized along y (the waves propagate along z, as shown in Figure 4.98). Thanks to the birefringence, the refractive index is different for the waves polarized along x (ordinary index n_o) and for the wave polarized along y (extraordinary index n_e). To fix ideas, in the case of SFG we have

$$n_e(T,\omega_3)\frac{\omega_3}{c} = n_o(T,\omega_2)\frac{\omega_2}{c} + n_o(T,\omega_1)\frac{\omega_1}{c} \tag{4.277}$$

To simplify, let us consider the particular case $\omega_1 = \omega_2 = \omega$ for which $\omega_3 = 2\omega$. In this case (SHG), Equation 4.277 requires $n_e(T,2\omega) = n_o(T,\omega)$ which can be satisfied by a proper choice of the crystal temperature, provided that $n_e(\omega) < n_o(\omega)$ (negative crystals). An elegant, alternative approach is represented by the quasi-phase matching (QPM) technique, in which several slices of non-linear medium are placed end-to-end, each slice being of length $\pi/|\Delta k|$, but with the signs of the coefficients $\chi^{(2)}$ alternating from one piece to the next. Essentially, the idea is to allow for a phase mismatch over some propagation distance, but to reverse the non-linear interaction at positions where otherwise the interaction would occur with the wrong conversion direction. In fact, *momentum* is conserved in QPM through an additional momentum contribution corresponding to the wavevector of the periodic structure (see for instance [323, 324] and references therein). In other words, phase matching is obtained if for some integer m, the following condition is satisfied

$$k_3 = k_1 + k_2 + m\frac{2\pi}{\Lambda} \rightarrow \frac{n(T,\lambda_i)}{\lambda_3} = \frac{n(T,\lambda_1)}{\lambda_1} + \frac{n(T,\lambda_2)}{\lambda_2} + \frac{1}{\Lambda} \tag{4.278}$$

where $n(T,\lambda)$ can be calculated by using the Sellmeier equation. Thus, given the pump wavelengths λ_1 and λ_2 (which sets λ_3 through energy conservation $\lambda_3^{-1} = \lambda_2^{-1} + \lambda_1^{-1}$, Equation 4.278 can be satisfied by properly choosing the temperature and the modulation period Λ of the crystal. When possible, first-order QPM ($m = 1$) is exploited, as the intensity of the generated field scales as $1/m^2$. In contrast to birefringent phase matching, by appropriate selection of Λ, QPM materials can be engineered for phase matching at any wavelengths within the transparency range of the crystal. This method enables a free choice of polarization of the interacting waves and the exploitation of the largest non-linear susceptibility component. Moreover, since QPM does not rely on birefringence, it can be used in isotropic materials with a high optical non-linearity.

The most popular technique for generating QPM non-linear crystals is periodic poling of ferroelectric materials such as lithium niobate (LiNbO$_3$), lithium tantalate (LiTaO$_3$), and potassium titanyl phosphate (KTP, KTiOPO$_4$) by ferroelectric domain engineering. Here, micro-structured electrodes are used to apply a strong electric field to the crystal for some time, so that the crystal orientation and thereby the sign of the nonlinear coefficient are permanently reversed solely below the electrode fingers. Typical poling periods range between 5 and 50 μm.

Starting from two cw pumping lasers, a SFG/DFG process can be used to realize a cw coherent radiation source operating at a higher/lower frequency (with respect to that of the pumping lasers). Therefore, *up-conversion* represents an effective way to extend the spectral coverage of laser sources to the ultraviolet region, whereas *down-conversion* performs the same task for the infrared. Another relevant second-order down-conversion non-linear optical process is that of optical parametric oscillation (OPO), which converts an input laser wave (the pump) into two output waves of lower frequency (idler and signal). In this case, the energy-conservation condition is expressed as $\omega_p = \omega_s + \omega_i$.

4.8.6.1 Sum frequency generation

As an example of SFG, we describe here the generation of cw ultraviolet radiation in periodically poled crystals [325]. Practical limitations to this approach are set by the availability of transparent materials at the considered wavelengths and by the ability to realize regular periodic structures. LiNbO$_3$ and KTP, widely used in the visible and IR range, are not transparent in the UV range. Lithium tantalate (LiTaO$_3$) is a valid alternative for UV generation, as its transparency window extends in the UV range, and has a nonlinear constant d_{33} similar to that of KTP. Also, LiTaO$_3$ is less sensitive to photo-refractive damage and green induced infrared absorption with respect to LiNbO$_3$. This latter feature makes LiTaO$_3$ more suitable than LiNbO$_3$ when high power densities are involved. However, the first-order QPM condition for efficient generation of UV sum frequency would require a poling period of about 2 μm. At present, such small periods cannot be easily realized, especially with a depth of at least several hundreds of micron, that is the typical beam dimension. Thus, a valuable option is to use a PPLT crystal with a poling period corresponding to a third-order QPM condition. The experimental apparatus is shown in Figure 4.99. The laser source is a semiconductor laser with an Ytterbium-doped fiber amplifier emitting up to 10 W of cw radiation around 1064 nm, with a linewidth of less than 300 kHz. In a first stage, second harmonic at 532 nm is generated. For this purpose, a temperature-controlled, 1%-MgO-doped, 30-mm-long, 1-mm-thick z-cut stochiometric MgO:LiTaO$_3$ crystal with a 7.97 μm period is used. The crystals' ends are antireflection coated for both fundamental and second harmonic. With the PPLT crystal an optimized SH power of 1.4 W with 8.5 W of pump power, with a resulting efficiency of 16.5% (0.65% W^{-1} cm^{-1}), is obtained. In the second stage, the outcoming infrared and green light interact in a second non-linear crystal (with waists of 90 μm for the fundamental beam and 60 μm for the second harmonic one). The latter crystal, identical to the first one except for the poling period, exploits third-order QPM which can be met with a period of about 6.5 μm, albeit at the cost of about a factor-of-ten decrease in efficiency. The generated UV radiation is monitored by a calibrated silicon photodiode and, in the most favorable case, a maximum UV power of 7 mW is measured. This is approximately one order of magnitude lower than what is expected for third-order QPM in PPLT, assuming $d_{33} = 13$ pm/V. Such discrepancy can be largely attributed to the unbalance in the duty cycle. In fact, the efficiency of m-th order QPM is proportional to $\sin^2(m\pi D)$, being D the duty cycle. Therefore, third-order QPM has three optimal values for D(1/6, 1/2, and 5/6), while the efficiency is null for $D = 1/3$ and 2/3. Indeed, the estimated mean duty cycle is very close to the worst cases.

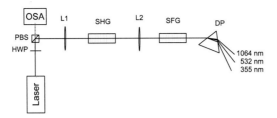

FIGURE 4.99
Schematic view of experimental setup. HWP, halfwave plate; PBS, polarizing beam splitter; OSA, optical spectrum analyzer; L, lens; DP, dispersive prism. SHG and SFG are temperature stabilized periodically poled crystals for second harmonic and sum frequency generation, respectively.

The above approach is particularly favorable when high spectral purity and tunability are required for the UV radiation. Indeed, it can exploit the robustness of high-performance and powerful lasers, such as solid state or fiber lasers, and naturally transfer, by virtue of the coherent character of non-linear optical processes, their advantageous spectral features to the final frequency. As we will see in Chapter 7, narrow-linewidth cw UV laser sources are crucial for high-resolution spectroscopy of atoms and for laser cooling and trapping.

4.8.6.2   Optical parametric oscillators

In some respects, an OPO is very similar to a laser: first, it makes use of a kind of laser resonator (but relying on optical gain from parametric amplification in a non-linear crystal rather than from stimulated emission); second, it exhibits a threshold for the pump power, below which there is only some parametric fluorescence. Let us start by considering a non-linear crystal illuminated by an intense wave (the pump) at frequency ω_3 and a weak wave (the signal) at ω_1; then, we assume that the phase-matching condition is satisfied for $\omega_2 = \omega_3 - \omega_1$ [166]. In this case, the signal is amplified while another wave, the idler, appears at ω_2. For a mathematical treatment, we rewrite Equation 4.273 and the corresponding ones for the fields $\mathcal{E}_2$ and $\mathcal{E}_1$

$$\frac{d\mathcal{E}_3}{dz} = i\frac{\omega_3}{n_3 c}\chi^{(2)}\mathcal{E}_1(z)\mathcal{E}_2(z) \tag{4.279}$$

$$\frac{d\mathcal{E}_2}{dz} = i\frac{\omega_2}{n_2 c}\chi^{(2)}\mathcal{E}_3(z)\mathcal{E}_1^*(z) \tag{4.280}$$

$$\frac{d\mathcal{E}_1}{dz} = i\frac{\omega_1}{n_1 c}\chi^{(2)}\mathcal{E}_3(z)\mathcal{E}_2^*(z) \tag{4.281}$$

Under the constant pump approximation, Equations 4.280 and 4.281 imply that

$$\frac{d^2\mathcal{E}_2}{dz^2} = \left(\frac{\chi^{(2)}|\mathcal{E}_3|}{c}\right)^2 \frac{\omega_1\omega_2}{n_1 n_2}\mathcal{E}_2 \equiv \gamma^2\mathcal{E}_2 \tag{4.282}$$

For the initial conditions $\mathcal{E}_2(z=0)$ and $\mathcal{E}_1(z=0) = \mathcal{E}_1(0)$, the solution is

$$\mathcal{E}_2(z) = i\sqrt{\frac{\omega_2 n_1}{\omega_1 n_2}}\frac{\mathcal{E}_3}{|\mathcal{E}_3|}\mathcal{E}_1(0)\sinh(\gamma z) \tag{4.283}$$

$$\mathcal{E}_1(z) = \mathcal{E}_1(0)\cosh(\gamma z) \tag{4.284}$$

Thus, as anticipated, the signal wave $\mathcal{E}_1$ is amplified, while a complementary wave $\mathcal{E}_2$, referred to as the idler, also appears. The usefulness of such a parametric amplification

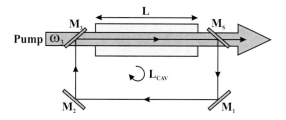

FIGURE 4.100
Layout of an optical parametric oscillator. The optical cavity can be resonant for just the signal beam or for both the signal and idler beams. (Adapted from [166].)

process mainly resides in its rather broad tunability range (obtained by adjusting the phase-matching condition). Next, focusing solely on the non-degenerate case (where signal and idler have different frequencies and the parametric interaction can amplify a signal field of any phase), we will consider the situation which originates when a parametric amplifier is placed inside a resonant cavity. The OPO is shown schematically in Figure 4.100. It consists of a nonlinear crystal of length L, in which the parametric interaction takes place, inserted within an optical cavity of length L_{cav}, assumed to be a ring cavity. This cavity is totally transparent at the wavelength of the pump beam. The two most typical configurations are:

- Singly resonant OPO: the cavity is transparent at the wavelength of the pump and idler beams. While M_1, M_2 and M_3 are perfectly reflecting mirrors, the output mirror M_s (transparent for the pump and idler) exhibits low transmission at the wavelength of the signal beam. Let's denote with R and T the reflection and transmission coefficients for the intensity at this mirror.

- Doubly resonant OPO: still transparent at the wavelength of the pump beam, the cavity is now resonant for both the signal and idler beam. The mirrors M_1, M_2 and M_3 are perfectly reflecting, while the output mirror M_s is weakly transmitting at frequencies ω_1 and ω_2.

In the following we restrict ourselves to the first case [166]. The behavior of a singly-resonant optical parametric oscillator (SR-OPO) is, in several but not all aspects, analogous to that of a homogeneously broadened laser. As for a laser, an oscillation is only set up if, for each round trip, the parametric gain dominates losses due to transmission at the output mirror. So, just as happens for the laser, there will be an oscillation threshold, governed here by the intensity of the pump beam. In principle, oscillation takes place on the cavity mode which experiences the highest gain, and only on that. Indeed, when oscillation starts, the pump power is depleted, the gain is lowered to the value of the loss of the cavity mode closest to the peak of the parametric gain curve, and the oscillation of other modes is thus inhibited [322]. Nevertheless, mainly due to mechanical instabilities of the cavity, frequent hops take place between different longitudinal modes among the tens (or hundreds) which lie within the gain profile [326]. Mode-hops can be successfully prevented increasing the free spectral range of the OPO resonator or making use of intra-cavity, wavelength-selecting elements, such as etalons, which restrict the number of cavity modes which can experience gain, thus letting the OPO operate at a single frequency [327]. Usually, low reflectivity (and hence low finesse) etalons are used, in order to reduce losses inside the cavity. Now, by virtue of Equation 4.284, the signal field leaving the crystal can be expressed as

$$E_1^{(+)}(z = L) = \mathcal{E}_1(0)e^{in_1\omega_1 L/c}\cosh\gamma L \qquad (4.285)$$

while the signal field that has reflected off the output mirror and propagated once around the cavity is given by

$$E_1^{(+)}(L_{cav}) = \sqrt{R}\mathcal{E}_1(0)e^{i\phi_1}\cosh\gamma L \qquad (4.286)$$

where $\phi_1 = [L_{cav} + (n_1 - 1)L](\omega_1 c)$ is the phase accumulated by the signal during the propagation over a complete cavity round trip. As for a laser, the conditions for steady-state oscillation are derived by imposing that, after one cavity round trip, the field is reproduced in both phase and amplitude; then, in complex value, we have $E_1^{(+)}(L_{cav}) = E_1^{(+)}(0) = \mathcal{E}_1(0)$, whereupon

$$\sqrt{R}e^{i\phi_1}\cosh\gamma L = 1 \qquad (4.287)$$

which implies the two real conditions

$$\cosh\gamma L = 1 \qquad (4.288)$$

$$(n_1 - L)L + L_{cav} = m\lambda_1 \qquad (4.289)$$

with m an integer. The first (second) condition expresses the equality of gain and loss (the resonant character of the cavity for the signal wave). For a low-intensity pump ($\gamma L \ll 1$) and a highly resonant cavity ($\sqrt{R} \simeq 1 - T/2$), the first condition yields the pump beam intensity at the oscillation threshold

$$|\mathcal{E}_3|^2 = \frac{n_1 n_2 c^2}{\omega_1 \omega_2} \frac{2\pi}{\mathcal{F}L^2(\chi^{(2)})^2} \qquad (4.290)$$

where $\mathcal{F} = 2\pi/T$ is the cavity finesse. Under typical experimental conditions, this is on the order of a few watts. When the pump exceeds this threshold, the energy transferred on each trip around the cavity from the pump to the signal and idler waves is no longer negligible; so, the pump wave amplitude diminishes during propagation, thus inducing a drop in the parametric gain (gain saturation phenomenon [322]). Due to the high value of the threshold intensity, development of effective continuous-wave, singly-resonant OPOs has been greatly facilitated by the advent of QPM ferroelectric crystals with a higher conversion efficiency.

In conclusion, in spite of the strong similarities discussed above, three major differences between the OPO and the laser must be emphasized:

- In the (singly-resonant) OPO, the emission wavelength is not determined by the level structure of the active medium, as it happens in lasers, but is rather related to the phase-matching conditions. These can be adjusted by changing the temperature or orientation of the crystal, so that tunability can be realized over a very broad spectral band (essentially limited only by the transparency window of the nonlinear medium) with a conversion efficiency (of the pump to the sum of the signal and idler waves) of up 90%.

- For an OPO system, gain saturation is caused by the pump depletion through energy transfer in parametric amplification whereas, in the laser case, it is due to the reduction of population inversion in amplification by stimulated emission;

- While the energy transfer from the pump mode to the signal and idler modes is practically instantaneous in the OPO, the excited laser level is able to store energy for an appreciable time length.

Linewidth of a cw SR-OPO

Let's consider an OPO singly resonant on the signal. The signal linewidth strongly depends on the stability of the OPO resonator (see below), while the spectral content of the idler, which constitutes the *ouput* of such an OPO, is substantially given by the convolution of the linewidth of the pump and of the signal [328].

When operating in free-running mode, the main elements which affect the width of the signal are essentially the crystal inside the cavity (in terms of its gain lineshape and characteristic dispersion curve) and the finesse of the cavity itself. This could be accounted for as follows [329, 330, 331].

$$\Delta\nu_s = \frac{\Delta\nu_c}{\sqrt{\langle n \rangle}} \tag{4.291}$$

where $\Delta\nu_s$ is the signal linewidth, $\Delta\nu_c$ is the single-pass bandwidth of the crystal (i.e. the width of the single-pass parametric gain curve) and $\langle n \rangle$ is the mean number of cavity transits (which, in turn, depends on the cavity finesse $\langle n \rangle = \mathcal{F}/\pi$), which can be evaluated starting from the reflectivity of the mirror coatings and of the crystal facets. For typical values $\Delta\nu_c = 500$ GHz and $\langle n \rangle = 100$, one gets $\Delta\nu_s = 50$ GHz, which is extremely high if the OPO has to be used for high-resolution spectroscopy.

In order to dramatically reduce the OPO linewidth it is necessary to lock the signal emission to a reference, e.g. a high-finesse cavity. In first approximation, the locking of the signal emission to a high-finesse optical cavity can be simulated considering the OPO resonator as stationary, i.e. as totally free from mechanical instabilities. In such an idealized view, an upper limit to the signal linewidth is given by the linewidth of the OPO resonator modes $\Delta\nu_{cav}$. Then, the linewidth of the idler is obtained by convolution of the pump and signal profiles. Therefore, for narrow-linewidth pump lasers (on the order of 10 kHz), $\Delta\nu_{cav}$ can be considered the maximum idler linewidth. Finally, it can be shown that the ultimate limit for the linewidth (FWHM) of a SR-OPO can be expressed as [332, 333]

$$\Delta\nu_s = \left(\frac{\kappa_s}{\kappa_i}\right)^2 \Delta\nu_p + 2\frac{h\nu_s}{P_s}\kappa_s^2 \tag{4.292}$$

where κ is the resonator decay rate (with $\kappa_i \gg \kappa_s$) and P is the output power. The first term depends on the spectral quality of the pump and on the finesse of the OPO cavity at signal and idler wavelengths; the second term is the so called quantum phase diffusion term and represents the OPO analog of the Schalow-Townes laser linewidth. For typical experimental parameters, the first term dominates. Being of the sub-Hz order, this latter value is practically negligible in the convolution with the pump linewidth for the final calculation of the idler linewidth.

Singly-resonant optical parametric oscillator for mid-infrared high-resolution spectroscopy

As an example of recent OPO systems, we present here a singly-resonant cw optical parametric oscillator, emitting more than 1 W in the 2.7-4.2 μm range, mainly developed for high-resolution molecular spectroscopy [334]. In this frame, wide tunability, narrow linewidth and high power are required for the source. In particular, the latter two properties are crucial for effectively probing sub-Doppler transition signatures, without the need for complicated experimental configurations, or even for trapping and manipulation of cold molecules. The experimental setup is shown in Figure 4.101. The OPO is pumped by a narrow-linewidth (FWHM=40 kHz at 1 ms) cw Yb-doped fibre laser, which seeds a Yb-doped fibre amplifier, delivering up to 10 W at 1064 nm. Laser frequency can be rapidly tuned over few GHz by changing the fibre cavity length through the action of a piezoelectric actuator, while a coarse and slow tuning can be achieved by changing the temperature of the laser cavity. The OPO cavity is in bow-tie configuration, singly resonating for the signal wavelength, with two plane mirrors and two curved ones (R.O.C.=100 mm). The optical path inside the cavity is 695 mm, corresponding to a longitudinal mode spacing of 430 MHz. All the cavity mirrors have high-reflectivity (HR) coating for the signal frequency ($R > 99.9\%$, between 1.4-1.8 μm). The curved mirrors have a high-transmission coating for the pump ($R < 2\%$) and the idler wavelengths ($R < 5\%$). The nonlinear crystal is a 50-mm-long periodically-poled sample of 5%-MgO-doped congruent lithium niobate, with seven poling periods ranging from 28.5 to 31.5 μm. The crystal end facets have anti-reflection coating for the pump ($R < 3\%$ at 1064 nm) and the signal ($R < 1\%$ in the range 1450-1850 nm), and high transmission coating for the idler ($R < 10\%$ in the range 2500-4000 nm). The crystal is temperature-stabilized with a precision better than 0.1°C and placed between the two curved mirrors, in correspondence of the smaller cavity waist. The value of this waist varies from 52 μm at 1.4 μm, to 59 μm at 1.8 μm. The amplifier output beam is coupled into the cavity through one of the curved mirrors and focused at the centre of the nonlinear crystal with a beam waist of 48 μm, so that pump and signal have approximately the same focusing parameter. An adjustable 400 μm-thick YAG etalon is placed between the two flat mirrors, in order to reduce mode-hop events. A Germanium (Ge) filter is used to separate the transmitted idler beam from residual pump and other spurious frequencies. The measured idler power as a function of the wavelength is shown in 4.102(a). The idler power is ~ 1 W, for wavelengths in the range 2.7-3.6 μm, while it decreases down to ~ 400 mW at longer wavelengths. The regular decrease at longer wavelengths is due to the inherent dependence on wavelength of the down-conversion efficiency. Scattering of the data can be partly attributed to small changes of the cavity losses, depending, in turn, on small inhomogeneities of the coatings at different wavelengths. Figure 4.102(b) shows the idler power as a function of the pump power at the wavelength of 2.93 μm, corresponding to the 31.5 μm-period grating at the temperature of 30°C. Experimental data have been fitted according to the empirical expression

$$P_i(P_p) = C \left[1 - \left(\frac{P_{\text{thr}}}{P_p} \right)^{\frac{1}{3}} \right] \tag{4.293}$$

where P_i and P_p represent the idler and pump power, respectively, P_{thr} is the OPO threshold, and C is a constant weakly depending on the cavity passive losses. As a result of the fit, the OPO threshold was found to be $P_{\text{thr}} = 3.1$ W.

The signal frequency is locked (by the PDH technique) to a high-finesse Fabry-Perot cavity consisting of two HR curved mirrors (R.O.C. = 1 m). The HR coating spans the same wavelength range of the OPO mirrors, providing high finesse ($\gtrsim 4000$) all over the OPO operating range. The reference cavity mirrors are glued on an invar spacer, held by a mechanical suspension and placed in a vacuum chamber for seismic and acoustic isolation. The chamber temperature is actively stabilized (< 0.1°C), reducing the cavity length fluctuations. For the PDH scheme, the signal beam leaking through one of the OPO cavity mirrors is coupled into a polarization-mantaining optical fibre and sent to a fibre-coupled electro-optic phase modulator (EOM), driven at 30 MHz by a local oscillator, and coupled into the reference cavity. A polarizing beam-splitter and a quarter waveplate deviate the beam reflected off the cavity to an InGaAs PIN photodiode. The ac-amplified signal from the detector is mixed with a reference signal from the local oscillator and low pass filtered. Such PDH signal is used as error signal for the servo-loop, whose output (correction signal) drives the OPO cavity PZT. The reference cavity is initially set to be nearly resonant with the signal frequency, by slightly changing the PZT voltage, then the servo-loop is switched on and operates. Once the locking was working, both the error and correction signals were acquired with an oscilloscope with on-board FFT routine, thus obtaining the power spectral density (PSD) for both signals. The correction signal PSD, within the servo bandwidth, represents the frequency noise of the free-running OPO, while the error signal

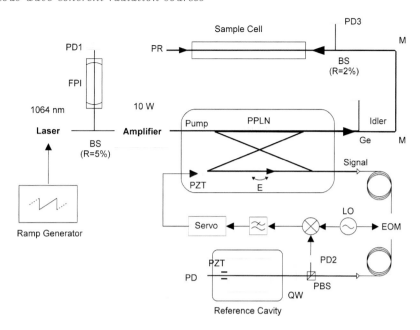

FIGURE 4.101
Experimental scheme of the OPO four-mirrors ring cavity and saturation spectroscopy setup. BS: beam splitter; E: YAG etalon; FPI: Fabry-Perot interferometer; Ge: germanium filter; LO: local oscillator; PBS: polarizing beam-splitter; PD: photodiode; PPLN: periodically-poled lithium niobate crystal; PZT: piezoelectric actuator; QW: quarter wave-plate.

PSD provides the in-loop residual frequency noise (Figure 4.103). The PDH spectrum out of the cavity resonance was also measured, as it sets the actual detection limit level. The servo-loop has a bandwidth of 4 kHz and is technically limited by the piezo resonances. At lower frequency (~ 100Hz) noise is reduced at the detection limit, while the structures visible up to 2 kHz are mostly due to mechanical resonances of the OPO cavity. In fact, in-loop noise is not the actual signal frequency noise, as it is the relative noise of the signal frequency with respect to the cavity resonance, i.e., it indicates how close the signal frequency is to the cavity resonance. A similar cavity showed an intrinsic frequency stability of 2×10^{-2} Hz/$\sqrt{\text{Hz}}$ around 1 kHz. However, for times longer than one second, a length drift of ~ 1 MHz/s was observed mainly due to the piezo relaxation. Thus, at least on a time scale longer than few ms, the in-loop PSD sets just a lower limit to the signal frequency noise, while for shorter times it can be reasonably assumed to be the actual frequency noise of the signal. With this assumption, from the in-loop PSD we calculated for the signal frequency a FWHM of ~ 70 kHz at 1 ms. This linewidth, together with the uncorrelated laser linewidth, determines for the idler a final linewidth of 80 kHz, quite comparable to typical linewidths attainable with DFG sources.

The spectroscopic performance of such an OPO was evaluated by carrying out sub-Doppler dip (phase-sensitive) detection (see next Chapter) on several transitions in the ν_1 ro-vibrational band of CH_3I (around 3.38 μm), resolving their electronic quadrupole hyperfine structure [335].

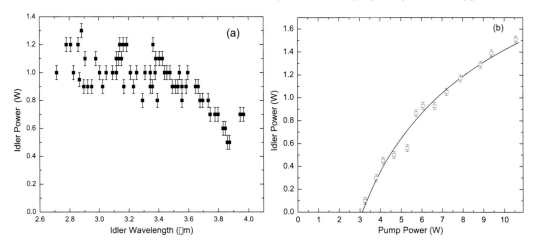

FIGURE 4.102

E(a) Experimental idler power as a function of the wavelength, from 2.7 μm up to 4 μm. (b) Idler power as a function of the pump power, for the 31.5-μm-period at the temperature of 30°C, corresponding to an emission wavelength of 2.93 μm. The continuous line represents data fitting by Equation 4.293.

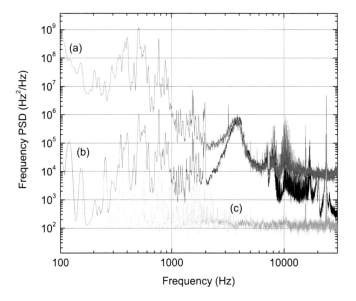

FIGURE 4.103

Power spectral densities concerning frequency stabilization of signal emission to the reference cavity. (a), PSD of the correction signal fed to the OPO piezo, corresponding to free-running signal noise; (b), PSD of the PDH signal corresponding to the in-loop residual frequency noise; (c), PSD of the out-of-resonance PDH, indicating the detection limit.

4.8.6.3 Difference frequency generators

Among all the recently developed IR laser sources, DFGs in periodically-poled non-linear (PPLN) crystals have proved to be the most reliable ones for high-resolution spectroscopic applications [336]. Indeed, due to the deterministic and coherent character of the DFG process, frequency stability and spectral purity properties, attainable for the visible/NIR pumping lasers, through well-established techniques, are automatically transferred to the MIR radiation. Although OPOs are also based on visible/NIR pumping lasers, their use in high-resolution spectroscopy has not yet taken off because complex setups are often necessary to obtain frequency-stable and single-mode behavior. On the contrary, OPO sources can be more attractive in applications where high power is required. Actually, the main drawback of DFGs is represented by their emission power, typically limited to a few mW.

DFG laser sources consist of two seed laser sources, namely a signal and a pump laser source, which are combined and focussed into a non-linear optical medium to generate an idler beam of the difference frequency of the respective seed lasers. The difference-frequency conversion efficiency has been investigated by Boyd and Kleinman, based on the electric field generated by two focused Gaussian beams, the dependence of the generation power on the focusing conditions, and the properties of the non-linear mixing material [337, 323]. In the case of two collinear Gaussian beams (with powers P_p and P_s at frequencies ω_p and ω_s, for the pump and signal beam, respectively) having identical confocal parameters $b = k_p w_p^2 = k_s w_s^2$ (here w is the beam waist), the DFG output power P_i, at the difference-frequency $\omega_i = \omega_p - \omega_s$, can be written as

$$P_i = \frac{256\pi^2}{c^3}\omega_i^2 \frac{d_{eff}^2}{n_i n_s n_p} \frac{h(\mu, \xi, \alpha)}{k_s^{-1} + k_p^{-1}} L P_p P_s e^{-\alpha L} \qquad (4.294)$$

here c is the speed of light in vacuum, n is the index of refraction, L is the crystal length, d_{eff} is the effective non-linear coefficient, and α is the absorption coefficient of the nonlinear optical medium at the DFG frequency. The units in Equation 4.294 are cgs and ε_0 was not factored out of d_{eff}. The subscripts s, p, i refer to the signal, pump, and idler (infrared), respectively. The focusing function $h(\mu, \xi, \alpha)$ involving walk-off and focused beam effects is given as (focusing point is assumed at the center of the crystal):

$$h(\mu, \xi, \alpha) = \frac{1}{4\xi} \int_{-\xi}^{+\xi} d\tau \int_{-\xi}^{+\xi} d\tau' \frac{e^{\frac{b\alpha}{4}(\tau-\tau')}}{1 + \tau\tau' - \frac{i}{2}(\tau - \tau')\left(\frac{1+\mu}{1-\mu} + \frac{1-\mu}{1+\mu}\right)} \qquad (4.295)$$

where $\mu = k_p/k_s$ and $\xi = L/b$ is the focusing parameter. The complex character of the $h(\mu, \xi, \alpha)$ function is only apparent: the integration, in fact, cancels the i dependence and yields a real function. Equation 4.294 shows typical features for a three-wave parametric mixing process

1. DFG power is proportional to the non-linear optical figure of merit, $d_{eff}^2/n_i n_s n_p$.

2. The output power varies linearly with the product of the input powers. In practice, the pumping power densities incident on the nonlinear optical crystal are usually limited by the laser induced damage of the crystal. A large optical damage threshold is thus highly desirable. Commercially available ferroelectric material-based QPM crystals, such as periodically poled LiNbO$_3$ (PPLN) or RbTiOAsO$_4$ (PPRTA), exhibit high laser induced damage thresholds and lead to mW CW DFG power by using high-power ($\sim$ W) pumping sources.

3. DFG power is proportional to the square of infrared frequency ω_i.

4. DFG power varies with the crystal length L in the case of Gaussian beam coupling, and reaches a maximum value with an optimum focusing parameter of $\xi \sim 1.3$. For example, optimum beam waists for near-IR pump and signal wavelengths inside the PPLN crystal typically range between 30 to 60 μm for 20 to 50 mm crystal

lengths.The h-function reduces to $h \sim \xi$ when using loose focusing parameter $\xi \ll 1$, which makes the DFG power proportional to L^2, as in the case of the plane-wave approximation.

Commercial PPLN crystals are available at 0.5 and 1 mm thickness. While the optimal focussing condition for conversion efficiency, as described in the above equations, is tied to the length of the crystal, it ignores the thickness and width of the crystal. The generated idler beam (with a beam waist that is determined by the signal and the pump beam) diverges much faster than the pump and signal beams, and hence the idler beam may clip causing diffraction and scattering noise if the crystal thickness is too thin or the crystal length too long. Thus, the most useful length of the crystal is determined by the combination of idler wavelength, PPLN crystal thickness, and focussing condition. Another consideration is absorption loss which occurs at wavelengths of 4.2 μm and longer in PPLN, limiting the useful range of DFG wavelengths up to 4.6 μm. In recent years, an alternative DFG crystal design utilizing a ridge type waveguide PPLN has become available, in which the conversion efficiency is no longer limited by diffraction or the focussing condition into a bulk crystal. In a waveguide, the cross section of the non-linear conversion is kept to the smallest possible guiding size over the length of the crystal, and thus is proportional to L^2. Waveguide PPLN crystals demonstrated conversion efficiencies of 100 times higher than bulk PPLN crystals, resulting in tens of mWs of DFG power. In particular, thanks to its resistance to photo-refractive damage, a QPM Zn:LiNbO$_3$ waveguide was recently used in conjunction with a high-power fiber amplifier as a pump source to realize a 3.4-μm DFG source with a tunability range of 10 nm and a maximum output power of 65 mW [338]. An updated survey of DFG sources is given in [337].

So far, the most sophisticated DFG source in terms of combined output-power and spectral characteristics was developed at Istituto Nazionale di Ottica. It is based on a special intra-cavity design [339] where the pumping sources are referenced to a Ti:Sa optical frequency comb synthesizer (OFCS) via a direct digital synthesis (DDS) scheme [340]. In essence, the intracavity setup enhances the idler output power up to 30 mW (at 4510 nm), while the DDS approach provides an intrinsic linewidth as low as 10 Hz. For a better understanding, we discuss the two apparatuses separately.

In the intracavity design, shown in Figure 4.104, the DFG non-linear crystal (a periodically poled MgO:LiNbO$_3$) is placed at the secondary waist position of a ring Ti:Sa laser, which is optically injected by an ECDL. The dashed box represents the baseplate of the Ti:Sa laser cavity, which is thermo-electrically stabilized within 1 mK around room temperature. The Ti:Sa gain crystal is water-cooled to efficiently remove the excess absorbed power from the 532 nm pump. The transmission value T of the output coupler (about 1.2% at 861 nm) was chosen as the best trade-off between a large power enhancement factor and a good coupling efficiency of the injecting ECDL. The former is maximum for $T = 0$, while the latter is maximum when T equals all other round trip losses, including the pump depletion in the DFG process. Indeed, these losses are experimentally quantified to about 2.4% at 10 W signal power. The 2-cm-length PP crystal, incorporating 7 channels with different poling periods (23.0, 23.1, 23.2, 23.3, 23.4, 23.5, 23.6 μm), is Brewster cut for λ_p. The pump, signal, and idler beams are angularly separated, due to dispersion at the output facet of the crystal. The idler beam is reflected out of the Ti:sapphire cavity by a gold mirror and then collimated by lens L3 ($f = 100$ mm). As already mentioned, the Ti:Sa laser is injection-locked by a fiber coupled ECDL with 838-863 nm tuning range, thus permitting tuning of the idler within the 3850-4540 nm range. However, the ECDL was operated constantly at the 861 nm wavelength (corresponding to an idler wavelength of 4510 nm) and 60 mW power throughout all measurements. An external 40-dB optical isolator

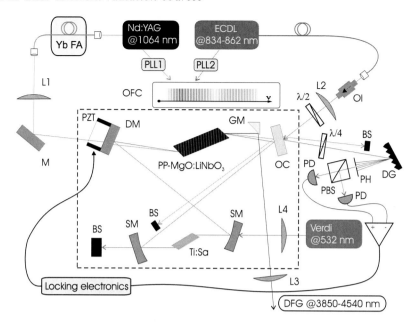

FIGURE 4.104
Layout of the intracavity DFG source. FA, fiber amplifier; ECDL, extended-cavity diode laser; OI, optical isolator; PLL, phase-locked-loop; OFC, optical frequency comb; L, lens; M, dielectric mirror; DM, dichroic mirror; GM, gold mirror; OC, output coupler; $\lambda/2$, half-wave plate; $\lambda/4$, quarter-wave plate; DG, diffraction grating; PBS, polarizing beam splitter; PH, pinhole; PD, photodiode; BS, beam stopper; SM, spherical mirror.

is required to avoid feedback effects from the Ti:Sa cavity, in addition to the internal one, having the same isolation level. A monolithic-cavity Nd:YAG laser seeds a Yb-doped fiber amplifier that can deliver a nominal power of 10 W, providing the signal radiation for the DFG process. The pump laser is phase-locked to the signal laser by use of the DDS scheme (see below). Lenses L1 ($f = 300$ mm) and L2 ($f = 250$ mm) perform an optimal matching of the Nd:YAG and ECDL beams to the Ti:Sa cavity mode. The cavity length is actively stabilized to keep it resonant with the injecting ECDL, by the polarization-based Hänsch-Couillaud technique. Proportional-integral processing electronics feeds the correction signal back to the PZT-mounted dichroic mirror, which is highly transmissive for the signal and highly reflective for the pump. A locking bandwidth as high as 8 kHz was achieved.

Now, let us discuss the DDS scheme with reference to Figure 4.105. Both pump and signal frequencies are beaten with the closest tooth of the OFCS (the corresponding integer orders N_p and N_s are measured by a wavemeter) and the respective RF beat notes $\Delta\nu_{pc}$ and $\Delta\nu_{sc}$ satisfy the following equations:

$$\Delta\nu_{pc} = \nu_p - N_p\nu_r - \nu_0 \tag{4.296}$$

$$\Delta\nu_{sc} = \nu_s - N_s\nu_r - \nu_0 \tag{4.297}$$

where $\nu_r \simeq 1$ GHz is the repetition rate and ν_0 is the OFCS carrier-envelope-offset (CEO), which is canceled from these beat notes by standard RF mixing. A low bandwidth ($\simeq 10$ Hz) phase-locked-loop (PLL) is used to remove the frequency drift of the Nd:YAG laser. A DDS circuit multiplies the $\Delta\nu_{sc} + \nu_0$ frequency by a factor N_p/N_s. A second PLL circuit with a wide bandwidth ($\simeq 2$ MHz) locks the $\Delta\nu_{pc} + \nu_0$ frequency to the DDS output by sending feedback corrections to the ECDL current and PZT voltage. The pump frequency is then $\nu_p = (N_p/N_s)\nu_s$, without any contribution from the OFCS parameters ν_0 and ν_r (at least at frequencies > 10 Hz). As a consequence, the absolute frequency ν_i of the generated idler radiation is given by the following equation:

$$\nu_i = \nu_p - \nu_s = \left(\frac{N_p}{N_s} - 1\right)\nu_s \qquad (4.298)$$

Therefore, the idler linewidth $\delta\nu_i$ can be expressed in terms of the signal linewidth $\delta\nu_s$, as follows:

$$\delta\nu_i = \left(\frac{N_p}{N_s} - 1\right)\delta\nu_s \qquad (4.299)$$

where, for all frequencies below 10 Hz, $\delta\nu_s$ traces the linewidth of the comb tooth around 1064 nm while, for all frequencies above 10 Hz, $\delta\nu_s$ coincides with the free-running Nd:YAG laser fluctuations. The accuracy of ν_i is only limited by the reference oscillator of the OFCS. In the original experiment this was a Rb/GPS-disciplined 10-MHz quartz with a stability of $6 \cdot 10^{-13}$ at 1 s and a minimum accuracy of $2 \cdot 10^{-12}$. From the above discussion it emerges that the attainable stability is ruled only by the signal laser. Of course, several approaches can be implemented to further enhance the signal frequency stability, e.g., by narrowing it onto Fabry-Perot cavities. In order to test the frequency stability of this source, a 1-m-long high-finesse cavity with maximum reflectivity at 4500 nm was built. The ZnSe plano-concave mirrors had a high-reflection coating on their concave surface (6 m radius of curvature) and an anti-reflection coating on the plane surface. Each mirror was measured to have 270 ppm losses (100 ppm absorption and 170 ppm transmission), corresponding to a finesse $\mathcal{F} = 11500$. The transmitted power through the resonant cavity was about 35% of the incident one and the achieved mode-matching was 86%. The mirror holders were separated by a three-bar invar structure which guaranteed a good passive thermal stability. The whole structure laid inside a vacuum chamber with a cantilever system damping mechanical vibrations in all directions. The vacuum conditions prevented frequency fluctuations due to pressure changes. A three PZT system was mounted on one mirror for fine cavity tuning. In order to use this cavity as frequency noise discriminator of the IR source, its passive frequency stability was accurately characterized. The cavity drift of about 1 kHz/s was measured from the linewidth of the cavity transmission averaged over long time scales (about 500 s) when illuminated with the OFCS-locked DFG source, having negligible drift in this time interval. At 100 Hz, the frequency noise induced by the PZT-driven electronics was measured to be one order of magnitude lower than the one of the IR source, and it decreased with a $1/f$ behavior up to 30 kHz. A resonance frequency of about 19 Hz and a damping time of about 5 s for the cantilever damped vibration system were measured by using an accelerometer. In Figure 4.106, the 19 Hz peak was assigned to the residual vibrational cavity noise at the cantilever resonance. Due to the narrow linewidth of this resonance, the noise amplitude fell by 15 dB at 20 Hz. For frequencies higher than 20 Hz, the vibrations were damped following a $1/f^2$ law in units of Hz/$\sqrt{\text{Hz}}$, as inferred by the solution of the differential equation for a damped harmonic oscillator forced by external vibration-induced white noise. To characterize the frequency noise of the DDS-based DFG source, the cavity length was tuned at a transmission corresponding to half of the peak value. Thus, the slope of the fringe side was used as a frequency-to-amplitude converter. The frequency noise spectral density recorded with a FFT spectrum analyzer is shown in Figure 4.106. The various lines highlight different behaviors of the spectrum: $1/f$ technical noise ($\nu < 2$ kHz), white noise ($\nu > 2$ kHz), cavity-cutoff region ($\nu > 10$ kHz), detector-cutoff region ($\nu > 400$ kHz). Following the above discussion about the cavity frequency stability, the cavity contribution to this noise can be considered negligible in the spectral

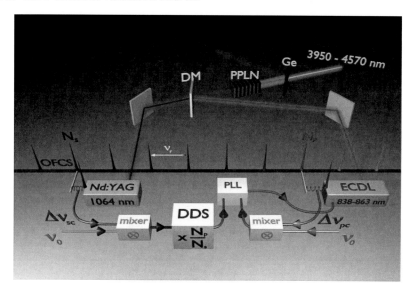

FIGURE 4.105
OFCS-referenced DFG infrared source. OFCS is used as a transfer oscillator to phase-lock the ECDL directly to the Nd:YAG. DM, dichroic mirror; Ge, germanium filter. See text for other acronyms.

range shown in the figure. From the power spectral density in the white-noise region an IR intrinsic linewidth of about 10 Hz can be inferred, while the time-integrated linewidth over 1 ms is about 1 kHz.

4.8.6.4 Tunable far-infrared radiation

Historically, frequency conversion by non-linear generation has been a key tool for synthesizing radiation, often continuously tunable over wide intervals, in regions where lasers did not directly emit. Non-linear difference frequency generation has been conveniently used for many years to generate cw radiation throughout the infrared region, till the millimetric range, where microwave sources had been available as coherent sources before the advent of the laser. The far-infrared (FIR) part of the electromagnetic spectrum spans the frequency range from 300 GHz to 10 THz (λ: 1 mm - 30 μm). In this spectral region rotational transitions of light molecules fall, which are of high interest for applicative research (mainly astrophysics and atmospheric physics) as well as for fundamental studies. Moreover, since absorption coefficients normally scale as either the square or the cube of frequency for $h\nu \ll kT$, FIR transitions are much more sensitive tools (compared to millimetric transitions) to detect neutral/ionized atoms/molecules. In spite of the very large scientific interest, high-resolution spectroscopic studies in the sub-millimetric started much later than in the neighboring microwave and infrared spectral regions, due to the lack of both adequate coherent sources and detectors. The introduction of optically pumped fixed-frequency FIR lasers and of liquid He cooled bolometric detectors (NEP in the order of $10^{-13} - 10^{-14}$ W/Hz$^{1/2}$) was the first step toward the development of high-resolution spectroscopy in this region. Wider tunability has then been achieved with the laser magnetic resonance (LMR) spectrometer. Such a technique only works on paramagnetic species, by tuning, via a magnetic field, the transition of interest into resonance with a FIR laser line. In this range, LMR is one of the most sensitive spectroscopic techniques, because it is intra-cavity in nature, with a minimum detectable absorption in the order of 10^{-10}. Nevertheless, the accuracy

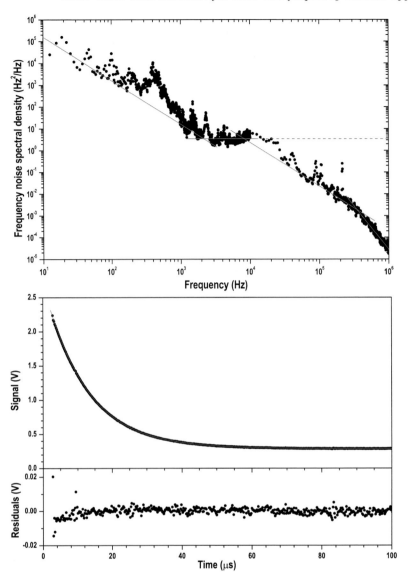

FIGURE 4.106

Frequency-noise spectral density for the DDS-based DFG source. Three spectra with different frequency spans (2 kHz, 10 kHz, 1 MHz) are stuck together and plotted in the same graph ($RBW = 3 \cdot 10^{-3} \times$ span).

of frequency measurements is limited to only several MHz due to the uncertainty in the extrapolation at zero magnetic field and the lack of reproducibility in the center of the FIR laser line. In the 80s, two metrological-grade techniques, based on difference frequency generation by non-linear conversion, were developed. The first of these is generation of microwave sidebands on the strongest FIR laser lines in Schottky diodes [341, 342, 343, 344]. It produces tunable FIR radiation up to about 3 THz (100 cm^{-1}) with 5 kHz linewidths and roughly 500 kHz accuracy. At present, with the recent improvement of harmonic multiplication from radio frequencies, it is only rarely implemented for FIR measurements. Thanks to the pioneering work of K.M. Evenson and co-workers, a significant improvement in the accuracy and very wide tunability were then obtained with the tunable far-infrared spectrometer (TuFIR) [345]. It is based on a synthesis of FIR radiation starting from infrared and microwave radiation mixed onto a non-linear device, namely the metal-insulator-metal (MIM) diode. In a later version [346], the TuFIR spectrometer produces radiation through non-linear mixing of three radiations (two from CO_2 lasers and one from a microwave source) in a third-order MIM diode [347]. The MIM diode generates microwave sidebands on the CO_2 difference frequency

$$\nu_{FIR} = \nu_1 - \nu_2 \pm \nu_{\mu w} \tag{4.300}$$

where ν_1 and ν_2 are the laser frequencies, and $\nu_{\mu w}$ is the microwave frequency. Figure 4.107 is a schematic of a third-order TuFIR spectrometer.

The radiation from lasers 1 and 2 is combined by means of a beam splitter and coupled onto the diode by a 25-mm focal length lens. The microwave radiation is coupled onto the diode by a bias-tee connected to the diode junction. The generated FIR, radiated from the diode's whisker in a long wire antenna pattern, is then collected and collimated by a 30-mm focal length off-axis section of a parabolic mirror. After passing through an absorption cell, the FIR is detected on a liquid He-cooled Si bolometer. The FIR radiation is frequency modulated (by frequency modulation of the CO_2 lasers) and detected with a lock-in amplifier. The far infrared frequency is tuned by scanning the microwave frequency. This is controlled by a personal computer, which also collects data from the lock-in amplifier. The standard frequency coverage of such a spectrometer is from 0.3 to 6 THz. The lower limit is set by the bolometer and the upper limit by the largest difference frequency between the two CO_2 lasers. Both CO_2 lasers are frequency stabilized to the 4.3 μm wavelength saturated fluorescence signal from low-pressure CO_2 cells (not shown in the figure) [348]. These frequencies have been measured to an absolute frequency with an uncertainty better than 5 kHz [349] [350]. Without using special locking techniques, a stability of about 25 kHz is obtained for each laser. The overall (statistics + systematic) frequency uncertainty of generated FIR radiation is thus 35 kHz. The spectrometer sensitivity is limited by the FIR power and the sensitivity of the detector. FIR powers up to a few hundred nW are generated with 150 mW from each laser and 6-10 dBm of microwave power applied to the MIM diode. For the best contacts, minimum detectable absorptions are around 10^{-4} in a 1-second integration time, corresponding to a minimum detection coefficient of 10^{-6} for

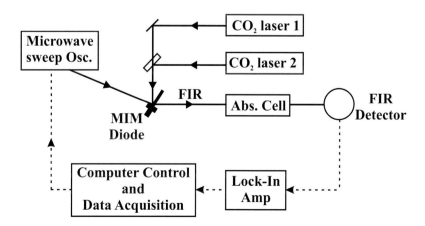

FIGURE 4.107
Schematic diagram of the tunable far-infrared spectrometer at LENS. The generated frequency is in the range of 0.3 to 6 THz.

a 1-meter long cell. The MIM diode consists of an electrochemically sharpened tungsten whisker (25-pm diameter and 3 to 7 mm long) contacting a metal base. The metal base has a naturally occurring thin oxide insulating layer. Both nickel and cobalt have been used as base materials, but cobalt is generally more consistent in the production of third-order FIR radiation.

With such a spectrometer, mainly operated at NIST, Boulder, Colorado, USA, at LENS, Florence, Italy, and later on in Japan [351], a large number of rotational and ro-vibrational transitions of light molecules and free radicals were measured with high accuracy [346, 352, 353, 354, 355, 356, 357, 358, 359, 360, 361, 362, 363]. In addition, atomic fine structure transitions could be measured [364, 365, 366], as well as transition dipole moments of molecules of atmospheric interest, with an accuracy up to 10^{-4} [367, 368]. In spite of the scarce reproducibility in the emission from the MIM diode, pioneering studies on the breakdown of the Born-Oppenheimer approximation could be performed and an analysis of sensitivity for such a spectrometer was performed [369]. Such a setup proved to be sensitive enough for the observation of tiny birefringence induced in a sample gas by a longitudinal magnetic field (Faraday effect), giving new insights in the achievable sensitivity of FIR spectrometers [370, 371]. Detailed studies of the limiting bandwidth of the TuFIR spectrometer showed that frequencies up to 10 THz could be generated, by using laser lines emitted from rare isotopes of CO_2 [372].

The MIM diode and present-day antennas

The MIM diode was for many years a key device for frequency metrology, being also used as a laser mixer in frequency chains, before the advent of combs [52]. Not so many studies could unveil its unique properties as a mixer, detector, and radiation generator, basically related to its ultrawide bandwidth. Notwithstanding, it is interesting to compare its properties with present-day antennas. Antennas find a very wide range of applications, including near-field optics and imaging (see [373] for a recent and intriguing review), with a need for scaling to nanometers, in order to operate in the visible range. An antenna is a structure that must efficiently convert incident radiation into localized energy (receiving antenna), and vice versa (transmitting antenna). The concept of antenna can be usefully scaled from radio frequencies up to optical frequencies, though the working principle changes. Indeed, at longer wavelengths (radio and microwave range) metals are perfect conductors and radiation cannot penetrate the metal at depths comparable with the antenna size (negligible skin depth). Instead, skin depth cannot be neglected at optical frequencies, where the antenna size is in the nanometer range. In this case, the description of radiation-antenna interaction is well explained by a propagating plasmon wave coupled to incident radiation. As a consequence, scaling of antenna size with wavelength differs from standard antenna theory, set for the radio-frequency range. Therefore, working principles of optical antennas depend on plasmonics and nanoptics and, ultimately, they are better described by quantum optics, similarly to atoms and molecules. Interestingly, MIM diodes were used for a long time in the whole range from radio to optical frequencies, without any significant change in their shape and mode of operation. Indeed, the longer straight part of it was considered useful for *transceiving* (receiving and transmitting) radio/micro/THz-waves while the tip was reputed good for shorter wavelengths. A first consideration, arising from present-day interpretation of antennas, is that the MIM diode could operate in a range (far/mid infrared) that is intermediate between the two extremes, radio and optical waves, where principles are more clear. Another point that is worth remarking is that its huge bandwidth is still unmatched in present devices (see, e.g., [374]), though they are very much specialized and benefit from micro/nano electro-optics for fabrication, as compared to hand-crafted fabrication procedures of MIM diodes. A possible origin of MIM unmatched wide bandwidth is possibly due to the hard-touch of the nanometer scale tungsten tip against the metal base, that reduces to a minimum the contact area and thus device capacitance. Finally, a re-visitation of the unique properties of the very simple MIM geometry in light of present-day physics could help in creating better transceiver devices for opto-electronics applications throughout the electromagnetic spectrum.

5

High-resolution spectroscopic frequency measurements

> Method is much, technique is much, but
> inspiration is even more.
> *Benjamin Nathan Cardozo*

A great building must begin with the
unmeasurable, must go through
measurable means when it is being
designed and in the end must be
unmeasurable.
Louis Isidore Kahn

Due to their unprecedented degree of monochromaticity and directionality, lasers have revolutionized the field of interferometry and spectroscopy. In the former branch, just think of the practical development of holography as well as of the ambitious experiments that are in progress to detect gravitational waves [375]. In the latter field, through the introduction of novel high-resolution and high-sensitivity interrogation techniques, lasers have allowed the study of atomic/molecular spectra with an unprecedented precision. In a nutshell, the resolution limit is now determined by the width of the spectral lines of the substance under investigation rather than by the instrumental width of the spectral apparatus. In turn, the implementation of more and more sophisticated spectroscopy-based frequency measurement schemes have improved the laser performance, eventually leading to the realization of optical standards. In this chapter we review the most advanced spectroscopic techniques for gaseous samples and the realization of standards using either absorption cells or effusive beams. Essentially, the same spectroscopic interrogation methods are also at the basis of frequency standards using samples of cold/trapped atoms and ions. These will be dealt with in Chapter 7.

5.1 Interferometric wavelength measurements

Interferometry is a vast branch of physics and a comprehensive treatment is beyond the scope of this book. In this section, we only discuss the basic principle of wavelength measurements. Indeed, the reader will surely remember from Chapter 1 that the current definition of the speed of light c, which is a pivot of modern frequency metrology, is based on an accurate, simultaneous measurement of the wavelength and frequency of a He-Ne laser. Even though the original experiment by K. Evenson et al. made use of a Fabry-Perot interferometer, here we focus on the Michelson interferometer (MI) which is at the heart of modern wavemeters.

In the basic MI (schematically shown in Figure 5.1) the light from a source, S, is divided by a 50% beam splitter oriented at 45 degrees to the beam. The transmitted beam travels to mirror M1 where it is back reflected to BS. 50% of the returning beam is deflected by 90 degrees at BS and it then strikes the screen, E (the other 50% is transmitted back towards the laser and is of no further interest here). The reflected beam travels to mirror M2 where it is reflected. Again, 50% of it then passes straight through BS and reaches the screen (the remaining 50% is reflected towards the laser and is again of no further interest here). The two beams that are directed towards the screen, E, interfere to produce fringes on the screen. For an incident plane wave, and denoting with s the distance from the source, at the screen we will have the superposition of the two following waves

$$E_1 = E_0 e^{i(-\omega t + ks_1 + \varphi_1)} \tag{5.1}$$

$$E_2 = E_0 e^{i(-\omega t + ks_2 + \varphi_2)} \tag{5.2}$$

where $k = \omega n/c$, n being the refractive index. Thus the total field can be expressed as

$$E_{tot} = E_1 + E_2 = E_0 e^{-i\omega t} e^{i(ks_1 + \varphi_1)} \left[1 + e^{i\Phi}\right] \tag{5.3}$$

where $\Phi = k\Delta s + \Delta\varphi$, with $\Delta s = s_2 - s_1 = 2d$ and $\Delta\varphi = \varphi_2 - \varphi_1$. Finally, the total intensity is

$$I_{tot} \equiv E_{tot} E_{tot}^* = 2I_0 \left(1 + \cos\Phi\right) = 4I_0 \cos^2\left(\frac{\Phi}{2}\right) \tag{5.4}$$

Hence, I_{tot} is maximum for $\Phi = 2m\pi$ and minimum for $\Phi = (2m+1)\pi$, $m = 0, 1, 2, \ldots$. This means that, if the two mirrors are perfectly parallel, then the whole illuminated area of the screen will be uniformly lit to a certain extent (including zero) for any value of Δs. With parallel incident light but slightly tilted mirrors (M1 or M2), instead, the interference pattern will consist of parallel fringes, which move into a direction perpendicular to the fringes as Δs is changed. To understand this, we consider the configuration in which M2 is tilted by the angle $\theta/2$ (see frame (b) of Figure 5.1). Waves are reflected normally by M1 and at an angle θ to the z-axis. Therefore, at the screen we will have

$$\begin{cases} E_1 = E_0 e^{-i\omega t} e^{i\varphi_1} e^{ik(2d+z)} \\ \\ E_2 = E_0 e^{-i\omega t} e^{i\varphi_1} e^{ik(x\sin\theta + z\cos\theta + \Delta\varphi/k)} \end{cases} \tag{5.5}$$

which provides

$$\begin{aligned} I_{tot} &= |E_1 + E_2|^2 = 4I_0 \cos^2\left[k\left(d - \frac{x}{2}\sin\theta + z\sin^2\frac{\theta}{2} - \frac{\Delta\varphi}{2k}\right)\right] \\ &= 4I_0 \cos^2\left[k\left(d - \frac{x}{2}\sin\theta + z_0\right)\right] \end{aligned} \tag{5.6}$$

where $z_0 = z\sin^2(\theta/2) - \Delta\varphi/2k$. Such intensity is maximum when

$$k\left(d - \frac{x_m}{2}\sin\theta + z_0\right) = m\pi \tag{5.7}$$

that returns

$$x_m = \frac{1}{\sin\theta}\left[2(z_0 + d) - \frac{m\lambda}{n}\right] \tag{5.8}$$

from which the fringe spacing is calculated as

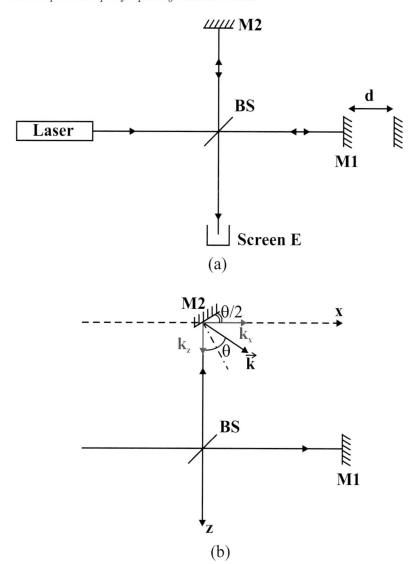

FIGURE 5.1
Basic layout of a Michelson interferometer.

$$\Delta x = x_m - x_{m+1} = \frac{\lambda}{n \cdot \sin \theta} \tag{5.9}$$

If d moves in time with uniform velocity v, $d = vt$, then the number N_{fr} of fringes that cross the center of the screen in a time T is given by

$$N_{fr} = \frac{\frac{dx_m}{dt}\big|_{t=T} \cdot T}{\Delta x} = \frac{2nd}{\lambda} \tag{5.10}$$

Therefore, the MI can be used for absolute wavelength measurements by counting the number of fringes when the mirror M1 is moved along a known distance d. Based on this principle, in the following we describe the realization of a compact, very accurate Michelson wavemeter [376]. The essential concept is illustrated in Figure 5.2. Two lasers, one of which has a known wavelength (reference laser), propagate simultaneously through the MI. While the relative lengths of the MI arms are varied by means of the travelling cornercubes, the interference fringes arising from both lasers are counted by the detectors. Therefore, as the mirror is moved by the distance d (which is the same for both lasers), the number of fringes counted for each of the two wavelengths is given by

$$N_R = \frac{2dn_R}{\lambda_R} \tag{5.11}$$

$$N_U = \frac{2dn_U}{\lambda_U} \tag{5.12}$$

where λ_R (λ_U) is the wavelength of the reference (unknown) laser and n_R (n_U) the refractive index of air at the reference (unknown) wavelength. This finally yields

$$\lambda_U = \frac{N_R}{N_U}\lambda_R\frac{n_U}{n_R} \tag{5.13}$$

or, equivalently,

$$\nu_U = \frac{N_U}{N_R}\nu_R\frac{n_R}{n_U} \tag{5.14}$$

Then, the reference and unknown fringes can be counted for a fixed time with a single conventional frequency counter (provided with an external reference for determining the counting time), configured to show directly the wavelength ratio. It is worth noting that, if the vacuum wavelength is required, the wavemeter should be operated, in principle, under vacuum conditions. Otherwise, correction for dispersion has to be applied a posteriori.

By virtue of its low cost and well-defined wavelength, a HeNe laser is typically used as the reference. In this case, the uncertainty in the final measurement can be expressed as

$$\left(\frac{\Delta\lambda}{\lambda}\right)^2 = \left(\frac{\Delta\lambda_{WC}}{\lambda}\right)^2 + \left(\frac{\Delta\lambda_{align}}{\lambda}\right)^2 + \left(\frac{\Delta\lambda_{HeNe}}{\lambda}\right)^2 + \left(\frac{\Delta\lambda_{count}}{\lambda}\right)^2 \tag{5.15}$$

where $\Delta\lambda_{WC}$, $\Delta\lambda_{align}$, $\Delta\lambda_{HeNe}$ and $\Delta\lambda_{count}$ are the uncertainties due to wavefront curvature, beam misalignment, HeNe wavelength, and counting resolution.

- Wavefront curvature in the laser beams will result in counting errors due to diffraction effects. The magnitude of this uncertainty, as estimated from simple diffraction theory, is given by

$$\left(\frac{\Delta\lambda_{WC}}{\lambda}\right) = \frac{\Delta\theta^2}{4} \tag{5.16}$$

where $\Delta\theta$ is the divergence of the beam. $\Delta\theta = 0.25$ mrad is a typical value for a HeNe laser.

- Relative misalignment of the reference and unknown beams will produce counting errors since the lengths traveled by the two beams will differ. This is a simple cosine error, which is, for small angles

$$\left(\frac{\Delta\lambda_{align}}{\lambda}\right) = \frac{1}{2}\left(\frac{\Delta x}{L}\right)^2 \tag{5.17}$$

where $\Delta x/L$ is the relative angular displacement. Typical values can be taken as $L = 0.5$ m and $\Delta x = 0.5$ mm.

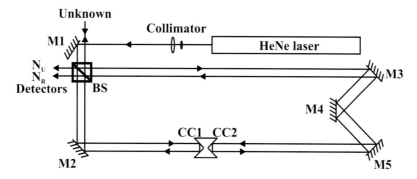

FIGURE 5.2

Wavemeter layout. (Adapted from [376].) The HeNe beam is reflected by mirror M1, then split at the non-polarizing beam splitter, BS. The transmitted beam reflects off M2 and then strikes the cornercube reflector, CC1, which sends the beam back parallel to its original path, with some transverse displacement. The beam reflected from BS travels via M3, M4, and M5 to CC2, where it also is reflected back parallel to its path. Mirror M4 is added for alignment purposes. The two beams, one from CC1 and the other from CC2, converge at the beamsplitter, producing two output beams. One is used to detect the interferences; the other is used as a guide to align the second laser. The motion of the cornercubes (which should be as much smooth and laterally stable as possible) is realized by mounting them on an air bearing. The latter consists of a cart through which compressed air flows; this forms a cushion which supports the cart over a track [376].

- The HeNe laser wavelength is uncertain due to the $\simeq 1.5$ GHz Doppler width of the Ne emission line (corresponding to a wavelength uncertainty of 2 pm), and to the unknown Ne isotope mixture used in the tube (amounting to 1 pm). Both uncertainties can be minimized by controlling the HeNe cavity temperature to obtain equal intensities of two operating longitudinal modes with orthogonal polarization [377].

- Concerning the uncertainty due to fringe counting, it is determined by the limited number of fringes counted during the measurement. For example, if 780000 ± 1 fringes are counted from the HeNe, we have $\Delta\lambda_{count}/\lambda = 1/780000 = 1.28 \cdot 10^{-6}$.

Summing all the uncertainties in quadrature, the final relative accuracy $\Delta\lambda/\lambda$ is on the order of few parts in 10^6. During the last 30 years, many variants of this measurement principle have been developed, achieving an accuracy of few tens of MHz. Major improvements included high-resolution counters, superior-performance optical and mechanical components, and augmented-stability reference lasers. However, the moving components within the interferometer arm necessitated elaborate mechanical and optical isolation of the wavelength meter from the actual experiment. More recently, wavelength meters (WLMs) based on solid-state Fizeau interferometers have taken the technological lead in precision wavelength measurement. They can achieve wavelength resolution down to the 10-MHz level without any moving parts. For highly accurate readout of the interference pattern, state-of-the-art CCD detectors are used. The solid-state etalon concept still relies on an accurate distance measurement in the first place (the distance between the mirror facets of the etalons has to be well characterized) yet the dependence of the measurement on the refractive index of the intermediate space is significantly reduced [378]. The measurement block is manufactured from solid components with well-known optical properties and mounted within a thermal isolation housing. The most precise Fizeau-type WLMs, with a resolution of 10^{-8}, require calibration with a frequency-stabilized reference laser. In commercial systems, these range from Zeeman-stabilized HeNe lasers at 473.612467 THz (fractional uncertainty $\pm 2 \cdot 10^{-8}$) to frequency-doubled Nd:YAG lasers locked against the iodine molecule absorption line (R56,32-0) at 563.259651965 THz (fractional uncertainty $\pm 2 \cdot 10^{-9}$).

5.2 Spectroscopic frequency measurements

Spectroscopy is a powerful tool to gain access to atomic and molecular physics, to test the foundations of quantum physical models, and is the cornerstone in research on fundamental constants. Insight and measurement precision both increase dramatically with resolution. In the case of gas-phase spectra, most often the first barrier that is encountered is represented by collision broadening. Usually this is overcome simply by using samples at lower pressures. But soon after, a more severe obstacle is hit, namely the Doppler effect. Well, laser spectroscopy has revolutionized this situation, currently allowing resolutions several orders of magnitude better than the Doppler limit. Indeed, many new high-resolution detection schemes have become feasible due to the high brightness, wide tunability, and narrow linewidth of lasers. Far from wishing to give a comprehensive review on laser spectroscopy techniques, for which the reader is referred to [379], after a short introduction on the basic principles of absorption laser spectroscopy, in the remainder of this chapter our discussion will focus on laser-based high-resolution spectroscopic techniques. Among these, several methods will be necessarily left out of the discussion, according to the following reasons. Despite having played a crucial role in fundamental physical studies, some approaches have been gradually either replaced by or evolved into more advanced ones. Here we are thinking of opto-galvanic spectroscopy [380], and coherent-spectroscopy techniques including level-crossing and quantum-beat schemes [381]. Although still in vogue, other techniques, including ionization spectroscopy with its particularly successful variant, namely resonance-enhanced multiphoton ionization (REMPI) spectroscopy [382]; velocity-modulation spectroscopy of molecular ions [383]; synchroton-radiation-based spectroscopy [384]; and laser-magnetic resonance spectroscopy [385], are today essentially a prerogative of Chemistry. Finally, some other techniques, above all the photo-acoustic one [386], are primarily used in applied spectroscopy for trace-gas detection. In short, we will deal solely with those spectroscopic techniques which do possess the potential for ultrahigh-resolution frequency measurements. In addition, although certainly fitting into the range of traditional high-resolution spectroscopic methods, a few particular schemes, like quantum-jump spectroscopy, double-resonance, and coherent population trapping, will be more naturally discussed in the frame of frequency standards (Chapter 7).

In practice, however, high resolution is most often related to high-sensitivity detection. Just as an example, if the gas pressure is lowered to reduce pressure broadening, less signal is available, which can be compensated for by employing longer path lengths or diminishing the measurement noise. So, we shall also deal with some (the most effective ones) of those which are usually referred to as high-sensitivity spectroscopic techniques. In fact, we will precisely start with them in order to devote, without any digression, the heart of this chapter to high-resolution spectroscopic techniques. It will be then clear to the reader that the distinction between the two categories is only of convenience and, in fact, a number of spectroscopic interrogation methods inherently possess the potential for both high-resolution and high-sensitivity detection. Indeed, the availability of a molecular-spectroscopy technique, able to combine the ultimate performance in terms of sensitivity, resolution, and frequency accuracy, can be crucial in many fundamental physical measurements.

5.2.1 Principles of absorption laser spectroscopy

As discussed in Chapter 4, the laser output very well approximates a polarized, monochromatic electromagnetic plane wave. Now, when the latter passes along the x direction through an absorbing gas, the change dI in the intensity depends in general in a non-linear way on the incident intensity

$$dI = -\Delta N\left(I\right) \cdot I \cdot \sigma \cdot dx \tag{5.18}$$

where σ is the absorption cross-section and the intensity-dependent population density difference can be written as a power series

$$\Delta N\left(I\right) = \Delta N_0 + \frac{d\left(\Delta N\right)}{dI}I + \frac{1}{2}\frac{d^2\left(\Delta N\right)}{dI^2}I^2 + \ldots \tag{5.19}$$

In this way, we can re-write

$$dI = -\left[\Delta N_0 I\sigma + \frac{d\left(\Delta N\right)}{dI}I^2\sigma + \ldots\right]dx \tag{5.20}$$

Here, the first term describes the linear absorption (Beer's law), while the second one reduces the absorption because $d\left(\Delta N\right)/dI < 0$ (indeed, if i is the lower level and f is the upper one, then $dN_i/dI < 0$ and $dN_f/dI > 0$). A classical experiment which points out the effects of the non-linear absorption consists of measuring the intensity I_{fl} of the fluorescence induced by a laser in a given sample as a function of the laser intensity I_L itself. In this case, one at first observes that I_{fl} increases linearly with I_L, but for higher laser intensities the growth becomes less than linear, as the absorption coefficient, and hence the relative absorption of the laser intensity, decreases. For even higher laser intensities, the fluorescence intensity eventually approaches a constant value (saturation), which is basically limited by the rate of the relaxation processes refilling the absorbing level. This saturation of the absorption can be used for Doppler-free spectroscopy, as will be outlined in a following section.

5.3 Frequency modulation spectroscopy

While in an ideal laser spectrometer, absorption signals would be affected only by the shot noise, in every actual situation, the measured noise level is several orders of magnitude higher. As already stressed in several places, such an excess noise is the result of contributions from many independent sources. Part of the excess noise affecting absorption signals has a random nature, and results in a Gaussian distribution of the output voltage around a mean value. Gaussian noise can be dramatically reduced by acquiring sequentially the signal and performing ensemble averaging. In this case, the noise level is expected to scale inversely proportional to the square root of the number M of averaged signals. However, since not all noise affecting the measurement system is Gaussian, it is not possible to average indefinitely. It is then crucial determining to which extent a quantity can be averaged successfully. Such analysis relies on the evaluation of the Allan variance as a function of the averaging time τ. Clearly, the τ value corresponding to the minimum of the Allan variance corresponds to the averaging time that provides the best stability of the mean value. Suppression of non-Gaussian, structured noise, such as, for instance, the $1/f$ noise, relies on completely different detection schemes, mostly based on heterodyne detection, which are outlined in the following. Heterodyne detection is an effective noise-reduction technique for spectroscopic applications, for it allows to move detection frequency away from DC, towards regions where the flicker noise is negligible, and at the same time to reduce the bandwidth over which the noise is detected. Heterodyne techniques basically rely on amplitude or frequency modulation of the probing radiation and coherent signal demodulation by phase-sensitive electronics. A periodic modulation of an optical frequency ν_0 is defined by a modulation frequency ν_m and an amplitude, or depth a

$$\nu(t) = \nu_0 + a \cos \nu_m t \tag{5.21}$$

Often, the modulation depth is expressed in terms of modulation index m, defined as the ratio between a and the width Γ of the absorption feature of interest. Different modulation-based detection schemes can be implemented, depending on the parameters ν_m, a and Γ [387].

5.3.1 Harmonic detection

The regime where both modulation depth and frequency are negligible with respect to the width of the absorption line to be detected (ν_m, $a \ll \Gamma$) is normally referred to as harmonic detection or derivative spectroscopy. Such a small-depth modulation results in a FM-AM conversion, as it can be seen by performing a Taylor expansion of the transmitted intensity (with $\Delta N_0 \equiv n$)

$$S(\nu) \equiv \frac{1}{nL} \frac{I_0 - I(\nu)}{I_0} = \sigma(\nu) \tag{5.22}$$

around ν_0:

$$S(\nu) \simeq \sigma(\nu_0) + \sigma'(\nu_0)(\nu - \nu_0) + \frac{\sigma''(\nu_0)}{2}(\nu - \nu_0)^2 + \frac{\sigma'''(\nu_0)}{6}(\nu - \nu_0)^3 + \dots \tag{5.23}$$

Since

$$\nu - \nu_0 = a \cos \nu_m t \tag{5.24}$$

then we have

$$S(\nu) \simeq \sigma(\nu_0) + \sigma'(\nu_0) a \cos \nu_m t + \frac{\sigma''(\nu_0)}{2} a^2 \cos^2 \nu_m t + \frac{\sigma'''(\nu_0)}{6} a^3 \cos^3 \nu_m t + \dots \tag{5.25}$$

which, recalling that

$$\cos^2 x = \frac{1 + \cos 2x}{2} \tag{5.26}$$

and

$$\cos^3 x = \frac{3 \cos x + \cos 3x}{4}, \tag{5.27}$$

yields

$$
\begin{aligned}
S(\nu) &\simeq \left[\sigma(\nu_0) + \frac{\sigma''(\nu_0)}{4} a^2 \right] + \left[\sigma'(\nu_0) a + \frac{\sigma'''(\nu_0)}{8} a^3 \right] \cos \nu_m t \\
&+ \frac{\sigma''(\nu_0)}{4} a^2 \cos 2\nu_m t + \frac{\sigma'''(\nu_0)}{24} a^3 \cos 3\nu_m t \\
&\simeq \sigma(\nu_0) + \sigma'(\nu_0) a \cdot \cos \nu_m t + \frac{\sigma''(\nu_0)}{4} a^2 \cdot \cos 2\nu_m t \\
&+ \frac{\sigma'''(\nu_0)}{24} a^3 \cdot \cos 3\nu_m t
\end{aligned} \tag{5.28}
$$

In the above relation one easily recognizes a term oscillating at frequency ν_m carrying information on the absorption cross section's first derivative, a term at frequency $2\nu_m$

carrying information on the absorption cross section's second derivative, and a term at frequency $3\nu_m$ carrying information on the absorption cross section's third derivative. These signals can be extracted by means of a lock-in amplifier. One of the advantages of derivative spectroscopy is that only noise components at ν_m or $2\nu_m$ are detected (whereas the whole detection bandwidth depends on the lock-in integration constant). These effects typically result in a one-order-of-magnitude enhancement in the signal-to-noise ratio (SNR) with respect to simple absorption spectroscopy. In addition, derivative detection offers the advantage of baseline flattening. For example, in second derivative spectra all the spectral components which are constant (baseline offset) and linear with wavelength (baseline slope) are eliminated. Equation 5.28, shows that the signals detected at frequency ν_m and $2\nu_m$ scale, respectively, as a and a^2; therefore, increasing modulation depth can lead to a further sensitivity enhancement. In this respect, the requirement $a \ll \Gamma$, necessary for the validity of the Taylor expansion, sets a limitation on the SNR improvement achievable with derivative spectroscopy.

5.3.2 Wavelength modulation spectroscopy

On the other hand, even under a deeper modulation, the absorption cross section remains a periodic function of time and, consequently, is liable for a Fourier expansion

$$\sigma(\nu_0 + a\cos\nu_m t) = \sum_{n=0} H_n(\nu_0)\cos n\nu_m t \tag{5.29}$$

$$H_n(\nu_0) = \frac{2}{\pi}\int_0^\pi \sigma(\nu_0 + a\cos\theta)\cos n\theta d\theta \tag{5.30}$$

Therefore, still it is possible to improve the SNR by increasing the modulation depth to values $a \gg \Gamma$. However, the signals detected at frequencies ν_m, $2\nu_m$, etc. can no longer be confronted with the terms of a Taylor expansion, but with the harmonics of the Fourier expansion as given by Equation 5.29. In literature, this regime ($\nu_m \ll \Gamma$, $a \gg \Gamma$) is referred to as wavelength modulation spectroscopy (WMS). Numerical evaluations of integrals defined by Equation 5.30 for different lineshape functions (either Lorentzian or Gaussian) can be found in [388]. These show that the amplitude of the WMS signal is actually not monotonic with the modulation depth, but different trends occur depending on the absorption profile under study.

5.3.3 Single- and two-tone frequency modulation spectroscopy

Frequency modulation (FM) spectroscopy relies on much faster modulation frequencies ($\nu_m \gg \Gamma$) [389, 390]. In this case, modulation results in the formation of two distinct sidebands, shifted of $\omega_m \equiv \pm 2\pi\nu_m$ with respect to the carrier $\omega_0 \equiv 2\pi\nu_0$. Now, the widely separated sidebands can individually interact with the absorption line of interest as the laser source is tuned through the spectral region. Alternatively, the entire line shape of the spectral feature can be scanned by tuning the radio frequency (ω_m) with the laser set at a fixed optical frequency (ω_0). In general, an external phase modulator can be used to convert the single-mode laser input into a *pure* frequency modulated optical spectrum with a low modulation index M. Figure 5.3 is a schematic of a typical experimental arrangement for FM spectroscopy.

As already shown during the discussion of the Pound-Drever-Hall technique, the phase-modulated output electric field can be expressed as

$$E_2(t) = E_0 e^{i(\omega_0 t + M\sin\omega_m t)} = E_0 e^{i\omega_0 t}\sum_{n=-\infty}^{+\infty} J_n(M)e^{in\omega_m t} \tag{5.31}$$

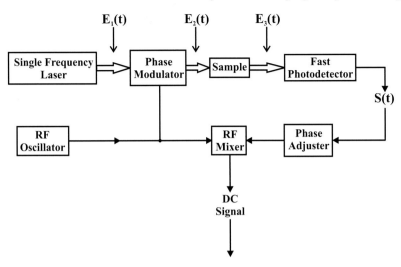

FIGURE 5.3
Typical experimental arrangement for FM spectroscopy.

where the expansion in a series of nth-order Bessel functions J_n characterizes the frequency components of the modulated light spectrum. Since $J_{-n}(M) = (-1)^n J_n(M)$, the lower-frequency components with odd n are 180 degrees out of phase from the upper sidebands. In the case of traditional FM spectroscopy, the modulation index is typically so small that the light spectrum consists essentially of the strong carrier and only one set of sidebands at $\omega_{\pm 1} = \omega_0 \pm \omega_m$, while higher-order sidebands are negligible. It is worth noting that we are implicitly assuming that the laser carrier linewidth is much smaller than ω_m, such that it does not enter critically into the theoretical formalism. For small M values we have $J_0(M) \simeq 1$ and $J_1(M) \simeq M/2$ whereupon

$$E_2(t) = E_0 e^{i\omega_0 t} \left[1 + \frac{M}{2} e^{i\omega_m t} - \frac{M}{2} e^{-i\omega_m t} \right] \tag{5.32}$$

We continue the theoretical analysis by assuming that the optical field defined by Equation 5.32 is passed through a sample that contains an absorption line of spectral width Γ. The effect of the sample on each frequency component $\omega_n = \omega_0 + n\omega_m$ can be characterized by the complex transmission function $T(\omega_n) \equiv T_n = \exp(-\delta_n - i\phi_n)$, where δ_n is the field amplitude attenuation and ϕ_n is the optical phase shift at ω_n. The transmitted optical field is then

$$E_3(t) = E_0 e^{i\omega_0 t} \left[T_0 + T_1 \frac{M}{2} e^{i\omega_m t} - T_{-1} \frac{M}{2} e^{-i\omega_m t} \right] \tag{5.33}$$

Since the photodetector electric signal $S(t)$ is proportional to the square modulus of $E_3(t)$, by assuming that $|\delta_0 - \delta_1|$, $|\delta_0 - \delta_{-1}|$, $|\phi_0 - \phi_1|$, $|\phi_0 - \phi_{-1}|$ are all small compared to 1, we get

$$S(t) = S_0 e^{-2\delta_0} [1 + (\delta_{-1} - \delta_1) M \cos \omega_m t + (\phi_1 + \phi_{-1} - 2\phi_0) M \sin \omega_m t] \tag{5.34}$$

This means that the in-phase $\cos \omega_m t$ component of the beat signal is proportional to the difference in loss experienced by the upper and lower sidebands, whereas the quadrature $\sin \omega_m t$ component is proportional to the difference between the phase shift experienced by the carrier and the average of the phase shifts experienced by the sidebands. The FM spectroscopy condition is achieved when ω_m is large compared with the spectral feature of interest such that only one sideband (let us say the upper one) probes the atomic/molecular line. In this case, the losses and phase shifts experienced by the carrier and lower sideband remain essentially constant. Thus $\delta_{-1} = \delta_0 = \bar{\delta}$ and $\phi_{-1} = \phi_0 = \bar{\phi}$ where $\bar{\delta}$ and $\bar{\phi}$ are the constant background loss and phase shift, respectively. If the quantities $\Delta\delta = \delta_1 - \bar{\delta}$ and $\Delta\phi = \phi_1 - \bar{\phi}$ are introduced, then Equation 5.34 further simplifies to

$$S(t) = S_0 e^{-2\bar{\delta}}[1 - \Delta\delta \cdot M \cos\omega_m t + \Delta\phi \cdot M \sin\omega_m t] \tag{5.35}$$

Thus the cosine component of the beat signal is now directly proportional to the absorption induced by the spectral feature, whereas the sine component is directly proportional to the dispersion induced by the spectral feature. The rf beat signal arises from a heterodyning of the FM sidebands with the carrier frequency, and thus the signal strength is proportional to the geometrical mean of the intensity of each sideband and the carrier. The null signal that results with pure FM light can be thought of as arising from a perfect cancellation of the rf signal arising from the upper sideband beating against the carrier with the rf signal from the lower sideband beating against the carrier. The high sensitivity to the phase or amplitude changes experienced by one of the sidebands results from the disturbance of this perfect cancellation.

The lineshapes of the FM spectroscopy signals depend critically on the ratio of sideband spacing to the width of the spectral feature. If, for example, the latter is Lorentzian, the dimensionless attenuation δ and phase shift ϕ can be expressed as

$$\delta(\omega) = \delta_{peak}\left(\frac{1}{R^2(\omega) + 1}\right) \tag{5.36}$$

$$\phi(\omega) = \delta_{peak}\left(\frac{R(\omega)}{R^2(\omega) + 1}\right) \tag{5.37}$$

with

$$R(\omega) = \frac{\omega - \Omega_0}{\Gamma/2} \tag{5.38}$$

where Ω_0 is the line center frequency and Γ the FWHM. The above three equations define $\delta_j = \delta(\omega_j)$ and $\phi_j = \phi(\omega_j)$ for each spectral component of the FM optical spectrum (again, the subscript $j = 0, \pm 1$ denotes the values at frequencies ω_0 and $\omega_0 \pm \omega_m$, respectively). Substitution of these results in Equation 5.34 gives a complete specification of the FM signal lineshape obtained when ω_0 or ω_m is scanned. If we define the following quantities

$$x_0 = \frac{\omega_0}{\Gamma/2} \qquad x_m = \frac{\omega_m}{\Gamma/2} \qquad X_0 = \frac{\Omega_0}{\Gamma/2} \qquad t_0 = \frac{x_0 - X_0}{x_m} \tag{5.39}$$

we can re-write the above expressions as

$$y_j \equiv \frac{\delta_j}{\delta_{peak}} = \frac{1}{1 + (t_0 + j)^2 x_m^2} \tag{5.40}$$

$$z_j \equiv \frac{(t_0 + j)x_m}{1 + (t_0 + j)^2 x_m^2} \tag{5.41}$$

with $j = 0, \pm 1$. In this way, referring to Equation 5.34, the absorption ($\cos\omega_m t$) signal is $S_{abs} \propto y_{-1} - y_1$, while the dispersion ($\sin\omega_m t$) signal is $S_{disp} \propto z_1 + z_{-1} - 2z_0$. A plot of such signals is shown in Figure 5.4.

Finally, we carry out a signal-to-noise analysis. For simplicity we consider the case where a purely absorptive feature is probed with a single isolated bandwidth and no background

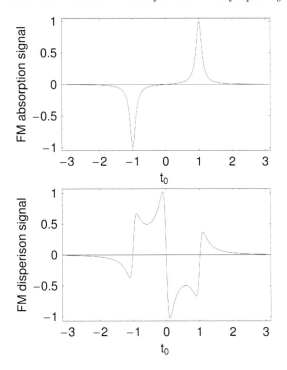

FIGURE 5.4

Typical absorption and dispersion signals in FM spectroscopy as a function of t_0 for $x_m = 10$ (see text).

absorption is present. Thus from Equation 5.35 with $\Delta\phi = \bar{\delta} = 0$, the slowly varying envelope $P(t)$ of the optical power incident on a photo-conductor (quantum efficiency η and gain g) is

$$P(t) = P_0(1 - \Delta\delta M \cos\omega_m t) \tag{5.42}$$

where P_0 is the total laser power. The generated current is $i(t) = \bar{i} + i_s(t)$ where the *dc* photocurrent is given by

$$\bar{i} = ge\eta\frac{P_0}{\hbar\omega_0} \tag{5.43}$$

while the beat-signal photocurrent is

$$i_s(t) = -ge\eta\frac{P_0}{\hbar\omega_0}\Delta\delta M \cos\omega_m t \tag{5.44}$$

Thus the rms power of the beat signal is

$$\overline{i_s^2(t)} = \frac{g^2 e^2 \eta^2}{2}\left(\frac{P_0}{\hbar\omega_0}\right)^2 \Delta\delta^2 M^2 \tag{5.45}$$

Since $1/f$ amplitude noise is insignificant at rf frequencies for single-mode lasers, the dominant noise sources are thermal noise and shot noise generated at the photodetector. The rms noise power is given by

$$\overline{i_N^2} = \overline{i_{SN}^2} + \overline{i_{TH}^2} \tag{5.46}$$

where

$$\bar{i}_{SN}^2 = 2eg\bar{i}\Delta f = 2g^2e^2\eta\left(\frac{P_0}{\hbar\omega_0}\right)\Delta f \qquad \bar{i}_{TH}^2 = \left(\frac{4k_BT}{R}\right)\Delta f \qquad (5.47)$$

where Δf and R represent the bandwidth and the input impedance of the detection electronics, respectively. The ratio of signal-to-noise is then

$$\frac{S}{N} = \frac{\overline{i_s^2(t)}}{\bar{i}_N^2} = \frac{\dfrac{g^2e^2\eta^2}{2}\left(\dfrac{P_0}{\hbar\omega_c}\right)^2\Delta\delta^2M^2}{2g^2e^2\eta\left(\dfrac{P_0}{\hbar\omega_0}\right)\Delta f + \left(\dfrac{4k_BT}{R}\right)\Delta f} \qquad (5.48)$$

from which it can be seen that it is always advantageous to increase either P_0 or M, to have η near unity, and to work with narrow-band detection electronics. For sufficiently high power, shot noise predominates over thermal one. In such conditions, we have

$$\frac{S}{N} = \frac{\overline{i_s^2(t)}}{\bar{i}_N^2} = \frac{\eta\left(\dfrac{P_0}{\hbar\omega_0}\right)\Delta\delta^2M^2}{4\Delta f} \qquad (5.49)$$

Actually achieving quantum-noise-limited detection sensitivity requires careful attention to experiment details. In this respect, the so-called residual amplitude modulation (RAM) plays a crucial role. Indeed, in practice, pure phase (or frequency) modulation is difficult to achieve, and the FM laser beam is amplitude modulated even when no sample is present. This happens either when the phase modulation is induced externally with an EOM or when it is obtained by modulating the current of a diode laser. The effect is that, besides the components at ω_0, $\omega_0 \pm \omega_m$ in Equation 5.32, a small component at ω_m may arise, resulting from a small imbalance in the amplitudes of the sidebands or a relative shift in phase which prevents the beat frequency from vanishing. This RAM can be detected by a photodiode and introduces a non-zero baseline. A comprehensive discussion on RAM, from both a theoretical and an experimental point of view, is given in [391]. In particular, a number of techniques to suppress RAM are also illustrated.

In some cases, however, FM scheme is limited by its severe technical requirements. In fact, since the condition $\omega_m \gg \Gamma$ must be accomplished, the modulation frequency is determined by the width of the spectral line under study. In the case of pressure broadened lines, for instance, modulations frequencies and, consequently, detectors as fast as few GHz are required. To overcome this limitation, an alteration of the described scheme can be adopted, namely the two-tone frequency modulation spectroscopy (TTFMS) [392]. In TTFMS two closely spaced modulation frequencies ω_1 and ω_2 are used (see Figure 5.5). These frequencies are both larger than the absorption linewidth, but their difference $\omega_1 - \omega_2 = \Omega$, at which the detection is performed, is relatively small (in the order of MHz), enough to lie within the detector bandwidth. In this way, one can also consider ω_1 and ω_2 as sampling essentially the same part of the absorption feature (which is typically broad with respect to Ω). Formally, the electric-field amplitude of the laser after modulation at these two distinct frequencies can be written as

$$E_{TT}(t) = E_0 e^{i(\omega_0 t + \beta_1\sin\omega_1 t + \beta_2\sin\omega_2 t)} \qquad (5.50)$$

where β_1 and β_2 are the FM indexes, and $\omega_1 = \omega_m + \Omega/2$ ($\omega_2 = \omega_m - \Omega/2$). Then we set $\beta_1 = \beta_2 \equiv \beta$, which is tantamount to taking the modulation power associated with ω_1

and ω_2 to be equal. Then, assuming $\beta \ll 1$, we get

$$E_{TT}(t) \simeq E_0 e^{i\omega_0 t} \left[1 + \frac{\beta}{2} e^{i\omega_1 t} - \frac{\beta}{2} e^{-i\omega_1 t} \right] \left[1 + \frac{\beta}{2} e^{i\omega_2 t} - \frac{\beta}{2} e^{-i\omega_2 t} \right]$$

$$= E_0 \left\{ e^{i\omega_0 t} - \frac{\beta^2}{4} e^{i(\omega_0 + \Omega)t} - \frac{\beta^2}{4} e^{i(\omega_0 - \Omega)t} \right.$$

$$\left. + \frac{\beta}{2} e^{i(\omega_0 + \omega_1)t} + \frac{\beta}{2} e^{i(\omega_0 + \omega_2)t} - \frac{\beta}{2} e^{i(\omega_0 - \omega_1)t} - \frac{\beta}{2} e^{i(\omega_0 - \omega_2)t} \right\} \quad (5.51)$$

This spectrum is shown pictorially in Figure 5.5, which is not drawn to scale, as the size of Ω is greatly exaggerated in order that the separate sidebands be discernible.

Thus, after interaction with the sample, described by the complex transfer function $T_j = \exp(\delta_j - i\phi_j)$, we have

$$E_{TTs}(t) = E_0 \left\{ T_0 \left[e^{i\omega_0 t} - \frac{\beta^2}{4} e^{i(\omega_0 + \Omega)t} - \frac{\beta^2}{4} e^{i(\omega_0 - \Omega)t} \right] \right.$$

$$\left. + \frac{\beta}{2} T_1 \left[e^{i(\omega_0 + \omega_1)t} + e^{i(\omega_0 + \omega_2)t} \right] - \frac{\beta}{2} T_{-1} \left[e^{i(\omega_0 - \omega_1)t} + e^{i(\omega_0 - \omega_2)t} \right] \right\} \quad (5.52)$$

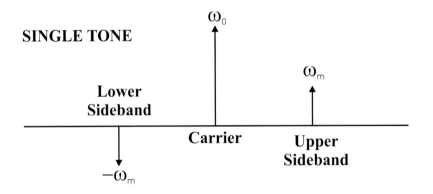

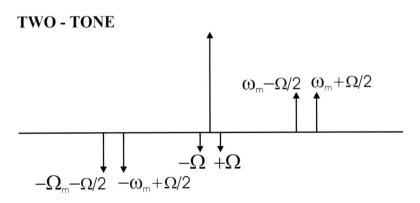

FIGURE 5.5
Spectral components of the laser after single and two-tone modulation.

The field is then incident on the photodetector that provides a signal $S_{TT}(t) \propto E_{TTs}E_{TTs}^{*}$. Since we extract the signal through narrow-band heterodyne detection at frequency Ω, we are interested only in the photocurrent output at Ω, although there will certainly be output at other frequencies. After some algebra, this is obtained as

$$S_{TTFM}(\text{at } \Omega) \propto \beta^2 \left[e^{-2\delta_1} + e^{-2\delta_{-1}} - 2e^{-2\delta_0} \right] \cos(\Omega t) \tag{5.53}$$

which, for small absorption, simplifies to

$$S_{TTFM}(\text{at } \Omega) \propto \beta^2 \left[2\delta_0 - \delta_1 - \delta_{-1} \right] \cos(\Omega t) \tag{5.54}$$

To illustrate these ideas, consider the effect of tuning the laser across an absorption line from below in frequency. First, the upper sideband pair will be absorbed. Thus δ_1 becomes non-zero, and a net negative signal results. Further scanning of the laser brings the absorption line onto the central group of sidebands and $2\delta_0$ becomes non-zero, and a net positive signal results. Finally, absorption of the lower sideband pair creates a nonzero δ_{-1}, and again a negative signal results. The recorded absorption spectrum will thus have a central positive peak and two symmetrically placed negative peaks with one half the height of the central peak. As shown in Figure 5.6, this behavior can be formally retrieved by use of Equation 5.40.

A typical experimental arrangement for TTFM spectroscopy is shown in Figure 5.7. Finally, it is worth noting that the TTFM signal is proportional to the square of the FM index rather than to the FM index as in conventional single-tone FM spectroscopy. Also, no phase information is encoded in the TTFM signal.

5.4 Magnetic rotation spectroscopy

Magnetic (Faraday) rotation (or magneto-optical effect) spectroscopy is a technique capable of enhancing the sensitivity of laser absorption experiments by taking advantage of the fact

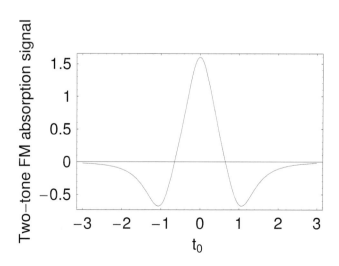

FIGURE 5.6
Typical TTFMS signal ($x_m = 2$).

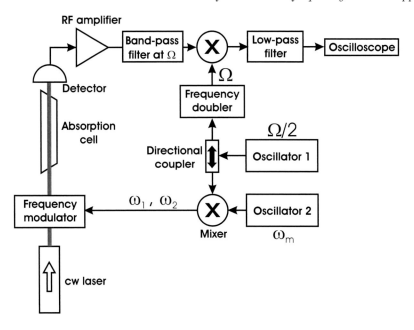

FIGURE 5.7
Typical experimental arrangement for TTFM spectroscopy.

that a transition in a paramagnetic molecule can alter the polarization state of incident linearly polarized light in the presence of a magnetic field. In such a configuration, the magnetic field breaks the magnetic (M_J) degeneracy of the rotational states (Zeeman effect). The resulting frequency shift of transitions is different for left-handed and right-handed circularly polarized light, giving rise to different refractive indices for these polarization components at a given radiation wavelength (circular birefringence). As a light beam, originally linearly polarized, propagates through the sample, this anisotropy leads to a rotation of the polarization axis. This magnetically induced birefringence in a longitudinal field and the related rotation of the polarization axis of linearly polarized light is called Faraday effect. This effect can be detected by means of two nearly crossed polarizers [393, 394, 395, 396]. In this way, laser amplitude noise is largely suppressed. Employing a static magnetic field in combination with a tunable laser, the sensitivity of direct absorption spectroscopy can be improved by 2-3 orders of magnitude. Another possibility is using an alternating magnetic field (modulated around $B = 0$) in combination with phase-sensitive detection (lock-in amplifier).

A typical experimental schematic is shown in Figure 5.8 [371]. In such an arrangement, continuous (dc) and sine-wave modulated (ac) magnetic fields can be applied (both with root-mean-square value equal to and different from zero). It is worth pointing out that, in the ac-modulated regime, the detected signal is in general due to a superposition of the Faraday and Zeeman modulation effects. Let us consider linearly polarized radiation that propagates along the z direction in a paramagnetic gaseous medium in the presence of a longitudinal magnetic field. The linearly polarized beam can be considered the superposition of right (σ_1) and left (σ_2) circular polarizations. For this reason, if the x direction is the polarization axis, the electric field can be written as

$$E(\omega) = \frac{E_0}{2}(\hat{x} + i\hat{y})e^{[i(\omega t - kz)]} + \frac{E_0}{2}(\hat{x} - i\hat{y})e^{[i(\omega t - kz)]} \qquad (5.55)$$

The externally applied magnetic field induces a difference in the refractive indices (n) and the absorption coefficients (α) for the σ_1 and σ_2 components. The difference $\Delta n = n^+ - n^-$ in the refractive indices introduces a rotation in the plane of polarization. Simultaneously, the originally linearly polarized radiation becomes elliptically polarized because of the difference $\Delta \alpha = \alpha^+ - \alpha^-$ in the absorption coefficients (magnetic circular dichroism). Thus, after interaction with the sample, the electric field becomes

$$E(\omega) = \frac{E_0}{2} e^{i\omega t} \left\{ \hat{x} \left[e^{\left(-ik^+L - \frac{\alpha^+}{2}L\right)} + e^{\left(-ik^-L - \frac{\alpha^-}{2}L\right)} \right] \right.$$
$$\left. + i\hat{y} \left[e^{\left(-ik^+L - \frac{\alpha^+}{2}L\right)} - e^{\left(-ik^-L - \frac{\alpha^-}{2}L\right)} \right] \right\} \tag{5.56}$$

where L is the interaction length. If the sample is placed between two nearly crossed polarizers, the transmitted field is given by

$$E(\omega) = \frac{E_0}{2} e^{i\omega t} e^{-ik^+L} e^{-\frac{\alpha^+}{2}L} \left\{ \left[1 + e^{\left(i\Delta kL + \frac{\Delta\alpha}{2}L\right)} \right] \sin\theta \right.$$
$$\left. + i \left[1 - e^{\left(i\Delta kL + \frac{\Delta\alpha}{2}L\right)} \right] \cos\theta \right\} \tag{5.57}$$

where $\Delta k = k^+ - k^- = \Delta n \omega / c$ and θ is the uncrossing angle between the polarizers. Then the transmitted intensity is evaluated as

$$I_t(\omega) = I_0 e^{-\frac{L}{2}(\alpha^+ + \alpha^-)} \left\{ \frac{1}{2} \left[\cosh\left(\frac{\Delta\alpha}{2}L\right) - \cos(\Delta kL + 2\theta) \right] + \xi \right\} \tag{5.58}$$

where I_0 is the incident intensity and ξ the polarizer extinction ratio. In the limit $\Delta kL \ll 1$ and $\Delta\alpha L \ll 1$ this reduces to

$$I_t(\omega) = I_0 e^{-\frac{L}{2}(\alpha^+ + \alpha^-)} \left[\sin^2\theta + \frac{\Delta kL}{2} \sin(2\theta) + \xi \right] \tag{5.59}$$

For small θ values then we get

$$I_t(\omega) \simeq e^{-\frac{L}{2}(\alpha^+ + \alpha^-)} I_0 \left[\theta^2 + \Delta kL\theta + \xi \right] \tag{5.60}$$

Thus we see that the Faraday signal is reduced by θ whereas the laser noise is suppressed by θ^2, which means that the SNR can be improved. In the above equation, the term which contains Δk is in general a complicated function of frequency because it depends on the Zeeman splitting associated with the transition [396]. In particular, this term is given by a summation of all the overlapping dispersion profiles that correspond to the $M_F + 1 \leftarrow M_F$ and $M_F - 1 \leftarrow M_F$ Zeeman components, where M_F is the component of the total angular momentum F in the field direction, which takes the values $M_F = F, F - 1, \cdots - F$. Assuming that all the $\Delta M_F = 1$ and $\Delta M_F = -1$ components are superimposed, an overlapping of only two dispersion curves can be considered.

As mentioned above, the Faraday effect has long been used to enhance sensitivity and selectivity in spectroscopic investigation of paramagnetic molecules, as it is an essentially zero-background technique like the related polarization spectroscopy technique (see later on in this chapter). However, this technique had mostly been applied in the visible and near infrared spectral regions where highly performing polarizing components can be used, with the aim to reduce the large amplitude fluctuations typical of tunable sources in this range,

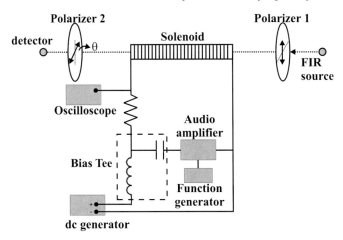

FIGURE 5.8
Schematic view of the experimental apparatus for observation of the magneto-optical (Faraday) effect by use of dc and sine-modulated ac magnetic fields.

such as cw dye and color-center lasers, with output powers of several hundred milliwatts. In 1997 an experiment at the European Laboratory for Nonlinear Spectroscopy, Florence, Italy first reported the observation of the Faraday effect in the far infrared range, at a wavelength around 85 micron, on a rotational transition of NO_2 with a tunable far-infrared spectrometer (see Chapter 4) [361]. The main disadvantage of this spectral region was that state-of-the-art polarizers in the FIR barely achieved extinction ratios of 10^{-3}. Moreover, because of the low-power levels available from TuFIR sources at that time (a few tens of nanowatts), detector noise (or thermal background noise) often limited the achievable sensitivity. Nevertheless, it was recognized that the far-infrared region is particularly convenient for magnetic rotation spectroscopy since Doppler-limited linewidths of molecular transitions are about 2 orders of magnitude narrower than in the visible range. Hence, the magnetic fields to rotate the polarization can be much lower. As just mentioned, the first demonstration of the Faraday effect in the far-infrared range was done on a rotational transition within the ground vibrational and electronic ground state of the NO_2 molecule, at 3514950.50 ± 0.04 MHz (as measured at zero magnetic field). To observe the effect, a collimated FIR beam was passed through a first polarizer (to fix the polarization axis) and then interacted with the gas sample, contained in a 2-m-long glass cell, and after the cell it was analyzed by a second polarizer. The cell was concentric to a plastic tube, around which two side-by-side solenoids, each 1-m-long, were wrapped. The inductance of each solenoid was about 3.5 mH, and peak currents of 0.5 A could be applied. Such current values generated a longitudinal flux density of 12.5 G when a temporally constant magnetic field was generated. Sinusoidal modulation was also applied to the magnetic field, using an audio amplifier. Peak-to-peak flux densities of 36 G were achieved at frequencies near 1 kHz. By combining together two polarizers, an extinction ratio up to $7 \cdot 10^{-4}$ could be achieved for the input and output of the cell. Varying the uncrossing angle θ between the two polarizers, a maximum of the S/N ratio was found for a large θ, close to $60°$, as shown in Figure 5.9. Actually, such a large uncrossing angle could be explained by the very low power used in that experiment and the predominant contribution of detector noise, at smaller angles, typically found to optimize the S/N in the visible, near-IR ranges. In that experiment, a mean value of the difference in the refractive index at the NO_2 line center for σ^+ and σ^- polarizations at a flux density of 36 G was measured to be $\Delta n = (5.90 \pm 0.07) \cdot 10^{-6}$. In a subsequent theoretical analysis, it was shown how differential detection schemes could be highly beneficial to increase, in

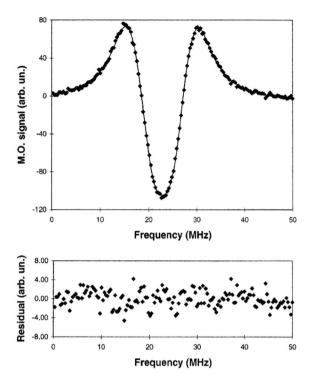

FIGURE 5.9
Top, experimental values and best-fit shape (curve) of the magneto-optical profile obtained for $\theta = 58°$, $B = 36$ G, $P_{NO_2} = 300$ mTorr. Bottom, the residual difference between the experimental values and the best-fit shape.

general, the sensitivity of far infrared coherent sources, like the tunable FIR spectrometer just described for magnetic rotation spectroscopy (see also Chapter 4) [369]. Later on, a thorough experimental and theoretical description of the continuous transition from a Zeeman effect regime to a Faraday (or magnetic rotation) effect regime was reported [371]. In this work, a key parameter to get increased detection sensitivity using a setup based on Magnetic Rotation (Faraday effect) emerges to be the coherent source power, given detection noise and optical bandwidth. Therefore, present availability of THz quantum cascade lasers could provide a boost for such spectroscopic techniques for detecting paramagnetic species.

5.5 Cavity-enhanced spectroscopy

The Lambert-Beer law shows that another valuable way to improve the SNR of an absorption signal is increasing the interaction path L between the sample and the probing radiation. The simplest technique to accomplish this task is using a multiple reflection cell (MRC). A MRC is a simple device consisting of two or more facing mirrors separated by a distance d. If the cell is properly aligned, the laser beam, coming from a hole in the input mirror, undergoes N_r reflections, each time traversing a different path before leaving the cell through an output hole. In this way, if the MRC contains an absorbing sample, the absorption pathlength is increased by a factor N_r without adding extra noise in the detected signal. The number of reflections is set by the cell geometry, and can reach a few hundreds, depending on the quality of mirrors. The main types of multiple reflection cells are the White [397] and Herriott type [398].

This multipass effect can be realized much more efficiently by placing the sample in an optical resonator. In fact, as already stressed, the use of an optical cavity leads to an enhancement of the effective absorption path-length proportional to the cavity finesse. The current state of the art for high reflectivity mirrors sets a limit on available cavity finesses in the range from 10^5 to 10^6. In the mid-infrared region mirrors with reflectivities of 99.97% are commercially available, providing path-length enhancements in the order of 10^4. However, from the experimental point of view, the intensity transmitted by an optical cavity is affected by an additional source of noise, introduced by the resonant nature of the signal. In fact, any relative frequency fluctuation between laser and cavity resonance results in a strong amplitude instability that affects the transmitted radiation. The sharper is the resonance linewidth (i.e., the higher is the finesse) the stronger is such FM-AM conversion. This places stringent requirements on the free-running frequency noise of the interrogating laser as well as the acoustic and seismic isolation required for the interferometer operation. Eventually, the same factor that enhances the absorption signal may affect its noise level, unless some refined technique is adopted to suppress this effect. Moreover, a better cavity finesse reduces the cavity mode line width, which makes the laser-cavity coupling more difficult. In the following, a few schemes are described, which successfully address these issues [336, 399].

5.5.1   Cavity-enhanced absorption spectroscopy

Cavity-enhanced absorption spectroscopy (CEAS) relies on a continuous coupling of laser light into the high-finesse cavity, which requires the laser frequency to be actively locked to one of the cavity modes. This can be accomplished either by electronic or optical feedback to the laser. In the former case, the well-established Pound-Drever-Hall (PDH) technique is used; however, any residual relative (between the laser and the cavity mode) frequency noise will evolve into amplitude noise in the transmitted light. Then, the molecular absorption profile is recovered by scanning the cavity free-spectral range. The general formula for the intensity transmitted from a cavity has been already given in Chapter 3. A more specific expression, corresponding to the case in which the laser frequency is locked to a cavity mode, will be provided in a following section, when dealing with cavity-enhanced saturation spectroscopy (see Equation 5.106). At the same time, while being highly efficient, electronic locking is complex, sensitive to external perturbations, and is difficult to use when the measurement instrumentation must be compact and robust (e.g., in environmental applications). As mentioned above, a second valuable option is represented by resonant optical feedback (OF) injection [400]. In this case, the laser-cavity coupling is arranged so as to permit a restrained return of the resonant intracavity field back to the laser, while avoiding reappearance of light directly reflected from the cavity input mirror. Essentially, for a wide interval of the OF level (defined as the ratio between the OF and the incident intensity) and a suitable OF phase (controlled by the distance between the laser and the optical cavity), the laser linewidth can be narrowed well below the cavity-mode one. As the very first result, on resonance, this maximizes the cavity injection efficiency. Moreover, when sweeping the laser across the cavity mode, such a narrowing effect is associated with a temporary laser-to-mode frequency locking. In this way, when performing fast frequency scans, one obtains a very accurate recording of the maxima of consecutive cavity modes. Furthermore, since the linewidth of the laser is well within that of the cavity mode, normalizing the transmitted signal to the incident light intensity authorizes the use of the cavity-transfer-function

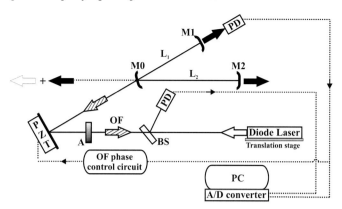

FIGURE 5.10
V-shaped cavity setup for OF-CEAS. Only a restrained part of the exiting resonant intra-cavity field is allowed to come back to the laser. At the same time, the realized optical feedback induces a laser linewidth reduction and provides an automatic laser-to-cavity-mode frequency locking. This elegantly overcomes the drawback of a noisy cavity output. The variable attenuator (A) adjusts the OF level, whereas the PZT-mounted steering mirror controls the phase of the OF field. The translation stage allows wider adjustment of the laser-cavity separation. The beam splitter (BS) is used to monitor the power incident on the cavity. (Adapted from [401].)

formalism to convert the cavity-enhanced spectra into absorption units. Finally, due to the regular comb structure of the cavity modes, a high degree of linearity is conferred to the frequency scale. This technique is called optical feedback cavity-enhanced absorption spectroscopy (OF-CEAS). A typical OF-CEAS setup, based on a V-shaped optical cavity, is shown in Figure 5.10. While the general principles of resonant optical feedback in diode lasers have already been outlined in Section 4.8.5.1, a more comprehensive discussion, specialized to the case of a three-mirror, V-shaped optical cavity, can be found in [401]. As demonstrated in this work, OF-CEAS is capable of reaching sensitivities in the $2.5 \cdot 10^{-8}$ $\mathrm{Hz}^{-1/2}$ range, limited by parasitic interference fringes.

5.5.2 Off-axis integrated cavity output spectroscopy

In the integrated cavity output spectroscopy (ICOS) technique, the cavity length (and hence its mode structure) and/or the laser frequency are dithered on a time scale, τ_{dith}, considerably faster than the typical temporal span for sweeping an absorption profile, τ_{sweep}. This has the effect of randomizing the input coupling of the light into the cavity. At the same time, the cavity output is integrated over a duration that is longer than τ_{dith}, but shorter than τ_{sweep} [402]. Also, the frequency dependence in the cavity transmission can be dramatically depressed implementing an off-axis configuration (OA-ICOS), where the laser beam is injected at an angle with respect to the main axis of the cavity, thus exciting a high density of transverse modes, whose spacing can practically equal their widths.

First investigated by Herriott, off-axis paths through optical resonators spatially separate the multiple reflections within the cavity until the re-entrant condition is satisfied, i.e., when the ray begins to retrace itself on the original path. This is dictated by the specific curvature r and spacing L of the mirrors forming the cavity. As shown in frame (a) of Figure 5.11, the multiple reflections appear on the mirrors as a series of spots in an elliptical pattern. The angle 2θ of a round-trip rotation is again purely determined by the geometry of the cavity and is given by $\cos\theta = 1 - L/r$. When $2m\theta = 2p\pi$, where m equals the number of

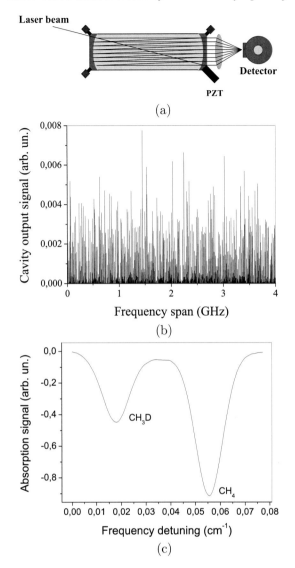

FIGURE 5.11

(a) Schematic of the off-axis cavity alignment evidencing the multiple-reflection beam path. If m equals the number of optical round-trip passes, the pattern becomes re-entrant for a cavity-effective free-spectral range $FSR = c/(2mL)$. In graph (b), a typical empty-cavity transmission spectrum is shown in the case of an on-axis FSR' of 166 MHz and an alignment with $m = 11$, so that $FSR = 15$ MHz. To wash out the cavity mode structure, both laser-frequency and cavity-length modulations are introduced and combined to time integration of the output signal. As an example, graph (c) also shows a typical sample absorption profile recorded with a DFG spectrometer, corresponding to ro-vibrational transitions of the CH_3D v_4 and CH_4 $v_2 + v_4$ bands, respectively, at 2960.617586 cm^{-1} and 2960.65530 cm^{-1}. The cavity was filled with pure methane in natural isotopic abundance at 100 mTorr pressure.

optical round-trip passes and p is an integer, the pattern becomes re-entrant, and the cavity effective free-spectral range (FSR) equals $c/2mL$. Actually, the sensitivity enhancement of the OA-ICOS method also depends on the ability to remove the cavity-resonance structure [403, 404]. This is in principle accomplished in the case of an effective FSR well below the laser linewidth $\Delta\Omega_L$, because the amount of energy coupled to the resonator virtually ceases to be a function of laser wavelength. However, in many experimental situations, the above condition is not met, and large-peak-intensity fluctuations are observed in the transmission spectrum, as shown in the example of frame (b) of Figure 5.11 for $FSR = 15$ MHz and $\Delta\Omega_L = 1$ MHz. Nevertheless, even in that case, if the width of the absorption feature of interest is much larger than the cavity effective FSR, both laser-frequency and cavity-length modulations can be introduced and combined to time integration of the output signal to efficiently flatten the cavity frequency response, while a sufficient number of data points are retained to define the absorption profile. The result of this experimental procedure can be appreciated in frame (c) of Figure 5.11 where a typical absorption line is shown for an absorbing gas in the cavity. The relationship between the laser tuning rate $G = f_{scan} \cdot \delta\omega$ (with f_{scan} as the frequency of the laser scan and $\delta\omega$ as the scan interval) and the cavity ring-down time τ also plays an important role for the stabilization of the cavity output. Indeed, the resonant energy build-up of the cavity can be efficiently suppressed when the laser is scanned through one effective FSR within the ring-down time. For a given value of $\delta\omega$, this sets a lower limit for the laser-frequency scan $f_{scan} \geq FSR/(\tau\delta\omega)$. Note that, at the same time, f_{scan} should also be chosen with a value much less than the characteristic cavity frequency $f_{cavity} = 1/(2\pi\tau)$, which describes the cavity-amplitude response. The optimum value for f_{scan} is chosen in the above range to maximize the signal-to-noise ratio while avoiding any distortion in the transmission spectrum. On the basis of this consideration, for all practical purposes, wavelength and electric-field phase information can be neglected in the description of the intracavity optical intensity, leading to a simplified theoretical model. Therefore, for any FSR-to-$\Delta\Omega_L$ ratio, to account for the presence of a cw laser injected into the cavity, we can simply add a constant source term to the standard differential equation used to describe the change of the intracavity power. This yields

$$\frac{dI}{dt} = \frac{c}{2L}[I_0 C_p T_m - 2I(1 - R_m)] \tag{5.61}$$

where I_0 is the incident intensity, C_p is a factor between 0 and 1 describing the amount of incident radiation coupled to the resonator, and T_m (R_m) is the mirror transmittivity (reflectivity). The steady-state transmitted intensity is found by multiplying the stationary solution of Equation 5.61 by the output mirror transmittivity $I_t = I_0 C_p T_m^2/[2(1 - R_m)]$. The presence of a weakly absorbing species in the cavity is taken into account by replacing the reflectivity with $R' = R_m \exp[-\alpha(\omega)P_s L] \simeq R_m[1 - \alpha(\omega)P_s L]$, and the output signal can be rewritten as

$$I_t(\omega) = \frac{I_0 C_p T_m^2}{2[(1 - R_m) + R_m \alpha(\omega)P_s L]} \tag{5.62}$$

with $\alpha(\omega)$ as the absorption coefficient of the selected transition and P_s as the absorber pressure. From Equation 5.62, the integrated absorbance (IA), namely, the area under the recorded absorption signal, can be calculated as

$$IA \equiv \int_{-\infty}^{+\infty} \frac{I_{t,\alpha=0} - I_t(\omega)}{I_{t,\alpha=0}} d\omega = \int_{-\infty}^{+\infty} \frac{\alpha(\omega)}{\alpha(\omega) + \dfrac{1}{P_s L_{eq}}} d\omega \tag{5.63}$$

where $L_{eq} = LR_m/(1 - R_m)$ is the effective absorption pathlength. The above formula for the quantity IA highlights the enhancement inherent to the OA-ICOS technique; indeed for very high L_{eq} values, the integrand tends to unity which yields, in turn, a large IA value.

The OA-ICOS method is, in general, limited by a low cavity transmission and a fluctuating coupling efficiency, and typically reach sensitivities in the 10^{-7} Hz$^{-1/2}$ range. The best results achieved so far with ICOS and OA-ICOS are $2 \cdot 10^{-10}$ Hz$^{-1/2}$ [405] and $2 \cdot 10^{-9}$ Hz$^{-1/2}$ [406], respectively.

5.5.3 Noise immune cavity-enhanced optical heterodyne molecular spectroscopy

All aforementioned techniques, however, do not make use, in their most common realizations, of any modulation methodology to reduce flicker noise. In this frame, wavelength modulation spectroscopy (WMS) has been combined with both cw-CEAS and OA-ICOS. In the former case, a sensitivity of $3 \cdot 10^{-11}$ Hz$^{-1/2}$ was obtained, but an extremely tight laser-to-cavity lock (1 mHz relative to the cavity) was mandatory [407]. In the latter case, an improvement in the detection sensitivity of a factor 20 was reported [408]. So far, the most rewarding approach to merge cw-CEAS with a modulation technique is represented by the noise-immune cavity-enhanced optical-heterodyne molecular-spectroscopy (NICE-OHMS) technique [407, 409, 399]. In this case, frequency modulation spectroscopy (FMS) is happily combined with CEAS to reach a close-to-shot-noise sensitivity in the 10^{-13} Hz$^{-1/2}$ range. The clever idea is to choose the modulation frequency exactly equal to the cavity free spectral range, such that all the components of the FM-triplet are transmitted through the cavity. Thus, any residual laser frequency noise will influence the three spectral components in an identical manner, and no conversion to amplitude noise will appear in the detected signal. In this sense, the technique is *immune* to laser frequency noise. The attractive feature is that, while taking advantage of the cavity enhancement, the benefits of FMS are fully exploited. Moreover, WMS can additionally be implemented in order to suppress leftover noise in the FMS background. Finally, like in cw-CEAS, the presence of counter-propagating high-intensity laser beams can be exploited to perform Doppler-free saturation spectroscopy.

Now let us discuss more quantitatively the case of Doppler-limited NICE-OHMS. Here, the three FM modes propagating inside the cavity interact with separate velocity groups of molecules. As a result, the interaction between the absorbers and the light can be described by addressing one mode at a time. The Doppler-broadened frequency modulated NICE-OHMS signal is obtained by simply multiplying Equation 5.34, describing the photodetector electric FM signal, by the cavity-enhancement factor $2\mathcal{F}/\pi$:

$$S^{FM,N-O} = \frac{2\mathcal{F}}{\pi} \{1 + [\delta(\omega_0 - \omega_m) - \delta(\omega_0 + \omega_m)]M \cos \omega_m t$$
$$+ [\phi(\omega_0 - \omega_m) + \phi(\omega_0 + \omega_m) - 2\phi(\omega_0)]M \sin \omega_m t\} \qquad (5.64)$$

where, as usual, $\mathcal{F}$ denotes the cavity finesse, ω_m the modulation frequency, and P_0 is the power of the laser beam. So, in the ideal case, where the cavity-enhancement effect applies only to the signal without introducing any extra noise, the sensitivity of FM spectroscopy will be augmented by the factor $2\mathcal{F}/\pi$. A generic schematic of the experimental setup for NICE-OHMS is shown in Figure 5.12 [399]. Three modulation frequencies are usually employed to optimize the performance of the spectrometer:

- A PDH scheme is implemented to lock the laser frequency against a longitudinal mode of the external cavity. For this purpose, the laser light is frequency modulated at ν_{PDH} (via an electro-optic modulator for instance) and the error signal derived in cavity reflection. By virtue of the noise-immune principle, the requisites for the PDH locking are looser, albeit the servo bandwidth still needs to be sufficiently high in order to realize an effective laser-to-cavity coupling (the needful bandwidth is basically set by the laser linewidth which, for high cavity finesse values, generally exceeds the cavity mode width). Depending on the specific type of laser used, the frequency correction signal is distributed among internal actuators and medium/high bandwidth external actuators (AOMs and/or EOMs), according to their respective frequency responses. After the PDH locking, scanning of the laser frequency along the absorption profile is accomplished, as usual, by tuning the cavity length. In this respect it is useful noting that, when acquiring Doppler-broadened spectra, high gain at low frequencies is needed (for the locking loop) so as to guarantee that the laser closely follows the large cavity scan.

- Then, a higher modulation frequency, ν_m, is also applied (typically through a second EOM) for NICE-OHMS detection. In order not to compromise the noise-immune principle as the FSR is varied during the cavity-length scan, such frequency is actively locked to the cavity FSR (via a second locking servo). Such active tracking of the cavity FSR is vital, particularly when scanning across a Doppler-broadened transition, where the FSR is varied by a significant fraction of the cavity-mode width. Conversely, this is less critical in Doppler-free detection, for which a shorter tuning range is needed. Commonly, the FSR-locking error signal is derived from cavity reflected light at the difference frequency $\nu_m - \nu_{PDH}$ (in the hundreds of megahertz range) by the so-called deVoe-Brewer technique [410]. However, it should be pointed out that, even in the presence of robust FSR tracking, due to the gas dispersion, the frequency distance between the cavity modes against which the three FM components are respectively locked changes in an asymmetric manner. As a consequence, the noise-immune condition is perfectly met only at resonance.

- In addition, cavity-length dithering (at frequency f_m) is usually accomplished in order to implement WMS. This helps to lessen the effects of low-frequency noise which inevitably comes into play through the RAM (both from the laser and the EOM) as well as via unwanted optical interference fringes.

Finally, the NICE-OHMS signal is detected in the cavity transmission, either in the FM or WM mode of detection. So far, the best detection sensitivity ($5 \cdot 10^{-13}$, 1-s integration time) with the NICE-OHMS technique was obtained in Doppler-free detection mode on C_2HD with a Nd:YAG laser (at 1064 nm) and an enhancement cavity of finesse 100000 ($FSR = 320$ MHz, 47-cm length)[409].

In spite of some indisputable technical hurdles, the NICE-OHMS technique offers unique advantages in terms of detection sensitivity and is becoming, for this reason, increasingly popular in the field of precision spectroscopy and optical frequency metrology.

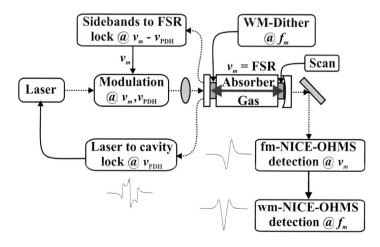

FIGURE 5.12
Schematic layout of a NICE-OHMS setup. (Adapted from [399].)

5.5.4   Cavity ring-down spectroscopy

Cavity ring-down spectroscopy (CRDS) relies on the fact that absorption is an additional loss in the cavity, and consequently changes its decay time, which is measured from the transient cavity transmission after an abrupt interruption of the injection [411, 412, 413, 414]. The absorbance is then evaluated through a comparison between the decay time of the transmitted laser light, respectively, in the presence and absence of the gas sample. Alternatively, the comparison can be made between the decay time in correspondence of two wavelengths, on- and off-resonance. While multiple longitudinal/transverse mode excitation, as realized when employing pulsed lasers, gives rise to multi-exponential decays (indeed each mode has its own characteristic ring-down time, RDT), excitation of a single cavity mode results, in principle, in a genuine single exponential process. Furthermore, use of pulsed lasers will more likely introduce unwanted interference effects in the light leaking out of the cavity, thus superimposing detrimental intensity modulations on the ring-down decay signal. As a result, the highest sensitivity will be achieved by use of a cw, narrow-linewidth laser. In the following we focus on such configuration. In this case, the ring-down event can be triggered either by breaking off the injected radiation or by rapidly scanning the laser over a cavity mode (or vice versa). The primary benefit of CRDS is that, in principle, it is not limited by the amplitude noise of the laser source, but only by the detection shot noise. In practice, however, drifts in the system between two consecutive measurements always prevent from achieving this ultimate limit and from averaging measurements over long times.

Figure 5.13 shows a generic schematic layout of the components making up a typical cw CRDS experiment. In the simplest scheme, the resonator length is continuously dithered (by an annular piezoelectric actuator mounted on one of the cavity mirrors) to bring the cavity mode frequencies into resonance with the laser beam. In this way, intensity will accumulate in the cavity (i.e., the cavity *rings up*); then, as a resonance builds up, a threshold detector triggers a large shift in the AOM frequency that rapidly brings the laser radiation out of resonance. If this switch-off is sufficiently fast (on the nanosecond timescale), a ring-down of the trapped light intensity will be recorded by a detector external to the cavity (obviously, the detector electrical bandwidth should be much greater than the inverse of the RDT). The average of many acquisitions, recorded by the oscilloscope, is then used to extract the

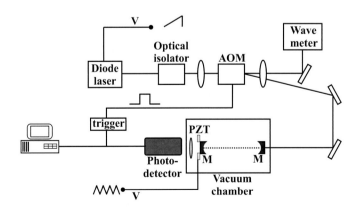

FIGURE 5.13

A schematic layout of a cw diode laser CRDS experiment. AOM=acousto-optic modulator; PZT=piezoelectric transducer; M=mirror. Also shown are the waveforms of applied voltages used for scanning the diode laser wavelength and modulating the cavity length. (Adapted from [413].)

RDT by means of a least-squares fitting routine; finally, hundreds of determinations of RDT yield the average cavity decay time $\bar{\tau}$ together with its standard error.

Next, let us derive an expression for the limiting detection sensitivity in CRDS [413]. To this aim, consider a cavity of length L which is excited by a laser beam of intensity I_{in}; then, after the first passage through the cavity, the intensity measured by the detector will be

$$I_0 = T^2 e^{-\alpha d} I_{in} \tag{5.65}$$

where T denotes the mirror intensity transmittivity, and α the frequency-dependent absorption coefficient of a sample filling a region of length d inside the cavity. The factor $e^{-\alpha d}$ expresses the one-pass intensity attenuation due to absorption and/or scattering by the homogeneous sample, according to the well-known Lambert-Beer law. For each successive round-trip, the intensity diminishes by an additional factor of $R^2 e^{-2\alpha d}$, with R being the mirror intensity reflectivity. Thus, after n round-trips, the intensity at the detector will be

$$I_n = [R \cdot e^{-\alpha d}]^{2n} I_0 = I_0 e^{-2n(\ln R + \alpha d)} \tag{5.66}$$

Next, by changing the discrete variable n to a continuous one, $t = 2Ln/c$, corresponding to the time spent in the cavity by the light to travel the distance $2Ln$, we can write

$$I(t) = I_0 e^{-\frac{ct}{L}(-\ln R + \alpha d)} \tag{5.67}$$

So, the cavity RDT, that is the time taken for the initial intensity to decrease by factor of e, is given by

$$\tau = \frac{L}{c(-\ln R + \alpha d)} \simeq \frac{L}{c(1 - R + \alpha d)} \tag{5.68}$$

which clearly shows that the RDT does not depend on the laser intensity (hence it is not affected by intensity fluctuations). In the absence of the sample, the RDT for an empty cavity is

$$\tau_0 = \frac{L}{c(1 - R)} \tag{5.69}$$

from which the mirror reflectivities can be accurately measured. If τ and τ_0 are experimentally recorded as functions of the laser frequency, the absorption spectrum of a sample within the cavity can be retrieved according to the following formula

$$\alpha(\nu) = \frac{L}{cd}\left(\frac{1}{\tau(\nu)} - \frac{1}{\tau_0}\right) \tag{5.70}$$

Finally, we can fix the limiting CRDS sensitivity as the minimum absorption coefficient, α_{min}, that can be detected in the limit of $\tau \to \tau_0$

$$\alpha_{min} = \frac{L}{cd\tau_0}\frac{\Delta\tau_{min}}{\tau_0} = \frac{1 - R}{d}\frac{\Delta\tau_{min}}{\tau_0} \tag{5.71}$$

where $\Delta\tau_{min}$ is the minimum detectable change in the cavity ring-down time. Equation 5.71 shows that the spectrometer sensitivity can be optimised either by increasing R or by increasing d, as well as by minimizing $\Delta\tau_{min}/\tau_0$ which represents the relative error of the cavity ring-down time measurements.

Within the above basic approach, sensitivities in the $\sim 10^{-8}$ Hz$^{-1/2}$ range can routinely be achieved. Higher sensitivities can be reached by actively locking the cavity and laser in resonance, which originates higher and more reproducible intra-cavity intensities and therefore improved signals. In the experiment by [415], a single-beam, dual-arm approach to CRDS was adopted to reduce shot-to-shot fluctuations and eliminate oscillations in the

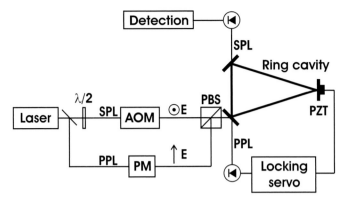

FIGURE 5.14
cw-CRDS setup employing a ring resonator locked to an external laser source. PPL=p-polarized light, SPL=s-polarized light, PM=phase modulator. (Adapted from [415].)

spectral backgrounds. More in detail, with reference to Figure 5.14, two orthogonal polarized beams are derived from the laser source (one s- and the other p-polarized with respect to the ring-down cavity mirrors) which see different mirror reflectivities at non-normal incidence (typically, dielectric mirrors have a lower reflectivity for PPL than for SPL). In this way, a single resonator can be simultaneously used as a low-finesse cavity for continuous locking to the probe laser (via the usual PDH technique) with one arm, and as a high-finesse cavity with the other arm, where the light is switched on and off (via the AOM) to perform CRDS measurements. Further improvement of this scheme, basically relying on simultaneous detection of an on- and off-resonance mode, allowed to push the detection sensitivity down to $4.2 \cdot 10^{-11}$ Hz$^{-1/2}$ [416].

An ac heterodyne technique in CRDS was also demonstrated achieving a detection sensitivity of $1.6 \cdot 10^{-10}$ at 1-s averaging [417]. In such configuration, two cavity modes, one probing the empty cavity and the other probing intracavity absorption, are excited simultaneously but with their intensities temporally out of phase, with one mode decaying and the other rising. Heterodyne detection between the two modes reveals the dynamic time constants associated with the empty cavity and the additional intracavity gas absorption.

5.5.4.1 Phase-shift (PS) CRDS

Phase-shift CRDS hinges on the principle that light trapped within a high-finesse optical cavity experiences a phase shift relative to light that bypasses the cavity [418]. In the basic PS CRDS scheme, a cw laser beam is sinusoidally intensity modulated at angular frequency Ω. Then, the time dependence of the light intensity entering the optical cavity, $I_{exc}(t)$, can be written as

$$I_{exc}(t) = I_0(1 + M \sin \Omega t) \qquad (5.72)$$

where M is the modulation depth. Since the light intensity inside the cavity also decays exponentially with the characteristic time $\tau(\nu)$, the intensity measured behind the cavity at time t is given by

$$I_{CRD}(t) = \frac{1}{\tau(\nu)} \int_{-\infty}^{t} I_0(1 + M \sin \Omega t) e^{-\frac{t-t'}{\tau(\nu)}} dt'$$

$$= I_0 \left\{ 1 + \frac{M}{\sqrt{1 + \Omega^2 \tau(\nu)^2}} \sin[\Omega t - \arctan \Omega \tau(\nu)] \right\} \qquad (5.73)$$

which means that the light transmitted through the cavity is phase shifted with respect to the incident beam by the amount ϕ given by

$$\tan\phi = -\Omega\tau(\nu) \tag{5.74}$$

Incidentally, the modulation depth decreases by the factor $1/\sqrt{1 + \Omega^2\tau^2}$. It follows therefore that the cavity ring-down time $\tau(\nu)$ can be determined in an intensity-independent way from a measurement of the ratio of the in-phase component to the out-of-phase component of the modulated light intensity that exits the optical cavity as a function of laser wavelength. For optimum detection sensitivity, Ω is selected such that $\Omega\tau(\nu)$ is around 1. Use of a lock-in amplifier enables measurement of the phase shift and determination of the ring-down time from which absorption coefficients can be derived in the usual way. PS CRDS is now finding increasing popularity in cavity-enhanced spectroscopy experiments making use of broadband light sources and in analytical applications of fiber-loop cw CRDS.

5.5.4.2 Saturated-absorption cavity ring-down spectroscopy

In principle, CRDS is not limited by amplitude noise of the laser source, but only by detection shot noise. However, variations of the empty-cavity decay rate always prevent us from achieving this ultimate limit and from averaging measurements over long times. The empty-cavity background could be subtracted by quickly switching the radiation frequency between nearby longitudinal cavity modes. Nevertheless, this method cannot be completely resolutive as different cavity modes are affected by uncorrelated fluctuations. To overcome this drawback, a novel spectroscopic technique, namely, saturated-absorption cavity ring-down (SCAR) was recently demonstrated [419]. Hinging on the decrease of the saturation level during each SCAR event, such a technique proved very effective in identifying and decoupling any variation of the empty-cavity decay rate. Saturation effects had already been observed in CRD experiments, but they had been either considered as a disadvantage to be avoided, or only used to get Lamb dips. Giusfredi et al., instead, developed and experimentally tested an effective model to take advantage of the SCAR effect in high-sensitivity and high-resolution spectroscopic detection. Let us assume that the gas interacts with intra-cavity radiation in a TEM$_{00}$ mode with a time-dependent intensity I and power P given by the following expressions

$$I(\rho, t) = I_0(t)\, e^{-2(\rho/w)^2} \tag{5.75}$$

$$P(t) = \frac{\pi w^2}{2} I_0(t) \tag{5.76}$$

where $\rho = \sqrt{x^2 + y^2}$ is the radial coordinate, $I_0(t) = I(\rho = 0,\, t)$ is the peak intensity on the cavity axis z, and w is the beam waist, assumed to be constant along z. We also define a time-dependent saturation parameter as

$$G(t) \equiv \frac{I_0(t)}{I_s} = \frac{P(t)}{P_s} \tag{5.77}$$

It is worth noting that we are neglecting the effects of the standing-wave light field inside the cavity and we are spatially averaging the different saturation levels of molecules interacting with light in node and anti-node positions. In the presence of inhomogeneous broadening due to a thermal Gaussian distribution of molecular velocities, the absorption coefficient is affected by saturation and obeys the following equation

$$\alpha(\rho, t) = \frac{\alpha_0}{\sqrt{1 + \dfrac{I(\rho, t)}{I_s}}} \tag{5.78}$$

where α_0 is the non-saturated value of α. Power attenuation due to gas absorption, along the z axis, can be expressed as [379]

$$\frac{dP}{dz}(t) = -2\pi \int_0^\infty \alpha(\rho, t) \, I(\rho, t) \, \rho d\rho = -\alpha_0 \frac{2P(t)}{1 + \sqrt{1 + \frac{P(t)}{P_s}}} \tag{5.79}$$

where the spatial integration over the transverse beam profile corresponds to a *local* approximation. That is correct at pressure regimes where the diffusion time of molecules is longer than the population relaxation time of molecular levels. Let us combine this continuous loss mechanism (following the Beer-Lambert law) with the discrete mirror losses (2 each cavity round-trip), and define the decay rates $\gamma_c = c(1-R)/l$ and $\gamma_g = c\alpha_0$ where R is the mirror reflectivity. Then, by using the speed of light relation $d/dz = (1/c)d/dt$, we get the following rate equation for the saturation parameter

$$\frac{dG}{dz}(t) = -\gamma_c G(t) - \gamma_g \frac{2G(t)}{1 + \sqrt{1 + G(t)}} \tag{5.80}$$

where $G(t=0) = G_0 = P_0/P_s$ is the saturation parameter value when the SCAR event starts, triggered by a threshold set on the signal detected in transmission. Numerical integration of Equation 5.80 can be preformed with the 4th-order Runge-Kutta algorithm within the CRD fitting procedure itself. Since the dynamic range of the decay curve exceeds 4 decades, it is useful to factorize G as follows

$$G(t) \equiv G_0 e^{-\gamma_c t} f(t; G_0; \gamma_c, \gamma_g) \tag{5.81}$$

whereupon the following differential equation is obtained

$$\frac{d}{dt} f(t) = -\gamma_g \frac{2f(t)}{1 + \sqrt{1 + G_0 e^{-\gamma_c t} f(t)}} \qquad f(0) = 1 \tag{5.82}$$

In Figure 5.15, SCAR measurements are compared with the model. The experimental data in this figure are the average of 3072 decay signals measured at a fixed frequency near the absorption peak of the chosen molecular transition ($03^31 - 03^30$ $R(50)$ transition of $^{12}C^{16}O_2$, recorded at room temperature and pressure of 50 μbar in a 1-m-long cavity formed by two 440-ppm-optical-loss high-reflectivity mirrors) over a 4-s time interval. The discrete noise values in the tail of the SCAR signal are due to the resolution of the digitizing oscilloscope. Almost flat residuals witness the validity of the theoretical model. A rough and intuitive explanation for the intrinsic ability of this technique to distinguish between empty-cavity and gas-induced decay rates can be given as follows. Three consecutive intervals can be recognized in the SCAR signal: zones A, B, and C carry information, respectively, on γ_c, $\gamma_c + \gamma_g$, and the detector offset due to thermal background (whose fitted value has been subtracted from the signal in figure). The A-B transition is marked by a slope change in the SCAR signal, which needs the condition $G_0 \gg 1$ to be well observable, as it is more evident in the f curves. As a consequence, a large detection dynamics is needed to measure the transition from strong-saturation to linear-absorption regime in the SCAR decay, before it falls below the noise level (B-C transition). The best choice for G_0 should give such three zones with similar durations.

As shown in Figure 5.16, the high-resolution performance of the SCAR technique was also tested by resolving the hyperfine structure of the ($00^01 - 00^00$) $R(0)$ transition of $^{17}O^{12}C^{16}O$ at natural abundance ($7.5 \cdot 10^{-4}$). Thanks to its ability to efficiently remove empty-cavity background losses, the SCAR technique has recently demonstrated to be a powerful tool for ultrahigh sensitivity spectroscopy. Indeed, radiocarbon detection was

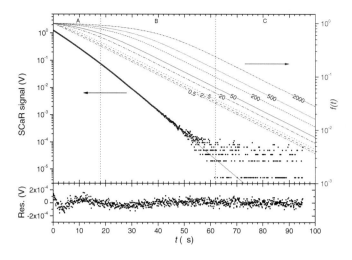

FIGURE 5.15
Comparison between experimental data and theoretical model for the SCAR technique. Experimental data points for the case $G_0 = 50$, the superimposed fit curve, and the residuals are plotted with left scale. Saturated decay functions f, simulated for different values of G_0 (labelling curves), are plotted with right scale.

achieved with a sensitivity to partial pressure of $^{14}CO_2$ in the part-per-quadrillion range [420]. This result simultaneously represents the lowest pressure of a gas phase simple molecule and a brand new, all-optical way for radiocarbon dating, directly challenging accelerator mass spectrometry (AMS) [421].

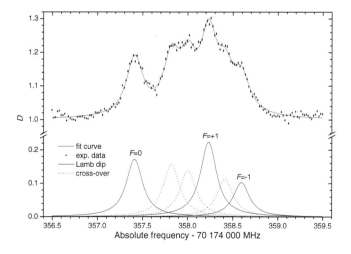

FIGURE 5.16
Sub-Doppler spectrum of the $(00^01 - 00^00)$ $R(0)$ transition of $^{17}O^{12}C^{16}O$ ($\nu_0 = 2340.765$ cm^{-1}, $S = 1.25 \cdot 10^{-22}$ cm), recorded with a pressure of 2 μbar. The dimensionless quantity D displaying the Lamb-dip features is defined as $D \equiv G_0 P_s / P_0$. The experimental data and a multi-Lorentzian curve fitting the 3 Lamb dips and the corresponding 3 crossovers are plotted.

5.6 Doppler-free saturation spectroscopy

Here begins our discussion on high-resolution spectroscopic techniques. Very soon after the advent of lasers, the first sub-Doppler spectroscopic technique (allowing the resolution of features on a frequency scale that is much smaller compared to Doppler width) was demonstrated. Also known as saturated absorption spectroscopy, this method was first observed by Lamb in the gain curve of a gas laser and then perfected by C. Bordé and T. Hänsch. It has revolutionized the field of high-resolution spectroscopy and is still widely employed. In a typical experimental setup (Figure 5.17), two counter-propagating laser beams derived from a single laser beam are sent through an atomic vapor cell (let us say along the z direction). The *pump* beam has a high intensity in order to make the gas transparent, while the transmittance of the *probe* beam (typically with a factor of ten smaller intensity) through the vapor is recorded by the detector and provides the actual spectroscopic signal. Let's start by considering the case where only the probe beam is allowed to go through the sample. Then, the usual Doppler-broadened absorption profile is recorded (we are assuming here that the gas pressure is sufficiently low to make the Lorentz broadening contribution negligible). Instead, if the pump beam is also sent through the vapor cell, a very narrow feature is observed in the signal at the resonance frequency $\nu = \nu_0$ of the atomic/molecular transition.

Qualitatively, the reason for this is the following: whenever atoms with non-zero velocity along the radiation propagation direction see the probe beam shifted into resonance due to the Doppler effect, at the same time the pump beam is shifted further away from resonance. By contrast, atoms at zero velocity do see both the pump and the probe beam into resonance. Then, if the photon flux in the pump beam is high enough, the ground is significantly depleted by the pump-induced absorption processes. As a result, the absorbance of the probe beam is reduced (compared to the case without pump beam) and a dip appears in its transmission spectrum.

The last comment of this qualitative introduction concerns the width of the Lamb dip which is, obviously, much narrower than the Doppler one. So, if the laser linewidth is small enough, the observed Lamb dip width can approach the natural linewidth of the atomic transition.

Next, let us try to derive a quantitative expression for such saturation-spectroscopy

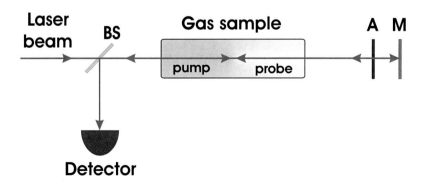

FIGURE 5.17
Illustration of the principle of Lamb-dip spectroscopy. The laser beam passes through the sample twice, using an arrangement of beam splitter (BS), mirror (M), and attenuator (A).

signals. For the sake of simplicity, let us consider atoms with only two internal states, the ground state $|g\rangle$ and the excited one $|e\rangle$. First, analyze the effect of the pump beam. In a two-level system, the populations in the two states obey the following equations

$$\begin{cases} \dot{N}_e = -\Gamma N_e + \sigma \Phi \left(N_g - N_e\right) \\ N_g + N_e = N = \text{const} \end{cases} \tag{5.83}$$

where the first term in the rate equation stems for spontaneous emission and the second term for stimulated processes, and $\Phi = I/h\nu$ is the incident photon flux. Then we have

$$\dot{N}_e = - \left(\Gamma + 2\sigma\Phi\right) N_e + \sigma\Phi N \tag{5.84}$$

whose steady-state solution is

$$\frac{N_e\left(\infty\right)}{N} = \frac{1}{2 + \dfrac{\Gamma}{\sigma\Phi}} \tag{5.85}$$

which gives

$$\frac{\triangle N_{pump}}{N} = \frac{N_g\left(\infty\right) - N_e\left(\infty\right)}{N} = 1 - \frac{1}{1 + \dfrac{\Gamma}{2\sigma\Phi}} \tag{5.86}$$

Now, if the absorption cross section has a Lorentzian profile with linewidth Γ and a Doppler-shifted resonance frequency

$$\sigma\left(\nu, v_z\right) = \sigma_0 \frac{\Gamma^2/4}{\left(\nu - \nu_0 + \dfrac{\nu_0}{c} v_z\right)^2 + \Gamma^2/4} \tag{5.87}$$

we have

$$\frac{\triangle N_{pump}}{N} = 1 - \frac{s}{1 + s + \dfrac{4\Delta^2}{\Gamma^2}} \tag{5.88}$$

where $\Delta = \delta + k v_z$, $\delta = \nu - \nu_0$, $s = \Phi/\Phi_{sat}$, $\Phi_{sat} \equiv I_{sat}/(h\nu) = \Gamma/\left(2\sigma_0\right)$. Then we can write

$$\frac{\triangle N_{pump}}{N} = 1 - \frac{s(\Gamma/2)^2}{\left(\nu - \nu_0 + \dfrac{\nu_0}{c} v_z\right)^2 + (\Gamma_s/2)^2} \tag{5.89}$$

with $\Gamma_{sat} = \Gamma\sqrt{1 + s}$. Next, let us consider the probe beam (and neglect interference effects between pump and probe). In this case, due to the opposite direction of propagation, the cross section is given by

$$\sigma_{probe}\left(\nu, v_z\right) = \sigma_0 \frac{\Gamma^2/4}{\left(\nu - \nu_0 - \dfrac{\nu_0}{c} v_z\right)^2 + \Gamma^2/4} \tag{5.90}$$

So the absorption coefficient, having the units of cm^{-1}, is given by

$$\alpha\left(\nu\right) = \int_{-\infty}^{+\infty} \sigma_{probe} \cdot \triangle N_{pump} \cdot dn\left(v_z\right) \tag{5.91}$$

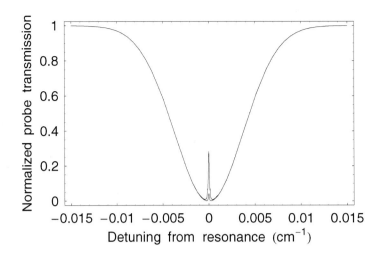

FIGURE 5.18

Normalized absorption coefficient according to Equation 5.91 as a function of the detuning $\nu - \nu_0$ for the following parameters: $T = 300$ K, $m = 30 \cdot 10^{-24}$ g, $\Gamma = 0.0001$ cm^{-1}, $\nu_0 = 3000$ cm^{-1}, and two different saturation parameters ($s = 0.1$ and $s = 1$).

where

$$dn\,(v_z) = n_0 \frac{e^{-\left(\frac{v_z}{v_p}\right)^2}}{v_p\sqrt{\pi}} dv_z \qquad (5.92)$$

is the fraction of atoms belonging to a certain velocity class, n_0 being the density of atoms in the vapor cell and $v_p = \sqrt{(2k_BT)/m}$ is the most probable velocity. The result of numerical integration of Equation 5.91 is shown in Figure 5.18 for typical parameters. The saturation dip appearing at the center of the Doppler(Voigt)-broadened profile has a FWHM approximately equal to Γ_{sat}. The simulation also shows that, in order to obtain an appreciable depth for such a dip, the laser intensity must be of the order of the saturation intensity.

To find an explicit expression for the saturation power

$$I_{sat} = \frac{h\nu\Gamma}{2\sigma_0}, \qquad (5.93)$$

one has to calculate the on-resonance cross section σ_0 in the frame of quantum mechanics. To this aim, we first observe that, for a polarized laser field (let us say along the x-direction), the on-resonance absorption cross section is proportional to $|x_{12}|^2$ where $x_{12} \equiv \hat{x} \cdot \boldsymbol{\mu}_{12}$ ($\boldsymbol{\mu}_{12} = \int \psi_2^* \boldsymbol{\mu}\psi_1 d\boldsymbol{r}$). Thus we can write

$$\sigma_0(x_{12}) = \alpha \cdot 3 \times \frac{|x_{12}|^2}{3} = \alpha \cdot 3 \times |\boldsymbol{\mu}_{12}|^2 = 3 \times \sigma_0(\boldsymbol{\mu}_{12}) \qquad (5.94)$$

where α is a proportionality constant and $|x_{12}|^2 = |\boldsymbol{\mu}_{12}|^2/3$ has been exploited (the factor $1/3$ arises from averaging of $|x_{12}|^2$ over all the possible spatial orientations of the atom). Next we can use Equation 4.72 (with $n = 1$) to express $\sigma_0(\boldsymbol{\mu}_{12})$ whereupon

$$\sigma_0(x_{12}) = 3 \times \frac{\lambda^2}{4\pi^2}\frac{A_{21}}{\Gamma} \qquad (5.95)$$

that, inserted into Equation 5.93, finally yields

$$I_{sat} = \frac{2\pi^2 hc\Gamma^2}{3\lambda^3\Gamma_{nat}} = \frac{\pi\varepsilon_0\hbar^2 c}{2} \frac{\Gamma^2}{|\boldsymbol{\mu}_{12}|^2} \tag{5.96}$$

where Equation 4.64 (with $g_2 = g_1$) has been used for $A_{21} \equiv \Gamma_{nat}$ in the last step. The above formula allows one to estimate the saturation intensity for a given transition, provided that the pertinent Γ and dipole moment are known. In this regard, the following clarification should be made: the actual value of Γ can be significantly larger than that of the natural linewidth. Most often, indeed, additional broadening mechanisms prevent observation of the natural linewidth for the detected Lamb dip. In other words, the value of Γ to be used in Equation 5.96 is the width of the Lorentzian profile resulting from the convolution of all homogeneous-broadening contributions Γ_i

$$\Gamma = \sum_i \Gamma_i = \Gamma_{nat} + \Gamma_{coll} + \Gamma_{transit} + \Gamma_{laser} + \dots \tag{5.97}$$

where the most common broadening mechanisms, namely the natural, the collisional, and the transit-time ones, have been highlighted. A possible non-negligible frequency jitter in the laser source has been considered too. Just to give an order of magnitude for saturation intensities in atoms, we consider the $5^2S_{1/2} - 5^2P_{3/2}$ transition (at 780.24 nm) in ^{87}Rb. In this case we have $\mu_{12} \sim 10^{-29}$ C·m and $\Gamma_{nat} \simeq 6$ MHz. So, with typical experimental parameters, $\Gamma \simeq \Gamma_{nat}$ and $I_{sat} \sim 1$ mW/cm^2 is found. This value is usually much lower than that for ro-vibrational molecular transitions, where weaker transition dipole moments occur. Moreover, in the case of molecules, the natural contribution Γ_{nat} in Equation 5.97 can be extremely low (few kHz or less), such that the saturation intensity is practically determined by the experimental parameters. Just as an example, for a reasonable width ($\Gamma = \Gamma_{coll} + \Gamma_{transit} + \Gamma_{laser} = 500$ kHz) and a relatively high transition dipole moment (on the order of 10^{-31} C·m), $I_{sat} \sim 100$ mW/cm^2 is found.

It is also useful to give here the explicit expressions for the various broadening contributions [422, 423]:

$$\Gamma_{nat} = \left(\frac{2\pi}{\lambda}\right)^3 \frac{|\boldsymbol{\mu}_{12}|^2}{3\pi\varepsilon_0\hbar} \tag{5.98}$$

$$\Gamma_{coll} = c_p p \tag{5.99}$$

$$\Gamma_{transit} = \sqrt{\frac{\ln 2}{\pi}} \frac{k_B T}{m} \frac{1}{w_0} \tag{5.100}$$

where c_p is the pressure-broadening coefficient (characteristic of the considered transition), p and T are the gas pressure and temperature, m the atomic/molecular mass, and w_0 is the radius of the Gaussian laser beam. Finally, recall that the width measured for the dip is $\Gamma_{sat} = \Gamma\sqrt{1+s}$.

In conclusion, three further observations are worthy of mention:

- Actually, since the simple theoretical requirements which led to Equation 5.91 can only be approximated in real experiments, the saturation contrast is much less pronounced and the dip linewidth is further broadened. Thus, the attainable experimental resolution is *degraded*. In particular, the following geometrical effects must be considered. Counter-propagating laser beams might be slightly misaligned, and the surface of constant phase of a beam is generally curved. Each of these deficiencies contributes a line broadening because a single molecule is effectively irradiated by two slightly different frequencies. Thus by simple geometry a misalignment of θ contributes a linewidth of

$$\sin\theta \cdot \Delta\nu_{Doppler} \tag{5.101}$$

Extending this geometry by integrating across the beam gives a curvature broadening of

$$\frac{\Delta\nu_{Doppler}}{\sqrt{2\pi R/\lambda}} \tag{5.102}$$

where R is the radius of curvature of the wave-front [424].

As an example of a real Doppler-free saturation spectrum, Figure 5.19 shows a Lamb-dip detection on the Doppler profile of the $R(4)$ lines around 3067 cm^{-1} performed with a DFG source according to a single-pass pump-and-probe scheme.

- When the atom has several sub-levels, either in the ground or excited state, other narrow, resonant (spurious) signals appear [166]. To fix ideas, let us consider the case with a single excited state, e, and two ground sub-levels, g and g', having an energy spacing less than the Doppler width. Now, if the pump beam is resonant on the transition $g \to e$, it will deeply alter the population of level e, which will affect the absorption experienced by the probe beam when it is tuned to either of the transitions $g \to e$ and $g' \to e$. As a result, a resonant change in the absorption will take place when the same velocity class interacts with both counter-propagating beams on either of the two transitions. This happens when the laser frequency ω is given by either $\omega = \omega_{g \to e}$, or $\omega = \omega_{g' \to e}$, or $\omega = (\omega_{g \to e} + \omega_{g' \to e})/2$. This latter is referred to as cross-over resonance. Thus, for atoms/molecules possessing a hyperfine structure, the saturated absorption signal generally exhibits a complicated form. The other side of the medal is that a dense grid of frequencies is available for laser frequency locking.

- The Doppler effect ultimately derives from time dilation, and is thus essentially relativistic, while the shift $\nu = \nu_0(1 + v/c)$, which is usually derived by simpler considerations, represents just the first-order correction for small velocities in collinear geometry. Higher-order corrections arise for all geometries, and even if the molecular velocity v is entirely perpendicular to the laser beam there will be a Doppler shift

$$\nu = \nu_0\sqrt{1 - \frac{v^2}{c^2}} \simeq \nu_0\left(1 - \frac{v^2}{2c^2}\right) \tag{5.103}$$

Clearly this cannot be removed by any counter-propagation method since it is independent of the sign of v. Nevertheless, it can be calculated and corrected for if the velocity distribution is known (see later on in this chapter). The additional linewidth contribution may be significant for lighter atoms but is generally negligible for heavier species (e.g., a few Hz for OsO_4).

5.6.1 Frequency locking to a Lamb dip

As already discussed in Chapter 4, saturated-absorption spectroscopy provides a narrow and stable reference for long-term frequency stabilization of a laser. A recent example of such an approach, shown in Figure 5.20, can be found in [425]. In that work, the frequency of a DFB quantum cascade laser (QCL) emitting at 4.3 μm was long-term stabilized to the Lamb-dip center of a CO_2 ro-vibrational transition by means of first-derivative locking to the saturated absorption signal. Also, thanks to the non-linear sum-frequency generation (SFG) process with a fiber-amplified Nd:YAG laser, the QCL mid-infrared radiation was linked to an optical frequency-comb synthesizer (OFCS) and its absolute frequency counted with a kHz-level precision and an overall uncertainty of 75 kHz. The saturation signal was

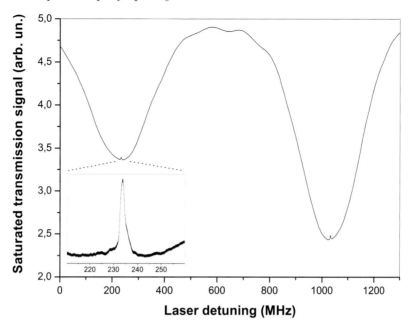

FIGURE 5.19

Lamb-dip detection on the Doppler profile of the R(4) lines at 3067.2344 cm^{-1} and 3067.2610 cm^{-1}, respectively. The cell was filled with pure CH$_4$ at 30 mTorr pressure. In the inset, the Lamb dip is shown on a 16-times magnified scale. The saturation contrast is nearly 4% while the FWHM is 2.5 ± 0.5 MHz. The spectra were averaged over several acquisitions, thus reducing the detection bandwidth down to about 1 kHz.

first observed in direct absorption, by continuously scanning the laser frequency with a slow (few Hz) saw-tooth modulation of the laser current (frame (a) of Figure 5.21).

A well contrasted Lamb dip (13% contrast factor) was obtained, exhibiting a HWHM of about 5 MHz. Wavelength modulation spectroscopy was then implemented at 40 kHz rate, with a modulation depth of about 3.5 MHz: this value was chosen in order to have the largest first-derivative signal without broadening the dip. The AC channel of the detector pre-amplifier was sent to an analog lock-in amplifier for demodulation. Frame (b) of Figure 5.21 shows a first-derivative saturated-absorption spectrum recorded in the same experimental conditions as in frame (a). The best fit of the experimental data was obtained with a convolution between two Gaussian functions (accounting for both the Doppler profile and the Lamb dip). The Lamb-dip HWHM resulting from the fit was about 5.3 MHz. Therefore, since the expected contributions of pressure and transit-time broadening to the Lamb-dip width (HWHM) were about 70 kHz and 100 kHz, respectively, the measured Lamb-dip width could be attributed to the free-running QCL frequency jitter. Then, the first-derivative saturation dip was used as the error signal to lock the QCL frequency to the transition center. To this aim, the lock-in output was processed by a PID controller, and the feedback signal sent to the modulation input of the current driver. Figure 5.22 shows a comparison between the frequency-noise power spectral density of the QCL in both locked and unlocked modes. A noise reduction exceeding 20 dB at frequencies lower than 30 Hz is obtained; the cutoff introduced by the lock-in time constant, set to its minimum available value of 1 ms, is also evident. The reduction of the slow frequency fluctuations is qualitatively presented in frame (b), where the lock-in output, in locked and unlocked conditions, is shown. Because of the small bandwidth, the loop is unable to reduce the laser high-frequency jitter.

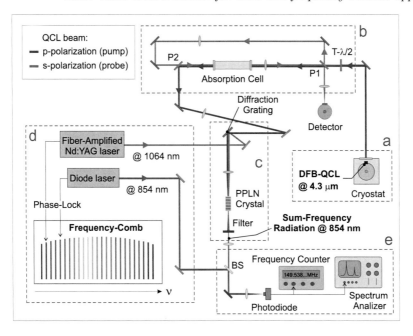

FIGURE 5.20 (SEE COLOR INSERT)
Schematic of the experimental setup for frequency locking to a Lamb dip. The main blocks of the apparatus are highlighted by dashed lines: the QCL housing and collimation (a), the saturation spectroscopy setup (b), the SFG assembly (c), the near-IR lasers phase-locked to the comb (d), and the beat-note detection and measurement (e). P1 and P2 are wire grid polarizers, $T - \lambda/2$ is a tunable half-waveplate and BS is a beam splitter. Despite the lack of mid-IR optical isolators, use of the crossed polarizers (P1 and P2) allows to have full control on the counter-propagating beams overlap and focusing, while avoiding undesired optical feedback to the laser.

Frequency stabilization techniques based on a combination of modulation and synchronous detection, like that just described, usually have better performance compared to schemes making use of direct detection. However, in certain applications, a modulated laser output may represent an obstacle. In that case, a side-locking (SL) technique can be adopted. We already mentioned in Chapter 4, that a modification of the traditional SL scheme, exhibiting first-order insensitivity to laser intensity fluctuations, has also been reported [180]. Now we have all the ingredients to understand such a method. Conventionally, the SL signal is generated by difference between the Doppler-broadened linear absorption and the saturated absorption signal, such that only the Doppler-free features survive on a flat background (see Figure 5.23). Then, a dc offset is added to the output of the differential amplifier in order to generate a zero-crossing signal at the input of the integrator, the value of such offset determining the frequency to which the laser is stabilized. With this approach, however, the fluctuations in laser power lead to changes in the frequency of the lock point. To overcome this drawback, one can exploit the different power dependencies of the Doppler-free saturated absorption spectrum and the Doppler-broadened background spectrum. Indeed, a wise combination of these components can make the intensity dependences in the error signal (ES) at the input of the integrator cancel, thus providing a first-order intensity insensitivity. It is worth noting that, working only for correlated intensity fluctuations in all three beams (pump, saturated probe, and unsaturated probe), such a technique is

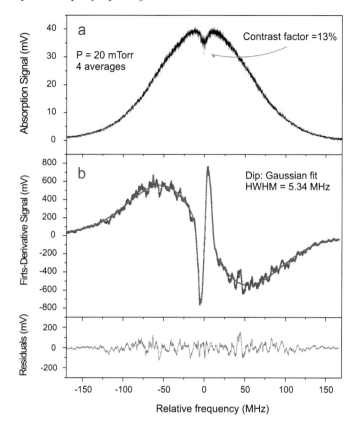

FIGURE 5.21
Saturation spectrum of the (01^11-01^10) P(30) CO_2 transition in direct-absorption (a) and first-derivative detection (b). The gas pressure in the cell was 20 mTorr (bringing to a relative absorption of about 10%) and the pump intensity interacting with the gas sample was about a factor 2 greater than the saturation level. The two traces have been recorded by a digital scope, with a sweeping time of about 0.2 seconds. The first-derivative signal was obtained by a lock-in amplifier with 1 ms integration time constant. The best fit and residuals are also shown.

ineffective against intensity changes arising from pointing fluctuations (caused, for instance, by fluctuations in the air refractive index or mechanical vibrations).

In principle, one could also apply this scheme to a Fabry-Perot resonator, in order to realize, by the simultaneous measurement of the incident and reflected (or transmitted) power, a power-independent lock to the side of a resonance.

5.6.2 Cavity-enhanced Doppler-free saturation spectroscopy

Fabry-Perot cavities represent a powerful tool for saturation of weakly absorbing samples. Indeed, if a cw laser is locked to a resonance frequency of a symmetrical cavity, according to Equation 3.39, the transmitted power is given by

$$P_t = M \frac{T^2}{(1-R)^2} P_0 \tag{5.104}$$

where P_0 is the incident laser power, T and R represent the power transmission and re-

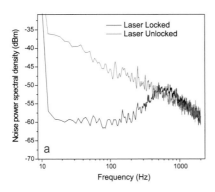
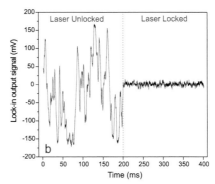

FIGURE 5.22

(a) QCL noise power spectral density in free-running and locking conditions, obtained by using the slope of a germanium-etalon transmission fringe as a frequency discriminator. (b) The lock-in output signal qualitatively shows the frequency fluctuations that can be compensated by the loop.

flection coefficients of the cavity mirrors, and M is the mode-match parameter that quantifies the coupling of the input power into the resonator. In the presence of a weakly absorbing gaseous medium, the transmitted power is calculated by replacing R with $R\exp[-\kappa(\nu)pL]$ (see Equation 3.53) in the above equation. Here $\kappa(\nu)$ is the absorption coefficient (i.e., the absorption cross-section expressed in $\mathrm{cm}^{-1}\mathrm{torr}^{-1}$), p the sample gas pressure (in torr), and L the cavity length. In case of weak absorption $(\exp[-\kappa(\nu)pL] \simeq 1-\kappa(\nu)pL)$, this procedure yields

$$P_t(\nu,p) \simeq \frac{P_t}{\left[1 + \dfrac{\kappa(\nu)pL_{eq}}{2}\right]^2} \tag{5.105}$$

where $L_{eq} = 2\mathcal{F}L/\pi$ is the effective interaction length, $\mathcal{F} \simeq \pi R(1-R)$ being the empty

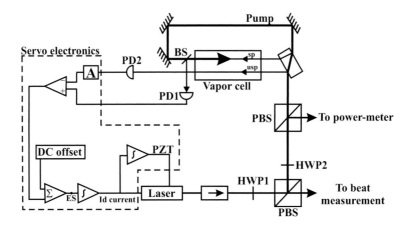

FIGURE 5.23

Experimental setup for power-insensitive side locking: PBS, polarizing beam splitter; BS, 50:50 beam splitter; HWP1, HWP2, half-wave plates; PD1, PD2, photodiodes; usp, saturated probe; usp, unsaturated probe; ECDL, extended-cavity diode laser. (Adapted from [180]).)

cavity finesse (see Equation 3.43). Therefore, the fractional change in the transmitted power, caused by the absorbing sample, is derived as

$$\frac{\Delta P_t(\nu_0)}{P_t} \equiv \frac{P_t(p=0) - P_t(\nu = \nu_0, p)}{P_t(p=0)} = 1 - \frac{1}{(1+x)^2} = \frac{x(2+x)}{(1+x)^2} \tag{5.106}$$

where $x = \kappa_0 p L_{eq}/2$ with $\kappa_0 \equiv \kappa(\nu_0)$, ν_0 being the line center frequency in cm^{-1}. Now, if saturation of absorption occurs, a narrow Lamb dip appears at the center of the Doppler-broadened profile. The change in the absorption cross section $\delta\kappa_0$ due to saturation leads to a variation of the cavity transmission which is given by

$$\frac{\delta P_t(\nu_0)}{P_t} = \frac{1}{P_t}\frac{dP_t(\nu_0, p)}{d\kappa_0}\delta\kappa_0 = \frac{pL_{eq}}{(1+x)^3}\delta\kappa_0 = \frac{P_c}{P_{sat}}\frac{x}{(1+x)^3} \tag{5.107}$$

where $\delta\kappa_0/\kappa_0 \simeq P_c/(2P_{sat})$ has been used in the last step and the saturation power is given by $P_{sat} = \pi w_0^2 I_{sat}$. Finally, the contrast G of the Doppler-free line with respect to the Doppler envelope can be calculated as

$$G \equiv \frac{\delta P_t(\nu_0)}{\Delta P_t(\nu_0)} = \frac{P_c}{P_{sat}}\frac{1}{(2 + \kappa_0 p L_{eq}/2)(1 + \kappa_0 p L_{eq}/2)} \tag{5.108}$$

It is worth noting that the amplitude of the saturation dip exhibits a different behavior as a function of the gas pressure, with respect to that of the line contrast. G decreases with increasing pressure. On the contrary, at very low pressures, when $\kappa_0 p L_{eq} \ll 1$ and the mean-free path of molecules is much greater than the beam radius, δP_t increases linearly with the pressure. Hence, out of the free-flight regime, when P_{sat} strongly depends on the pressure, δP_t starts to decrease [426]. A typical experimental setup for cavity-enhanced saturation spectroscopy is shown in Figure 5.24 [427]. In that work, 5 μW of DFG radiation at 4.25 μm were coupled to a confocal FP cavity (finesse $\simeq 550$) to record Lamb dips of weak transitions in the fundamental ro-vibrational band of CO_2 up to the $J = 82$ level. Frequency locking of the IR radiation to the FP cavity was achieved with a PDH scheme. For this purpose, phase modulation of the generated IR beam was obtained by transferring, through the non-linear process, the phase modulation imposed on the DFG signal beam at 1064 nm (provided by a Nd:YAG laser) by an electro-optic modulator (driven at only 3.6 MHz to match the limited 4-MHz bandwidth of the liquid-nitrogen-cooled InSb detector). The error signal from the feedback electronics was sent to the current driver of the master diode laser (MDL), which optically injected the pump slave diode laser. For detection of the transmission signal, the cavity length (and hence the frequency) was scanned by sending a voltage ramp to the cavity's three piezoelectric transducers to perform both direct-absorption and first-derivative recording. In the latter case, the cavity was dithered with a low-frequency (< 1 kHz) sinusoidal modulation added to the ramp. The signal was then demodulated by a lock-in amplifier.

5.6.2.1 Doppler-free NICE-OHMS

As already mentioned, Lamb dips can be observed with NICE-OHMS due to the presence of high-intensity counter-propagating waves inside the cavity [428, 399]. Since three modes propagate in each direction in the cavity, sub-Doppler dispersion signals occur at detunings ($\Delta\omega$) of 0 (mainly when the two carriers interact with molecules with zero axial velocity), $\pm\omega_m/2$ (when one carrier and one sideband address a common velocity group of molecules), and $\pm\omega_m$ (when the two lower or the two upper sidebands interact with the same group of molecules). As the FM detection scheme is insensitive to the carrier absorption, the center peak, which has the largest amplitude in the dispersion signal, is instead missing in the absorption recording. For an extensive investigation of the shape and size of sub-Doppler

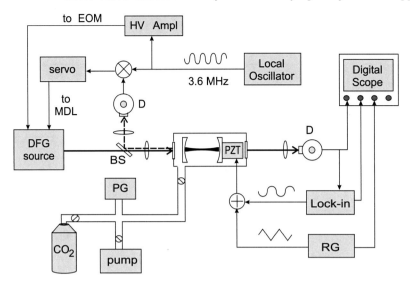

FIGURE 5.24
Optical setup for cavity-enhanced saturation spectroscopy: EOM, electro-optic modulator; D, InSb detector; BS, beam splitter; PG, pressure gauge; PZT, piezoelectric transducer; RG, ramp generator; HV, high voltage; MDL, master diode laser.

NICE-OHMS signals, the reader is referred to [429]. Here we just mention that in the first demonstration, the NICE-OHMS technique was exactly applied to obtain the saturated signal of a C_2HD overtone transition at 1064 nm [430]. In that case, the obtained sensitivity was $1.2 \cdot 10^{-10}$ at 1 s averaging. Moreover, selection of slow molecules gave a linewidth four times better narrower than the room-temperature transit-time limit. More recently, fiber-laser-based sub-Doppler NICE-OHMS was carried out on C_2H_2 at 1531 nm up to saturation degrees of 100 [428]. The apparatus is shown in Figure 5.25. While the two rf modulation are usually applied through separate, consecutive EOMs, here a fiber EOM is used to create both the 20-MHz sidebands for the PDH lock and the 379.9-MHz sidebands for the NICE-OHMS detection. Locking of the FM modulation frequency to the cavity FSR is accomplished by the deVoe-Brewer technique, while scans over the sub-Doppler signals are performed by applying a low-frequency (40-100 mHz) ramp to the cavity input PZT. Detected in cavity transmission, the signal is then demodulated at 379.9 MHz with a double-balanced mixer. A phase shifter in the reference arm is used to set the FM detection phase to dispersion phase by maximizing the sub-Doppler signal. In addition, in order to suppress low-frequency noise, a WM dither at 125 Hz is directed to the cavity output PZT and the signal is further demodulated with a lock-in amplifier. Finally, after proper adjustment of the gain and phase, both the scan and the dither are fed forward to the laser PZTs in order to remove some load from the locking servo. Peculiar examples of strongly saturated NICE-OHMS signals are displayed in Figure 5.26.

5.7 Doppler-free polarization spectroscopy

Polarization spectroscopy (PS) was first reported in 1976 by Wieman and Hänsch as a useful Doppler-free method offering a considerably better signal-to-noise (S/N) ratio in

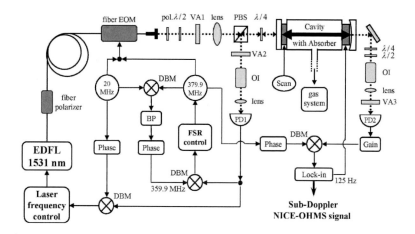

FIGURE 5.25
Experimental setup for sub-Doppler NICE-OHMS. The following legend holds: EDFL=Erbium-doped fiber laser, EOM=electro-optic modulator, pol.=free space polarizer, $\lambda/2$=half-wave plate, VA=variable attenuator, PBS=polarizing beam splitter cube, $\lambda/2$=quarter-wave plate, OI=optical isolator, PD=photodetector, DBM=double balanced mixer, Phase=phase shifter, Gain=separate gain stage, BP=bandpass filter, nodes ($\bullet$)=power splitters/combiners. The dotted lines correspond to the free-space laser beam path. (Courtesy of [399].)

comparison with standard saturation spectroscopy [431]. In a typical PS setup, a strong pump beam and a weak probe beam with different polarizations and counter-propagating through the target sample are tuned to the desired optical transition (Figure 5.27). The optical pumping induced by the polarized pump beam induces a birefringence in the medium and a consequent detectable polarization change in the weak probe beam. In fact, PS can be regarded as a kind of saturation spectroscopy, where the change in the complex refractive index is proportional to the pump intensity.

We now derive expressions for the amplitude and shapes of the polarization signals obtained with PS [432, 433, 434]. First of all, the z axis is fixed by the propagation direction

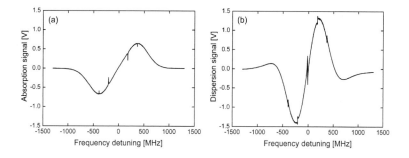

FIGURE 5.26
Typical sub-Doppler FM-NICE-OHMS absorption (a) and dispersion (b) features over the Doppler-broadened envelopes. The experiment used 500 ppm of C_2H_2 at 20 mTorr intracavity pressure and an intracavity power of 4.6 W, corresponding to a saturation degree of 50 for the carrier and 1.5 for the sidebands. (Courtesy of [399].)

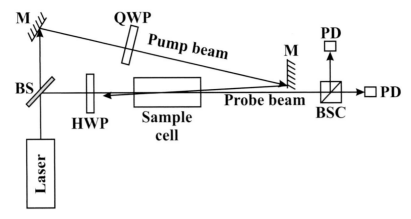

FIGURE 5.27
Experimental setup for Doppler-free polarization spectroscopy. The beam splitter (BS) picks off a fraction of light coming from the laser to form the probe beam. Then, the nearly counter-propagating pump and probe beams overlap in the vapor cell. A half-wave plate (HWP) is used to rotate the plane of polarization of the probe with respect to the axis of the beam splitting cube (BSC); a quarter-wave plate (QWP) is used to make the pump circularly polarized. M=mirror, PD=photodetector. (Adapted from [432].)

of the probe beam. Then, we can write the initial state, which is linearly polarized in a plane at an angle ψ with respect to the x axis, in terms of the circular polarization vector basis (σ^+ and σ^- components)

$$\boldsymbol{E} = \begin{pmatrix} E_x \\ E_y \end{pmatrix} = E_0 \begin{pmatrix} \cos\psi \\ \sin\psi \end{pmatrix} = E_0 \left\{ \frac{e^{-i\psi}}{2} \begin{pmatrix} 1 \\ i \end{pmatrix} + \frac{e^{+i\psi}}{2} \begin{pmatrix} 1 \\ -i \end{pmatrix} \right\} \tag{5.109}$$

In propagating through the cell of length L, these two components undergo differential absorption and dispersion due to the gas and cell windows, so that the probe-beam electric field after the cell is given by

$$\boldsymbol{E}(L) = E_0 \left\{ \frac{e^{-i\psi}}{2} \begin{pmatrix} 1 \\ i \end{pmatrix} e^{-ik_+ L} e^{-\alpha_+ L/2} e^{-ik_{w+}l} \right.$$
$$\left. + \frac{e^{+i\psi}}{2} \begin{pmatrix} 1 \\ -i \end{pmatrix} e^{-ik_- L} e^{-\alpha_- L/2} e^{-ik_{w-}l} \right\} \tag{5.110}$$

with

$$k_{\mp} = \frac{\omega}{c} n_{\mp} \tag{5.111}$$

$$k_{w\mp} = \frac{\omega}{c} n_{w\mp} \tag{5.112}$$

where α^+ and n^+ (α^- and n^-) are the absorption coefficient and the refractive index experienced by the σ^+ (σ^-) components. The refractive indices of the window (thickness l) are also in general complex $n_{w\mp} = (1/l)\left[b_{R\mp} - i(c/\omega)b_{I\mp}\right]$. Then, if we define

$$\Omega_1 = \frac{\omega}{2c}(\Delta n \cdot L + \Delta b_R) \tag{5.113}$$

$$\Omega_2 = \frac{L}{4}\Delta\alpha + \frac{1}{2}\Delta b_I \tag{5.114}$$

$$n = \frac{1}{2}(n_+ + n_-), \quad \alpha = \frac{1}{2}(\alpha_+ + \alpha_-), \quad b_R = \frac{1}{2}(b_{R+} + b_{R-}),$$

$$b_I = \frac{1}{2}(b_{I+} + b_{I-}), \quad \Delta n = n_+ - n_-, \quad \Delta \alpha = \alpha_+ - \alpha_-,$$

$$\Delta b_R = b_{R+} - b_{R-}, \quad \Delta b_I = b_{I+} - b_{I-} \tag{5.115}$$

we get

$$\boldsymbol{E}(L) = \frac{E_0}{2} e^{-i\frac{\omega}{c}(nL + b_R)} e^{-\left(\frac{\alpha L}{2} + b_I\right)} \begin{pmatrix} e^{-i\psi} e^{-i\Omega_1} e^{-\Omega_2} + e^{i\psi} e^{i\Omega_1} e^{\Omega_2} \\ i e^{-i\psi} e^{-i\Omega_1} e^{-\Omega_2} - i e^{i\psi} e^{i\Omega_1} e^{\Omega_2} \end{pmatrix} \tag{5.116}$$

In conventional PS, the probe beam is horizontally polarized (the angle ψ is equal to zero) while the analyzer is a linear polarizer slightly misaligned from the vertical such that a change in signal is detected when the plane of polarization of the probe is rotated (the pump beam is always circularly polarized). If ϑ is the misalignment angle of the polarizer relative to the vertical axis (y axis), the transmitted electric field is

$$E_t = E_x \sin \vartheta + E_y \cos \vartheta \tag{5.117}$$

whereupon, by use of Equation 5.116, we get

$$I_t = I_0 e^{-(\alpha L + 2b_I)} |\sin \vartheta \cos \Phi + \cos \vartheta \sin \Phi|^2 \tag{5.118}$$

where $\Phi \equiv \Omega_1 - i\Omega_2$. When plotting this expression for different ϑ values, two important drawbacks can be identified: first of all, the need of keeping ϑ very small, which implies a very weak signal; secondly, the unavoidable presence of an offset, comparable or even larger than the amplitude of the PS signal itself, and correlated to the laser intensity fluctuations.

To overcome these limitations, the following double-balanced variant can be used. The probe beam is decomposed into horizontal and vertical components by the beam splitting cube and the difference in their intensity gives the PS signal

$$
\begin{aligned}
I_{signal} &= I_x - I_y = E_0^2 e^{-(\alpha L + 2b_I)} \cos(2\psi + 2\Omega_1) \\
&= E_0^2 e^{-(\alpha L + 2b_I)} (\cos 2\psi \cos 2\Omega_1 - \sin 2\psi \sin 2\Omega_1)
\end{aligned} \tag{5.119}
$$

which, in the hypothesis $\Omega_1 \ll 1, \psi = \pi/4$ (which maximizes the dispersive component), simplifies to

$$I_{signal} = -I_0 e^{-(\alpha L + 2b_I)} \left(\Delta n \cdot L \cdot \frac{\omega}{c} + \Delta b_R \cdot \frac{\omega}{c} \right) \tag{5.120}$$

with $I_0 = E_0^2$. Now we assume that the laser frequency is scanned across a single resonance and that the spectral profile of the difference in absorption is a (power broadened) Lorentzian

$$\Delta \alpha = \frac{\Delta \alpha_0}{1 + \xi^2} \tag{5.121}$$

where $\Delta \alpha_0$ is the maximum difference in absorption at the line center and $\xi = (\omega_0 - \omega)/\Gamma_s$ is the scaled detuning in units of half the saturated linewidth Γ_s. According to the Kramers-Kronig dispersion relation we also have

$$\Delta n = \frac{2(\omega_0 - \omega)}{\Gamma_s} \frac{c}{\omega} \frac{\alpha}{2} = \frac{c}{\omega} \Delta \alpha_0 \frac{\xi}{1 + \xi^2} \tag{5.122}$$

whereupon we finally obtain

$$I_{signal} = -I_0 e^{-(\alpha L + 2b_I)} \left(\Delta \alpha_0 \frac{\xi}{1 + \xi^2} \cdot L + \Delta b_R \cdot \frac{\omega}{c} \right) \tag{5.123}$$

which, for optically isotropic windows ($\Delta b_R = 0$), reduces to

$$I_{signal} = -I_0 e^{-(\alpha L + 2b_I)} \left(\Delta\alpha_0 \frac{\xi}{1+\xi^2} L \right) \tag{5.124}$$

i.e., we obtain a signal that is dispersion shaped, being the derivative of the sub-Doppler linewidth. This remains true for a wide range of angles, centered around $\psi = \pi/4$ at which the maximum amplitude is obtained. Finally we note that the finite extinction ratio of the polarisers ($\eta \sim 10^{-6}$ to 10^{-8}) can be accounted for by writing

$$I_d = I_0 e^{-(\alpha L + 2b_I)} \left[\eta + \Delta\alpha_0 L \frac{\xi}{1+\xi^2} \right] \tag{5.125}$$

In a more recent realization, a double-balanced PS experiment has been carried out with a quantum cascade laser at 4.3 μm on a hot-band CO_2 transition [433]. The experimental apparatus is shown in Figure 5.28, while typical sub-Doppler PS signals are shown in Figure 5.29. The balanced technique, in particular, allows to obtain sharp spectra with much higher signal-to-noise ratios with respect to standard saturation spectroscopy. Because of the absence of any modulation onto the laser, such signals are particularly suited for an efficient frequency stabilization of the laser to the molecular line.

5.8   Doppler-free two-photon spectroscopy

This technique was first proposed by Chebotaev and co-workers [435, 436, 166]. Consider a two-quantum atomic/molecular transition in the field formed by two counter-propagating

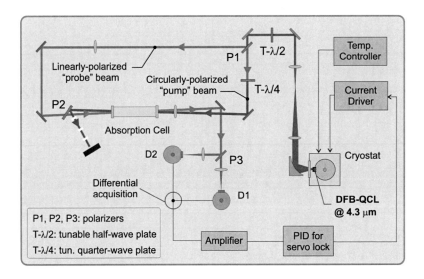

FIGURE 5.28
Experimental setup for double-balanced PS with a QCL laser source. The wire-grid polarizers P1 and P3 are used as polarizing beam splitters. The rotation angle ϑ of the analyzer polarizer P3 is adjusted with respect to P2 in order to obtain the desired signal: $\vartheta \simeq 0°$ (P2 and P3 almost crossed) corresponds to the standard PS sugnal, while $\vartheta \simeq 45°$ is used for balanced detection. D1 and D2 are identical InSb detectors.

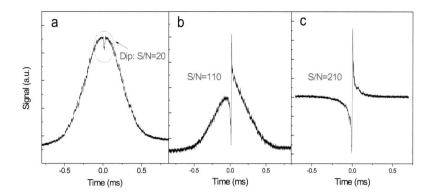

FIGURE 5.29
Lamb-dip detection on the $(01^11-01^10)P(30)$ ro-vibrational transition of CO_2 at 2310.5062 cm^{-1}. The S/N ratio on the sub-Doppler feature increases when upgrading from (a) the single-pass standard saturation spectroscopy to (b) the standard polarization spectroscopy, up to (c) the double-balanced technique. The three traces were acquired in identical experimental conditions.

(along the x-direction) laser beams of the same frequency ω (Figure 5.30). Then, for a particle moving with velocity $\boldsymbol{v}$, the frequencies of the two travelling waves in the atom's rest frame are, respectively, given by $\omega - \boldsymbol{k}_1 \cdot \boldsymbol{v} = \omega(1 - v_x/c)$ and $\omega - \boldsymbol{k}_2 \cdot \boldsymbol{v} = \omega(1 + v_x/c)$. As a result, if the two-photon transition is driven by a double excitation in a single travelling wave, the resonance condition depends on the atomic velocity as

$$\omega_f - \omega_i = 2\omega \left(1 - \frac{v_x}{c}\right) \tag{5.126}$$

In this case, for a given value of ω, a single velocity class is excited, which corresponds to a resonance with a Doppler-broadened lineshape. Conversely, if the atoms absorb one photon from each of the counter-running waves, the condition of two-photon resonance is given by

$$\omega_f - \omega_i = \omega \left(1 - \frac{v_x}{c}\right) + \omega \left(1 + \frac{v_x}{c}\right) = 2\omega \tag{5.127}$$

In this type of resonance, all particles, regardless of velocity, take part in the two-photon absorption, which originates a sharp increase in the absorption signal. Qualitatively, the latter is thus the sum of a wide Doppler profile, corresponding to two-quantum absorption from a unidirectional wave, and a narrow Lorentzian resonance, arising from two-quantum absorption by all particles for which $\omega_f - \omega_i = 2\omega$. As we will derive below, the amplitude contrast of such a Lorentzian peak roughly equals the ratio of the Doppler width to the homogeneous width.

The two main advantages of two-photon spectroscopy with respect to saturation spectroscopy can be summarized as follows:

- All the atoms contribute to the absorption, irrespective of their velocity, whereas in saturation spectroscopy only a small fraction of the atoms produces the narrow resonance.

- The width of the resonance peak is unattached by the wavefront curvature as the two photons are absorbed simultaneously at the same point in space (recall that, in order to obtain narrow resonances in saturation spectroscopy, the wave vector should have rigorously the same direction along the whole cross section of the standing wave). As a

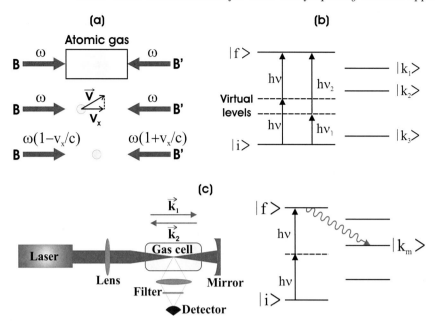

FIGURE 5.30
Doppler-free two-photon spectroscopy scheme. (a) An atomic vapor interacts with two counter-propagating laser beams of the same frequency. A picture is given in both the lab frame and in the atom's rest frame. (b) Two-photon transition with two equal or different photons. (c) Experimental arrangement for Doppler-free two-photon absorption with fluorescence detection. (Adapted from [166] and [379].)

result, light beams with huge cross sections can be employed, thus dramatically reducing the transit-time broadening. Indeed, the two-photon transition probability is independent from the orientation of atomic velocities about the standing light wave. Therefore, transit broadening can effectively be diminished by guaranteeing that the particles move along the standing wave rather than across it.

The quantitative treatment of two-photon absorption in an atomic (molecular) vapor relies on a quantum mechanical approach. According to second-order perturbation theory, for a stationary atom subjected simultaneously to two laser fields, the transition rate (probability per unit time) for two-photon absorption is found to be [322]

$$R_{i \to f}^{(2)} = \left| \sum_k \left(\frac{\mu_{fk}\mu_{ki}/2\hbar^2}{\omega_{ki} - \omega_1} + \frac{\mu_{fk}\mu_{ki}/2\hbar^2}{\omega_{ki} - \omega_2} \right) \right|^2 \frac{E_1^2 E_2^2 \Gamma}{[\omega_{fi} - (\omega_1 + \omega_2)]^2 + \gamma^2/4} \qquad (5.128)$$

where E_1 (ω_1) is the electric field amplitude (angular frequency) associated with laser beam 1, μ_{fk} (μ_{ki}) is the dipole matrix element (having the units of C·m) between the states f and k (k and i), $\omega_{ki} = \omega_k - \omega_i$, $\omega_{fi} = \omega_f - \omega_i$, and $\Gamma = \Gamma_i + \Gamma_f$ is the sum of the widths of the initial and final levels. The above sum extends over all real atomic levels k that are connected by allowed one-photon transitions with the initial level i. Assuming that a single level dominates the sum, Equation 5.128 reduces to

$$R_{i \to f}^{(2)} = \left| \frac{\mu_{fk}\mu_{ki}/2\hbar^2}{\omega_{ki} - \omega_1} + \frac{\mu_{fk}\mu_{ki}/2\hbar^2}{\omega_{ki} - \omega_2} \right|^2 \frac{I_1 I_2 \gamma}{[\omega_{fi} - (\omega_1 + \omega_2)]^2 + \Gamma^2/4} \qquad (5.129)$$

where $I_1 = E_1^2$ and $I_2 = E_2^2$. Next, for an atom moving with velocity $\boldsymbol{v}$, Equation 5.129 is readily generalized to

$$R_{i \to f}^{(2)} = \left| \frac{1}{\omega_{ki} - \omega_1 - \boldsymbol{k_1} \cdot \boldsymbol{v}} + \frac{1}{\omega_{ki} - \omega_2 - \boldsymbol{k_2} \cdot \boldsymbol{v}} \right|^2$$
$$\cdot \frac{\mathcal{S} I_1 I_2 \gamma}{[\omega_{fi} - (\omega_1 + \omega_2) - \boldsymbol{v} \cdot (\boldsymbol{k_1} + \boldsymbol{k_2})]^2 + \Gamma^2/4} \tag{5.130}$$

where $\boldsymbol{k_1} = \omega_1/c\hat{k}_1$ and $\mathcal{S} = |\mu_{fk}\mu_{ki}/2\hbar^2|^2$. Now, let us consider two counter-propagating beams generated from the same laser beam ($\omega_1 = \omega_2 = \omega$ and $I_1 = I_2 = I$) and let us distinguish between the three possible transition rates:

1. The first photon (i.e., the photon connecting level i to the virtual level) is absorbed from laser beam 1 and the second photon (i.e., the photon connecting the virtual level to level f) from laser beam 2. This event is non-distinguishable from that in which the first photon is absorbed from laser beam 2 and the second photon from laser beam 1. Thus, from Equation 5.130 with $\boldsymbol{k_1} = -\boldsymbol{k_2} = \boldsymbol{k}$ one has

$$\mathcal{R}_1 = \left| \frac{1}{\omega_{ki} - \omega - \boldsymbol{k} \cdot \boldsymbol{v}} + \frac{1}{\omega_{ki} - \omega + \boldsymbol{k} \cdot \boldsymbol{v}} \right.$$
$$\left. + \frac{1}{\omega_{ki} - \omega + \boldsymbol{k} \cdot \boldsymbol{v}} + \frac{1}{\omega_{ki} - \omega - \boldsymbol{k} \cdot \boldsymbol{v}} \right|^2 \frac{\mathcal{S} I^2 \Gamma}{(\omega_{fi} - 2\omega)^2 + \Gamma^2/4}$$
$$= \left| \frac{1}{\omega_{ki} - \omega - \boldsymbol{k} \cdot \boldsymbol{v}} + \frac{1}{\omega_{ki} - \omega + \boldsymbol{k} \cdot \boldsymbol{v}} \right|^2 \frac{4 \mathcal{S} I^2 \Gamma}{(\omega_{fi} - 2\omega)^2 + \Gamma^2/4} \tag{5.131}$$

2. Both the photons are absorbed from the laser beam 1. Now, Equation 5.130 with $\boldsymbol{k_1} = \boldsymbol{k_2} = \boldsymbol{k}$ yields

$$\mathcal{R}_2 = \left| \frac{1}{\omega_{ki} - \omega - \boldsymbol{k} \cdot \boldsymbol{v}} \right|^2 \frac{4 \mathcal{S} I^2 \Gamma}{(\omega_{fi} - 2\omega - 2\boldsymbol{k} \cdot \boldsymbol{v})^2 + \Gamma^2/4} \tag{5.132}$$

3. Both the photons are absorbed from the laser beam 2. Now, Equation 5.130 with $\boldsymbol{k_1} = \boldsymbol{k_2} = -\boldsymbol{k}$ yields

$$\mathcal{R}_3 = \left| \frac{1}{\omega_{ki} - \omega + \boldsymbol{k} \cdot \boldsymbol{v}} \right|^2 \frac{4 \mathcal{S} I^2 \Gamma}{(\omega_{fi} - 2\omega + 2\boldsymbol{k} \cdot \boldsymbol{v})^2 + \Gamma^2/4} \tag{5.133}$$

For non-resonant transitions ($|\omega_{ki} - \omega| \gg \boldsymbol{k} \cdot \boldsymbol{v}$), all the above denominators ($\omega_{ki} - \omega + \boldsymbol{k} \cdot \boldsymbol{v}$) can be approximated as ($\omega_{ki} - \omega$), whereupon

$$\mathcal{R}_{tot} = \mathcal{R}_1 + \mathcal{R}_2 + \mathcal{R}_3 = \mathcal{B}(\omega) \cdot \left\{ \frac{4}{(\omega_{fi} - 2\omega)^2 + \Gamma^2/4} \right.$$
$$\left. + \frac{1}{(\omega_{fi} - 2\omega - 2\boldsymbol{k} \cdot \boldsymbol{v})^2 + \Gamma^2/4} + \frac{1}{(\omega_{fi} - 2\omega + 2\boldsymbol{k} \cdot \boldsymbol{v})^2 + \Gamma^2/4} \right\} \tag{5.134}$$

with

$$\mathcal{B}(\omega) \equiv \frac{4 \mathcal{S} I^2 \Gamma}{|\omega_{ki} - \omega|^2} = \frac{4 \mathcal{S} I^2 \Gamma}{|\omega_{fi}/2 + \delta - \omega|^2} \tag{5.135}$$

where the frequency difference between the virtual level and the intermediate level k has been introduced. Then, when the frequency 2ω is scanned around ω_{fi} by a small amount (a few Γ), for typical values of the detuning δ, one has $\delta \gg (\omega_{fi}/2 - \omega)$ such that $\mathcal{B}(\omega)$ is, in effect, constant $\mathcal{B}(\omega) = 4\mathcal{S}I^2\Gamma/|\delta|^2 \equiv \mathcal{B}$. Therefore, Equation 5.134 can be re-written as

$$\mathcal{R}'_{tot} \equiv \frac{\mathcal{R}_{tot}}{\mathcal{B}} = \frac{4}{(\omega_{fi} - 2\omega)^2 + \Gamma^2/4}$$

$$+ \frac{1}{(\omega_{fi} - 2\omega - 2\boldsymbol{k} \cdot \boldsymbol{v})^2 + \Gamma^2/4} + \frac{1}{(\omega_{fi} - 2\omega + 2\boldsymbol{k} \cdot \boldsymbol{v})^2 + \Gamma^2/4}$$

$$= \frac{4}{(\omega_{fi} - 2\omega)^2 + \Gamma^2/4}$$

$$+ \frac{1}{(\omega_{fi} - 2\omega - \frac{2\omega}{c}v_z)^2 + \Gamma^2/4} + \frac{1}{(\omega_{fi} - 2\omega + \frac{2\omega}{c}v_z)^2 + \Gamma^2/4} \tag{5.136}$$

where the z direction has been assumed coincident with that of propagation of the laser beam. With the introduction of the variable $t = \omega_{fi} - 2(\omega/c)v_z \equiv \omega_{fi} - 2kv_z$, integration over the velocity distribution

$$\frac{e^{-(v_z/v_p)^2}}{\sqrt{\pi}v_p} dv_z \tag{5.137}$$

yields

$$\frac{1}{\sqrt{\pi}} \int_{-\infty}^{+\infty} \mathcal{R}'_{tot} e^{-y^2} dy = \frac{4}{(\omega_{fi} - 2\omega)^2 + \Gamma^2/4}$$

$$+ \frac{2\sqrt{\pi}}{kv_p\gamma} \int_{-\infty}^{+\infty} \frac{1}{\pi} \frac{\Gamma/2}{(t - 2\omega)^2 + \Gamma^2/4} \cdot e^{-\left(\frac{t - \omega_{fi}}{2kv_p}\right)^2} dt$$

$$\simeq \frac{4}{(\omega_{fi} - 2\omega)^2 + \Gamma^2/4} + \frac{2\sqrt{\pi}}{kv_p\Gamma} e^{-\left(\frac{2\omega - \omega_{fi}}{2kv_p}\right)^2} \tag{5.138}$$

where, in the limit of very small Γ, the Lorentzian representation of the Dirac delta function has been used in the last step. Equation 5.138 represents the superposition of a Doppler-broadened background and a narrow Lorentzian profile. The peak of the Doppler profile, however, amounts only to the fraction

$$\eta = \frac{2\sqrt{\pi}}{kv_p\gamma} \frac{\gamma^2}{16} \simeq \frac{\Gamma}{2\Gamma_D} \tag{5.139}$$

of the Lorentzian peak height, $\Gamma_D = 4kv_p \ln 2$ being the FWHM of the Doppler profile. Just as an example, for $\Gamma = 2$ MHz and $\Gamma_D = 1$ GHz, the Doppler-free signal is about 1000 times higher than the peak of the Doppler-broadened background. To conclude this elementary discussion, frame (c) of Figure 5.30 shows an experimental arrangement for Doppler-free two-photon absorption with fluorescence detection. Other, more sophisticated schemes will be described in the remainder of this chapter.

5.9   Second-order Doppler-free spectroscopy

As already mentioned in Section 5.6, a more intransigent limitation is posed by the second-order Doppler effect (SODE). In principle, there are basically two ways to suppress the SODE influence. The first one is connected with cooling/trapping of particles and will be dealt with in Chapter 7. The second one relies on optical selection of cold particles in a gas by means of the inhomogeneous saturation in a standing light wave [437, 438]. Under transit-time conditions ($\Gamma\tau_0 \ll 1$, where Γ is the homogeneous linewidth and $\tau_0 = a/v_0$ the transit-time for a particle with an average thermal velocity v_0 crossing the light beam of radius a), the main contribution to the saturation resonance is provided by slow particles whose transverse velocity is much less than an average thermal one. Equivalently, the main contribution to the saturation resonance is determined by cold particles with an effective temperature $T_{eff} \sim (\Gamma\tau_0)^2 T_0$, with T_0 being the gas temperature. Therefore the SODE shift of a resonance is small $\delta \simeq (\Gamma\tau_0)^2 \Delta_0$ at $\Gamma\tau_0 \ll 1$ ($\Delta_0 = -(1/2)(v_0/c)^2\omega_0$). Thus, the saturation lineshape approaches, in essence, that for a single stationary particle.

The expression for the resonance shape under the transit-time conditions with allowance for SODE can be approximated as [439]

$$\alpha(\Omega) = \frac{\alpha_0\kappa\beta^2}{4}\Gamma^2 \int_0^\infty \frac{W(u)du}{\gamma^2 + [\Omega + (1/2)(u/c)^2\omega_0]^2} \tag{5.140}$$

where $W(u) = (u/v_0^2)\exp(-u^2/v_0^2)$ is the Maxwellian distribution of particles over the transverse velocities u, $\kappa = (2dE/\hbar\Gamma)^2$ denotes the saturation parameter, $2E$ represents the field amplitude, d the dipole matrix element of the absorbing transition, $\gamma = \Gamma + u/a$, α_0 is the unsaturated absorption coefficient, $\beta = \Gamma\tau_0$, and $\Omega = \omega - \omega_0$. The physical sense of Equation 5.140 is plain, i.e., the resulting resonance shape is a superposition of separate resonances, each located at the frequency $\omega = \omega_0 - (1/2)(u/c)^2\omega_0$ with half-width $\gamma = \Gamma + u/a$.

In a saturated absorption spectroscopy experiment, an effective method to select slow absorbers out of the thermal distribution consists of diminishing the laser intensity. In this way, intuitively, just the slow atoms lagging enough in the laser field will significantly contribute to the saturation signal. At the same time, the use of slow atoms/molecules also increases the interrogation time of the laser radiation with the particles and therefore leads to a reduced transit-time broadening. The drawback is that such a scheme only exploits a little fraction of the molecules in the thermal velocity distribution, which strongly reduces the saturation signal, thus necessitating the use of an ultrastable laser spectrometer. Within this approach, the first observation of SODE-free optical resonances with a width of 50 Hz was reported in [438]. The experiment was carried out in methane on the $F^{(2)}P(7)v_3$ absorption line (at 3.39 μm) with a special spectrometer based on the use of a He-Ne laser with a telescopic beam expander (TBE) inside the cavity. This permitted recording the resonance at a pressure of 10^{-6} torr and lower, and making an effective selection of cold particles. With this method, a relative spectral linewidth of less than $3 \cdot 10^{-13}$ was observed. Such an approach has been applied successfully also on OsO_4 and SF_6 at 10 μm [440].

Besides this saturated absorption-based technique, slow-molecule detection was also carried out in a (Doppler-free) two-photon scheme [441]. More recently, an extremely refined implementation of the latter scheme was applied to the Hydrogen 1S-2S transition [442] and will be the object of a more detailed discussion (see Section 5.12.5). As already mentioned, however, the most effective method to reduce the detrimental influence of high velocities is to cool the absorbers. For example, atoms and ions can be cooled by laser radiation if they have a strong cycling transition; in the case of molecules, instead, due to the lack of cycling transitions, several alternative cooling/trapping techniques are emerging. These aspects will be discussed in Chapter 7.

5.10   Sub-Doppler spectroscopy in atomic/molecular beams

Another valuable option to overcome Doppler broadening consists of reducing the velocity distribution of the atoms/molecules by producing them in collimated beams. As we will see in a short while, although the mean beam velocity may be quite high, like in supersonic beams, the spread of velocities is strongly reduced. As a result, the effective temperature for spectroscopy may be as low as a few Kelvins (in other words, the beam molecules possess poor relative kinetic energy). Moreover, when probing low-divergence beams in a transverse geometry, both Doppler broadening (originating from a spread of velocities in the laser propagation direction) and shifts (due to the beam velocity) are heavily suppressed.

As schematically shown in frame (a) of Figure 5.31, a molecular beam is created by letting a gas escape (via a small orifice) from a high-pressure vessel into an evacuated chamber. If the mean free path in the gas reservoir is large compared to the hole size, then molecules will every now and then leak without undergoing collisions. The resulting effusive beam is characterized by the fact that distributions over velocities and internal degrees of freedom (rotation, vibration) are the same as in the vessel. In such beams, the most probable velocity of molecules is typically a few hundred meters per second, while a small fraction of them will have speeds around ten meters per second. These slower molecules can be filtered out by means of curved electrostatic guides or mechanical selectors [443]. Higher fluxes of slow and internally colder molecules can be obtained when the effusive beam is formed by means of the buffer gas-cooling technique (see below).

By contrast, if the mean free path is shorter than the orifice diameter, which is achieved either for higher vessel pressures or with larger holes, molecules will frequently collide while escaping. As we will discuss below, this will produce adiabatic cooling of all degrees of freedom. In other words, all the energy originally available to the molecule in the reservoir is converted, in the expansion region, into kinetic energy (directed flow), i.e., a supersonic beam of internally cold molecules is created.

5.10.1   Effusive beams

Operation at thermal equilibrium and the absence of collisions imply that an effusive beam contains a well-defined equilibrium distribution of internal states. This is an important advantage over many other beam sources, which are much less well defined in this respect. For this reason, in order to point out the main features of sub-Doppler spectroscopy on atomic/molecular beams, we focus here on effusive beams, following the treatment by [379]. This starts with the density of molecules impinging on the area A of the aperture having speed between v and $v + dv$

$$n(v) = n_0 \frac{4}{\sqrt{\pi}} \frac{v^2}{v_p^3} e^{-(v/v_p)^2} dv \tag{5.141}$$

where n_0 is the total density of molecules. As a consequence, the far-field density in the effusive flow beam as a function of polar coordinates is given by

$$n(v, r, \theta) = \frac{n(v)}{4\pi} d\Omega = n_0 \pi^{-3/2} \frac{v^2}{v_p^3} e^{-(v/v_p)^2} \frac{A \cdot \cos\theta}{r^2} dv$$

$$\equiv C \frac{v^2}{r^2} e^{-(v/v_p)^2} \cos\theta dv \tag{5.142}$$

where $d\Omega$ is the solid angle element under which the area A of the source is seen

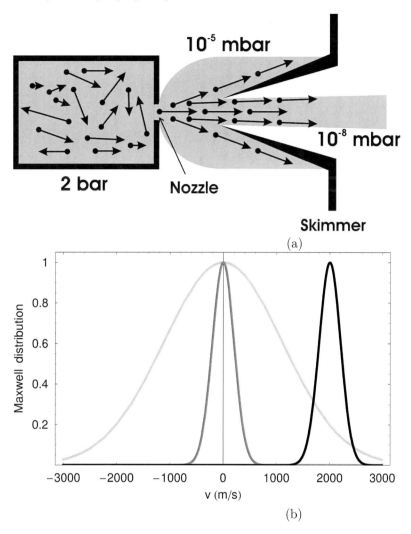

FIGURE 5.31

Principle of a supersonic expansion. (a) A high-pressure gas is expanded via an orifice into vacuum, molecular velocities being represented by the arrows. The randomness in direction and magnitude of the thermal static gas in the vessel is converted into directed flow after the nozzle. (b) Velocity distribution for helium in a static gas at room temperature (light gray), a static gas at 10 K (gray), and a supersonic expansion at 10 K with a mean forward velocity of 2 km/s (black): when a static gas is cooled, the velocity distribution narrows, but the peak remains centered at zero velocity; in a supersonic expansion, instead, besides the narrowing in the velocity distribution, the maximum shifts to a non-zero value, as the flow is preferentially in one direction. (Adapted from [445].)

from a given point (with polar coordinates θ and r), 4π is the total solid angle, and $C = (n_0\pi^{-3/2}A)/v_p^3$. With reference to Figure 5.32, we choose the molecular beam axis as the z-axis, place the slit in the y direction, and consider a laser beam propagating along the x direction. Molecules effusing from the reservoir through the orifice (area A) into the vacuum chamber pass through a narrow slit (width b) at a distance d from A before reaching

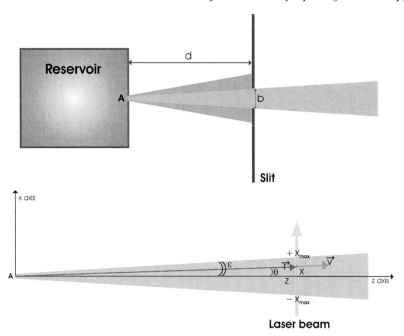

FIGURE 5.32
Schematic experimental arrangement for laser excitation spectroscopy with reduced Doppler width in a collimated molecular/atomic beam.

the interaction region with the laser beam. The quantity $b/2d = \tan \epsilon \simeq \varepsilon$ is referred to as the collimation angle.

Since the vectors $\boldsymbol{r}$ and $\boldsymbol{v}$ are parallel to each other, the following relations hold: $x = r \sin \theta$ and $v_x = v \sin \theta$ which imply $v_x = (x/r)v$ $(dv_x = (|x|/r)dv)$. Thus, we can write

$$n(v_x, x) = C \frac{z}{|x|^3} v_x^2 e^{-(rv_x/xv_p)^2} dv_x \equiv \tilde{n}(v_x, x)dv_x \tag{5.143}$$

Now, when the laser beam travels along the x-direction through the molecular beam, its power decreases as

$$P(\omega) = P_0 \exp\left[-\int_{x_1}^{x_2} \alpha(\omega, x)dx\right] \simeq P_0 \left[1 - \int_{x_1}^{x_2} \alpha(\omega, x)dx\right] \tag{5.144}$$

where small absorption has been assumed and the absorption coefficient is given by

$$\alpha(\omega, x) = \int_{-\infty}^{+\infty} \tilde{n}(v_x, x)\sigma(\omega, v_x)dv_x \tag{5.145}$$

such that

$$\frac{P_0 - P(\omega)}{P_0} = \int_{-\infty}^{+\infty} \left[\int_{x_1}^{x_2} n(v_x, x)\sigma(\omega, v_x)dx\right] dv_x \tag{5.146}$$

As usual, the cross section $\sigma(\omega, v_x)$ is expressed by the Lorentzian formula

$$\sigma(\omega, v_x) = \sigma_0 \frac{\Gamma^2/4}{(\omega - \omega_0 - \frac{\omega_0}{c}v_x)^2 + \Gamma^2/4} \tag{5.147}$$

Inserting Equation 5.147 and Equation 5.143 into Equation 5.146 we obtain

$$\frac{P_0 - P(\omega)}{P_0} = \frac{\sigma_0 \Gamma^2}{4} Cz \int_{-\infty}^{+\infty} \left[\int_{-x_{max}}^{+x_{max}} \frac{(1/|x|^3) v_x^2 e^{-(rv_x/xv_p)^2}}{(\omega - \omega_0 - \frac{\omega_0}{c} v_x)^2 + \Gamma^2/4} dx \right] dv_x \qquad (5.148)$$

that, with the change of variable $\omega_0' = (\omega_0/c)v_x$, becomes

$$\frac{P_0 - P(\omega)}{P_0} \simeq C' \int_{-\infty}^{+\infty} \left[\int_0^{+x_{max}} \frac{(\omega_0'^2/x^3) \exp\left[-\frac{c^2}{\omega_0^2 v_p^2} \frac{z^2}{x^2} \omega_0'^2 \right]}{(\omega - \omega_0 - \omega_0')^2 + \Gamma^2/4} dx \right] d\omega_0' \qquad (5.149)$$

where $z/r = \cos\theta \simeq 1$ has also been exploited and

$$C' = \frac{\sigma_0 \Gamma^2}{2} Cz \frac{c^3}{\omega_0^3} \qquad (5.150)$$

Analytical integration of Equation 5.149 over x provides

$$\frac{P_0 - P(\omega)}{P_0} = C'' \int_\infty^{+\infty} \frac{e^{-\frac{c^2 \omega_0'^2}{\omega_0^2 v_p^2 \varepsilon^2}}}{(\omega - \omega_0 - \omega_0')^2 + \Gamma^2/4} d\omega_0' \qquad (5.151)$$

where $x_{max} = z\varepsilon$ has been used and

$$C'' = C' \frac{\omega_0^2 v_p^2}{c^2 z^2} \qquad (5.152)$$

Finally, with the change of variable $\omega' = \omega_0 + \omega_0'$ we get

$$\frac{P_0 - P(\omega)}{P_0} = C'' \int_\infty^{+\infty} \frac{e^{-\frac{c^2 (\omega' - \omega_0)^2}{\omega_0^2 v_p^2 \varepsilon^2}}}{(\omega - \omega')^2 + \Gamma^2/4} d\omega' \qquad (5.153)$$

which represents a Voigt profile, but with the Doppler width reduced by the collimation ratio of the beam $\varepsilon = b/(2d)$. For example, a collimation ratio of 0.01 corresponds to a reduction of the Doppler width by a factor 100.

5.10.2 Supersonic beams

As anticipated, when the mean free path of the molecules is much smaller than the size of the orifice, a supersonic free jet is formed by allowing the gas to expand into a vacuum. While referring to [444] for a detailed description of the supersonic expansion, here we briefly discuss only those aspects which are of interest in the context of precision spectroscopy [445, 379]. Conservation for the total energy (of a mole of mass m) before (E_0) and after (E_f) the expansion (in the z-direction) requires

$$E_0 = U_0 + p_0 V_0 + \frac{1}{2} m u_0^2 = E_f = U_f + p_f V_f + \frac{1}{2} m u_f^2 \qquad (5.154)$$

where $U = U_{trans} + U_{rot} + U_{vib}$ represents the internal energy, pV the potential energy, and $(1/2)mu^2$ the kinetic-flow energy associated with the mean forward velocity u along

z. Next we can safely assume $u_0 \ll u_f$ (the gas is, in fact, at thermal equilibrium in the vessel) and $p_f \ll p_0$ (the gas expands into vacuum), whereupon

$$U_0 + p_0 V_0 = U_f + \frac{1}{2} m u_f^2 \qquad (5.155)$$

This equation shows that, if most of the energy $U_0 + p_0 V_0$ is converted into the $(1/2) m u_f^2$ term, then a beam with small internal energy U_f will be produced. In other words, the enthalpy necessary to establish the directed mass flow is gradually furnished by the enthalpy which was associated with the random thermal motion in the static gas (upper frame of Figure 5.31). The cooling that is produced directly, as the internal energy is decreased during the supersonic expansion, is that of the translational motion. The velocity distribution is then described by a modified Maxwellian

$$n(v_z) \propto \exp \left[-\frac{m(v_z - u_f)^2}{2 k_B T_{trans,f}} \right] \qquad (5.156)$$

around the flow velocity u_f. The distribution width, fixed by the translational temperature $T_{trans,f}$, defines the relative energy of collisions in the expanding gas. From a spectroscopic point of view, this represents an important advantage: indeed, in spite of the high velocity of the molecular flow, the spread of velocities within the beam is strongly reduced. Moreover, at least in the early stages of the expansion, collisions between molecules occur, which tend to cool the vibrational and rotational degrees of freedom too (in a sense, the vibrating and rotating molecule finds itself in a translationally cold bath). This results in a compression of the population distribution $n(v, J)$ into the lowest vibrational and rotational levels, which drastically simplifies the molecular absorption spectrum and greatly enhances the strength of transitions starting from low rotational levels. However, the three internal degrees of freedom are cooled to different extents. Indeed, non-translational cooling stops when the density becomes too low to allow a significant number of collisions (this happens in the downstream expansion, where the translational temperature decreases but the molecular density also does). Therefore, only degrees of freedom that equilibrate rapidly with the translational bath will be extensively cooled: since the cross sections for collisional energy transfer $\sigma_{vib-trans}$ or $\sigma_{vib-rot}$ are generally much smaller than $\sigma_{rot-trans}$ and $\sigma_{rot-trans} \ll \sigma_{trans-trans}$, then we deduce $T_{trans} < T_{rot} < T_{vib}$.

With regard to spectroscopic applications, the last relevant issue is that of condensation. Fortunately, this is a slow process in a supersonic expansion inasmuch as, while cooling requires only two-body collisions, it needs at least three-body collisions to constitute condensation nuclei. Ultimately condensation or formation of clusters will take place, thus fixing the minimum temperature attainable for a supersonic free jet. A well-established strategy to counteract this drawback consists of mixing a small quantity of the molecule of interest with a large amount of a carrier gas, such as helium or other rare gases, and then expanding the mixture to form a jet. In this way, most of the collisions will occur between carrier gas atoms, which considerably weaken the inter-atomic forces.

We close this section by pointing out that supersonic beams are most often operated in a pulsed mode, which relaxes pumping requirements thus allowing use of millimeter-sized orifices. In such beams, densities as high as 10^{13} molecules cm^{-3} at translational temperatures below 1 K can be realized, the mean forward velocity being determined by the temperature and pressure of the source as well as by the mass of the carrier gas. For example, the final velocity of molecules seeded in a room-temperature Kr (Xe) expansion is about 440 m·s^{-1} (330 m·s^{-1}) [443]. High-flux and low-temperature properties, however, are not the only desirable characteristics for a spectroscopic experiment. Indeed a low molecular speed is crucial to greatly enhance the interrogation time and hence the ultimate resolution in spectroscopic frequency measurements. For this reason, such seeded pulsed beams often just represent ideal starting points for subsequent Stark, Zeeman, and optical deceleration experiments. These techniques will be discussed in the first sections of Chapter 7 which are devoted to trapping and cooling of atoms, ions, and molecules.

5.10.3 Buffer-gas-cooling

An alternative effective tool to produce either cold, dense stationary gases or high-flux beams of cold atoms and molecules is the buffer-gas-cooling (BGC) method [446]. With reference to Figure 5.33, the cooling process of the desired molecular species A, at initial temperature T_i, takes place inside a cryogenic cell filled with He buffer gas at low temperature T_b and density n_{He}. After a characteristic number of collisions N_{coll}, the translational temperature T of A comes arbitrarily close to equilibrium with the buffer gas, such that $T \simeq (1 + \epsilon)T_b$ when $N_{coll} = -k \cdot \ln(\epsilon T_b/T_i)$. Here $k = (m_A + m_{He})^2/(2m_a m_{He})$ and m_A (m_{He}) is the mass of A (He). Rotational degrees of freedom are also cooled during these collisions (obviously, only the ground vibrational state is populated in this temperature regime). Then, a beam of A is formed by allowing both He and A particles to exit the cell into a high-vacuum environment via a small hole of diameter d. In order to obtain the highest flux of cold, slow A molecules, several relevant parameters must be taken into account. First, the number of cold A particles in the beam depends on both n_{He} and the cell geometry. During thermalization, a particle of A typically travels a distance $R \simeq N_{coll}/(n_{He} \cdot \sigma_t)$, where σ_t is a thermally averaged cross section for elastic collisions. Hence, for a cell with distance R_h from the entry point to the hole, the particles of A will be efficiently thermalized before exiting only if $R < R_h$. Secondly, the forward velocity v_f of the thermalized beam of A particles is determined by the ratio d/λ where $\lambda = 1/(n_{He} \cdot \sigma_c)$, σ_c being the elastic cross section of cold A-He collisions. In the effusive limit ($\lambda \gg d$), v_f will be given approximately by the thermal velocity of cold A particles, i.e., $v_f^{eff} \simeq v_A = \sqrt{2k_b T/m_A}$. By contrast, when $\lambda \ll d$, the A particles will become entrained in the outward flux of He, so that $v_f^{entr} \simeq v_{He} = \sqrt{2k_B T/m_{He}}$. Since $m_A \gg m_{He}$ for most molecular species of interest, v_f is much smaller in the effusive limit than for an entrained beam.

From the above considerations, one can recognize that the conditions for efficient thermalization and for a slow beam are in conflict: thermalization is most efficient for n_{He} above a threshold value, while effusive flow demands that n_{He} be less than a different one. The best compromise is obtained when R_h and d are chosen so that such thresholds coincide $n_{He}^{-1} = d\sigma_c = (R_h \sigma_t)/N_{coll}$. This corresponds, of course, to a partial entrainment of A in the helium flow and, therefore, to a forward velocity which lies between v_f^{eff} and v_f^{entr}. Finally, one also wishes to maximize the number N_A of A molecules in the beam. Monte Carlo simulations of the beam formation process show that N_A increases rapidly (approximately $\propto n_{He}^3$) up to a critical value of n_{He} (n_{He}^{cr}), above which it remains roughly constant at a value given by $N_A^{max} = N f_{max}$, where N is the total number of A particles inside the buffer gas cell and $f_{max} = d^2/(8R_h^2)$ is the ratio between the area of the hole and that of a hemisphere at radius R_h. In light of the above discussion, the properties of the beam (flux, forward velocity, and temperature) are simultaneously optimized according to the following procedure. Just to fix ideas, let us consider the specific case of fluoromethane molecules, where the quantity $1/n_{He}^{cr}\sigma_c$ sets the hole diameter d to the mm order. Moreover, assuming $T_b = 3$ K and $\epsilon = 0.1$, we have $N_{coll} \simeq 40$; since σ_c and σ_t are of the same order of magnitude, (typically $\sigma_c \geq \sigma_t$), this implies that $R_h \simeq dN_{coll}$ is of the order of a few cm. This eventually returns $f_{max} = 1/(8N_{coll}^2) \simeq 10^{-4}$, which ensures a relatively intense flux starting from a typical value of N. In a real experiment, once the geometrical parameters (d, R_h) have been

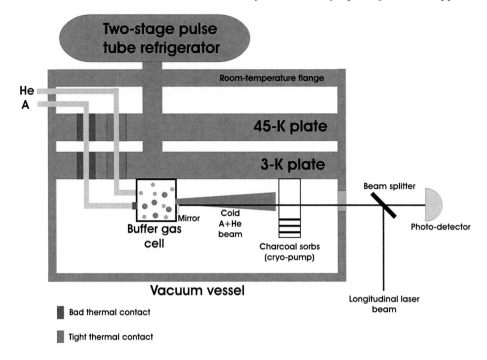

FIGURE 5.33 (SEE COLOR INSERT)
Buffer-gas-cooling setup. The buffer gas cell, mounted in a stainless-steel vacuum vessel, is a gold-coated copper box, with the top face attached to the 3-K plate of a cryostat based on a commercial pulse-tube cooler. The exit hole is centered on one side face. Buffer gas continuously flows into the cell through a narrow tube that is thermally anchored to the 3-K plate too. In this way, both the cell walls and the He gas are typically at $T_b = 3$ K. Outside of the buffer-gas cell, good vacuum is maintained in the beam region by means of charcoal sorbs which pump away (pumping speed of about 1000 l/s) the helium and non-guided molecules.

fixed, the optimum buffer gas density is found by scanning n_{He} around the value n_{He}^{cr}. The produced molecular beam can be monitored using laser absorption spectroscopy. In particular, the absorption signal coming from the longitudinal (relative to the molecular beam) probe laser beam is fitted to a distribution of the form $f_l(v) \propto \exp[-m_A(v-v_l)^2/2k_BT_l]$ to extract the flux, the forward velocity v_l and temperature T_l (recall that, rather than forming an effusive beam, the A molecules are partially entrained in the helium flow). For transverse temperature, $f_t(v) \propto \exp[-m_Av^2/2k_BT_t]$ can be used for fitting. It is experimentally found that the BGC process guarantees complete thermalization in the sense that longitudinal, transverse, and rotational temperature coincide with each other. At the end of such BGC stage, a continuous CH_3F beam with a flux of 10^{13} molec/s, $v_l = 100$ m/s, and $T = 3$ K can be obtained.

In conclusion, we refer the reader to [447] for a comprehensive review on the different BGC flow regimes as well as to other physical/technical considerations.

In principle, any type of spectroscopy that has been used to study static gases may be applied to investigate samples cooled in a supersonic expansion. In a *classical* scheme, saturation spectroscopy is performed simply by letting a laser beam cross the molecular beam perpendicularly and then be reflected on itself by means of a mirror located on the opposite side of the molecular beam. The most sophisticated scheme relies, however, on two-photon Ramsey fringes. This will be described in the following section.

5.11   Ramsey fringes

This method was developed by Ramsey to reduce systematic effects such as time-of-flight broadening and ac Stark shifts in rf experiments. The basic idea is to replace one single long interaction region by two spatially separated driving fields oscillating phase coherently, i.e., with a fixed relative phase (Figure 5.34). The passage through the first interaction zone leaves an atom in a coherent superposition of ground state and excited state. During the time-of-flight T between the two regions the atomic wave function oscillates freely with the transition frequency. In the second interaction region, depending on the relative phase of the atom-field system, the atom is either excited or de-excited. As a function of the driving frequency, Ramsey interference fringes in the excitation probability can be observed [448].

Now let us go into details of this method by following the original theoretical treatment [449]. Consider a molecular system which at time t_1 enters a region where it is subjected to an oscillatory perturbation which induces transitions between two molecular eigenstates p and q with energies W_p and W_q, respectively. Assume that the perturbation V is of such a form that

$$V_{pq} = \hbar b e^{i\omega t}, V_{qp} = \hbar b e^{-i\omega t}, V_{pp} = 0, V_{qq} = 0 \tag{5.157}$$

where b represents the Rabi frequency. Then, if

$$\psi(t) = C_p(t)\psi_p + C_q(t)\psi_q \tag{5.158}$$

the Schrödinger time dependent equations reduce to

$$i\hbar \dot{C}_p(t) = W_p C_p(t) + \hbar b e^{i\omega t} C_q(t) \tag{5.159}$$

$$i\hbar \dot{C}_q(t) = W_q C_q(t) + \hbar b e^{-i\omega t} C_p(t) \tag{5.160}$$

If at time t_1, C_p and C_q had the values $C_p(t_1)$ and $C_q(t_1)$, respectively, then, as it can

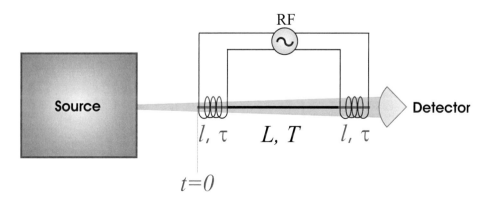

FIGURE 5.34
Schematic setup for a Ramsey in the RF domain.

be readily confirmed by direct substitution, solution of the above equations at time $t_1 + T$ is

$$C_p(t_1 + T) = \left\{ \left[i\cos\theta \, \sin\frac{aT}{2} + \cos\frac{aT}{2} \right] C_p(t_1) \right.$$

$$\left. - i\sin\theta \, \sin\frac{aT}{2} e^{i\omega t_1} \, C_q(t_1) \right\} e^{i\left(\frac{\omega}{2} - \frac{W_p + W_q}{2\hbar}\right)T} \quad (5.161)$$

$$C_q(t_1 + T) = \left\{ -i\sin\theta \, \sin\frac{aT}{2} e^{-i\omega t_1} \, C_p(t_1) \right.$$

$$\left. + \left[-i\cos\theta \, \sin\frac{aT}{2} + \cos\frac{aT}{2} \right] C_q(t_1) \right\} e^{-i\left(\frac{\omega}{2} + \frac{W_p + W_q}{2\hbar}\right)T} \quad (5.162)$$

where

$$\cos\theta = (\omega_0 - \omega)/a, \quad \sin\theta = 2b/a \quad (5.163)$$

$$a = \left[(\omega_0 - \omega)^2 + (2b)^2 \right]^{1/2}, \quad \omega_0 = (W_q - W_p)/\hbar \quad (5.164)$$

In the particular case $b = 0$ we have

$$C_p(t_1 + T) = e^{-\frac{iW_p T}{\hbar}} C_p(t_1) \quad (5.165)$$

$$C_q(t_1 + T) = e^{-\frac{iW_q T}{\hbar}} C_q(t_1) \quad (5.166)$$

Now, consider a molecule on which the perturbation acts while the molecule goes a distance l in time τ after which it enters a region of length L and duration T in which b is zero, after which it is again acted on by the perturbation for a time τ. Then, if t is taken to be zero when the first perturbation begins to act, and if $C_p(0) = 1, C_q(0) = 0$, Equations 5.161 and 5.162 with $t_1 = 0$ and $T = \tau$ yield

$$C_p(\tau) = \left[i\cos\theta \sin\frac{a\tau}{2} + \cos\frac{a\tau}{2} \right] e^{i\left(\frac{\omega}{2} - \frac{W_p + W_q}{2\hbar}\right)\tau} \quad (5.167)$$

$$C_q(\tau) = -i\sin\theta \, \sin\frac{a\tau}{2} e^{-i\left(\frac{\omega}{2} + \frac{W_p + W_q}{2\hbar}\right)\tau} \quad (5.168)$$

Next, Equations 5.165 and 5.166 with $t_1 = \tau$ give

$$C_p(\tau + T) = e^{-\frac{iW_p T}{\hbar}} C_p(\tau) \quad (5.169)$$

$$C_q(\tau + T) = e^{-\frac{iW_q T}{\hbar}} C_q(\tau) \quad (5.170)$$

Finally, Equations 5.161 and 5.162 with $t_1 = \tau + T$ and $T = \tau$ provide

$$C_p(2\tau + T) = \left\{ \left[i\cos\theta \, \sin\frac{a\tau}{2} + \cos\frac{a\tau}{2} \right] C_p(\tau + T) \right.$$

$$\left. - i\sin\theta \, \sin\frac{a\tau}{2} e^{i\omega(\tau+T)} \, C_q(\tau + T) \right\} e^{i\left(\frac{\omega}{2} - \frac{W_p + W_q}{2\hbar}\right)\tau} \quad (5.171)$$

$$C_q\left(2\tau+T\right)=\left\{-i\sin\theta\ \sin\frac{a\tau}{2}e^{-i\omega(\tau+T)}\ C_p\left(\tau+T\right)\right.$$

$$+\left[-i\cos\theta\ \sin\frac{a\tau}{2}+\cos\frac{a\tau}{2}\right]C_q\left(\tau+T\right)\right\}e^{-i\left(\frac{\omega}{2}+\frac{W_p+W_q}{2\hbar}\right)\tau} \tag{5.172}$$

The elimination of the intermediate values of the C's gives $C_q\left(2\tau+T\right)$ and hence the final probability that the system changes from state p to q

$$P_q \equiv \left|C_q\left(2\tau+T\right)\right|^2$$

$$= 4\sin^2\theta\ \sin^2\frac{a\tau}{2}\left[\cos\frac{\delta T}{2}\cos\frac{a\tau}{2}-\cos\theta\sin\frac{\delta T}{2}\sin\frac{a\tau}{2}\right]^2 \tag{5.173}$$

where

$$\delta = \omega_0 - \omega = 2\pi(\nu-\nu_0) \tag{5.174}$$

In the immediate vicinity of the resonance ($\cos\theta \simeq 0$, $\sin\theta \simeq 1$) this simplifies to

$$P_q \simeq \frac{1}{2}\sin^2 a\tau\left[1+\cos\delta T\right] \tag{5.175}$$

From this one finds that optimal excitation of the atom is achieved for two interactions with $a\tau=\pi/2$, i.e., $\pi/2$ pulses. Equation 5.175 predicts, as a function of the detuning δ, fringes with a periodicity equal to $1/T$, the FWHM of the central fringe being $\delta\nu = 1/2T$. Hence, the achievable resolution with Ramsey excitation improves with increasing separation between the applied fields. Obviously, if there is a phase difference $\Delta\Phi = \Phi_2 - \Phi_1$ between the two interacting zones, then 5.175 has to be modified to

$$P_q \approx \frac{1}{2}\sin^2 a\tau\ \left[1+\cos\delta T+\Delta\Phi\right] \tag{5.176}$$

This means that phase differences in general shift the center of the Ramsey structure by the amount $\nu_0' - \nu_0 = \Delta\phi/(2\pi T)$. As a consequence, for the operation of precise frequency standards a few methods have been devised to keep such phase shifts as low and as constant as possible (see Chapter 7).

The transition probability expressed by Equation 5.173 applies to only a single molecular velocity, since times T and τ are inversely proportional to the velocity. Hence, Equation 5.173 must be averaged over the distribution of molecular velocities occurring in the molecular beam. By virtue of Equation 5.141, one can write

$$P_{q,av} = \frac{(4v_p/\sqrt{\pi})\int_0^{+\infty} y^3 e^{-y^2} P_q dy}{(4v_p/\sqrt{\pi})\int_0^{+\infty} y^3 e^{-y^2} dy} = 2\int_0^{+\infty} y^3 e^{-y^2} P_q dy \tag{5.177}$$

where $y = v/v_p$. By insertion of Equation 5.173 into Equation 5.177 and subsequent numerical computation, one obtains the behavior shown in Figure 5.35 for two different temperature values. From such plot two main features can be recognized: 1) in general, the fringe contrast decreases gradually as we move away from the center frequency; 2) the lower is the temperature the higher is the amplitude and the number of fringes recorded in the same frequency interval. As a result, a significant improvement in the signal-to-noise ratio, and hence in resolution of the frequency measurement, is expected when using a cold molecular beam.

The extension of Ramsey's idea to the optical spectral region seems quite obvious if the RF fields are replaced by two phase-coherent laser fields. However, the transfer from

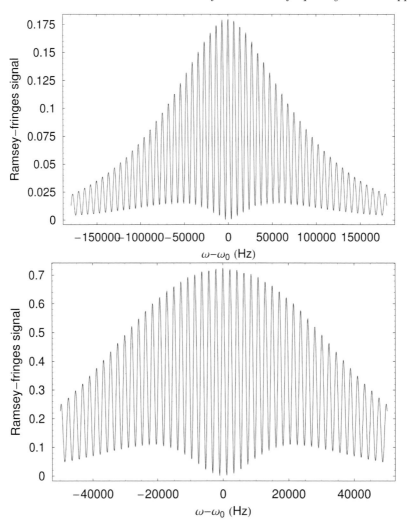

FIGURE 5.35

Ramsey fringes averaged over the distribution of molecular velocities (according to Equation 5.177) for $L = 50$ cm, $l = 1.5$ cm, $b = 8.83$ kHz, $16 \times 1.67 \cdot 10^{-27}$ Kg, and for two different temperature values: $T = 300$ K (upper frame) and $T = 30$ K (lower frame).

the RF region, where the wavelength λ is larger than the field extension l, to the optical range where $\lambda \ll l$, causes some difficulties. In this case, even if the two optical light fields oscillate phase coherently, the fringe pattern washes out, since the phase of a running wave varies in space on the scale of the optical wavelength. In the optical domain, this method must be associated with a sub-Doppler technique in order to avoid scrambling of the fringe pattern. The use of three- and four-zone configuration [450] imposes severe conditions on parallelism and equidistances that, in practice, limit the distance between zones and hence the ultimate resolution. By contrast, in the case of two-photon spectroscopy with two photons being absorbed from opposite directions in a standing wave, the phase is space independent [451, 452]. With reference to Figure 5.36, the quantitative description starts from the probability amplitude $c_n^{(2)}(t)$ for the atom to be in level n at time t as derived in the two-photon-absorption theory [453]. Thus, after the interaction (in a time $\tau = l/v$) with the laser beam in the first zone we can write

$$c_n^{(2)}(\tau) = \sum_m \frac{\mu_{nm}\mu_{mg}|E|^2}{\hbar^2(\omega_{mg} - \omega)} \cdot \frac{e^{i(\omega_{ng} - 2\omega)\tau} - 1}{\omega_{ng} - 2\omega} \equiv D_{ng}(\omega) \cdot \frac{e^{-i\Delta\tau} - 1}{\Delta} \quad (5.178)$$

where ω (E) represent the angular frequency (the electric field) associated with the laser beam, μ_{nm} (μ_{mg}) is the dipole matrix element corresponding to the $n \to m$ ($m \to g$) transition, and $\Delta = \omega_{ng} - 2\omega$. The transit (in a time $T = L/v$) through the free-field region followed by the passage through the second interaction zone is instead described by

$$c_n^{(2)}(\tau, T) = e^{-i\Delta T} e^{-\gamma T} \cdot D_{ng}(\omega) \left[\frac{e^{-i\Delta\tau} - 1}{\Delta} \right] \quad (5.179)$$

where $1/\gamma$ is the spontaneous lifetime of the upper level, which accounts for the fact that part of the excited molecules decay before they reach the second zone. Then the final probability amplitude for the atom to be in level n is obtained as

$$\left| c_n^{(2)}(\tau) + c_n^{(2)}(\tau, T) \right|^2 = \frac{2D_{ng}^2(\omega)}{\Delta^2}(1 - \cos\Delta\tau)\left(1 + e^{-2\gamma T} + 2e^{-\gamma T}\cos\Delta T\right)$$

$$\simeq \mathcal{B}(\omega)\left\{ \mathcal{A} + \mathcal{C}\cos\left(2\pi\frac{\nu - \frac{\nu_{ng}}{2}}{P}\right) \right\} \quad (5.180)$$

where $\mathcal{B}(\omega) = D_{ng}^2(\omega)\tau^2$, $\mathcal{A} = 1 + e^{-2\gamma T}$, $\mathcal{C} = 2e^{-\gamma T}$, and $\Delta\tau \ll 1$ has been used in the last step. Therefore, the transmission signal will consist of periodical fringes, with period $P = v/(2L)$ and contrast $\mathcal{C}$, which are superimposed on the broader (Doppler-free) two-photon background signal $\mathcal{B}(\omega)$ arising from the absorption in one single zone. Equation 5.180 generalizes Equation 5.173, or Equation 5.175 to the optical domain. Therefore, further averaging over the distribution of molecular velocities is needed to obtain the actual signal. Again, this will result in a gradual attenuation of the side fringes. It is worth noting that, in the derivation of the final probability amplitude for the RF-domain case, the spontaneous lifetime of the upper level was not considered (see Equations 5.169 and 5.170) because in the RF domain much smaller γ values usually arise, such that $\gamma T \ll 1$.

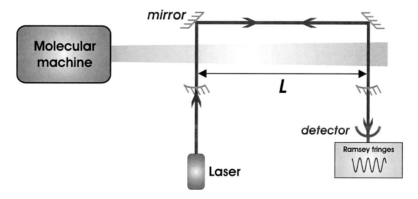

FIGURE 5.36
Each of the interaction tracts corresponds to a laser standing wave. The condition that the relative phase between them is fixed is automatically fulfilled by generating both standing waves inside the same Fabry-Perot cavity.

In conclusion, the ingenious idea of the separated-fields Ramsey excitation scheme is in the possibility of tailoring the transit-time broadening and the resolution independently from each other. Indeed, the former can be augmented by shortening the interaction time at each single zone (so that a large fraction of the atoms can contribute to the signal), whereas the overall measurement resolution is determined by the much longer time-of-flight between the two interaction zones. For these reasons, as we will discover in the next chapters, the Ramsey interrogation scheme is at the basis of atomic/molecular fountains as well as of sophisticated experiments employing cold molecular beams. Moreover, if the atomic state being studied decays spontaneously, the separated oscillatory fields method permits the observation of narrower resonances than those anticipated from the lifetime and the Heisenberg uncertainty principle, provided the two separated oscillatory fields are sufficiently far apart; only states that survive sufficiently long to reach the second oscillatory field can contribute to the resonance. This method, for example, has been used by Lundeen et al. in precise studies of the Lamb shift [454].

We close this section by describing a very refined experiment in the frame of two-photon optical Ramsey fringes [455]. With reference to Figure 5.37, the supersonic SF_6 beam has a forward velocity of 400 m/s and the distance between the two interaction zones is 1 m, corresponding to a fringe periodicity of 200 Hz. The transition of interest is P(4) E^0 in the v_3 band (Figure 5.38). A folded, U-shape cavity is used to provide the two phase-coherent stationary waves of the Ramsey spatial interferometer. A carbon dioxide laser (at 28.4 THz) is used to excite two-photon Ramsey fringes on the molecular beam. Such a laser (Ramsey laser) is offset phase locked to a second CO_2 laser (reference laser) which is stabilized, in turn, against a 2-photon transition in SF_6 (FWHM 40 kHz). Then, AOM1 is needed to tune the excitation beam to the desired frequency (76 MHz below the r level). Moreover, the signal transmitted from the U cavity is used for locking the cavity resonance onto the excitation frequency. However, to increase the signal strength, the Ramsey fringe signal is detected via stimulated emission in a one-photon transition (from the upper energy level to the intermediate ro-vibrational level) in a separate Fabry-Perot cavity. To this aim, AOM2 is used to bring the detection beam into resonance with the $e \to r$ transition. In this way, a fringe signal (reflecting the periodic behavior of the population in level e) is recorded by detector 2, as the frequency of the excitation beam is scanned (in steps) across the two-photon resonance. Finally, the frequency measurement is ultimately referenced to the Cs primary standard. For this purpose, a comb is created by a sum-frequency generation (SFG) process (in a non-linear crystal) between the reference CO_2 laser and a mode-locked Ti:Sa fs laser (see next chapter). The resultant frequency comb can be expressed as $f_q^{SFG} = qf_r + f_0 + f(CO_2)$. This SFG comb overlaps the high-frequency part of the initial comb, and the beat notes $f_q^{SFG} - f_p = f(CO_2) - (p-q)f_r$ are obtained, which are insensitive to f_0. A large number of (p,q) pairs gives the low-frequency beat note $f(CO_2) - mf_r$, which is used to phase lock the mth harmonic of the repetition rate to the CO_2 laser frequency. The repetition rate is simultaneously compared to a 100 MHz or 1 GHz frequency reference, and the error signal is returned to the CO_2 laser via a servo loop of bandwidth of 10-100 mHz. The reference is based on a combination of a hydrogen maser and a cryogenic oscillator controlled with a Cs fountain (LNE-SYRTE signal), and is transferred to the laboratory as an amplitude modulation on a 1.5 μm carrier, via 43 km of optical fiber. The phase noise added by the fiber introduces an instability of a few 10^{-14} for a 1 s integration time, reducing to around 10^{-15} over 1000 s. These are figures for passive transfer, but can be improved more than 10 times when the fiber noise is compensated. All radio frequency oscillators in the system are also referenced to the LNE-SYRTE signal. In this experiment, the mean value of about 500 individual measurements was $\nu(SF_6, P(4)E^0 = 28412764347320.26 \pm 0.79)$ Hz, where the uncertainty ($2.8 \cdot 10^{-14}$ as a fractional value) was the standard deviation.

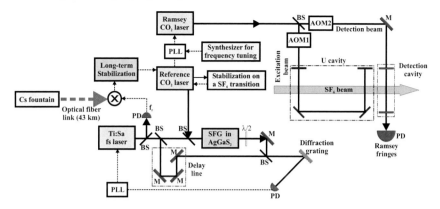

FIGURE 5.37

Detailed experimental setup for two-photon optical Ramsey fringes. (Adapted from [455].)

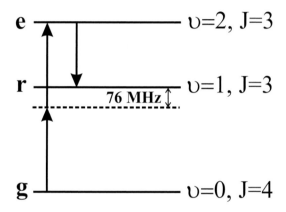

FIGURE 5.38

Schematic of the three levels involved in the P(4) E^0 resonance. (Adapted from [451].)

5.12 Laser frequency standards using thermal quantum absorbers

In its simplest realization, an optical frequency standard consists of a narrow-linewidth laser which is made to interact with a sample of atoms/ions/molecules that possess appropriate absorption lines. When the laser frequency is scanned across a given resonance, an absorption feature is detected. The latter is then converted (by means of a suitable stabilization circuit) into an error signal, which is eventually used to keep the laser frequency at the line center. With this in mind, it is quite obvious that a weak dependence of the line-center frequency on external fields as well as a high quality factor are highly desirable properties for a reference atomic/molecular transition. Finally, it is worth stressing that such a frequency-stabilized laser can be viewed as an optical standard only if its frequency is determined in relation to the primary Cs atomic clock.

Over the years, optical frequency standards at various frequencies (from the blue to the infrared) have been developed in several laboratories, many of them operating with the absorbing gas confined in a cell.

5.12.1 Iodine-stabilized lasers

Molecular iodine offers several distinct advantages as an optical frequency standard. First, it is characterized by a rich spectrum of narrow absorption lines in the visible range. Second, being rather heavy, its velocity is relatively low at room temperature. Last, by properly adjusting the temperature of the absorption cell, one can control the vapor-pressure-mediated collision-induced shifts [456]. For iodine standards based on He-Ne lasers (at $\lambda \simeq 543$, 612, 633 and 640 nm), the fractional uncertainty of output frequency can be as low as $2.5 \cdot 10^{-11}$ with an output power of about 100 μW. As shown in Figure 5.39, standards with higher power levels can be achieved using frequency-doubled Nd:YAG lasers (532 nm) [456]. Here, about 100 mW of radiation at 532 nm are passed through a first polarizing beam splitter (the deflected beam being available for experimental use). PBS2 then divides the laser beam into a pump (saturating) beam and a probe beam, and a modulation-transfer Doppler-free saturation scheme is implemented. For this purpose, an AOM shifts the frequency $\omega/2\pi$ of the pump beam by $\delta/2\pi = 80$ MHz. Also, the rf signal driving the AOM is switched on and off at a frequency of 23 kHz. As a result, (deflected and frequency-shifted) saturating pump beam is speedily chopped. Afterwards, the pump beam passes through the cell and periodically saturates the absorption of those iodine molecules whose Doppler-shifted transition frequency coincides with the frequency of the pump beam. Finally, the saturated absorption is probed by a counter-propagating laser beam of frequency $\omega/2\pi$. Now, since the superposition of these two counter-propagating beams gives rise to a *walking wave* structure, where the nodes and anti-nodes move with a velocity $c \cdot \delta/(2\pi)$, saturated absorption mechanism applies to iodine molecules that move with this same velocity component (and thus experience a standing wave). Correspondingly, the laser frequency $\omega_L/2\pi$ at the center of the observed saturation dip is Doppler-shifted by an amount of $\delta/2$ (to first order) and the transition frequency of the molecule at rest is given by $\omega_0/2\pi = (\omega_L + \delta/2)/2\pi$. Moreover, phase-sensitive detection is implemented. To this aim, the phase of the probe beam is modulated at 5 MHz such that, when the laser frequency is tuned across the molecular resonance, an intensity modulation at that frequency occurs, which is detected by the photodetector (PD) and demodulated by means of the double balanced mixer (DBM). In turn, this latter signal is phase sensitively detected by a lock-in, which is driven with the chopping frequency (23 kHz). With this scheme, frequency offsets generated by a residual linear absorption are strongly suppressed and only the saturated absorption contributes to the error signal. As a result, the achieved stability can be as low as a few 10^{-11}. Tunable lasers were also used to take advantage from molecular iodine reference lines in the green. In particular, a solid state tunable source emitting around 541 nm wavelength and having a continuous tunability range of about 1.2 THz was set up. An extended cavity DBR (distributed Bragg reflector) diode laser emitting cw around 1083 nm was amplified, in a Yb-doped fiber amplifier, and frequency doubled in a PP-KTP nonlinear crystal [225]. With this source, frequency locked onto sub-Doppler I_2 lines, a frequency stability of 4 parts in 10^{13} (300 s integration time) for the diode laser frequency was achieved and used as reference for precise measurements of atomic helium fine structure transitions around 1083 nm [226].

5.12.2 Acetylene-stabilized lasers

With its two isotopomers, $^{12}CH_2$ and $^{13}CH_2$, the acetylene molecule provides a very dense grid of ro-vibrational overtone transitions in the telecom spectral range (approximately between 1.515 and 1.553 μm). Potentially, any of these lines can be used for laser frequency stabilization. Nevertheless, due to the weakness of such overtone transitions, high laser intensities are needed in order to generate Doppler-free saturated absorption signals. This can be accomplished either by employing high-power erbium lasers in direct-excitation

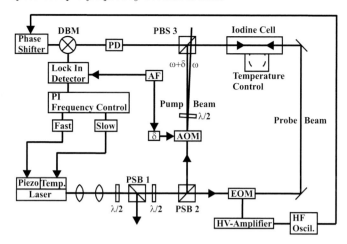

FIGURE 5.39
Layout of an iodine-stabilized Nd:YAG laser. (Adapted from [456].)

schemes [457], or by resorting to cavity-enhanced methods [458]. In this regard, for instance, application of the NICE-OHMS technique may considerably improve the stability of acetylene-stabilized diode lasers, which is currently almost an order of magnitude worse than that of iodine-stabilized He-Ne lasers [456].

5.12.3   Methane-stabilized lasers

Because of its spherical-top symmetry, which provides it with a spectrum only weakly influenced by any spurious external field, the CH_4 molecule has always played a crucial role in frequency metrology. Moreover, overlapping with the 3.39 μm line of the He-Ne laser, the $F_2^{(2)}$ component of the ν_3 $P(7)$ transition has historically been used as a reference frequency for such laser. Eventually, the celebrated simultaneous measurement of the wavelength and frequency of this laser has led to the current definition of the speed of light. Typically, a CH_4-stabilized He-Ne laser relies on a 60-cm-long laser resonator with an intracavity CH_4 absorption cell (approximately 30 cm in length) [456]. The FWHM of the obtained Doppler-free saturation signal (about 300 kHz) is mainly determined by transit-time broadening. By expanding the diameter of the laser beam, this was strongly reduced up to resolve the hyperfine structure and even the 1.3-kHz recoil splitting in the saturation signal [459]. As already discussed, however, this scheme is powerless against the shift and broadening caused by the second-order Doppler effect, which can only be reduced by the use of slow molecules. With the method described above (see Section 5.9), a relative spectral linewidth of less than $3 \cdot 10^{-13}$ was observed [438].

5.12.4 OsO_4-stabilized lasers

Possessing a number of lines with favorable characteristics in the 10-μm-wavelength range (in particular, the isotopomers $^{188}OsO_4$, $^{190}OsO_4$, and $^{192}OsO_4$ have no hyperfine structure), OsO_4 has widely been used for frequency stabilization of CO_2 lasers [460]. Various configurations have been realized so far, ranging from large external absorption cells [440], to optical resonators containing the molecular gas [462, 460], as well as to molecular beams [463]. With the best OsO_4-stabilized systems, a fractional frequency stability of a few $10^{-15}/\sqrt{\tau/s}$ and a reproducibility of $2 \cdot 10^{-13}$ have been reported [456].

Most often, atomic/molecular samples for frequency standards are prepared in beams rather than in absorption cells. As already discussed, the main advantage of this configuration is that a transversal excitation geometry can be adopted, which strongly suppresses first-order Doppler effects (see Section 5.10). Moreover, the excitation and detection zones can be spatially separated, which affects in a beneficial way the signal-to-noise ratio. Last but not least, the refined Ramsey-fringes technique, discussed in Section 5.11, can be implemented.

5.12.5   Atomic hydrogen standard

For atomic beams, the hydrogen standard is clearly the most advanced one. Recently, the measurement of the $1S - 2S$ transition frequency in atomic hydrogen has been further improved via two-photon spectroscopy on a 5.8 K beam to a fractional frequency uncertainty of $4.2 \cdot 10^{-15}$. Such progression mainly arose from an improved stability of the spectroscopy laser, and a better determination of the main systematic uncertainties. Giving here a detailed description of this experiment will be quite instructive, as it puts together practically all the concepts presented so far in this chapter anticipating, at the same time, some others contained in future chapters. With reference to Figure 5.40, coherent radiation at 243 nm (13 mW) is obtained by frequency doubling twice (within two resonant cavities) the laser radiation coming from a tapered amplifier injected, in turn, by an ECDL master oscillator emitting at 972 nm. A linewidth of less than 1 Hz and a fractional frequency drift of $1.6 \cdot 10^{-16}$ s^{-1} are obtained by locking the master laser to a high-finesse ULE cavity. Then, a fiber-laser frequency comb (250-MHz repetition rate) phase coherently links the cavity frequency to an active hydrogen maser which is, in turn, referenced to a mobile cesium fountain atomic clock. Cycle slip detection is also applied as described in [464].

Emerging from a cooled copper nozzle, the atomic hydrogen beam then enters the differentially pumped excitation region ($10^{-5}/10^{-8}$ mbar). The latter is shielded by a grounded Faraday cage from electric stray fields. Excitation of the 1S-2S transition is accomplished by two counter-propagating photons in the 243-nm standing laser wave formed by the enhancement cavity. Then, for detection, excited, metastable atoms are quenched via the $2P$

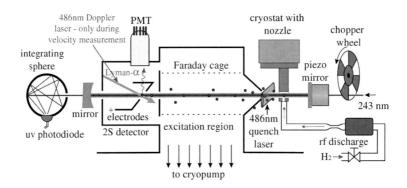

FIGURE 5.40
Schematic of the beam apparatus for an improved measurement of the hydrogen 1S-2S transition frequency. A standing laser wave at 243 nm (between gray mirrors) excites the sharp 1S-2S transition in a collinearly propagating cold thermal beam of atomic hydrogen. The 2S state is detected after quenching with a localized electric field which releases a Lyman-α photon. (Courtesy of [442].)

state by an electric field (10 V/cm) and the emitted 121-nm photons are collected by a photomultiplier tube (PMT). The intracavity power is monitored by measuring the cavity transmission be means of a photodiode connected to an integrating sphere. For spectroscopy, an external magnetic field of 0.5 mT is applied to separate the magnetic components and use the transitions $(1S, F = 1, m_F = \pm 1) \rightarrow (2S, F' = 1, m_F = \pm 1)$ whose Zeeman shifts of the ground and excited state almost completely cancel. The hyperfine centroid frequency is obtained by adding $\Delta f_{HFS} = +310712229.4(1.7)$ Hz as calculated from the experimental results for the $1S$ and $2S$ hyperfine splittings. The recorded spectra are shown in Figure 5.41 for different detection delays τ (see below). At each frequency point, Lyman$-\alpha$ photons are counted for 1 s. In principle, for such light atoms and high frequencies, the recoil shift is quite large. Indeed, due to the recoil of the atom when it emits or absorbs a photon, there is a frequency shift given by $h\nu^2/2Mc^2$ where M is the mass of the atom. For example, it amounts to 12.6 MHz for the Lyman$-\alpha$ line. Obviously, it can be corrected for, usually with a very low residual fractional uncertainty ($3.6 \cdot 10^{-18}$ in the experiment under description). Thus, the two main systematic effects to be compensated for are the ac Stark shift (or light shift) and the second-order Doppler effect.

The former is caused by the interaction of the incident radiation with the non-resonant states of the atom. The effect of far-detuned laser light on the atomic levels can be treated as a perturbation in the second order of the electric field, i.e., linear in terms of the field intensity. As a general result of second-order time-independent perturbation theory for non-degenerate states, an interaction (Hamiltonian $\mathcal{H}_1$) leads to an energy shift of the i-th state (unperturbed energy $\mathcal{E}_i$) that is given by

$$\Delta E_i = \sum_{j \neq i} \frac{|\langle j|\mathcal{H}_1|i\rangle|^2}{\mathcal{E}_i - \mathcal{E}_j} \tag{5.181}$$

For an atom interacting with laser light, the interaction Hamiltonian is $\mathcal{H}_1 = -\hat{\mu} \cdot \boldsymbol{E}$ with $\hat{\mu} = -e\boldsymbol{r}$ representing the electric dipole operator. For the relevant energies $\mathcal{E}_i$, one has to apply a *dressed-state* view, considering the combined system *atom plus field* [465]. In its ground state the atom has zero internal energy and the field energy is $n\hbar\omega$ according to the number n of photons. This yields a total energy $\mathcal{E}_i = n\hbar\omega$ for the unperturbed state. When the atom is put into an excited state by absorbing a photon, the sum of its internal energy $\hbar\omega_0$ and the field energy $(n-1)\hbar\omega$ becomes $\mathcal{E}_j = \hbar\omega_0 + (n-1)\hbar\omega = -\hbar\Delta_{ij} + n\hbar\omega$. For a two-level atom, the interaction Hamiltonian is $\mathcal{H}_1 = -\mu E$ and Equation 5.181 simplifies to [466]

$$\Delta E = \pm\frac{|\langle e|\mu|g\rangle|^2}{\Delta}|E|^2 = \pm\frac{3\pi c^2}{2\omega_0^3}\frac{\Gamma}{\Delta}I(\boldsymbol{r}) \tag{5.182}$$

for the ground and excited state (upper and lower sign, respectively). We have used the relation $I = 2\varepsilon_0 c|\tilde{E}|^2$ and the expression for the dipole matrix element. This effect is thus proportional to the incident light intensity and is generally in the range of several hundreds of kHz in usual experiments involving cw lasers. Finally, the dc Stark effect is also worthy of mention: a static electric field may shift and broaden the line, by mixing levels of opposite parity. In a typical experiment, it is very difficult to eliminate stray electric fields altogether, residual amplitudes lying in the range of 10 mV cm^{-1}. Similarly, stray magnetic fields are responsible for the dc Zeeman effect.

Coming back to the current experiment, in order to correct for the ac Stark effect, a double-pass acousto-optic modulator (AOM) in zero-th order is placed in front of the enhancement cavity to quickly alter the power level under otherwise identical conditions. Then, the transition frequency is extrapolated to zero intensity. However, a small quadratic contribution must be taken into account before applying such linear extrapolation, the asso-

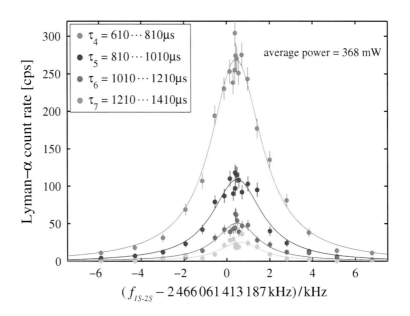

FIGURE 5.41 (SEE COLOR INSERT)

Single scan line profile for different delays along with Lorentzian fits. With the detection delay τ, an upper limit is set on the atoms' velocity, which reduces the signal accordingly. The full width at half maximum (FWHM) is about 2 kHz. (Courtesy of [442].)

ciated fractional uncertainty amounting to $0.8 \cdot 10^{-15}$. Moreover, the fractional uncertainties related to the dc Stark and Zeeman effects are 0.4 and 0.38 $\cdot 10^{-15}$, respectively.

The Doppler effect, due to the velocity v of the atoms, is cancelled to first order thanks to the two-photon excitation scheme. The remaining second-order Doppler (SOD) shift $(-v^2 f_{1S-2S}/2c^2)$ is compensated as follows. First, the excitation light is chopped at 160 Hz which allows time-of-flight resolved detection of atoms excited to the 2S state. Evaluating only 2S counts recorded at a certain delay τ after the light has been switched off by the chopper wheel allows the fastest atoms to escape. This samples the slow tail of the velocity distribution and removes most of the SOD; in other words, the delay τ between blocking the 243 nm radiation and the start of photon counting sets the upper limit for the atomic velocity of $v < l/\tau$, where l is the distance between the nozzle and detector. Then, with the help of a multichannel scaler all photons are counted and sorted into 12 adjacent time bins $\tau_1 = 10...210$ μs,... $\tau_{12} = 2210...2410$ μs. Therefore, for each scan of the laser frequency over the hydrogen 1S-2S resonance, up to 12 spectra are obtained, measured with different delays. Then, to correct for the SOD shift, an elaborate theoretical model is used to fit all the delayed spectra of one scan simultaneously with a set of 7 parameters. The result of the fitting procedure is the 1S-2S transition frequency for a hydrogen atom at rest. The overall fractional uncertainty in the SOD correction amounts to $\sigma/f_{1S-2S} = 2.0 \cdot 10^{-15}$.

The other two dominant contributions in the total uncertainty budget come from statistics $(2.6 \cdot 10^{-15})$ and lineshape model $(2.0 \cdot 10^{-15})$. Summarizing all corrections and uncertainties, the value $f_{1S-2S} = 2466061413187035(10)$ Hz is found for the $1S - 2S$ hyperfine centroid frequency. Other uncertainty sources, which were less significant in the above measurement, but are relevant to the operation of frequency standards, will be mentioned in Chapter 7.

5.12.6 Calcium standard

Other suitable candidates are represented by Ca and Mg. Indeed, the intercombination transitions of alkaline earth atoms, like those shown in Figure 5.42, are celebrated examples of references for optical frequency standards. Just to give an idea, Mg and Ca exhibit natural linewidths as low as 0.04 kHz and 0.4 kHz, respectively. Additionally, in both cases, the frequencies of the $\Delta m_J = 0$ transitions are almost insensitive to electric and magnetic fields.

So far, most attention has focused on Ca atomic beams. Early work started in 1979 when Barger et al. obtained a resolution as low as 1 kHz [467]. Later, such 657-nm intercombination transition was extensively investigated at various labs worldwide, and a transportable beam standard was also constructed for dissemination purposes [468]. The key components of such a standard were a pre-stabilized diode laser system with a linewidth below 2 kHz and a miniaturized thermal (effusive) calcium beam apparatus. A separated-field (optical Ramsey) excitation scheme was used to resolve spectroscopic structures of a width below 20 kHz, which were used to stabilize the laser frequency to the center of the intercombination line. The characterization of the stability and reproducibility of the standard was investigated in comparison with a stationary standard based on laser-cooled Ca atoms resulting in a relative frequency uncertainty of $1.3 \cdot 10^{-12}$ [456].

We conclude this chapter by dealing with two other important, vast spectroscopy branches, namely Fourier Transform Infrared (FTIR) spectroscopy and Raman spectroscopy. Although not commonly employed as ultrahigh-resolution techniques, these undoubtedly deserve some consideration, particularly in connection with two modern, sophisticated applications. Indeed, both Raman and FTIR spectroscopy are experiencing a renewed interest thanks to the advent of modern optical frequency comb synthesizers based on fs lasers (see Chapter 6) [469, 470, 471]. Moreover, Raman scattering is at the basis of advanced cooling schemes (refer to Chapter 7).

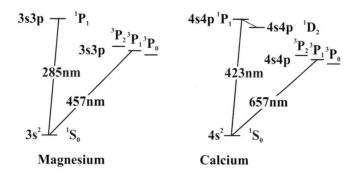

FIGURE 5.42
Partial energy diagrams of the alkaline earth metals magnesium and calcium. (Adapted from [456].)

5.13   Fourier transform spectroscopy

Fourier Transform Infrared (FTIR) spectrometry was originally developed to overcome the limitations encountered with dispersive instruments in recording infrared molecular spectra over wide spectral intervals. Indeed, the main difficulty was the slow scanning process and a method for measuring all of the infrared frequencies simultaneously, rather than individually, was needed. So a solution was developed which employed an interferometer. The latter is indeed able to produce a unique type of signal (interferogram) which has all of the infrared frequencies *encoded* into it. In this way, the signal can be measured very quickly, usually on the order of a few seconds. Thus, the time element per sample is reduced to a matter of a few seconds rather than several minutes. With reference to Figure 5.43, to measure a spectrum with an FT spectrometer, a Michelson interferometer is illuminated with a white or polychromatic source of radiation and the movable mirror is translated over a distance $-x_{max}, +x_{max}$ which depends on the desired resolution. The output signal is passed through a sample, and the resulting interferogram signal is received by an IR detector. The signal produced by the detector is sampled at certain increments of x. Then, a means of *decoding* the individual frequencies is needed, which can be accomplished through the Fourier Transform (via computer) [472, 473].

Formally, an interferogram for a polychromatic source which consists of frequencies from $0 \to \overline{\nu}_m$ is

$$I(x) = \int_0^{\overline{\nu}_m} I(\overline{\nu})[1 + \cos(2\pi\overline{\nu}x)]d\overline{\nu}$$

$$= \frac{1}{2}I(0) + \int_0^{\overline{\nu}_m} I(\overline{\nu})\cos(2\pi\overline{\nu}x)d\overline{\nu} \tag{5.183}$$

With many different wavelengths present, such interferogram exhibits the following fea-

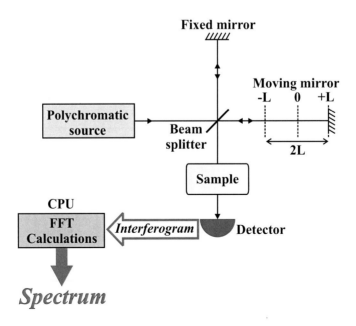

FIGURE 5.43
Schematic diagram of a Michelson interferometer configured for FTIR ($L = x_{max}/2$).

tures. Ideally, it is symmetrical about $x = 0$; when $x = 0$ the interference between all of the frequencies is constructive, giving rise to a central maximum; instead, for $x = \infty$ the frequencies combine both constructively and destructively and the net contribution due to the integral in Equation 5.183 vanishes. Thus $I(\infty) = I(0)/2$ and Equation 5.183 becomes

$$F(x) = \int_0^{\bar{\nu}_m} I(\bar{\nu}) \cos(2\pi\bar{\nu}x)d\bar{\nu} \tag{5.184}$$

where $F(x) = I(x) - I(\infty)$. Note that $F(x)$ is symmetric about $x = 0$ because cosine is an even function. Now let us take the Fourier transform of $F(x)$

$$\int_{-\infty}^{+\infty} F(x)e^{2\pi i f x}dx = \int_{-\infty}^{+\infty} F(x)\cos(2\pi f x)dx$$

$$= \int_0^{\bar{\nu}_m} I(\bar{\nu}) \left[\int_{-\infty}^{+\infty} \cos(2\pi\bar{\nu}x)\cos(2\pi f x)dx \right] d\bar{\nu}$$

$$= \frac{1}{2}\int_0^{\bar{\nu}_m} I(\bar{\nu}) \left\{ \int_{-\infty}^{+\infty} \left[e^{2\pi i(\bar{\nu}+f)x} + e^{2\pi i(\bar{\nu}-f)x} \right] dx \right\} d\bar{\nu}$$

$$= \frac{1}{2}\int_0^{\bar{\nu}_m} I(\bar{\nu})[\delta(\bar{\nu}+f) + \delta(\bar{\nu}-f)]d\bar{\nu} = \frac{I(f)}{2} \tag{5.185}$$

where $I(-f) = 0$ has been exploited. Finally, the spectrum is obtained as

$$I(f) = 2\int_{-\infty}^{+\infty} F(x)e^{2\pi i f x}dx = 4\int_0^{+\infty} F(x)\cos(2\pi f x)dx \tag{5.186}$$

The spectrum $I(f)$ of the sample under investigation is precisely recovered by computing Equation 5.186 (in practice, two measurements of $I(f)$, with and without the sample, are required). In practice, however, the interferogram is from $-x_{max}$ and $+x_{max}$, not $-\infty$ to $+\infty$. Such limited mirror travel manifests into a loss of information in recovering $I(f)$. The relationship between resolution $\delta_{\bar{\nu}}$ and mirror displacement x_{max} may be derived as follows. Consider a monochromatic wave illuminating the interferometer. From Equation 5.184 we get

$$F_{mon}(x) = I(\bar{\nu}_1) \cos(2\pi\bar{\nu}_1 x) \tag{5.187}$$

In this case, according to 5.186 and the above considerations, the spectrum is practically determined as

$$I_{mon}(f) = 2\int_{-x_{max}}^{+x_{max}} I(\bar{\nu}_1)\cos(2\pi\bar{\nu}_1 x)e^{2\pi i f x}dx$$

$$= 2\int_{-\infty}^{+\infty} I(\bar{\nu}_1)\cos(2\pi\bar{\nu}_1 x)e^{2\pi i f x}rect(x)dx$$

$$= 2F.T.[F_{mon} \cdot rect]$$

$$= 2\int_{-\infty}^{+\infty} F.T.[F_{mon}](u) \cdot F.T.[rect](y-u)du \tag{5.188}$$

where the rectangular function

$$rect(x) = \begin{cases} 1 & -x_{max} \le x \le +x_{max} \\ 0 & \text{elsewhere} \end{cases} \tag{5.189}$$

has been introduced and the convolution theorem used in the last step. Since

$$F.T.[rect] = \int_{-\infty}^{+\infty} rect(x)e^{2\pi i f x}dx = 2x_{max}sinc(2\pi f x_{max}) \tag{5.190}$$

and

$$F.T[F_{mon}] = \int_{-\infty}^{+\infty} I(\overline{\nu}_1)\cos(2\pi\overline{\nu}_1 x)e^{2\pi i f x}dx$$

$$= \frac{1}{2}\int_{-\infty}^{+\infty} I(\overline{\nu}_1)\left[e^{2\pi i(f+\overline{\nu}_1)x} + e^{2\pi i(f-\overline{\nu}_1)x}\right]dx$$

$$= \frac{1}{2}I(\overline{\nu}_1)[\delta(f+\overline{\nu}_1) + \delta(f-\overline{\nu}_1)]$$

$$= \frac{1}{2}I(\overline{\nu}_1)\delta(f-\overline{\nu}_1)] \tag{5.191}$$

(the unphysical negative frequency has been discarded in the last step), Equation 5.188 becomes

$$I_{mon}(f) = 2\int_{-\infty}^{+\infty} x_{max}sinc(2\pi u x_{max})I(\overline{\nu}_1)\delta(f-\overline{\nu}_1-u)du$$

$$= I(\overline{\nu}_1)2x_{max}sinc[2\pi(f-\overline{\nu}_1)x_{max}] \equiv I(\overline{\nu}_1)\cdot ILS \tag{5.192}$$

where the instrumental lineshape $ILS = 2x_{max}sinc[2\pi(f-\overline{\nu}_1)x_{max}]$ has been introduced. Now we observe that the *side lobes* of the *sinc* function drop off 22% below zero and that such unacceptable ringing in the spectrum is closely related to the sharp edges of the rectangular function. Thus, an apodization procedure is needed, that is choosing a gentler aperture function. The most common one is the following triangular aperture

$$tri(x) = \begin{cases} 1 - \dfrac{|x|}{x_{max}} & -x_{max} < x < +x_{max} \\ 0 & \text{elsewhere} \end{cases} \tag{5.193}$$

whose Fourier transform of $tri(x)$ is

$$F.T.[tri(x)] = \int_{-x_{max}}^{+x_{max}} \left(1 - \frac{|x|}{x_{max}}\right)e^{2\pi i f x}dx$$

$$= \frac{\sin(2\pi f x_{max})}{\pi f} - \frac{2}{x_{max}}\int_0^{x_{max}} x\cos(2\pi f x)dx$$

$$= x_{max}sinc^2(\pi f x_{max}) \tag{5.194}$$

This is characterized by the absence of negative side lobes, an increased linewidth, and small-size positive lobes. In this case, a monochromatic line $\overline{\nu}$ gives the following spectrum

$$I'_{mon}(f) = I(\overline{\nu}_1)ILS' \tag{5.195}$$

with $ILS' = x_{max}sinc^2[\pi(f-\overline{\nu}_1)x_{max}]$. The relation between resolution $\delta_{\overline{\nu}}$ and mirror scan $2L$ may be derived by considering an interferometer to be illuminated with two monochromatic sources, $\overline{\nu}_1$ and $\overline{\nu}_2$, where $\overline{\nu}_2 - \overline{\nu}_1 = \delta_{\overline{\nu}}$. In order to resolve $\overline{\nu}_1$ and $\overline{\nu}_2$, one generally uses the Rayleigh criterion, which is fulfilled when the central maximum of $ILS'(f-\overline{\nu}_2)$ falls upon the first zero of $ILS'(f-\overline{\nu}_1)$. Thus, one obtains

$$\delta_{\overline{\nu}} = \frac{1}{x_{max}} = \frac{1}{2L} \tag{5.196}$$

where the last step follows from the fact that, when the moving mirror travels between $-L$ and L, the optical path difference varies between $-x_{max} = -2L$ and $x_{max} = 2L$. For example, if $2L$ is 10 cm, then $\delta_{\bar{\nu}} = 0.1$ cm^{-1}. However, the above resolution assumes that a perfectly collimated light beam propagates throughout the interferometer. In practice, due to divergence of the input source, paths travelled by the rays at the edge of the beam differ from those travelled by the axial ones. This means that, for some value of x, the axial and extreme rays will interfere destructively at all frequencies, such that no further increase in resolution originates from ulterior mirror displacement. The maximum tolerable beam divergence without degrading resolution is formulated as the maximum half-angle of the beam, $\gamma_{max} = \sqrt{\delta_{\bar{\nu}}/\nu_{max}}$ (with ν_{max} being the highest frequency in the spectrum). The necessary reduction in the beam divergence, as set by ν_{max}, is undertaken by inserting an aperture into the output beam from the interferometer.

Another important issue concerns the number of data points to be sampled from an interferogram in order to retrieve the spectrum with full information (within the limit of resolution). According to the Nyquist criterion, if the spectrum extends from $0 < \nu \leq \nu_{max}$, this is given by $2\nu_{max}/\delta_{\bar{\nu}}$. For example, if a spectrum is to be measured with a resolution of 1 cm^{-1}, and if the highest frequency in the spectrum is 5000 cm^{-1}, then 10000 data points must be sampled on each side of the $x = 0$ point of the interferogram. According to the Cooley-Tukey fast Fourier transform algorithm, the data are digitized in equal increments of path difference Δx [474]. Rather than collecting data by the *step and integrate* procedure, as in early systems, modern instruments use a *rapid scan* approach, in which the infrared radiation is modulated (in the kHz frequency range), and many interferograms are acquired and averaged. State-of-the-art, commercially available FTIR systems may exhibit spectral resolutions as low as 30 MHz.

We close this pedagogic discussion by mentioning two celebrated advantages associated with FTIR with respect to dispersive instruments. The first one, known as Jacquinot's or the throughput advantage, arises from the fact that, due to the absence of grating and slit between the source and the detector, the energy loss in a FT spectrometer is much reduced compared to a dispersive instrument. This means that interferometric spectrometers permit the observation of spectra from very weak sources and indeed they are widely utilized in astronomical observations.

The Felgett advantage can be understood as follows. Consider a spectrum of width $\Delta\bar{\nu}$ to be measured with resolution $\delta\bar{\nu}$, which corresponds to a number of elements

$$M = \frac{\Delta\bar{\nu}}{\delta\bar{\nu}} \tag{5.197}$$

Now, if a grating/prism instrument is used, each of them is severally observed for a time T/M, with T being the time spent to acquire the whole spectrum. Thus, since the noise is proportional to $\sqrt{T/M}$ in an element of width $\delta\bar{\nu}$ (in the infrared region the noise is indeed random and independent of the signal level), we can write

$$\left(\frac{S}{N}\right)_{grating} \propto \frac{T/M}{\sqrt{T/M}} = \sqrt{T/M} \tag{5.198}$$

For an interferometer, instead, the signal coming from all the elements is accepted at the same time, such that the signal in one element is $\propto T$, while the noise is again $\propto \sqrt{T}$. As a result, for the signal-to-noise in an interferometer, we get

$$\left(\frac{S}{N}\right)_{interferometer} \propto \frac{T}{\sqrt{T}} = \sqrt{T} \tag{5.199}$$

Finally, the Felgett advantage can be expressed as

$$\frac{(S/N)_{interferometer}}{(S/N)_{grating}} = \sqrt{M} \tag{5.200}$$

While an updated and comprehensive treatment of FTIR spectroscopy can be found in [475], specific applications in the emerging field of FTIR spectroscopy based on optical frequency combs will be discussed in the next chapter.

5.14   Raman spectroscopy

First discovered by C.V. Raman in 1928, this spectroscopic method relies on inelastic scattering of light by molecules. The laser radiation interacts with the vibrational, rotational or electronic modes in the molecular system, resulting in the energy of the photons being shifted up or down (Figure 5.44).

In a first type of interaction (Stokes process), following the collision between a laser photon of energy $\hbar\omega_1$ and a molecule in the initial level E_i, a photon with lower energy, $\hbar\omega_s$, is detected, while the molecule is found in a higher-energy level E_f

$$\omega_s = \omega_1 - \omega_{fi} \quad \text{with} \quad \omega_{fi} \equiv \frac{E_f - E_i}{\hbar} \tag{5.201}$$

Conversely, if the incoming photon is scattered by a molecule in an excited state, superelastic scattering may happen, where the excitation energy is relocated to the scattered photon, which now possesses a higher energy than the incident photon (anti-Stokes process):

$$\omega_{as} = \omega_1 + \omega_{if} \quad \text{with} \quad \omega_{if} \equiv \frac{E_i - E_f}{\hbar} \tag{5.202}$$

A third type of interaction, referred to as elastic Rayleigh scattering, is possible for a molecule with no Raman-active modes (see below). In this case, after absorbing a photon of frequency ω_1, the excited molecule returns back to the same basic vibrational state and emits light with the same frequency ω_1.

In order to understand the physical origin of Raman scattering, we now consider an approach in which both the electromagnetic radiation and the material system are treated

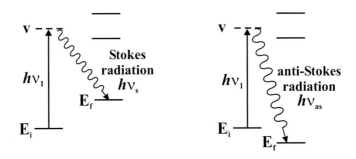

FIGURE 5.44
Schematic level diagram of Raman scattering.

classically [476]. The electric field (associated with the laser) distorts the electron cloud of the molecule, thereby creating an induced electric dipole moment which, in turn, will emit, i.e., scatter, EM radiation. Our target is to derive the frequency-dependent, linear induced-electric-dipole vector ($\boldsymbol{\mu}_{ind}$) of a molecule, by exploiting the relationship

$$\boldsymbol{\mu}_{ind} = \boldsymbol{\alpha} \boldsymbol{E} \tag{5.203}$$

where $\boldsymbol{E} \equiv (E_x, E_y, E_z)$ is the electric field vector associated with the incident monochromatic plane wave of frequency ω, and $\boldsymbol{\alpha}$ denotes the polarizability tensor of the molecule. In general, this latter quantity will be a function of the nuclear coordinates and hence of the molecular vibrational frequencies. Therefore, the frequency dependence of the induced-electric-dipole vector can be retrieved by inserting into Equation 5.203 the frequency dependence of $\boldsymbol{E}$ and $\boldsymbol{\alpha}$. For the sake of simplicity, the scattering system will consist of just one molecule which is space-fixed in its equilibrium configuration: the molecule does not rotate, but the nuclei are free to vibrate about their equilibrium positions. By denoting with $Q_k, Q_l, ...$ the normal coordinates of vibration associated with the molecular vibrational frequencies $\omega_k, \omega_l, ...$, then we can expand each component α_{ij} of the polarizability tensor in a Taylor series as follows

$$\alpha_{ij} = (\alpha_{ij})_0 + \sum_k \left(\frac{\partial \alpha_{ij}}{\partial Q_k}\right)_0 Q_k + \frac{1}{2} \sum_{k,l} \left(\frac{\partial^2 \alpha_{ij}}{\partial Q_k \partial Q_l}\right)_0 Q_k Q_l + ... \tag{5.204}$$

where the summations are over all normal coordinates, and the subscript '0' indicates that the derivatives are taken at the equilibrium configuration. Moreover, we shall neglect the terms involving powers of Q higher than the first (electrical harmonic approximation). Next, let's focus our attention on one normal mode, Q_k, whereupon we can rewrite Equation 5.204 as

$$(\alpha_{ij})_k = (\alpha_{ij})_0 + (\alpha'_{ij})_k Q_k \tag{5.205}$$

where the quantities

$$(\alpha'_{ij})_k = \left(\frac{\partial \alpha_{ij}}{\partial Q_k}\right)_0 \tag{5.206}$$

are the components of the derived polarizability tensor. Since Equation 5.206 holds for all tensor components, we can write

$$\boldsymbol{\alpha}_k = \boldsymbol{\alpha}_0 + \boldsymbol{\alpha}'_k Q_k \tag{5.207}$$

Under the assumption of simple harmonic motion, the time dependence of Q_k is given by

$$Q_k = Q_{k0} \cos(\omega_k t + \delta_k) \tag{5.208}$$

where Q_{k0} represents the normal coordinate amplitude and δ_k a phase factor. Then, Equation 5.208 and Equation 5.207 can be combined to yield the time dependence of the polarizability tensor resulting from the k-th molecular vibration:

$$\boldsymbol{\alpha}_k = \boldsymbol{\alpha}_0 + \boldsymbol{\alpha}'_k Q_{k0} \cos(\omega_k t + \delta_k) \tag{5.209}$$

Now we introduce $\boldsymbol{E} = \boldsymbol{E}_0 \cos \omega_1 t$ and 5.209 into Equation 5.203 to obtain

$$\boldsymbol{\mu}_{ind} = \boldsymbol{\alpha}_0 \boldsymbol{E}_0 \cos \omega_1 t + \boldsymbol{\alpha}'_k \boldsymbol{E}_0 Q_{k0} \cos(\omega_k t + \delta_k) \cos \omega_1 t \tag{5.210}$$

which by a little algebra can be re-formulated as

$$\boldsymbol{\mu}_{ind} = \boldsymbol{\mu}_{Ray}(\omega_1) + \boldsymbol{\mu}_{Ram,s}(\omega_1 - \omega_k) + \boldsymbol{\mu}_{Ram,as}(\omega_1 + \omega_k) \tag{5.211}$$

where

$$\boldsymbol{\mu}_{Ray}(\omega_1) = \boldsymbol{\alpha}_0 \boldsymbol{E}_0 \cos\omega_1 t \tag{5.212}$$

corresponds to Rayleigh scattering, and

$$\boldsymbol{\mu}_{Ram,s}(\omega_1 - \omega_k) = (1/2)\boldsymbol{\alpha}'_k \boldsymbol{E}_0 Q_{k0} \cos[(\omega_1 - \omega_k)t - \delta_k] \tag{5.213}$$

$$\boldsymbol{\mu}_{Ram,as}(\omega_1 + \omega_k) = (1/2)\boldsymbol{\alpha}'_k \boldsymbol{E}_0 Q_{k0} \cos[(\omega_1 + \omega_k)t + \delta_k] \tag{5.214}$$

correspond to the Stokes and the anti-Stokes component of Raman scattering, respectively. So far we have considered only the induced dipole moment; if the molecule also possesses a permanent dipole moment $\boldsymbol{\mu}_{perm}$, this depends on the nuclear displacements of the vibrating molecule as well. For small displacements from the equilibrium position, this quantity can also be expanded into a Taylor series

$$\boldsymbol{\mu}_{perm} = (\boldsymbol{\mu}_{perm})_0 + \sum_k \left(\frac{\partial\boldsymbol{\mu}_{perm}}{\partial Q_k}\right)_0 Q_k + ... \tag{5.215}$$

which for just one normal mode reduces to

$$\boldsymbol{\mu}_{perm} = (\boldsymbol{\mu}_{perm})_0 + \left(\frac{\partial\boldsymbol{\mu}_{perm}}{\partial Q_k}\right)_0 Q_k \tag{5.216}$$

Thus the total dipole moment for the molecule is

$$\begin{aligned}
\boldsymbol{\mu}_{tot} = \boldsymbol{\mu}_{perm} + \boldsymbol{\mu}_{ind} \;=\;& (\boldsymbol{\mu}_{perm})_0 + \boldsymbol{\alpha}_0 \boldsymbol{E}_0 \cos\omega t \\
& + \left(\frac{\partial\boldsymbol{\mu}_{perm}}{\partial Q_k}\right)_0 Q_{k0} \cos(\omega_k t + \delta_k) \\
& + \frac{1}{2}\boldsymbol{\alpha}'_k \boldsymbol{E}_0 Q_{k0} \cos[(\omega - \omega_k)t - \delta_k] \\
& + \frac{1}{2}\boldsymbol{\alpha}'_k \boldsymbol{E}_0 Q_{k0} \cos[(\omega + \omega_k)t + \delta_k]
\end{aligned} \tag{5.217}$$

where the third term describes the IR spectrum. Now we realize that:

- A vibrational mode is infrared active if the derivative of at least one component of the derived dipole moment vector with respect to the normal coordinate Q_k, taken at the equilibrium position, is non-zero: $(\partial\boldsymbol{\mu}_{perm}/\partial Q_k)_0 \neq 0$.

- The vibrational mode is Raman active if at least one component of the derived polarizability tensor with respect to the normal coordinate Q_k, taken at the equilibrium position, is non-zero: $\boldsymbol{\alpha}'_k \neq 0$.

- Since the classical equilibrium polarizability tensor $\boldsymbol{\alpha}_0$ always has some non-zero components, all molecules exhibit Rayleigh scattering.

To illustrate these concepts, in the following we consider a few simple cases. Let's start with a homonuclear diatomic molecule, A2, which has just one mode of vibration. Due to the symmetry of the electron distribution, such a molecule has no permanent dipole moment in the equilibrium position; also, this symmetry does not change as the internuclear separation is varied, so that the dipole remains zero during a vibration: as a result, the derivative is zero and the vibration is infrared inactive. Concerning Raman activity, we can represent the non-zero polarizability of the molecule by an ellipsoid having (at equilibrium) one principal axis along the bond direction and the other two principal axes at right angles to the bond direction. In general, this polarizability ellipsoid is characterized by three components; in the

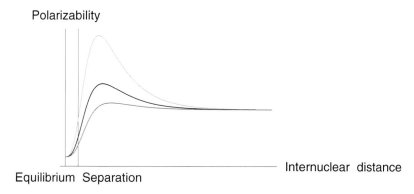

FIGURE 5.45

Qualitative plots of α (black), $\alpha_{\parallel}$ (light gray), and $\alpha_{\perp}$ (gray) as functions of the internuclear distance, for a diatomic homonuclear molecule. (Adapted from [476].)

case of a σ bond, however, these reduce to two components: the polarizability along the bond ($\alpha_{\parallel}$) and the polarizability at right angles to the bond ($\alpha_{\perp}$). Then, for a given internuclear separation, the mean polarizability is defined by $a = (1/3)(\alpha_{\parallel} + \alpha_{\perp})$, while the anisotropy by $\gamma = (\alpha_{\parallel} - \alpha_{\perp})$. In order to calculate how these quantities vary with the internuclear separation (in the neighborhood of the equilibrium position), one has to resort to quantum mechanics. Shown in Figure 5.45, the result is that the vibrations of A2 diatomic molecules are Raman active. We next consider the example of a heteronuclear diatomic molecule (AB), again having just one mode of vibration. The above arguments for the polarizability variations also apply to this case, so that the vibration will be Raman active. Then, let's address the issue of infrared activity. We first observe that, due to the asymmetry in the electron distribution, the AB molecule necessarily possesses a permanent dipole moment. The typical behaviour of the dipole moment component along the bond direction against the internuclear distance is shown in Figure 5.46, whereas the components at right angles to the bond direction are, of course, always zero. As a general result, the maximum dipole moment occurs at an internuclear separation different from the equilibrium distance, which implies that the derivative at the equilibrium position is non-zero: the vibration in AB molecules is thus infrared active. The A2 and AB cases are summarized in Figure 5.47.

Finally, Figure 5.48 also shows the polarizability and dipole moment variations in the neighbourhood of the equilibrium position for a linear ABA molecule.

In conclusion, we should mention that, although the classical theory honestly describes the frequencies $\omega_1 \pm \omega_k$ of the Raman lines, it suffers from three major limitations which can be overcome only within a quantum mechanical treatment: first, it does not give the correct line strengths; second, it is not applicable to molecular rotations (actually, in a classical frame, you cannot assign specific discrete rotational frequencies to molecules); third, it cannot yield information as to how $\boldsymbol{\alpha}'_k$ is related to the properties of the scattering molecule.

To complete this introductory discussion, we deal with a generic concept/experimental setup for measurement of Raman scattering [477]. Since scattered light leaves the sample in all directions, the detectors may be placed at any angle. Figure 5.49 shows the coordinate system for a typical 90-degree scattering arrangement. The incident light propagates along the X direction, while the scattered light is detected along the Y direction. Raman spectra are measured using incident light polarized in the Z direction. The two polarization components of the scattered light are then resolved by means of a polarization analyzer located

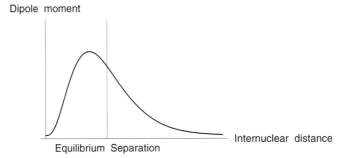

FIGURE 5.46
Dipole moment as a function of internuclear distance in a diatomic molecule AB. The specific form of the plot, and hence the magnitude and sign of the derivative, varies from one molecule to another. (Adapted from [476].)

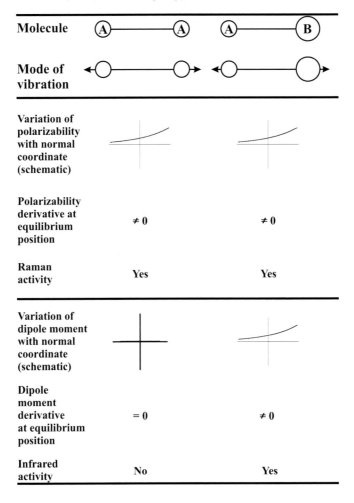

FIGURE 5.47
Illustration of vibrational Raman and infrared activities for an A2 and an AB molecule. (Adapted from [476].)

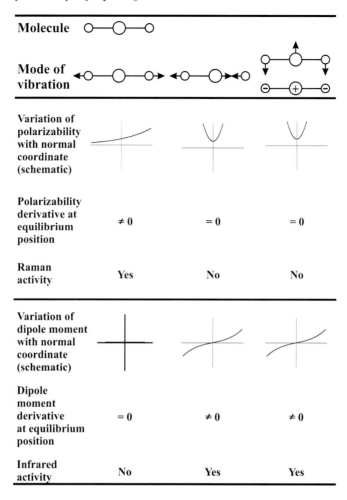

Molecule		
Mode of vibration		
Variation of polarizability with normal coordinate (schematic)		
Polarizability derivative at equilibrium position	$\neq 0$	$= 0$	$= 0$
Raman activity	Yes	No	No
Variation of dipole moment with normal coordinate (schematic)		
Dipole moment derivative at equilibrium position	$= 0$	$\neq 0$	$\neq 0$
Infrared activity	No	Yes	Yes

FIGURE 5.48

Vibrational Raman and infrared activities for a linear ABA molecule. (Adapted from [476].)

between the sample and the detector. The polarized (depolarized) spectrum is measured by observing the scattered light polarized in the Z (X) direction, $I_{pol} \equiv I_{\parallel}$ ($I_{dep} \equiv I_{perp}$). I_{pol} and I_{dep} are in turn proportional to the differential cross sections of the scattering process. These are obtained by projecting the polarizability tensor onto the polarization directions $\hat{e}_{sc}$ and $\hat{e}_1 \equiv (0, 0, 1)$ of the scattered and incident radiation (in cgs units)

$$\left(\frac{d\sigma_{if}}{d\Omega}\right)_{sc,1} = \frac{\omega_1 \omega_{sc}^3}{c^4} \left| \hat{e}_{sc} \begin{pmatrix} (\alpha_{XX})_{if} & (\alpha_{XY})_{if} & (\alpha_{XZ})_{if} \\ (\alpha_{YX})_{if} & (\alpha_{YY})_{if} & (\alpha_{YZ})_{if} \\ (\alpha_{ZX})_{if} & (\alpha_{ZY})_{if} & (\alpha_{ZZ})_{if} \end{pmatrix} \cdot \begin{pmatrix} 0 \\ 0 \\ 1 \end{pmatrix} \right|^2 \tag{5.218}$$

where the transition polarizability tensor having elements such as $(\alpha_{XX})_{if}$, $(\alpha_{XY})_{if}$, etc. connects the initial and final rotational and/or vibrational states, and

$$\hat{e}_{sc} = \begin{cases} (1, 0, 0) & \text{dep} \\ (0, 0, 1) & \text{pol} \end{cases} \tag{5.219}$$

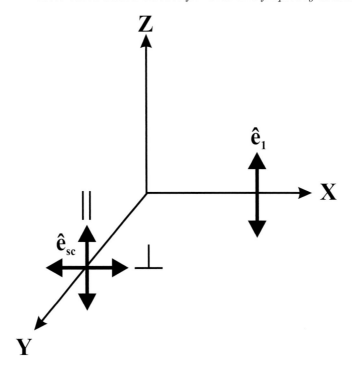

FIGURE 5.49
Ninety-degree Raman scattering geometry. Uppercase letters XYZ indicate the space-fixed coordinate system, i.e. the lab reference frame, whereas lowercase letters refer to the body-fixed coordinate system, i.e. a coordinate system fixed in the molecule. (Adapted from [477].)

So, in the 90 degree scattering arrangement, we have

$$\left(\frac{d\sigma_{if}}{d\Omega}\right)_{sc,1} = \frac{\omega_1\omega_{sc}^3}{c^4} \cdot \begin{cases} |(\alpha_{ZX})_{if}|^2 & \text{dep} \\ |(\alpha_{ZZ})_{if}|^2 & \text{pol} \end{cases} \tag{5.220}$$

Note that, since polarizabilities are normally expressed in CGS units of cm^3, the cross section 5.220 has, as it must, the units of a surface. On the other hand, the elements of the transition polarizability tensor are conveniently expressed according to the Kramers-Heisenberg-Dirac formalism in the molecule frame of reference [477]

$$(\alpha_{\rho\sigma})_{if} = \frac{1}{\hbar}\sum_n \left[\frac{\langle i|\mu_\rho|n\rangle\langle n|\mu_\sigma|f\rangle}{\omega_1 + \omega_{nf} + i\Gamma_n} - \frac{\langle i|\mu_\sigma|n\rangle\langle n|\mu_\rho|f\rangle}{\omega_1 - \omega_{nf} - i\Gamma_n}\right] \tag{5.221}$$

where ρ and σ are directions x, y, z in the molecule frame. In Equation 5.221, the sum extends over all molecular levels $|n\rangle$ having a width Γ_n accessible by single-photon transitions from the initial state $|i\rangle$. Now the problem is to convert molecule-frame tensor elements to the necessary lab-frame components. In the conventional 90° scattering arrangement, we require α_{ZZ} for the polarized spectrum and α_{ZX} for the depolarized one. We are interested here in a sample of randomly oriented molecules, as in the gas or liquid phase. Then, it can be shown that [477]

$$|\alpha_{ZZ}|^2 = \frac{1}{3}\Sigma^0 + \frac{2}{15}\Sigma^2 \tag{5.222}$$

$$|\alpha_{ZX}|^2 = \frac{1}{6}\Sigma^1 + \frac{1}{10}\Sigma^2 \tag{5.223}$$

where the invariants Σ^0, Σ^1 and Σ^2 are written in terms of the molecule-frame components of the polarizability:

$$\Sigma^0 = \frac{1}{3}|\alpha_{xx} + \alpha_{yy} + \alpha_{zz}|^2 \tag{5.224}$$

$$\Sigma^1 = \frac{1}{2}\left[|\alpha_{xy} - \alpha_{yx}|^2 + |\alpha_{xz} - \alpha_{zx}|^2 + |\alpha_{yz} - \alpha_{zy}|^2\right] \tag{5.225}$$

$$\Sigma^2 = \frac{1}{2}\left[|\alpha_{xy} + \alpha_{yx}|^2 + |\alpha_{xz} + \alpha_{zx}|^2 + |\alpha_{yz} + \alpha_{zy}|^2\right]$$
$$+ \frac{1}{3}\left[|\alpha_{xx} - \alpha_{yy}|^2 + |\alpha_{xx} - \alpha_{zz}|^2 + |\alpha_{yy} - \alpha_{zz}|^2\right] \tag{5.226}$$

Now we have all the ingredients to calculate Equation 5.220, provided that the molecule-frame components $(\alpha_{\rho\sigma})_{if}$ of the polarizability have been previously evaluated starting from the corresponding wavefunctions. In conclusion, the intensity of a Raman line at the Stokes or anti-Stokes frequency is proportional to the cross section defined by Equation 5.220 times the product between the population density in the initial level $N(E_i)$ and the incident laser intensity I_1

$$I_{\text{spont. Raman}} \propto \left(\frac{d\sigma_{if}}{d\Omega}\right)_{sc,1} \cdot N(E_i) \cdot I_1 \tag{5.227}$$

Since, at thermal equilibrium, the population density $N(E_i)$ follows the Boltzmann distribution, the intensity of the Stokes lines, for which the initial state may be the vibrational ground state, is usually quite larger than that of the anti-Stokes lines, for which the molecules must have initial excitation energy. Even with strong pump laser beams, the intensity of spontaneously scattered Raman light is often very weak (indeed the scattering cross sections in spontaneous Raman spectroscopy are typically on the order of 10^{-30} cm^2). Nevertheless, according to Equation 5.221, the Raman scattering cross section raises significantly if the laser frequency ω_1 matches a transition frequency ω_{nf} of the molecule (resonance Raman effect). Under such conditions, one can achieve an enhancement of 10^6, but at the expense of increasing the background due to fluorescence. Since both fluorescence and Raman scattering are emitted isotropically, normally the fluorescence cannot be suppressed easily. To overcome this problem, several techniques have been developed including the Inverse Raman effect, the Hyper-Raman effect, the Raman-induced Kerr effect, Surface-enhanced Raman scattering, Coherent anti-Stokes Raman spectroscopy (CARS), and Stimulated Raman scattering (SRS). Since we are only interested in high-resolution Raman spectroscopy of gases with cw laser sources, we can restrict our discussion to these two latter techniques.

5.14.1 Coherent anti-Stokes spectroscopy

Among the different variants of Raman spectroscopy, coherent anti-Stokes Raman spectroscopy (CARS) has probably the most general utility and offers several distinct advantages [478]. In the CARS scheme, two relatively high-power laser beams, at frequencies ω_l and ω_s, are focused in the sample under investigation. As a result of mixing the two lasers, a coherent beam, resembling a laser beam at frequency $\omega_{as} = 2\omega_l - \omega_s$, is generated in the medium (see Figure 5.50), the efficiency of such conversion process depending critically upon the presence of molecular resonances at a frequency $\omega_l - \omega_s$. A CARS spectrum is then retrieved by recording the intensity variation of the beam at ω_{as}, as $\omega_l - \omega_s$ is swept over the molecular resonance Conversion of the two laser beams into the anti-Stokes component at

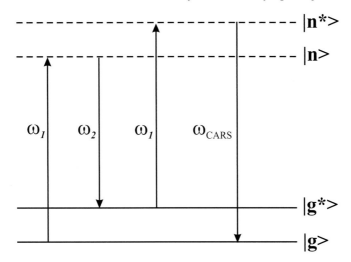

FIGURE 5.50
The energy diagram of coherent anti-Stokes Raman spectroscopy: $\omega_1 \equiv \omega_l$, $\omega_2 \equiv \omega_s$, $\omega_{CARS} \equiv \omega_{as}$. The solid lines represent molecular energy levels and the dashed lines indicate virtual levels.

$\omega_{as} = 2\omega_l - \omega_s$ is a direct consequence of the non-linear dielectric properties of the material. As already discussed, the macroscopic polarization $\boldsymbol{P}$ of a medium in an applied electric field $\boldsymbol{E}$ may generally be expressed as a power series

$$P_i = \sum_j \chi_{ij}^{(1)} E_j + \sum_j \sum_k \chi_{ijk}^{(2)} E_j E_k + \sum_j \sum_k \sum_l \chi_{ijkl}^{(3)} E_j E_k E_l + ... \qquad (5.228)$$

where $\chi^{(i)}$ denotes the dielectric susceptibility tensor of rank $i+1$ associated with the i-order of the electric field. In the scope of Raman spectroscopy, the first-order term (which is the only relevant term under low intensity fields) describes spontaneous Raman scattering, the second-order one is responsible for the hyper-Raman effect, while the third-order term is related to CARS. In an isotropic medium, such as a gas, inversion symmetry exists in its macroscopic dielectric properties, so that the lowest order non-linearity which may be present is due to the third-order susceptibility. In what follows, we shall assume the same direction for all the involved electric fields so as to treat each field as a scalar quantity. Then, using complex notation, we express the electric field at frequency $\omega_i = 2\pi c/\lambda_i$ as

$$E_i(\omega_i) = \frac{1}{2}\left[\mathcal{E}_i e^{i(k_i z - \omega_i t)} + c.c.\right] \qquad (5.229)$$

where $\mathcal{E}_i$ is the amplitude, $k_i = \omega_i n_i/c$ is the momentum vector and n_i the index of refraction at ω_i. In the CARS case, only two frequencies at $\omega_1 \equiv \omega_l$ and $\omega_2 \equiv \omega_s$ are introduced; consequently, only those terms in which the polarization varies by $\omega_3 = 2\omega_1 - \omega_2$ are retained. Thus we have

$$P^{(3)} = \frac{1}{8}[3\chi^{(3)}(-\omega_3, \omega_1, \omega_1, -\omega_2)]\mathcal{E}_1^2\mathcal{E}_2^* e^{i[(2k_1 - k_2)z - (2\omega_1 - \omega_2)t]} + c.c. \qquad (5.230)$$

where the usual Bloembergen notation is adopted such that $\chi^{(3)}(-\omega_a, \omega_b, \omega_c, \omega_d)$ represents the susceptibility for the process in which $\omega_a = \omega_b + \omega_c + \omega_d$. By substitution of Equation 5.230 into Maxwell's equations, the gain equation for plane waves is obtained as

$$\frac{d\mathcal{E}_{as}}{dz} = -\frac{i\pi\omega_{as}}{2cn_{as}}[3\chi^{(3)}]\mathcal{E}_l^2\mathcal{E}_s^* e^{i(2k_l - k_s - k_{as})z} \qquad (5.231)$$

where the previous subscripts 1, 2, and 3 have been replaced by l, s, and as to symbolize the laser frequency, the Stokes frequency, and the anti-Stokes frequency, respectively. By integrating Equation 5.231 and using the intensity $I_i = (c/8\pi)|\mathcal{E}_i|^2$, we get

$$I_{as} = \left(\frac{4\pi^2\omega_{as}}{n_{as}c^2}\right)^2 |3\chi^{(3)}|^2 I_l^2 I_s L^2 \left[\frac{\sin(\Delta kL/2)}{\Delta kL/2}\right]^2 \tag{5.232}$$

where L is the length over which the beams are mixed through the sample and $\Delta k = 2k_1 - k_s - k_{as}$ represents the momentum mismatch resulting from the fact that, due to dispersion in the medium, propagating waves move in and out of phase. Since dispersion in gases is usually pretty small, phase matching over moderate path lengths is readily attained (of course, this is not the case for condensed media). Next, we want to find an explicit expression for $\chi^{(3)}$ [478, 322]. To this aim we first observe that, in the presence of a vibrational resonance, the total susceptibility associated with the process $2\omega_l - \omega_s = \omega_{as}$ is the sum of a frequency-dependent resonant part and a nearly frequency-independent non-resonant part

$$\chi^{(3)} = \chi^{(3)}_{res} + \chi^{(3)}_{NR} \tag{5.233}$$

It is precisely the presence of the $\chi^{(3)}_{NR}$ term that limits the sensitivity of the technique, as Raman spectra can be recorded only to an extent that $\chi^{(3)}_{res}$ exceeds $\chi^{(3)}_{NR}$. An expression for $\chi^{(3)}_{res}$ can be derived following a classical approach. We start by assuming that the vibrational mode can be described as a simple harmonic oscillator of resonance frequency ω_v and damping constant Γ. We also relate the molecular polarizability α to a bond stretching coordinate q (i.e., the deviation of the internuclear distance from its equilibrium value)

$$\alpha = \alpha_0 + \left(\frac{\partial\alpha}{\partial q}\right)_0 q + \dots \tag{5.234}$$

where the subscript 0 denotes the equilibrium position. In this way, the force exerted on the oscillator by a field because of this polarizability is given by

$$F = \frac{1}{2}\left(\frac{\partial\alpha}{\partial q}\right)_0 \langle E^2\rangle \tag{5.235}$$

where the angular brackets denote a time average over an optical period. The equation of motion for the simple damped harmonic oscillator is thus

$$\ddot{q} + \Gamma\dot{q} + \omega_v^2 q = \frac{1}{2m}\left(\frac{\partial\alpha}{\partial q}\right)_0 \langle E^2\rangle \tag{5.236}$$

where

$$E = E_l + E_s = \frac{1}{2}\left[\mathcal{E}_l e^{i(k_l z - \omega_l t)} + \mathcal{E}_s e^{i(k_s z - \omega_s t)} + c.c.\right] \tag{5.237}$$

Since we are only interested in the time-varying part of $\langle E^2\rangle$, we can write

$$\langle E^2\rangle = \frac{1}{2}\mathcal{E}_l\mathcal{E}_s^* e^{i(Kz - \Omega t)} + c.c. \tag{5.238}$$

where only the term oscillating at the lowest frequency has been retained and $K = k_l - k_s$ and $\Omega = \omega_l - \omega_s$. Now by adopting for Equation 5.236 the following trial solution

$$q(t) = \frac{1}{2}\left[q(\Omega)e^{i(Kz - \Omega t)} + c.c.\right] \tag{5.239}$$

we find the amplitude of the molecular vibration as

$$q(\Omega) = \frac{(1/2m)(\partial\alpha/\partial q)_0 \mathcal{E}_l \mathcal{E}_s^*}{\omega_v^2 - \Omega^2 - i\Omega\Gamma} \tag{5.240}$$

Now, since the polarization of the medium is by definition given by

$$\begin{aligned}
P(t) &\equiv N\left(\frac{\partial\alpha}{\partial q}\right)_0 q(t)E(t) \\
&= \frac{N}{4}\left(\frac{\partial\alpha}{\partial q}\right)_0 \left[q(\Omega)e^{i(Kz-\Omega t)} + c.c.\right] \\
&\quad \cdot \left[\mathcal{E}_l e^{i(k_l z - \omega_l t)} + \mathcal{E}_s e^{i(k_s z - \omega_s t)} + c.c.\right]
\end{aligned} \tag{5.241}$$

where N is the molecular number density, the component of the polarization oscillating at $\omega_{as} = 2\omega_l - \omega_s$ is

$$\begin{aligned}
P_{as}(t) &= \frac{N}{4}\left(\frac{\partial\alpha}{\partial q}\right)_0 q(\Omega)\mathcal{E}_l e^{i(2k_l - k_s)z} e^{-i(2\omega_l - \omega_s)t} + c.c. \\
&= \frac{1}{8}\frac{\dfrac{N}{m}\left(\dfrac{\partial\alpha}{\partial q}\right)_0^2}{\omega_v^2 - \Omega^2 - i\Omega\Gamma}\mathcal{E}_l^2 \mathcal{E}_s^* e^{i(2k_l - k_s)z} e^{-i(2\omega_l - \omega_s)t} + c.c.
\end{aligned} \tag{5.242}$$

where Equation 5.240 has been used in the last step. Comparison of Equation 5.242 with Equation 5.230 then yields

$$3\chi_{res}^{(3)}(\omega_{as}) = \frac{N}{m}\left(\frac{\partial\alpha}{\partial q}\right)_0^2 \frac{\Delta_j}{\omega_v^2 - (\omega_l - \omega_s)^2 - i\Gamma(\omega_l - \omega_s)} \tag{5.243}$$

where $N\cdot\Delta_j$ represents the difference in population between the lower and upper state for a particular transition j, Δ_j accounting for the statistical distribution, i.e. $\Delta_j = 1(0)$ in the limit of zero (infinite) temperature. In the vicinity of a resonance, we have $\omega_v - (\omega_l - \omega_s) \ll \omega_v$ and Equation 5.243 simplifies to

$$3\chi_{res}^{(3)}(\omega_{as}) \simeq \frac{A}{2[\omega_v - (\omega_l - \omega_s)] - i\Gamma} \tag{5.244}$$

with $A = (N/m)(\partial\alpha/\partial q)_0^2(1/\omega_v)$. Inserting the square modulus of Equation 5.244 into Equation 5.232, under phase matching conditions, the intensity of the beam at ω_{as} can be expressed as

$$I_{as} = \left[\frac{4\pi^2 \omega_{as} L N}{m\omega_v n_{as} c^2}\left(\frac{\partial\alpha}{\partial q}\right)_0^2\right]^2 \frac{1}{4[\omega_v - (\omega_l - \omega_s)]^2 + \Gamma^2} I_l^2 I_s \tag{5.245}$$

from which one can see that the strength of the CARS signal increases, in particular, with the square of the molecular density N and with the product $I_l^2 I_s$. As anticipated, the intensity I_{as} of the beam at frequency $\omega_{as} = 2\omega_l - \omega_s$ changes as $\omega_l - \omega_s$ is swept around the molecular resonance at ω_v, giving rise to a Lorentzian profile. However, if the non-resonant susceptibility $\chi_{NR}^{(3)}$ is non-negligible, it is the quantity $|\chi_{res}^{(3)} + \chi_{NR}^{(3)}|^2$ that must be inserted into Equation 5.232. This results in a distortion of the Lorentzian lineshape. Moreover, as it is apparent from the above derivation, Equation 5.245 only incorporates, via Γ, the effect of homogeneous broadening mechanisms. Actually, Doppler broadening can be accounted for by replacing Equation 5.244 with

$$3\tilde{\chi}_{res}^{(3)}(\omega_{as}) \simeq A\sqrt{\frac{m}{2\pi k_B T}} \int_{-\infty}^{+\infty} \frac{e^{-\frac{mv_z^2}{2k_B T}}}{2[\omega_v(1 + \frac{v_z}{c}) - (\omega_l - \omega_s)] - i\Gamma} dv_z \qquad (5.246)$$

where v_z is the velocity component along the direction of propagation (z) of the beam at ω_{as} [479].

In conclusion, the advantages of CARS can be summarized as follows [379]:

- As suggested by the presence of a resonant denominator in Equation 5.245, the intensity of the anti-Stokes signal is by far larger than in spontaneous Raman spectroscopy.

- While in spontaneous Raman spectroscopy molecular lines are Doppler-broadened by the thermal motion of the scattering molecules, in CARS experiments the ultimate spectral resolution is virtually dictated by the bandwidth of the two incident lasers and the molecular intrinsic linewidth.

- Since the anti-Stokes wave forms a highly collimated beam, the detector can be far away from the interaction region, which considerably diminishes any spontaneous background signal.

A typical experimental arrangement to generate CARS signals with cw lasers is shown in Figure 5.51. Such an apparatus was used to measure the $Q(2)$ vibrational line in D_2 around 2987 cm^{-1} [480]. The source providing the ω_l frequency was an argon laser at 514 nm (5 W), while the one providing ω_s was a tunable, single-mode dye laser near 607 nm (50 mW). The two laser beams were combined, using the constant deviation prism P1, and focused into the cell with the lens L1. The three emerging beams, Ar, dye, and anti-Stokes ($\lambda_{as}^{-1} = 2\lambda_l^{-1} - \lambda_s^{-1} \simeq 445$ nm), were collimated by L2 and dispersed by the second prism P2. The use of the various diaphragms, prims P2, narrow-band filter F reduced the parasitic light at the CARS wavelength to a point below the dark count of the photomultiplier. With this setup, the center frequency of the line under investigation was measured for different gas pressure values. The extrapolated zero-pressure frequency was then 2987.237 ± 0.001 cm^{-1}.

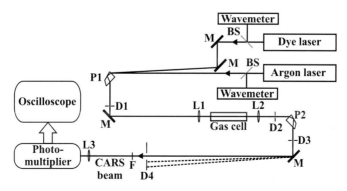

FIGURE 5.51
Schematic layout of the apparatus for cw CARS in gases.

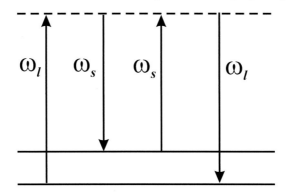

FIGURE 5.52
Energy diagram of stimulated Raman scattering.

5.14.2   Stimulated Raman scattering

Stimulated Raman scattering (SRS) is a third-order non-linear $\chi^{(3)}$ process too. In this case, however, the role of the laser at ω_2 is played by the Stokes wave generated by the pump laser (at ω_l) itself (Figure 5.52). The physical origin of SRS can be grasped in terms of the interactions illustrated in Figure 5.53 [322]. In the upper frame, the molecular vibration is shown to modulate the refractive index of the medium at frequency ω_v, which impresses frequency sidebands onto the laser field. In the lower frame, the Stokes field at frequency $\omega_s = \omega_l - \omega_v$ is shown to beat with the laser field, which produces a modulated intensity, $I(t) = I_0 + I_1 \cos(\omega_l - \omega_s)t$. This latter coherently excites the molecular oscillation at $\omega_l - \omega_s = \omega_v$. These two processes strengthen each other: the lower-frame interaction gives rise to a stronger molecular vibration which, by the upper-frame interaction, originates a larger Stokes field; in turn, this produces a stronger molecular vibration. Such a parametric interaction can be quantitatively described within the same exact formalism introduced above for CARS. Typically, SRS is a very strong scattering process: more than 10% of the energy of the incident laser beam is converted into the Stokes frequency, the emission occurring in a narrow cone in the forward and backward directions.

The most straightforward form of SRS is that in which a linearly polarized pump beam (at $\omega_l \equiv \Omega$) induces a Raman gain (loss) at the Stokes (anti-Stokes) frequency ω which is measured by a probe beam linearly polarized either parallel or perpendicular to the pump (inverse Raman effect is the terminology commonly used to designate the induced loss at

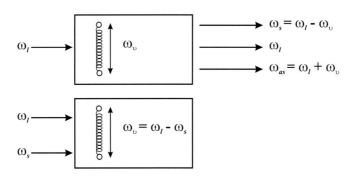

FIGURE 5.53
Interpretation of stimulated Raman scattering. (Adapted from [322].)

the anti-Stokes frequency). The basic setup for cw SRS is shown in Figure 5.54 [481]. The pump beam is modulated prior to being focused through the sample, thus producing a modulated Raman gain at the Stokes frequency which is synchronously (at the pump beam modulation frequency) detected by the probe beam $P(\omega)$. Then, the Raman spectrum is obtained by scanning the spectral difference frequency between the pump and probe. In the case of Gaussian pump and probe beams with identical beam parameters which are focused collinearly through a sample in exact coincidence, it can be shown that the fractional power gain $\delta P(\omega)/P(\omega)$ at the probe frequency ω induced by a pump beam at frequency Ω is given approximately by

$$\frac{\delta P(\omega)}{P(\omega)} = \frac{96\pi^2 \Omega \omega}{nc^3} \Im[\chi_{(3)}^{iijj}(-\omega, \omega, \Omega, -\Omega)]P(\Omega) \qquad (5.247)$$

where n is the refractive index at ω and $\chi_{(3)}^{iijj}$ is the third-order non-linear susceptibility. From the above expression, the linear dependence of the SRS signal on $\chi^{(3)}$ and the pump power $P(\Omega)$ can be appreciated, which contrasts with the quadratic dependence of CARS. Therefore, SRS techniques offer significantly stronger output signal power levels for low-density applications using low-power cw sources.

Although SRS is a powerful technique per se, some variants have been developed to provide enhanced capabilities. In this respect, an example is given by the optically heterodyned polarization technique. Here the beam polarizations are no longer parallel or orthogonal to each other and a polarization analyzer is added at the exit of the sample. In essence, such scheme combines the SRS principle with the SNR enhancement inherent to polarization spectroscopy. Just as an example, let us consider a linearly polarized pump. In this case, one can choose the pump field E_Ω along the x axis with the probe analyzing polarizer oriented at 45 degrees transmitting along the $x - y$ diagonal. Then, the probe is made to transmit a small component through the polarizer by the insertion of a HWP into the probe beam path prior to the sample. In this way, a linearly polarized probe yields $\Im[\chi^{(3)}]$, whereas an elliptically polarized probe gives $\Re[\chi^{(3)}]$. In both cases, the output signal is still linear in $\chi^{(3)}$.

In the very end, we refer the reader to [482] for an updated and extended review on high-resolution Raman spectroscopy of gases.

Another important application of SRS is in the field of Raman lasers. Here, starting from a tunable pump laser at frequency ω_L, intense coherent radiation sources at frequencies $\omega_L \pm n\omega_v$ ($n = 1, 2, 3, \dots$) can be realized, a striking difference with respect to standard lasers lying in the absence of population inversion. An enticing feature is that essentially any Raman laser wavelength can be attained with a proper choice of the pump wavelength, provided that both wavelengths fall in the transparency window of the Raman gain medium. The latter is usually based on either an optical fiber [483], or a solid-state bulk crystal [484, 485], or a silicon waveguide on a chip [486], or a silica toroidal microcavity [487], or a

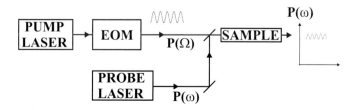

FIGURE 5.54
Schematic experimental setup for cw SRS.

gas [488]. In all cases, low threshold powers are possibly achieved with the aid of high-finesse laser resonators.

In general, one major drawback of these devices is that, since Raman amplification works only at high intensities, other, undesired nonlinearities (e.g. Kerr effect or four-wave mixing) come into play, which eventually prevent from achieving narrow-linewidth operation.

6

Time and frequency measurements with pulsed laser systems

6.1 Introduction

Exploiting the high peak intensities characteristic of short laser pulses to excite multiphoton processes or to drive highly nonlinear phenomena (and generate new wavelengths, for example), while maintaining the high resolution characteristic of CW sources to investigate very narrow spectral structures of atoms or molecules, is one of the forbidden dreams of the laser experimentalist. Unfortunately, the two conditions of short pulse duration and high spectral resolution normally appear in striking contrast, since short pulses invariably correspond to broad spectral bandwidths that limit the frequency resolution to the inverse of the pulse duration (see Figure 6.1a).

However, this is not the end of the story because several techniques have been devised in relatively recent years to overcome this apparent limit and, surprisingly, it is exactly by using very short pulses (in the form of a laser frequency comb) that the highest precisions have been achieved in frequency and time measurements.

A glimpse of the mechanisms that can be used to beat the frequency limitations connected to the use of short laser pulses can be gained from a simple example: if pairs of time-delayed and phase-locked pulses (like those generated by splitting a single laser pulse by means of a Michelson interferometer) are used, a simple Fourier transformation shows that the corresponding spectrum preserves a broad bell-shaped envelope, but also acquires a sinusoidal modulation with a spectral period given by the inverse of the temporal separation τ between the two pulses (see Figure 6.1b). In the ideal case, where the two time-delayed pulses with electric fields $E_1(t)$ and $E_2(t + \tau)$ are perfectly phase-locked, their combined spectrum is easily found as:

$$I_t(\omega, \tau) = I_1(\omega) + I_2(\omega) + 2\sqrt{I_1(\omega)I_2(\omega)} \cos(\omega\tau) \qquad (6.1)$$

which reduces to

$$I_t(\omega, \tau) = 2I(\omega)(1 + \cos(\omega\tau)) \qquad (6.2)$$

in the case of equal pulse intensities $I(\omega) = I_1(\omega) = I_2(\omega)$.

It is this $1/\tau$ fringe period that now sets the instrumental resolution and allows one, in principle, to investigate very fine spectral features if a long time delay τ is available. In principle, the spectral resolution $\Delta\nu$ achievable by these methods just depends on the maximum time delay τ between the exciting pulses as $\Delta\nu \approx \frac{1}{\tau}$. If the delay is introduced by translating an optical delay line of a distance L (like in one arm of a Michelson interferometer), the delay is $T = \frac{2L}{c}$ where c is the speed of light, and the resolving power achievable with a given mirror displacement is thus

$$R = \frac{\nu}{\Delta\nu} \approx \frac{2L}{\lambda}, \qquad (6.3)$$

i.e., it simply corresponds to twice the number of wavelengths (λ) scanned by the mirror movement.

One of the first demonstrations of high-resolution spectroscopy of multiphoton atomic transitions with pairs of ultrashort pulses dates back to 1996, when it was showed that it was indeed possible to measure line splittings (the hyperfine separation of the $8S_{\frac{1}{2}}$ state in cesium in that case) in a two-photon transition with a spectral resolution much better than that given by the single-pulse spectral width [489]. Actually, this is based on the same principle of Fourier Transform Spectroscopy, that normally uses broad-bandwidth cw sources to perform medium-to-high resolution studies in the medium and far infrared; one of the advantages of employing this technique with short pulses is that one can use their high peak intensities to move to different spectral regions or to investigate new transitions involving two or more photons.

The idea of using a pair of phase-locked pulses in order to achieve a better spectral resolution can also be extended by the use of longer sequences of equally time-delayed and phase-locked pulses. The spectrum that one obtains in this case still exhibits the broad bandwidth related to the short pulse duration, but is now modulated in a sharper and sharper fashion as longer pulse sequences are used (see Figure 6.1c). In the case of N equally spaced (by a delay τ) and phase-locked pulses, the textbook solution for the corresponding spectrum is given by the expression:

$$I(\omega, \tau) = I_0(\omega)\left(\frac{\sin N\omega\tau/2}{\sin \omega\tau/2}\right)^2 \qquad (6.4)$$

and the spectral interference pattern is the well-known array of intense and sharp interference maxima at $\omega_n = n\frac{2\pi}{\tau}$, with some small residual modulations in between. In the ideal limit of an infinite train of phase-locked and equally-spaced pulses of duration τ_p, the resulting spectrum essentially consists of a *comb* of infinitely sharp lines, equally separated by a frequency spacing corresponding to the inverse of the inter-pulse period and extended over a frequency range inversely proportional to τ_p.

The advantages of such a peculiar spectral distribution are evident: this spectral comb can be used as a precise ruler to measure unknown frequency intervals in a relatively simple way. By locking two laser lines to two different *teeth* of the comb, and by counting the integer number of interposed teeth, one can immediately obtain the unknown frequency gap, if the separation between the teeth is well known. A mode-locked laser is a natural way for generating such an ideally infinite sequence of time-delayed pulses with a well-defined phase relationship [490]. Its spectrum (given by the set of equally-spaced longitudinal modes of the cavity) is a broad comb of frequencies with a mode separation equal to the measurable and controllable pulse repetition rate (see Figure 6.1c). The largest frequency gap that can

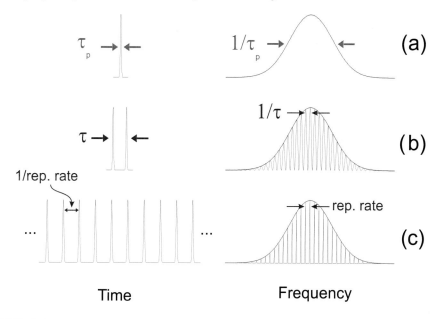

FIGURE 6.1
A single laser pulse has a frequency bandwidth which scales with the inverse of its duration τ_p (a). If one uses a pair of phase-locked pulses, delayed by a time τ, the resulting spectrum maintains the broad envelope of width $1/\tau_p$ but with a sinusoidal modulation of spectral period $1/\tau$ (b). This width sets the new instrumental resolution. If one uses an infinite sequence of pulses, locked in phase and equally delayed in time of a time interval τ, the spectrum breaks up in a *comb* of very narrow lines (the *teeth*) equally spaced by a frequency interval $1/\tau$ (c).

be bridged with such a comb is determined by the inverse of the pulse duration τ_p (if the pulse is Fourier transform-limited), but it can be widely extended if non-linear interactions are used to broaden the spectrum.

In this chapter we will discuss the basics of ultrashort pulse generation by means of the mode-locking phenomenon and we will then concentrate on the development of the new tool of Optical Frequency Comb Synthesizers (OFCSs) that have brought about a revolution in the way we measure frequency and time (see the 2005 Nobel lectures of T. W. Hänsch and J. Hall [52, 53]). We will then discuss the extension of precision frequency measurements based on ultrashort laser pulses and OFCSs to new spectral ranges, which has recently paved the way to the exploration of a rich and almost untouched territory.

6.2 Theory of mode locking

The emission of coherent radiation from a laser is governed by a subtle interplay between the gain spectrum of the active medium and the resonant frequencies of the laser cavity itself. As we have seen before, continuous-wave (CW) lasers emitting light at a single, well-defined frequency that is highly stable in time are normally used for precision spectroscopy. To achieve this single-frequency emission big efforts are usually made to force their oscillation on a single transverse and longitudinal mode of the cavity. While the transverse mode distribution can be narrowed by placing spatial filters in the cavity, the longitudinal mode selection is usually performed by shortening the cavity or introducing frequency-selective losses (spectral filters) that allow only one mode to experience sufficient gain. If such spectral filters are removed and the emission profile of the gain medium is wide enough, several longitudinal modes of the cavity can simultaneously oscillate and the output intensity of the laser is no longer constant with time.

Assuming that emission takes place only in the fundamental TEM$_{00}$ transverse mode, the laser can thus emit on many longitudinal modes (see Figure 6.2a), whose frequencies ν_m satisfy the condition

$$\nu_m = \frac{mc}{2nL} \tag{6.5}$$

where m is a positive integer, c is the speed of light, L is the length of the cavity, and n is an effective average refractive index. Here we initially consider that the laser elements are not dispersive, i.e., that the refractive index does not depend on the frequency; then, the different longitudinal modes are equally spaced with a constant separation $\nu_r = \nu_{m+1} - \nu_m = c/2nL$.

The electric field of a laser that oscillates on M adjacent longitudinal modes of frequency $\omega_m = 2\pi\nu_m = \omega_p + 2\pi m\nu_r$ centered about a mode of frequency $\omega_p = 2\pi p\nu_r$ (with p a large positive integer) with the same amplitude ε_0 can thus be written as

$$\tilde{E}^+(t) = \frac{1}{2}\varepsilon_0 e^{i\omega_p t} \sum_{m=(1-M)/2}^{(M-1)/2} e^{i(2m\pi\nu_r t + \phi_m)}, \tag{6.6}$$

where m ranges from $(1 - M)/2$ to $(M - 1)/2$ and ϕ_m is the phase of mode m. In general, this gives rise to a field that periodically repeats itself with a period corresponding to the cavity round-trip time $\tau_r = 1/\nu_r$ (see Figure 6.2b). For random relative phases ϕ_m among the different longitudinal modes, as is normally the case for a free-running laser, their emission adds up incoherently and the result is a noisy output with an average intensity that equals the sum of the intensities of the individual modes. Actually, when the laser operates in this regime a competition exists among the different modes to be amplified by stimulated emission, and this causes big fluctuations in the relative phases and amplitudes of the modes, which explain the big fluctuations of the instantaneous output intensity. If, on the contrary, all the modes have the same phase ϕ_0 (or a fixed phase separation exists between successive modes, such as $\phi_{m+1} - \phi_m = \alpha$), then their emission will periodically add up coherently, resulting in a sequence of intense and short bursts of light (see Figure 6.2c).

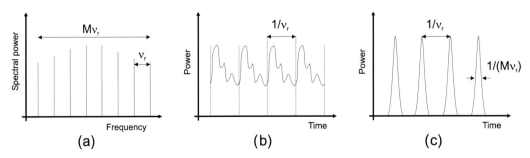

FIGURE 6.2
(a) Emission spectrum of laser oscillating on a set of many equidistant longitudinal modes; (b) corresponding time evolution of the laser power in the case of random phase variations among the different longitudinal modes; (c) mode-locked emission in the time domain.

In this case the sum in Equation 6.6 can be easily evaluated and the total laser field is

$$\tilde{E}^+(t) = \frac{1}{2}\varepsilon_0 e^{i\phi_0} e^{i\omega_p t} \frac{\sin(M\pi\nu_r t)}{\sin(\pi\nu_r t)} \tag{6.7}$$

that, for large M, corresponds to a train of well-isolated pulses spaced by $\tau_r = 1/\nu_r$ and whose duration is approximatively given by $\tau_P \approx 1/(M\nu_r)$. It is straightforward to see that the minimum pulse duration is thus given by the inverse of the laser bandwidth, and that the maximum of the field amplitude at the peaks is M times that of a single mode $\epsilon_{peak} = M\epsilon_0$. The peak intensity now grows with the square of the number of modes and can thus become much larger than the single-mode emission if many modes are made to lase coherently.

Setting a constant phase relationship among the different longitudinal modes of the cavity is what is usually referred to as *mode locking*. If the conditions for mode locking are met, then the laser emission consists of a regular train of very short pulses with high peak intensity. The purpose of locking the modes is indeed to organize their competition in such a way that the relative phases stay constant. Equivalently, in the time domain it corresponds to forcing the output intensity of the laser to consist of a periodic series of pulses resulting from the shuttling back and forth of a wave packet within the laser cavity.

In the following section we will review some of the most important mechanisms currently used to achieve mode locking and the techniques to control the dispersion in the laser cavity as to obtain the shortest output pulses.

6.3   Mode-locking mechanisms and dispersion compensation schemes

Mode locking in a laser requires a mechanism that results in higher net gain for a train of short pulses, compared to cw operation. This can be done by inserting in the cavity an active element, such as an acousto-optic modulator, or, passively, by including a saturable absorber (real or effective). Passive ML yields the shortest pulses because the self-adjusting mechanism becomes more effective than active ML, which can no longer keep pace with the ultrashort time scale associated with pulses as short as tens or even just a few fs. In real saturable absorption, which occurs in dyes or semiconductors, the shortest pulse width is limited by the finite response associated with the relaxation time of the excited state. On the other side, effective saturation absorption typically relies on the non-linear refractive index of some materials, together with spatial effects or interference to produce a higher net gain for more intense pulses. In this case, the ultimate limit on minimum pulse duration is basically due to an interplay among the ML mechanism, group velocity dispersion, and net gain bandwidth. Currently, the generation of ultrashort optical pulses is dominated by the Kerr-lens and the non-linear polarization-rotation mode-locking mechanisms, described in the next two sections.

6.3.1   Ti:sapphire lasers and Kerr-lens mode locking

The primary reason for using Ti:sapphire is its enormous gain bandwidth (700-1000 nm), which allows the generation of spectra of approximately 100 THz bandwidth directly in the oscillator, corresponding to ultrashort pulses with durations of approximately 10 fs. Moreover, the Ti:sapphire crystal can also serve as the nonlinear material for mode locking through the Kerr effect which manifests itself as an increase of the nonlinear index at

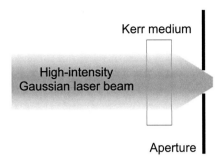

FIGURE 6.3
Kerr-lens mechanism: at high optical intensities, a Gaussian index profile is generated in the crystal; this acts as a lens and focuses the beam. Therefore, only high intensities are fully transmitted through the aperture, while low intensities experience losses. Since short pulses produce higher peak powers, mode-locked operation is encouraged.

increasing optical intensity
$$n(x, y) = n_0 + n_2 I(x, y). \tag{6.8}$$

Since the transverse intensity profile of the intracavity beam is Gaussian, a Gaussian index profile is created in the Ti:sapphire crystal, which makes the latter equivalent to a lens. As a consequence, the beam tends to focus, the focusing increasing with the optical intensity. Together with a correctly positioned effective aperture, this effective lens can act as a saturable absorber, i.e., high intensities are focused and hence are fully transmitted through the aperture, while low intensities experience losses. Since short pulses produce higher peak powers, they experience lower losses, making mode-locked operation favorable (see Figure 6.3).

This mode-locking mechanism has the advantage of being simple and essentially instantaneous, but has the disadvantages of not being self-starting and of requiring a critical misalignment from optimum cw operation. Spectral dispersion in the Ti:sapphire crystal due to the variation of the index of refraction with wavelength will result in temporal spreading of the pulse each time it traverses the crystal. At these wavelengths, sapphire displays normal dispersion, where longer wavelengths travel faster than shorter ones. In order to prevent the pulses from spreading, the overall group velocity dispersion (GVD) experienced by them in a cavity round trip has to be minimized. So, to counteract the Ti:S normal dispersion, prism sequences are normally used to provide adjustable negative dispersion [491]. It is also possible to generate anomalous dispersion by using the so-called chirped mirrors [492]. These have the disadvantage of less adjustability, if used alone, but they allow shorter cavity lengths and give additional control over higher order dispersion. If used in combination with prisms they allow one to produce pulses even shorter than those achieved using prisms alone [493].

The Ti:sapphire ML laser can deliver several watts of average output power but, because it is a free-space laser system, the cavity must be carefully engineered for good stability. The compactness and cost of the system are generally limited by the necessary high-power ($>$ 5 W) solid-state pump lasers, usually frequency-doubled cw Nd:YVO$_4$ lasers operating at 532 nm. The laser cavity itself can be built in an extremely compact framework, providing the highest repetition rate of all comb sources. Recently, a 10-GHz Ti:sapphire comb was demonstrated [494].

6.3.2 Fiber-based lasers and nonlinear-polarization-rotation mode-locking

Two main kinds of fiber lasers, the Erbium- and the Ytterbium-doped ones, are currently used for metrological-related applications based on their mode-locked operation.

Er-doped fiber lasers [495, 496, 497, 498, 499] owe their success to the widespread use of Er:fiber amplifier technology in optical communication systems [500]. Therefore, Er:fiber lasers may now be built from compact, inexpensive, and extremely reliable industrial components such as the 980-nm grating-stabilized pump laser diodes and fiber-optic couplers, splitters, and multiplexers. The cavity of a fiber laser consists mainly of a closed beam path that makes these systems inherently stable. Their emission wavelength of approximately 1.55 μm also allows the use of highly nonlinear fibers (HNFs) for spectral broadening [501] that, as we will see in the following, is a necessary requisite for obtaining a stable frequency comb. In contrast to PCFs and related fibers with air holes, these highly nonlinear silica fibers consist of solid cladding and core (diameter ≈ 4.0 μm), which allows them to be spliced onto standard communication fibers with low loss. Consequently, a higher level of integration is possible for Er:fiber systems, which improves the overall stability. Modern Er:fiber ML lasers provide repetition rates as high as 250 MHz, average powers of several hundred milliwatts, and excellent long-term stability; they are also cost-effective, compact, and extremely user-friendly (turnkey operation).

Yb-doped femtosecond fiber lasers have the same basic design as Er:fiber systems and exhibit similar advantages such as turnkey operation, compactness, and high intrinsic stability. Their shorter operation wavelength of approximately 1.03 μm, however, does not allow them to employ as many industrial components and requires the use of PCFs or similar fibers for spectral broadening. Yb:fiber ML lasers are mostly recognized for their excellent high-power capabilities. Due to the small pump defect of the laser transition (976-nm pump, 1.03-μm emission) and the availability of fibers with extremely high doping concentrations, amplified Yb:fiber systems exhibit tens of watts of output powers [502, 503] and repetition rates can be as high as 1 GHz. Therefore, these systems are an ideal choice for transferring the advantages of ML laser operation into the ultraviolet [504, 505, 506] and infrared [507] regions via nonlinear frequency conversion.

Fiber-based fs ML lasers are based on different mechanisms. One approach is additive pulse mode-locking (APM) whose working principle is illustrated in Figure 6.4 [508]. The fiber is contained in a resonator which has the same round-trip time as the laser resonator and is coupled to it by a semi-transparent dielectric mirror. Pulses returning from the fiber resonator interfere with those pulses which already are in the main laser resonator. For proper adjustment of the resonator lengths, there is constructive interference near the peak of the pulses, but not in the wings, because the latter have acquired different nonlinear phase shifts in the fiber. As a result, the peak of a circulating pulse is enhanced, while

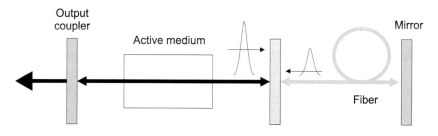

FIGURE 6.4
Schematic of the additive-pulse mode-locking mechanism.

the wings are attenuated. The APM technique makes it possible to obtain rather short pulses without using very special components and can work in different wavelength regions. However, the resonator length adjustment is rather critical, questioning the practicability of the technique for commercial products.

Therefore, mode locking in fiber lasers is usually achieved by using nonlinear polarization in a fiber. When an intense optical pulse propagates in an optical fiber which is not polarization mantaining, a nonlinear (i.e., intensity-dependent) change to some elliptical polarization state is produced. The physical cause of this effect is related to self-phase and cross-phase modulation as well as to some uncontrolled birefringence in the fiber. Thus, if the pulses afterwards pass a polarizer, the power throughput actually depends on the optical peak power. A typical mode-locking configuration contains some fiber polarization controller or waveplates, which can be adjusted so that the maximum transmission (minimum loss) at the polarizer occurs for the highest possible optical intensity [509]. Figure 6.5 shows the whole setup for a highly stable, frequency-controlled mode-locked erbium fiber laser [510].

The cavity is unidirectional and an all-fiber design is obtained by using a WDM coupler for pump light coupling as well as a 90/10-fiber coupler for output coupling. Moreover, a zero-dispersion cavity is obtained by balancing the dispersion from the positive and negative fibers. A single polarization is selected by the intracavity polarizing isolator and the polarization controllers on each side of the isolator are used to optimize polarization evolution for optimum mode-locked operation. The described nonlinear-polarization-rotation mode-locking (NPRM) mechanism is actually related to APM. Indeed, in this case two polarization modes coupled to each other through the intensity-dependent birefringence are interferometrically combined at a polarizing element, where interference occurs, as in APM.

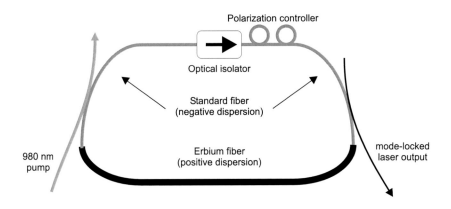

FIGURE 6.5
Experimental setup for a stretched pulse Er-doped fiber ring laser [509]. A typical mode-locking configuration contains fiber polarization controllers, which can be adjusted so that the maximum transmission (minimum loss) at the polarizer occurs for the highest possible optical intensity. Moreover, a zero-dispersion cavity is obtained by balancing the dispersion from the positive and negative fibers.

6.4 Optical frequency comb synthesis from mode-locked lasers

As we have seen above, an ideal mode-locked laser emits a train of equally-spaced pulses with a period τ_r , which corresponds to a *comb* of modes in the spectral domain whose spacing $\nu_r = 1/\tau_r$ is constant. However, differently from the ideal situation described in Section 6.2, dispersion from active and passive elements inside the cavity causes the refraction index not to be a constant across the laser gain bandwidth and, therefore, the group and phase velocities v_g and v_p of the propagating field to differ. Consequently, after the pulse envelope has undergone each full round trip in the laser cavity, the field phase front has accumulated a time delay that can be expressed as

$$\tau_{CEO} = 2L\left(\frac{1}{v_g} - \frac{1}{v_p}\right) \tag{6.9}$$

and the phase of the wave at the carrier frequency ν_c has acquired a shift from the pulse envelope of

$$\phi_{CEO} = 2\pi\nu_c\tau_{CEO}. \tag{6.10}$$

Since an additional phase ϕ_{CEO} (CEO is for carrier-envelope offset) is acquired by all pulses exiting the laser cavity (see Figure 6.6) at a repetition rate $\nu_r = 1/\tau_r$, this constant (at least in principle) phase slippage causes an additional frequency contribution $\nu_0 < \nu_r$, given by

$$\nu_0 = \frac{\phi_{CEO}}{2\pi\tau_r} \tag{6.11}$$

that has to be taken into account in the determination of the optical frequencies of the spectral comb lines emitted by a mode-locked laser, which then results in

$$\nu_m = \nu_0 + m\nu_r. \tag{6.12}$$

The spectrum of a "perfect" mode-locked laser is shown in Figure 6.6, and is thus composed of a series of equally-spaced lines (the *teeth* of the comb) whose frequencies are perfectly defined by the knowledge of the mode index m, the offset frequency ν_0, and the repetition rate of the laser ν_r. Since both these frequencies lie in the rf domain, the large (of the order of $10^5 \div 10^6$) integer mode index m can act as a multiplying wheel to establish a direct link between the rf and the visible range, thus giving access to the absolute determination of optical frequencies if one is able to accurately measure and control both ν_0 and ν_r.

In a "real" mode-locked laser, neither ν_0 nor ν_r are strictly constant because of vibrational motions of the mirrors and fluctuations in the pump power; several techniques have then been devised in the recent years to accurately measure and stabilize them.

6.4.1 Comb stabilization

For the comb generated by a ML laser to be useful as a precise reference for absolute optical frequencies, control of its spectrum, i.e., the absolute position and spacing of the comb lines, is necessary. Although measurement of ν_r and ν_0 are in principle sufficient to determine an absolute optical frequency, it is generally preferable to use the measurements in a feedback loop to actively stabilize or lock one or both of them to suitable values. In terms of the above description of the output pulse train, this means control of the repetition rate, ν_r, and the pulse-to-pulse phase shift, ϕ_{CEO}, which may be achieved by making appropriate adjustments to the operating parameters of the laser itself.

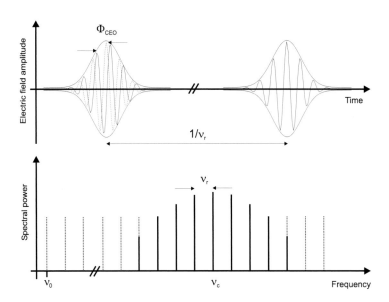

FIGURE 6.6
Scheme of a series of ultrashort laser pulses emitted from a mode-locked laser. Successive pulses are exact replicas of each other apart from a phase factor ϕ_{CEO}. This translates in the frequency domain to a comb-like structure of narrow spectral modes separated by the laser repetition frequency $\nu_r = 1/\tau_r$ and with an offset from zero frequency given by ν_0.

Some experiments only require control of the repetition rate, $\nu_r = 1/\tau_r = v_g/L$, where v_g is the round-trip group velocity and L is the cavity length. The simplest way of locking ν_r is therefore by adjusting L . Mounting either end mirror in the laser cavity on a translating piezoelectric actuator (PZT) easily achieves this. The actuator is then typically driven by a phase-locked loop that compares ν_r or one of its harmonics to an external clock. Locking the comb spacing alone is sufficient for measurements that are not sensitive to the comb position such as measurement of the difference between two laser frequencies ν_{l1} and ν_{l2}.

However, many experiments require the control of both ν_r and ν_0. To do so, both the round-trip group delay and the round-trip phase delay must be controlled. Since both depend on the cavity length, a second parameter in addition to it must be controlled. In an ideal situation, an orthogonal control of ν_r and ν_0 would be desirable to allow the servo loops to operate independently. If this cannot be achieved, one servo loop will have to correct changes made by the other or, if necessary, orthogonalization can be achieved by either mechanical design, or by electronic means.

A mode-locked laser that uses two intracavity prisms to produce the negative group velocity dispersion necessary for Kerr-lens mode-locking provides an additional knob to adjust the comb parameters. By using a second PZT to slightly tilt the mirror at the dispersive end of the cavity about a vertical pivot, one can introduce an additional phase shift proportional to the frequency distance from the central one, which displaces the pulse in time and thus changes the round-trip group delay (see Figure 6.7). In the frequency domain this corresponds to introduce a change of the length of the cavity that depends linearly on the frequency. This leads to changes in both ν_r and ν_0 but leaves the central frequency mode on the pivot axis constant.

In addition to tilting the mirror after the prism sequences, the difference between the group and phase delays can also be adjusted by changing the amount of glass in the cavity

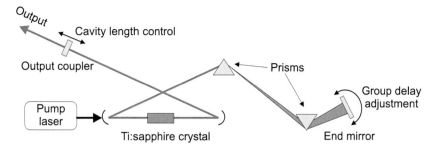

FIGURE 6.7
Schematic of a typical Kerr-lens mode-locked Ti:sapphire laser with controls for both repetition rate and offset frequency stabilization.

or adjusting the pump power. The first approach changes the difference between the group delay and phase delay due to dispersion in the glass but has the disadvantage of mechanical limitations that prevent rapid response time in a servo loop. On the other hand, changing the pump power can be achieved very rapidly by means of acousto- or electro-optical modulators. Its main effect is that of changing the power of the intracavity pulse, which has been empirically shown to change the pulse-to-pulse phase due to nonlinear effects.

Both ν_r and ν_0 have to be precisely measured in order to control them in a feedback loop. Measurement of ν_r is straightforward; one simply detects the pulse train with a fast photodiode. Although it can range from tens of MHz to several GHz, ν_r is typically around 100 MHz and its measurement poses no particular problem. On the other hand, measuring ν_0 is not as simple. An elegant solution is available in the case where the optical spectrum spans an octave in frequency, i.e., the highest frequencies are a factor of 2 larger than the lowest frequencies. If one uses a second harmonic crystal to frequency double a comb line, with index n, from the low-frequency portion of the spectrum, it will have approximately the same frequency as the comb line on the high-frequency side of the spectrum with index $2n$ (see Figure 6.8). Measuring the heterodyne beat between these yields a difference frequency

$$2\nu_n - \nu_{2n} = 2(n\nu_r + \nu_0) - (2n\nu_r + \nu_0) = \nu_0 \tag{6.13}$$

which is just the sought offset frequency. More details about the stabilization of a femtosecond frequency comb will be given later in the context of absolute frequency measurements.

Thus, an octave-spanning comb spectrum enables simple measurement of ν_0 by means of such a so-called f-to-2f interferometer, but such a broad spectrum is not readily available from a mode-locked laser oscillator (although octave spanning lasers have been recently demonstrated). Therefore, additional external mechanisms normally have to be used to substantially broaden the spectrum. One possible way to achieve such extreme spectral broadening relatively simply is through white-light continuum generation. Provided that the laser pulses are intense enough, focusing them into a suitable transparent material results in the generation of a white-light continuum that contains wavelengths ranging from the IR to the near UV. Although continuum generation is a complex issue involving changes in the temporal and spatial beam characteristics, the dominant process and the starting mechanism leading to spectral superbroadening is the self-phase modulation of the pulse, which is due to the intensity-dependent refractive index of the medium. In any material with a third-order nonlinear susceptibility $\chi^{(3)} \neq 0$ and an instantaneous response, the refractive index depends on the intensity $I(t)$ of the propagating field as

$$n(t) = n_0 + n_2 I(t) \tag{6.14}$$

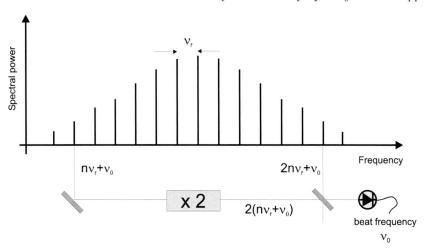

FIGURE 6.8
Conceptual scheme for the measurement of the offset frequency ν_0 of a mode-locked laser whose spectrum spans more than an optical octave.

So, after propagation through a short length L of such a medium, the field at a carrier frequency ν_0 experiences a time-dependent shift in the instantaneous frequency that is equal to

$$\Delta\omega(t) = -\omega_0 n_2 L/c \frac{dI(t)}{dt}, \tag{6.15}$$

resulting in red detuning at the leading edge of the pulse and blue detuning at the trailing edge. A 100-fs pulse focused to a peak intensity of 10^{14} W/cm^2 in a 1-mm-thick medium with a typical nonlinear index coefficient of 10^{-16} cm^2/W will give rise to a continuum pulse with a frequency excursion that is comparable to the frequency of the carrier itself.

The generation of the continuum is, however, the result of a very complex interplay among competing processes, and the exact characteristics of the output field appear strongly dependent on the exact initial conditions of the interaction and hardly predictable. In particular, one is led to expect that the white-light pulses produced by phase-locked pump pulses have lost any precise phase relationship in the generation process. Considering that phase coherence among successive pulses in the train is an essential ingredient for the generation of a broadband frequency comb, such white-light pulses may, at first glance, appear inadequate for this purpose.

Unexpectedly, in 1997, a simple experiment [511] proved that supercontinuum generation could preserve the phase coherence of the pulses and set the basis for extension of the frequency comb to a full octave. However, at that time pulses intense enough to observe this effect could only be produced in amplified systems at a kHz repetition frequency, too dense a frequency grid to be used in frequency space. It was necessary to wait until 1999 before a group from Bell Labs [512] reported the massive spectral broadening of relatively low power fs pulses in a photonic crystal fiber (PCF). These strongly-guiding fibers are made of an array of air holes that confine the light to a pure silica core region embedded within the array. The large refractive index contrast between the pure silica core and the "holey" cladding, and the resultant strong nature of the optical confinement, allows the design of fibers with very different characteristics to those of conventional ones. Here, a very small core size of 1μm leads to increased nonlinear interaction of the guided light with the silica. At the same time the very strong waveguide dispersion substantially compensates the material dispersion of the silica at wavelengths below 1 micron. This gives an overall

GVD which can be zero around the central wavelength of standard Ti:sapphire lasers. As a result, short optical pulses travel further in these fibers before being dispersed, which further increases the nonlinear interaction and allows very broad spectra to be generated at relatively low peak powers. Soon it was discovered that supercontinuum could also be obtained in tapered fibers [513].

With the availability of photonic-crystal and tapered fibers, broad frequency combs could be easily generated and it became straightforward to set up a frequency chain measuring the interval between an optical frequency ν and its second harmonic. Due to the availability of the PCF this was first demonstrated in J. Hall's group in Boulder [514, 515] and shortly afterwards in T. W. Hänsch's group in Garching [516].

6.4.2 Measurements with a frequency comb

As the frequency comb has developed over the last decade, two general approaches have emerged for its application as an ultraprecise measurement tool. In the first case, the comb serves simply as a frequency ruler against which a cw laser is calibrated and measured. It is the cw laser that then performs the spectroscopy. This kind of applications will be described first in what follows. The second general approach employs the frequency comb to directly probe atomic and molecular samples. This approach will be described later, followed by a review of the most recent and interesting applications of frequency combs to a variety of scientific fields.

6.4.2.1 Measuring frequency differences

The simplest use of the frequency comb produced by a mode-locked laser is the measurement of the frequency difference between two optical sources, typically, single-frequency lasers. If the absolute frequency of one of the two sources is known, this yields an absolute measurement of the other. This scheme was first demonstrated by Kourogi [517] with an optical frequency comb generator based on the addition of frequency sidebands to a CW laser in a periodically-modulated optical cavity, but the introduction of optical combs based on mode-locked lasers has largely extended the range of measurable frequency differences. Since the absolute frequency reference is constituted by one of the cw laser sources, the comb is just used as a precise frequency ruler in this case; therefore, an accurate knowledge of its repetition rate ν_r, determining the tooth frequency spacing, is all that is needed. A simplified scheme of this kind of experiments is shown in Figure 6.9. The first example of this measurement scheme was the determination of the absolute frequency of the D_1 line in atomic cesium, performed in 1999 by Hänsch's group at MPQ [518, 519]. Here, the 18.4 THz frequency gap between a diode laser stabilized on the saturated cesium line and the fourth harmonic of a methane-stabilized He-Ne laser was bridged by locking these two emission lines to two teeth of a femtosecond frequency comb. If the a priori uncertainty on the unknown transition frequency is smaller than roughly half the separation between two comb teeth, then determining the integer number of teeth between the two laser lines leads to a precise measurement of the frequency under study. Shortly after this initial experiment, the group of J. Hall at JILA demonstrated the possibility of widening the measurable frequency gap by using a mode-locked laser with shorter pulse duration and further broadening its spectrum by self-phase-modulation in a single-mode optical fiber [520]. This allowed them to bridge the 104 THz frequency gap between a Ti:Sapphire laser locked to a two-photon transition of ^{85}Rb at 778 nm and a iodine-stabilized Nd:YAG laser at 1064 nm, thus improving the precision of the available frequency standard.

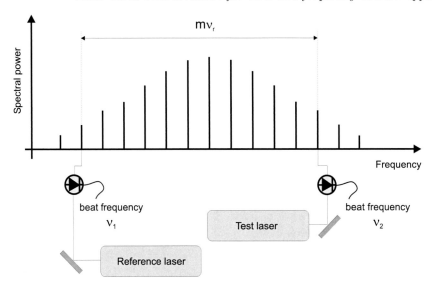

FIGURE 6.9
Conceptual scheme for the measurement of the *large* difference between two laser emission frequencies. The two laser outputs are beated against the closest "teeth" of a mode-locked laser, whose repetition rate ν_r is accurately measured or controlled. The frequency of the test laser is easily measured as $\nu_x = \nu_{ref} + m\nu_r \pm \nu_1 \pm \nu_2$; where ν_{ref} is the frequency of the reference laser, and the unambiguous identification of the integer m relies on a sufficiently accurate estimate of the unknown frequency ν_x.

6.4.2.2 Measuring absolute frequencies with an octave-spanning comb

The experiments described above showed the great potential of femtosecond frequency-combs for high-precision frequency measurements. However, it was soon realized that one could also use them for directly comparing the optical frequencies to the microwave primary standard without intermediate steps. Depending on the available bandwidth from the mode-locked laser, one can use different schemes to this purpose. The first realization of a direct radio-to-optical frequency conversion using a femtosecond laser relied on a rather complex scheme based on the comparison of different optical harmonics of the same fundamental laser frequency [521]. It allowed to reference the frequency of a methane stabilized He-Ne laser to an integer multiple of the cesium clock that controlled the mode spacing. By doing this, every other frequency in the setup, including every mode of the comb, was known with the precision of the cesium clock. Radiation at 486 nm, corresponding to the 7th harmonic of the He-Ne laser, was then used to measure the absolute frequency of the hydrogen 1S-2S two-photon resonance, occurring at the fourth harmonic of this wavelength. The outcome was one of the most precise measurements of an optical frequency at those times, providing a transition frequency value as accurate as 1.9 parts in 10^{14}, limited by the reproducibility of the hydrogen spectrometer.

Much simpler schemes can be adopted if a spectrum spanning more than one optical octave is available. As shortly discussed above, different methods can now achieve this goal with simple and compact setups. With an octave-spanning frequency comb, the frequency of a single-frequency laser can be determined by measuring the frequency interval between the laser and its second harmonic, i.e., $2\nu - \nu = \nu$, where ν is the frequency of the single-frequency laser (see Figure 6.10). This technique requires a minimum effort to control the

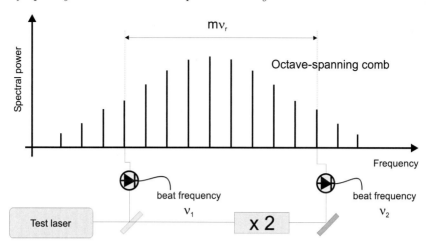

FIGURE 6.10
Conceptual scheme for the absolute measurement of an unknown laser frequency with an octave-spanning frequency comb from a mode-locked femtosecond laser. The difference between the fundamental laser frequency and its second harmonic (that is, the laser frequency itself) can be measured with the method described in the previous section.

comb, as only the comb spacing ν_r needs to be controlled. In principle, the comb spacing does not have to be locked, it only needs to be measured.

This technique was utilized at JILA [515] to make one of the first direct microwave to optical measurements using a single mode-locked laser. In this experiment the octave-spanning frequency spectrum was obtained by spectral broadening in a piece of microstructured optical fiber. Here, the frequency difference between the 1064 nm line of a Nd-YAG laser stabilized to molecular iodine transition and its second harmonic was measured. It was thus possible to achieve the first measurement of the absolute frequency of an optical transition starting directly from a microwave Cs clock.

6.4.2.3 Absolute optical frequency synthesizer

As we have briefly shown above, the absolute frequency of the m-th tooth of a femtosecond frequency comb is fully determined by Equation 6.12. Therefore, by measuring and controlling the comb offset frequency ν_0 and repetition rate ν_r, both in the microwave region and directly linkable to the Cs time-frequency standard, one can directly synthesize millions of absolutely known optical frequencies corresponding to the individual comb teeth. Such an absolute optical frequency synthesizer can be used as an absolute frequency scale against which every laser line falling within the ML laser bandwidth can be measured.

Let us now look in more detail to a typical scheme for implementing a self-referencing optical frequency synthesizer based on a titanium-sapphire mode-locked laser and on the f-2f method. The laser spectrum is first broadened to more than one optical octave with a microstructure fiber. After the fiber, a dichroic mirror separates the infrared ("red") part from the green ("blue"). The former is frequency doubled in a non-linear crystal and reunited with the green part to create a wealth of beat notes, all at ν_0. The number of contributing modes is given by the phase matching bandwidth $\Delta\nu_{pm}$ of the doubling crystal and can easily exceed 1 THz. To bring all these beat notes at ν_0 in phase, so that they all add constructively, an adjustable delay in the form of a pair of glass wedges or cornercubes is used. It is straightforward to show that the condition for a common phase of all these beat

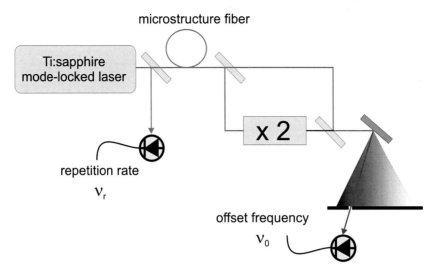

FIGURE 6.11
Schematic view of the setup for a self-referenced frequency comb. The repetition rate ν_r is simply measured with a photodiode, while an f-2f interferometer measures the offset frequency ν_0 after octave spectral broadening of the mode-locked laser pulses. The feedback mechanisms to measure and stabilize the laser ν_r and ν_0 are not shown here for simplicity.

notes is that the green and the doubled infrared pulse reach the photodetector at the same time. The adjustable delay allows to compensate for different group delays, including the fiber. In practice the delay needs to be correct within $c\Delta\nu_{pm}$ which is 300 μm for $\Delta\nu_{pm} = 1$ THz. Outside this range a beat note at ν_0 is usually not detectable. The maximum signal-to-noise ratio when measuring ν_0 is obtained for equal intensities reaching the detector within the optical bandwidth that contributes to the beat note [522]. In practice this condition is most conveniently adjusted by observing the signal-to-noise ratio of the beat note with a radio frequency spectrum analyzer. A grating is also frequently used to prevent the extra optical power, that does not contribute to the signal but adds to the noise level, from reaching the detector. Typically only a very moderate resolution is sufficient and it is usually not necessary to use a slit between the grating and the photodetector. When detecting the beat note as described above, more than one frequency component is obtained. Observing the beat notes between frequency combs, not only the desired component $k = 2m - m' = 0$ is registered, but all positive and negative integer values of k contribute, up to the bandwidth of the photodetector. This leads to a set of radio frequency beat notes at $k\nu_r \pm \nu_0$ for $k = ..., -1, 0, +1, ...$ In addition, the repetition rate ν_r and its harmonics will most likely give the strongest components. A low-pass filter with a cutoff frequency of $0.5\nu_r$ selects exactly one beat note at ν_0. The design of such a filter may be tricky, mostly depending on how much stronger the repetition rate signal exceeds the beat note at ν_0, and for this reason it is desirable to work at high repetition rates. In addition, a larger repetition rate concentrates more power in each mode further improving the beat notes with the frequency comb.

Once ν_0 and ν_r have been measured, it is necessary to control and stabilize them. As discussed above, several knobs can be used for controlling these two frequencies, and stabilization is usually performed by mixing the detected frequencies with radio frequency references. In this way the frequency difference is generated, low-pass filtered, and with appropriate gain sent back to the proper actuators. When properly designed, such phase-

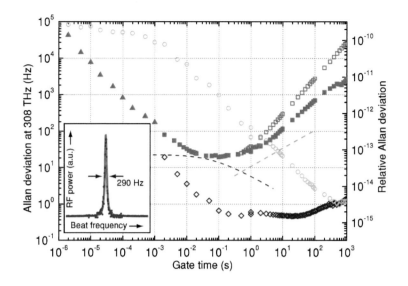

FIGURE 4.39

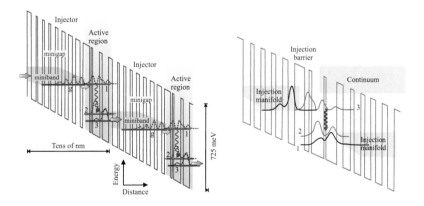

FIGURE 4.93

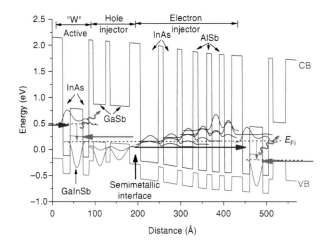

FIGURE 4.97

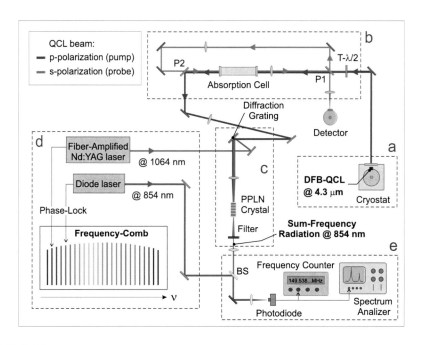

FIGURE 5.20

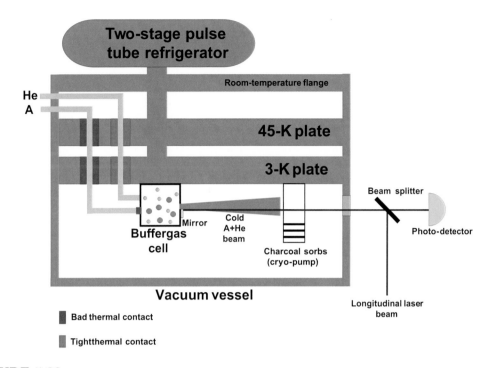

FIGURE 5.33

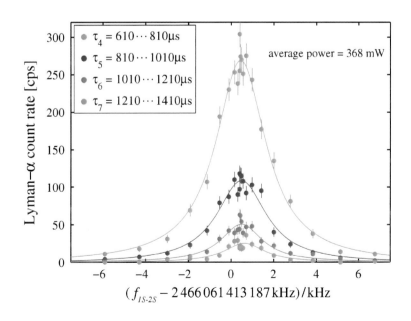

FIGURE 5.41

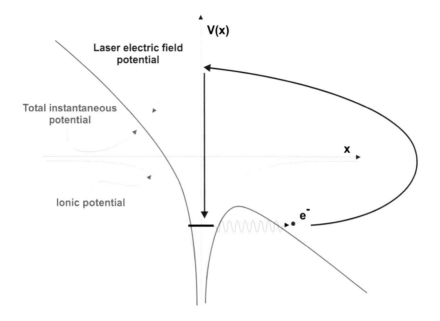

FIGURE 6.21

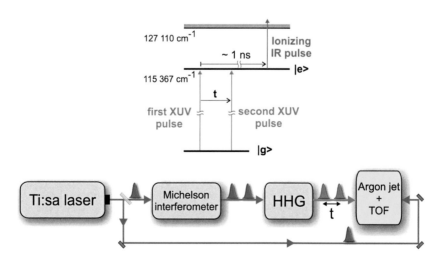

FIGURE 6.25

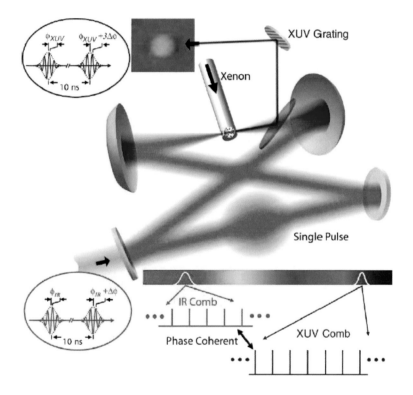

FIGURE 6.30

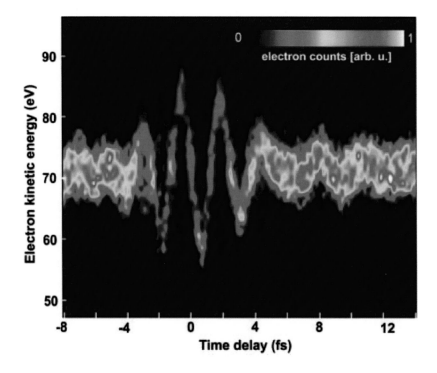

FIGURE 6.32

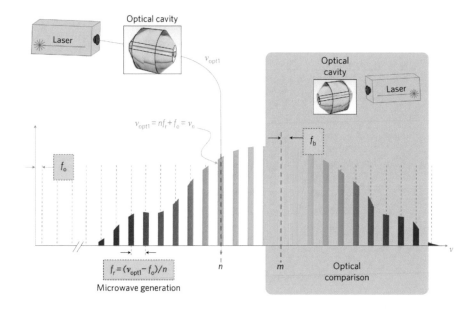

FIGURE 7.15

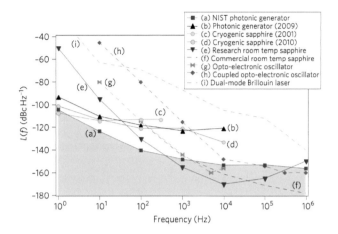

FIGURE 7.18

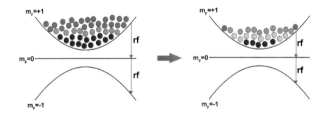

FIGURE 7.23

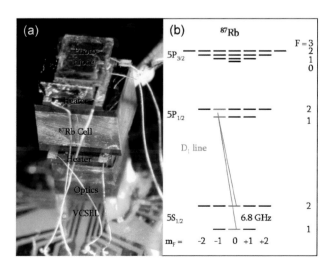

FIGURE 7.51

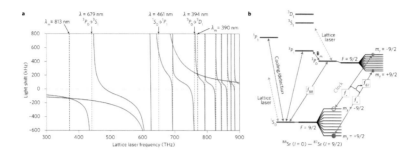

FIGURE 8.5

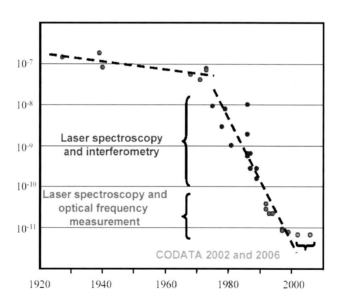

FIGURE 8.8

locked loops force one oscillator, the repetition rate or the offset frequency, to stay in phase with another, the radio frequency reference. In particular, since for averaging times shorter than acoustic vibrations of several ms period, a typical free-running titanium-sapphire laser shows better phase stability than a high-quality synthesizer, it is important to use a moderate servo bandwidth (of a few 100 Hz at most) for phase locking the repetition rate. Stabilizing the carrier envelope frequency, even though it generally requires faster electronics, does not have the stability and accuracy issues that enter via the repetition rate due to the large multiplication factor m in Equation 6.12. Any fluctuation or inaccuracy in ν_0 just adds to the optical frequencies rather than in the radio frequency domain where it is subsequently multiplied by m.

In general, measuring the frequency of an unknown cw laser at ν_L by means of a stabilized frequency comb involves the creation of a beat note ν_b with the comb itself. For this purpose one needs a good spatiotemporal mode match between the beam of the cw laser and that of the comb. One usually reflects out a small portion of the frequency comb around ν_L by means of a dichroic beam splitter and mixes it with the cw laser output at the right intensity ratio for an optimum signal-to-noise ratio. The frequency of the cw laser is then given by

$$\nu_L = m\nu_r \pm \nu_0 \pm \nu_b \qquad (6.16)$$

The signs may be determined by introducing small changes to one of the frequencies with a known sign while observing the sign of changes in another frequency. For example the repetition rate can be increased by picking a slightly different frequency of the reference oscillator. If ν_L stays constant we expect ν_b to decrease (increase) if the "+" sign ("−" sign) is correct. The other quantity that needs to be determined is the mode number m. If the optical frequency ν_L is already known to a precision better than the mode spacing, the mode number can simply be determined by solving the corresponding Equation 6.16 for m and allowing for an integer solution only. A coarse measurement could be provided by a wave meter, for example, if its resolution and accuracy is trusted to be better than the mode spacing of the frequency comb. If this is not possible, at least two measurements of ν_L with two different and properly chosen repetition rates may leave only one physically meaningful value for ν_L. This technique for measuring absolute frequencies using a self-referenced comb was first demonstrated at JILA by Jones et al. [514] for the measurement of the frequency of a 778 nm single-frequency laser locked to a two-photon transition in ^{85}Rb (see Figure 6.12).

6.4.2.4 Direct frequency-comb spectroscopy

The previous examples essentially use a frequency comb to enhance a conventional spectroscopy setup by providing a frequency axis for the laser sources. In a different approach, combs can be used to interrogate a sample directly. Interestingly, this is the route that was launched in the late 1970's to explore pairs of comb modes and multiphoton transitions in atomic systems [490, 523]. Direct frequency-comb spectroscopy may bring a number of advantages to particular applications, thanks to the inherent accuracy and spectral purity of the comb modes and to the wide spectral coverage, all in a collimated single-spatial-mode beam with teeth that can be coupled to a matched cavity for long effective path lengths [15]. However, the large quantity of OFCS modes often results in a very low power per mode, limiting sub-Doppler spectroscopy to atomic (or molecular) beams or cooled systems. Moreover, if used in a transmission scheme, the presence of many non-resonant modes can be detrimental to the detected signal-to-noise ratio.

The most straightforward scheme to implement involves a frequency comb illuminating a sample and the detection of fluorescence from an excited state while either ν_r or ν_0 is scanned. In this configuration, the frequency-comb is like a multimode laser with narrow-

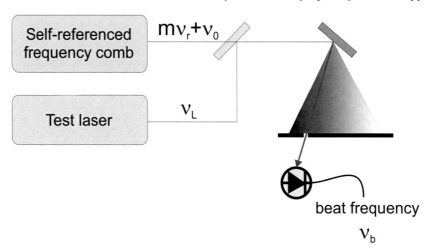

FIGURE 6.12
Simplified view of a typical scheme for absolute frequency measurements with a self-referenced optical frequency comb. The emission of the test laser is superposed to that from the comb, and the appropriate part of the common spectrum is detected with a photo-diode to measure the beat frequency ν_b. The unknown laser frequency ν_L is then obtained from Equation 6.16.

linewidth modes at precisely known optical frequencies and a spectral coverage of hundreds of terahertz. As a result, all the well-known cw-laser spectroscopic techniques can be used with the additional advantage of the absolute frequency calibration against primary standards and of the spectral coverage of different atomic or molecular structures with a single laser source.

This basic approach has been implemented with alkali atoms in a magneto-optical trap (MOT), beam, and cell [524, 525, 526, 527]. With cold atoms, or propagation orthogonal to an atomic beam, Doppler broadening can be greatly reduced, yielding uncertainties in the determination of transition frequencies that begin to rival the best measurements presented with cw laser spectroscopy. The first single-photon DFCS experiment was the measurement of the 5S-5P transition (at 384 THz) excited by a 100-MHz Ti:S comb in a ^{87}Rb optical molasses [528]. Here, the ν_r and ν_0 parameters of the comb were conveniently chosen as to only excite this particular Rb transition. Similar experiments have been performed to demonstrate EIT with a comb [529, 530], and the D1 and D2 lines of cesium at 900 nm and 850 nm, respectively, were measured with a 1-GHz Ti:Sa comb in a thermal beam [531, 526, 532].

Another interesting application of single-photon DFCS spectroscopy was the frequency measurement of a forbidden transition, like the P1 calcium clock transition at 657 nm [533]. In this work, the two main limitations for this spectroscopic technique were overcome: the OFCS power was enhanced and its linewidth narrowed. The first was achieved by amplifying the comb radiation by a comb-injected slave diode laser [534]. With such an enhanced source, together with long interrogation times, thanks to the use of cold Ca atoms, saturation spectroscopy, photon recoil splitting, and Ramsey-Bordé fringes were observed. For OFCS narrowing, a short-term, highly stable cw fiber laser was used as a comb reference oscillator instead of the usual RF one.

If the comb mode separation ν_r is smaller than the Doppler profile of the transition in a room temperature sample, comb excitation becomes velocity selective and transparency holes can be probed by scanning a separate cw laser [535, 536].

Counterpropagating beams can also provide reduced Doppler broadening when multi-photon transitions are being investigated [527, 537]. Similarly to the single-photon DFCS spectroscopy, multi-photon transitions can be driven by using OFCSs. Here, two typical situations can occur: multiphoton transitions with an intermediate atomic or molecular resonance, and those in which the pulse spectrum is far detuned from an intermediate resonance. Example of the former is an experiment of multiphoton spectroscopy of the 5D and 7S hyperfine manifolds of rubidium atoms via the resonant 5S-5P single-photon transition [525, 528]. In this case, the multiphoton comb spectroscopy can be viewed as a two-mode cw laser with frequencies resonant with the intermediate transitions. The hyperfine spectrum was recorded and frequency measured by changing the ν_r frequency with an accurate choice of its initial value to get the proper resonant frequency and to discriminate between multi-photon and single-photon transitions. Vice versa, multiphoton DFCS of the 8S, 9S, and 7D hyperfine manifolds of cesium [538] is a demonstration of multiphoton comb spectroscopy without intermediate resonant transition. In this case, instead of matching the comb frequencies with the two intermediate single-photon allowed transitions by means of accurate control of the frequency comb parameters, the motion of the atoms is used for Doppler-velocity selection of atoms simultaneously excited by two comb modes with an energy equal to the considered two-photon atomic transition. With these experiments, frequencies and hyperfine coefficients of the Cs states involved in the measured transitions were determined with an accuracy of 50-200 kHz.

At first glance it may seem that a two-photon transition requires high power that is generally not available in a single comb mode. However, it is straightforward to see that in this case the modes can sum up pairwise such that the full power of the frequency comb contributes to the transition rate, as initially proposed by Ye. F. Baklanov and V. P. Chebotayev [523]. Suppose the frequency comb is tuned such that one particular mode, say near the center, is resonant with the two-photon transition. This means that two photons from this mode provide the necessary transition energy. In this case, also the combination of modes symmetrically placed with respect to the central one are resonant, as they sum up to the same transition frequency. In fact, all modes contribute to the transition rate in this way. The same applies if the two-photon resonance occurs exactly halfway between two modes, as schematically shown in Figure 6.13. It can be shown [523] for unchirped pulses that the total two-photon transition rate is the same as if one would use a continuous laser with the same average power.

One disadvantage of fluorescence detection is that multiple comb elements can also simultaneously interact with different transitions for the same value of ν_r and ν_0. Thus, the detected fluorescence may arise from two (or many) levels which share the same excited state, rendering the individual transitions indistinguishable.

Multiplexing with a VIPA spectrometer

Other techniques exist, however, that circumvent this problem by directly measuring the power and/or phase of individual comb teeth that have interacted with the atomic or molecular gas. This is perhaps the most interesting application of DFCS, because it also allows the parallel detection of a huge number of transitions in the comb bandwidth for a mixture of different atomic or molecular species.

Implementing this kind of highly-multiplexed frequency-comb spectroscopy requires a detection system that is broadband, yet highly resolving in the spectral domain. Simple grating spectrographs can provide a resolution of several tens of gigahertz and bandwidths of hundreds of gigahertz to several terahertz; however, frequency combs typically exhibit a ν_r that rarely exceeds 1 GHz and bandwidths of several hundred terahertz. This means that a grating-based spectrometer falls short in both categories. An alternative multi-channel detection technique is therefore needed which can discriminate the comb teeth.

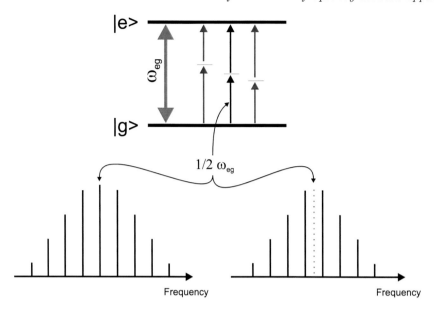

FIGURE 6.13

Direct frequency-comb spectroscopy of a two-photon transition. All the symmetrically placed pairs of modes contribute to the excitation if the two-photon transition is resonant with one of the comb modes if it occurs exactly halfway between two comb modes.

The most promising spatial detector for comb modes is the grating-VIPA (virtually imaged phased array) detector which uses two dispersive elements (grating and plane-parallel etalon) acting in two orthogonal spatial directions. The VIPA provides ≈ 1 GHz resolution in the visible spectral range (≈ 500 MHz at 1550 nm) [539, 540, 541]. When combined with a lower dispersion grating in the orthogonal spatial direction, 5–10 THz of bandwidth can be captured in a single measurement taking a few milliseconds. Briefly, the comb light is focused in a line and illuminates the VIPA etalon at a small angle. As a result of the multiple interference inside the VIPA, different frequencies exit from the etalon at different angles. This pattern is repeated for each free-spectral range of the VIPA, but different orders can be separated by dispersion in the orthogonal direction by a standard grating. The resulting output is detected by a CCD array where each single OFCS mode is clearly distinguishable from the others. Frequency calibration as well as details on the indexing of the numerous detected modes are described in [540].

This technique has been successfully employed for molecular fingerprinting and trace gas detection in the visible, near-infrared, and mid-infrared spectral regions. When combined with a multipass cell or enhancement cavity, minimum detectable absorption below 1×10^{-9} cm^{-1} has been achieved along with sensitivities to concentrations below 10 ppb for some common gases [539, 541]. In particular, Thorpe and co-workers demonstrated highly efficient CRDS, in which a Ti:S OFCS was coherently coupled to a high-finesse Fabry-Perot optical cavity [539]. In this configuration, more than 100000 optical comb teeth, each coupled to a specific longitudinal cavity mode, undergo ring-down decays when the cavity input is shut off. Sensitive intracavity absorption information is then simultaneously available across 100 nm in the visible and NIR regions. In this way, real-time, quantitative measurements were made of the trace presence, the transition strengths and linewidths, and the population redistributions due to collisions and temperature changes for molecules such as O_2, NH_3

and H_2O. In a similar experiment fiber-comb-based CRD spectroscopy was used to record a 100-nm bandwidth spectra of NH_3, C_2H_3, and CO around 1500 nm [542].

Vernier spectrometer

At low tooth spacings where the VIPA approach cannot be used, or if one wants to avoid the use of a hard-to-align VIPA element altogether, a different scheme can be applied. Gohle and co-workers first developed it by performing DFCS in conjunction with a high-finesse optical resonator [543]. In this experiment, ν_r is intentionally mismatched to the cavity FSR in such a way that only every $x - th$ comb mode is on resonance. This results in an effective filtering of the original comb so that, for sufficiently large x, it can be resolved with a simple grating spectrograph (see Figure 6.14). This situation may be realized by choosing $\nu_r/\nu_{FSR} = x/(x+1)$ or $\nu_r/\nu_{FSR} = x/(x-1)$, as in a Vernier scale. As the resonator length is tuned, the next set of $x\nu_r$-spaced comb modes is brought into resonance and so on, until the resonator round-trip length has been scanned by a wavelength.

By recording several spectra over the entire sweep, a final spectrum may be reconstructed that exhibits both a sampling period given by the comb spacing (i.e., ν_r) and the final spectral resolution limited by the width of an individual comb line.

Fourier transform spectrometer

As we have seen above, Fourier transform spectroscopy (FTS) can be used to measure complex broadband optical absorption or emission from molecules at high resolution by measuring the laser spectrum via time-domain interference [544]. Commonly used FT spectrometers employ thermal light sources to cover 5–50000 cm^{-1} with resolutions as low as 0.001 cm^{-1}. Acquisition time is a major shortcoming of FTS because the resolution scales directly with the range of the scanning arm; higher spectral resolution thus requires longer

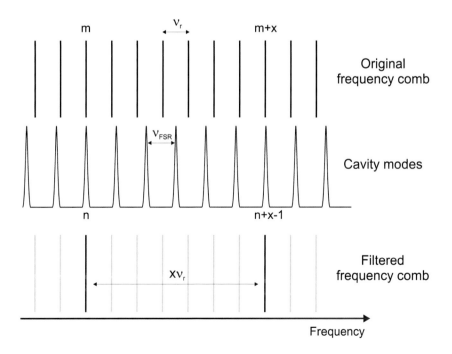

FIGURE 6.14
Scheme of a Vernier detection approach for a comb-cavity system. With a proper mismatch, the cavity filters the comb emission to contain only lines separated by $x\nu_r$. By sweeping ν_r, one can tune every comb line onto resonance sequentially.

scans and lower acquisition times. An important advantage of FTS lies in the possibility of recording broad optical spectra, but it requires a broadband light source if absorption is to be measured. Good spatial mode-matching of the waves traveling in two arms of the interferometer is required to obtain an interference signal with significant contrast. Unfortunately, most white-light sources such as tungsten lamps do not possess sufficient spatial coherence for good mode matching unless a diffraction-limited emission cone is used, which greatly reduces the intensity of the source. Therefore, thermal sources require a considerable averaging time to achieve a sufficiently good signal-to-noise ratio. Replacing the thermal source with a frequency comb (to act as a diffraction-limited white-light source), a technique known as frequency comb Fourier transform spectroscopy, can overcome this problem and immediately provides increased brightness and reduces the averaging times [545, 546]. Furthermore, the scheme may be directly extended to the mid-infrared region with suitable detectors.

However, an even greater potential was recently unveiled [471]. Because the comb is a pulsed laser source with a very precise repetition rate, an effective modulation/lock-in detection scheme that samples exactly at ν_r can be used. Unlike usual modulation frequencies, which are in the kilohertz range, ν_r is typically 100 MHz and above, which virtually eliminates any low-frequency noise. In addition to the brightness factor (a decrease of approximately tenfold in averaging time) the lock-in detection may add another factor of 10^5 in improvement [471]. In addition, the quadrature components of the lock-in signal can also be measured, which carry information on the dispersion of the sample, whereas traditional FTSs can only measure absorption.

An interesting interpretation of the FTS data is based on the Doppler shift that the light bouncing off the mirror, which is moving at a speed v, experiences. This generates a new frequency comb with a Doppler shifted mode spacing of $(1(v/c))\nu_r$. Of course, rather than Doppler-shifting a copy of the frequency comb, one could start out with two different frequency combs that operate with slightly different repetition rates. This allows recording to be done without moving parts, scanning to be much faster, and, even more importantly, it can provide very accurate calibration by controlling ν_0 and ν_r of the two combs rather than trying to determine the velocity v. This is exactly the idea behind the multi-heterodyne detection scheme described below.

Multi-heterodyne detection

This approach employs two frequency combs with their repetition rates slightly detuned from each other (e.g., by ≈ 1 kHz). One of the combs serves as a reference, while the second acts as a probe, which passes through the sample [547, 548, 549]. Data are acquired from the heterodyne beat between the two combs using a high-speed detector and digitizer. The basic operation can be understood in both the time and frequency domains. In the former, the slight detuning of the combs results in multiple heterodyne beats between individual pairs of modes from each comb. This is the time domain equivalent to the pulse trains from the two combs walking through each other, similar to what could be achieved with a scanning delay line, although with no moving parts. The digitized data are similar to that acquired from the conventional scanning Michelson interferometer of a Fourier transform spectrometer, and the complex spectrum (amplitude and phase) is obtained via Fourier transformation. With mutual stabilization of the frequency combs, sequential interferograms can be obtained and averaged, resulting in high signal-to-noise over broad spectral bandwidths with high spectral resolution [549]. While two frequency combs are required, this approach has the advantage of employing a single point detector. It is also compatible with cavity enhancement, in which case a minimum detectable absorption $\approx 1 \times 10^{-8}$ cm^{-1} has been demonstrated [550].

6.4.2.5 Other measurement schemes and applications

Combs in astronomy

Spectroscopy has always been one of the most important tools for studying the universe. Current astronomical spectrographs have sufficient short-term stability to measure the extremely small Doppler shifts resulting from planets orbiting distant stars or the possible slow change in the cosmological red-shift. However, even the best spectrographs are optomechanical systems that require periodic calibration, which is currently performed using atomic vapor lamps. Now, in much the same way as the frequency-stabilized laser has replaced the krypton lamp for the realization of the meter, broad-bandwidth frequency combs with large mode spacing may begin to replace such conventional discharge lamps and absorption cells for astronomical spectrographs requiring the highest level of calibration. Such broadband combs indeed possess the required properties of high accuracy, great spectral range, and nearly perfect long-term stability.

A typical application is the detection of extrasolar planets from the changing recoil velocity (and the consequent Doppler shift of spectral lines) of their star during the orbital period. These recoils velocities are extremely small unless a massive planet in close orbit is considered. For example, the variation imposed by the Earth on the velocity of our Sun has an amplitude of only $v_e=9$ cm/s with a period of one year. To detect such a tiny modulation in other stars, where it is also superimposed with a center-of-mass motion of typically hundreds of km/s and with the motion of the Earth around the Sun, a relative Doppler shift of $v_e/c = 3 \times 10^{-10}$ needs to be measurable. Converted to visible radiation, this requires a resolution of 150 kHz and the same reproducibility after half the orbital time. While this may be many orders of magnitude less precise than present optical frequency standards and clocks, there are other factors involving both the astronomical instrumentation and the frequency comb that make such precision challenging. Spectral lines from atoms and ions from interstellar clouds and the surface of stars are subject to strong line broadening due to collisions and the Doppler broadening of typically several GHz due to their thermal motion. Given these rather broad lines, the required spectral or velocity resolution can be obtained only by using the statistics of many lines observed simultaneously. It thus requires a very broad-bandwidth spectrometer with extremely small irregularities in the calibration curve. A frequency comb appears to be the optimum tool for this kind of calibration task, both in terms of providing an equidistant dense calibration and for allowing long-term reproducibility that derives from the possibility to reference to a precise clock (see scheme of Figure 6.15). In this case even a simple GPS disciplined rubidium clock suffices for the required 3×10^{-10} reproducibility to detect Earth-like extrasolar planets. However, for the comb to be useful for this application its repetition rate should be high enough such that individual comb teeth can be resolved by the spectrograph. In a typical echelle-type astronomical spectrograph, this amounts to a mode spacing of several tens of gigahertz in the visible or near infrared. Moreover, such large mode spacing is required over bandwidths of hundreds of nanometers, which is an extremely difficult problem, particularly in the visible portion of the spectrum. Some promising avenues towards this goal include the direct generation of high-repetition-rate self-referenced frequency combs, such as a recently demonstrated 10 GHz Ti:sapphire [551], mode filtering of lower repetition rate sources [552, 553, 554, 555], or the generation of broadband combs via parametric means in microresonators (see later) [556, 557, 558, 559]. Several groups are moving forward with calibration efforts using narrower bandwidth combs, and preliminary results appear promising [560, 561, 562].

Other possible approaches to astronomical applications include the direct heterodyne detection with a frequency comb. In this scheme the star light is mixed with the frequency comb on a fast photodetector producing a radio frequency spectrum identical to the optical

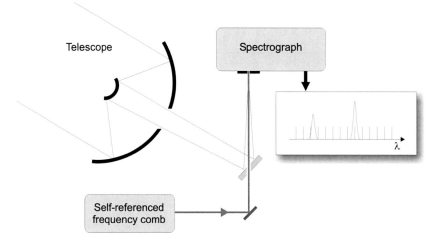

FIGURE 6.15
Simplified view of a possible arrangement for the use of a femtosecond frequency comb for the calibration of spectral lines from astronomical sources. The set of comb modes acts as an extremely broad-bandwidth and high-reproducibility frequency ruler that avoids frequent re-calibration of the spectrometer.

spectrum but shifted by the optical frequency. Signal processing can then be done with a radio frequency spectrum analyzer rather than with an optical spectrometer. This type of detection is known for producing shot noise limited signals and is used to demonstrate noise levels below the shot noise limit with squeezed light. The frequency comb can provide a large number of optical local oscillators to shift any optical component within its span to the radio frequency domain.

Frequency/time transfer

With the new generations of ultrastable clocks and frequency standards, the possibility of distributing time and frequency information to remote locations without loss of accuracy or precision has become an important topic. For example, applications such as long-baseline radio astronomy or particle acceleration and free-electron lasers require stringent timing over extended distances. The phase coherence and broad optical bandwidths of frequency combs can be exploited for frequency transfer and timing synchronization across fiber networks.

For frequency transfer, a coherent optical frequency comb has been used to translate the optical clock frequency to a phase-coherent 1.5 μm cw laser. This laser signal can then be transmitted over long distances of a Doppler-compensated optical fiber to a remote frequency comb, where its frequency is translated back to either a second optical frequency or the RF domain, depending on the application [563, 564]. Fractional frequency uncertainties of 10^{19} are achievable over large distances, allowing for the faithful dissemination and comparison of state-of-the-art optical clocks. The demonstrated performances thus show that the various technologies of phase-stabilized cw lasers, frequency combs, and fiber links can be successfully incorporated into a fiber network capable of preserving the stability and phase noise of current optical sources while bridging large distances and traversing the optical spectrum.

Time transfer, as opposed to frequency transfer, is a more challenging task that has yet to be demonstrated in a comb-based system, mainly due to fiber dispersion. However, timing synchronization across a local network has indeed been demonstrated at sub-10-fs levels by exploiting the precise timing of the pulse train from a frequency comb source [565].

Stable microwave sources

The initial drive for the development of self-referenced frequency combs was to multiply the frequency of RF clocks up to the optical domain. However, frequency combs can also be operated in the reverse direction, dividing the optical domain down to the RF regime.

As already mentioned, a cw laser oscillator stabilized to a high-finesse optical cavity (resonant Q approaching 10^{11}) is one of the lowest phase noise electromagnetic oscillators available in any frequency range [209]. One can then use this ultrastable cw oscillator as the reference for a frequency comb. A self-referenced frequency comb is phase-locked to a continuous-wave laser that is itself phase-locked to a high-finesse optical cavity. The low phase noise properties of the cw laser oscillator can thus be transferred to all the elements of the optical frequency comb, including the repetition rate and its harmonics.

In the limit of perfect phase-locking, the comb's repetition rate is a fraction of the continuous-wave laser frequency, with the phase noise given by the intrinsic thermal cavity noise. In the simplest case, a fast photodetector at the output of the stabilized frequency comb will generate photocurrent at frequencies equal to the spacing of the comb modes [566, 567]. For example, in the case of a 1 GHz frequency comb, the photocurrent output from the photodiode is made up of tones at 1,2,3,4.... GHz, up to the cutoff frequency of the diode. It will thus generate a microwave signal with a close-in phase noise lower than that achieved with room-temperature dielectric RF oscillators. This kind of signal can then be used to synthesize other frequencies or waveforms with low-phase noise in the microwave, millimeter-wave, or terahertz domains. These low-phase-noise microwaves can support sensitive Doppler RADAR and precision microwave interferometry measurements.

Waveform generation

A broad array of uniformly and widely spaced comb modes with milliwatt-level power would be valuable for applications in the field of microwave photonics and optical arbitrary waveform generation. In this case one can envision using line-by-line amplitude and phase control of the individual teeth of a frequency comb for the generation of arbitrarily synthesized optical waveforms with low phase noise and timing jitter (see Figure 6.16). When combined with high-speed photodetection, the manipulation of comb modes spaced by tens of gigahertz can also be used to generate arbitrary RF waveforms that are challenging, if not impossible, to synthesize with conventional techniques involving digital-to-analog converters. Demonstrations of different pulse shaping architectures using un-stabilized frequency combs have been carried out [568, 569], and the additional aspect of line-by-line pulse shaping with a low-phase-noise comb could add enhanced capabilities for applications in signal processing, secure communications, radar, and imaging, to name a few. Coherent LIDAR is also a motivation behind the generation and metrology of arbitrary optical waveforms [570]. The ability to generate arbitrary optical waveforms would permit coherent excitation of atoms or molecules well beyond today's demonstrations, which have so far been limited by the relatively simple comb structure available to Raman or two-photon transitions.

Ranging

Another important application of frequency combs is related to the precise measurement of distances, which can be easily converted to a time or frequency determination. In fact, length is a basic physical quantity and its precise measurement is of fundamental significance in science and technology. The ability to determine the absolute distance to an object (ranging) is important for applications such as large-scale manufacturing and future space satellite missions involving tight formation flying, where fast, accurate measurements of distance are critical for maintaining the relative pointing and position of individual satellites

Today, optical interferometers are commonly used to measure distances with an accuracy better than an optical wavelength; indeed, in extreme cases, such as gravitational wave detection, which calls for ultrasensitivity, the accuracy can be many orders of magnitude below the wavelength. Generally speaking, laser ranging determines the phase shift

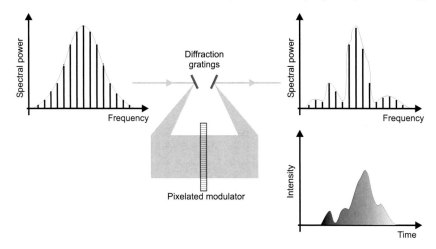

FIGURE 6.16
Scheme of a possible setup for arbitrary waveform generation based on the amplitude and phase modulation of the individual modes of a frequency comb.

of a signal after it has traveled a certain distance. Conventional laser interferometry techniques use continuous-wave lasers to measure the phase of optical wavelengths to attain subwavelength-resolution over ranges exceeding a few tens of meters. But measurements are limited to relative range changes (it is not possible to determine directly the absolute value of the distance being measured), meaning that this type of approach yields a small ambiguity range (range window) of just half the wavelength used. It also means that to measure any distance at all, an interferometer mirror must be moved along the entire distance while fringes are counted. Time-of-flight techniques, on the other hand, measure distance through pulsed or rf-modulated waveforms. Such systems offer large measuring ranges but poorer resolution (50–100 μm). In pursuit of the best of both worlds (larger ambiguity range and good resolution) multiple wavelengths can be combined to generate a longer "synthetic wavelength." Unfortunately, this requires either a tunable laser source or multiple laser sources to generate stable, accurate optical wavelengths over a wide spectral range. As such, the accuracy of absolute distance measurements is greatly affected by the individual precision of the wavelengths used.

As with spectroscopy, frequency combs can thus be used to support these conventional laser ranging approaches by functioning as a precise spectral ruler. Examples include the synthetic wavelength interferometer, operating with a sequence of RF harmonics of the pulse repetition rate. Using the principle of multi-wavelength interferometry, the frequency of a single cw laser can be consecutively locked to a sequence of selected modes of a stabilized optical comb and this scheme was used in determining the absolute length of gauge blocks with an overall calibration uncertainty of 15 nm [571]. In order to overcome the drawback of time-consuming consecutive measurements at different wavelengths, a different scheme was adopted by Schuhler et al. [572]. By frequency stabilizing two lasers (emitting around 1.3 μm) to a fiber-based optical frequency comb, they realized a two-wavelength source permitting the generation of synthetic wavelengths from tens of micrometers to several meters. By generating synthetic wavelengths as small as 90 μm with an accuracy better than 0.2 ppm and with a phase accuracy better than $2\pi/200$, the system allowed researchers to reach a nanometer accuracy.

However, frequency combs can also be used directly for distance measurements instead of just as spectral rulers. An interesting alternative for absolute length metrology at arbitrary distances with a resolution better than an optical fringe was proposed by Ye [573]. The approach relies on the consideration that a phase-stabilized mode-locked fs laser can provide both incoherent, time-of-flight information and coherent, fringe-resolved interferometry. In fact, an ultrafast pulse train allows determination of an absolute distance by use of time-of-flight information as in the incoherent measurement approach. However, a phase-stabilized femtosecond pulse train also allows one to build optical interference fringes when pulses traversing different arms of an interferometer are allowed to overlap and interfere, given a proper adjustment of the mode-locked laser. Essentially one can thus create a white-light interferometry condition at any desired length difference between the two arms in the interferometer.

More recently, researchers at the Delft University of Technology and Korea Advanced Institute of Science and Technology have combined three distance-measuring techniques in one instrument [574]: spectrally resolved interferometry (SRI) to reach nanometer-scale resolution, but with a nonambiguity range of about 1.5 mm; time-of-flight (TOF), which measures arbitrarily long absolute distances with a resolution of about 7.5 mm; and synthetic wavelength interferometry (SWI), which has a resolution of about 0.16 mm and a nonambiguity range of greater than 7.5 mm. In this way, SWI could link the high resolution of SRI to the absolute measuring capability of TOF. Later, this approach was extended to the precise measurement of long (tens of meters) path differences, only limited by the uncertainties in the refractive index of air [575]. Here, an 815-nm ML Ti:sapphire laser is used as the frequency-comb source, with both repetition frequency and offset frequency referenced to a cesium atomic clock. The frequency-comb source has a pulse duration of 40 fs and a repetition rate of 1 GHz, which corresponds to a pulse-to-pulse distance of 30 cm. After collimation, the source is sent to a Michelson interferometer, with one part of the beam (the short reference arm) reflected by a corner cube mounted on a piezoelectric transducer, and the other part of the beam (the long measurement arm) reflected by another corner cube, which can be moved along a 50-m-long measurement bench. A correlation function is measured by overlapping the beams on an avalanche photodiode (APD) while the PZT is modulated.

Building on these works, Coddington et al. [576] recently demonstrated a new comb-based LIDAR technique. Their approach combines the advantages of both time-of-flight and interferometric approaches to provide absolute distance measurements simultaneously from multiple reflectors and at low power. They use a pair of stabilized broadband, fiber-based, femtosecond-laser frequency combs with pulse trains of slightly different repetition rates (100.021 and 100.016 MHz). One comb acts as the "signal" source and samples a distance path defined by reflections off a target and reference plane; the other acts as a broadband local oscillator and recovers range information in an approach equivalent to linear optical sampling. In this way, pulse time-of-flight information can be obtained, yielding 3-μm distance precision with a 1.5-m ambiguity range in 200 μs. Through the optical carrier phase, the measurement accuracy improves to better than 5 nm in 60 ms, and through the radio frequency phase the ambiguity range can be extended to 30 km, potentially providing ranging with an accuracy of 2 parts in 10^{13} at long distances.

These techniques share the common aim of extending the subwavelength precision of the interferometric principle at long range by making use of the frequency comb of femtosecond lasers. In a recent experiment at KAIST [577], a completely non-interferometric approach based on the time-of-flight principle using femtosecond comb lasers was used for long distance ranging. In this case the arrival times of a train of reference pulses are compared to that of the same train reflected from a distant object by means of a nonlinear optical cross-correlation technique. The relative delay between one pulse of the reference train and one of the object train is finely locked to zero by adjusting the laser repetition rate ν_r. This allows one to determine the distance as an integer multiple of the cavity length with direct traceability to the RF time standard. In comparison to some of the aforementioned interferometric techniques, this simple time-of-flight method offers nanometer precision at a higher sampling rate and with less post-processing computation. The measuring capability is also well maintained at long range, with neither periodic ambiguity nor coherence limitation in the measurable distance. A sub-10 nm measurement precision in vacuum can thus be achieved with this method at integration times as short as 10 ms, regardless of the extent of the distance to be measured.

6.4.3 Relevant properties of a mode-locked laser for frequency-comb applications

Although Ti:sapphire-based frequency combs were the first to be spectrally broadened to an octave and self-referenced, the last decade has witnessed an expansion of the frequency comb technology to new femtosecond lasers with some distinct advantages over the original Ti:sapphire. Primarily, erbium- and ytterbium-based fiber-based lasers, which are directly pumped by laser diodes, have emerged as low-cost, compact, and robust alternatives. At the same time, the search for the ideal frequency comb continues and each of these new different lasers has its own benefits and disadvantages. However, in choosing the appropriate frequency comb for a given application, there are several key properties that need to be considered.

Spectral Bandwidth and Coverage The first issue to be addressed is probably the spectral range to be covered. At the moment, the wavelength range from 400 nm to nearly 2200 nm is well covered by a combination of Ti:sapphire-, Er:fiber-, and Yb-based lasers. Although Ti:sapphire is the only one capable of directly generating octave-spanning spectra [578, 579, 580], all the others employ nonlinear optical fibers to obtain the necessary spectrum for self-referencing.

Frequency and Amplitude Noise In the ideal case, the frequency comb should add no noise in excess of the reference oscillator that controls ν_r and ν_0. With reasonable control electronics, this is typically not an issue on time scales greater than ≈ 0.01 s, as the comb faithfully reproduces the reference in accordance with Equation 6.12. However, on time scales from $2/\nu_r$ up to ≈ 1 ms, the noise properties of different frequency combs can vary substantially. Frequency noise that arises from temperature and acoustically driven fluctuations can usually be overcome with good mechanical design and well-designed control servos; however, amplitude and frequency noise coming from a noisy pump laser or from fundamental noise within the laser (amplified spontaneous emission, ASE) can be more challenging to eliminate [581, 582]. Generally speaking, noise on the pump laser will be transferred to both fluctuations in the amplitude and frequency of the comb modes up to Fourier frequencies corresponding to the characteristic gain dynamics (typically ≈ 5–10 kHz in Er:fiber [582], ≈ 0.5–1 MHz in Ti:sapphire [583]). Femtosecond lasers with high intracavity power and short pulses will have less ASE-induced frequency noise on the comb. In this regard, Ti:sapphire and other solid-state lasers with cavity losses of a few percent should have an advantage over fiber lasers where losses can exceed 50%. External to the mode-locked laser itself, the main source of amplitude and phase noise comes from spectral broadening in nonlinear fibers. Both technical and fundamental noise (from ASE and photon shot noise) can be amplified in the nonlinear fiber, such as the microstructured fiber and the highly nonlinear fiber (HNLF) [584, 585, 586]. In some cases, this can lead to optical spectra nearly devoid of the original comb structure, although such decoherence is generally less of a factor when short pulses (e.g., <50 fs) are used to pump a short nonlinear fiber having zero dispersion near the central wavelength of the femtosecond laser [585].

Repetition Rate For most of the traditional frequency metrology experiments, it is desirable to have the highest practical repetition rate at which an octave spectrum is obtained. This is typically in the range of a few hundred megahertz up to a few gigahertz. Given a fixed average power, a higher repetition rate provides more power per frequency mode. The development of broad bandwidth combs with mode spacings larger than 10–50 GHz is very challenging but very valuable for emerging applications in microwave photonics as well as in the calibration of astronomical spectrographs, as described above. Generation in this case is challenging with conventional mode-locked laser-based frequency combs due to the necessity of a short cavity length. Moreover, because the pulse energy scales inversely with repetition rate, the nonlinear spectral broadening at rates of many tens of gigahertz is much less effective than at 100 MHz repetition rate. To date, the highest repetition rate for which octave-spanning spectra and self-referencing has been achieved is 10 GHz with a ring Ti:sapphire laser [551]. Along the lines of higher repetition rates, a fascinating development of the past few years was the parametric generation of frequency combs in highly nonlinear microresonators (see later).

For applications that employ the comb directly for spectroscopy, the best repetition rate is more difficult to define. Lower rates ($\leq$100 MHz) with higher energy per pulse are desirable for high spectral resolution and when nonlinear frequency conversion is needed to get to mid-infrared or ultraviolet wavelengths. On the other hand, gigahertz repetition rates enable the direct resolution of individual modes and can permit nonlinear spectroscopy with an individual mode [587].

Some tunability (1–5%) of the repetition rate is also desirable in the determination of the mode index.

Size, Weight, and Power A definite and important trend for frequency comb sources is towards smaller, more efficient, more robust, and less expensive sources. Compelling applications in spectroscopy, length measurement, waveform synthesis, and optical atomic clocks will ultimately require frequency combs that can operate in real-world conditions outside the research lab. Such environments require not only robustness, but also the overall power usage is an important factor. It is clear that the currently most significant reduction in power usage comes with the direct diode-pumped Er and Yb fiber-based systems, which are about 10 times more efficient than Ti:sapphire.

6.4.4 Microresonator-based frequency combs

In view of the above considerations, for many interesting applications of frequency combs it seems advantageous to further reduce their footprint and at the same time to increase the repetition rate into the frequency range above 10 GHz. A new optical frequency comb generation principle has recently emerged that goes exactly in this direction by using parametric frequency conversion in high quality factor microresonators [556]. This approach provides access to high repetition rates in the range of 10 to 1000 GHz, may lead to a new generation of combs that enable planar integration, and may permit a direct link from the radio frequency (RF) to optical domain on a chip.

The general scheme is based on the interaction between a continuous-wave pump laser with the modes of a monolithic microresonator via the Kerr nonlinearity. Thanks to the small volume of the resonator, light can be highly confined so that its intensity and nonlinear interaction with the medium are enhanced. The ultrahigh Q-factor (proportional to the photon cavity-storage time, and in excess of 10^8 in some cases) results in long interaction lengths and can lead to extremely low thresholds (submicrowatt power level) for nonlinear optical effects.

An important class of microresonators are whispering gallery mode (WGM) resonators (such as microdisks, microspheres, microtoroids, or microrings) which confine light by total internal reflection around the perimeter of an air-dielectric interface. The optical WGM resonances correspond to an integer number of optical wavelengths around the microres-

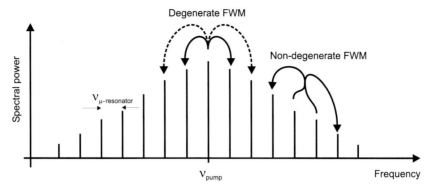

FIGURE 6.17
Generation of an equidistant set of frequency modes by four-wave mixing in a microresonator pumped by a cw laser. Both degenerate and non-degenerate processes concur in producing new sidebands spaced by the cavity free spectral range.

onator's perimeter. For resonators made of materials that exhibit inversion symmetry, the elemental nonlinear interaction is third-order in the electric field, which gives rise to the process of parametric four-wave mixing (FWM). This, because of energy conservation, converts two pump photons into signal and idler photons at frequencies that are equally spaced with respect to the pump [588]. If the signal and idler frequencies coincide with optical microresonator modes, the parametric process is enhanced, resulting in efficient sideband generation. In WGM resonators momentum is also intrinsically conserved when signal and idler modes are symmetrically located with respect to the pump mode. The initially generated signal and idler sidebands can interact with each other and produce higher-order sidebands by non-degenerate FWM, which ensures that the frequency difference of pump and first-order sidebands is exactly transferred to all higher-order sidebands. Thus the successive, cascaded FWM to higher orders intrinsically leads to the generation of phase-coherent sidebands spaced by the cavity FSR, that is, an optical frequency comb (Figure 6.17). Dispersion, the variation of the free spectral range of the cavity with wavelength, ultimately limits this conversion process and leads to a finite bandwidth of the comb generation process, because the cascaded FWM becomes much less efficient when the generated sidebands do not coincide with the cavity mode spectrum.

Optical comb generation with a microresonator thus fundamentally requires a high-Q cavity with small mode volume that is made from a material with a third-order nonlinearity and low dispersion. Toroidal microresonators [589] were the first system in which optical frequency comb generation was demonstrated [556]. They consist of a small silica toroidal WGM and can attain Q-factors in excess of 10^8. Highly efficient coupling into these planar devices can be achieved by using the evanescent field of tapered optical fibers (see Figure 6.18). Using tapered fiber coupling of 100 mW pump power at 1550 nm, a 375-GHz repetition rate frequency comb was first attained with a spectral bandwidth exceeding 350 nm, and the uniformity of the mode spacing was shown to be better than 1 part in 10^{17} [556]. This broad bandwidth was possible thanks to the intrinsic dispersion compensation in silica in the 1550-nm region. Pumping around this wavelength has recently led to the creation of broad frequency combs that exhibit more than a full octave span in wavelength [590].

Millimeter-scale crystalline resonators, made by polishing a cylindrical blank, can also feature exceptional Q-factors that exceed 10^{10}. Input and output coupling of light in this case is achieved with evanescent prismatic couplers, and optical frequency combs with a mode spacing as low as 12 GHz have been demonstrated [557]. This frequency is low enough to be

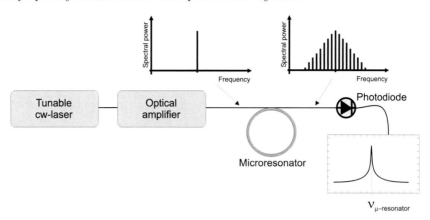

FIGURE 6.18
The amplified output of a tunable cw laser is coupled to a microresonator by means of the evanescent field of a tapered optical fiber. If the laser frequency coincides with one of the cavity modes, a broad frequency comb can be generated, as witnessed by the appearance of a peak at the resonator free spectral range in the detected transmitted signal.

directly detected using a photodetector, and such microresonators have been employed to generate microwave signals having high spectral purity [591, 592]. Fabry-Perot fiber-based cavities can equally give rise to optical comb generation [559] by using an interplay of the third-order nonlinearity and Brillouin scattering.

Optical frequency combs have also been generated in more compact silicon photonic circuits that integrate both resonator and waveguide on the same chip. In particular, optical frequency comb generation in integrated silicon nitride (SiN) resonators fabricated using approaches compatible with widespread complementary metal-oxide semiconductor (CMOS) technology has been recently demonstrated [593]. Notwithstanding a significantly smaller Q-factor, that is partially compensated for by the tight confinement of the field inside the microresonators, SiN exhibits a third-order nonlinearity that is approximately one order of magnitude larger than that of silica or crystalline materials such as CaF_2 or MgF_2. The main benefit of this approach is that it is fully planar and offers considerable flexibility, such as access to dispersion engineering through suitable resonator coatings. Moreover, this approach allows direct integration of the waveguide on the same chip-scale platform providing a means to fabricate a compact packaged device. Using an integrated SiN resonator, optical frequency comb generation has recently been demonstrated in this manner [594]. In addition, parametric comb generation has been demonstrated with another planar integrated resonator using a doped silica glass (Hydex) [558] that is also CMOS compatible and exhibited even higher optical Q-factors ($> 10^6$).

An important aspect of the optical frequency comb in metrology is the stabilization of the comb repetition rate and carrier envelope frequency. Although microresonators do not feature any moveable parts, ν_0 is directly accessible by varying the pump laser frequency and the repetition rate ν_r can be varied by use of the intensity-dependent round-trip time of the cavity. Due to both thermal effects (i.e., the change of refractive index due to heating by absorbed laser power) and the Kerr nonlinearity, power variations of the laser are converted to variations in the effective path length of the microresonator and therefore change the mode spacing of the frequency comb [591]. By using electronic feedback on both laser frequency and laser power, a microresonator Kerr comb with a mode spacing of 88 GHz has been fully frequency stabilized, with an Er:fiber laser-based frequency comb serving as reference. This latest result has been achieved by mitigating some of the problems that still affect microresonator-based frequency combs. In fact, due to the small volume, many noise processes (such as thermo-refractive and thermo-elastic noise) are enhanced in microresonators compared to more standard cavities. Phase stabilization is only viable when the noise on the measured ν_0 and ν_r beat notes is sufficiently low to be compensated with servo-control techniques. Moreover, especially for the planar waveguide and microtoroid systems discussed above, the reduction of ν_r to a readily measured frequency below 100 GHz may be an additional important issue.

Solving these problems will probably give access to compact frequency combs that have the interesting possibility of working also with different materials. Semiconductors such as InP, Ge, SiN, or Si are transparent in the mid-IR. Hence, microresonator-based mid-IR combs could enable a new generation of optical sensing devices in the important spectroscopic "molecular fingerprinting" regime, with a multichannel generator and spectral analysis tools residing on the same chip.

Besides the possible applications described above, the extremely high repetition rate available to microresonator-based frequency combs combined with a high power per mode (> 1 mW) could find interesting applications in high-capacity telecommunications. The advantage of the optical comb generator is that it can simultaneously generate hundreds of telecommunication channels from a single low-power chip source. Thus, a single laser source and a microresonator-based comb can in principle replace the individual lasers used for each channel in telecommunications.

6.5 Extension of OFCSs into novel spectral regions

6.5.1 High-order laser harmonics and extensions to the XUV

Current amplified pulsed laser systems can reach peak intensities of the order of $10^{15} - 10^{20}$ Wcm^{-2} and, under these conditions, a perturbative approach is no longer adequate to model the nonlinear response of materials to radiation. As a matter of fact, at a peak intensity of 3.5×10^{16} Wcm^{-2}, the electric field of the laser corresponds to the field which binds an electron to the proton in the ground state of an hydrogen atom: the assumption that successive terms in the expansion of the medium polarization in powers of the incident field get progressively smaller with the order cannot hold. The most impressive consequence of the breakdown of the perturbative approach is the deviation from the expected exponential decay of the intensity of successive harmonic orders. The appearance of a so-called plateau, a region of the spectrum where several harmonics are generated with almost constant efficiency, is the characteristic feature of high-order harmonic generation (HHG) in gases [595, 596].

Here, short and intense laser pulses are typically focused in a pulsed gas jet to produce coherent radiation in the extreme ultraviolet (XUV) and soft X-rays, a wavelength range (1-100 nm) where the lack of coherent sources has greatly limited the possibility of spectroscopic investigation, so far.

As a consequence of the isotropy of the gaseous medium, only odd-order harmonics are generated, and one finds that the region of efficient production of high harmonics is limited by a so-called cutoff. The most energetic photons generated in the process possess a cutoff energy E_{cutoff}, which is experimentally found to follow the simple law:

$$E_{cutoff} \approx I_P + 3.2 U_P \tag{6.17}$$

here I_P is the ionization potential of the atoms in the gas jet, and U_P is the ponderomo-

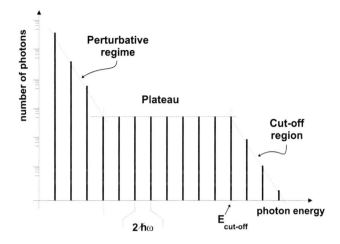

FIGURE 6.19
Schematic view of the typical spectrum of high-order harmonics. Only odd-order harmonic photons are generated, with successive harmonics separated by twice the energy of the pump laser. The first harmonics follow the predictions of a perturbative approach, with intensities exponentially decreasing with the order. Then a plateau is formed and extends up to E_{cutoff}.

tive energy, which corresponds to the mean kinetic energy acquired by an electron in one cycle of oscillation in the laser field:

$$U_P = \frac{e^2 E^2}{4m\omega^2} \tag{6.18}$$

with e and m the electron charge and mass, respectively, E the amplitude of the laser field, and ω its frequency [597]. It is evident that high laser intensities and high ionization potentials are required in order to reach very short wavelengths. Indeed, rare gas atoms and ultrashort pulses of very high peak intensity are normally used to generate high harmonics. A further advantage of using hard-to-ionize gases and short pulses is given by the requirement that neutral atoms survive long enough to experience the full peak pulse intensity instead of being completely ionized at the leading edge of the pulse itself. If $W(I)$ is the atomic ionization rate and τ_p the pulse duration, then the maximum effective intensity useful to produce harmonic photons is the so-called saturation intensity I_s, approximately given by $W(I_s)\tau_p = 1$; $W(I)$ being a monotonously increasing function of I, higher saturation

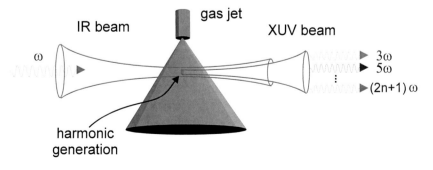

FIGURE 6.20
High-order harmonic generation in a gas jet.

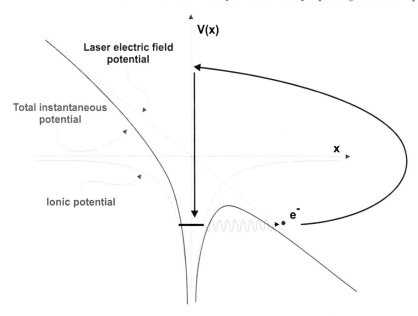

FIGURE 6.21 (SEE COLOR INSERT)
The three-step model for HHG. Electrons are first tunnel-ionized by the intense laser field, then they are accelerated during an optical cycle and finally recombine with the ion, thus releasing the accumulated kinetic energy in the form of harmonic photons.

intensities can be reached with shorter laser pulses and, of course, with light noble gases having higher ionization potentials.

6.5.1.1 Generation and properties of high-order laser harmonics

The standard picture of the high-order harmonic generation (HHG) process can be easily visualized from a semi-classical perspective in a single-atom approach (see Figure 6.21): every half optical cycle of the laser pulse, electrons undergo tunnel ionization through the potential barrier formed by the atomic potential and by the electric field potential of the laser itself; after being accelerated in the ionization continuum by the field, they may come back to the ion core and recombine to emit harmonic photons that release the accumulated kinetic and ionization energy [598].

Some of the most relevant features of harmonic radiation can be simply obtained by using this single-atom picture and solving the classical equations of motion for an electron, freed with zero velocity at different moments in an oscillating electric field. Depending on the phase of the field at the moment of ionization, the electron can either escape from the parent ion or oscillate in the field and come back to the original position after a fraction of the optical period T. Only electrons ionized within $T/4$ after each field maximum contribute to harmonic emission, by recombining with the ion and converting their kinetic and potential energy into photons (see Figure 6.22).

The experimentally found cutoff law can be simply obtained by noting that the maximum return kinetic energy (brought back by electrons which escape slightly after each field maximum every half optical cycle) corresponds to about $3.17\,U_p$. For the generation of harmonics with photon energies below the cutoff limit, there are always two different trajectories (corresponding to two different release times within each half cycle) that the electron can follow in the continuum such that it returns to the ion core with the correct energy.

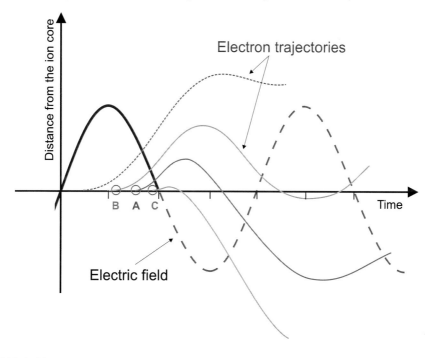

FIGURE 6.22

Schematic representation of the main electron trajectories involved in the HHG process (only the first half optical cycle is considered for simplicity, but things repeat with a T/2 periodicity). Electrons which are ionized when the absolute value of laser field amplitude is growing just escape from the ion and do not contribute to the process, while electrons released after the peak can come back to the core position and recombine to emit harmonic photons. Electrons ionized in A return with the maximum kinetic energy (about 3.17 U_p) and generate the most energetic photons in the cutoff region. Electrons released in B and C follow different trajectories in the continuum but return with the same kinetic energy, thus both contributing to the generation of the same harmonic.

These two trajectories, although giving rise to photons with the same energy, may correspond to quite different permanence times of the electrons under the influence of the laser field, and some of the spatial and temporal characteristics (phase-matching relations, defocusing and frequency modulation) of the emitted harmonic radiation will thus be strongly affected [599, 600].

Harmonic emission during a single pump pulse is thus seen to be composed of a train of XUV photon bunches, essentially released twice every optical cycle of the laser field. This temporal structure translates into the characteristic spectral structure of high-order harmonics, with peaks corresponding to the odd orders, separated by twice the laser frequency. The semiclassical model [598] also offers a simple explanation for other experimentally observed characteristics of harmonic radiation. A linear polarization of the laser is needed to make sure that the electron oscillating in the field may hit the ion and recombine with it when it comes back after the oscillation in the continuum; experiments have shown that just a small degree of ellipticity is sufficient to completely inhibit the generation of harmonics. It is also clear that, although the expressions for E_{cutoff} and U_p seem to favor long-wavelength pump pulses for the generation of high harmonics, actually, the longer time that it takes for the electron to return to the ion (proportional to the pump laser wavelength) implies a significant spreading of the electron wavepacket during its oscillation and a final lower overlap with the ion wavefunction, which in turns causes the probability of recombination to decrease.

6.5.1.2 Ramsey-type spectroscopy in the XUV with high-order harmonics

In recent years there has been an increasing interest towards high spectroscopic resolution in the vacuum ultraviolet (VUV) and extreme ultraviolet (XUV) regions, where important investigations involving several atomic and molecular species can be performed [601]. A broad and simple tunability in all this spectral range and extending to the soft X-rays is only achieved by synchrotron facilities, which, when used in combination with the best monochromators available, can provide radiation with a bandwidth as small as $0.5 \div 1$ cm^{-1} [602]. Apart from their limitations in spectral resolution, synchrotron sources suffer from some lack of accessibility, and tabletop-size sources would be much more desirable for a more widespread use. On the other hand, laser-based narrow-band cw sources hardly exist at such short wavelengths [603].

In this context harmonic sources look like a very appealing alternative but the extremely broad bandwidth associated with their short pulse duration seems to prevent their use for spectroscopy. In fact, even if some low-to-medium-order harmonics can be generated with pump pulses in the nano- and picosecond range [604, 605, 606], allowing one to keep a good spectral resolution for selected applications, higher-order harmonics can only be created at intensities above 10^{13} W/cm^2 by ultrashort laser pulses; and a 100-fs pulse is already characterized by a spectral width in the THz range.

A way to overcome this limit is with the application of the two-pulse technique described above to the harmonic radiation, for example by splitting and delaying the XUV pulses by means of a Michelson interferometer before sending them to the samples under study. It is interesting to note that, according to Equation 6.3 and in the case of XUV radiation with wavelengths in the 30-100 nm range, just a few mm of mirror displacement can achieve the same resolving power of the best synchrotron monochromators. Unfortunately, the use of this technique with HOH pulses is far from straightforward, mainly because good interferometers cannot be built to work in the XUV due to the lack of suitable optics. This problem can be solved by moving the pulse-splitting stage in the path of the laser beam, in order to create two phase-locked and time-delayed pump pulses that would then generate equally phase-related XUV pulses. Of course, in order for this technique to work, one has to make sure that the phase lock is preserved in the generated pulses: if the process of HOH generation were an incoherent one, no phase relationship could be preserved between the XUV pulses and such a scheme for spectroscopy with harmonics would be useless.

The mutual phase coherence between the harmonic pulses produced by phase-locked pump pulses was clearly demonstrated in a series of experiments performed in Florence and Lund at the end of the 90's, where the stable interference pattern between two secondary XUV pulses showed that the harmonic generation process could preserve the memory of the pump pulse phases [607, 608, 599]. Not only was the phase not scrambled in the process, but it was also demonstrated that a negligible frequency chirp was imparted to the XUV pulses. Nice, stable, and highly contrasted interference fringes indicated that the generation process was not as phase-destructive as it was initially thought, and showed that harmonic generation could become a suitable source for XUV interferometry [609] and high-resolution two-pulse spectroscopy.

Successively, other experiments performed in Florence and Paris [610, 611] directly demonstrated the appearance of the modulated spectrum described in Equation 6.2 by using collinear XUV pulses as produced by time-delayed pump pulses exiting an unbalanced Michelson interferometer.

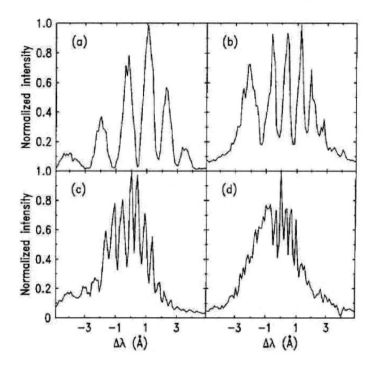

FIGURE 6.23

Experimental spectra of harmonics 11 (a), 15 (b), 19 (c), and 23 (d) generated by two laser pulses delayed by 120 fs and focused at 2×10^{14} Wcm^{-2} [611]. According to expectations, the spectra exhibit the broad envelope of the single pulse, with a superposed sinusoidal modulation showing fringes with a period $\delta\lambda = \lambda^2/c\tau$.

Here, the temporally integrated signal $I(\omega_s, t)$ observable at the exit slit of an XUV spectrometer centered at ω_s as a function of the delay t can be obtained by considering the harmonic field as the sum of two identical and temporally separated pulses. If $F(\omega - \omega_s)$ is the transmission function of the monochromator filter, which we assume to be symmetric and much narrower than the single-pulse spectrum $I_0(\omega)$, one obtains:

$$I(\omega_s, t) \propto I_0(\omega_s)\left(1 + \tilde{F}(t)\cos(\omega_s t)\right) \tag{6.19}$$

where $\tilde{F}(t)$ is the Fourier transform of $F(\omega)$, normalized to have $\tilde{F}(0) = 1$, and corresponds to the visibility of the resulting interference fringes. Measuring the fringe visibility as a function of the delay thus simply yields the Fourier transform of the filter transmission function. Expression 6.19 shows that two short time-delayed pulses that would not normally interfere due to their temporal separation may be forced to overlap again as a result of the broadening of their temporal profile introduced by the spectral filtering. If the filtering action of the monochromator is replaced by the narrow resonance of an atomic system, this technique may then allow the study of its spectral characteristics with an unprecedented resolution for this wavelength region. This approach is at the basis of Ramsey-type and Fourier transform spectroscopic schemes in the XUV with high-order harmonic pulses.

Ramsey-type spectroscopy with ultrashort laser pulses [612, 613] had been demonstrated in applications to bound state spectroscopy with optical sources in single-photon as well as in multiphoton and multistep transitions [489, 614]. Its general idea is similar to what was first introduced by N. Ramsey in 1950 [449] to face the transit-time-limited interaction between the molecules in a jet and a microwave field. In that case a greater effective interaction time was obtained by using two widely spaced interaction zones. In an analogous way one may think to extend the interrogation time characteristic of an ultrashort harmonic field by using pairs (or sequences) of collinear, phase-coherent, and time-delayed pulses. Consider the sinusoidally modulated two-pulse spectrum depicted in Figure 6.1b and imagine that a narrow atomic transition is placed somewhere under the broad spectral envelope. If the delay between the incoming pulses is varied, the fringe pattern moves and the transition will get in and out of resonance in a sinusoidal fashion. Of course, when the delay gets so long that the period of the spectral fringes becomes comparable or smaller than the transition linewidth, then the contrast of the excitation modulation will tend to decay. From a different point of view, one can consider the first short pulse as inducing a coherence in the two-level system and creating a dynamical polarization of the medium. The induced polarization oscillates at the transition frequency with a decaying amplitude during the dephasing time. The second pulse, depending on its phase with respect to the polarization oscillation, can enhance or destroy the residual system excitation. As a result, any excitation-related observable exhibits interference fringes when varying the delay between the two pulses. In more complex excitation schemes, involving more than two interacting states, the modulation of the fringe pattern and the appearance of beat notes can give information on the energy separation of nearby levels and on their lifetimes. In principle, the same information can also be retrieved with a Fourier transform spectroscopy (FTS) scheme, by directly measuring the integrated transmission of the pulse pair through the absorbing sample as a function of the interpulse delay. By a Fourier transformation of the resulting interferograms, the spectrum of the transmitted radiation can be retrieved.

High-order harmonics are ideal candidates to study the spectral characteristics of high-lying excited states with one-photon transitions, and Ramsey and FTS schemes have been recently demonstrated.

An experiment in Saclay [615] showed that it is possible to measure the spectral profile of harmonics by employing a FTS setup and measuring the interferogram produced by pairs of XUV pulses as a function of their delay. This illustrated the potential of the technique but was not followed by any real spectroscopic application, mainly because of the small absorption of the XUV radiation by a gaseous sample at low density which results in undetectable changes in the harmonic spectral profile.

Ramsey schemes have in principle a much higher sensitivity since the signal comes from directly measuring an atomic excitation, which, in the case it results in ionization, can have a unit detection efficiency. The first test of Ramsey-type spectroscopy with harmonics was performed on a pair of krypton autoionizing states, resonant with the ninth harmonic of a Ti:sapphire laser [616, 617]. The quantum interference manifested itself as a fringe pattern in the electron/ion signal caused by the ionization versus the delay between the pulses, with a fringe spacing given by the atomic transition period [618], in this case about 0.29 fs (see Figure 6.24). On the other hand, the modulation of the fringe contrast on the scale of the state lifetime, amounting to hundreds of fs, reflected the decay of the autoionizing states and the beating of the two transition frequencies.

More recently, the same Ramsey-type scheme based on pump pulse splitting has been successfully extended to the investigation of bound highly-excited states in Ar [619]. In this case, much longer level lifetimes (and correspondingly much narrower transition linewidths, although limited by Doppler broadening) allowed to put to a stringent test the spectral resolution limits of the original two-pulse XUV Ramsey scheme. Since the upper level lay close to the first ionization threshold of the atom, excited electrons could be promoted to the continuum by a further absorption of a delayed single IR photon at the fundamental wavelength (see Figure 6.25).

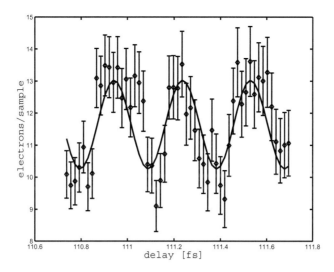

FIGURE 6.24

First Ramsey fringes in the excitation of an autoionizing transition in the XUV by means of a pair of time-delayed high-order harmonic pulses. The fringe spacing corresponds to the 0.29-fs period of the atomic transition.

However, if accurate measurements of transition frequencies are to be performed, all the atomic quantum interference fringes have to be accurately followed over a long time interval. While this is the norm for standard FTS in the visible and IR regions, following this approach for Ramsey spectroscopy in the XUV is far from straightforward and almost technically impossible. In fact, since these measurements are normally performed in an ion/electron counting regime, they are intrinsically very slow and one has to face long acquisition times that pose severe constraints on the overall system stability. One can overcome these constraints and make measurement times substantially shorter, by acquiring fringes only over a limited subset of randomly-chosen delay intervals in the whole delay range T that is needed for a given target spectral resolution.

It is worth examining in some detail a few different cases: let us assume that we are dealing with a single resonant atomic transition of negligible linewidth compared to the inverse of the time delay. The simplest case is depicted in Figure 6.26a where the atomic interference fringes are acquired while spanning the delay t in a single time window between t_0 and $t_0 + \delta t$: this is the case of a typical measurement like the one depicted in Figure 6.24. The square modulus of the Fourier transform (FT) of the Ramsey signal is a sinc^2-shaped curve centered at the atomic transition frequency and with a spectral width $\approx 1/\delta t$ (right column of Figure 6.26a).

If the Ramsey signal is acquired while the delay t of the second pulse is sequentially spanned in two δt–long windows starting at t_0 and $t_0 + T$ (T being accurately measured on the scale of the atomic oscillation period), the above spectral curve is further modulated by a $\cos^2$ term of period $1/T$ (see Figure 6.26b). This high-frequency modulation opens the possibility of high-resolution spectroscopy. However, if $T \gg \delta t$, many different spectral maxima are eligible as the real atomic frequency and the identification is highly ambiguous unless a previous determination of the frequency is available with an uncertainty $< 1/2T$ [620].

This ambiguity can also be solved by sampling the Ramsey fringes while scanning the delay of the second pulse in several time windows of width δt, either regularly spaced, or

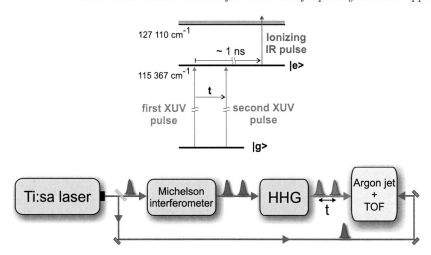

FIGURE 6.25 (SEE COLOR INSERT)
Scheme of a setup for Ramsey spectroscopy of highly-excited bound states in argon [619].
A single IR pump pulse is split int two delayed replicas in a highly-stable Michelson interferometer. After high-order harmonic generation, the two XUV pulses are used to excite Ar
atoms in a jet. A successive IR pulse probes the system excitation by ionizing atoms in the
upper excited level (see atomic level scheme above). Ramsey fringes are observed in the ion
signal as a function of the delay between the exciting pulses.

with a random sampling of the delay intervals. The random approach has various advantages
in this sense, as different sinusoidal modulations with incommensurable periods rapidly
cancel the satellite comb peaks (see Figure 6.26c), leading to a substantial reduction of the
acquisition time. Moreover, the width of the surviving tooth, corresponding to the atomic
resonance, approaches the theoretical resolution limit $(1/T)$ also in this case [621].

In principle, this scheme imposes to scan and accurately measure the delay t over all the
measurement windows. Anyhow, as far as one is concerned with a well-isolated resonance,
a remarkable experimental simplification can be obtained by just accurately monitoring the
relative delay of distinct pairs of measurements. In other words, the basic two-time-window
acquisition schematically depicted in Figure 6.26b can be repeated several times for a set
of random delays spanning the whole range up to T, as in Figure 6.26d.

Then, the square moduli of the FTs corresponding to different pairs can be simply
combined by taking their product. The advantage of this approach is that this high spectral
resolution can be achieved by combining a very limited number of short acquisition pairs
at random but accurately known relative delays, instead of performing a very demanding
interferometrically calibrated single acquisition of Ramsey fringes over a long delay of several
picoseconds. This corresponds to a large reduction of the acquisition time, while preserving
the final spectral resolution and with the fundamental experimental advantage of combining
data from independent measurement runs. In the bottom inset of Figure 6.27 a zoom of
the frequency region around the resonance is shown for each of the 20 pairs of Ramsey
measurements. After a few acquisitions the resonance peak is correctly identified and its
width becomes progressively narrower as more pairs at longer relative delay are added [622].

Another approach to Ramsey-type spectroscopy with harmonics has been recently developed by the group of Eikema and Ubachs in Amsterdam [623] and uses the amplification
of successive pulses from a mode-locked, phase-stabilized, laser oscillator to generate two
or more accurately delayed harmonic pulses. In this case, the time separation between the
driving pump pulses for harmonic generation is provided by the accurately known repetition

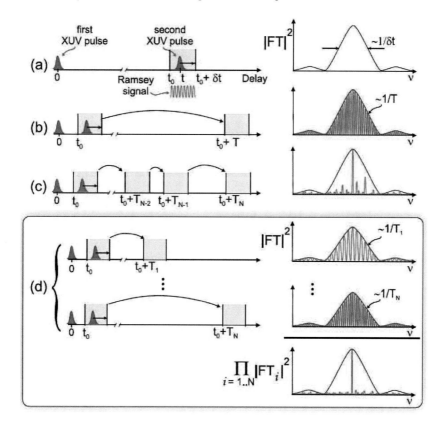

FIGURE 6.26
Different possibilities for scanning the relative delay between the excitation pulses in a Ramsey experiment with harmonics. The corresponding spectral information achievable by Fourier-transforming the Ramsey signals is shown in the right panels. See the text for details.

rate of a stabilized femtosecond frequency comb laser, instead of being introduced by splitting a single amplified laser pulse in a Michelson interferometer. More in detail, electro-optic modulators are used to select two or more (consecutive or not) pulses from the mode-locked pulse train. These pulses are then amplified in a Ti:sapphire nonsaturating amplifier up to an energy of about 15-25 μJ per pulse, while their spectrum is limited to about 0.5-0.7 nm in order to excite just a single transition of interest.

In the first experiment of this kind [623] a two-photon transition at 212.55 nm in krypton was excited by producing the necessary radiation by fourth-harmonic generation of the amplifier output at 850.2 nm through sequential frequency doubling in two beta-barium borate (BBO) crystals (see Figure 6.28). Sequences of two or three consecutive pulses of the mode-locked train were used and the excited state population was then probed by a delayed ionization pulse at 532 nm.

The inter-pulse delay T was scanned by changing the comb laser repetition frequency around 75 MHz. With a single pulse, the excitation probability is clearly constant. With two pulses, a clear cosine oscillation with a high contrast is observed, while three-pulse excitation gives a pulse-like structure appearing with narrower peaks compared to two-pulse excitation. Also in this case, the unambiguous identification of the mode corresponding to the true position of the resonance is not trivial and requires in principle a preliminary

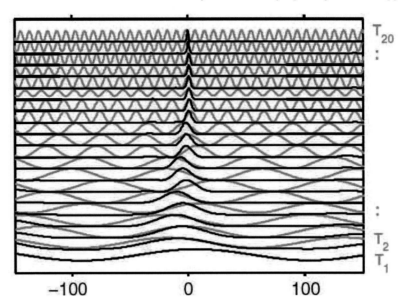

FIGURE 6.27
Effect of increasing the number of randomly delayed Ramsey scans on the retrieved spectral signal. The more scans are added at incommensurable delays, the less satellite peaks appear in the spectrum, until a single narrow resonance is clearly left. Absolute frequency measurements in the XUV are thus possible [622].

determination of the transition frequency with a sufficient accuracy. One can obviate to this problem as illustrated above, by performing scans at different delays and then looking for coincident peaks. Here, one can perform different sets of measurements at different laser repetition rates of the mode-locked oscillator. In the experiment of reference [623] three different frequencies were used to correctly identify the two-photon resonance and determine its absolute frequency with an uncertainty of 3.5 MHz, one order of magnitude better than previous measurements based on single nanosecond laser pulses.

In a successive experiment the same group used 125 nm pulses to excite a bound state in xenon which was subsequently probed by ionization with an additional delayed visible pulse [624]. Radiation at the desired wavelength was produced by third harmonic generation in a gas cell of the frequency-doubled output from a Ti:sapphire amplifier at 750 nm. In this case two electro-optic pulse pickers were used to selectively amplify either one pulse, two pulses separated by an integer number of oscillator round-trip times, or a pulse train consisting of up to six pulses, from a mode-locked frequency-comb laser. Once again, the time delay between them is scanned on an attosecond time scale by varying the repetition frequency ν_r of the frequency comb oscillator, while keeping the phase difference between successive pulses ϕ_{CEO} fixed. If two pulses are used with their separation varied in steps corresponding to an increasing number of oscillator round-trip times, cosine oscillations of decreasing period are observed, as shown in Figure 6.29b, so that the precision in the determination of the maxima is increased. On the other hand, by adding more pulses to the first two, the frequency resolution can be also increased, as is shown in Figure 6.29a: in the limit of adding an infinite number of pulses and infinite transition lifetime, sharp modes would emerge, resembling the original frequency comb structure.

The above two experiments did not really use the process of high-order harmonic generation to move comb-assisted Ramsey spectroscopy to the XUV. This was achieved more

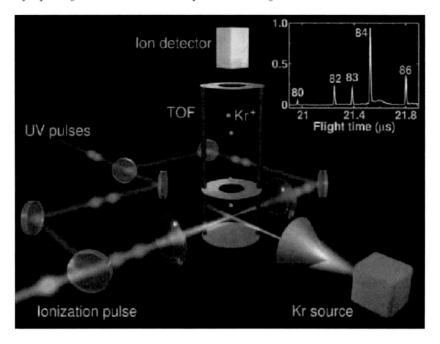

FIGURE 6.28
Schematic view of the experimental setup for the Ramsey-like spectroscopy of krypton at 212 nm [623]. A phase-stabilized oscillator is used here (after amplification) to provide an accurate control of the inter-pulse delay between the two exciting pulses.

recently by the Amsterdam group through the amplification of a pair of subsequent pulses from an IR frequency-comb laser with a double-pulse parametric amplifier to the milli-joule level. These energetic pump pulses could then be used for standard HHG in a krypton jet to efficiently produce radiation at the 15th harmonic at 51.5 nm [625]. The pairs of XUV pulses were used to excite the transition from the $1s^2$ ground state to the $1s5p$ excited state of ^{4}He and state selective ionization was then performed with an additional pulse at 1064 nm. Like in previous experiments, the delay between the XUV pulses was varied with attosecond precision by changing the repetition frequency of the mode-locked oscillator, while widely different repetition rates were used to unambiguously identify the correct resonance. Such measurements led to a new determination of the ground state ionization energy for ^{4}He of $h\times 5\,945\,204\,212(6)$ MHz, with an improvement in the accuracy of almost one order of magnitude compared to previous measurements.

The two different approaches to Ramsey spectroscopy in the XUV based on: (1) the pulse-splitting technique by a Michelson interferometer, and (2) the amplification of successive pulses from a stabilized mode-locked laser have different merits and drawbacks. In the comb-based scheme the time delay between successive pulses is accurately known and controlled via the repetition frequency of the laser, but the amplification process is known to introduce additional phase shifts that have to be accurately monitored and taken into account for the correct determination of the resonance frequency. On the other hand, the approach based on pulse-splitting owes its accuracy to the accuracy of the path displacement in the interferometer and therefore significant stability problems have to be faced, especially when long time delays (high spectral resolutions) are needed. The latter approach also allows in principle a continuous scan of the delay over long intervals and has no minimum delay to start from, as is the case for comb-based Ramsey spectroscopy. The minimum delay for the

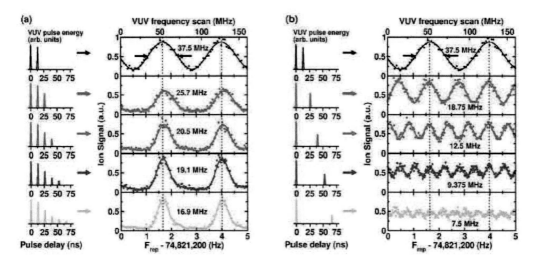

FIGURE 6.29
Ionization signal from Xe atoms following different schemes of excitations by UV pulses at 125 nm [624]. See the text for details.

latter is set by the inverse of the maximum repetition rate of the mode-locked oscillator and this technique is therefore intrinsically limited to the excitation of states whose lifetimes are longer than the comb interpulse delay, currently around 10 ns.

6.5.1.3 Cavity-enhanced XUV frequency combs

The above Ramsey-type schemes for precision spectroscopy in the XUV essentially rely on the frequency conversion of just a pair of delayed and mutually phase coherent pulses. A significant step forward would be made if the entire (ideally infinite) train of phase-locked pulses emitted from a stabilized mode-locked laser could be directly converted to the XUV. In this case one could translate the whole comb structure characteristic of the visible or near-IR laser spectrum to the XUV. If the original comb tooth frequencies were given by the usual form of Equation 6.12, then the XUV comb structure of the k-th harmonic should become:

$$\nu_h = h\nu_r + k\nu_0 \tag{6.20}$$

where the integer index h now spans an interval of k times larger values than the index m in Equation 6.12, and also the offset frequency ν_0 of the pump comb is multiplied by k when moved to the XUV. It is clear that obtaining this kind of result would allow to extend all the benefits and spectroscopic techniques discussed above for the frequency comb femtosecond lasers in the visible and near-IR to a new, and almost unexplored, spectral region.

Unfortunately, the original frequency comb structure is normally lost due to the reduction of the pulse train repetition rate required to actively amplify single pulses to the energies required for the HHG process to take place. In fact, the energy of pulses directly coming from a mode-locked laser oscillator is usually in the nJ range, while at least several μJs are needed for harmonic generation. Amplifying the energy of ultrashort pulses to this range implies reducing their repetition rate in a CPA scheme, typically in the kHz regime. The resulting frequency comb would thus be far too dense to be useful. It would for example require a preliminary knowledge of the investigated resonance frequencies at unavailable precisions. The mode spacing would also be too narrow for direct frequency comb spectroscopy which requires the separation of the modes, i.e., the repetition rate, to be larger than the observed transition linewidth.

To obviate to this situation a new approach to high order harmonic generation has been recently developed that allows the direct generation of XUV radiation at the high repetition rates of the original mode-locked laser oscillators. It consists of coherently coupling the pulse train emitted from the laser to an external enhancement cavity where the nonlinear process of harmonic generation can take place. Cavity-enhanced nonlinear frequency conversion is a standard technique when used in combination with CW laser emission, and has allowed impressive conversion efficiencies due to the resonant enhancement of the laser field circulating in the cavity. However, coupling the femtosecond pulse train to an enhancement cavity is not as straightforward: instead of just locking the monochromatic laser line to a single cavity mode, here all the longitudinal modes of the cavity have to be simultaneously resonant with each mode ν_m that originates from the laser. This can be accomplished with a resonator of appropriate length and zero group velocity dispersion, whose offset frequency ν_0 is matched to the laser. Rephrased into time domain language, coherent pulse addition in a resonator occurs if (1) the round-trip time of the pulse is the same as the period of the laser, (2) the pulse envelope does not change its shape during one round-trip, and (3) the carrier phase shift with respect to the envelope of the circulating pulse is the same per round trip as the pulse to pulse change from the laser. If these conditions are met, then the pulses arriving from the mode-locked cavity and those circulating in the cavity always meet at the cavity input coupler with the right timing and the right relative phase to coherently add their amplitudes in the cavity by constructive interference.

For increasing cavity finesse, the stored pulse undergoes an increasing number of round-trips and therefore its energy is progressively enhanced. At the same time, also, dispersion compensation becomes more critical, but it can be achieved with appropriately designed chirped mirrors.

If a low-density gas target is now placed in a position corresponding to a focus of such a resonator, it can act as the nonlinear medium for HHG. In contrast to the usual HHG schemes, the power that is not converted into the XUV after a single pass through the medium is "recycled" and can contribute in subsequent passes, so that higher total conversion efficiencies than for conventional schemes can be expected. In the first experiments of cavity-enhanced harmonic generation from a frequency comb mode-locked laser performed at Garching [626] and at JILA [627], the peak intensities in the cavity focus were in the range of 5×10^{13} W/cm^2, sufficient for the generation of high harmonics up to the 7th [627] or 15th [626] order.

As the HHG process is extremely nonlinear, it has been a significant concern that frequency comb coherence could not be maintained through HHG. If the harmonic generation process is coherent, i.e., if it does not disrupt the precise phase relationships between the successive pulses in the generated train of XUV pulses, then one may expect that radiation at each harmonic frequency is further broken into a comb structure with teeth at frequencies simply given by Equation 6.20. Indeed, it is a prerequisite that coherence is maintained between successive pulses in the train as this sets the frequency comb linewidth to be less than the repetition frequency. This comb linewidth determines the ultimate frequency resolution for experiments using an XUV frequency comb.

The fact that the phase coherence is preserved to a certain degree in the HHG process had already been demonstrated in early experiments [599], and all the Ramsey-type spectroscopic schemes with harmonics cited above critically rely on this property. However, showing that the phase coherence is well preserved over the timescale of single pulses or over a pair of successive pulses is not a guarantee that an accurate phase lock is established among all the pulses in the XUV train, so the existence of an XUV frequency comb has not been fully demonstrated yet.

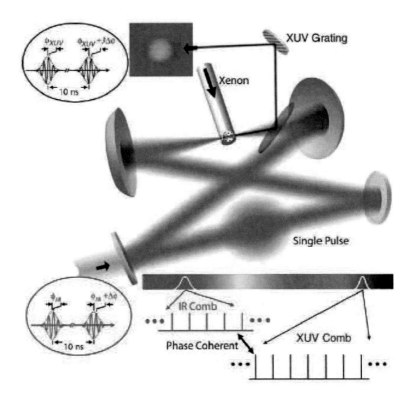

FIGURE 6.30 (SEE COLOR INSERT)
First experimental scheme for the cavity-enhanced generation of high-order harmonics from a phase-stabilized oscillator [627]. See the text for details.

Both the first two experiments [626, 627] showing the possibility of producing cavity-enhanced high-order harmonics also successfully tested the preservation of the comb coherence. This was done only up to the third harmonic by performing a beat experiment between the radiation produced by the gaseous medium in the cavity and that generated in a more conventional low-order harmonic generation in nonlinear crystals. The presence of clear and narrow beat lines in the temporally overlapped heterodyne signals demonstrated that the full temporal coherence of the original near-IR comb had been faithfully transferred to the third harmonic in the HHG process. The third harmonic is however generated in a fully perturbative regime and this demonstration thus cannot be directly translated to higher orders, where highly non-pertubative processes take place.

A successive experiment in Garching [628] showed that it is possible to produce much higher power at XUV wavelengths by using a lower repetition rate mode-locked laser. In this case, the use of a frequency comb at 10.8 MHz repetition rate and a long (13.9 m) enhancement cavity allowed the production of up to the 19th harmonic and achieved a power in the μW range for harmonics in the plateau. Such power levels are now of a sufficient order of magnitude as to be able to excite two-photon XUV transitions like the 1S-2S one in hydrogen-like He$^+$ at 60 nm. The group of J. Ye at JILA [504] also investigated the production of an XUV frequency comb by the up-conversion of the pulse train from a high-power mode-locked femtosecond fiber laser at 1070 nm. In this case below-threshold harmonics (that is, with photon energies below the ionization potential of the atom) from the 7th to the 13th order were generated and analyzed, and a direct measurement of the pulse-to-pulse coherence for the 7th harmonic at 153 nm was performed. Very recently, the same group reported the generation of extreme-ultraviolet frequency combs, reaching wavelengths of 40 nanometers [506]. These combs were now powerful enough to observe single-photon spectroscopy signals for both an argon transition at 82 nm and a neon transition at 63 nm, thus confirming the combs coherence in the extreme ultraviolet. Moreover, the absolute frequency of the argon transition was determined by direct frequency comb spectroscopy. The resolved ten-megahertz linewidth of the transition, which is limited by the temperature of the argon atoms, is unprecedented in this spectral region and places a stringent upper limit on the linewidth of individual comb teeth.

6.5.1.4 Attosecond pulses

The possibility of controlling the offset frequency ν_0 and thus stabilizing the carrier envelope phase ϕ_{CEO} of a mode-locked laser [522, 514] has opened new opportunities for high-intensity laser physics. Setting the offset frequency to zero, which can be readily accomplished with the technique of self-referencing, it is possible to obtain a train of pulses that have a fixed phase of the carrier with respect to the envelope. This means that all the pulses in the train have exactly the same electric field, even though the value that the carrier envelope phase actually assumes upon stopping its pulse-to-pulse slippage is unknown.

The precise value of ϕ_{CEO} does not make a big difference for long pulses that comprehend many optical field cycles under the pulse envelope, but may give substantially different results in experiments performed with few-cycle pulses [629], if they are used to drive processes which depend on the electric field in high order [630]. High-order harmonic generation is one of such cases. Let us examine what happens when trying to generate cutoff harmonics with two ultrashort pump pulses having the extreme configurations of ϕ_{CEO} that correspond to so-called "cosine" or "sine" pulses (see Figure 6.31). In the first case a peak of the field oscillation corresponds to the envelope maximum, while in the second case a zero of the field coincides with the peak of the envelope. It is easy to observe that the maximum value of the electric field is different in the two cases. Since generation of cutoff harmonics only takes place close to the peak of the pulse envelope, it may happen that, in the "cosine" configuration of the carrier envelope phase, only the central field peak may be sufficiently intense to start the process. Therefore, a single burst of high-energy, cutoff harmonic photons will be produced for each pump pulse. Due to the extreme temporal localization of the event, much shorter than the pump optical cycle, the duration of the XUV burst is in the attosecond (10^{-18} s) regime. Stabilizing ϕ_{CEO} to zero to produce trains of few-cycle "cosine" pump pulses are therefore the conditions for generating single, isolated, attosecond pulses.

In the opposite "sine" case, on the other hand, the two field maxima close to the envelope peak may not be sufficiently intense to generate bursts of cutoff harmonic photons altogether. Otherwise, they might produce two lower-intensity, attosecond bursts in the same pump pulse, separated by a delay corresponding to a roughly half optical cycle.

A clear signature of these effects is obtained by measuring the harmonic spectra produced by these different pulses [631, 632, 633]. While in the "cosine" pulse case single attosecond pulses produce a continuum spectrum in the cutoff region, in the "sine" case attosecond XUV pulses are produced in time-delayed pairs and thus produce a characteristic sinusoidal spectral interference with a spectral period equal to the inverse of their half-cycle temporal delay.

Detecting and stabilizing the carrier envelope phase not only allowed the production of attosecond pulses for the first time, but it also allowed to completely recover the electric-field transients of ultrashort pulses. Such a direct measurement uses the crosscorrelation between single attosecond pulses generated the way described above with the driving pulses. The

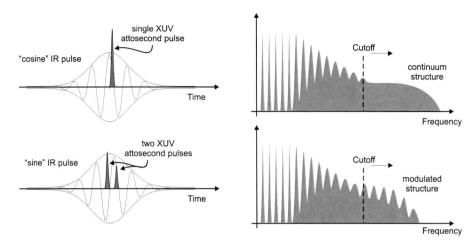

FIGURE 6.31
Generation of single attosecond pulses from a few-cycle IR pump pulse. The absolute phase of the driving pulses is crucial in the outcome of this highly nonlinear process. A "cosine" pulse can produce a single attosecond pulse in the cutoff region of the spectrum. A "sine" pulse can produce a pair of lower-intensity attosecond pulses with a delay corresponding to a half optical cycle. This is made evident by the sinusoidal structure of the spectrum in the cutoff region.

attosecond pulses are used to ionize atoms and the liberated electrons are accelerated by the instantaneous field of the ultrashort infrared pulses. By changing the delay between the pulses, the electric-field transient can be sampled with a temporal resolution on the attosecond scale [634]. Figure 6.32 shows the result of such a measurement.

6.5.2   Mid- and far-infrared OFCSs

Most ro-vibrational transitions of simple molecules fall in the IR region of the spectrum. Accurate frequency measurements of these transitions not only allow a better knowledge of the energy level structure of these molecules, but also provide a "natural" grid of frequency references for a variety of applications including environmental monitoring, astrophysics spectroscopy, and telecommunications, among others. Direct comb synthesis in the IR is therefore extremely useful for absolute frequency metrology of molecular spectra. In this frame, a few schemes have been devised for the extension of OFCSs to the IR region.

Broadening the spectrum of fs mode-locked lasers through highly nonlinear optical fibers has already succeeded in extending combs up to a wavelength of 2.3 μm [635]. However, so far, the approach which has produced the most relevant results in molecular spectroscopy relies on OFCS-assisted DFG IR laser sources.

The metrological performance of optical frequency comb synthesizers (OFCSs) has been transferred to the mid-IR, around 4.2 μm, by phase-locking the pump sources of a difference frequency generation (DFG) scheme to the teeth of a near-IR comb [636]. With this apparatus the frequency of some CO_2 transitions in the antisymmetric stretching vibrational mode around 4.3 μm was measured [422]. Sub-Doppler spectroscopy on these transitions was also performed by coupling the low DFG radiation power into a high-finesse Fabry-Perot cavity in order to enhance the laser intensity to the CO_2 saturation values [637, 638]. An important feature of comb-referenced DFG sources is the possibility of realizing very narrow

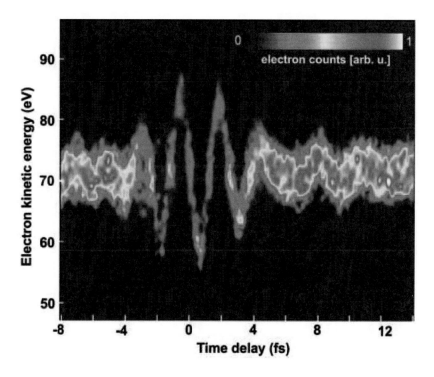

FIGURE 6.32 (SEE COLOR INSERT)
Snapshot of the electric field of a few-cycle IR pulse as measured by a streaking technique using single attosecond probe pulses [634].

linewidth sources which are useful both in high-precision and high-sensitivity applications. Indeed, as the DFG pump and signal lasers fall in the OFCS coverage range, phase-locking their frequencies to the nearest teeth of the comb gives a very narrow idler linewidth only limited by the excess phase noise between the two comb teeth due to the propagation of the repetition rate phase-noise to the optical frequencies.

Such a method, however, precludes the possibility of direct comb referencing for laser sources emitting directly in the mid-IR and not based on frequency mixing, such as quantum cascade lasers. More recently, a 270-nm span frequency comb at 3.4 μm has been realized by difference frequency generation between two spectral peaks emitted by a single uniquely designed Ti:Sa fs laser [639]. A composite frequency comb ranging from the violet to the mid-IR (0.4-2.4 μm) has also been obtained from a phase-controlled fs Ti:Sa laser and a synchronously pumped optical parametric oscillator [640].

A more flexible and direct approach for the creation of mid-IR combs has been demonstrated in Naples [641]. It is essentially based on a DFG process between a continuous-wave (cw) laser source and a near-IR fiber-based OFS. The pump beam is generated by an external-cavity diode laser (ECDL) emitting in the range 1030-1070 nm and is amplified by an Yb-doped fiber amplifier which delivers up to 0.7 W, preserving the linewidth of the injecting source (less than 1 MHz). The signal beam originates from a near-IR OFCS covering the 1-2 μm octave; a fraction of its output covering the 1500-1625 nm interval is fed to an external Er-doped fiber amplifier. The amplified comb beam has an average power of 0.7 W and spans from 1540 to 1580 nm (with a 100 MHz spacing), corresponding to nearly 50000 teeth in the frequency domain. Concerning the intensity and frequency noise spectral density, simultaneous amplification of many optical carriers does not introduce any appreciable additional noise in the process, owing to the slow gain dynamics which makes the amplifier practically immune from crosstalk effects. The two pumping laser beams are finally combined onto a dichroic mirror and focused into a periodically poled lithium-niobate (PPLN) crystal for DFG. Then, the mid-IR comb centered near 3.3 μm, with a bandwidth of 180 nm (5 THz) and an overall power of about 5 μW, is detected with a liquid-nitrogen-cooled HgCdTe detector. By tuning the 1-μm laser wavelength, and the quasi-phase-matching conditions of the crystal, the center frequency of the DFG comb can be tuned from 2.9 to 3.5 μm.

Such a scheme can be similarly applied to different IR regions, provided that nonlinear crystals with suitable transparency and conversion efficiency are used. Actually, use of advanced fiber-based devices, which benefit from the continuous progress in telecom technology, may give additional advantages for the realization of more and more effective setups.

The first direct frequency comb referencing of a 3-μm DFG source [642] was also achieved in Naples with a scheme similar to the previous one (Figure 6.33). However, in this experiment, the Er amplifier was also simultaneously seeded by a narrow-linewidth diode laser emitting in the 1520-1570 nm interval. As a consequence, two DFG processes take place simultaneously in the PPLN crystal, and both a cw source and a 3-μm comb arise from the interaction. The beat-note signal recorded at 3-μm by the fast MCT detector is used to phase-lock the cw coherent radiation to its closest tooth in the mid-IR and then use it for gas-cell spectroscopy. In this way, absolute frequency metrology of molecular vibration spectra was demonstrated by tuning the comb mode spacing across the Doppler-broadened absorption profile of a CH_4 ro-vibrational transition.

A similar scheme was also used to realize the first absolute frequency measurement performed using a quantum cascade laser (QCL) referenced to an OFCS [643]. For this purpose, a 4.43 μm QCL was used for producing near-IR radiation at 858 nm by means of sum-frequency generation (SFG) with a Nd:YAG source in a PPLN crystal. The absolute frequency of the QCL source was then measured by detecting the beat note between the sum frequency and a diode laser at the same wavelength, while both the Nd:YAG and the diode laser were referenced to the OFCS. Doppler-broadened line profiles of $^{13}CO_2$ molecular transitions were recorded with such an absolute frequency reference. Even better accuracies in the line center determination were achieved by means of sub-Doppler saturation spectroscopy performed with the same OFCS-assisted QCL spectrometer [425]. In this case, the QCL was frequency locked to the saturation dip while the beat note frequency between the diode laser and the SFG radiation was counted, yielding an accuracy 20 times better than the Doppler-limited measurements.

The mentioned spectroscopic works also address the issue of developing new effective molecular IR clocks. Such clocks have the advantage that they can be implemented with simple sub-Doppler spectroscopic techniques in gas cells or molecular beams, offering the opportunity to construct less expensive, transportable and stable oscillators. On the other hand, the mid-IR clock quality factor is intrinsically lower than in the optical-UV region, but, nonetheless, direct comb referencing represents the simplest and the most direct way to work in the crucial molecular "fingerprint" region.

Mid-IR clock oscillators can be developed by using the ro-vibrational transitions of spherical top molecules as CH_4 [639, 644], OsO_4 [645] and SF_6 [646]. As for the UV/visible clock transitions, OFCSs have been used to measure either the absolute transition frequency or to characterize the stability performance of these molecular-based clocks. The CO_2 laser locked to the OsO_4 ro-vibrational transition at 28 THz (10.5 μm) has been used for a long time in frequency chain synthesis as an intermediate oscillator to bridge the visible and microwave regions. Thanks to a femtosecond Ti:Sa OFCS, the CO_2/OsO_4 frequency has been measured with an accuracy of 10^{-12} and with a stability of 3×10^{-14} in 1 s [645]. The same group has proposed to use a CO_2 laser that is frequency locked to a ro-vibrational transition of SF_6 as another possible reference clock at 10 μm [646]. In both cases, the same experimental approach was used to link the IR frequencies to the Cs primary standard: a SFG nonlinear up-conversion process involving the CO_2 laser and visible radiation generates radiation still in the visible, whose frequency can be measured against the OFCS. In the CO_2/OsO_4 experiment, the SFG signal laser was an 852 nm OFCS-phase-locked diode laser, and the generated 788 nm SFG pump was used to phase-lock another diode laser at this wavelength. In this way, the beat-note SNR between the OFCS and the diode laser at 788 nm was increased. A more elegant approach was adopted in the CO_2/SF_6 experiment. Here, the SFG signal laser was the low-frequency part of the comb, giving rise to a pump comb coinciding with the high-frequency part of the originary OFCS. The SNR value of the beat note between the generated and the originary comb around 788 nm is increased, due to the collective contribution of the many comb pairs involved in the frequency counting. The main advantages of this scheme are that only one phase-lock loop is needed (to control the OFCS repetition rate frequency against the microwave Cs primary standard) and that the beat note is CEO free. Such a technique was first introduced to develop a CEO-free clockwork by using a CH_4-stabilized He-Ne laser at 3.39 μm as an OFCS oscillator [639, 644]. A 10^{-13} frequency stability at 1 s was measured for this He-Ne/CH_4 OFCS with an absolute frequency reproducibility of 10^{-12} at the same time scale. Finally, it is worth giving a short overview on progress in the development of frequency combs in the far infrared (FIR) spectral window (lying between the mid-IR and the microwave regions), roughly covering the 10 THz-0.3 THz range. Also for FIR, as in the other spectral windows discussed above, the importance of combs is related both to the very precise frequency referencing of other FIR sources and to their direct use as broadband sources. As already discussed in Chapter 4, laboratory coherent sources in this region generally have low power, ranging from the nanowatt to the few-mW range; the latter figure applies to the most recent cw quantum cascade lasers, whereas the lower powers typically afflict generation schemes based on frequency synthesis that, for this reason, are often limited in terms of spectral coverage at higher FIR frequencies. In this framework, comb generation in the microwave range (i.e., below 300 GHz) proved successful [647]. On the other hand, referencing of cw QCLs to fiber-based NIR combs was explored either by using current modulation in photo-conductive antennas [648] or by using electro-optic crystals [649]. However, these approaches suffered from low efficiency and required most of the QCL power for referencing to the comb, thus hindering its use for experiments. Moreover, these two approaches did not allow to generate a free-propagating THz comb, similarly to what was demonstrated in other spectral regions. The latter drawback has been overcome in a very recent experiment, where an air-propagating comb, covering the 0.1-6 THz spectral range, was created by difference frequency generation among teeth of a near-IR comb in a surface-waveguide lithium niobate crystal [296]. In this experiment, phase-locking of a QCL laser, emitting around 2.5 THz, to a comb tooth was also demonstrated, paving the way to a number of unpredictable metrological applications in the FIR range.

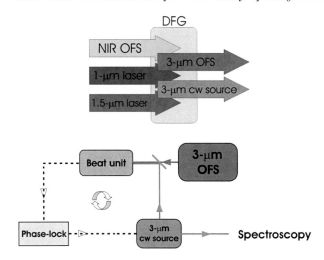

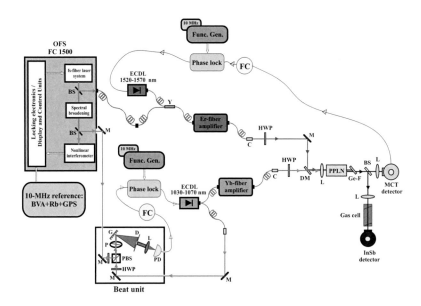

FIGURE 6.33

Upper Frame: Concept scheme for direct phase-locking a cw laser to a 3-μm comb. Two simultaneous DFG processes in a PPLN crystal are exploited to produce an OFS and a coherent source at 3 μm. Then, the detected beat-note signal is used to phase-lock the probe radiation directly to the MIR comb, while a second beam is allowed to record molecular spectra with an absolute frequency scale. Lower Frame: detailed experimental setup. First, the DFG pump source at 1-μm is phase-locked to the FC1500 comb. Then, the signal beams are obtained by simultaneous amplification of a cw 1.5-μm laser and the 1.5-μm portion of the NIR OFS. The MIR beat note is used to phase lock the 1.5-μm laser, and therefore the 3-μm probe radiation, to the MIR OFS, while reflection by a CaF$_2$ window is used for gas-cell spectroscopy. The following legend holds: ECDL=external cavity diode laser, G=diffraction grating, M=mirror, HWP=half wave plate, PBS=polarizing beam splitter, P=linear polarizer, D=iris diaphragm, L=lens, PD=InGaAs photo-detector, DM=dichroic mirror, Ge-F=germanium filter, FC=frequency counter, BS=beam splitter, C=fiber collimator, Y=fiber splitter.

7

Frequency standards

> Even a broken clock is right twice a day.
> *Stephen Hunt, The Court of the Air*

I love that you get cold when it's 71 degrees out.
from the movie "When Harry Met Sally..."

7.1 General features of frequency standards and clocks

Today, clocks and oscillators are vital devices in countless scientific and technological applications, encompassing sophisticated tests of fundamental physical theories on one side as well as synchronization of communication systems on the other one. There are by now many diverse realizations, ranging from common wristwatch quartz oscillators to ultra-refined primary frequency standards. In spite of this wide assortment, both in terms of cost and performance, all oscillators can be characterized within the same formalism. This is also true for time and frequency dissemination. Indeed, transmitting time over dial-up telephone lines with an uncertainty of about 1 ms hinges on the same principles that are the foundation of the 1-ns-uncertainty, satellite-based time broadcasting. Let us start by summarizing a few considerations scattered in previous chapters and putting some order in the nomenclature [650].

A **frequency standard** is a stable oscillator used for frequency calibration or reference. It generates a fundamental frequency with a high degree of accuracy and precision (sometimes, higher-order harmonics are also used as references). The oscillator, in turn, consists of two elements: a **generator** producing periodic signals and a **discriminator** controlling the output frequency. If the discriminator is actively oscillating and the output frequency of the device is fixed by a resonance in its response, then we speak of a self-sustained oscillator and, consequently, of **Active Frequency Standard**. Shining examples are represented by pendulum and quartz-crystal oscillators, as well as lasers and masers. Conversely, if the frequency-selective feature of the discriminator is probed by a tunable generator whose frequency is then locked to the peak of the discriminator response function (via a feedback loop), then we speak of **Passive Frequency Standard**. Paradigmatic examples are offered by cesium, rubidium, and optical atomic standards.

Then, a **standard clock** comprises: 1) a frequency standard; 2) a device to count off the cycles of the oscillation emitted by the frequency standard; 3) a display to output the result.

While some common properties of standards and clocks, certain relevant frequency-noise processes affecting their performance, as well as the statistics tools used for their characterization, were extensively discussed from a general point of view in Chapter 2, here we present the most relevant specific systems, which can be divided into four main categories:

- Microwave dielectric oscillators. This class basically comprises quartz and cryogenic sapphire oscillators.

- A second category of microwave oscillators is represented by photonic microwave oscillators based on whispering gallery mode (WGM) resonators and ultrastable microwaves generated via optical frequency division.

- Microwave atomic standards and clocks. These include, in particular, active hydrogen masers, cesium and rubidium fountains as well as microwave ion standards.

- Optical atomic clocks. These include trapped-ion clocks and neutral atom optical lattice clocks. Actually, this topic will be dealt with in the next chapter. Indeed, representing the cutting-edge technology in the field of absolute frequency metrology, optical atomic clocks are not yet well established as frequency standards and research is underway at a number of laboratories worldwide to investigate the stability and accuracy of different systems. In addition, hinging on a unique mix of the finest ingredients, introduced one by one in each chapter, optical atomic clocks capture the essence of the whole book, in a perfect reunion with the first chapter.

For a better understanding of these two latter items, a preliminary discussion on cooling/trapping of atoms and ions will be also provided in this chapter. Then, we will conclude by addressing the issue of time and frequency dissemination, with special emphasis on the emerging transfer techniques based on optical means.

7.2 Quartz oscillators

Piezoelectricity is the main feature of a crystal which is used as an oscillator. Essentially, it is the appearance of an electric potential difference (EPD) across certain faces of a crystal when a mechanical strain (bending, shear, torsion, tension, or compression) is exerted on it. In a crystal with a non-symmetrical unit cell (UC), by virtue of the applied strain, the ions in each UC are displaced, thereby inducing the UC electric polarization. Due to the regularity of the crystalline structure, these effects accumulate, eventually determining the appearance of an EDP between definite facets of the crystal. Conversely, when an external voltage is applied across the crystal, the latter undergoes mechanical movement (Figure 7.1). In this case, the ions in each UC are dislocated by electrostatic forces, leading to a mechanical deformation of the entire crystal. Among the various crystalline substances which would lend themselves to be used in a frequency reference, by virtue of its many favorable characteristics, quartz has become the most popular [95, 65, 59]. It has a formula unit composition of SiO_2. Alpha-quartz (characterized by a trigonal crystal system), which is thermodynamically stable up to 573 °C, is the most common polymorph of the silica

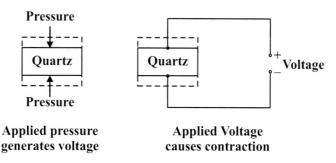

FIGURE 7.1
Illustration of piezoelectricity principles. (Adapted from [95].)

minerals. Indeed it offers two unique advantages: first, it can be grown with high purity and then easily machined; second, it is characterized by high stiffness and elasticity, which allows to fabricate plates with high-frequency and high-Q mechanical oscillations. Most physical properties of a quartz are anisotropic, as one can sense by the complicated shape of the macroscopic quartz crystal (Figure 7.2). As a consequence, alterations during the growth process which affect anisotropy will induce crystal imperfections. Early diagnosis, both in terms of crystal orientation and possible flaws, is accomplished by the use of polarized light, X-rays, and chemical etching. The major axis of quartz growth is not anisotropic to light and is, as such, called the optic axis. In literature, it is labelled as the Z axis in an orthogonal X, Y, Z coordinate system. The choice of specific axes and angles at which the crystal is cut to realize a plate of quartz material is of the utmost importance in determining the physical and electrical parameters of the oscillator. The most common cuts, conventionally referred to as AT, BT, etc., are shown in Figure 7.3. These are characterized by different trade-offs between electrical-mechanical capability and temperature coefficient. For example, AT-cut crystals exhibit extremely small temperature coefficients, whereas SC-cut ones offer excellent stress compensation. As already mentioned, when the obtained plate is subjected to a voltage alternating at appropriate frequency, the crystal will start vibrating and produce a steady signal, the mode of vibration depending on the specific cut. Just as an example, an extensional (thickness shear) mode is excited for an X (AT) cut. Different vibration modes are illustrated in Figure 7.4. Both harmonic and non-harmonic signals and overtones may be realized in such a vibration set-up. While unwanted non-harmonic signals are suppressed by preparing highly polished and properly shaped quartz pieces, harmonic overtones are highly desirable, as they permit the realization of higher-frequency references using the same cut. For example, a 10-MHz quartz oscillator can be attained as the 5th overtone of an AT-cut element. The final value of the center oscillation frequency depends crucially on the precision with which the plate thickness is controlled. For this reason, fine adjustment of the center frequency is often undertaken *a posteriori* by covering the quartz plate with gold layers (to give an idea, a monolayer of gold typically changes the frequency by 2 parts in 10^7).

Now, application of the quartz oscillator as a reference requires including it as a tuned circuit in a suitable feedback arrangement, such that the mechanical vibrations stabilize the oscillator's frequency. To better understand this, let us start by considering the electric equivalent circuit of a quartz crystal unit (Figure 7.5) [65]. The $L - C$ series describes the swapping between the mechanical energy stored in the crystal elastic deformation and the electric energy stored in the capacitor, while the resistance R account for dissipation of the energy from the oscillation to the thermal energy (both in the crystal itself and in the mounts). Finally, C_0 represents the static capacitance of the electrodes and the leads. The

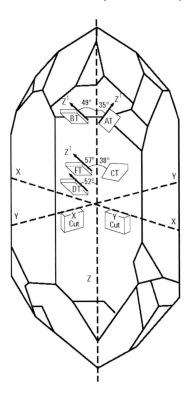

FIGURE 7.2
Axis orientation in a doubly terminated quartz crystal. (Adapted from [59].)

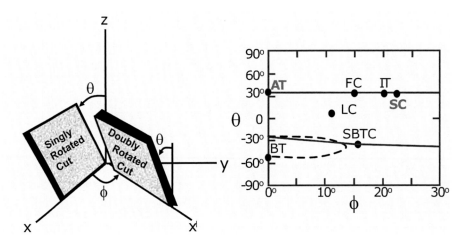

FIGURE 7.3
Orientations of singly and doubly rotated cuts (left) and angles of a few significant cuts (right). (Adapted from [59].)

values of these components are determined by the crystal parameters (basically the cut, the size, and the shape) and give, in turn, the quartz resonance frequency. Using the Laplace

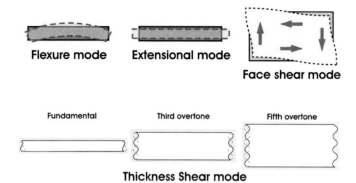

FIGURE 7.4
Different vibrational modes in a quartz crystal. (Adapted from [59].)

transform $(s = j\omega)$, the impedance of the equivalent circuit can be determined as

$$Z(s) = \frac{s^2 + (R/L)s + \omega_s^2}{sC_0\left(s^2 + (R/L)s + \omega_p^2\right)} \qquad (7.1)$$

where

$$\omega_s = \frac{1}{\sqrt{LC}} \qquad (7.2)$$

is the series resonant frequency and

$$\omega_p = \sqrt{\frac{C + C_0}{LCC_0}} \simeq \omega_s\left(1 + \frac{C}{2C_0}\right) \qquad (7.3)$$

is the parallel resonant frequency ($C_0 \gg C$). In practice, for small R/L values, ω_s (ω_p) is a zero (a pole) of $Z(s)$. Generally, the series resonance is a few kilohertz lower than the parallel one. Just as an example, from typical values of a 4-MHz quartz ($L \simeq 100$ mH, $C \simeq 0.015$ pF, $C_0 \simeq 5$ pF, $R \simeq 100\ \Omega$) we have $\nu_s = (1/2\pi)\omega_s \simeq 4.109$ MHz, $\nu_p = (1/2\pi)\omega_p \simeq 4.115$ MHz and $(1/2\pi)(R/L) \simeq 159$ Hz, such that $\omega_s, \omega_p \gg (R/L)$ is satisfied. Crystals below 30 MHz are generally operated between series and parallel resonance, which means that the crystal appears as an inductive reactance in operation. Crystals above 30 MHz (up to > 200 MHz) are generally operated at series resonance where the impedance appears at its minimum and equal to the series resistance (see Figure 7.6). As explained in Chapter 2, the quartz resonator must be inserted into an amplifier circuit to form an oscillator. In this configuration, a fraction of energy is fed back to the crystal, thus inducing

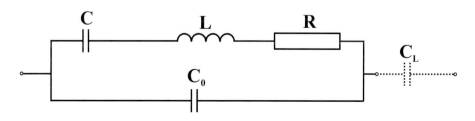

FIGURE 7.5
Electrical equivalent circuit for a quartz crystal unit. (Adapted from [65].)

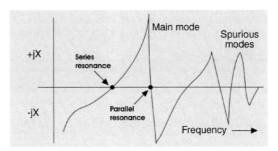

FIGURE 7.6
Reactance (i.e., the imaginary part of the impedance) plot for a quartz crystal, typical of any overtone.

vibrations which tend to stabilize the generated frequency at the resonance value [59]. A simplified amplifier feedback circuit is shown in Figure 7.7. The series capacitance, tunable by an applied voltage, is used to slightly change the phase of the feedback so as to finely tune the oscillator frequency. Such a Voltage Controlled Crystal Oscillator (VCXO) represents a good compromise between the good intrinsic stability of a crystal oscillator (a few parts per million) and the tunability. At this point, it should be well known that, at the frequency of oscillation, the closed-loop phase shift is $\phi = 2n\pi$. After initial energizing, the only signal present in the circuit is noise. Then, only the noise component, whose frequency fulfils the above phase condition, propagates around the loop with increasing amplitude until the amplifier gain is reduced by non-linearities of the active elements (*self-limiting*). At steady state, the closed-loop gain is equal to 1. If a phase perturbation $\Delta\phi$ occurs, the frequency must shift by $\Delta\nu = -\nu_0(2Q_L)^{-1}\Delta\phi$ to maintain the $2n\pi$ phase condition, where Q_L is the loaded Q of the crystal in the network.

The quartz plate mounting is also paramount to attain elevated Q-factors, as well as to minimize the cross-coupling between different vibrational modes and external stress. In this respect, a critical point is represented by the stress induced by deposition of the metallic electrodes on the quartz plate, which constitutes a major source of *aging*. This is smartly averted by the so-called BVA (Boitier a Vieillissement Ameliore, Enclosure with Improved Aging) structure, where the electrodes are on ancillary plates with a few-micron gap to the vibrating quartz. This also enables the attainment of a higher Q-value, which is not deteriorated through damping introduced by the electrode material.

Moreover, due to dependence of the crystal unit's frequency on temperature, high immunity to temperature fluctuations must be achieved in most demanding applications. This can be better accomplished by use of an oven controlled crystal oscillator (OCXO), in which the crystal and other temperature sensitive components are in a stable oven which is adjusted to the temperature (usually $\gtrsim 80°C$) where the crystal's frequency vs. temperature has zero slope. Temperature compensated crystal oscillators (TCXOs) and microcomputer compensated crystal oscillators (MCXOs) are also often employed, albeit their performances are quite lower.

Inspired by [651], in the following we analyze the factors that most significantly affect both the frequency accuracy and stability of quartz oscillators.

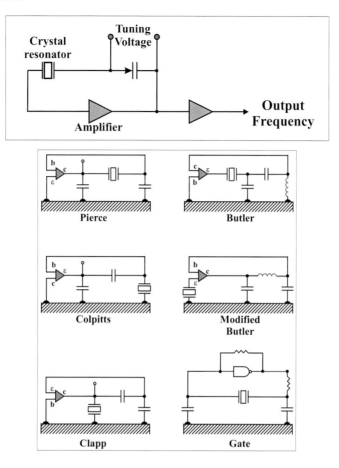

FIGURE 7.7
Schematic of a quartz oscillator. Among the various types, the Pierce, the Colpitts, and the Clapp circuits are the most widely used. They only differ by the location of the rf ground points. In the Butler circuits, the emitter current is the crystal current. The gate oscillator is a Pierce-type which uses a logic gate plus a resistor instead of the transistor. (Adapted from [59].)

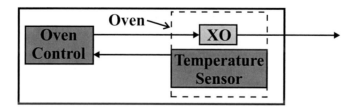

FIGURE 7.8
Oven controlled crystal oscillator. (Adapted from [59].)

7.2.1 Factors affecting crystal oscillator frequency accuracy

Temperature

Although crystals with different cut angles exhibit different frequency-temperature characteristics, a quite general feature is the cubic dependence on temperature. Also, the zero temperature coefficient point is usually changed by varying the angle between crystal wafer and crystal axis. Just as an example, in a wide temperature range (between -55 and $105°$ C), the relative frequency change of AT-cut crystals can be kept as low as $\pm 2 \cdot 10^{-5}$ with a proper angle processing.

Aging

In literature, the change in the crystal resonator frequency caused by the operational time is referred to as aging. Most often this fractional variation of the vibrational frequency $(\Delta f/f)$ is monotonic, albeit in some cases the aging rate can reverse sign over time. Aging essentially results from: thermal gradient effects (in this respect, the behavior of SC-cut OCXOs is much better than that of AT-cut OCXOs); pressure release effects; the increase or decrease of the mass of the crystal polar plates (respectively due to gas absorption or decomposition); changes in the crystal structure.

Drive level

At high-precision levels, the oscillator frequency also depends on the alternating current, i, which flows across the crystal, according to the following equation [651]

$$\Delta f/f \simeq k i^2 \tag{7.4}$$

where k is a constant related to the specific crystal. When such drive electric current is too large, the aging performance and the long-term frequency stability will be worse. Conversely, with a too small drive level, the crystal electric current may be dominated by the noise, which adversely affects the short-term frequency stability. Thus, a compromise must be found. As an example, the driving level is less than 70 μA in 5-MHz high-precision crystal oscillators.

Retrace and thermal hysteresis

The so-called retrace error is nothing but the frequency variation that occurs when power is removed from an oscillator (for several hours) and then re-applied on it. In this case, indeed, the oscillator frequency tends to stabilize at a slightly different value. Particularly, the retrace error plays an important role in OCXOs. Moreover, due to lattice defects and stress relief in the mounting structure, the frequency-temperature characteristic of a quartz oscillator does not retrace itself exactly upon temperature cycling (thermal hysteresis).

Frequency pushing and pulling

While frequency pushing (expressed in MHz/V) just represents the sensitivity of the oscillator output frequency to the supply voltage, frequency pulling is a measure of the frequency change caused by a non-ideal load. By influencing the oscillator circuitry, both of them modify the phase or amplitude of the signal reflected into the oscillator loop, which, in turn, changes the oscillator frequency.

Tuning port reference voltage drift

Since both tuning range and sensitivity strongly depend on the tuning port reference voltage, any drift in the latter will also perturb the oscillator frequency accuracy.

7.2.2 Factors affecting frequency stability

Oscillator tuning port noise

In light of the above remark, by disturbing the tuning sensitivity, tuning-port noise can also affect the oscillator stability.

Reference source noise

Of course, if the quartz oscillator is linked to a reference one, the source noise is somehow transferred to the output frequency.

Power supply noise

This affects both frequency and phase, determining cycle-to-cycle jitter.

Vibration-induced noise and other external sources

Needless to say, due to the crystal sensitivity to acceleration, random and/or periodic mechanical vibrations can introduce a considerable amount of phase noise in a high-performance quartz. Other detrimental factors include spurious electric and magnetic fields, ambient pressure and humidity, gas permeation, etc.

7.2.3 State-of-the-art ultrastable quartz oscillators

Today, the most advanced quartz oscillators (see, for instance, OCXO8607 model from Oscilloquartz S.A.) rely on the technique of housing a state-of-the-art BVA, SC-cut, 3rd-overtone crystal resonator and its associated oscillator components in double oven technology (operating temperature range -30 to $+60$ ° C). Standard output (7-dBm-level sine waveform) frequencies are 5 and 10 MHz (harmonics < -40 dBc and spurious < -70 dBc) with a phase noise ($BW = 1$ Hz) of -150 dBc at 1 kHz and the following frequency stability ($\Delta f/f$) features (http://www.oscilloquartz.com/):

- Long-term stability (aging after 30 days of continuous operation) of $4 \cdot 10^{-9}$/year;

- Stability over temperature range $\leq 2 \cdot 10^{-10}$ peak to peak;

- Stability versus power supply of $5 \cdot 10^{-11}$ ($V_{cc} \pm 10\%$);

- Stability versus load changes of $2 \cdot 10^{-11}$ (50 $\Omega \pm 10\%$);

- Short-term stability (Allan deviation) $\sigma_y(\tau) = 1 \cdot 10^{-13}$ (1-30 s);

- g sensitivity $< 5 \cdot 10^{-10}/g$

Characterization of a few other ultrastable quartz oscillators can be found in [652].

7.3 Cryogenic sapphire oscillators

As we will see later on in this chapter, fractional frequency stabilities of passive atomic standards are now close to the $10^{-14}/\sqrt{\tau}$ level (with τ being the measurement time), only restricted by the number of interrogated atoms. This exacts an interrogation oscillator with a short-term stability better than 10^{-14}, which, as just discussed, cannot be afforded by current quartz technology. In this regard, ultrastable cryogenic microwave oscillators have demonstrated a short term frequency stability in the range 10^{-14} to a 10^{-16} [653]. Such oscillators are based on resonators exhibiting very high electrical Q-factors (in excess of 10^8), which can only be reached in super-conducting cavities or sapphire resonators operating at low temperatures in liquid-helium-cooled systems. Recently, great efforts have been made to exempt the need for liquid helium and realize more compact, portable liquid-nitrogen-cooled systems. For the sake of truth, we have to mention here that room-temperature sapphire oscillators for research applications exhibit comparable phase-noise performance [654]. However, this very mature technology can be now considered at its end and will hence be omitted. By contrast, the younger technology of cryogenic sapphire oscillators (CSOs) is still open to strong improvement. For this reason, in the following we focus solely on CSOs [653].

These have been built by several groups around the world since the late 1980s. Monocrystalline sapphire (Al_2O_3) is characterized by uniaxial anisotropic complex permittivity. At microwave frequencies the permittivity parallel (perpendicular) to the crystal's c-axis is 11.5 (9.4). Typically, the cavity design consists of a sapphire cylinder (or disk), with the crystal c-axis aligned to the cylinder axis within $1°$, surrounded co-axially by a metallic shield (made from copper, silver, or superconducting material) to hinder rf radiation losses.

When this metallic shield is realized as a coating on the sapphire surface, then the resonator can be regarded as a vacuum cavity filled with dielectric. In this case, the theoretical description is practically that for a vacuum cavity (see Section 3.1), leading to transverse-magnetic ($TM_{m,n,q}$) or transverse-electric ($TE_{m,n,q}$) modes. In such a configuration, resonators operate on low order modes (i.e., low azimuthal numbers) and the Q-factor is determined by the surface resistance (even for superconductive coatings).

In a second arrangement, the sapphire element is suspended in a metallic can at some distance from it, as illustrated in Figure 7.9 [655, 656]. Here, rf energy is coupled into the resonator via magnetic loop probes, thus creating an electro-magnetic field distribution mainly confined within the sapphire. An adequate theoretical treatment reveals that in this configuration the resonator modes are in general an hybrid between TM and TE [657]. In literature, a mode with a dominant axial (z-direction) electric-field dependence is denoted as an $E_{m,n,p+\delta}$-mode (quasi-$TM_{m,n,p+\delta}$ or $WGH_{m,n,p+\delta}$), whereas one with a dominant axial magnetic-field dependence is indicated as a $H_{m,n,p+\delta}$-mode (quasi-$TE_{m,n,p+\delta}$ or $WGE_{m,n,p+\delta}$). Here m, n, and p are the number of azimuthal, radial, and axial variations and δ is a number slightly less than 1. Physically, this corresponds to the whispering gallery mode (WGM) configuration, where the electromagnetic energy resides in the sapphire component, but close to the dielectric/vacuum interface through a physical mechanism not different from total internal reflection in optical systems. The higher the azimuthal number, the more the modes become WG-like. Cryogenic WGM resonators typically use modes on the order of ten. The unloaded electrical quality factor for such a shielded dielectric resonator can be expressed as [653]

$$Q = \frac{1}{R_s\Gamma^{-1} + p_\varepsilon \tan\delta + p_\mu\chi''} \tag{7.5}$$

where p_ε and p_μ denote the mode electric and magnetic filling factors (that are a measure of the respective field energies stored in the dielectric to the total stored energy), $R_s\Gamma^{-1}$ is the metallic-shield loss, $\tan\delta \equiv \varepsilon''/\varepsilon'$ is the tangent loss of the dielectric material (ε' and ε'' being the real and imaginary part of the relative dielectric constant, respectively), and χ'' is the imaginary part of the ac susceptibility originating from paramagnetic impurities. For a carefully designed resonator, the unloaded-cavity Q-factor is determined by the tangent loss, such that it is on the order of 100000 at ambient temperature, higher than 10 millions at 77 K, and up to 10^9 at the liquid-helium temperature.

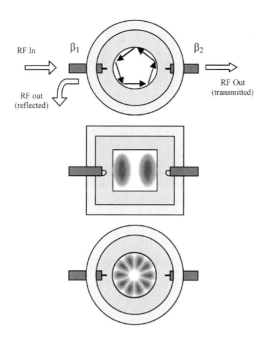

FIGURE 7.9
Basic sapphire-loaded cavity design. The sapphire cylinder (typical size: diameter 3-5 cm, height 1-3 cm) is suspended in the metal cavity. Radio frequency energy is coupled into and out of the resonator via magnetic loop probes. Top and bottom illustrations indicate, respectively, the total internal reflection and field pattern for the $TE_{5,5,\delta}$ mode. (Courtesy of [655].)

A major difficulty encountered with sapphire WG-mode resonators is represented by the high sensitivity to thermal fluctuations, which originates from the heavy temperature dependence of the sapphire permittivity. To overcome this drawback, a few temperature stabilization and compensation schemes have been then worked out. In essence, the resonator structure has to incorporate some perturbation mechanism to compensate for the sapphire permittivity thermal sensitivity. In this way, the modified resonator will exhibit a turnover temperature, T_0, where the thermal sensitivity as a whole vanishes at the first order. However, since the perturbation inevitably degrades the resonator Q-factor, the turnover temperature is in practice restricted to $T_0 \leq 80$ K. Different techniques including paramagnetic, dielectric, or mechanical compensation have been implemented so far [658, 659].

- Thermo-mechanical compensation - The sapphire resonator is made of two disks slightly spaced by a copper (or silver) piece. Since the dilatation coefficient of the copper is higher than that of the sapphire, any increase in the resonator temperature causes a corresponding extension in the relative height of the gap. In this way, the mean value of the relative permittivity seen by the electric field (which is essentially axial for a WGH mode) diminishes, which represents the opposite behavior of the natural sapphire relative permittivity variation.

- Dielectric compensation - Since sapphire has a positive temperature coefficient of permittivity (TCP), thermal compensation can be attained by combination with a dielectric having a negative TCP, like rutile (TiO_2). In such a scheme, the resonator consists of two thin rutile rings placed on the two flat surfaces of the sapphire puck. Within this approach, a turnover temperature of 53 K and a Q-factor of the order of 10^7 can be obtained.

- Paramagnetic compensation - Another method relies on doping the sapphire crystal with paramagnetic Ti^{3+} ions. The effect of the resulting temperature-dependent magnetic susceptibility on the resonator frequency is opposed to those caused by the sapphire permittivity sensitivity. With a 0.1% by weight Ti^{3+} ions concentration, the turnover temperature is in the range of 20-77 K, depending on the WG mode.

Furthermore, WG modes can be perturbed by a number of low-Q spurious modes (these are, for instance, other WG modes as well as the usual empty-cavity modes perturbed by the presence of the sapphire) that can degrade the resonator performance. To realize an effective modal selection, without resorting to a sharp bandpass filter in the loop, the so-called open-cavity technique has been demonstrated (Figure 7.10) [659]. In this configuration, only the metal caps are retained, while the cylinder is replaced with a microwave absorber. Such assembly is then inserted in a vacuum chamber whose internal walls are overlaid with a microwave absorber. Now, the spurious resonances are no longer confined, thus being completely removed from the spectrum. By contrast, the high-order WG modes are nearly unaffected by the absorber. Indeed, the Q-factor measured for them is still higher than $2 \cdot 10^8$. Concerning the achievable flicker floor, for a resonator operating on a high-order ($m = 15$) WGH mode at 10.959 GHz, a value of $\sigma_y(\tau) = 7 \cdot 10^{-15}$ (Allan deviation) was observed (as measured against a microwave synthesizer driven by a hydrogen maser), the cryogenic oscillator instability showing up at $\tau > 100$ s [659]. A typical oscillator loop is shown in Figure 7.11. Two coaxial cables link the cooled resonator to the sustaining circuit at room temperature. Due to the large temperature gradients experienced by these cables, which manifests in a strong fluctuation of their electrical length, a significant phase drift originates along the oscillator loop. To compensate for this and make the oscillation frequency equal to the resonator frequency, a Pound servo is implemented. To this aim, the loop signal is phase modulated by a voltage phase shifter at a frequency ν_{mod} higher than the resonator bandwidth. A circulator (C) is used to direct the signal reflected by the resonator to a tunnel diode. Operating as a quadratic detector, the latter provides a voltage that is synchronously demodulated at ν_{mod} in a locking amplifier. The reflected signal comprises the residual carrier at ν_{osc} and two sidebands at $\nu_{osc} \pm \nu_{mod}$. When the oscillating frequency ν_{osc} just equals the resonator one ν_0, the residual carrier is sucked by the resonator and only the two sidebands survive. These latter are then mixed in the diode which gives as output a voltage, for which the a.c. component is at $2\nu_{mod}$: the demodulated signal is zero. Conversely, if $\nu_{osc} \neq \nu_0$, the diode voltage is modulated at ν_{mod}: the demodulated signal is proportional to $\nu_{osc} - \nu_0$. The locking amplifier output signal is integrated and sent back to the VCPS bias stage. In this way, the phase fluctuations originating in the loop are corrected in real time in the loop bandwidth. Finally, in order to mitigate the detrimental effects due to microwave power dissipation and radiation pressure inside the resonator, the power injected into the latter is stabilized by a standard power control comprising a quadratic detector as the sensor and a voltage controlled attenuator (VCA) in the loop.

Another interesting experimental issue is related to the difficulty of an accurate characterization of the phase noise of a CSO. Indeed, high-resolution phase noise measurements would require two identical phase-locked CSOs. Unfortunately, due to mechanical tolerances, two resonators machined from the same high-purity crystal typically may differ by 100 kHz-1 MHz, while the sapphire resonator bandwidth is only on the order of 10 Hz. Figure 7.12 shows a setup which overcomes this limitation [660]. It was used to characterize the phase noise of a first CSO (CSO1) placed on a specially designed cryo-cooler against a

FIGURE 7.10
Picture of the opened cavity WG resonator. (Courtesy of [659].)

second CSO (CSO2) which was cooled in a large liquid-helium dewar. Both CSOs operated on a whispering gallery mode at 9.99 GHz, their frequencies differing by 745 kHz (at 6 K). Also, both CSOs used a Pound frequency stabilization and a power servo, as described above. Then, the two CSO signals were mixed. The resulting 745-kHz beat note was first amplified and then compared to the signal derived from the frequency division of a 95-MHz signal coming from a low-noise RF synthesizer. The latter was eventually phase locked on the beat-note signal. The PLL acted on the varactor of a 10 MHz VCO used as reference for the RF synthesizer.

We close this section by analyzing the fundamental limitations to the frequency stability of a sapphire oscillator [653]. An oscillator can be regarded as a resonator whose losses originate from an amplifier with a noise temperature T_N. By virtue of the Townes-Schawlow formula, the thermal noise energy, $k_B T_N$, is responsible for the following fractional frequency

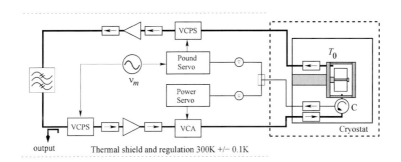

FIGURE 7.11
Typical circuit implemented for a WG-mode CSO. Only the resonator is cooled, whereas the sustaining circuit is outside the cryostat. The bold line corresponds to the oscillator loop, while the thin lines refer to the electronic controls. VCPS=voltage controlled phase shifter, VCA=voltage controlled attenuator. (Courtesy of [659].)

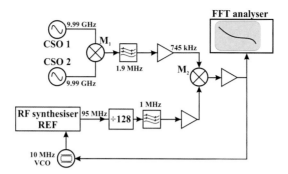

FIGURE 7.12
Setup for measuring the extremely low phase noise of a CSO. (Adapted from [660].)

fluctuations

$$\frac{\Delta f}{f_T} = \sqrt{\frac{k_B T_N}{2PQ^2\tau}} \tag{7.6}$$

where τ is the measurement integration time and P the power. Another important fluctuation term is related to radiation pressure. Indeed, both its amplitude and fluctuations increase as P grows. The resulting fluctuating mechanical deformation degrades the resonator frequency stability which can be characterized by [653]

$$\frac{\Delta f}{f_P} = \frac{1}{f}\frac{\partial f}{\partial P}\sqrt{\frac{Phf}{\tau}} \tag{7.7}$$

where h is Planck's constant. Since the above two terms have inverse power dependences, there will be an optimum power level which minimizes their sum

$$P_{opt} = \sqrt{\frac{A_n}{2}}\left(Q\frac{1}{f}\frac{\partial f}{\partial P}\right)^{-1} \tag{7.8}$$

where $A_n = k_B T_N/(hf)$ is the amplifier noise number. The corresponding minimum frequency fluctuation is given by

$$\left(\frac{\Delta f}{f}\right)_{min} = 2^{3/4}A_n^{1/4}\sqrt{\frac{h}{2\pi\tau}\frac{1}{f}\frac{\partial f}{\partial U}} \tag{7.9}$$

where $(1/f)\partial f/\partial U$ is on the order of 10^{-6} J^{-1} for sapphire, $U = PQ/(2\pi f)$ representing the energy stored in the resonator. So, if the amplifier has ideal quantum-limited performance ($A_n = 1$), we find the limit $(\Delta f/f)_{min} \simeq 10^{-20}/\sqrt{\tau}$. However, measured stabilities are orders of magnitude above this quantum limit for several reasons. First, the resonator is driven by an amplifier with $A_n \gg 1$. Second, due to technical power instabilities, radiation pressure fluctuations are much larger than assumed in Equation 7.7. Some other technical sources of frequency instability are discussed in [656]. As a result, for integration times τ between a few seconds and about one hundred seconds, the best sapphire oscillators constructed so far can reach a flicker floor in the Allan deviation at $\sigma_y(\tau) \simeq 3 \cdot 10^{-16}$ [653, 656].

7.4 Photonic microwave oscillators based on WGM resonators

We have already seen that a laser locked to a narrow resonance of a high-finesse cavity represents an ordinary source of stable, narrow-linewidth optical signals. In this respect, being characterized by small size, high transparency windows, and narrow resonances, whispering gallery mode (WGM) resonators may offer a unique tool for laser stabilization. Precisely in this frame, the thermodynamic (fundamental) limitations of the frequency stability of WGM resonators have been evaluated in [661, 662].

Here we discuss, instead, the possibility of direct generation of high-stability microwave signals using such resonators. Photonic microwave oscillators commonly rely on the generation and subsequent demodulation of polychromatic light to provide a suitable beat-note signal. Among the different possible approaches, one effective scheme is based on excitation of hyperparametric oscillations in WGM resonators [663]. While usual parametric oscillations hinge on a $\chi^{(2)}$ nonlinearity (involving three photons) and impose phase matching conditions between far separated optical frequencies, hyperparametric oscillations rely on a $\chi^{(3)}$ nonlinearity (involving four photons: two pump, signal, and idler) and require phase matching conditions among nearly degenerate optical frequencies. Here, the signal and idler optical sidebands grow from vacuum fluctuations at the expense of the pumping wave. In the current specific configuration, such a four-photon process, $\hbar\omega+\hbar\omega \to \hbar(\omega+\omega_M)+\hbar(\omega-\omega_M)$, is triggered by an external pumping source at ω, with ω_M being determined by the free spectral range of the WGM resonator $\omega_M \simeq \Omega_{FSR}$. This latter frequency, obtained through demodulation of the oscillator output via a fast photodiode, exactly represents the desired microwave signal. Its spectral purity is obviously improved by increasing the Q-factor of the resonator, which also decreases the pumping threshold value for the oscillation (down to few microwatts).

Next, let us discuss the phase noise associated with such a system. With reference to Figure 7.13, the amplitude of the phase-modulated electric field at the exit of the resonator can be expressed as

$$E_{out} = e^{i\omega_0 t} \left[A + B e^{i\omega_M t} e^{i\phi(t)} - B e^{-i\omega_M t} e^{-i\phi(t)} \right]$$
$$= e^{i\omega_0 t} \left\{ A + 2iB \sin[\omega_M t + \phi(t)] \right\} \tag{7.10}$$

where A (B) is the amplitude of the carrier (each sideband), ω_M the distance between each generated sideband and the pumping light, and $\phi(t)$ is the oscillator phase noise. The field amplitude at the photodiode is then

$$E_{PD} = E_{out} e^{-\alpha} r + E_{LO} e^{i\omega_0 t} e^{i\psi_0} \sqrt{1 - r^2} \tag{7.11}$$

where $e^{-\alpha}$ is the amplitude loss at the output coupler (Cp2), r the amplitude transmittivity of the splitter Sp2, and E_{LO} (ψ_0) the amplitude (phase) of the local oscillator, that is the electric field propagating through the delay line. If we choose $\psi_0 = \pi/2$, then the DC optical power on the photodiode is

$$P_{PD} = P_A e^{-2\alpha} r^2 + P_{LO}(1 - r^2) \simeq P_{LO}(1 - r^2) \tag{7.12}$$

where P_A is the power of optical carrier escaping the resonator and P_{LO} is the power of the local oscillator. Similarly ($\psi_0 = \pi/2$), by neglecting the terms quadratic in B, the AC photocurrent in the photodiode is given by

$$j = 4\mathcal{R} E_{LO} r \sqrt{1 - r^2} e^{-\alpha} B \sin[\omega_M t + \phi(t)] \tag{7.13}$$

where $\mathcal{R}$ is the responsivity of the photodiode. Hence, the demodulated time-averaged microwave power is

$$P_{mw} = 8\rho r^2 \mathcal{R}^2 P_{LO} P_B e^{-2\alpha}(1 - r^2) \tag{7.14}$$

where ρ is the resistance at the output of the photodiode. Thus, the generated microwave power is proportional to the power of the optical sidebands. Finally, the phase noise of the oscillator can be characterized by a modified Leeson formula

$$S_\phi = \left[1 + \frac{\eta P_{mw}}{2\rho\mathcal{R}^2 P_{PD}^2} \frac{P_{PD}}{P_B} \frac{\gamma^2}{f^2}\right] \frac{2\hbar\omega_0}{\eta P_{PD}}$$

$$\simeq \left[1 + 4\eta e^{-2\alpha} r^2 \frac{\gamma^2}{f^2}\right] \frac{2\hbar\omega_0}{\eta P_{LO}(1 - r^2)} \tag{7.15}$$

where γ is the HWHM of the loaded WGM resonance and η is the quantum efficiency of the detector. The above equation shows that (i) the phase noise power spectral density is independent from the efficiency of the parametric process; and (ii) the phase noise corner frequency can be greater than the HWHM of the loaded WGM resonance. These conclusions are confirmed by the experimental observations, as shown by the typical phase noise plot of Figure 7.14. Indeed, the phase noise floor agrees with shot-noise-limited operation and the corner frequency is determined by the linewidth of the WGM resonances. Apart from increasing the resonator quality factor, the oscillator performance could be considerably improved through the generation of multiple harmonics. In this respect, a significant noise reduction in the generated microwave signal would be achieved through the creation of an optical comb. Indeed, assuming that the comb has N equidistant, phase-locked optical harmonics, the phase noise of the beat note between any two neighboring harmonics is N times lower than that of the optical frequencies. However, the number of generated optical harmonics is limited by the presence of stimulated Raman scattering (SRS), which adds noise to the generated microwave signal. Starting right after the hyperparametric oscillations (when the power of the first signal and idler sidebands are several percents of the overall optical power entering the resonator), SRS is particularly efficient because modes corresponding to Stokes light (generated in different mode families uncoupled from the fiber couplers) typically have higher quality factors. To overcome this drawback, WGM resonators of a proper shape as well as special geometrical/spectral dampers could be employed so as to decrease the SRS efficiency. In turn, this would eventually lead to an increase in the number N without extra noise.

Since N is inversely proportional to the square root of the finesse of the resonator and the phase diffusion of the microwave signal is inversely proportional to the finesse, one can guess that the noise properties will still be improved if the finesse of the resonator is increased.

7.5 Generation of ultrastable microwaves via optical frequency division

A promising alternative approach to generate high-frequency stability microwaves exploits high-Q optical resonators in conjunction with all-optical frequency division. Indeed, as discussed in more places, low absorption and scattering in the optical domain can provide 10^{11}-level Q-factors in a Fabry-Perot cavity. In addition, if well-isolated, this resonator will exhibit average length fluctuations in the order of ~ 100 attometer on a 1 s timescale. As a result, a fractional frequency instability as low as $2 \cdot 10^{-16}$ (for averaging times of 1-10

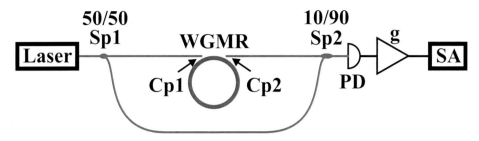

FIGURE 7.13
Experimental setup of a whispering gallery photonic hyperparametric microwave oscillator. Light from a YAG laser is split at Sp1, one part being sent into the CaF_2 WGM resonator (toroidal shape with a diameter of several millimiters and thickness in the range of several hundred microns) through coupler Cp1. The resonator loaded Q factor is on the order of 10^9, the intrinsic Q exceeding 10^{10}. The other part is sent to a fiber delay line. Then, splitter Sp2 is used to combine the light coupled out of the resonator (via Cp2) with that propagated through the delay line. In this way, about 90% of light from the resonator goes to the fast photodiode (PD). The output microwave signal is pre-amplified with an amplifier (g) and directed to a microwave spectrum analyzer (SA). (Adapted from [663].)

s) can be attained for a continuous-wave (cw) laser by stabilizing it against such a cavity. Transfer of this stability to a microwave signal is precisely the topic of this section [664]. Figure 7.15 illustrates the principle of such a photonic oscillator.

In essence, the frequency stability of the high-performance cw laser, that is the optical reference at ν_{opt1}, is transferred to the repetition rate ($f_r \simeq 0.1 - 10$ GHz) of a self-referenced femtosecond laser frequency comb. For this purpose, the nth comb element is phase locked to ν_{opt1}, while simultaneously stabilizing the offset frequency f_0. In this way, the optical cavity stability is transmitted to the OFC mode spacing $f_r = (\nu_{opt1} - f_0)/f_r$. Then, the stabilized pulse train is detected by a fast photodiode, which produces photocurrent at frequencies equal to f_r and its harmonics (up to the cutoff frequency of the photodiode). Since the phase noise level obtained for such microwave signals is lower than that available from commercially available microwave references, accurate characterization requires comparing two similar, fully independent systems (Figure 7.16). In this apparatus, the first (second) optical divider comprises an octave-spanning 1 GHz Ti:sapphire femtosecond laser which is phase locked to a cavity-stabilized cw laser at 578 nm, ν_{opt1} (1070 nm, ν_{opt2}). In each system, the stabilized pulsed output illuminates a high-speed photodiode, thus producing a microwave signal at 1 GHz and harmonics up to ~ 15 GHz. Then, the two independent 10-GHz tones, selected by bandpass filters, are mixed after proper amplification. Finally, the mixer output is analyzed to determine the relative frequency and phase fluctuations. Phase noise data are presented in Figure 7.17. In conclusion, with this kind of photonic oscillator, a 10 GHz signal has been demonstrated with an absolute instability of $\leq 8 \cdot 10^{-16}$ at 1 s of averaging. Such performance is comparable to that produced by the best microwave oscillators, but without the need for cryogenic temperatures. Also, Figure 7.18 shows a comparison with other leading microwave generation technologies in the 10 GHz range.

Such phase-coherent division retains the fractional frequency instability, while reducing the phase fluctuations by a factor of $\sim 5 \cdot 10^{14}$ (500 THz/10 GHz).

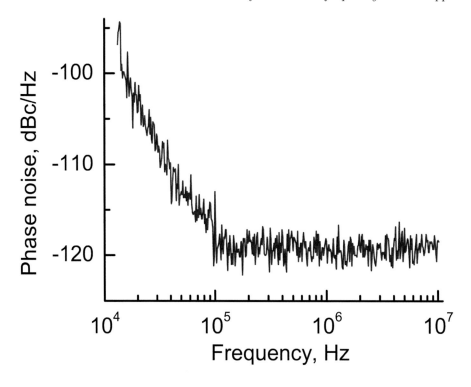

FIGURE 7.14
Typical phase noise of a WGM-based oscillator with a floor of -126 dBc at 300 kHz, corresponding to a HWHM of the resonance of $\simeq 115$ kHz and a delay line length of ~ 115 m. For a total optical power on the detector of 6 mW, the power of the generated microwave signal leaving the detector is -46.1 dBm. After the microwave amplifier (characterized by a noise figure of 7 dB and a gain of 43.1 dB), a power of -3 dBm is reached. (Courtesy of [663].)

7.6   Trapping and cooling of neutral atoms

While by now the world of cold atoms represents *per se* an exciting and vast field of contemporary physics, here, inspired by a few specialized reviews [665, 666, 667, 668, 669], we only discuss those aspects which are strictly necessary for understanding frequency standards and clocks.

As already extensively discussed, two primary restrictions arise in spectroscopic studies which use room-temperature gaseous samples, due to the high thermal velocities (on order of hundreds of m/s) of the atoms/molecules. First, Doppler effects cause displacement and broadening of the spectral lines. Second, the ultimate attainable spectral resolution is set by the limited observation time. The wish to decrease motional effects in spectroscopy and atomic clocks was a major motivation for the cooling of both neutral atoms and ions. Indeed, although thermal atomic/molecular velocities can be reduced by refrigeration (as the square root of temperature), any gas in equilibrium (other than spin-polarized atomic hydrogen) would be condensed at temperatures corresponding to values below 1 m/s. So, a change in the cooling paradigm was needed. This was provided by the advent of laser

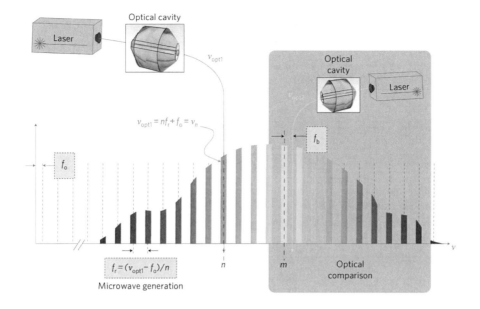

FIGURE 7.15 (SEE COLOR INSERT)
Principle of phase-coherent division of an ultrastable optical signal to the microwave domain through an optical frequency comb (OFC). First, the nth comb element is phase locked to ν_{opt1} (while simultaneously stabilizing the offset frequency f_0), which transfers the optical cavity stability to the OFC mode spacing $f_r = (\nu_{opt1} - f_0)/f_r$. In the case of a high-fidelity optical divider, the sub-hertz optical linewidth of the reference laser is translated into a microhertz linewidth on f_r. (Courtesy of [664].)

cooling and trapping of neutral atoms. In this light, starting from the initial proposals dating back to the seventies, several methods were devised which enabled reaching lower and lower temperatures, down to the observation of quantum degeneracy. In the following, the most successful schemes are presented.

7.6.1 Optical molasses

Due to conservation of total momentum, when an atom with mass m and velocity $\boldsymbol{v}$ absorbs or emits a photon with frequency ν, its velocity varies by the amount $v_r = h\nu/mc = h/(m\lambda)$, that is the recoil velocity. While this velocity change is generally small, if several photons are scattered, a sizeable change in the atomic speed can result. Although the absorption of each photon is followed by a spontaneous emission event, due to randomness of the emission direction, spontaneous emission does not have, on average, a net affect on the atomic velocity. For a more quantitative description of the phenomenon, let us consider the simple model of a two-level atom interacting with a plane light wave propagating in the direction $\hat{n}$ with frequency ν_L and intensity I [665]. Let E_g and E_e be the energies of the lower and upper state of the atom, respectively ($E_e - E_g = h\nu_A$). The momentum exchange between the radiation field and the atom manifests itself in a force (referred to as spontaneous force or radiation pressure) acting on the latter, $\boldsymbol{F}_{sp}$. The latter is given by the momentum of a single photon times the number of absorption-emission cycles in the unit of time, R. In turn, R is the product between the rate of spontaneous emission, $1/\tau$,

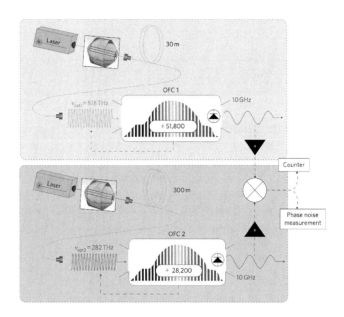

FIGURE 7.16
Experimental setup for the generation and characterization of the 10 GHz low-noise microwaves. In each independent system, stable light from the cavity is transmitted to the OFC through an optical fiber. After phase-locking the combs against their respective optical references, the two independent 10-GHz signals are mixed. The obtained mixed-down product is characterized via frequency and phase-noise measurements. In order to evaluate the optical stability of the OFCs and the cw lasers, the beat signal f_b between the second stabilized cw laser ν_{opt2} and a tooth of the comb independently stabilized by ν_{opt1} was also measured. (Courtesy of [664].)

and the probability of occupation of the excited level. Since the latter can be expressed as

$$N'_e \equiv \frac{N_e}{N_g + N_e} = \frac{1}{2} \frac{I/I_0}{1 + (I/I_0) + (4/\Gamma^2)[\delta - (\nu_L/c)\boldsymbol{v} \cdot \hat{n}]^2} \quad , \tag{7.16}$$

we have $R = (1/\tau)N'_e$ and $\boldsymbol{F}_{sp} = \hat{n}(h\nu_L/c)R$, where N_e (N_g) is the number of atoms in the excited (ground) state, $\delta = \nu_L - \nu_A$, I_0 is the saturation intensity, and $\Gamma = 1/(2\pi\tau)$ is the natural width of the atomic resonance. So we have

$$\boldsymbol{F}_{sp} = \hat{n}\frac{h\nu_L}{c}\frac{1}{2\tau}\frac{I/I_0}{1 + (I/I_0) + (4/\Gamma^2)[\delta - (\nu_L/c)\boldsymbol{v} \cdot \hat{n}]^2} \tag{7.17}$$

Now, as contained in the original proposal by T.W. Hänsch and A.L. Schawlow in 1975, an effective cooling configuration is realized by irradiating a moving atom with counter-propagating laser beams of the same frequency ν_L, which is tuned slightly below ν_A [670]. Indeed, by virtue of the Doppler shift, the atom will absorb preferentially photons moving opposite to its velocity, thus slowing down; the kinetic energy of the atoms is dissipated and converted into energy of the electromagnetic field (Figure 7.19). By neglecting stimulated emission ($I/I_0 \ll 1$) and using the above equation, the total force experienced by the atoms can be written as the sum of the forces exerted by each of the laser beams

$$F_{sp} = \frac{h\nu_L}{2c\tau}\left[\frac{I/I_0}{1 + (4/\Gamma^2)[\delta - (v/c)\nu_L]^2} - \frac{I/I_0}{1 + (4/\Gamma^2)[\delta + (v/c)\nu_L]^2}\right] \tag{7.18}$$

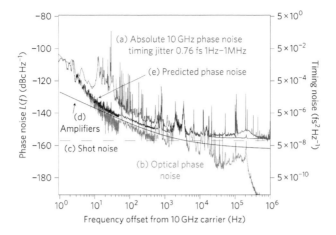

FIGURE 7.17

Measured phase-noise spectrum for a single 10-GHz photonic oscillator (a) and a single optical reference (b) scaled to 10 GHz as determined from f_b. Contributions from shot noise and amplifiers are also plotted. Curve (e), that is the sum of curves (b), (c), and (d), then yields the estimated achievable phase noise. (Courtesy of [664].)

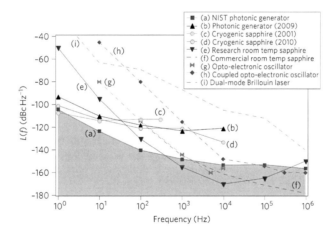

FIGURE 7.18 (SEE COLOR INSERT)

Trends for single-sideband phase noise in several leading microwave-generation technologies in the 10 GHz range. Spurious tones are neglected for all data. (a), result from the work described in the text. (b), previous Er:fiber and Ti:sapphire optical frequency divider results. (c),(d), cryogenic sapphire oscillators. (e), research room-temperature sapphire oscillator. (f), commercial room-temperature sapphire oscillator. (g),(h), long-fiber (g) and coupled (h) opto-electronic oscillators. (Courtesy of [664].)

which, in the limit of small velocities ($v < \Gamma c/\nu_L$), becomes

$$F_{sp}(v) = 16\pi h \frac{\nu_L^2}{c^2} \frac{\delta}{\Gamma} \frac{I/I_0}{\left[1 + (2\delta/\Gamma)^2\right]^2} v \equiv -\alpha v \tag{7.19}$$

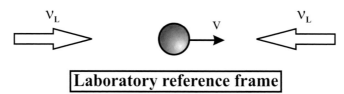

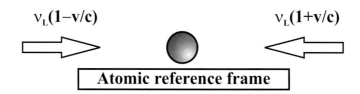

FIGURE 7.19
Principle of 1-D cooling in optical molasses.

with

$$\alpha = 16\pi h \frac{\nu_L^2}{c^2} \frac{\delta}{\Gamma} \frac{I/I_0}{\left[1 + (2\delta/\Gamma)^2\right]^2} \tag{7.20}$$

Thus, for small v values, the spontaneous force is a linear function of velocity as in a viscous medium. This suggested the name optical molasses for this cooling configuration. Such a scheme is readily generalized to 3-D by employing three orthogonal pairs of laser beams (each made of two counter-propagating beams with opposite circular polarizations). The first experimental demonstration was provided by S. Chu in 1985 with sodium atoms [671].

Concerning the minimum achievable temperature, two counteracting effects must be considered. The first one is the cooling effect produced by the viscous force: the rate at which energy is removed by cooling is $\dot{E}_{cool} = \boldsymbol{F} \cdot \boldsymbol{v} = -\alpha v^2$. The second is a heating process which is imputable to the discrete, random character of the exchange in linear momentum between the atom and the electromagnetic field. In other words, while the average viscous force reduces to zero the average atomic velocity, its fluctuations are responsible for a mean square velocity different from zero. Since for an atom having zero average speed the absorption probabilities of a photon in the $+z$ and $-z$ directions are equal, each absorption event represents a $\hbar \boldsymbol{k}_L$-length step in the atomic momentum random-walk, with equal probabilities of positive and negative steps. The same is true for spontaneous emission. Eventually, each fluorescence cycle represents two steps in a random-walk. Thus, we will have for the mean square momentum

$$\frac{d\langle P^2 \rangle}{dt} = 2\left(\frac{h\nu_L}{c}\right)^2 R' \tag{7.21}$$

where R' is obtained by R always in the limit of small atomic velocity, $I/I_0 \ll 1$, and assuming that the on-resonance saturation parameter in the two counter-propagating beams is $2I/I_0$

$$R' = \frac{1}{2\tau} \frac{2I/I_0}{1 + (2\delta/\Gamma)^2} \tag{7.22}$$

So the heating rate is

$$\dot{E}_{heat} \equiv \frac{1}{2m}\frac{d\langle P^2\rangle}{dt} = \frac{1}{m}\left(\frac{h\nu_L}{c}\right)^2\frac{1}{\tau}\frac{I/I_0}{1+(2\delta/\Gamma)^2} \qquad (7.23)$$

When the two effects balance ($\dot{E}_{heat} = \dot{E}_{cool}$), we will have

$$\dot{E}_{heat} = \dot{E}_{cool} = \alpha v^2 \qquad (7.24)$$

which gives

$$k_B T = mv^2 = \frac{h\Gamma}{4}\left(\frac{\Gamma}{2\delta} + \frac{2\delta}{\Gamma}\right) \qquad (7.25)$$

The minimum temperature is then obtained for $\delta = \Gamma/2$ and is given by

$$k_B T_D = \frac{h\Gamma}{2} \qquad (7.26)$$

That is the so-called Doppler limit. Typical values are 240 μK for sodium, 120 μK for cesium and 140 μK for rubidium.

In the context of laser-cooled atoms, another meaningful reference temperature is the recoil one, T_{rec}. This corresponds to the fact that, due to the discrete character of the exchange of momentum, the spread in the atomic energies cannot fall below the energy associated with the single photon recoil. Thus we can define

$$\frac{1}{2}k_B T_{rec} = \frac{1}{2m}\left(\frac{h\nu_L}{c}\right)^2 \qquad (7.27)$$

Just as an example, that is 200 nK for cesium. Soon after the demonstration of the first optical molasses, accurate time-of-flight-based temperature measurements, first carried out by W. Phillips and co-workers, pointed out a temperature value significantly smaller than the expected Doppler limit and not far from the recoil limit [672]. This surprising result was explained by J. Dalibard and C. Cohen-Tannoudji as a consequence of the combination of multilevel atoms, polarization gradients, light shifts, and optical pumping [673]. How these cooperate to produce sub-Doppler laser cooling is illustrated below with a simple model [667]. Since the laser detuning typically employed in an optical molasses is not much larger than Γ, both differential light shifts and optical pumping transitions will occur for distinct ground-state Zeeman sublevels. Moreover, due to the spatial modulation of the laser polarization, such light shifts and optical pumping rates will depend, in a correlated manner to each other, on position. To fix ideas, let us consider the laser configuration of Figure 7.20a, where two plane waves (having the same frequency and intensity) counter-propagate along the z axis, with orthogonal linear polarizations. Then, it is easy to see that the polarization of the resulting total field converts from σ^+ to σ^- and vice versa every $\lambda/4$, being elliptical or linear in between. Next, let us assume an angular momentum $J_g = 1/2$ for the atomic ground state (and $J_e = 3/2$ for the excited state), such that the two Zeeman sublevels $M_g = \pm 1/2$ will experience different light shifts, as a function of the laser polarization; in other words, their Zeeman splitting will exhibit spatial modulations of period $\lambda/2$. Also, optical-pumping transfers between the two sublevels, $M_g = -1/2 \to M_g = +1/2$ for a σ^+ polarization and $M_g = +1/2 \to M_g = -1/2$ for a σ^- polarization, will originate from real processes of absorption (of photons by the atom) followed by spontaneous emission; once again, the optical pumping rates are spatially modulated with a period $\lambda/2$. With the proper detuning sign, and here we come to the heart of the affair, optical pumping always transfers atoms from the higher to the lower Zeeman sublevel, but there is something else: the finite value of the optical pumping time introduces a time lag between internal and external variables. As a result, an atom which is moving to the right, starting from the bottom of a valley (for example in the state $M_g = -1/2$ at a place where the polarization is σ^+), can climb up the Zeeman potential hill and reach the top before absorbing a photon. Once on the peak, the probability for the atom to be optically pumped in the other sublevel is maximum. Thus, the atom finds itself again in the bottom of a valley, and so on (double arrows of Figure 7.20b. Essentially, reminiscent of the myth of Sisyphus, the atom is continuously forced to climb up successive potential hills, dissipating its kinetic energy into potential energy in the meantime, without ever having a chance of going down and recovering kinetic energy. The potential energy gained at the expense of kinetic energy is dissipated in the spontaneous Raman anti-Stokes process (the spontaneously emitted photon has an energy higher than the absorbed laser photon). In each Sisyphus cycle, the total energy E of the atom is diminished by an amount on the order of U_0, with U_0 being the depth of the optical potential wells of Figure 7.20b. This happens until E drops below U_0, in which case the atom stays trapped in the well. Therefore, the Sisyphus cooling mechanism provides temperatures T_{Sis} such that $k_B T_{Sis} \simeq U_0$. In conclusion, since the light shift $U_0 \propto \hbar I/\delta$ is much smaller than $\hbar\Gamma$ at low laser intensity, much lower temperatures can be reached compared to the Doppler cooling case. Actually, when U_0 becomes on the order of the recoil energy $E_{rec} = (1/2)k_B T_{rec}$, the Sisyphus cooling is contrasted by the heating caused by the recoil due to the spontaneously emitted photons (which increases the kinetic energy of the atom) and hence ceases functioning. As a consequence, as confirmed by a full quantum theory of Sisyphus cooling as well as by experiments, the lowest achievable temperatures are on the order of a few E_{rec}/k_B.

However, as already discussed, optical molasses are most often realized with counter-propagating laser beams having opposite circular polarizations. Sub-Doppler temperatures are reached in this configuration too. The mechanism, theoretically explained in [673], is similar to the one considered above (this time, however, it only works for $J \geq 1$ in the ground state).

Furthermore, it is in fact possible to deceive the single-photon recoil limit and realize

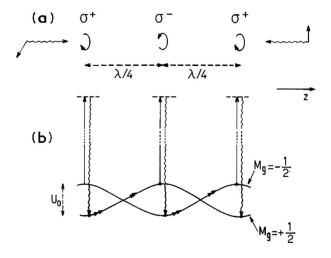

FIGURE 7.20
Principle of Sisyphus cooling. (Courtesy of [667].)

atomic temperatures T lower than T_{rec}. Two schemes have proved particularly effective to enter such sub-recoil laser cooling regime: velocity selective coherent population trapping (VSCPT) and Raman cooling [667]. In both of them, the idea is to make the photon absorption rate depend on the atomic velocity v and vanish for $v = 0$: $R_{abs}(v = 0) = 0$. In this way, the absorption of light is quenched for an atom with $v = 0$, which prevents spontaneous re-emission and hence the associated random recoil. As a result, ultracold atoms with $v \simeq 0$ are defended from the *detrimental* effects of the light. Conversely, atoms with $v \neq 0$ undergo the usual cycles of absorption followed by spontaneous emission: the corresponding random walk in v space eventually transfers them to the $v \simeq 0$ dark state, where accumulation can occur. Concerning the condition $R_{abs}(v = 0) = 0$, in the VSCPT case it is realized by exploiting destructive quantum interference between different absorption amplitudes, whereas Raman cooling relies on proper sequences of stimulated-Raman and optical-pumping pulses.

7.6.2 Magneto-optical traps

While providing viscous damping, the optical molasses is not a trap: the atoms are indeed free to diffuse around, thus eventually escaping from the interaction region. First proposed by J. Dalibard, the magneto-optical trap (MOT) overcomes this drawback by combining the usual molasses configuration with a quadrupole magnetic field. In the most popular realization, the three orthogonal pairs of laser beams intersect in the zero point of the magnetic field generated by an anti-Helmholtz pair (i.e., by two identical parallel coils with currents flowing in opposite directions). The working principle is illustrated by the 1-D scheme of Figure 7.21 in the simple case of an atom with a $J = 0$ ($J = 1$) ground (excited) state. The magnitude of the magnetic field is proportional to the distance from the trap center: $B_z(z) = bz$. The Zeeman frequency shifts corresponding to the $m_J \neq 0$ excited states are given by $\Delta\nu(z) = \Delta E(z)/h = \pm(g_J\mu_B/h)bz \equiv \kappa z$, where g_J is the Landé factor of the excited state and μ_B the Bohr magneton. This introduces a spatially dependent term into the detuning

$$\delta(z) = \delta \mp (v/c)\nu_L \mp \kappa z \qquad (7.28)$$

Qualitatively, if $\nu_L < \nu_A$, an atom, which is located to one side of the $z = 0$ position, will preferentially absorb photons from the laser beam coming from that direction, thus experiencing a restoring force towards the center of the trap. Once again, in the low-intensity limit, the force acting on an atom can be expressed as the sum of the forces exerted by each of the beams. In this case, by insertion of Equation 7.28 into Equation 7.18, in the limit of small v and κz values, one obtains

$$F_{MOT} = \frac{h\nu_L}{2c\tau}\frac{I}{I_0}\left[\frac{1}{1 + (4/\Gamma^2)[\delta - (v/c)\nu_L - \kappa z]^2}\right.$$
$$\left. - \frac{1}{1 + (4/\Gamma^2)[\delta + (v/c)\nu_L + \kappa z]^2}\right]$$
$$\simeq 16\pi h\frac{\nu_L}{c}\frac{\delta}{\Gamma}\frac{I/I_0}{\left[1 + (2\delta/\Gamma)^2\right]^2}\left(\frac{\nu_L}{c}v + \kappa z\right) \qquad (7.29)$$

where the viscous term is responsible for dissipation of kinetic energy and the elastic one provides the additional confinement. Essentially, the motion of an atom in a MOT is that of a damped harmonic oscillator. In conclusion, it is worth pointing out that, given the relevant figures for the experimentally realized magnetic field gradients ($\nabla \boldsymbol{B}$), the magnetic force $\boldsymbol{F}_m = -\boldsymbol{p}_m \cdot \nabla \boldsymbol{B}$ due to direct interaction with the atomic magnetic moment ($\boldsymbol{p}_m$) is negligible in a MOT.

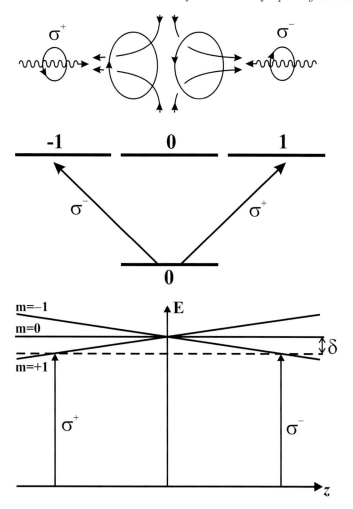

FIGURE 7.21
Principle of 1-D magneto-optical trapping: schematic experimental setup together with transition scheme. The spatially dependent term introduced in the detuning by the quadrupole magnetic field is also represented. (Adapted from [665].)

Atoms can be collected in a MOT either from a slowed atomic beam or from the room-temperature vapor in a cell. In the context of atomic frequency standards, to procure a fine duty cycle between preparing and probing the sample, it is advisable to load a huge number of atoms in a little time. In this regard, the number N_{MOT} of atoms trapped in a MOT can be expressed by the rate equation

$$\frac{dN_{MOT}}{dt} = R_c - \frac{N_{MOT}}{\tau_{MOT}} - \beta N_{MOT}^2 \qquad (7.30)$$

where R_c denotes the rate of capture, τ_{MOT} the average time spent by an atom in the MOT, and the term containing the coefficient β represents the loss rate due to collisions among the trapped atoms. Then, in the case of a vapor-cell trap, by neglecting the last term in Equation 7.30, the steady-state number of confined atoms is given by $N_{MOT,s} = \tau_{MOT} R_c$. Next, we have to express the quantities τ_{MOT} and R_c [665]. The former is evaluated as the loss rate due to collisions between the trapped atoms and the hot background-vapor atoms:

$\tau_{MOT}^{-1} = n\sigma\sqrt{3k_BT/m}$, where n is the density of the atoms in the vapor, T the temperature of the cell, and σ the collisional cross section. The latter, defined as the rate at which atoms with a velocity smaller than the capture velocity of the trap (v_c) enter the volume V determined by the intersection of the laser beams, can be expressed as

$$R_c = \frac{nV^{2/3}v_c^4}{2}\left(\frac{m}{k_BT}\right)^{3/2} \tag{7.31}$$

In turn, v_c represents the maximum velocity for which an atom can be arrested within the distance fixed by the laser beam diameter d: $v_c = \sqrt{dv_{rec}/\tau_{exc}}$, where v_{rec} is the recoil velocity and τ_{exc} the excited-state lifetime. Finally, we obtain

$$N_{MOT,s} = \tau_{MOT}R_c = \frac{\pi}{\sqrt{6}}\frac{d^4v_{rec}^2}{\tau_{exc}^2\sigma}\left(\frac{m}{2k_BT}\right)^2 \tag{7.32}$$

where $V = (\sqrt{\pi}d)^3$ has been exploited. Just as an example, considering a Cs MOT ($\tau_{exc} = 30$ ns, $\sigma = 2\cdot10^{-13}$ cm^2, $v_{rec} = 3.5$ mm/s) with $T = 300$ K and $d = 1$ cm, one finds $N_{MOT,s} \simeq 5\cdot10^9$ atoms. In practice, the maximum achievable value for $N_{MOT,s}$ is restricted by the βN_{MOT}^2 loss term, which can no longer be neglected above a critical value of trapped atoms. Similar considerations apply to the maximum attainable atomic density, which is additionally limited by the repulsive forces between trapped atoms caused by re-absorption of scattered photons.

While dense samples ($\rho \geq 10^{10}$ atoms/cm^3) of $N \gg 10^7$ atoms are easily cooled in a MOT, the maximum phase-space density is in the range $\rho\Lambda_{dB}^3 \sim 10^{-5} - 10^{-4}$, with $\Lambda_{dB} = h/\sqrt{2\pi mk_BT}$ being the thermal de Broglie wavelength.

Incidentally, we mention that, besides the above radiation-pressure traps, optical dipole traps are also widely used. These rely on the dipole force exerted on an atom by nearly-resonant light with a non-uniform intensity $I(\boldsymbol{r})$. In this case, rather than the scattering of photons due to spontaneous emission, processes of absorption-stimulated emission are exploited [465]. In the dressed-atom picture, such dipole force results from the spatially-varying shift of atomic levels induced by a light field having a non-uniform intensity profile: for a laser frequency lower (higher) than the atomic resonance frequency, the atoms will be pulled towards high-intensity (low-intensity) regions. In the former (latter) case, the detuning $\delta = \nu_L - \nu_A$ is negative (positive) and we speak of red (blue) dipole trap. For a sufficiently large detuning, the depth of a dipole trap varies as $I(\boldsymbol{r})/\delta$, while the photon scattering rate is proportional to $I(\boldsymbol{r})/\delta^2$. Since no cooling mechanism is present in a dipole trap, only atoms colder than its depth can be loaded into it; their lifetime is precisely limited by the heating associated with the photon scattering rate. In its simplest realization, a dipole trap consists of a single, red-detuned, focused laser beam. Here, three-dimensional confinement is obtained at the focus, where the light field intensity exhibits an absolute maximum. In blue-detuned dipole traps, atoms are repelled by the light walls and confined in regions where light is virtually absent. Such an optical bottle beam trap can be realized, for instance, with a strongly focused blue detuned laser beam, which passes through a computer-generated circular π phase hologram displayed on a spatial light modulator. The most alluring features of optical dipole traps are the following: first, atoms can be confined in extremely small volumes; second, contrary to the case of magnetic trapping, the atoms are not polarized, thus enabling experiments where mixed-spin systems are studied; third, fast on/off switching is feasible; last but not least, they can be employed with atoms for which magnetic trapping is not possible.

A comprehensive, tutorial review on optical dipole traps is given by [466], whereas an updated reference on their use in the frame of all-optical Bose-Einstein condensation (see below) is represented by [674].

7.6.3   Bose-Einstein condensation

Albeit brief, this treatment cannot ignore the greatest achievement so far in the field of ultracold neutral atoms, namely the Bose-Einstein condensation (in Bose gases) and its counterpart in Fermi gases [675]. An additional motivation to provide a short *excursus* on this topic is that some schemes, originally developed to produce and probe BECs, also proved to be, more in general, valuable tools for creating cold atomic samples with improved phase-space densities (like those often employed in present-day frequency standards) and for characterizing more accurately their relevant parameters (number of atoms, temperature, ...), respectively.

After a MOT stage, two additional stages are needed, that are magnetic trapping and evaporative cooling.

7.6.3.1   Magnetic trapping

Essentially, magnetic traps are inhomogeneous magnetic field configurations, $\boldsymbol{B}(\boldsymbol{r})$, endowed with a local minimum, such that atoms with a non-zero magnetic moment $\boldsymbol{\mu}$ experience an interaction energy given by $W = -\boldsymbol{\mu} \cdot \boldsymbol{B}$ and hence a force $\boldsymbol{F} = \boldsymbol{\nabla}(\boldsymbol{\mu} \cdot \boldsymbol{B})$. In principle, the direction of this force depends on the relative magnetic moment/field orientation. However, the treatment is dramatically simplified if the atomic motion is adiabatic (i.e., the atom does not change Zeeman sublevel). In that case, one can introduce a local potential defined by the product between the atomic magnetic moment and $|\boldsymbol{B}|$. Therefore, atoms which are in low-field-seeking (lfs) states (that are states whose energy increases with $|\boldsymbol{B}|$) are acted on by a restoring force towards the minimum region of $|\boldsymbol{B}|$, thus being trapped. As a consequence of the Wing theorem, which basically states that, in a region devoid of charges and currents, the strength of a static magnetic field can have local minima but no local maxima, exclusively atoms in lfs states can be trapped. It is worth pointing out that, in order to remain in a low-field-seeking state, atoms have to retain their spin orientation relative to the magnetic field. This is only possible if the rate of change of the magnetic field in the reference frame of the moving atom $\omega_F = \boldsymbol{v} \cdot \boldsymbol{\nabla}(\boldsymbol{B}/B)$ is much smaller than the Larmor frequency $\omega_L = \mu B/\hbar$, $\boldsymbol{v}$ being the velocity of the atom. This cannot be achieved, for instance, in the center of a magnetic quadrupole trap where $B = 0$. Thus, magnetic traps with an offset field must be used, such that the Larmor frequency is always sufficiently large (always $B \neq 0$).

Compared to MOTs, where the photon recoil limit hampers achieving of ultralow temperatures, magnetic traps permit, by virtue of their conservative character, the atomic confinement at much lower temperatures. The two main drawbacks are:

- Magnetic traps are much shallower than MOTs. For typical potential depths ($\Delta B \sim 100$ G) atoms with $\mu \simeq \mu_B$ can be trapped only if $k_B T < \mu_B \Delta B \simeq 10^{-25}$ J, corresponding to a temperature of about 10 mK. This means that it is necessary to pre-cool the atoms (precisely in a MOT, for instance) before loading them into a magnetic trap.

- Just due to the conservative nature, a separate mechanism is needed to chill the magnetically confined atoms. In this sense, the most effective approaches are represented by evaporative cooling (see next section) and sympathetic cooling via elastic collisions with another cold species in the trap.

Among the several configurations of magnetic traps that have been used in BEC experiments, we focus here on the Ioffe-Pritchard (IP) trap whose elements and current arrangement are schematically shown in Figure 7.22. In its original geometry, this comprises four straight bars that generate a linear quadrupole field plus two end so-called pinch coils which provide the axial confinement (along the z direction). By using cylindrical coordinates, close to the origin (that is the trap center) the field components are

$$B_z = c_1 + c_3(z^2 - \rho^2/2) + ... \tag{7.33}$$

$$B_\rho = -c_3 z\rho + c_2\rho\cos(2\phi) + ... \tag{7.34}$$

$$B_\phi = -c_2\rho\sin(2\phi) + ... \tag{7.35}$$

where the c_1, c_2, and c_3 coefficients can be calculated starting from the expressions for the field generated by a coil and a straight conductor. Then we can write

$$|\boldsymbol{B}|^2 = c_1^2 + 2c_1 c_3 z^2 + [c_2^2 - c_1 c_3 - 2c_2 c_3 z\cos(2\phi)]\rho^2 + c_3^2(z^4 + \rho^4/4) + ... \tag{7.36}$$

showing that, with a suitable choice of the experimental parameters, the origin can be made a minimum in any direction. The associated potential is harmonic with a nonzero bias field (corresponding to the c_1 term) at the origin

$$W = \mu\left[c_1 + c_3 z^2 + \frac{1}{2}\left(\frac{c_2^2}{c_1} - c_3\right)\rho^2\right] \tag{7.37}$$

A bias field of a few Gauss is typically sufficient to suppress Majorana spin flips. In conclusion we mention that, among the different variants of the IP trap, particularly effective is the so-called *cloverleaf* trap [668].

7.6.3.2 Evaporative cooling

This technique relies on the selective removal of those trapped atoms whose energy is higher than the average energy per atom and on the subsequent re-thermalization of the sample through collisions. Since this process diminishes the average energy of the atoms remaining in the trap, the new equilibrium state corresponds to a lower gas temperature. Evaporative cooling can be made more efficient by forcing its proceeding rate. Rf-induced evaporation is the most popular manner to accomplish this. Here, rf radiation is used to flip the atomic spin so as to convert the attractive confinement force into a repulsive one, which expels the targeted atoms from the trap (Figure 7.23). The energy-selective character of this method arises from the fact that the resonance frequency is proportional to the magnetic field, and hence to the potential energy of the atoms. With regard to transitions between magnetic sublevels m_F, the resonance condition for the magnetic field strength B is $|g|\mu_B B = \hbar\omega_{rf}$, where g is the Landé g−factor. Since the trapping potential is given by $m_F g\mu_B [B(r) - B(0)]$, only atoms which have a total energy $E > \hbar|m_F|(\omega_{rf} - \omega_0)$ will evaporate (ω_0 is the rf frequency which induces spin flips at the bottom of the trap). A detailed discussion of the

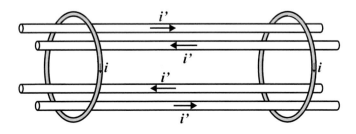

FIGURE 7.22
Scheme of a Ioffe-Pritchard trap.

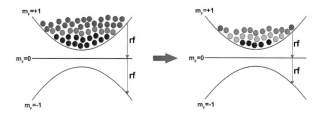

FIGURE 7.23 (SEE COLOR INSERT)
Principle of rf-induced evaporative cooling. Atoms populate the trap according to their energies, which are described by a Boltzmann distribution. The most energetic atoms can travel further from the center of the trap than the less energetic atoms can, and are more likely to interact with the rf field. The latter induces transitions between the trapped $m_F = 1$ state to the untrapped $m_F = -1$ one. After this expulsion, the remaining atoms will collide and re-thermalize, forming another Boltzmann distribution at a lower average energy. The high-energy tail of the Boltzmann distribution contains a disproportionate share of the total energy of the cloud, so the removal of just a small number of hot atoms can significantly impact the average energy of the cloud. By gradually reducing the frequency of the rf field, one can lower the temperature until a Bose-Einstein condensate is formed.

general features of evaporative cooling is given in [676]. Here we only mention that an efficient evaporative cooling requires the rate of elastic collisions to greatly exceed that of collisions which are responsible for atom losses from the magnetic trap. These can be distinguished into collisions with the background gas and inelastic collisions. The former ones can be reduced by decreasing the pressure of background gas; ultrahigh vacuum chambers are indeed used in the experiments. The latter ones can be either binary collisions, such as dipole and spin relaxation, or three-body recombination. Since evaporative cooling relies on collisions, a high density is needed, which makes inelastic collisions unavoidable. Therefore, for a specific atomic species, the efficiency of such process crucially depends on how the relevant collisional parameters compare to each other. The success of BEC experiments on alkali atoms is precisely due to the fact that they are characterized by fairly large elastic cross sections.

By summarizing, the *standard* procedure to achieve BEC can in fact be divided into three main steps [677, 678]:

- **Laser cooling and trapping of atoms** - This is usually accomplished by a double-MOT apparatus in which the atoms are loaded into a first MOT from the vapor and then transferred (by pushing laser beams) into a second, high-vacuum chamber.

- **Magnetic trapping** - A delicate step in this stage is the initial adiabatic compression of the confined atoms. Indeed, just after loading, the atoms are adiabatically compressed by raising the trap curvature. This induces an increment in the temperature and density of the atoms, with the phase-space density remaining constant. The associated gain in the rate of elastic collisions is vital to realize the proper re-thermalization conditions for the subsequent evaporative cooling process.

- **Evaporative cooling** - When optimized, this process can provide a dramatic reduction of the gas temperature with an enhancement of the phase-space density by 5-6 orders of magnitude, thus allowing to reach the condition for BEC (phase-space density ~ 1).

7.6.3.3 Probing a BEC

Most often, cold atomic samples are characterized by measuring the power emitted by the fluorescent light. Alternatively, one can analyze the absorption or the phase shift suffered by the light when transmitted through the sample. In the specific case of BEC, one of the latter two approaches is adopted [668, 665].

Absorption Imaging

In this scheme, the atomic cloud is illuminated with a short pulse of resonant light and the shadow cast by the sample is imaged onto a CCD camera. However, due to the high optical density, quantitative measurements cannot be performed directly in the trap. In practice, at the end of the evaporative cooling, the trapping fields are switched off and the absorption image is recorded after several milliseconds of ballistic expansion of the cloud. This also relaxes spatial resolution requirements for the optical system. Then, from a single image, information can then be extracted on the density and the temperature of the atoms in the trap. In practice, however, a sequence of three pictures is acquired for each cycle: the first picture of the atomic cloud is followed by a picture of the laser probing beam itself (i.e., without any atoms present); finally, a background frame, without the probe beam, is recorded. This is obviously a destructive observation method because each atom scatters several photons while being probed, which heats the gas. Experiments are therefore performed by repeated cycles of loading, evaporation, and probing.

When a weak (saturation effects are negligible) probe laser beam passes through an atomic cloud along the z direction, it experiences both absorption and a phase shift. By neglecting the unessential $e^{i\omega t}$ term, these can expressed as

$$E(z) = E_i e^{-n''kz} e^{in'kz} \equiv t E_i e^{i\varphi} \tag{7.38}$$

where n' and n'' represent, respectively, the real and imaginary part of the sample refractive index. Thus, assuming as usual a Lorentzian shape for the absorption interaction, we have

$$t \equiv e^{-n''kz} = \exp\left(-\frac{\tilde{n}\sigma_0}{2}\frac{1}{1+\delta^2}\right) \tag{7.39}$$

and

$$\varphi \equiv n'kz = -\frac{\tilde{n}\sigma_0}{2}\frac{\delta}{1+\delta^2} \tag{7.40}$$

where σ_0 is the resonant cross section ($\sigma_0 = 3\lambda^2/(2\pi)$ for a two-level atom), $\delta = (\omega - \omega_0)/(\Gamma/2)$ is the laser detuning in half linewidths, and $\tilde{n}(x,y) = \int n(x,y,z)dz$ is the so-called column density. Since photodetectors are not sensitive to phase, the absorption image shows the spatial variation of t^2 from which, in turn, that of $\tilde{n}(x,y)$ can be easily retrieved. The above derivation assumes that light enters and exits the cloud at the same (x,y) coordinate (thin-lens approximation).

Dark-ground and Phase-contrast Imaging

In dark-ground imaging, a collimated probe beam propagates through a weakly absorbing sample and the coherently scattered light is imaged onto a camera. Instead, the probe beam is blocked after passing the sample by means of a small opaque object located at the position where the beam comes to a focus (Figure 7.24a). Since the probe light field emerging from the sample (E_f) can be separated into the scattered and unscattered radiation, we can write

$$E_f \equiv t E_i e^{i\varphi} = E_i + \Delta E \tag{7.41}$$

which, blocking the unscattered light (E_i), gives the dark-ground signal

$$\langle I_{dg} \rangle = \frac{1}{2} |E_f - E_i|^2 = I_i[1 + t^2 - 2t \cos \varphi] \tag{7.42}$$

For small φ values, such signal is quadratic in φ. Compared to absorption imaging, the dispersive approach has two significant advantages. First, since the information is contained in φ, the probe frequency can be tuned sufficiently far from resonance in order to interrogate dense atomic clouds directly in the trap. Second, it can be a non-destructive detection method.

Phase-contrast imaging can be thought of as a homodyne detection technique where the unscattered light represents the local oscillator and interferes with the scattered radiation. This is accomplished by shifting, via a *phase plate* (i.e., an optical flat with a small bump or dimple in the center), the phase of the unscattered light by $\pm \pi/2$ in the Fourier plane of the imaging lens (Figure 7.24(a)). Then, the intensity of a point in the image plane is given by

$$\langle I_{pc} \rangle = \frac{1}{2} \left| tE_i e^{i\varphi} + E_i \left(e^{\pm i \frac{\pi}{2}} - 1 \right) \right|^2 = I_i \left[t^2 + 2 - 2\sqrt{2}t \cos \left(\varphi \pm \frac{\pi}{4} \right) \right] \tag{7.43}$$

For small φ one obtains

$$\langle I_{pc} \rangle = I_i[t^2 + 2 - 2t \pm 2t\varphi] \tag{7.44}$$

which is linear in φ. Thus, phase-contrast imaging should be preferred to dark-ground imaging for small signals.

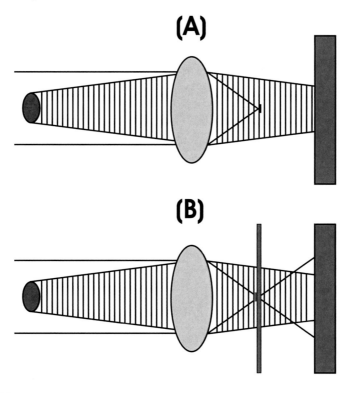

FIGURE 7.24
Principle of dark-ground (a) and phase-contrast (b) imaging. (Adapted from [668].)

Once the column density has been experimentally determined with one of the above techniques, a suitable fitting function can be used to extract the temperature T and the total number of atoms N. Just to fix ideas, let us consider the case of absorption imaging of a thermal cloud above the critical temperature for condensation. After a time t of ballistic expansion, the density profile can be expressed as

$$n(x,y,z,t) = \frac{N}{(2\pi)^{3/2}\sigma_x(t)\sigma_y(t)\sigma_z(t)}e^{-\frac{x^2}{2\sigma_x^2(t)}}e^{-\frac{y^2}{2\sigma_y^2(t)}}e^{-\frac{z^2}{2\sigma_z^2(t)}} \qquad (7.45)$$

such that $\int n(x,y,z,t)dxdydz = N$. By integrating Equation 7.45 along the direction of propagation of the laser beam, we then obtain the column density

$$\tilde{n}(x,y,t) = \frac{N}{2\pi\sigma_x(t)\sigma_y(t)}e^{-\frac{x^2}{2\sigma_x^2(t)}}e^{-\frac{y^2}{2\sigma_y^2(t)}} \qquad (7.46)$$

which can be fitted to $\tilde{n}_{exp}(x,y,t)$ in order to extract N, $\sigma_x(t)$ and $\sigma_y(t)$. Then, these two latter quantities can be used to deduce the radii in the trap $\sigma_x(t=0)$ and $\sigma_y(t=0)$, eventually yielding the temperature T via the equipartition principle

$$\frac{3}{2}k_BT = m\left[\frac{1}{2}\omega_x^2\sigma_x^2(t=0) + \frac{1}{2}\omega_\rho^2\sigma_\rho^2(t=0)\right] \qquad (7.47)$$

Note that for traps with cylindrical symmetry, like those most commonly used, one has $\omega_y = \omega_z \equiv \omega_\rho$, which obviously implies $\sigma_y(t) = \sigma_z(t) = \sigma_\rho(t)$. The relationship between $\sigma_i(t)$ and $\sigma_i(t=0)$ depends on the system under investigation. For a classic ideal gas it is given by [679]

$$\sigma_i(t) = \sigma_i(t=0)\sqrt{1+\omega_i^2t^2} \qquad (7.48)$$

In the case of a BEC at "$T=0$", the column density is more adequately fitted with an inverted Thomas-Fermi parabola and similar expressions can be derived for the ballistic expansion [680].

In conclusion, as already mentioned, the production of ultracold ensembles of atoms has revolutionized the field of atomic and optical physics generating, in addition, much interest among researchers in other, traditionally disjoint fields. In particular, the aforementioned realization of all-optical BECs has played a crucial role in the advancement of research in the field of quantum degenerate gases. Indeed, the advantage of the all-optical approach is that magnetic fields can then be used to control the interactions between the atoms (making attractive interactions repulsive, for example) with the powerful Feshbach-resonance technique [681]. Moreover, optical traps have allowed ytterbium to be condensed in recent years [682]. Ytterbium is notable because it is the only atom with two valence electrons to be condensed so far. As we will see later on, generation of a BEC in Yb is also helpful in optical lattice clocks using even isotopes of alkaline-earth-like atoms. By now, BECs on microelectronic chips [683] are also routinely produced. A quite complete list of ongoing research directions including Optical lattices; Quantum gases in low dimensions; Disordered and designed potentials; Ultracold Molecules; Long-range and dipolar interactions; Precision measurements with quantum gases; Atom lasers; Quantum magnetism; Non-equilibrium phenomena; Quantum simulation; Few-body systems; and Hybrid quantum systems, can be found in the Book of Abstracts related to the Conference, *Bose-Einstein Condensation 2011 Frontiers in Quantum Gases* [684]. In the very end, it is worth mentioning that BEC of photons in an optical microcavity was also observed [685].

7.7   Cold stable molecules

Just as the achievement of quantum degeneracy in Bose and Fermi gases has opened unimaginable scenarios in fundamental and applied Physics, so the impact of creating cold (1 mK - 1 K) and ultracold ($<$ 1 mK) molecules is expected to lead to discoveries reaching far beyond the focus of traditional molecular science. As an example, a BEC of polar molecules would represent a quantum fluid of strongly and anisotropically interacting particles and thereby elucidate the link between BECs in dilute gases and in dense liquids. Furthermore, while presenting complexities and challenges for experimental control, the extra degrees of freedom available in molecules offer unique opportunities for the exploration and manipulation of exotic quantum phases [686]. Unfortunately, the cooling techniques so far developed for atomic species cannot be readily applied to molecules, because of the lack of a closed set of cycling transitions, due primarily to the absence of strict selection rules between vibrational levels. At present, ultracold molecules can be obtained solely in the form of dimers, basically via magneto-association or photo-association of alkali atoms close to quantum degeneracy. However, a huge drawback of such indirect techniques is that they only work with molecules whose constituent atoms can be laser cooled and trapped. Thus, for the conceivable future, many chemically relevant species (hydrides, nitrides, oxides, fluorides, etc.) will be still excluded. Meanwhile, a number of direct, more versatile approaches have been demonstrated to bring stable molecules into the cold temperature regime [687, 688]. Among these, the most successful methods are represented by buffer-gas cooling (BGC) and Stark deceleration. In the former scheme, already discussed in Chapter 5, both translational and rotational degrees of freedom of the desired molecular species are cooled via elastic collisions with a helium buffer gas in a cryogenic cell. Then, a molecular beam can be formed by allowing the molecules to exit the chamber via an orifice. Continuous buffer-gas beams are now routinely produced in a number of laboratories with low temperature (around 3 K), low velocity (100 m/s), and high intensity (10^{14} s^{-1}·sr^{-1}). The latter technique, invented by G. Meijer [443], relies on Stark-effect-based deceleration of a molecular beam via time-varying inhomogeneous electric fields and is presented into more detail in the following. As such a method operates only with polar molecules, a more recent variant, named as optical Stark deceleration (OSD), has also been introduced [689]. In this case, the change in the molecular velocity is caused by the optical dipole force induced by use of far-off-resonant pulsed optical fields in the 10^{12} W/cm^2 range.

7.7.1   Stark decelerator

In order to describe the behavior of neutral polar molecules in an external inhomogeneous electric field $\boldsymbol{E}$, let us first recall that the interaction energy and the force associated with a molecular dipole $\boldsymbol{\mu}$ are, respectively, given by $W = -\boldsymbol{\mu} \cdot \boldsymbol{E}$ and $\boldsymbol{F} = \nabla(\boldsymbol{\mu} \cdot \boldsymbol{E})$. If the molecular motion is adiabatic (i.e., the molecule does not change the Stark sublevel) it can be described by a local potential given by the molecular dipole times $|\boldsymbol{E}|$. In particular, in the presence of a minimum of $|\boldsymbol{E}|$, the so-called low-field-seeking (lfs) states (that are states whose energy increases with increasing electric field) experience a restoring force towards the minimum region. To better understand the working principle of a Stark decelerator (SD), let us consider the specific case of a symmetric top molecule, like CH$_3$F. In this case, rotational states are labelled by three quantum numbers: J, the total angular momentum quantum number; K, describing the projection of the vector $\boldsymbol{J}$ onto the molecular axis; and M, representing the projection of $\boldsymbol{J}$ onto the local electric field vector. Also, since the Stark deceleration process does not enhance the phase-space density, the beam entering the

SD is typically prepared via supersonic expansion to cool both the internal and external molecular degrees of freedom. By virtue of such rotational cooling, the CH_3F beam entering the Stark decelerator consists almost exclusively of molecules that are in states with $J = 0$ or 1. Among these, the most populated lfs states are $|JKM\rangle = |100\rangle, |11-1\rangle, |1-11\rangle$. In its original version, the Stark decelerator consists of a longitudinal array of electric field stages, separated by a distance L (Figure 7.25). Each stage is formed by two parallel r-radius cylindrical hardened steel rods, spaced d apart (L, r, and d are on the mm order). One of the rods is connected to a positive and the other to a negative switchable high-voltage (HV) power supply. Alternating stages are connected to each other. At a given time, the even-numbered stages are switched to the HV difference and the odd-numbered ones are grounded. In this way, molecules in lfs states that approach the plane of the first pair of electrodes experience the increasing electric field as a potential hill, and will lose kinetic energy on its upward slope. When leaving the high-field region, however, the molecule would regain the same amount of kinetic energy, due to the acceleration on the downward slope of the hill. This can be avoided by switching off the electric field when the molecule has reached a position that is close to the top. At the same time, the pairs of electrodes that were grounded are switched to HV difference. Consequently, the molecule will find itself anew in front of a potential hill and will again lose kinetic energy when climbing it. By repeating this process many times, the velocity of the molecule can be reduced to an arbitrarily low value. As the electric field close to the electrodes is higher than that on the molecular beam axis, molecules in lfs states will experience a force focusing them towards it. This occurs, however, only in the plane perpendicular to the electrodes. In order to focus the molecules in both transverse directions, the electrode pairs that make up one deceleration stage are alternately positioned horizontally and vertically. By contrast, high-field-seeking (hfs) states are deflected from the molecular beam axis and will eventually be lost. In this respect, it is crucial that the molecules stay in the lfs state throughout the decelerator. If they came in zero electric field, projection onto hfs states might occur. As already discussed in the case of magnetic trapping of atoms, these so-called Majorana transitions are prevented by guaranteeing that the electric field never drops below a certain minimum value. This also ensures that the Stark splitting exceeds that corresponding to the highest-frequency component of the rf radiation, which originates from the fast switching of the electric fields. After exiting the SD, the decelerated molecular packets can be either directly utilized for experiments or further manipulated in a variety of elements. Prominently, after deceleration to a standstill, they can be loaded into traps or storage rings for confinement of several seconds.

Since the first demonstration in 1999, a variety of SDs have been designed and built, including decelerators based on wire electrodes [690], integrated on a chip [691], and travelling wave decelerators [692].

Output beams from either BGC or SD setups can be eventually loaded into electric/magnetic traps for collisional studies [693] or serve as the underpinning for further cooling stages to approach the μK threshold. These include cavity-assisted laser cooling [694], electrostatically remixed magneto-optical trapping [695], and Sisyphus-type opto-electrical cooling [696]. Such schemes could ultimately lead to superior samples of ultracold molecules when combined with followup techniques such as evaporative or sympathetic cooling [697]. Besides representing indispensable initial sources for any cooling route, molecular samples provided by BGC (SD)-based machines already have the potential to dramatically influence a variety of research domains encompassing precise control of chemical reactions, study of novel dynamics in low-energy collisions, and ultrahigh-resolution molecular spectroscopy. In

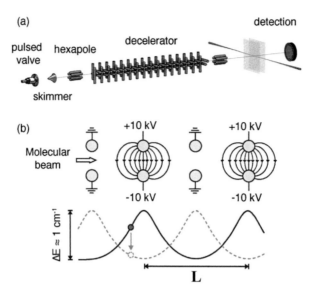

FIGURE 7.25
Experimental setup and principle of Stark deceleration. (Courtesy of [688].)

this latter frame, potentially all the techniques discussed in Chapter 5 for molecular-beam spectroscopy, two-photon Ramsey fringes above all, can be applied.

7.8   Trapping and cooling of ions

The seemingly obvious opportunity of trapping a charged particle by exploiting a mere electrostatic inhomogeneous field ($\boldsymbol{E}(\boldsymbol{r})$) to create a restoring force ($\boldsymbol{F} = q\boldsymbol{E}$) towards a given point O in space is unfortunately impeded by the Gauss equation for electrostatics $\nabla \cdot \boldsymbol{E} = 0$, which implies $\nabla \cdot \boldsymbol{F} = 0$ (Earnshaw theorem). To elude this, one must resort to an association of static magnetic and electric fields (Penning trap) or, alternatively, to a time-dependent inhomogeneous electric field (radio frequency trap or Paul trap).

7.8.1   Paul traps

Among the different configurations which are able to provide stable confinement, the one corresponding to a time-averaged trapping potential which is harmonic and symmetric is the most utilized [698, 699, 700]. A typical three-electrode configuration is shown in Figure 7.26, comprising two endcaps and a ring electrode whose surfaces are infinite hyperboloids of revolution described by

$$\frac{r^2}{r_0^2} - \frac{z^2}{z_0^2} = \pm 1 \tag{7.49}$$

where r_0 is the inner radius of the ring electrode, z_0 is half the distance between the two endcaps, and the $+$ and $-$ sign correspond to the ring electrode and endcap, respectively. Application of the time-varying voltage $\Phi_0 = U_0 - V_0 \cos \Omega t$ to the electrodes will result in

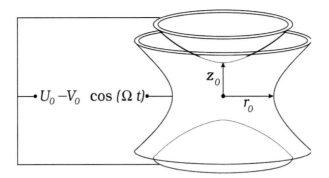

FIGURE 7.26
Electrode structure for a Paul-type trap. (Adapted from [700].)

the following potential

$$\Phi(r, z, t) = \frac{U_0 - V_0 \cos \Omega t}{r_0^2 + 2z_0^2}(r^2 - 2z^2) \tag{7.50}$$

Thus, the ion is stable (unstable) in the axial (radial) direction for half the cycle and vice versa for the other half. However, by selecting proper values for certain parameters, stability in all three dimensions can be obtained. To find out such stability conditions, let us first derive the electric field strength $\boldsymbol{E} = -\nabla\Phi$ as

$$E_r = -\frac{\Phi_0}{d_0^2}r \qquad E_\varphi = 0 \qquad E_z = \frac{2\Phi_0}{d_0^2}z \tag{7.51}$$

where $2d_0^2 \equiv r_0^2 + 2z_0^2$ has been introduced. It is worth mentioning that several experiments use $r_0 = \sqrt{2}z_0$. Next, we can write down the equations of motion for an ion of mass m and charge e as

$$\ddot{r} + \frac{e}{md_0^2}(U_0 - V_0\cos\Omega t)\,r = 0 \tag{7.52}$$

$$\ddot{z} - \frac{2e}{md_0^2}(U_0 - V_0\cos\Omega t)\,z = 0 \tag{7.53}$$

If the above equations are parametrized by defining the dimensionless quantities

$$\xi \equiv \frac{\Omega t}{2} \quad a_r = \frac{4eU_0}{md_0^2\Omega^2} \quad q_r = \frac{2eV_0}{md_0^2\Omega^2} \quad a_z = -2a_r \quad q_z = -2q_r \tag{7.54}$$

then a set of Mathieu's differential equations is found

$$\frac{d^2w_i}{d\xi^2} + (a_i - 2q_i\cos 2\xi)\,w = 0 \qquad i = r, z \tag{7.55}$$

The values of a and q for which the solutions are stable (i.e., oscillating with limited amplitude rather than exponentially growing) simultaneously along the z direction and in the radial plane, which is an obvious requisite for 3-D confinement, are extracted by drawing (on the same set of axes) a composite plot of both the stability boundaries: the overlap regions correspond to 3-D trapping. Among these, being the most practical, the one close to the origin (a and q values much less than 1) is almost exclusively used for ion confinement [699]. As the Mathieu equation is nothing but a differential equation with periodic coefficients, the general form of the stable solutions is dictated by Floquet's theorem [701]

$$w_i\left(\xi\right) = A e^{i\beta_{w_i}\xi} \sum_{n=-\infty}^{+\infty} C_{2n} e^{i2n\xi} + B e^{-i\beta_{w_i}\xi} \sum_{n=-\infty}^{+\infty} C_{2n} e^{-i2n\xi} \tag{7.56}$$

Here, the real-valued characteristic exponent β_{w_i} and the coefficients C_{2n} are functions of a_i and q_i only and do not depend on initial conditions, whereas A and B are arbitrary constants that may be used to satisfy boundary conditions or normalize a particular solution. Without loss of generality, with the initial condition $A = B$, we have

$$w_i\left(\xi\right) = 2A \sum_{n=-\infty}^{+\infty} C_{2n} \cos\left[\left(2n + \beta_{w_i}\right)\xi\right] \tag{7.57}$$

By inserting this formula into Equation 7.55 and defining

$$D_{2n} = \frac{a_i - \left(2n + \beta_{w_i}\right)^2}{q_i} \tag{7.58}$$

one obtains the recursion relations

$$C_{2n} = \frac{C_{2n-2}}{D_{2n}} + \frac{C_{2n+2}}{D_{2n}} \tag{7.59}$$

Numerical values for these coefficients can be deduced by truncating the continued fractions after the desired accuracy is reached. In the case $a_i, q_i \ll 1$, the lowest-order approximation to the ion trajectory corresponds to $C_{\pm 4} = 0$. In this case one gets for $n = 0$

$$C_0 = \frac{C_{-2}}{D_0} + \frac{C_2}{D_0} \tag{7.60}$$

and for $n = 1$

$$C_2 = \frac{C_0}{D_2}, \qquad C_{-2} = \frac{C_0}{D_{-2}} \tag{7.61}$$

which combined together give

$$D_0 = \frac{1}{D_{-2}} + \frac{1}{D_2} \tag{7.62}$$

This equation translates into the following equation for β_{w_i}

$$\frac{a_i - \beta_{w_i}^2}{q_i} = q_i \left[\frac{1}{a_i - \left(\beta_{w_i} - 2\right)^2} + \frac{1}{a_i - \left(\beta_{w_i} + 2\right)^2} \right] \tag{7.63}$$

which, for $4 \gg a_i, 4\beta_{w_i}, \beta_{w_i}^2$, yields

$$\beta_{w_i} \simeq \sqrt{a_i + \frac{q_i^2}{2}} \tag{7.64}$$

Note that under this approximation $C_2 = C_{-2} \simeq -q_i/4C_0$. Therefore we have

$$w_i\left(\xi\right) = 2A \sum_{n=-1}^{+1} C_{2n} \cos\left[\left(2n + \beta_{w_i}\right)\xi\right]$$

$$= 2A \left\{ C_0 \cos\left(\beta_{w_i}\xi\right) + C_2 \cos\left[\left(2 + \beta_{w_i}\right)\xi\right] + C_2 \cos\left[\left(-2 + \beta_{w_i}\right)\xi\right] \right\}$$

$$\simeq 2A C_0 \cos\left(\beta_{w_i}\xi\right) \left\{ 1 + \frac{2C_2}{C_0} \cos\left(2\xi\right) \right\} = 2A C_0 \cos\left(\beta_{w_i}\xi\right) \left\{ 1 + D_0 \cos\left(2\xi\right) \right\}$$

$$= 2A C_0 \cos\left(\omega_i t\right) \left\{ 1 - \frac{q_i}{2} \cos\left(\Omega t\right) \right\} \tag{7.65}$$

where

$$\omega_i = \sqrt{a_i + \frac{q_i^2}{2} \frac{\Omega}{2}} \equiv \beta_{w_i} \frac{\Omega}{2} \tag{7.66}$$

In essence, this is the motion of an oscillator of frequency ω_i whose amplitude is modulated with the trap's driving frequency Ω (the driven excursions are $180°$ out of phase with the driving field and a factor $q_i/2$ smaller than the amplitude of the secular motion). Since it is assumed that $\beta_{w_i} \ll 1$, the oscillation at ω_i, referred to as the secular motion or macromotion, is slow compared with the superimposed fast micromotion at Ω. For typical experimental parameters ($\Omega = 100$ MHz, $U_0 = 1$ V, $V_0 = 500$ V, $d_0 = 0.001$ m, $m = 200 \cdot 10^{-27}$ Kg), we have $q_r^2/2 = 0.0032$ and $a_r = 0.00032$ such that $\beta_r = \sqrt{a_r + q_r^2/2} \simeq q_r/\sqrt{2}$ and $\beta_z = \sqrt{a_z + q_z^2/2} \simeq \sqrt{2} q_r$, hence $\omega_r \simeq \omega_z/2$. Most often, in order to observe ion spectra in the Lamb-Dicke regime (see below) as well as to implement laser cooling techniques, it is desirable to have a very small trap, where the single ion is strongly confined and the secular frequencies are elevated. For a given mass to charge ratio, this can be accomplished either by increasing V_0 or by reducing d_0. In the former case, an obvious limitation is represented by the material used for the electrodes (a too large value of V_0 will trigger unwanted electric sparks). In the latter case, the difficulty in machining miniature traps restricts the d_0 value to around ~ 200 μm. As a consequence, secular frequencies on the order of tens of MHz have been reported so far.

In conclusion, a few features are worth being highlighted:

- By virtue of the large difference in the frequencies ω_i and Ω, the slow motion at frequency ω_i can be safely considered as separate, while time averaging over the fast oscillation at Ω. For $q_i^2/2 \gg a_i$ and $r_0^2 = 2z_0^2$, this corresponds to the time-averaged pseudo-potentials of depth D_z and D_r given by

$$eD_z = \frac{1}{2} m\omega_i^2 z_0^2 \quad \Rightarrow \quad D_z = \frac{eV_0^2}{4mz_0^2\Omega^2} \tag{7.67}$$

$$eD_r = \frac{1}{2} m\omega_i^2 r_0^2 \quad \Rightarrow \quad D_r = \frac{eV_0^2}{4mr_0^2\Omega^2} = \frac{eV_0^2}{8mz_0^2\Omega^2} \tag{7.68}$$

- Equation 7.56 shows that the motional spectrum contains the frequencies $(\beta_{w_i} + 2n)\Omega/2$, with n being an integer (fundamental frequencies correspond to $n = 0$). This can be experimentally demonstrated by exciting a given motional frequency by an additional (weak) radio-frequency (rf) field applied to the electrodes: at resonance, some ions may leave the trap, providing a signal proportional to the number of trapped ions.

- When trying to maximize the density of trapped ions, one naturally tends to increase the potential depth D_i so as to confine the highest number of particles. Actually, the oscillation amplitudes correspondingly increase, which results in an augmented ion loss for higher q_i in a trap of a given size. Thus, a compromise must be found.

- Ideally, once stored in a Paul trap, the ions would remain there for an infinitely long time. In practice, the lifetime is set by collisions with the neutral background molecules. Thus, operation in an ultrahigh vacuum is necessary. At pressures around 10^{-8} Pa, storage times of many hours are routinely obtained, whereas lifetimes of many months can be achieved at extremely low pressures (10^{-14} Pa), as realized by cryopumping. Under certain conditions, however, the storage time can benefit from higher background pressures. Indeed, it has been found that, when the mass of the ion exceeds that of the neutral atom/molecule, ion-neutral collisions tend to damp the ion motion, thus increasing the lifetime in the trap.

- While in thermal equilibrium at high temperatures, a cloud of trapped ions is described by a Gaussian density distribution (averaged over a period of the micromotion), as the temperature is gradually lowered, the cloud diameter contracts until a homogeneously charged sphere is obtained. As a rule of thumb, the average kinetic energy of the confined ions amounts to $1/10$ of the potential depth.

- Thus far we have ignored the strong interaction between the ions arising from their mutual Coulomb repulsion. If the ions' kinetic energies are small compared to the energy of such interaction, then the particles will be arranged in quasi-crystalline structures. For example, a few ions may be aligned in the nodal line of a linear quadrupole trap (see below), just like pearls on a string. For higher numbers of ions (up to 10^5), instead, more complex structures like helices can take place. Also, due to the collective oscillations of the ions, new motional frequencies appear [65].

- Another significant issue for single-ion traps is the displacement of the equilibrium point of the confined ion away from the trapping node. This may be caused by stray static electric fields of varied origin as well as by misalignment of the trap electrodes. As a consequence, since the micromotion extent increases with the distance from the field node, unpleasant Doppler and Stark shifts can occur. This effect can be minimized by applying proper trim voltages to additional static compensation electrodes, which are situated close to the main trap structure, so as to bring the ion location back to the nodal point [700]. The most effective method to monitor the micromotion level consists of measuring the Doppler modulation of the trapped-ion absorption rate [702]. Indeed, this latter manifests itself in a modulation of the observed fluorescence inasmuch as the radiative lifetime for the upper level of the cooling transition is much shorter than the rf period. As a result, a strong correlation exists between the detected photon arrival time (PAT) and the rf drive phase, which can be exploited for very precise adjustments of the ion position: in practice, this is accomplished by measuring the change in the modulation amplitude and phase of the PAT relative to the trap drive rf potential as the trim potential is tuned.

7.8.1.1   Linear Paul traps

Although the quadrupolar hyperbolic configuration provides the best approximation to a harmonic potential over a wide trap volume, this structure is not well suited for easy viewing of single-ion fluorescence or for the injection of laser excitation beams [700]. For this reason, most single-ion traps use a different electrode geometry. Among the various configurations, a very useful one is the linear rf trap, shown schematically in Figure 7.27 [701]. Here, a common rf potential is applied to the dark electrodes; the other electrodes are held at rf ground through capacitors connected to ground. The lower right portion of the figure shows the $x - y$ electric fields from the applied rf potential at an instant when the rf potential is positive relative to the ground. A static electric potential well is created (for positive ions) along the z axis by applying a positive potential to the outer segments (gray) relative to the center segments (white). This design is characterized by the presence of an approximate field nodal line along the central axis. Then, if the axial potential is rendered pretty weak compared to the x, y ones, two or more confined ions will line up along the trap axis (in contrast with a conventional Paul trap, where only one ion can be free of micromotion, as this condition only exists at the center of the trap). This can be useful for addressing individual ions with laser beams, each having very low kinetic energy. While this scheme has been successfully implemented for microwave transitions, where localization to a length-scale below that of the incident microwave radiation is easily obtained, an adequate degree of confinement and cooling is much more stringent in the case of optical transitions.

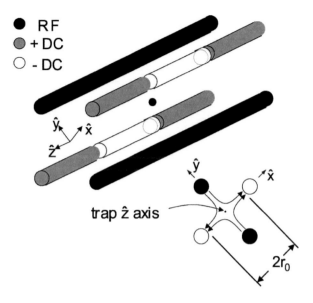

FIGURE 7.27
Schematic of a linear rf Paul trap. (Courtesy of [701].)

7.8.2 Penning traps

The Penning trap makes use of the same electrode arrangement as in the rf Paul trap, but it only exploits static electric and magnetic fields [703]. The first component is a homogeneous magnetic field B along the z axis, which forces a particle of charge e and mass m on a circular orbit (around the z axis) by the Lorentz force. This two-dimensional harmonic motion (radial trapping) is referred to as cyclotron motion and has a frequency of $\omega_c = |eB|/m$. However, this magnetic field does not confine the ions along the z axis. Then, an additional electric quadrupole field is superimposed; it is applied to the ring with respect to the endcaps and is described by the potential

$$U = U_0 \frac{2z^2 - r^2}{2d_0^2} \tag{7.69}$$

that is Equation 7.50 with $V_0 = 0$ (recall that the radial coordinate r must appear in the potential as it does for the potential to satisfy the Laplace equation: $\nabla^2 U = 0$). In this way, the axial z motion is a bound, harmonic oscillation provided that $eU_0 > 0$. As a result, the ions follow a complex orbit composed of three independent harmonic eigenmotions. The classical equations of motion of the ion are

$$m\frac{d^2\boldsymbol{r}}{dt^2} = e\boldsymbol{E}\left(\boldsymbol{r}\right) + e\frac{d\boldsymbol{r}}{dt} \times \boldsymbol{B} \tag{7.70}$$

that is

$$m\ddot{x} = e\left(E_x + \dot{y}B\right) \tag{7.71}$$

$$m\ddot{y} = e\left(E_y - \dot{x}B\right) \tag{7.72}$$

$$m\ddot{z} = eE_z \tag{7.73}$$

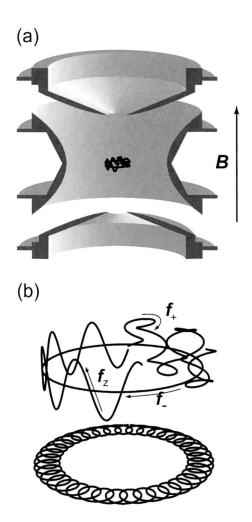

FIGURE 7.28
Cross section through a Penning trap (a) with the resulting motion of the ions (b). Here $f_z \equiv \omega_z$, $f_- \equiv \omega_m$, and $f_+ \equiv \omega_c'$. (Courtesy of [704].)

where $\boldsymbol{E} \equiv -\boldsymbol{\nabla}U \equiv (U_0/d_0^2)\,(x, y, -2z)$. The axial motion is decoupled from the magnetic field. It is a simple harmonic motion

$$\ddot{z} + \omega_z^2 z = 0 \tag{7.74}$$

with frequency

$$\omega_z^2 = \frac{2eU_0}{md_0^2} \tag{7.75}$$

Typically, the quadrupole potential represents a relatively weak addition (relative to the magnetic field) in the sense that $\omega_z \ll \omega_c$. For the $x - y$ components we have

$$\ddot{x} = \frac{\omega_z^2}{2}x + \omega_c \dot{y} \tag{7.76}$$

$$\ddot{y} = \frac{\omega_z^2}{2}y - \omega_c \dot{x} \tag{7.77}$$

By adding the first of the two above equations and i times the second one, and using the complex quantity $w = x + iy$ we have

$$\ddot{w} = \frac{\omega_z^2}{2} w - i\omega_c \dot{w} \tag{7.78}$$

which can be solved by $w = w_0 e^{-i\omega t}$, yielding the quadratic equation

$$\omega^2 - \omega\omega_c + \frac{\omega_z^2}{2} = 0 \tag{7.79}$$

with the real solutions

$$\omega_m = \frac{\omega_c}{2} - \sqrt{\frac{\omega_c^2}{4} - \frac{\omega_z^2}{2}} \quad \text{magnetron frequency} \tag{7.80}$$

$$\omega_c' = \frac{\omega_c}{2} + \sqrt{\frac{\omega_c^2}{4} - \frac{\omega_z^2}{2}} \quad \text{reduced cyclotron frequency} \tag{7.81}$$

Thus the radial motion is the epicyclic superposition of two uniform circular motions. For the above roots to be real, the condition $\omega_c^2 > 2\omega_z^2$ must be satisfied, which is equivalent to $|e|B^2/m > 4|U_0|/d_0^2$. It is worth noting that the sum of these two radial frequencies corresponds exactly to the free cyclotron frequency $\omega_c = \omega_m + \omega_c'$, while $\omega_c^2 = \omega_c'^2 + \omega_m^2 + \omega_z^2$. Both relations can be used to determine the cyclotron frequency. Typical experimental values are a few Volt for U_0 and a few Tesla for B, r_0 and z_0 being on the order of a few mm (as for Paul traps); for $U_0 = 5$ V, $d_0 = 1$ cm, $B = 1$ T, and $m = 200 \cdot 10^{-27}$, the following typical values are found for the various frequencies: $\omega_c = 800$ kHz, $\omega_z = 283$ kHz, $\omega_m = 26$ kHz, and $\omega_c' = 774$ kHz. In other words, for normal trapping parameters, it is always true that $\omega_m \ll \omega_z \ll \omega_c' \simeq \omega_c$. There are distinct differences between the magnetron motion and the cyclotron and the axial motion. The latter represents a harmonic oscillation and consequently the energy is swapped between its kinetic and potential part. Due to the high velocity and the small radius, the energy of the cyclotron motion is mainly a kinetic one, whereas the nature of the magnetron motion is essentially potential energy.

7.8.3 Trap loading

Since particles can be confined in a trap only if their kinetic energy is lower than the potential well barriers, injection of defined-energy ions into a fixed-height trap is not possible. Thus, the simplest loading technique consists of creating the ions inside the confining volume by photon or electron ionization of neutral atoms [705]. However, this approach is hardly applicable to low-abundance isotopes or antiparticles that are generated in accelerator facilities. In this case, within the time of the ions' passage through the trap region, it is often necessary to rapidly raise the potential barrier or, conversely, diminish the particles' kinetic energy by an effective cooling method [65].

7.8.4 Ion cooling techniques

The prime reason to cool trapped ions is to enhance their storage time. Indeed, during the loading procedure, the ions are not created exactly at the trap center; thus, they gain kinetic energy when falling towards it. Moreover, when produced by crossing an electron beam with an atomic one, the ions are hot just after generation. Thus, without cooling, the ions are readily lost as they can collide with other particles (e.g., background gas molecules) and surpass the potential well [699]. Also, cooling is of utmost importance in quantum information processing and high resolution spectroscopy. In the former application, cooling schemes are implemented to prepare the ion in its ground vibrational state before logic operations are performed. In the latter case, Doppler effects are heavily suppressed; just to give an idea, for a well depth of 20 eV, the mean kinetic energy of confined ions is around $\sim 20/10 = 2$ eV, whereupon the fractional frequency shift due to the second-order Doppler effect for an ion of mass number 200 is on the order of $\Delta\nu/\nu = -(mv^2/2)/(mc^2) \simeq$ 2eV$/(200 \cdot 0.94$ GeV$) \simeq 10^{-11}$, which is clearly unacceptable for a frequency standard.

There are various cooling schemes available. These include resistive cooling, buffer-gas cooling, radiative cooling, laser cooling, and sympathetic cooling. For the first three methods the reader is referred to [705], whereas laser cooling and sympathetic cooling will be discussed into a certain detail here.

7.8.4.1 Laser cooling

When dealing with laser cooling of trapped ions, two opposite regimes can be identified: the heavy-particle and the fast-particle limit [699].

In the former, the spontaneous transition probability greatly exceeds the ion oscillation frequency, such that laser cooling resembles that of a free particle: the lowest achievable temperature is posed by the Doppler limit.

In the latter regime, where the opposite inequality relationship holds between the ion oscillation frequency and the spontaneous transition probability, motional sidebands appear in the emission spectrum. This gives rise to a different cooling mechanism called sideband cooling: the ion energy can be progressively reduced by tuning the laser to one of the lower sidebands. When such scheme is modified so as to become a two-photon process, the so-called Raman sideband cooling technique is realized.

The three laser cooling schemes mentioned above (applicable to both ions trapped in the Paul and Penning traps) are described in more detail in the following [699].

Laser cooling to the doppler limit

If the ion has a suitable energy-level structure, one can make use of the laser cooling techniques developed for neutral atoms. As already shown, the lowest achievable temperature using Doppler cooling is given as

$$T_{min} = \frac{\hbar\Gamma}{2k_B} \tag{7.82}$$

where Γ is the natural linewidth of the cooling transition (recall that, in order to reach this temperature, the cooling laser must be detuned by half the linewidth below resonance: $\Delta\omega = -\Gamma/2$). Typical values for T_{min} are in the mK range. In many cases, the kinetic energy reduction accomplished by Doppler cooling is enough to depress second-order Doppler shifts below the 10^{-17} level as well as to enter the Lamb-Dicke regime, where the ion motion amplitude is below the wavelength of the probe laser.

Resolved sideband cooling

Being bound by a harmonic potential, the discrete energy levels of the ion are described as

$$E = \hbar\omega_i\left(\frac{1}{2} + n_{\omega_i}\right) \tag{7.83}$$

where n_{ω_i} represents the vibrational quantum number and ω_i is the vibrational oscillation frequency of the trapped ion. Next, let us denote with ω_0 the rest frequency of the ion cooling transition. Then, as viewed in the laboratory, the absorption and emission spectrum for an atom oscillating in its confining well has resolved components at ω_0 and $\omega_0 \pm m\omega_i$ where m is a positive integer, provided that $\omega_i \gg \Gamma$. Therefore, as shown in Figure 7.29, a modified energy-level structure will originate for the ion, where each electronic level has a discrete ladder of energies corresponding to different vibrational quantum numbers. Obviously, such structure can only be observed if the laser irradiating the ion has a linewidth smaller than the separation of the energy levels and if the recoil energy satisfies the same condition. As a result, sideband cooling can be undertaken by tuning a laser to the first red sideband such that

$$\omega_L = \omega_0 - \omega_i \tag{7.84}$$

where ω_L is the laser frequency and ω_0 coincides with the difference in energy between the ground and the first electronic excited state. In this way, the ion will be lifted up to the first excited electronic state but lowered by one vibrational number; when the ion decays back into the ground electronic state, the vibrational quantum number will be conserved: this process diminishes the overall vibrational quantum number by one (the zero change in the vibrational number can be understood in analogy with the Franck-Condon principle for molecular transitions, where the strongest transition is the one that maximizes the overlap between the initial and final wavefunction). By iteratively tuning the laser to the red sideband, the ion is eventually optically pumped down to its ground vibrational state. In other words, on average, each scattered photon reduces the ion vibrational quantum number n_{ω_i} by 1, such that $\langle n_{\omega_i} \rangle \ll 1$ is obtained and the particle spends most of its time in the ground-state level of its confining potential. When $\langle n_{\omega_i} \rangle \ll 1$, T is no longer proportional to $\langle n_{\omega_i} \rangle$ but depends logarithmically on $\langle n_{\omega_i} \rangle$

$$k_B T = \frac{\hbar \omega_i}{\ln\left(1 + \dfrac{1}{\langle n_{\omega_i} \rangle}\right)} \tag{7.85}$$

Such a cooling scheme was first demonstrated on a single Hg^+ ion confined in a Paul trap (Figure 7.30) with an oscillation frequency of $\omega_i/2\pi = 2.96$ MHz [706]. First, Doppler cooling was performed by tuning a 194-nm laser onto the $^2S_{1/2} \rightarrow ^2 P_{1/2}$ transition. Then, resolved sideband cooling was carried out. For this purpose, the 194-nm radiation was switched off and the narrow $^2S_{1/2} \rightarrow ^2 D_{5/2}$ transition ($\Gamma \sim 1$ Hz) was driven using a 281.5-nm laser (note that the criterion $\omega_i \gg \Gamma$ is largely fulfilled). On the other hand, this also corresponds

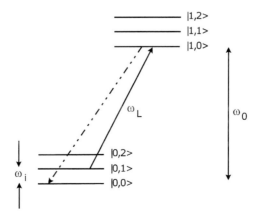

FIGURE 7.29
Modified energy levels in the resolved sideband limit. (Adapted from [699].)

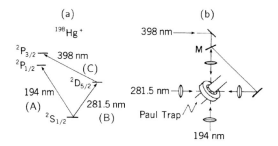

FIGURE 7.30
(a) Optical transitions involved in the Hg II sideband cooling experiment. (b) Geometrical arrangement of the laser beams, the 194-nm fluorescence being detected orthogonal to the figure plane. Mirror M reflects (transmits) the 194-nm radiation (398-nm). (Courtesy of [706].)

to a poor sideband scattering rate (on the order of six photons per second); to enhance this, a 398-nm laser was employed in order to couple the $^2D_{5/2}$ level to the $^2P_{3/2}$ one. Only the y- and z- degrees of freedom were cooled, where the values of $\langle n_y \rangle, \langle n_z \rangle = 0.051$ were reached. The latter value was estimated by measuring the asymmetry in the absorption spectrum (indeed, as $\langle n \rangle$ approaches zero, the lower sideband height is substantially reduced because there is no vibrational energy level that it can couple to); then, the ratio of the amplitudes of the lower to upper sidebands yields

$$\langle n_{\omega_i} \rangle = \frac{1}{\sqrt{1 - \dfrac{S_L}{S_U}}} - 1 \tag{7.86}$$

where S_L and S_U represent the absorption strength on the lower and upper motional sideband, respectively. With the value $\langle n_{\omega_i} \rangle = 0.051$ Equation 7.85 returns a temperature of the ion below 50 μK, corresponding to a situation where the ion is in the lowest level of the trapping potential for about 95% of the time. More recently, a similar experiment was performed using a single Ca$^+$ ion in a Paul trap [707].

Raman sideband cooling

Unfortunately, the criterion $\omega_i \gg \Gamma$ is met only in very few ions. In all other cases, Raman sideband cooling comes to the rescue. Here, a stimulated Raman transition between two ground-state sublevels is exploited to compensate for the absence of narrow single-photon transitions. To go into details, let us consider the system shown in Figure 7.31 whose levels are labelled as $|J, n\rangle$, with J (n) representing the internal (vibrational) energy levels. Then, initially in the $|1, n\rangle$ level, the ion is addressed with two laser beams whose frequencies are $\omega_{L1} = \omega_{R1} - \Delta$ and $\omega_{L2} = \omega_{R2} - \Delta + \omega_i$ where $\omega_{R1} = \omega_{0,n} - \omega_{1,n}$ and $\omega_{R2} = \omega_{0,n} - \omega_{2,n}$. The difference $\omega_{L1} - \omega_{L2}$ is such that, in each stimulated Raman transition, the ion is sent into the $|2, n - 1\rangle$ level, ceding one vibrational quantum of energy. The cooling process is then finalized by means of a third laser which addresses the transition between $|2, n - 1\rangle$ and the short-lived $|aux\rangle$ level. The resulting spontaneous Raman transition delivers the ion to the $|1, n - 1\rangle$ level. Over many transitions, on average, the vibrational quantum number is conserved by the character of the spontaneous Raman process. Therefore, by iterating this process, the ion is optically pumped into the ground vibrational state. Raman sideband cooling was first demonstrated by Monroe and co-workers in 1995 on a single Be$^+$

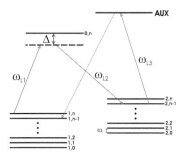

FIGURE 7.31
Illustration of the Raman sideband cooling scheme. (Adapted from [699].)

ion confined in a Paul trap, leading to the values $\langle n_x \rangle$, $\langle n_y \rangle$, $\langle n_z \rangle$ =0.033, 0.022, 0.029 [708].

7.8.4.2 Sympathetic cooling

When laser cooling is hardly applicable due to, e.g., an unfavorable energy level structure, one can resort to the so-called sympathetic cooling method. Here, the ion of interest is cooled through collisions with an auxiliary species that is accessible to laser cooling. Originally carried out to cool different isotopes of the same ionic species (^{26}Mg$^+$ and ^{26}Mg$^+$ isotopes) by laser cooled ^{24}Mg$^+$ ions [709], sympathetic cooling has also been accomplished between different species. In the first realization, ^{198}Hg$^+$ ions were sympathetically cooled in a Penning trap by the use of laser cooled ^{9}Be$^+$ ions [710]. More recently, sympathetic cooling of a single trapped ^{174}Yb$^+$ ion by a Bose-Einstein condensate of ^{87}Rb was observed [711].

7.8.5 Spectroscopy of trapped particles in the Lamb-Dicke regime

As first pointed out by Dicke, Doppler-free spectroscopy is feasible if the spatial excursions of an atom are constrained to less than the wavelength λ of the probed transition [712]. While in the rf and microwave domain such a Lamb-Dicke criterion is readily met by virtue of the large radiation wavelengths, an extremely tight confinement of the particles is required at shorter wavelengths, which most often introduces other perturbations that are harmful to the line narrowing. In this frame, a happy haven is represented by ion traps where, simply undergoing a harmonic motion, the particles are virtually constrained without disturbance. For example, the Lamb-Dicke condition can be safely fulfilled in the optical domain for a single, laser-cooled ion that is tightly confined in a miniature rf Paul trap (Coulomb repulsion between ions makes it difficult, but not impossible, to meet the Lamb-Dicke criterion when more than one ion is in the trap).

In order to better understand the spectral narrowing attainable by this method, let us transform the electric field vector into the rest frame of the atom [713]

$$E(\boldsymbol{r}(t), t) = \boldsymbol{E}_0 \cos(\omega t + \boldsymbol{k} \cdot \boldsymbol{r}(t) + \phi) \tag{7.87}$$

where $|\boldsymbol{k}| = 2\pi/\lambda$. Now, if $|\boldsymbol{r}(t)| \ll \lambda$, then $\boldsymbol{k} \cdot \boldsymbol{r}(t) \to 0$ regardless of the relative direction of motion. So, the atom begins to accumulate (lose) phase during the first half of its cycle, but before gaining (losing) π radians and losing coherence with the field, it reverses its direction and returns to the original phase setting. Hence, the only significant response of the atomic system occurs at $\omega = \omega_0$. Concerning the second-order Doppler shift, when the ion is laser-cooled to about 1 mK (that is near the Doppler-cooling limit for most ions using a strongly allowed transition), it amounts to few parts in 10^{18}. Moreover, since the ion is confined in a region of nearly zero field, other shifts in the resonance frequency owing to electric and magnetic components could be at the same level. In spite of these nice features, however, the signal-to-noise ratio from a sample consisting of just one ion is generally poor. To overcome this drawback, the so-called electron shelving technique (EST) has been demonstrated [714]. Such a scheme can be understood as follows. Consider an ion that has both a strongly allowed and a weakly allowed transition sharing the ground state (see Figure 7.32). Next, suppose that, once trapped and cooled, the ion is initially in the ground state; then, radiation at a frequency near the weak resonance is pulsed on, possibly causing a transition to the long-lived upper level. After that, light with a frequency near that of the strong transition is pulsed on. If the atom has made a transition in the previous step, no fluorescence will be observed; otherwise, fluorescence will be observed at an easily detectable level. Essentially, the idea is that detecting the presence or absence of fluorescence from the strongly allowed transition (which can scatter as many as 10^8 photons per second) is much more manageable than detecting the one photon that is eventually emitted when the metastable state decays. More importantly, a transition with a narrow natural linewidth can be detected with nearly unit efficiency by the presence or absence of the strong fluorescence. In other words, the achievable signal-to-noise ratio is limited only by the quantum statistical fluctuations in making the weak transition.

In the EST, the line profile of the clock transition is retrieved from the number of quantum jumps detected in a given number of interrogation periods, as the probe laser frequency is swept across the resonance. Stabilization against the reference transition is then obtained by repeatedly stepping the probe laser frequency between the two estimated half-intensity points and measuring the imbalance in the quantum jump rate; from such imbalance a correction signal for the probe laser can also be derived [715].

With this powerful technique, J. Bergquist et al. were able to obtain the fully resolved recoilless optical resonance and motional sidebands of the narrow $S - D$ transition at $\lambda = 282$ nm on a single, laser-cooled $^{198}Hg^+$ ion confined in a miniature rf Paul trap. Each single-photon transition to the electric-quadrupole-allowed metastable D state was detected with nearly unit efficiency by monitoring for the presence (no transition made) or absence (weak transition made) of fluorescence from the strongly allowed $S - P$ transition at 194 nm. The fractional resolution of this spectrum already exceeded $3 \cdot 10^{-11}$ [716]. More recently, the spectral linewidth observed for this transition, limited by the frequency fluctuations in the probe laser, was less than 80 Hz, corresponding to a fractional resolution of less than 10^{-13}. Laser cooling to the zero-point energy of the trap's harmonic well reduced the spatial excursions of a trapped ion to less than 2.5 nm [706].

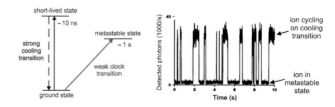

FIGURE 7.32

In the electron shelving technique, a single ion is detected by monitoring the fluorescence signal from the strong cooling transition. Each time the ion is excited to the upper (metastable) state of the clock transition, the fluorescence disappears until the ion decays back to the ground state. (Courtesy of [715].)

7.9 Microwave atomic standards

Having finished this digression on cooling/trapping of atoms/ions, we now attack atomic frequency standards. At first sight, these would appear exempt from the many factors that afflict the performance of artifact-based standards, as the transition frequency is fixed, in principle, solely by the atomic structure. While this is true to a great extent, the details of the interaction between the atoms and the probe as well as the physics/electronics interplay somehow influence the output frequency, thus misting the clear line of demarcation with a frequency standard defined by an artifact [650].

7.9.1 Metrological properties of the active hydrogen maser

An example of an active atomic standard is provided by the Active Hydrogen Maser (AHM). While the general working principle was already discussed in Chapter 4, here we focus on more technical aspects as well as on the frequency accuracy and stability. In the following we will neglect Passive Hydrogen Masers which are characterized by a poorer metrological performance.

7.9.1.1 Maser design

A modern, high-performance (commercial) system can be schematically described as follows [717]. First of all, molecular hydrogen is supplied (under electronic servo control) by a compact storage bottle; molecules are then dissociated into atoms in a discharge bulb. The atoms emerge from such source through a collimator represented by a small elongated hole. Then, a magnetic state selector prepares an atomic beam in the proper quantum state and directs it towards a Teflon-coated quartz storage bulb. The latter is surrounded by a right cylindrical microwave cavity (working on the TE_{011} mode), resonant at the hydrogen transition frequency, where the maser action is stimulated. The microwave signal from the cavity is coupled (via a small loop connected with a coaxial cable) to a low-noise heterodyne receiver (mixer + high-resolution frequency synthesizer) whose output is sent to a phase-locked loop. This latter eventually locks a voltage controlled crystal oscillator (VCO) to the maser output (Figure 7.33). To further enhance short-term stability, a higher-quality crystal oscillator is also employed. Temperature-controlled multipliers, dividers and buffer amplifiers provide several isolated outputs at standard frequencies. In order to minimize systematic perturbations of the maser output frequency, some precautions must be taken:

- Sputter-ion pumps preserve high-vacuum conditions and remove the hydrogen in excess;

- The cavity is magnetically shielded and isolated from external temperature variations;

- The *C-Field* (i.e., the internal magnetic field) is controlled by means of an axial magnetic field coil wound on the inside of the first shield;

- An automatic frequency control system is incorporated to hold the cavity at a constant frequency relative to the hydrogen emission line (see below).

7.9.1.2 Frequency shifts

Generally, due to several effects, the actual frequency of a hydrogen maser does not coincide exactly with that of the ground state splitting of the unperturbed hydrogen atom [65, 7, 718]. The first three contributions that we shall consider are common to all types of atomic frequency standards, whereas the last three are specific to the hydrogen maser.

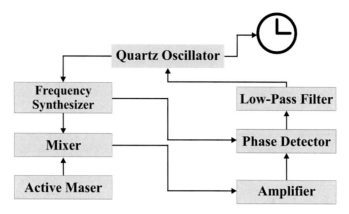

FIGURE 7.33
Block diagram of an active hydrogen maser. (Adapted from [3].)

Second order Zeeman effect (SOZE) - For weak magnetic fields, the clock transition frequency exhibits a quadratic Zeeman effect

$$\Delta\nu_{B^2} = 2.7730 \cdot 10^{-1} \mathrm{Hz} \left(\frac{B}{\mu\mathrm{T}}\right)^2 \tag{7.88}$$

corresponding to a fractional shift of $\Delta\nu_{B^2}/\nu_0 \simeq 2 \cdot 10^{-12}$ for a typical value of $B = 0.1$ μT. Frequency shifts induced by changes in the magnetic field within the storage bulb are ordinarily controlled by employing multiple layers of permalloy. For better performance, however, active compensation is needed. In one valuable approach, leakage of the external magnetic field through the outer shielding layer is sensed by a magnetometer and feedback is provided via a coil wound around the next inner layer. In practice, a fractional uncertainty below 10^{-13} is obtained in the value of this SOZE shift.

Second order Doppler effect - The second order (SO) Doppler shift is given by

$$\frac{\Delta\nu_{SO}}{\nu_0} = -\frac{(3/2)k_B T}{mc^2} \tag{7.89}$$

where the numerator represents the average kinetic energy of the hydrogen atoms which thermalize by impact at the storage cell wall temperature, T. For a typical operational temperature $T = 313$ K, this shift amounts to $-4.3 \cdot 10^{-11}$ and keeping the bulb temperature constant within 0.1 K leads to a fractional uncertainty of $1.4 \cdot 10^{-14}$.

Cavity pulling - As already mentioned, cavity pulling occurs when the microwave cavity eigenfrequency is not tuned exactly to the atomic resonance (see Equation 4.34). To reduce this effect, hydrogen masers are equipped with one or another method of automatic tuning, leading to an uncertainty less than 10^{-14} (see next section).

Wall shift - In an active maser, the e.m. field inside the resonant cavity induces in the hydrogen atom a dipolar magnetic moment (DMM) which then maintains the e.m. field itself. Collision with the storage bulb wall modifies the phase of the induced DMM. This originates a frequency shift of the form $\Delta\nu_w = K/D$, where K is a constant depending on the coating properties and the temperature and D is the bulb diameter. As an example, in the case of FEP Teflon 120, at $T = 313$ K, we have $K \simeq -0.4$ Hz·cm; for $D = 15$ cm this yields $\Delta\nu_w/\nu_0 \simeq -2 \cdot 10^{-11}$.

Spin-exchange collisions - As already mentioned, collisions between hydrogen atoms in the *clock states* cause both a frequency shift and a broadening of the atomic resonance. In turn, such additional broadening may give rise to a frequency shift via the cavity pulling. As we will see in a short while, suitable auto-tuning cavity methods can almost completely compensate for these effects, thus reducing the residual spin-exchange incidence down to a few times 10^{-13}.

Magnetic-inhomogeneities relaxation - The magnetic field at the storage bulb may contain undesired radial components. Accordingly, atoms moving through such a field may experience a radiation spectrum at the frequency corresponding to transitions of the $\Delta m_F = \pm 1$ type. This induces relaxation among the populations of the energy levels and influences the maser coherence. However, by properly designing the solenoid and the magnetic shields, this kind of relaxation can be suppressed below the 10^{-13} level for $B = 0.1~\mu$T. A crucial parameter is the size of the neck hole in the shields: the smaller it is, compatible with the required pumping speed, the more one reduces the probability of magnetic-inhomogeneity relaxation.

7.9.1.3   Automatic tuning of the resonant cavity

As already explained, a strict limitation is imposed on the cavity tuning frequency, if the oscillation frequency has to stay constant. Whatever the adopted scheme, an error signal is derived that is proportional to the detuning of the cavity relative to the value which affords an oscillation frequency coinciding with the atomic transition one. Then, feedback is provided by some transducer coupled to the cavity. In the following we only discuss the cavity frequency-switching servo, which provides the best metrological performance [719].

Cavity frequency-switching servo

With reference to Figure 7.34a, let us denote with f_0 the reference for the cavity servo, namely the hydrogen emission line. The cavity frequency is switched continuously between the two resonance values with a difference f_w. If the cavity average frequency, f_c, deviates from f_0, then a modulation voltage Δv is impressed on the maser output signal, which is next sent to the servo electronics in order to remove the cavity frequency offset. The design of the cavity frequency control system is shown in Figure 7.34b. The modulation period generator (MPG) biases with a square-wave modulation voltage a varactor (i.e., a voltage-variable capacitor diode) connected in series with a coupling loop within the maser cavity. This produces the two frequencies between which the cavity is switched. A signal from the MPG is also sent to the synchronous detector (demodulator and bandpass filter circuit). A second coupling loop delivers the maser signal from the cavity to the receiver. Here, after proper amplification and filtering, the modulation signal is detected and sent to the other input of the demodulator. The output of the latter is an *up* or *down* control signal which is passed to the cavity frequency control register. This is a digital integrator whose output voltage biases a second varactor in another coupling loop, thus controlling the average frequency of the cavity. Figure 7.34b also shows the servo which independently controls the cavity temperature. The above cavity frequency-switching method gives three distinct advantages: first, its operation needs no other stable frequency references; second, unlike conventional automatic spin-exchange tuning, it requires no beam intensity switching, which preserves the maser short term stability (phase noise); third, it enables the maser to deliver long-term stability normally attributed to the most stable of cesium atomic standards.

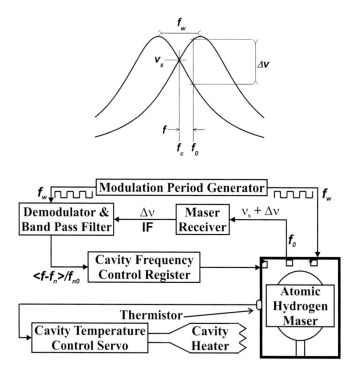

FIGURE 7.34
Principle and system block diagram of cavity frequency-switching servo. (Adapted from
[719].)

7.9.1.4   Frequency stability

The frequency instability of an active hydrogen maser, as characterized by the Allan deviation, exhibits distinct trends for different sampling times τ [65, 7]. For $\tau < 0.1$ s, it is basically determined by that of the quartz oscillator servo-controlled by the atomic resonance. The short-term instability (between 0.1 and 10 s) can be described by a $1/\tau$ behavior, whereas the medium-term one (for τ greater than about 10 s) is well represented by a $1/\sqrt{\tau}$ dependency. For longer times (between 1000 and 10000 seconds), the Allan deviation reaches a flicker floor (below the 10^{-15} level). After that, primarily due to the cavity drift, the frequency instability rises again. Both the flicker-floor level and the subsequent increase are significantly improved by the cavity frequency-switching method discussed above.

The medium-term instability basically originates from the perturbation induced by the thermal radiation field on the e.m. field built up by the stimulated emission. Since such disturbance contributes with random phase to the excited mode of the resonator, the resulting frequency noise is white and the corresponding Allan deviation is given by

$$\sigma_y(\tau) = \frac{1}{Q_{at}} \sqrt{\frac{k_B T}{2 P_{at}}} \frac{1}{\sqrt{\tau}} \tag{7.90}$$

where Q_{at} is the quality factor of the atomic resonance and P_{at} the power dissipated in the cavity.

In the short-term regime, additional white phase noise, mainly arising from fluctuations in the resonator length and in the phase of the electronic circuit (the amplifier above all),

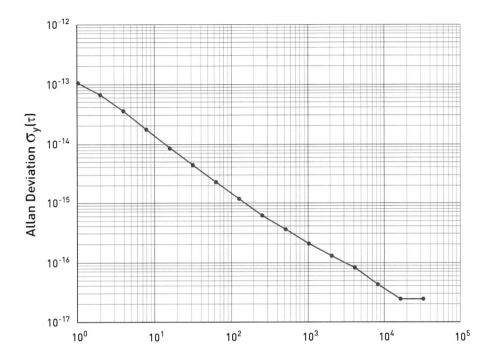

FIGURE 7.35
Frequency stability plot of a high-performance, commercial active hydrogen maser with
auxiliary output generator [717].

contributes to the signal. The associated Allan deviation can be expressed as

$$\sigma_y(\tau) = \sqrt{\frac{3k_B T f_h}{\pi^2 \nu_0^2 P_{at}} \left(1 + F\frac{P_{at}}{P_r}\right)\frac{1}{\tau}} \tag{7.91}$$

where P_r is the power received by the amplifier, F the noise factor of the latter, and
the cut-off frequency f_h defines the bandwidth of the equipment employed to measure the
frequency fluctuations.

The Allan deviation of a high-performance, commercial active hydrogen maser (AHM)
is shown in Figure 7.35. By virtue of their excellent stability for periods from about 10 s to
a day, AHMs are generally used as so-called *flywheels* for Cs cold-atom fountains (see next
section).

7.9.1.5 Cryogenic hydrogen masers

Since both the short- and medium-term frequency instabilities increase with the temperature, several efforts have been made to cool the resonant cavity down to few Kelvin [7].
A further motivation has been provided by the observation that recombination and relaxation of the hydrogen atom are strongly lessened on a sub-Kelvin helium film. In the same
configuration, furthermore, while the hyperfine resonance broadening induced by collisions
between hydrogen atoms is three orders of magnitude lower than at room temperature (at
equal densities), the frequency shift due to collisions on the film is of the same order of magnitude as in room-temperature-operated masers. This can notably enhance both the atomic
quality factor and the power dissipated in the cavity, which would contribute to reducing

the maser frequency instability as well. On the basis of these considerations, a fractional instability in the 10^{-18} range was predicted for a maser at 0.5 K. However, this assumed that previous experimental findings and theoretical results would still apply at cryogenic temperatures. Unfortunately, this did not prove true. First, at 0.5 K, the helium pressure causes a drastic reduction in the mean free path of the hydrogen atoms (down to 1 cm); as a consequence, the atoms no longer average out the spatial variations in the magnetic fields. Second, the relative velocities of the hydrogen atoms are smaller at low temperatures, which increases the collision time between the atoms themselves, thus leading to additional frequency shifts. These problems, in conjunction with the technical hurdles encountered to practically implement cryogenic masers, have relegated their use almost exclusively to basic research [720, 721, 7].

7.9.2 Cesium clocks

Several properties made cesium a good choice as the atomic resonance source for a primary frequency standard [6, 5]. Similar to mercury, cesium is a soft, silvery-white ductile metal that becomes a liquid at about 28.4 °C. Being fairly heavy (133 amu), cesium atoms move at a relatively slow speed (about 130 m/s at room temperature); this allows cesium atoms to stay in the interaction zone longer than hydrogen atoms, for example, which travel at a speed of about 1600 m/s at room temperature. Cesium also possesses a relatively high hyperfine frequency (9.2 GHz) when compared to rubidium (6.8 GHz) and hydrogen (1.4 GHz). The only stable (natural abundance 100%) isotope ^{133}Cs has a nuclear spin $I = 7/2$ which, combined with the total spin $J = 1/2$ of the electron shell, gives rise to the two hyperfine states $F = I + J = 4$ and $F = I - J = 3$. In a magnetic field $\boldsymbol{B}$, these split into 16 components (Figure 7.36). The Cs clock utilizes the transition with the lowest sensitivity to magnetic fields $|F = 4, m_F = 0\rangle \equiv |4, 0\rangle \leftrightarrow |F = 3, m_F = 0\rangle \equiv |3, 0\rangle$. The frequency of the corresponding levels is given by the Breit-Rabi formula [163]

$$\nu\big|_{F=I\pm\frac{1}{2},m_F\rangle} = \frac{-\nu_0}{2\,(2I+1)} + \frac{\mu_B g_I m_F B}{h} \pm \frac{\nu_0}{2}\sqrt{1 + \frac{4m_F x}{(2I+1)} + x^2} \tag{7.92}$$

with

$$x = \frac{(g_J - g_I)\,\mu_B B}{h\nu_0} \tag{7.93}$$

$g_J \simeq 2$ and $g_I \simeq -0.0004$ being the appropriate fine structure Landé factor and the nuclear g-factor, respectively. From this, one can see that the $|F = 4, m_F = 0\rangle$ state is a low-field-seeking (lfs) state, while the $F = 3, m_F = 0\rangle$ one is a high-field-seeking (hfs) state. Then, the frequency separation between such levels is given by

$$\Delta\nu_{|4,0\rangle\leftrightarrow|3,0\rangle}(B) \simeq \nu_0 + \frac{\nu_0}{2}x^2$$

$$= \nu_0 + 4.2745\cdot 10^{-2}\ \text{Hz}\left(\frac{B}{\mu\text{T}}\right)^2 \tag{7.94}$$

where ν_0 (9192631770 Hz) is the hyperfine separation in zero field between the states $F = I + 1/2$ and $F = I - 1/2$ (Figure 7.37).

7.9.2.1 Cesium-beam frequency standards

Figure 7.38 schematically illustrates the setup for a cesium beam frequency standard, which can be traced back to the seminal work by Rabi and Ramsey [6, 722, 65]. The atoms effuse into a high vacuum chamber through a nozzle from an oven, heated to a temperature of

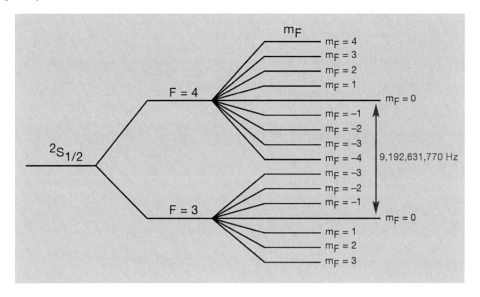

FIGURE 7.36
Schematic diagram of the cesium clock transition. (Adapted from [6].)

about 100 °C or higher. The small energy separation between the $F = 3$ and $F = 4$ states implies that such levels are about equally populated in the thermal beam. In the state-preparation region, an inhomogeneous field, generated by a Stern-Gerlach magnet, spatially separates atoms in the various m_F states, and atoms in one of the ground-state levels ($|3, 0\rangle$ or $|4, 0\rangle$) are directed towards the Ramsey cavity. However, due to the velocity spread in the atomic beam, this separation is imperfect, so some atoms in other m_F states are mixed in with the ground-state atoms that go through the cavity. Obviously, this state-

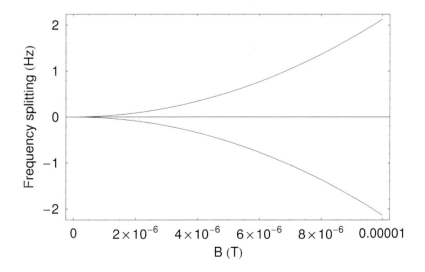

FIGURE 7.37
Frequency splitting for the $|4, 0\rangle$ (upper curve) and $|3, 0\rangle$ (lower curve) states in ^{133}Cs. The transition frequency between these two states is used to define the second. The frequency splittings are defined here as $\nu(|4, 0\rangle, B) - \nu(|4, 0\rangle, B = 0)$ and $\nu(|3, 0\rangle, B) - \nu(|3, 0\rangle, B = 0)$.

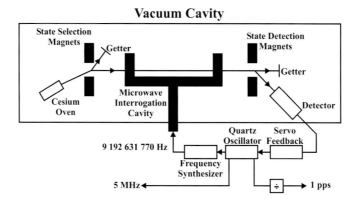

FIGURE 7.38
Schematic setup of an early cesium beam frequency standard. (Adapted from [6].)

preparation method entails the rejection of most of the atoms entering the system. While inside the Ramsey cavity, the cesium beam is exposed to the output signal from a quartz-based microwave synthesizer whose frequency is tuned to the desired atomic resonance ($\nu_0 = 9192631770$ Hz). In this way, some of the atoms will change their magnetic state. In the state-detection region, a second, identical Stern-Gerlach magnet is employed to channel to a hot-wire (Langmuir-Taylor type) detector only those atoms that have been stimulated by the microwave field to the other ground, $m_F = 0$ level. Then, the central feature in the Ramsey fringe pattern, with its maximum at ν_0, is used to stabilize the VCXO frequency against the atomic transition one. For this purpose, the signal from the detector is sent to a servo feedback circuit which steers the quartz oscillator frequency so as to maximize the number of atoms reaching the detector itself. Finally, standard output frequencies, such as 1 Hz, 5 MHz, and 10 MHz, are derived from the locked quartz oscillator and used as reference signals.

Inside the Ramsey cavity, a constant magnetic field (denoted as C field) is applied to separate energetically the otherwise degenerate magnetic sub-levels, thus allowing the selective excitation of the clock transition. It is worth pointing out that, in order to excite the $\Delta m_F = 0$ magnetic dipole transition, the Ramsey field must oscillate parallel to the C field axis (so as to have zero component of angular momentum along it), which fixes the mutual orientation of the microwave cavity and the coils generating the C field. Figure 7.39 shows the typical arrangement in a primary laboratory standard. Concerning the C field value, it has to be large enough to isolate the clock transition from the other ones; however, for a larger value, the clock frequency is affected to a bigger extent by fluctuations in the magnetic field itself: typically, a value in the μT range is used. In addition, due to the dependence of the clock-transition frequency on the magnetic field, efficient shielding against the ambient magnetic field (and its associated fluctuations) is also implemented. Furthermore, as a consequence of the second-order Zeeman effect experienced by the atoms in the C field region, the center frequency of the Ramsey feature generally deviates from the unperturbed transition frequency. Thus, to guarantee that the VCXO output frequency represents the exact SI value, the value of the applied C field must be accurately known.

Different thermal-cesium-beam devices were developed over an approximate 40-year period (1959-1998) and served as national primary frequency standards. Based on this design, Cs atomic clocks are also commercially available from several manufacturers. One of the most significant improvements made to thermal beam frequency standards consisted of using narrow-linewidth lasers for state selection and detection, thus replacing the magnets and

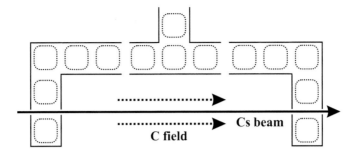

FIGURE 7.39
Cross section of a typical Ramsey resonator showing the C field (generated by a solenoid) and the magnetic field lines of the standing rf wave. (Adapted from [65].)

the hot-wire detector found in the earlier standards. Indeed, using lasers in place of magnets offers several distinct advantages. First, instead of merely filtering out atoms that are in the wrong energy state, as it happens in magnetic selection, the lasers optically pump as many atoms as possible into the desired energy state, which generates a much stronger signal. An example of such improved configuration is represented by the NIST-7 standard, now residing in the NIST museum in Boulder. In spite of the reduced Ramsey cavity length (just 155 cm) and the not so narrow resonance width (65 Hz), NIST-7 eventually reached an accuracy of $5 \cdot 10^{-15}$ (below a nanosecond per day), nearly 20 times better than its predecessor. In the most widely adopted approach, as illustrated in Figure 7.40, 852.355-nm-wavelength laser radiation is tuned to excite atoms from the $F = 4$ to the $F' = 3$ state. After about 30 ns, these excited atoms decay back (by spontaneous emission) into the $F = 4$ and $F = 3$ hyperfine-split ground states; since only atoms in the $F = 4$ state are excited repeatedly, after a few cycles, all population is optically pumped into the $F = 3$ state. Moreover, if the configuration shown in Figure 7.41 is implemented, where the optical-pumping beam is perpendicular to the atomic one, then the excitation is not velocity selective. As opposed to the restricted angle of acceptance in the magnetic selector, this gives rise to an atomic beam that has a higher number of atoms and is more homogeneous spatially. Finally, the absence of intense magnetic gradients associated with the Stern-Gerlach selectors, strongly suppresses Majorana transitions. At the output of the Ramsey cavity, atoms in the $F = 4$ state can be detected by monitoring the fluorescence signal originating from the $F = 4 \rightarrow F' = 5$ laser excitation and the subsequent spontaneous decay. By virtue of the quantum mechanical selection rule $\Delta F = 0, \pm 1$, this is a cycling transition (i.e. atoms excited to the $F' = 5$ level can decay only back to the $F = 4$ state), so that the excitation-followed-by-emission process can take place many times. As a result, even for a low detection probability, a fair number of fluorescence photons is readily obtained, thus enabling to detect virtually each excited atom. A drawback related to optically-pumped frequency standards may be represented by the presence of spurious light from the pumping/detection region along the interrogation path of the atomic beam. Via the ac Stark effect, such radiation can introduce a light shift, which is experimentally investigated by changing the laser power.

The various sources of frequency shift in Cs standards will be discussed directly in the frame of fountain-based ones.

7.9.2.2   Cesium fountain clocks

The short-term stability of NIST-7 was typically $\sigma_y(\tau) = 7 \cdot 10^{-13} \tau^{-1/2}$, meaning that frequency measurements with uncertainties near 10^{-15} (one standard deviation) could be made

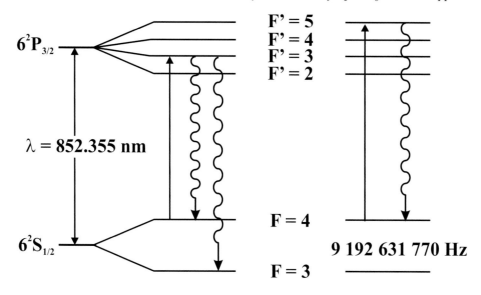

FIGURE 7.40
Optical pumping for state preparation in a Cs atomic clock.

in about 10 hours. As mentioned above, the central spectral feature in the Ramsey pattern had a linewidth $\Delta\nu$ of about 65 Hz corresponding to $Q_{at} \equiv \nu_0/\Delta\nu \simeq 1.5 \cdot 10^8$. However, the performance of optically-pumped thermal-beam standards was still fundamentally limited by the short interaction time (about 10 ms) resulting from the high velocity of the cesium atoms. First introduced by Zacharias in the 1950's, the concept of a fountain standard aimed precisely at increasing the interaction time. The original idea was to implement the usual Ramsey's two-pulse interaction scheme in a vertical Cs beam standard with just one Ramsey zone [6]. In such a configuration, slow atoms from the Cs oven would go through the interaction zone while travelling upward, stop and invert their direction under the gravity influence, and eventually go through the same interaction zone while travelling downward.

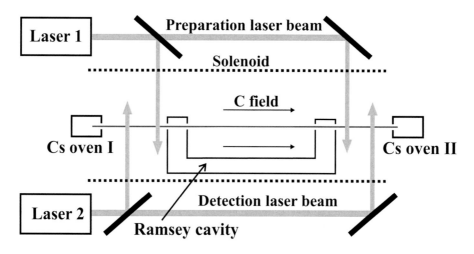

FIGURE 7.41
Setup of a Cs clock with optical state selection and detection. (Adapted from [65].)

With a ballistic flight of only a meter upwards, the interaction time would be enhanced to nearly 1 s, corresponding to a very narrow Ramsey resonance (<1 Hz). Unfortunately, Zacharias' device did not work. Actually, all of the slow atoms were scattered out of the beam by the fast atoms that overtook them. About thirty years later, Zacharias' premature intuition was revived by Steven Chu at Stanford University. As we will see in a short while, application of the laser cooling/trapping techniques discussed in previous sections played a crucial role to overcome the mentioned difficulties. Chu's group built fountains first with sodium and then with cesium atoms, although neither device was employed as a primary frequency standard. Researchers at the Bureau National de Metrologie - Systemes de Reference Temps Espace (BNM-SYRTE) in Paris later realized the first fountain-based primary frequency standard. Since then, several other metrology laboratories worldwide have constructed laser-cooled Cs fountain standards.

Figure 7.42 shows schematically the key elements of fountain standards [723, 6, 163, 65]. Four main tasks must be accomplished on the atoms:

1. They must be cooled and stored in a source, and then ejected upwards;

2. They must be optically pumped into the lower of the two ground-state hyperfine substates ($F = 4$);

3. They must frankly trace their vertical trajectories in a weak uniform magnetic field (the C field), interacting with the Ramsey field only at the beginning and end of their flight;

4. Population in each of the two hyperfine sublevels ($F = 4$ and $F = 3$) must be monitored to derive the frequency stabilization signal.

The first two task are performed in the laser-cooled source, while the phase-coherent Ramsey excitation is carried out sequentially in time, as the atoms go through the same microwave cavity on their way upwards and then downwards. Finally, the transition probability measurement is accomplished optically hinging on the detection of resonance fluorescence from the cycling $F = 4 \leftrightarrow F' = 5$ transition. The entire fountain cycle takes around 1 s. In the following, a more detailed description is given.

In the cold-atom source, six independent laser beams are used to form either a MOT or optical molasses (OM), in which Cs atoms from a $\approx 10^{-8}$-torr-pressure vapor are captured and cooled. When the starting point is a MOT, the quadrupole field is switched off after few hundreds of milliseconds, giving way to a subsequent OM stage where the atoms are further cooled. Conversely, when starting directly with an OM, the lin⊥lin polarization configuration is used. This reduces the number of cooled atoms by roughly a factor of ten, but gives three significant advantages: first, as a result of the increased size, the atomic cloud is less dense, which considerably reduces the collisional frequency shift; second, a larger cloud realizes a more efficient filling of the microwave cavity aperture, which originates a more homogeneous trajectory distribution thus improving compensation of transverse cavity phase gradients; third, in contrast to MOTs, the atomic cloud size is independent on the number of loaded atoms, which strongly simplifies the determination of the collisional frequency shift. After the MOT/OM stage, the atomic cloud is launched upward at 3-4 m/s. This is conveniently accomplished in a moving molasses where the frequency ν_1 (ν_2) of the downward (upward) pointing vertical laser beam is red(blue)-detuned by $\delta\nu$ with respect to the frequency ν utilized to cool the atoms in the OM. More in detail, by denoting with z the upward direction, the superposition of two counter-propagating waves takes the form

$$E(z,t) = E_0 e^{i[(\omega+\delta\omega)t - kz]} + E_0 e^{i[(\omega-\delta\omega)t + kz]}$$
$$= 2E_0 e^{i\omega t}\cos(\delta\omega t - kz) = 2E_0 e^{i\omega t}\cos(2\pi\delta\nu t - 2\pi z/\lambda) \quad (7.95)$$

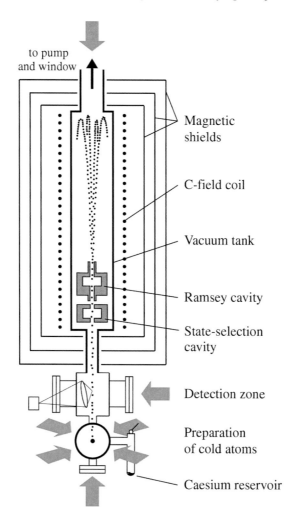

to pump
and window

Magnetic
shields

C-field coil

Vacuum tank

Ramsey cavity

State-selection
cavity

Detection zone

Preparation
of cold atoms

Caesium reservoir

FIGURE 7.42
Simplified setup of an atomic fountain clock. (Courtesy of [724].)

whereupon the atoms move upwards with the speed of the phase fronts given by $v = z/t = \lambda \cdot \delta\nu$. During this launch, an optimized polarization-gradient cooling scheme is also carried out, where the intensity of the cooling lasers is ramped down within a millisecond to about 0.5 mW·cm^{-2}, accompanied by a simultaneous increase of the detuning $(10\Gamma - 12\Gamma)$. Summarizing, in typically 500 ms, 10^5-10^9 atoms are collected, launched upwards and cooled down to below 1 μK. Then, the laser light is completely extinguished to avoid disturbing the cesium atoms along their ballistic-flight path. The *ball* of launched atoms (about 1 cm in diameter) is almost exclusively in the $F = 4$ ground state, but all m_F sublevels are populated. State selection is then performed in a microwave state-preparation cavity by use of a π-pulse at 9.192 GHz which drives the $|4, 0\rangle \rightarrow |3, 0\rangle$ transition, followed by a short optical blast which removes the other $F = 4$ atoms. Then, the remaining Cs atoms, all in the $|3, 0\rangle$ state, enter a cylindrical TE_{011} microwave cavity. The latter is typically made of copper with a Q of $\simeq 10000$ and tuned to 9.192 GHz. In passing up through such cavity, the atoms experience a first $\pi/2$ pulse. After reaching apogee about 1 m above the microwave cavity, the atomic clouds begins to fall due to gravity. On the way down, the atoms go through the cavity a second time (about 0.5 s after their first passage) and experience a second Ramsey $\pi/2$ pulse.

The detection region is generally located above the cooling one. The population N in each of the $|3,0\rangle$ and $|4,0\rangle$ states is measured according to the following steps:

- In their fall, the atoms cross a transverse standing wave formed by retro-reflecting on a mirror a σ^+-polarized, $\simeq$1-mW/cm^2-intensity probe beam, typically tuned half a natural linewidth below the $(^6S_{1/2} \ F = 4) \longrightarrow (^6P_{3/2} \ F' = 5)$ cycling transition. In such a standing-wave configuration, unidirectional light pressure is suppressed, which avoids accelerating the atoms. The resulting fluorescence-light pulse (a few milliseconds of duration) is recorded by a low-noise photodiode whose time-integrated signal is proportional to the population in the $|F = 4, m_F= 0\rangle$ state. A collection efficiency of about 0.8% is typically realized, corresponding to a detection of few hundred photons per atom;

- Then, these atoms are pushed away by the radiation pressure of the travelling wave that is realized by blocking the retro-reflected probe beam;

- After few ms, atoms in the $|F = 3, m_F= 0\rangle$ state cross two superimposed laser beams. The first one, resonant with the $(^6S_{1/2} \ F = 3) \longrightarrow (^6P_{3/2} \ F' = 4)$ transition, rapidly pumps the atoms into the $F = 4$ state. Then, the second beam, having the same parameters as the aforementioned probe beam, gives rises to a second fluorescence pulse which, detected by a second photodiode, now yields the population in the $|F = 3, m_F= 0\rangle$ state.

Thus, in each fountain cycle, the signals from the two photodetectors are combined to yield the ratio

$$p \propto \frac{N\,(F = 4)}{N\,(F = 3) + N\,(F = 4)} \tag{7.96}$$

Being normalised to the total number of atoms, such signal p is largely independent of shot-to-shot fluctuations in the atom number. It is used for the frequency stabilization according to the following procedure. The synthesizer output frequency, feeding the Ramsey cavity, is square-wave modulated from shot to shot of the fountain cycle, such that the atoms are probed alternately on each side of the central Ramsey fringe. The resulting two transition probabilities, p_{left} and p_{right}, are then compared to each other. If $p_{left} \neq p_{right}$, then correction is applied to the synthesizer frequency. This gives the frequency deviation of the fountain relative to reference. A typical Ramsey pattern for NIST-F1 is shown in Figure 7.43, where the upper frame corresponds to the envelope while the lower one displays an expanded section around the central Ramsey fringe. The linewidth of the latter is around 1 Hz corresponding to $Q_{at} \simeq 10^{10}$. It is worth noting that the large number of fringes originates from the narrower velocity distribution.

We close this section with a few technical considerations:

- In order to depress the extent of cold-atom loss via collisions with background gas as well as the degree of detected stray fluorescence, pressure must be kept below 10^{-10} Torr in the microwave, free-flight and detection regions;

- A highly homogeneous C field (defining the quantization axis) of about 1 mG is needed; this is accomplished by means of a solenoid plus a set of compensation coils which surround the microwave and free-flight region. Environmental magnetic fluctuations are mitigated by a first mu-metal layer, wrapped around the whole experiment, plus 3 extra layers surrounding the free-flight zone. In addition, an active servo system, driving a set of coils inside the outermost shield, is also implemented. As a result, the residual fluctuations in the interrogation region are kept below 0.1 μG;

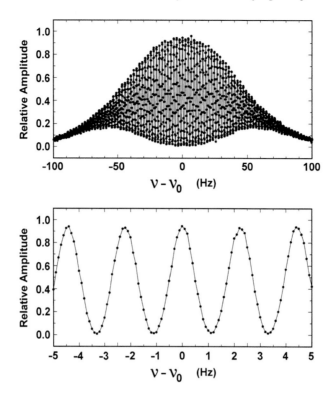

FIGURE 7.43
Typical Ramsey pattern for an atomic fountain clock. (Courtesy of [722].)

- Commonly referred to as the local oscillator (LO), the interrogation field feeding the Ramsey cavity is synthesized by a low-phase-noise frequency chain. The ultimate metrological performance of an atomic fountain is inherently related to the phase noise, spectral purity and phase stability of this LO. With state-of-the-art (BVA) quartz oscillators, the typical frequency-stability of a Cs fountain is limited to $10^{-13}\tau^{-1/2}$ (with τ being the averaging time in seconds), which makes the achievement of the 10^{-16}-accuracy goal a real challenge (actually, a single measurement would imply an averaging time of longer than one week). This limitation was overcome by the realization of an advanced flywheel oscillator consisting of an ultrastable cryogenic sapphire oscillator in conjunction with an H maser [725]. Here, starting from the low phase noise 11.932-GHz output signal of the CSO, a 100-MHz tunable signal was synthesized that was weakly phase-locked to the H-maser output signal with a time constant of about 1000 s. This latter time value corresponded to the intersection between the two Allan deviation trends (Figure 7.44). Then, by proper multiplication and mixing of the CSO frequency with the frequency of a synthesizer, the 9.2 GHz microwave signal for the interrogation was generated. Implementing this system as a local oscillator for a Cs cold atom fountain, a record frequency stability of $1.6 \cdot 10^{-14}\tau^{-1/2}$ was obtained. While providing a great improvement in the fountain short-term stability, CSOs require unwieldy and expensive liquid helium refills approximately every 30 days; also, at each refill, almost a day is needed to restore optimum performance of the CSO. An alternative solution to ensure fully continuous operation would consist of employing a pulsed-tube cryocooler. However, as already discussed in Section 7.5, another valuable option is represented by the utilization of an ultra-low-noise, optical-frequency-comb-based microwave generator. Atomic fountains using such optically-derived flywheel oscillators

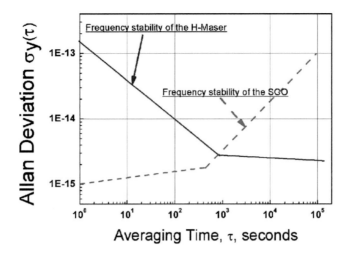

FIGURE 7.44
Trend lines for the Allan deviation of the CSO and the H maser used to realize a flywheel with superior performance. (Courtesy of [725].)

have already reached short-term stabilities in the medium-low $10^{-14}\tau^{-1/2}$ range [726], limited by quantum projection noise (see next Section).

7.9.2.3 Uncertainty budget in a Cs fountain clock

Statistical uncertainties

Generally, the relevant noise contributions are adequately described by white-noise processes, so that the Allan standard deviation $\sigma_y(\tau)$, proportional to $\tau^{-1/2}$, can be used for a quantitative analysis. In a well-designed fountain clock, the dominant noise sources are [724]:

1. Quantum projection noise - It originates from the alternate operation on the left and right sides of the central Ramsey fringe, where the transition probability is neither 0 nor 1. In fact, according to quantum mechanics, if an atomic system is prepared in a linear superposition $|\psi\rangle = \alpha |a\rangle + \beta |b\rangle$ of the two states $|a\rangle$ and $|b\rangle$, except when α or $\beta = 0$, the outcome of a measurement indicating whether the system is in $|a\rangle$ or $|b\rangle$ cannot be predicted with certainty (one can only state that the probability of finding the system in $|a\rangle$ is $|\alpha|^2$ with $|\alpha|^2 + |\beta|^2 = 1$) no matter how accurately the state has been prepared. Such unavoidable source of measurement fluctuations was named "quantum projection noise", as it can be regarded as deriving from the random projection of the state vector onto one of the states compatible with the measurement process [727];

2. Photon shot noise - This arises from the statistical detection of a high number of photons per atom;

3. Electronic detection noise - It is intrinsic to the electronic detection process, generally accomplished by a photodetector in conjunction with a transimpedance amplifier;

4. Local oscillator noise - This stems from a down-conversion process of the LO frequency-noise components due to the non-continuous probing of the atomic

transition frequency. This is known as the Dick effect which can be understood as follows [163]. Since the interrogation must of necessity be interrupted during part of each measurement cycle, in this lapse, the LO servo loop receives no corrective feedback signal, which implies that the control on the LO is discretely periodic. This slow sampling rate *aliases* the high-frequency noise of the LO to lower frequencies that are close to the signal, thus introducing spurious frequency shifts in the standard. The extent of degradation depends on the specific interrogation scheme and has been calculated for different systems and parameters [728]. In most advanced frequency standards, the Dick effect represents a serious issue for the ultimate achievable stability; for this reason, local oscillators must be selected, whose noise properties match the adopted interrogation scheme. Of course, the degradation of the stability due to the Dick effect decreases when the duty cycle of interrogation time versus cycle time approaches 100%. This means that a continuous-beam fountain clock would be practically immune from the Dick effect. A continuous fountain poses, however, a number of technical and experimental challenges [729].

Then, the Allan standard deviation of the relative frequency fluctuations $y(t)$ for an atomic fountain can be expressed as [724]

$$\sigma_y(\tau) = \frac{1}{\pi Q_{at}} \sqrt{\frac{T_c}{\tau}} \sqrt{\frac{1}{N_{at}} + \frac{1}{N_{at}\epsilon_c n_{ph}} + \frac{2\sigma_{\delta N}^2}{N_{at}^2} + \gamma} \tag{7.97}$$

where one-to-one correspondence exists between the different terms under the square root and the noise contributions listed above. Here τ is the measurement time (expressed in s); $T_c \simeq 1$ s is the fountain cycle duration ($\tau \gg T_c$); $Q_{at} = \nu_0/\Delta\nu \simeq 1 \cdot 10^{10}$ is the atomic quality factor (with $\Delta\nu$ being the width of the Ramsey fringe and ν_0 the Cs hyperfine frequency); N_{at} is the number of detected atoms; n_{ph} the average number of photons scattered per atom; ϵ_c the photon collection efficiency; $\sigma_{\delta N}$ represents the uncorrelated rms fluctuations of the atom number per detection channel (incidentally, note that the factor 2 in the third term arises from the fact the number of atoms in the $|F = 4, m_F = 0\rangle$ and $|F = 3, m_F = 0\rangle$ the states are measured separately); γ is the contribution from the interrogator oscillator. When detecting high numbers of photons per atom, using state-of-the-art low-noise electronic components and employing a low-noise cryogenic sapphire oscillator, the noise contributions respectively associated with the second, third and fourth term can be neglected. In this case, the limit set by the quantum projection noise can be reached. This was demonstrated in [730] where, by replacing the quartz oscillator with a CSO, an instability of $4 \cdot 10^{-14}(\tau/s)^{-1/2}$ range was achieved with $N_{at} = 6 \cdot 10^{15}$.

Systematic uncertainties

The frequency outputted by a fountain deviates from the unperturbed Cs transition frequency ($\nu_0 = 9192631770\,\text{Hz}$) by the sum of all systematic shifts. By virtue of the symmetry of the fountain geometry, the low velocities of the atoms, and the narrow linewidth, such difference is typically on the order of 1 mHz (for comparison, it amounts to about 2 Hz for a thermal-beam primary standard). Its precise evaluation as well as the estimate of the corresponding uncertainty represent major difficulties when operating a fountain as a primary clock. Just to give an idea of the complexity of the matter, we spend a few words for each of the most significant frequency-shifting sources [722, 6, 724, 65].

Second-order Zeeman effect - The C field inherently causes a frequency shift due to the second-order Zeeman effect. According to Equation 7.94, for the typical C-field strength of 0.1 μT, a fractional frequency shift of $5 \cdot 10^{-14}$ is obtained. Experimentally, such shift is accurately evaluated first by measuring the frequency of the $|4, 1\rangle \to |3, 1\rangle$ magnetic field-sensitive transition and then applying the found value to correct for the shift in the $|4, 0\rangle \to |3, 0\rangle$ clock transition (since such a correction asks for the determination of $\langle B^2 \rangle$, which equals $\langle B \rangle^2$ only in the case of a really constant C field, an excellent homogeneity is required for the latter). Due to the high number of Ramsey fringes, a delicate issue of this procedure is the uncertainty associated with the identification of the central fringe on the magnetically sensitive transitions. In the most valuable approach, the Ramsey pattern around the central fringe is repeatedly measured for the $m_F = 1$ transition while launching to various heights. In this way, the Ramsey fringes interfere constructively on the central one, losing coherence away from it. Even mis-assigning 1 full fringe (considered unlikely), a fractional frequency error in the low 10^{-16} range is obtained.

In the case of NIST-F1, for example, magnetic-field fluctuations of about 10^{-12} T exist in the C-field region; the resulting frequency shift, in the order of 10^{-18} for the $m_F = 0$ clock transition, can thus be ignored. Therefore, the overall quadratic-Zeeman-shift uncertainty is dominated by the central-fringe-location one; this is conservatively assigned the value of $3 \cdot 10^{-16}$, corresponding to the aforementioned misassignment of one whole fringe.

Cavity-related shifts: residual first-order Doppler and cavity pulling - If the atomic trajectories were perfectly vertical, frequency shifts induced by inevitable cavity phase variations (both axial and radial) would be entirely cancelled, as the interaction between each atom and the Ramsey field would occur once with velocity $+v$ (upwards) and once later with $-v$ (downwards). In particular, transverse residual thermal velocities as well as possible misalignments in the launching direction give rise to a spread in the trajectories between the first and the second passage; this eventually results in a residual first-order Doppler frequency shift, usually referred to as distributed cavity phase (DCP) shift. By carefully designing the Ramsey cavity and accurately calculating its actual phase distribution, such DCP shift has been reduced to the high 10^{-17} range, but with an associated uncertainty still on the medium-low 10^{-16} order [726].

As in hydrogen masers, the cavity pulling effect is due to the interference inside the microwave resonator between the field radiated by the input coupler and the field radiated by the atomic magnetic dipoles, when the atoms pass through the cavity. This interference induces a time-dependent phase shift between the field inside the resonator and the signal delivered by the interrogation oscillator and thus a shift of the clock frequency. An approximate expression for the latter is given by [731, 732]

$$\Delta \nu_{pull} \equiv \nu - \nu_0 \simeq K N_{at} \frac{\nu_0(\nu_C - \nu_0)}{(\nu_C - \nu_0)^2 + \nu_0^2/4Q_C^2} \tag{7.98}$$

where N_{at} is the number of atoms crossing the resonator, K represents an amplitude coefficient whose magnitude depends on the microwave power (typically on the order of a few 10^{-16} Hz), and the other symbols have the same meaning as in Equation 4.34. Being much larger for the Rabi-pedestal part of the lineshape than for the Ramsey-fringe part, the cavity-pulling frequency shift is accurately evaluated by measuring the microwave power dependency of the offset between the Rabi pedestal and the Ramsey fringe for all seven Zeeman components of the Cs hyperfine transition [733]. Ultimately, while having the same physical origin as in a hydrogen maser, the cavity pulling effect is usually negligible in an atomic fountain, where the operating conditions of the Ramsey cavity are far from maser oscillation. Also, by virtue of the proportionality to N_{at}, it is automatically corrected for when the collisional shift correction is applied (see below).

Microwave lensing - This shift originates from the modification of the atomic motion due to the interaction between the standing-wave field in the microwave cavity and the atoms themselves [726]. In analogy with the optical-spectroscopy case of an atom which interacts with a laser beam propagating along the z direction, such shift can be expressed (in energy) as $(\hbar k)^2/(2m)$, where $\hbar k$ is the momentum of the absorbed photon and m the atomic mass. Transferring to the microwave domain and considering the ^{133}Cs clock transition, the corresponding fractional frequency shift is $1.5 \cdot 10^{-16}$. However, this naive picture overestimates the effect. Actually, since the plane wave is not infinite (in the transverse directions), the discrete recoil experienced by the atom in the z direction is less than $\hbar k$ [734]. Also, the magnetic dipole force that arises in the transverse directions acts as a lens to focus (or defocus) the atomic wave packets; this produces a frequency shift that is not discrete, but varies linearly with the field amplitude and strongly depends on the atomic state detection. When sophisticated theoretical calculations are carried out in this frame, the microwave-lensing fractional frequency shift is indeed found to be in the high 10^{-17} range.

Second-order Doppler effect (SODE) - In principle, the atomic velocity distribution can be retrieved from the Fourier transform of the observed Ramsey pattern, as this includes the contributions of all atoms with their different speeds. However, in contrast to thermal beam clocks, the mean quadratic velocity of the atoms in a typical fountain clock is such that the fractional SODE shift is on the order of 10^{-17}. Thus, time dilation and the associated uncertainty can be safely neglected.

Microwave leakage - This effect originates from unintended interactions occurring between the atoms and stray microwave radiation existing near the outside of the Ramsey cavity. Such an extraneous field has a complex, uncontrolled conformation, mainly consisting of both a standing-wave and a travelling-wave component. The resulting frequency shift is unlikely to be disentangled from other systematic sources (e.g. the distributed cavity phase) and exhibits a complicated dependence on various fountain parameters, the microwave power above all. So far, typical associated uncertainties are in the low 10^{-16} range, as evaluated by comparison with specially developed theoretical models [735]. More recently, a novel experimental approach was implemented, based on the use of low-phase-transients switchable microwave synthesizers. These not only have inherently low leakage, but also can be effectively switched off as the atoms come out from the Ramsey cavity [726].

Neighboring transitions - Consisting of up to 16 ground-state sublevels, the ^{133}Cs structure allows, in principle, the coupling of several levels apart from those involved in the clock transition (CT). Although the resulting neighbouring transitions ideally have a symmetrical amplitude distribution (relative to the CT center), an asymmetric overlap may arise, in practice, due to the way the atomic trajectories fit within the magnetic field lines. This causes a distortion in the observed intensity signal, whose maximum is displaced from the true resonance frequency. In particular, Rabi pulling originates from an overlap of the clock line with the wings of the adjacent $F = 3, m_F = 1 \leftrightarrow F = 4, m_F = 1$ and $F = 3, m_F = -1 \leftrightarrow F = 4, m_F = -1$ lines. Ramsey pulling stems, instead, from the contributions of transitions with $\Delta m_F = \pm 1$. Potentially, frequency shifts also originate from the so-called Majorana transitions, which may occur between different magnetic substates of the same hyperfine transition ($\Delta F = 0, \Delta m_F \neq 0$), close to zero crossings of the magnetic field. Commonly, thanks to the state-selection process, the impact of these frequency pulling effects is reduced to well below 10^{-16}.

Electronics - Subtle errors arising from modulation distortion and integrator offsets as well as from switching transients, round-off and aliasing may affect the performance of an even carefully devised digital servo system, eventually leading to a frequency shift. Spurious frequency components and phase noise in the microwave Ramsey field may further worsen the situation. Refined models for the sensitivity to all superpositions of amplitude-modulation (AM) and phase-modulation (PM) noise as well as methods for measuring these effects have been developed [736, 737]. Again, the present uncertainty is of a few 10^{-16}.

Light shift - The interaction between the laser light and the atoms during their ballistic flight implies a frequency shift via the ac Stark effect. To overcome this drawback, at the proper stage of the fountain operation, the laser light is frequency detuned far from the resonance and additionally blocked by mechanical shutters. As already discussed in previous occasions, such shift is experimentally evaluated by measuring its dependence on the laser light intensity. In this way, the associated uncertainty is typically reduced to the very low 10^{-16} region.

Background gas collisions - For the typical vacuum pressures in the ballistic flight region (in the low 10^{-7} Pa range), the effect of residual gas collisions is evaluated to be well below 10^{-16}.

Blackbody radiation - The blackbody shift results from the interaction between the Cs atoms and the thermal radiation emanated by the walls of the 300-K vacuum enclosure. In spite of being quite large ($\sim 2 \cdot 10^{-14}$), this shift can be corrected with an uncertainty as low as $2 \cdot 10^{-16}$ (corresponding to an uncertainty of 1 K in the thermal radiation temperature) by resorting to suitable theoretical models. In the case of the Cs atomic clock, the fractional ac stark shift caused by the blackbody radiation has been calculated as

$$f_{bb} = -1.573(3) \cdot 10^{-14} \text{Hz} \left(\frac{T}{300 \text{ K}} \right)^4 \times \left[1 + 0.014 \left(\frac{T}{300 \text{ K}} \right)^2 \right] \qquad (7.99)$$

for a vacuum enclosure at temperature T. Going beyond the current limit will require either a huge amount of theoretical work to gain a deeper insight into the blackbody shift, or the realization of a cryogenic vacuum system ($T = 77$ K) to reduce it by two orders of magnitude. Work is in progress at NIST to accomplish this latter option [738].

Gravitational frequency shift - As discussed later on in this chapter, a major outcome of general relativity is that the frequencies of two non-local clocks are shifted apart by the amount $\Delta \nu = \nu \Delta \Phi / c^2$, where $\Delta \Phi$ denotes the difference between the values of the gravitational potential at the two respective locations. When referenced to the rotating geoid, such difference can be approximated as $\Delta \Phi \simeq g \Delta h / c^2$, where g is the local acceleration due to Earth's gravitation and Δh is the (small) height change on the surface of Earth (i.e. higher clocks run faster). In the case of cesium, this shifts the clock frequency by approximately 10^{-16} per meter (relative to its elevation above sea). At present, the uncertainty of this correction is insignificant ($3 \cdot 10^{-17}$), but is likely to become increasingly important in future, more accurate standards.

Cold collision shifts - A leading uncertainty source is represented by the frequency shift caused by collisions among the cold atoms in the cloud. At the temperatures under consideration, the collision physics strongly simplifies up to be described by just one parameter, namely the scattering length a associated with the s-wave process. Then, if we denote with $|\alpha\rangle$ and $|\beta\rangle$ the two internal states between which Ramsey transitions are induced, the mean-field shift is given by [739]

$$\Delta \nu = -\frac{2\hbar}{m} \cdot \sum_j n_j (1 + \delta_{\alpha,j})(1 + \delta_{\beta,j})(a_{\alpha,j} - a_{\beta,j}) \qquad (7.100)$$

where the j index refers to the atomic Zeeman substate $|F, m_F\rangle$, n_j are the partial densities (since these are both space and time dependent, the measured frequency shift appear as an average over these variables), the Kronecker symbol accounts for quantum statistics and $a_{\alpha j}$ ($a_{\beta j}$) is the scattering length associated with the binary collision between one atom in the internal state $|\alpha\rangle$ ($|\beta\rangle$) and the other one in the state $|j\rangle$. The trouble is particularly severe in the case of cesium whose collisional cross-section is unusually high at the cloud temperatures used in fountains. This forces one to keep the Cs density at a relatively low level; the residual shift is then evaluated by measurements at different densities followed by zero-density extrapolation. Currently, such shift can be as high as $4.0 \cdot 10^{-16}$ with an uncertainty in the correction of $1.0 \cdot 10^{-16}$. Several schemes have been proposed to further suppress this effect. In principle, one could even more reduce the atomic cloud density and hence the collision rate. However, this would degrade the signal-to-noise ratio and the short-term stability (according to Equation 7.97). In a smart alternative approach, theoretically developed at Istituto Nazionale di Ricerca Metrologica (INRM), as many as 10 clouds of laser-cooled Cs atoms are launched in rapid succession in such a way that the first atomic *ball* has the highest apogee, the second one has its apogee just below the first, and so on, thus ensuring that all the trajectories intersect in the detection region. While preserving the final number of detected atoms and hence the ultimate signal-to-noise ratio, this technique effectively lowers the average Cs density by about a factor of 10. Indeed, a reduction in the density-shift uncertainty down to $3 \cdot 10^{-17}$ is expected with this approach [6]. However, preliminary experimental tests (performed at NIST with seven *balls*) have revealed that, due to the various launch heights and travel times, the correction of systematic effects might be different for each *ball* in the sequence, thus requiring further consideration.

Conceptually, an even simpler solution would be to select another element having a lower collisional cross section. As we will see in a short while, this has motivated the construction of rubidium fountain clocks.

In conclusion, the relevant contributions to systematic uncertainty are on the order of a few 10^{-16} or less. Then, the square root of the sum of squares of the individual (independent) contributions yields the total systematic uncertainty. A comparison between the seven primary Cs fountain clocks can be found in [724]. As already pointed out, the superior stability of the SYRTE fountains basically stems from the cryogenic sapphire oscillator available there. A more updated survey on atomic fountains at LNE-SYRTE is given in [726].

7.9.3 Rubidium clocks

Rubidium has one stable isotope, ^{85}Rb, with a natural abundance of 72% and another, very long living isotope, ^{87}Rb, with a natural abundance of 28%. Having a larger hyperfine splitting in its ground state (6.83 GHz), this latter is more attractive for metrological purposes. The energies of the hyperfine levels of ^{87}Rb in a magnetic field are shown in Figure 7.45. Again, the transition between the states with only a weak quadratic dependence on the magnetic field, $|F = 2, m_F = 0\rangle \leftrightarrow |F = 1, m_F = 0\rangle$, is selected as the clock transition. In this case, the Breit-Rabi formula yields

$$\Delta\nu_{B^2} = \simeq 5.74 \cdot 10^{-2} \text{Hz} \left(\frac{B}{\mu T}\right)^2 \qquad (7.101)$$

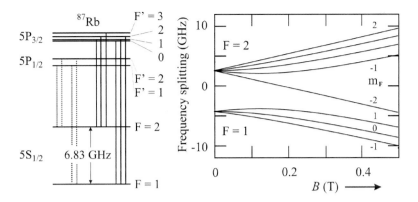

FIGURE 7.45
Left: hyperfine structure of the ground state and the first excited electronic states of ^{87}Rb.
Right: energies of the hyperfine states of ^{87}Rb in a magnetic field.

7.9.3.1 Rb fountain clocks

As anticipated, the construction of ^{87}Rb fountains was motivated by the expectation of a lower collisionally induced fractional frequency shift, that is one of the dominant contributions in the accuracy budget of Cs fountains. Indeed, early measurements performed on rubidium fountains confirmed that the collisional shift of the ground hyperfine transition is from 30 to 35 times lower for ^{87}Rb than for ^{133}Cs [739], thus opening new perspectives for improving the accuracy of cold-atom microwave frequency standards.

The working principle is the same as for a Cs-based fountain. Thus, *mutatis mutandis*, experimental setups for Rb fountains are essentially identical to those constructed for cesium [740]. An updated, comprehensive survey on the design, operating parameters, and accuracy evaluation of a Rb fountain is given in [732]. This refers to the fountain at NPL, which is characterized by three distinctive features:

- A double-MOT system is used, where a magneto-optical source provides an intense and continuous flux of cold ^{87}Rb atoms which is then loaded into the main MOT. In such a way, this latter can get to contain a high number of ^{87}Rb atoms, without the vacuum chamber being contaminated by the thermal rubidium vapour;

- The temperature of the fountain interrogation region is tuned and stabilized with an uncertainty of 0.1 K;

- The fountain uses a Ramsey cavity of high loaded quality factor $Q_c \simeq 28500$ in conjunction with an atomic sample of large transverse size, which suppresses the frequency uncertainty due to the distributed phase.

The fractional frequency accuracy of this Rb fountain was estimated to be $3.7 \cdot 10^{-16}$, while the frequency stability, limited essentially by noise in the local oscillator (the 6.8-GHz synthesizer with a stability of $1.5 \cdot 10^{-13}\tau^{-1/2}$), was measured to be $7 \cdot 10^{-16}$ after one day of averaging $(2 \cdot 10^{-13} \ \tau^{-1/2})$. When compared with the corresponding figures in best Cs fountain clocks $(3 \cdot 10^{-16}$ and $3 \cdot 10^{-14} \ \tau^{-1/2})$, these numbers suggest that there is still enough room for improvement. Furthermore, accurate investigations have revealed that the dominant contribution (at room temperature) in the Rb case is represented by the blackbody radiation shift rather than the collisional shift, contrary to what was thought at the beginning. In this sense, Rb fountain clocks have lost their key advantage over Cs fountains.

7.9.3.2 Lamp-based Rb cell standards

Apart from sophisticated fountains, optically-pumped, gas-cell rubidium microwave standards represent a well-established type of (transportable) secondary frequency reference, which provides competitive frequency stabilities up to medium-term timescales ($\sim$ days) in compact volumes around 0.25 to 2 liters.

In the traditional arrangement [65, 7], shown in Figure 7.46), the light form a small (volume less than 1 cm^3, which keeps the Rb atoms in the Lamb-Dicke regime) discharge lamp (containing 1 mg of ^{87}Rb together with a noble gas such as krypton) passes through a filter cell (containing ^{85}Rb and a noble gas such as argon at a pressure of around 10^4 Pa) and is then directed to the resonance cell (volume of a few cm^3, operational temperature around 80 °C) containing ^{87}Rb and a low-pressure buffer gas (about 10^3 Pa) that is a mixture of nitrogen and rare gases. Due to a favorable coincidence between the spectra of ^{87}Rb and ^{85}Rb, behind the filter cell, the radiation contains only the components that can excite transitions from the $F = 1$ ground state. In this way, the atoms enter one of the P states, from which relaxation into the two ground-state levels occurs either by fluorescence or by collisions with nitrogen atoms. After a few cycles, virtually all ^{87}Rb atoms are optically pumped into the $F = 2$ state; this makes the resonance cell transparent to the filtered light, thus maximizing the light intensity reaching the photodetector. Then, as the microwave radiation is applied, the transition $F = 2 \rightarrow F = 1$ in the ground state is stimulated and the $F = 1$ level is thereby re-populated. As a consequence, a fraction of the incident filtered light is absorbed, which slightly reduces the level of the photodetector signal. The resulting absorption feature is used to generate a feedback signal that tunes a crystal oscillator (VCXO) so as to keep the microwave frequency from the synthesiser at the atomic resonance. The presence of the buffer gas in the resonance cell has several different reasons. First, the resulting dilution reduces the mean free path between two collisions below the wavelength, thus suppressing Doppler broadening of the hyperfine resonance. At the same time, the rate at which atoms collide with the cell walls is also diminished, which reduces the associated relaxation and frequency shift (this also prevents the blackening of the glass cell due to the high chemical activity of alkali metals). The buffer gas pressure is such that diffusion velocities of rubidium atoms are extremely small (on the order of 1 cm/s), whereupon they

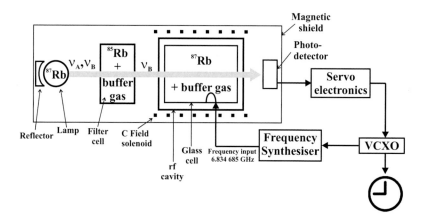

FIGURE 7.46

Scheme of a lamp-based Rb frequency standard. Typically, the microwave cavity sustaining the TE$_{111}$ (or TE$_{011}$) mode exhibits a loaded quality factor of $Q_c \simeq 400$. Very compact designs can be obtained by utilizing a magnetron-type microwave resonator and combining in a single cell the functions of the filter and the absorption cells. (Adapted from [65].)

can be practically regarded as stationary during their relaxation time (about 1 ms). As a drawback, collisions between the rubidium atoms and the buffer-gas constituents induce a frequency shift in the resonance line. For a single-component buffer gas, this is typically on the order of 10^{-7} (in fractional form), but it can be considerably lowered by use of suitable mixtures; this also curbs frequency shifts resulting from temperature fluctuations ($\simeq 10^{-9}$ K). Apart from such collisional shift and other sources that are common to all atomic clocks, a peculiar frequency shift is related to the double-resonance interrogation scheme, where two transitions sharing the same energy level are simultaneously excited (see next Section); also referred to as light shift, inasmuch as connected with optical pumping, this amounts to a few Hz.

Concerning frequency instability, high-performance, lamp-based devices exhibit an Allan deviation of $\sigma_y(\tau) \geq 4 \cdot 10^{-12}\tau^{-1/2}$ for 1 s $< \tau <$ 1000 s. After reaching a flicker floor (between 10^{-12} and 10^{-13}, for $\tau >$ 1000 s), $\sigma_y(\tau)$ increases again due to the influence of systematic effects. These include the aforementioned chemical reactivity between rubidium and glass as well as changes in the buffer gas composition and pressure (caused by diffusion through the walls). As a result, in the long term (for τ greater than $10^3 - 10^4$ s), rubidium clocks drift by a few times 10^{-11} per month. For this reason, as illustrated later on in this Chapter, rubidium clocks are often disciplined by global positioning systems (GPSs).

7.9.3.3 Laser-based Rb cell frequency standards

A clear limit of lampbased standards is represented by the fact that a troublesome background signal originates from the broad-band light that does not directly contribute to optical pumping. This latter can be more effectively accomplished by replacing the lamp+filter system with a laser. In this respect, two schemes are presented here [741], which also allow us to introduce in a more natural fashion two interrogation techniques that were excluded from Chapter 5, namely double-resonance (DR) and coherent population trapping (CPT) spectroscopy.

Rb clocks based on double-resonance spectroscopy

The first DR experiment is illustrated in Figure 7.47 [742, 743]. Mercury atoms in an external constant magnetic field, $\boldsymbol{B}_0$, are irradiated with linearly polarized light, which drives a π transition to the $m_J = 0$ level of the $^3\mathrm{P}_1$ excited state. The emission from these atoms is also linearly polarized π light. Now, by application of a high-frequency (rf or microwave), perpendicular (relative to $\boldsymbol{B}_0$) magnetic field, $\boldsymbol{B}_1$, one can induce transitions $\Delta m = \pm 1$ (within the $^3\mathrm{P}_1$ excited state), thus populating the Zeeman substates $m = 1$ and $m = -1$. What one eventually detects is the emission from these levels (occurring in a direction perpendicular to that of the π emission), that is circularly polarized σ light. The essence of such double-resonance technique (double excitation with light and rf/microwave radiation) is that $\Delta m = \pm 1$ transitions between Zeeman substates can be measured with extremely high detection sensitivity, as the rf/microwave quanta with small energies are detected via the much more energetic light quanta.

Optical pumping is a closely related technique. Its principle may be conveniently illustrated referring to the sodium D lines, e.g. the transition from the $^2\mathrm{S}_{1/2}$ ground state to the $^2\mathrm{P}_{1/2}$ excited state. By application of a static magnetic field, both terms are split into Zeeman components $m_J = \pm 1/2$ (see Figure 7.48). Now, if the pumping light is circularly polarized, let's say σ^+, only transitions from the $m_J = -1/2$ ground state to the $m_J = +1/2$ excited state can take place, which populates exclusively the $^2\mathrm{P}_{1/2}, m_J = +1/2$ level. On the other hand, emission from this state can occur either as σ^+ light, leading to the $^2\mathrm{S}_{1/2}, m_J = -1/2$ initial state, or as π light, leading to the $^2\mathrm{S}_{1/2}, m_J = +1/2$ state. The

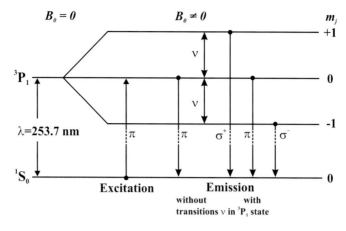

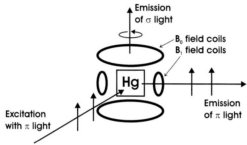

FIGURE 7.47

Double-resonance method (concept and experimental arrangement) in the case of Hg atoms. One pair of coils produces the constant field B_0, while the other one generates the high-frequency field B_1. (Adapted from [741].)

overall effect of this pumping cycle on the ground state is to increase the population of the $m_J = +1/2$ terms at the cost of those with $m_J = -1/2$. Then, relaxation processes (caused by collisions among the Na atoms or with the cell walls) or microwave-induced transitions (named electron spin resonance transitions) from the upper to the lower ground-state sublevel can be detected optically, namely through the change in the intensity of the absorption from $^2S_{1/2}, m_J = -1/2$ to $^2P_{1/2}, m_J = +1/2$.

In Figure 7.49 the block diagram of a laser-pumped, double-resonance Rb atomic frequency standard is shown [744]. In this work, two main issues are addressed in order to master the short-term-stability and the long-term frequency drifts, respectively:

1. the realization of a laser pump-light source with a sufficient frequency and intensity stability;

2. the control, via a fine-tuning of the cell's buffer gas content (Ar+N$_2$), of both the temperature coefficient of the resonance cell and the light-shift effects under conditions of laser pumping.

The laser head consists of a 780-nm-wavelength DFB laser diode (linewidth of 4.5 MHz, output power of few tens of microwatts, relative intensity noise @300 Hz of $7 \cdot 10^{-14}$ Hz^{-1}, frequency modulation noise @300 Hz of 4 kHz/$\sqrt{\text{Hz}}$, and power stability better than 0.1 %/day at fixed environmental conditions), frequency-stabilized (by FM modulation techniques) to the $F = 1 \rightarrow \{F' = 0, F' = 1\}$ crossover sub-Doppler saturated-absorption line

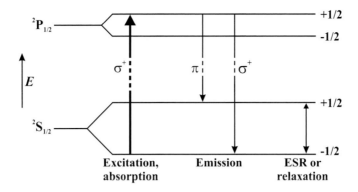

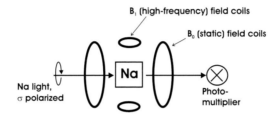

FIGURE 7.48

Optical pumping scheme for Na atoms. Only atoms in the ground state $m_j = -1/2$ absorb the σ^+ light with which the sample is irradiated. π transitions occurring in emission from the excited state lead to an increase in the population of atoms with $m_j = +1/2$. With the high-frequency field, transitions from $m_j = +1/2$ to $m_j = -1/2$ are induced, increasing the number of atoms which are able to absorb the pumping light. (Adapted from [741].)

(derived from a small reference cell) of the ^{87}Rb D_2 ($5S_{1/2} \to 5P_{3/2}$) transition. In this way, an efficient optical pumping of the $5S_{1/2}$ ground-state population into the $F = 1$ hyperfine term can be achieved for the ^{87}Rb atoms contained in the resonance cell encased in the clock module. A constant magnetic field applied to this cell lifts the Zeeman degeneracy and isolates the hyperfine ground-state transition ($F = 1, m_F = 0 \to F = 2, m_F = 0$), while two shields strongly suppress fluctuations in the ambient magnetic field. Also, a telescope is used in order to expand the laser beam to the cell diameter, so as to sample a maximum number of atoms at low light intensity. Then, the clock transition (at 6.83 GHz) is probed by applying (via a TE_{011} magnetron-type resonator) a microwave field. This latter is provided by a high-quality microwave synthesiser with a phase-noise at 6.8 GHz of -112 dBc/Hz @300 Hz (6.8 GHz carrier); the microwave frequency is then locked to the centre of the obtained DR line (linewidth of 467 Hz, contrast of 35%) using frequency modulation of the microwave and lock-in detection techniques. With this setup, the authors could measure an intensity-light-shift coefficient of $-1.9 \cdot 10^{-12}/\%$, a frequency-light-shift coefficient of $2.2 \cdot 10^{-17}/\%$ and a temperature coefficient $< 6.6 \cdot 10^{-12}/$K. Ultimately, a short-term frequency stability of $4 \cdot 10^{-13} \tau^{-1/2}$ for 1 s $< \tau <$ 1000 s was achieved, with a medium- to long-term stability reaching the $1 \cdot 10^{-14}$ level at 10^4 s.

Chip-scale Rb clocks based on coherent-population-trapping spectroscopy

In recent years, the coherent population trapping (CPT) phenomenon has attracted much interest in view of realizing an optically-pumped passive frequency standard using alkali metal atoms. As opposed to the DR technique, the CPT scheme involves exciting the atoms

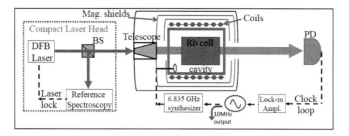

FIGURE 7.49
Schematic of the setup for the double-resonance Rb clock. (Courtesy of [744].)

exclusively by means of coherent optical radiation fields. The resulting advantage is twofold: first, no microwave cavity is requested for the excitation of the hyperfine transition; second, when the mutually coherent optical fields are derived from the same laser by suitable frequency modulation, a significant reduction of the light shift can be achieved [745].

While CPT is best understood within the density matrix approach [166], here we only give a simplified physical picture referring to the so-called Λ system shown in Figure 7.50 [746]. Let's assume for simplicity that the excited state $|3\rangle$ decays into the two long-lived ground states, $|1\rangle$ and $|2\rangle$, at an equal rate $\Gamma/2$, while the ground states decay into each another at the rate γ, such that $\Gamma \gg \gamma$. Next, let's denote with Ω_1 (Ω_2) the Rabi frequency corresponding to the incident light field detuned by δ_1 (δ_2) from the $|1\rangle - |3\rangle$ ($|2\rangle - |3\rangle$) atomic transition. Then, the atom-light interaction Hamiltonian can be written as

$$\hat{H}_{int} = \Omega_1|3\rangle\langle 1| + \Omega_2|3\rangle\langle 2| + h.c. \tag{7.102}$$

Now, among the infinite possible superposition states for an atom in the ground state, let us consider the following one

$$|nc\rangle = \frac{1}{\sqrt{\Omega_1^2 + \Omega_2^2}}(\Omega_2|1\rangle - \Omega_1|2\rangle) \tag{7.103}$$

In such case, one obtains

$$\langle 3|\hat{H}_{int}|nc\rangle = 0 \tag{7.104}$$

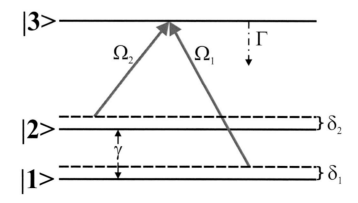

FIGURE 7.50
A three-level atom-light system in Λ configuration, illustrating the CPT concept. (Adapted from [746].)

i.e. the atom cannot be excited by the given combination of incident light fields (by contrast, for any other superposition state, there is always a finite excitation probability). This implies that $|nc\rangle$ represents a *dark* state where atoms accumulate: indeed, while the spontaneous emission from the excited state tends to populate it, no process is capable of exciting the system out of it. In other words, once in $|nc\rangle$, the atom no longer interacts with the applied field, i.e. the system becomes transparent. In particular, the atom is no longer subject to optical pumping cycles and no longer fluoresces. Experimentally, as the laser frequency difference is scanned, the fluorescence (transmission) spectrum displays a *dark* (*bright*) line in a narrow region around the difference frequency between the levels $|1\rangle$ and $|2\rangle$. It can be shown that the FWHM of such *dark* (or *bright*) line is approximately given by [747]

$$FWHM_{dark} = 2(\gamma + \Omega_R^2/\Gamma) \tag{7.105}$$

where we have supposed for simplicity $\Omega_1 = \Omega_2 \equiv \Omega_R$.

The CPT phenomenon is precisely at the basis of chip-scale atomic clocks [748]. The micro-fabricated frequency references being developed at the National Institute of Standards and Technology (NIST) make use of miniature cells (1 mm³ volume) that are filled with ^{87}Rb by reacting BaN$_6$ and ^{87}RbCl in a controlled environment. Besides ^{87}Rb, such reaction produces alkali atoms, Ba, Cl, and N$_2$. Most of this nitrogen is pumped off before sealing the cell, while the vacuum chamber (including the interior of the cell) is backfilled with 24 kPa of neon and 11 kPa of argon. Thanks to the presence of this buffer gas, the CPT effect takes place over a width which, no longer masked by the atom transit time across the laser beams (on the order of several hundred kHz, depending on the actual physical arrangement), is greatly reduced due to the Dicke effect and mainly ruled by ground state relaxation and laser power broadening (Equation 7.105). This situation gives rise to linewidths of the same magnitude as those obtained in the observation of magnetic resonance lines by means of standard radio frequency techniques. Then, the cell is integrated with heaters, an optics assembly, a 795-nm-wavelength vertical-cavity surface-emitting laser (VCSEL), and a photo-detector to form a functional physics package for an atomic clock of volume 12 mm³, consuming 195 mW of power at an ambient temperature of 22 °C. The atomic hyperfine resonance is excited by use of CPT spectroscopy on the D_1 lines (Figure 7.51). For this purpose, the two phase-coherent circularly-polarized light fields are created by modulating the VCSEL current at half the frequency of the ground-state hyperfine splitting, thus producing (on the optical carrier) two first-order sidebands separated by 6.8 GHz. As usual, a longitudinal magnetic field is applied to lift the degeneracy of the Zeeman sub-states. Then, the CPT resonance is detected by measuring the transmitted optical power by use of phase-sensitive detection with a lock-in amplifier. Finally, a feedback loop stabilizes the modulation frequency of 3.4 GHz onto the center of the CPT resonance. For this ^{87}Rb clock, a fractional frequency instability of $4 \cdot 10^{-11}\tau^{-1/2}$ was achieved for integration times between 1 s and 10 s, with a residual long-term drift ($\tau > 50$ s) of $-5 \cdot 10^{-9}$/day. This represented a significant improvement compared to a previous micro-fabricated clock exciting the D_2 transition in Cs [262].

7.9.4 Microwave ion clocks

Trapped ions can provide reference frequencies in the microwave or in the optical domain. In the former case, mostly magnetic dipole transitions between hyperfine components in the ground state are used. We will not dwell much on this topic, as research on ion frequency standards is by now focusing on optical transitions (see next chapter), albeit work is still in progress in a few laboratories in developing compact and rugged ion clocks for metrological applications including space operation (see for instance [749]).

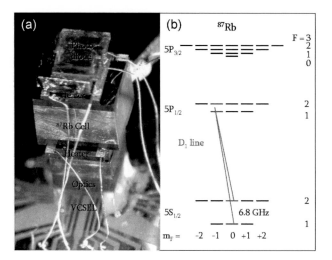

FIGURE 7.51 (SEE COLOR INSERT)
Picture of the ^{87}Rb D_1 chip-scale atomic clock and level diagram for CPT spectroscopy. The two optical transitions (D_1 radiation at 795 nm), represented by the arrows, form the Λ system in the case of σ^- polarization. These transitions connect the ground state levels $F = 1, m_F = 0$, $F = 2, m_F = 0$ (of interest in the implementation of a frequency standard) to a common level $F' = 2, m_F = -1$ of the excited state. (Courtesy of [748].)

Most microwave laboratory ion standards are based on laser-cooled ions in a linear Paul trap [750]. Recent work has concentrated mainly on ^{199}Hg$^+$ and ^{171}Yb$^+$, with some work on ^{111,113}Cd$^+$ in a conventional Paul trap (there has also been work on laser-cooled ^{9}Be$^+$ ion clouds in a Penning trap). The relevant energy levels for Hg II and Yb II are shown in Figure 7.52 along with the optical pumping transitions. In particular, a stable and accurate

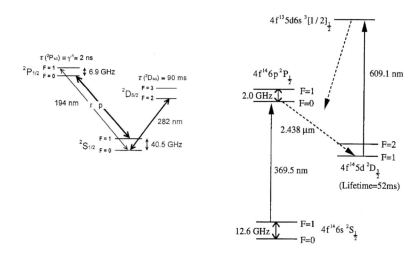

FIGURE 7.52
Energy levels for ^{199}Hg$^+$, ^{171}Yb$^+$ for use as microwave frequency standards (from [751] and [752]).

frequency standard based on the 40.5 GHz ground-state hyperfine transition in Hg II was demonstrated [751]. The ions were confined in a cryogenic linear Paul (rf) trap and laser cooled to form a linear crystal. With seven ions and a Ramsey interrogation time of 100 s, the fractional frequency stability was $3.3(2) \cdot 10^{-13} \, \tau^{-1/2}$ for measurement times $\tau < 2$ h. The ground-state hyperfine interval was measured to be 40507347996.84159(14)(41) Hz, where the first number in the parentheses was the uncertainty due to statistical and systematic effects, and the second was the uncertainty in the frequency of the time scale to which the standard was compared. Improvements in accuracy to the 10^{-16} level were envisaged following the development of a smaller trap requiring lower voltage and operated at a lower frequency, resulting in a strong reduction of the ac Zeeman shift from any asymmetric currents in the trap rods. Additionally, more ions could be trapped and individually monitored to reach the projection noise limit. A few experimental details are worthy of mention. With reference to Figure 7.52, first, the ions were cooled with both beams p and r for approximately 300 ms. Next, beam r was blocked for about 60 ms to optically pump the ions into the $^2S_{1/2}, F = 0$ level. Both beams were then blocked during the Ramsey microwave interrogation period, which consisted of two $\pi/2$ microwave pulses of duration $t_R = 250$ ms separated by the free precession period $T_R = 100$ s. Transitions to the $F = 1$ state were detected by reapplying only beam p until the ion was pumped optically into the $F = 0$ state ($\simeq 15$ ms), while the number of detected scattered photons (typically about 150 per ion) was counted. This process completed one measurement cycle. The microwave frequency was synthesized from a low-noise quartz oscillator locked to a reference hydrogen maser.

Microwave frequency standards based on the 12.6 GHz ground-state hyperfine transition in Yb II have been under development at the National Measurement Institute, Australia, for many years. Using a laser-cooled ion cloud ($\sim 10^4$ ions are cooled to below 1 K in a linear trap), the transition frequency was measured in 2001 to an accuracy of 8 parts in 10^{14}, limited by the homogeneity of the magnetic field due to the stainless-steel vacuum chamber. The short-term stability was estimated to be $5 \cdot 10^{-14} \, \tau^{-1/2} \text{Hz}^{-1/2}$. Since this experiment had no cryo-cooling of the vacuum walls, when the cooling laser was switched off for 12.6 GHz microwave interrogation, the ion heating rate was somewhat faster than the Hg II work at NIST (measurements showed that after 20 s the temperature had risen to about 3 K). Cooling and optical pumping were carried out with a 369 nm laser. The presence of the metastable $^2D_{3/2}, F = 1$ state can cause problems for efficient cooling of this ion: during cooling on the $^2S_{1/2}, F = 1 \rightarrow^2 P_{1/2}, F = 0$ cycling transition, the ion can decay into the metastable level and thus no longer be able to be cooled using a 369-nm laser. Thus, a clearing laser at 609 nm is used to return the ion to the ground state. Later, a new chamber in the alloy CrCu was realized, which is non-magnetic and has good vacuum properties. Uncertainties associated with field inhomogeneity in this improved vacuum system were below 1 part in 10^{15}. Also, other systematic uncertainties such as AC Zeeman shift, microwave imperfections, and pressure shifts were reduced to permit operation in the 10^{-15} accuracy range [753].

7.10 Time transfer and frequency dissemination

Nowadays, apart from cutting-edge scientific and technological implications, an astonishing and increasing number of daily applications rely on the dissemination of frequency and time signals with a varied range of precision and accuracy requirements. Just to mention a few, these include: transportation management, navigation and communications; industrial processes, agriculture, and finance; utility, surveillance, emergency, and environmental services. Whatever the method used to transfer time/frequency in a specific context, the ultimate attainable precision/accuracy depends not only on the performance of the various means and devices utilized to send/receive the signals, but also on the overall care that is paid to calibrate the employed instrumentation and to develop the physical model which is at the basis of the transmission/comparison procedure. Just to give an idea of the prickly aspects involved in this matter, the accuracy of frequency standards and clocks used in the most demanding scientific applications is by now such that relativistic effects must be taken into account in time and frequency comparisons.

We start this section by addressing the issue of the realization of time scales. While the reader is referred to [7, 3] for a comprehensive discussion, only a brief account is given here, following [650, 65]. After that, we will deal with the subject of transmitting time and frequency information.

7.10.1 Realization of time scales

According to the definition of the SI second given in Chapter 1, this should be realized with an unperturbed Cs atom in free space; also, the observer should be at rest relative to the atom itself. In this way, the *proper* time of the clock is measured, which is independent of any conventions with respect to coordinates or reference frames. In practical timekeeping, however, clocks at different locations are compared utilizing time signals, which necessitates the introduction of coordinate frames. In fact, being defined in terms of the SI second as carried out on the rotating geoid, the so-called International Atomic Time (TAI) is a coordinate (rather than a proper) time scale. To appreciate the importance of such distinction, just think of a *perfect* cesium clock orbiting around the earth in a global positioning system (GPS) application. Then, to compensate for the frequency offset with respect to TAI, an observer on the earth must apply all the corrections extensively discussed so far, including the gravitational redshift, the first- and second-order Doppler shifts, etc.

7.10.1.1 Realization of TAI

TAI is computed retroactively by the International Bureau of Weights and Measures (BIPM), based on data furnished by a worldwide network of approximately 50 timing laboratories, for a total of about 400 atomic clocks (either primary or commercial). Schematically, the procedure is the following (see Figure 7.53). Every five days, each laboratory measures the time differences between each of its clocks, TA(k), and its own laboratory time scale, designated UTC(lab). Approximating the universal coordinated time scale UTC (to be discussed below), the UTC(lab) scales (about 60 at present) are evaluated starting from selected ensembles clocks. Moreover, in order to compare data from different laboratories, following a protocol defined by the BIPM, each laboratory also measures the time differences between UTC(lab) and GPS time (it is worth pointing out that GPS satellites are used only as transfer standards, and the satellite clocks drop out of the data). Then, all time-difference data are transmitted to the BIPM which, using the so-called ALGOS algorithm, computes their weighted average to generate an intermediate time scale referred to as echelle atomique libre (EAL). After that, the length of the EAL second is compared with the SI second as realized by primary clocks in some major timing laboratories, and a correction is applied so as to steer the duration of the EAL scale unit. The resulting scale is TAI.

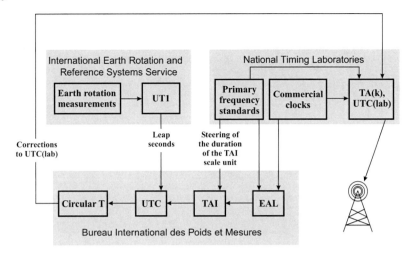

FIGURE 7.53
Simplified scheme for the realization of TAI and UTC. (Adapted from [65].)

7.10.1.2 Coordinated universal time

Unfortunately, the length of the day defined using $24 \times 3600 = 864000$ TAI seconds was shorter (by about 0.03 ppm) than the length of the day defined by the UT1 time scale (the latter is derived by observing and modelling the rotation of the earth around its polar axis), and the times of the two scales were destined to diverge more and more over the years. Therefore, since our daily life and astronomical navigation are both governed by the earth's rotation, the Universal Coordinated Time (UTC) was adopted in 1972. Derived from TAI, UTC is made to approximately follow UT1 ($|\text{UTC}(t) - \text{UT1}(t)| < 0.9$ s) by insertion of leap seconds. Hence, while their scale units coincide, UTC and TAI differ by an integer number n of seconds which depends on the earth's irregular angular velocity

$$\text{UTC}(t) = \text{UT1}(t) - n \tag{7.106}$$

Since UTC is derived from TAI, it is an atomic time scale too. A leap second is occasionally inserted in UTC worldwide at the same epoch, the time interval between two leap seconds being provided by the International Earth Rotation and Reference Systems Service (IERS) on the basis of astronomical observations of Earth's rotation. Finally, the differences [UTC-UTC(lab)] are published periodically in BIPM Circulaire T and it is recommended that [UTC-UTC(lab)]<100 ns; thus, the local UTC(lab) have to be steered to follow UTC.

7.10.2 Transmitting time information

In spite of their intimate relationship, time and frequency differ substantially in how they are distributed and hence in the overall uncertainty budget associated with the dissemination process. Obviously, uncertainties in the delay in the transmission channel between the clock and the user influence time distribution in a direct way, directly contributing to the error budget. Conversely, frequency is a time interval rather than an absolute time, such that the uncertainty in its transfer is essentially dictated by temporal fluctuations in the transmission delay rather than by its absolute magnitude. In other words, accurate knowledge of the time delay (either by modelling or measurements) of a channel is needed when carrying time information, whereas just stability is requested for such delay when transmitting frequency information. Starting with time transfer, in the following we will give a brief account on a couple of methods. But first we have to enunciate a few key concepts and definitions, following the report by [754].

- As already mentioned, in special and general relativity there are two types of time, proper and coordinate. Measured by an unperturbed (i.e., insensitive to environmental conditions, gravity, and accelerations) clock accompanying the observer, proper time, τ, is invariant in any coordinate change. Also, since the SI second is defined exclusively in terms of the Cs atom radiation periods (without indication of a specific gravitational potential, or state of motion), any observer can realize it as the unit for proper times. On the other hand, proper time cannot describe phenomena in extended domains, in which cases coordinate time, t, must be employed. In addition to spatial coordinates, (x_1, x_2, x_3), this constitutes the fourth dimension, $x_0 = ct$, of the space-time reference system. In a given reference system, coordinate time represents a univocal way of dating. In metrology coordinate time cannot be measured, but only computed, the relation between proper time of an observer and coordinate time being provided by the metric.

As well known, in special relativity, for instance, the infinitesimal interval between two events is given by $ds^2 = -c^2 d\tau^2 = -c^2 dt^2 + x_1^2 + x_2^2 + x_3^2$, such that

$$c^2 \left(\frac{d\tau}{dt}\right)^2 = c^2 - \left[\left(\frac{dx_1}{dt}\right)^2 + \left(\frac{dx_2}{dt}\right)^2 + \left(\frac{dx_3}{dt}\right)^2\right] \tag{7.107}$$

from which the relationship between proper and coordinate time depends on the velocity v of the clock according to the celebrated formula

$$\frac{d\tau}{dt} = \sqrt{1 - \frac{v^2}{c^2}} \equiv \frac{1}{\gamma} \tag{7.108}$$

The situation is a bit more complex in general relativity, where the metric takes into account the surrounding masses and energy. In this case, the gravitational field close to the Earth is conveniently dealt with by defining a coordinate system having the origin of the space axes at the Earth's center of mass and which does not rotate with respect to the most distant observable bodies of the Universe. The metric tensor of such a *non-rotating geocentric coordinate system* can be expressed as [754]

$$ds^2 = -c^2 d\tau^2 = -\left(1 - \frac{2U}{c^2}\right) dx_0^2 + \left(1 + \frac{2U}{c^2}\right) [dx_1^2 + dx_2^2 + dx_3^2] \tag{7.109}$$

where $U = U_E + U_T$ is the sum of the Newtonian gravitational potential U_E generated by the Earth (vanishing at infinity) and the Newtonian tide-generating potential U_T of external bodies (vanishing at the geocentre), both taken with a positive sign. At the level of frequency uncertainty of 10^{-14}, a satisfactory approximation to the gravitational potential of the Earth U_E is provided by

$$U_E = \frac{GM_E}{r} + J_2 GM_E a^2 \frac{1 - 3\sin^2 \phi}{2r^3} \tag{7.110}$$

where $r = (x_1^2 + x_2^2 + x_3^2)^{1/2}$, ϕ is the geocentric latitude (the geocenter being O, the geocentric latitude of a point P is the angle between OP and the equatorial plane; this is positive towards the North), $GM_E = 3.986004418 \cdot 10^{14}$ m^3/s^2 is the product of the gravitational constant and the mass of Earth, $J_2 = 1.082636 \cdot 10^{-3}$ is the quadrupole moment coefficient of Earth, and $a = 6378136.5$ m is the equatorial radius of Earth. Next, let us consider the effect of the tidal potential U_T. Increasing with geocentric distance, this is of order 10^{-15} at the distance of geostationary satellites. Its value is obtained by summing the contributions of all celestial bodies A (in practice, the Sun and the Moon). The lowest-order term in the expression for U_T is found as

$$(U_T)_1 = \frac{1}{2} \sum_A Q_{ij}^A x^i x^j \qquad (7.111)$$

where the notation of summation on repeated indexes has been used and with

$$Q_{ij}^A = \frac{GM_A}{r_{EA}^3} \left(3 \frac{r_{EA}^i r_{EA}^j}{r_{EA}^2} - \delta_{ij} \right) \qquad (7.112)$$

Here, r_{EA}^i are the differences of coordinates, in a coordinate system centered at the barycentre of the solar system, between the center of mass of the Earth and the body A, r_{EA} is the coordinate distance between these points, and δ_{ij} is the Kronecker delta. All the values appearing in Equation 7.112 can be found in astronomical ephemerides. The next-order term for U_T gives rise to frequency corrections of order 10^{-18} on the ground and $2 \cdot 10^{-16}$ at the altitude of geostationary satellites [755].

By means of Equation 7.109 and defining the coordinate speed of the clock (with respect to the centre of Earth) as

$$v = \frac{\sqrt{dx_1^2 + dx_2^2 + dx_3^2}}{dt} , \qquad (7.113)$$

the relationship between proper and coordinate time can be now calculated as (the subscript nrg stands for non-rotating geocentric)

$$\frac{d\tau}{dt} = 1 - h_{nrg}(t) \qquad (7.114)$$

with

$$h_{nrg}(t) = c^{-2} \left[U(t) + \frac{v^2(t)}{2} \right] + \mathcal{O}(c^{-4}) \qquad (7.115)$$

Thus, more in general, we can always write

$$\frac{d\tau}{dt} = 1 - h(t, x_1, x_2, x_3) \qquad (7.116)$$

with $h(t, x_1, x_2, x_3)$ given as a power series of $1/c$, its actual form depending on the specific case under consideration. Also, since on a given world-line $x_i = x_i(t)$, we may enlighten the notation as

$$\frac{d\tau}{dt} = 1 - h(t) \qquad (7.117)$$

For many applications, it is convenient to adopt a coordinate system which rotates with the Earth. This is obtained from the non-rotating coordinate system implied by Equation 7.109 through a spatial rotation and is valid at the same level of approximation. When using this coordinate system, one obtains

$$h_{rg} = c^{-2} \left[\tilde{U}_g + \Delta \tilde{U}(t) + \frac{V^2(t)}{2} \right] + 2c^{-2} \omega \frac{dA_E}{dt} \qquad (7.118)$$

where X_i is the triplet of spatial coordinates in the geocentric coordinate system rotating with Earth (the X_3-axis is close to Earth's rotation axis), $V_i = dX_i/dt$ (V is the modulus of $\boldsymbol{V}$), $\omega = 7.292115 \cdot 10^{-5}$ rad/s is the angular velocity of rotation of the earth, $\tilde{U}_g = 6.26368575 \cdot 10^7$ m^2/s^2 is the constant potential in the geocentric rotating coordinate system at the level of the rotating geoid (including the potential of the centrifugal force), $\Delta\tilde{U}$ is the difference of gravitational potential in the geocentric rotating coordinate system (including the potential of the centrifugal force) between a specified location and the geoid (note that $\Delta\tilde{U}$ is negative above the geoid), and A_E is the area of the equatorial projection of the surface swept out by a vector with origin at the Earth's center of mass and terminus at a moving point (it is measured in the rotating coordinate system and is positive in eastward motion). The last term in Equation 7.118 results from the Sagnac effect, originating from the fact that the clocks on Earth co-rotate with the same angular velocity such that their velocity is dependent on the latitude. The explicit expressions for the potential terms, corresponding to an uncertainty in normalized frequency not exceeding 10^{-18} (which may be necessary for current advanced metrological applications), follow from rather complex evaluations [756]. At an uncertainty level of 10^{-16}, the numerical expression for $\Delta\tilde{U}/c^2$ is

$$\Delta\tilde{U}/c^2 = -1.08821 \cdot 10^{-16} \, b/\text{m} - 5.77 \cdot 10^{-19} \, b/\text{m} \sin^2\phi$$
$$+1.716 \cdot 10^{-23} \, (b/\text{m})^2 \tag{7.119}$$

where b/m represents the altitude above the geoid expressed in meters. As just mentioned, the above expression is accurate to better than 10^{-16} when b is less than 15 km and known with an uncertainty of less than 1 m.

- The convention applied for the synchronization of two clocks is as follows. Two events are said to be coordinate simultaneous when they have the same time coordinate in some specified coordinate system, whatever their space coordinates x_i. Then, two clocks A and B are said to be coordinate synchronized during a time interval $t_1 \leq t \leq t_2$, when their readings, $\tau_A(t)$ and $\tau_B(t)$, considered as coordinate simultaneous events, remain equal during this interval.

- Let us denote with ν_C the proper frequency of a standard clock C, i.e., the frequency derived from the proper second at the location of the standard itself. In advanced metrological applications, this must be regarded as a quantity varying with time. Here we are particularly interested in the dependence on the coordinate time t. Thus, by designating with the constant $\nu_{C,0}$ the proper frequency stated by the builder, it is customary to define the proper normalized frequency deviation as

$$y_C(t) = \frac{\nu_C(t) - \nu_{C,0}}{\nu_{C,0}} \tag{7.120}$$

that is a dimensionless quantity close to zero. Then, Equation 7.117 generalizes to

$$\frac{d\tau_C}{dt} = \frac{d\tau_C}{d\tau}\frac{d\tau}{dt} = [1 + y_C(t)][1 - h(t)] \tag{7.121}$$

7.10.2.1   Portable clocks

According to Equation 7.121, when the time is delivered from the point P (time t_1) to the point Q (time t_2) by means of a transportable clock C, the difference in the clock readings, which arises due to the transport, is

$$\tau_C(t_2) - \tau_C(t_1) = \int_{W(C)} [1 + y_C(t)][1 - h(t)]dt \tag{7.122}$$

which, in the limit $y_C \to 0$, reduces to

$$\tau_C(t_2) - \tau_C(t_1) = (t_2 - t_1) - \int_{W(C)} h(t)dt \qquad (7.123)$$

where $W(C)$ is the world-line of clock C and the appropriate expression for $h(t)$ is used. Such technique was adopted by many timing laboratories in the past. In practice, the limitation of this approach stems from the finite frequency stability of the portable clock (as it is transported along the path) as well as by uncertainties in various relativistic corrections that may have to be applied.

7.10.2.2 Global positioning system

At present, the US Global Positioning System (GPS) is the main tool used for time transfer between remote sites, albeit a second system, named Global Navigation Satellite System (GLONASS), is operated by Russia. Additional systems, currently in their development stage, are planned like the European Galileo system or the Chinese COMPASS. Here we focus on the GPS system following the treatment given in [757, 65] to which the reader is referred for a more comprehensive discussion. As shown in Figure 7.54, the nominal GPS system is a constellation of at least 24 satellites orbiting around the Earth in 6 planes: approximately circular, these orbits are inclined 55° to the equator around which they are spaced at a 60° separation (the mean distance of a satellite to the Earth's center of mass is about 26600 km). Atomic clocks (cesium, rubidium) are operated on board of each

FIGURE 7.54
The GPS satellite constellation.

satellite. The nominal output frequency of the clocks is $f_0 = 10.23$ MHz, from which the carrier frequencies on the L1 and the L2 band, $f_1 = 1575.42$ MHz and $f_2 = 1227.60$ MHz, are derived (by multiplying f_0 by 154 and 120, respectively). Being below 2 GHz, these microwave signals can be received without any directional antenna. Prior to transmission, the two carriers are phase modulated with pseudo-random noise (PRN) binary codes. In more detail, a 1.032-MHz-chip-rate coarse/acquisition (C/A) code is used for f_1, while a 10.23-MHz-chip-rate precision (P) code modulates both f_1 and f_2. Being unique for each satellite, these continuously repeated codes are used for identification and tracking purposes through a correlation process performed by a receiver. An additional 50-bits-per-second binary code, referred to as the navigation message, is superimposed to the PRN-codes by an exclusive-or process. It contains information about the status of both the individual satellite and the overall constellation, the satellites' positions (ephemeris data), the offset of the satellites' clocks relative to a common reference timescale (the GPS system time), as well as information about the ionosphere. The satellites are governed by a ground segment comprising several monitor stations, a few uplink stations, and a master control station (located at Colorado Springs) where, on the basis of the calculated offset between the individual clocks and the system time, suitable clock steering commands are sent to the satellites. The system time, derived from all clocks in the satellites and the ground stations, is kept in close agreement with the time scale of UTC (United States Naval Observatory); it is the master control station that assesses the performance of the various clocks in each satellite and decides which of them are used for the signal generation. Since the transmitting power of the GPS satellites is below 40 W, the signal is hidden in the thermal noise on the Earth. This requires a correlation process for the signal reception and demodulation. For this purpose, an oscillator internal to the receiver generates a local copy of the PRN-code, which is then electronically time shifted and multiplied with the incoming antenna signal to provide the correlation function. If the received satellite PRN-code coincides with the replica signal, then the correlation function is at a stable maximum and the receiver tracking loops can lock to the carrier frequencies of the satellite. Obviously, such a process necessitates a local replica of the received carrier-frequency too; also, this replica must be frequency shifted to compensate for the Doppler effect suffered by the received satellite carrier frequency. In this way, the receiver can measure the phase of the received signal with an ambiguity of multiples of cycles of the carrier frequency. In modern receivers, this is accomplished with a precision of better than 1% of the wavelength.

A Global Navigation Satellite System (GNSS) receiver determines its local position on Earth by simultaneously comparing the signals with time stamps from different satellites with its local clock. The distance between the user U at the coordinates X, Y, Z and the i-th satellite of known position x_i, y_i, z_i is measured by the time delay Δt_i elapsed from the transmission to the reception of the signal. Thus, if the clock in the user's receiver and the one on board the satellite were perfectly synchronized, the true range from a first satellite could be calculated as $R_1 = c \cdot \Delta t_1$. Following the same procedure, a second satellite straight provides the U position in the plane containing the two satellites and the user, as one of the two intersection points between the circles with respective ranges R_1 and R_2. Three-dimensional location in space necessitates a third satellite. However, one of the most significant error sources is the GPS receiver's clock. Due to the very large value of c, the evaluated distances from the GPS receiver to the satellites are extremely sensitive to errors in the GPS receiver clock; for example an error of one microsecond (1 μs second) corresponds to an error of 300 meters. In order to avoid using an extremely accurate and expensive GPS receiver clock, so as to build inexpensive, mass-market GPS receivers, a fourth satellite is instead used. Let us assume, for instance, that the time T_U of the user's clock advances the system time T_{GNSS} from the satellites by $\delta t_u = T_U - T_{GNSS}$. Then, the ranges computed from the apparent time differences between the satellite clocks and the user clock are referred to as *pseudoranges* $P_i = R_i + c \cdot \delta t_u$. Now, if four different pseudoranges P_i are measured, one can write down a set of four equations

$$(x_1 - X)^2 + (y_1 - Y)^2 + (z_1 - Z)^2 = (P_1 - c \cdot \delta t_u)^2 \tag{7.124}$$

$$(x_2 - X)^2 + (y_2 - Y)^2 + (z_2 - Z)^2 = (P_2 - c \cdot \delta t_u)^2 \tag{7.125}$$

$$(x_3 - X)^2 + (y_3 - Y)^2 + (z_3 - Z)^2 = (P_3 - c \cdot \delta t_u)^2 \tag{7.126}$$

$$(x_4 - X)^2 + (y_4 - Y)^2 + (z_4 - Z)^2 = (P_4 - c \cdot \delta t_u)^2 \tag{7.127}$$

containing four variables: the three spatial coordinates X, Y, Z and the offset of the user clock δt_u. This can be solved for the unknowns by algebraic or numerical methods. Then, in practice, the code-measured pseudorange for one satellite $"i"$ can be expressed as

$$C_i = \sqrt{(x_i - X)^2 + (y_i - Y)^2 + (z_i - Z)^2} + c \cdot \delta t_u + \Delta_C + \varepsilon \tag{7.128}$$

where all physical influences on the signal path and the satellite clock are lumped into the delay Δ_C, and ε describes the noise in the measurement process. In turn, the Δ_C term can be modelled as

$$\Delta_C = \delta_{ion} + \delta_{tro} + \delta_{tide} + \delta_{rel} + \delta_{mul} \tag{7.129}$$

where δ_{ion} (δ_{tro}) represents the signal delay caused by the ionosphere (troposphere) re-fractivity, δ_{tide} characterizes changes related to the deformation of the Earth's surface (e.g., by the gravitational potential of the Moon and the Sun, ocean tides, . . .), δ_{rel} incorporates relativistic effects, and δ_{mul} describes multipath delays induced by signal reflections at buildings, trees, and hard grounds nearby the receiver's antenna as well as diffraction effects. According the desired precision degree, more or less each of the above delays has to be removed by suitable physical modelling [757].

While in positioning applications one is interested in determining the coordinates of the receiver, in the framework of time and frequency comparisons the quantity of concern is the receiver clock offset, δt_u (here, in principle, the signals from just one satellite would be suffi-cient, as the receiver's antenna position is fixed and well known). In this case, measurements must be referenced to an external timescale, physically provided by a 1 pulse-per-second (PPS) signal. A typical system used in a metrology laboratory consists of the receiver itself, usually located in a temperature-stabilized environment, and an antenna which is situated outside, preferably at a site which minimizes multipath and diffraction effects (Figure 7.55). First, the GPS measurements are linked to the time of an internal oscillator. This latter is synchronized to an external frequency or, alternatively, to the frequency derived from the satellite's signals (dashed lines). Then, an internal 1 PPS signal is generated by a time interval counter (TIC) and compared to the external 1 PPS signal. Since the GNSS mea-surements are referenced to the internal oscillator too, this cancels out when combining TIC and GNSS data. Modern state-of-the-art time and frequency receivers usually incorporate geodetic receiver circuit boards. Originally intended for precise positioning, these also out-put a 1 PPS signal derived from the internal clock. Moreover, they are capable to track the C/A-code as well as the P-code on both frequencies and to measure the carrier-phase. Finally, the boards are integrated together with the TIC and a computer which combines the data.

In conclusion, we should mention that a few other techniques exist to compare the readings of two distant clocks, based on the transfer of electromagnetic signals with radio frequencies. Among these, two-way satellite time and frequency transfer (TWSTFT), here omitted for brevity, represents the most accurate tool [650, 758].

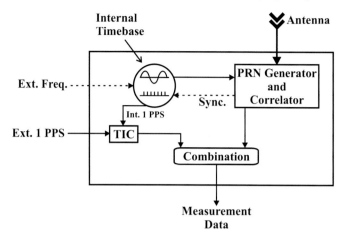

FIGURE 7.55
Simplified schematic of a basic time and frequency transfer receiver. (Adapted from [757].)

7.10.3 Frequency transfer

Before addressing the state-of-the-art techniques for frequency transfer both in the microwave and optical domain, we stop to describe in the following a paradigmatic, widely used scheme for absolute optical frequency measurements, which contains many of the elements described in this and the previous chapter.

7.10.3.1 A democratic absolute frequency chain

So far, the most refined optical frequency measurements accomplished with femtosecond combs have relied on local cesium standards or hydrogen maser references. However, the latter are available only at national laboratories or large institutions. Conversely, being much less expensive and bulky, high-quality BVA quartz oscillators, Rb clocks, and GPS timing receivers are widespread. One can construct with them an absolute (traceable to the Cs primary standard) frequency chain, as illustrated in Figure 7.56. The offset f_0 and repetition rate f_r of the optical frequency comb synthesizer (OFCS) are both stabilized against a 10-MHz BVA quartz oscillator. The latter, in turn, is phase-locked to a rubidium clock, which is eventually phase-locked to the primary Cs standard via the GPS system (most often, a commercial GPS-disciplined rubidium reference is directly employed).

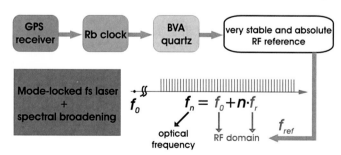

FIGURE 7.56
Absolute chain for optical frequency measurements consisting of an OFCS stabilized against a quartz-rubidium-GPS reference.

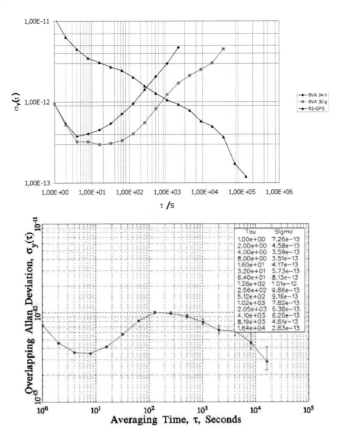

FIGURE 7.57

(a) Allan deviation plots for the quartz oscillator (squares), and for the GPS-disciplined rubidium reference (triangles). (b) Allan deviation plot for the quartz-rubidium-GPS chain reference (PLL time constant=100 s).

As typical Allan deviation plots show (Figure 7.57a), the quartz oscillator exhibits excellent stability on the $1 - 10^3$ s time scale, whereas the GPS-disciplined rubidium oscillator performs best over longer periods. By phase locking the two with a time constant ranging from 10^2 to 10^3 s, a reference characterized by the quartz short-term stability and the GPS long-term stability is obtained, whose Allan deviation plot is shown in Figure 7.57b. As a result, when measuring a frequency which is phase-locked to such a chain reference, averages can be performed over extremely prolonged acquisition periods, greatly improving precision.

7.10.3.2 Dissemination of microwave frequency standards

Now, we resume the thread of discussion. In light of the above considerations, comparisons between distant clocks can be carried out employing either two-way satellite time transfer systems or a satellite link, like a global positioning system (GPS). In the latter case, a resolution at one day of a few 10^{-14} (10^{-15}) is achieved with a commercial (geodesic) receiver. Alternatively, indirect comparisons can be accomplished utilizing transportable clocks (with a present record accuracy of slightly below 10^{-15} for the Cs fountain) as transfer oscillators between fixed high-stability clocks.

Apart from these more traditional systems, optical fibers are being experimented as valuable tools to establish local networks for the dissemination/comparison of time and frequency standards, both in the microwave and optical domains. This approach offers two main advantages: first, environmentally isolated fibers are much more stable than open-air paths, particularly at short time scales; second, active stabilization of the fiber-optic distribution channels can also be exploited to ameliorate the stability of the transmitted standard itself [759].

In the microwave domain, signals in the form of amplitude modulation of an optical carrier can be transmitted through optical fibers. The principle of such scheme is illustrated in Figure 7.58, which refers to a transfer between two laboratories in Paris, namely LPL and SYRTE [760]. In this experiment, a few sections of 1.55-μm telecommunication single-mode optical fibers were interconnected to form a 43-km-total-length link. This latter was fed with a DFB laser diode emitting at 1.55 μm. By modulating the diode junction current with the 100-MHz reference signal, a synchronous intensity modulation of the laser output was produced, to be then detected at the distant end of the optical link. In order to fully characterize the frequency transfer, two independent correction systems were implemented. In the first one, the rf signal detected at SYRTE after one round trip in the same fiber was compared to the reference one. The extracted perturbation value, $2\Delta\psi$, was used to compensate one-path perturbations, $\Delta\psi$, by phase shifting the 100-MHz signal in the opposite sense. This provided a corrected 100-MHz signal at LPL, then sent back to SYRTE for evaluation via the second fiber. A second compensation system was necessary for such transfer. Implemented at LPL, this latter acted directly on the optical path. A correction signal was generated by comparing the phase of the 100 MHz arriving from SYRTE and the 100 MHz modulation after one round-trip in the second fiber. Fast fluctuations were compensated by stressing a 15-m optical fiber wrapped around a cylindrical piezoelectric actuator (implemented at the input of the second optical fiber), while slow perturbations were compensated by heating a 1-km optical fiber spool. Also, in order to raise isolation between channels, the forward and return beams were modulated at different frequencies, 1 GHz and 100 MHz, respectively. Within this approach, a transfer instability of $5 \cdot 10^{-15}$ at 1 s and $2 \cdot 10^{-18}$ after 1 day was achieved.

7.10.3.3 Optical frequency transfer

Sharing with optical clocks the inherent advantage of a higher spectral resolution, a direct transfer of the optical carrier itself would provide better stability. As an illustrative example (Figure 7.59), we consider here a fiber link that was recently setup to bridge almost 70 km geographical distance between PTB and LUH in Germany [761]. The link consists of two parallel fibers (F1 and F2) in the same underground cable which, for characterization purposes, are connected to each other forming a 146-km loop with the input and remote ends located in the same lab at PTB. The light source is a stabilized fiber laser (at 1542 nm) with a fractional frequency instability of $2 \cdot 10^{-15}$ (1 s) and a coherence length exceeding 10000 km. Only 3 mW optical power is injected into the fiber link to keep below the threshold for stimulated Brillouin scattering. A bidirectional erbium-doped fiber amplifier (EDFA) partially compensates the 43 dB single-pass optical attenuation. Mechanical perturbations and temperature changes inducing optical path length fluctuations are compensated with the commonly used fiber-noise-cancellation scheme described in Section 4.7.3. Here, the whole fiber link is regarded as one arm of a Michelson interferometer, which is phase stabilized against a reference arm. Derived from a double-pass signal (in-loop, at PD1), a correction signal controls the frequency and phase of an acousto-optic modulator (AOM1) at the fiber input; AOM2 serves to distinguish light reflected by the Faraday mirror (FM) at the remote end from spurious reflections along the fiber. The stabilized link is characterized by

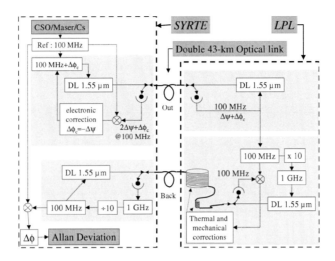

FIGURE 7.58
Schematic of the optical link compensation for dissemination of microwave frequency standards. (Courtesy of [760].)

measuring a beat signal between outgoing laser light and light at the remote end (at PD2). With this system, the achieved Allan deviation is $3 \cdot 10^{-15}/\tau$; after 30000 s, the relative uncertainty for the transfer is at the level of $1 \cdot 10^{-19}$.

In a similar experiment, the phase coherence of an ultrastable optical frequency reference was fully preserved over actively stabilized fiber networks of lengths exceeding 30 km (from JILA to NIST); using frequency combs at each end of the coherent-transfer fiber link, a heterodyne beat between two independent ultrastable lasers, separated by 3.5 km and 163 THz, achieved a 1-Hz linewidth [762].

More recently, it was demonstrated that the structure of an optical frequency comb,

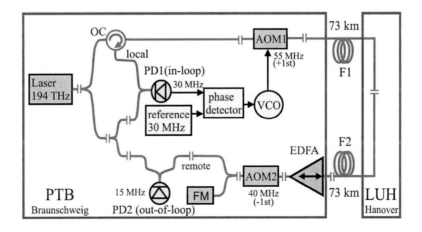

FIGURE 7.59
Schematic setup of active fiber noise compensation for a 146 km fiber link. OC, optical circulator; VCO, voltage-controlled oscillator. (Courtesy of [761].)

transferred over several km of fiber, can be preserved with a fractional accuracy better than $3 \cdot 10^{-18}$, a level compatible with the best optical frequency references currently available. The stability of the mode spacing after optical-microwave conversion was also tested and found to be preserved to better than $2 \cdot 10^{-15}$ at 1 s and $4 \cdot 10^{-17}$ for averaging times greater than 1000 s [763]. The experimental setup implemented to cancel the environmentally induced fiber noise is shown in Figure 7.60. Both the repetition rate $f_r = 100$ MHz and the offset f_0 of the optical frequency comb (from a commercial 1.56 μm amplified erbium-doped mode-locked fiber laser generating sub-150 fs optical pulses) are stabilized to a 10 MHz signal from a hydrogen maser. Being situated in a different part of the laboratory, the laser source is connected to the experimental setup via approximately 10 m of SMF-28 fiber, which broadens the pulse duration to about 17 ps before it enters the first, 90:10 power splitter. The 10% output is employed for monitoring purposes (providing the local pulse train), whereas the 90% output is used to propagate the comb over a 7.7-km-length (spooled) single-mode fiber (SMF) towards the receiver end; here, a portion is returned to the transmitter end, again via SMF. In order to accurately compare the signal that has travelled 7.7 km with that injected at the input of the fiber, both the SMF ends are placed in the same laboratory. The forward and backward travelling pulse trains are separated by means of optical circulators (CIR1, CIR2), while a dispersion compensating fiber module (DCF) recompresses the pulses to a duration below 100 ps; in addition, a free-space delay line is adjusted to ensure appropriate temporal overlap between the local and the returned pulse train. The returned frequency comb is combined with the original one after the latter has been frequency shifted by $f_{AOM} = 104$ MHz (via an acousto-optic modulator). Among the notes resulting from the beating between their optical modes, the one at lowest frequency (4 MHz) is amplified and phase compared with a maser-referenced synthesizer. This generates an error signal which, after suitable integration, is applied to two fiber stretchers and a thermally-controlled fiber spool to compensate for fast (up to a few kHz) and slow phase fluctuations, respectively. The precision with which the whole optical-frequency-comb

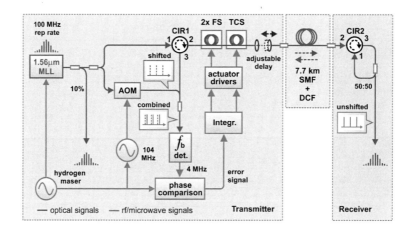

FIGURE 7.60

Experimental setup for dissemination of an optical frequency comb over fiber. The detection stage (f_b det.) comprises a photodiode, a filter and cascaded amplifiers. MLL: mode-locked laser; CIR: circulator; FS: fiber stretchers; TCS: thermally controlled spool; Integr.: integrator; AOM: acousto-optic modulator; SMF: single-mode fiber; DCF: dispersion compensating fiber. (Courtesy of [763].)

structure is preserved across the several-km-scale fiber transmission is tested by measuring the mode spacing f_r and the frequency of a selected optical mode (Figure 7.61).

In the former case, the 80th harmonic (8 GHz) of f_r at the receiver end of the fiber is phase compared with that detected directly at the output of the laser using a microwave mixer. By integrating the power spectral density of the phase noise fluctuations between 1 Hz and 100 kHz, as measured with an FFT analyser, a timing jitter less than 17.5 fs is calculated. In particular, the effectiveness of noise cancellation can be appreciated by noting that the phase noise at 1 Hz offset from the carrier is âĹŠ91 dBc/Hz, very close to the noise measured when the 7.7-km SMF and the DCF are replaced by a 2 m fiber and an attenuator set to provide the same overall loss. The measurement of the frequency stability is accomplished by converting the output voltage of the microwave mixer (logged every 0.5 s with a digital voltmeter) into phase changes and then evaluating the corresponding Allan deviation.

In the latter case, a selected optical comb mode is beaten, before and after the fiber link, against a continuous-wave (CW), 542-nm-wavelength laser stabilized to a ULE cavity. Since changes in the frequency of the CW laser are common mode, any difference observed between the two resulting 35-MHz beat frequencies (the broadband noise of which is filtered by means of 200-kHz-bandwidth tracking oscillators) originates from the fiber noise. The power spectral density of the phase fluctuations between the two beat frequencies, and hence of the transferred optical mode, is measured with a digital phase detector (with a linear range extending over 256π) followed by an FFT analyser. The timing jitter evaluated from 1 Hz to 100 kHz is 5.2 fs corresponding to a phase change of approximately 2π. The frequency stability calculated from the phase data is $4 \cdot 10^{-17}$ at 1 s and approximately $2 \cdot 10^{-18}$ for timescales of a few thousand seconds (corresponding to a timing jitter smaller than 10 fs). The mean frequency offset between the two beats corresponds to a transfer accuracy for the optical mode frequency of $2.6 \cdot 10^{-18}$. This clearly shows that, due to the noise introduced in the optical-to-microwave conversion, the accuracy of the transferred repetition rate is worse than that of the optical modes.

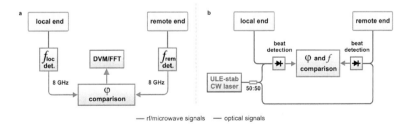

FIGURE 7.61

Experimental layout for measuring phase noise and frequency stability of the mode spacing (a) and transferred optical modes (b). The detection stages denoted as f_{loc} and f_{rem} comprise a fast photodiode, a narrow bandpass filter and microwave amplifiers. (Courtesy of [763].)

8

Future trends in fundamental physics and applications

The long unmeasured pulse of time moves
everything. There is nothing hidden that
it cannot bring to light, nothing once
known that may not become unknown.
Sophocles

Whatever has been said, narrated, drawn,
reality represents much more than all this.
Johann Wolfgang von Goethe

8.1 Optical atomic clocks

As discussed in the previous chapter, with the present frequency stability of $6 \cdot 10^{-16}$ and accuracy of $1.1 \cdot 10^{-15}$, microwave frequency standards based on fountains are now in their adulthood. Also, within a few years, further affordable betterments could push the stability into the low 10^{-16} range (over one-day averaging time) and the accuracy to $1 \cdot 10^{-16}$. We have also learnt that, when the quantum projection noise dominates over technical noise sources, the instability of an oscillator which is locked to an atomic transition with frequency ν_0 and linewidth $\Delta\nu$ can be expressed as

$$\sigma(\tau) = \frac{\Delta\nu}{\pi\nu_0} \sqrt{\frac{T}{N\tau}} \tag{8.1}$$

where T is the cycle time required to perform a single determination of the line center frequency and N denotes the number of detected atoms [764]. Note that this formula is valid for averaging times $\tau > T$. Also, a signal contrast of 100% and the absence of dead time (between the cycles) are assumed. Now, while the processes that limit the linewidth of an atomic transition in the microwave region are essentially comparable to those in the optical domain, a 5-orders-of-magnitude enhancement is gained by moving from the operating frequency of the cesium primary standard to optical frequencies. Thus, potentially, a dramatic reduction in the instability can be obtained by establishing a frequency standard on an optical transition. At risk of becoming pedantic, before going on, we insist in elucidating the principle of operation of a passive atomic frequency standard (Figure 8.1) [715].

The core is represented by an atomic reference with resonant response centered at frequency ν_0 and is interrogated by a suitable local oscillator of frequency ν. Then, an absorption signal arises when ν is tuned around ν_0, from which an error signal is derived. Via a servo control loop, the latter is used to accord the frequency ν of the local oscillator so as to hold it as close as possible to the frequency ν_0 of the reference transition. Finally, such

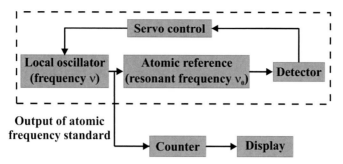

FIGURE 8.1
Schematic layout of a passive atomic frequency standard. (Adapted from [715].)

frequency standard is converted into a clock, by counting the cycles of the stabilized local oscillator.

In an optical atomic clock, the three main ingredients specialize as follows (see Figure 1.9):

1. The local oscillator is an ultranarrow linewidth laser. Indeed, in order not to degrade the great potential offered by the natural linewidths of clock transitions in optical frequency standards, few hertz or less, it is mandatory to diminish the probe laser linewidth to a comparable level. As extensively discussed in Chapter 4, this is usually accomplished by using the Pound-Drever-Hall (PDH) stabilization technique to lock the probe laser to a high finesse ultrastable optical cavity. As discussed in Chapter 3, this commonly consists of a pair of identical concave mirrors optically contacted on to the ends of an ultralow expansion (ULE) glass spacer. With a typical resonator length of 10 cm and a finesse of about 200000, a cavity resonance linewidth of 5-10 kHz is obtained. As a result, provided that the fidelity of the lock to the cavity is high enough, the frequency stability of the probe laser will be limited by that of the cavity resonance rather than by its own noise. As this kind of ULE cavity is not tunable, an acousto-optic modulator (AOM) is needed to shift the probe laser frequency into coincidence with some cavity resonance (sometimes the AOM drive frequency is also controlled in such a way as to compensate for the first-order residual drift of the ULE cavity). In this configuration, linewidths in the range 0.4-1 Hz and fractional frequency instabilities of a few 10^{-16} (for averaging times of a several seconds) have been reported by a number of laboratories. As already examined in Chapter 3, the most afflictive fluctuations in the ULE cavity length arise from low-frequency (below 100 Hz) seismic and acoustic vibrations which couple to cavity (via the support structure) and tend to deform it. Thus, mechanical designs with reduced sensitivity to vibrations are of utmost importance and many successful efforts have been made in this sense. Then, once the temperature is well constrained at the zero thermal expansion point of the ULE spacer and the effects of vibrations are efficiently suppressed, the cavity frequency stability will be dictated by dimensional changes due to thermal fluctuations in the mirror substrates and their coatings. In conclusion, it has been demonstrated that the performance achieved with state-of-the-art optical local oscillators is indeed very close to the thermal noise limit. Nevertheless, in the near future one may expect to further reduce the thermal noise limit through an attentive choice of cavity materials and geometry as well as the operating temperature;

2. Authoritative candidates for the atomic reference are either transitions in single laser-cooled trapped ions or transitions in cold neutral atoms. These will be presented separately in the two next sections.

3. The counting device is provided by a femtosecond optical frequency comb [765]. In an optical clock, the carrier-envelope offset frequency f_0 is detected using the usual self-referencing technique. Then, a first phase-locked loop (PLL) forces $f_0 = \beta f_r$ (e.g., by controlling the pump power of the femtosecond laser). In addition to f_0, a second heterodyne beat f_b is measured between an individual comb element $f_m = f_0 + m f_r$ and the optical standard laser oscillator that is locked to the clock transition frequency f_{at}. Similarly, a second PLL forces $f_b = \alpha f_r$ (e.g., by changing the cavity length of the femto-second laser with a piezo-mounted mirror). The constants α and β are integer ratios which can be implemented with frequency synthesizers that use $f_r/100$ as a reference. In this way, every element of the comb, as well as their frequency separation f_r, is phase-coherent with the laser locked to the atomic standard. In formulas

$$\begin{cases} f_b = f_{at} - (f_0 + m f_r) \equiv \alpha f_r \\ f_0 = \beta f_r \end{cases} \quad \Rightarrow \quad f_r = \frac{f_{at}}{m + \alpha + \beta} \qquad (8.2)$$

Thus, $f_r = f_{at}/(m+\alpha+\beta)$ is the countable microwave output of the clock, which is readily detected by illuminating a photodetector with the broadband spectrum from the frequency comb.

At this point, a crucial issue is testing how well the femtosecond comb transfers the optical frequency standard stability to other spectral regions. This task is accomplished by comparing the outputs from two independent combs that are locked to the same optical reference [766, 715]. In the optical domain, such comparison is carried out by looking at the heterodyne beat signal between the two combs. In the microwave domain, instead, the pulse train from each comb is directed onto a photodetector. The two resulting signals are then combined into an electronic mixer whose output provides the input of a conventional rf counter. In both cases, any instability due to the optical reference is common to the two combs and thus cancels out during the comparison. These experiments revealed that, while in optical-heterodyne comparisons the excess fractional frequency instability introduced by the comb is on the order of $2 \cdot 10^{-17}$ at 1 s (averaging down to the 10^{-19} range in a few thousand seconds), microwave signals extracted by photodetection have significantly worse stability. This suggests that excess noise may be introduced by the photodetection process.

8.1.1 Trapped ion optical clocks

As discussed in the previous chapter, a single laser-cooled ion confined in a Paul-type trap offers some favored properties to realize an optical frequency standard. First, entering the Lamb-Dicke regime strongly suppresses Doppler effects (broadening and shifts) that are related to the ion motion relative to the probing radiation. Second, in a cryogenic environment, perturbations from blackbody radiation and atomic collisions are both extremely small. Third, the ion storage time can last for months, thus considerably increasing the probe interaction time [767]. We also described an effective detection scheme, known as the electron shelving technique, by which the narrow reference transition can be detected with nearly 100% efficiency.

TABLE 8.1
Wavelength and theoretical linewidth of the clock transition in most advanced trapped-ion optical frequency standards currently under development in several labs.

Ion	Clock Transition	λ (nm)	$\Delta\nu$ (Hz)	Lab	$\delta\nu/\nu$ (10^{-15})
^{27}Al$^+$	^{1}S$_0$-^{3}P$_0$	267	0.008	NIST	0.65
^{40,43}Ca$^+$	^{2}S$_{1/2}$-^{2}D$_{5/2}$	729	0.14	Innsbruck, NICT, Marseilles	2.4
^{88}Sr$^+$	^{2}S$_{1/2}$-^{2}D$_{5/2}$	674	0.4	NPL, NRC	3.8
^{115}In$^+$	^{1}S$_0$-^{3}P$_0$	237	0.8	MPQ, Erlangen	180
^{171}Yb$^+$	^{2}S$_{1/2}$-^{2}D$_{3/2}$	436	3.1	PTB, NPL	3.2
^{171}Yb$^+$	^{2}S$_{1/2}$-^{2}F$_{7/2}$	467	$\sim 10^{-9}$	PTB, NPL	20
^{199}Hg$^+$	^{2}S$_{1/2}$-^{2}D$_{5/2}$	282	1.8	NIST	0.65

Note: Last column reports the fractional uncertainty measured for the clock transition relative to the SI second. (Adapted from [768].)

A number of potentially suitable transitions in different ions have been recognized for the accomplishment of an optical frequency standard. Each of them does possess clear advantages, but it is not free from drawbacks. These can be either related to some key clock-transition parameters, sensitivity to environmental perturbations above all, or to more technical hurdles in the implementation of the experimental setup. So far, much progress has been made for the ions reported in Table 8.1 [768]. Just as an example, partial term diagrams showing the laser wavelengths needed to cool and probe the trapped ion are also displayed in Figure 8.2 for ^{88}Sr$^+$, ^{171}Yb$^+$, and ^{199}Hg$^+$.

While Ca$^+$, Sr$^+$, Yb$^+$, and Hg$^+$ exhibit alkali-like atomic structure, ^{27}Al$^+$ and ^{115}In$^+$ are more similar to alkaline earth elements.

In the former case, the ^{2}D states are metastable, decaying to the ^{2}S$_{1/2}$ ground state via an electric quadrupole transition. Interestingly, the ^{171}Yb$^+$ ion also possesses a low-lying, 6-year-lifetime ^{2}F$_{7/2}$ state which can decay to the ground state only via an electric octupole transition. Although considerably harder to drive (compared to the quadrupole clock transition), the latter offers the potential advantage of noticeably longer interrogation times. Recently, the Physikalisch-Technische Bundesanstalt (PTB) group was successful in measuring the unperturbed transition frequency of such electric octupole transition, ^{2}S$_{1/2}(F = 0)$-^{2}F$_{7/2}(F = 3)$, as 642121496772645.15(52) Hz with a fractional uncertainty of $7.1 \cdot 10^{-17}$ [769].

In the latter case, the clock transition is the strongly spin-forbidden ^{1}S$_0$-^{3}P$_0$ one. Such a captivating transition has no electric quadrupole shift together with a small blackbody Stark shift. As a counterpart, an evident experimental challenge is represented by the construction of laser sources at the deep UV wavelengths required for cooling and probing. For instance, the cooling wavelength for ^{27}Al$^+$ falls around 167 nm. To overcome this drawback, the ^{27}Al$^+$ ion is co-trapped with an auxiliary ^{9}Be$^+$ ion that can be cooled at the more convenient wavelength of 313 nm. Then, the Coulomb interaction couples the two ions, leading to sympathetic cooling of the ^{27}Al$^+$ ion. In this scheme, due to the absence of cooling fluorescence from ^{27}Al$^+$, the usual electron shelving interrogation cannot be applied, but the clock-transition information can nevertheless be mapped back to the ^{9}Be$^+$ ion for readout [770]. This approach, called quantum-logic spectroscopy, will be detailed in Section 8.9.3.1.

As shown in Table 8.1, the uncertainty of the best absolute frequency measurement, that for ^{199}Hg$^+$ [771], is dominated by the uncertainty of the cesium microwave primary frequency standard used as the reference. Meanwhile, other standards are approaching this limit. This clearly demonstrates the urgent need to directly compare optical frequency standards without intermediate microwave references. For this purpose, a femtosecond optical frequency comb can be used. The principle of such a direct comparison is illustrated in

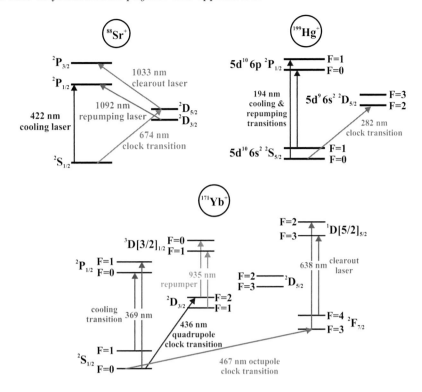

FIGURE 8.2
Partial term diagrams showing the laser wavelengths required for operation of an optical clock based on ^{88}Sr$^+$, ^{171}Yb$^+$, and ^{199}Hg$^+$. (Adapted from [768].)

Figure 8.3 for the ^{199}Hg$^+$ and ^{27}Al$^+$ optical standards [772]. In that experiment, the ratio of the two clock transition frequencies was determined with a relative uncertainty of $5.2 \cdot 10^{-17}$.

In terms of frequency stability, notable results have been achieved with the comparison between the ^{199}Hg$^+$ and ^{27}Al$^+$ standards [772]. Here, a combined fractional frequency instability of $4 \cdot 10^{-15}\tau^{-1/2}$ has been measured for averaging times up to 2000 s, representing an upper limit to the instability of each standard. Another remarkable achievement concerns the ^{171}Yb$^+$ quadrupole standard, where a frequency stability of $9 \cdot 10^{-15}\tau^{-1/2}$ has been observed in the comparison of two similar systems [773]. More recently, an optical clock with a fractional frequency inaccuracy of $8.6 \cdot 10^{-18}$ was constructed, based on quantum logic spectroscopy of an Al$^+$ ion [774]. In that experiment, a simultaneously trapped Mg$^+$ ion served to sympathetically laser cool the Al$^+$ ion and detect its quantum state. Then, the frequency of the $^1S_0 - {}^3P_0$ clock transition was compared to that of a previously constructed Al$^+$ optical clock with a statistical measurement uncertainty of $7.0 \cdot 10^{-18}$. The two clocks exhibited a relative stability of $2.8 \cdot 10^{-15}\tau^{-1/2}$, and a fractional frequency difference of $-1.8 \cdot 10^{-17}$, consistent with the accuracy limit of the older clock.

It has become apparent by now that the best cesium microwave primary frequency standards have been surpassed by the most advanced ion optical frequency standards. Indeed, based on the electric quadrupole transitions in ^{199}Hg$^+$, ^{88}Sr$^+$, and ^{171}Yb$^+$, secondary representations of the second were adopted in 2006 by the International Committee for Weights and Measures (CIPM) [775].

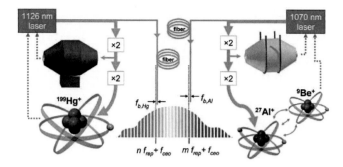

FIGURE 8.3

Frequency ratio measurement for the comparison of ^{199}Hg$^+$ and ^{27}Al$^+$ optical clock frequencies. The fourth harmonic of a 1126-nm laser is locked to the clock transition of ^{199}Hg$^+$, the second harmonic being used for pre-stabilization against an ULE Fabry-Perot cavity (which narrows the laser linewidth down to 1 Hz). The same holds for the 1070-nm laser which, driving the spectroscopy transition in the ^{27}Al$^+$ ion, is then locked to the detection transition in the co-trapped ^{9}Be$^+$ ion, according to the quantum-logic spectroscopy scheme. After this, the two laser frequencies are compared by means of a femtosecond comb: the beat-note frequency $f_{b,Hg}$ ($f_{b,Al}$) of the mercury (aluminum) clock laser with the $n(m)$th comb tooth, the comb carrier-envelope offset, f_{ceo}, and repetition rate, f_{rep}, enter the frequency ratio measurement. (Courtesy of [772].)

8.1.1.1 Systematic frequency shifts

In the following we survey the most important systematic frequency shifts that contribute to the overall uncertainty budget for a trapped-ion optical frequency standard. While only a brief, general discussion on the various shift sources is given here, the reader is referred to [768, 764, 715] for a detailed report on the extent to which each of them affects the metrological performance of all the ions listed in Table 8.1.

Zeeman shifts

For the odd isotopes of alkali-like ions, like ^{199}Hg$^+$ and ^{171}Yb$^+$, $m_F = 0 - m_F = 0$ components exist that are field-independent to first order. Concerning the second-order Zeeman shift, operation in μT-range magnetic fields should reduce the corresponding fractional uncertainty to the 10^{-17}-10^{-18} level, depending on the specific ion species. When using even isotope ions, like ^{88}Sr$^+$ and ^{40}Ca$^+$, a first-order Zeeman shift is instead present, which is cancelled out by over-and-over-again and consecutively probing two Zeeman components that are symmetrically placed about the centroid of the multiplet. Finally, the linear magnetic-field dependence of the ^{1}S$_0$-^{3}P$_0$ transitions in ^{115}In$^+$ and ^{27}Al$^+$ is removed in a similar manner.

Electric quadrupole shift

In many trapped-ion optical frequency standards, the dominant systematic frequency shift is represented by the electric quadrupole shift of the clock transition frequency, caused by the interaction between the electric quadrupole moments of the pertinent atomic states with any residual electric field gradient present at the trapped-ion position. As already mentioned, for the ^{1}S$_0$-^{3}P$_0$ transitions in ^{115}In$^+$ and ^{27}Al$^+$ such shift is zero because both the upper and lower clock transition levels have null angular momentum and hence no quadrupole moment. For the other ions being explored as optical frequency standards, the electric quadrupole shift is wholly attributable to that of the upper clock transition level, as the spherically symmetric $^2S_{1/2}$ state has no quadrupole moment. Even with low-flux ion loading techniques and effective 3-D micromotion compensation, unwanted field gradients within the trap can readily produce quadrupole shifts of several hertz. Providentially, a keen method to erase this shift has been demonstrated, which entails taking the average of measurements for several different Zeeman components [776, 715].

Second-order Doppler shift

As already explained, first-order Doppler shift is eliminated by laser cooling the trapped ion into the Lamb-Dicke regime. However, residual thermal motion and micromotion give rise to second-order Doppler shifts. In the former case such shift is given by [768]

$$\left(\frac{\Delta\nu_{Doppler}}{\nu}\right)_{thermal} = -\frac{3k_B T}{2Mc^2} \tag{8.3}$$

where M denotes the ion mass and the ion temperature T is close to the Doppler cooling limit $T_{min} = \hbar\gamma/(2k_B)$, γ being the natural linewidth of the cooling transition. Then, the corresponding fractional frequency shift will be in the range 10^{-18}-10^{-20}, depending on the mass.

The second-order Doppler shift related to the micromotion is instead much more pronounced. As a consequence, accurate minimization of the 3-D micromotion is mandatory to keep this effect as low as the part in 10^{17} level. If the trapping-field angular frequency satisfies the condition $\Omega \gg \gamma$, then the micromotion can be monitored by measuring the between the scattering rate R_1 (R_0) when the laser is tuned to the first sideband (carrier) [764]. Then, in the low intensity limit, the micromotion contribution to the second-order Doppler shift can be expressed as [764]

$$\left(\frac{\Delta\nu_{Doppler}}{\nu}\right)_{micromotion} \simeq -\left(\frac{\Omega}{\omega\cos\phi}\right)^2 \tag{8.4}$$

where ω is the angular frequency of the transition and ϕ is the angle between the probe laser beam and the direction of the micromotion. The above equation shows that the micromotion contribution scales quadratically with the clock transition wavelength.

Stark shifts

The main Stark-shift sources in the clock-transition frequency are [764]

- Motionally induced exposure of the ion to electric fields. Indeed, both the micromotion and the thermal motion expose the ion to a non-zero rms electric field, whose magnitude and constancy can be further influenced by stray charge within the trap. When the ion is cooled to the Doppler limit, again the micromotion contribution dominates, thereby requiring careful compensation to reach the 10^{-18} level.

- Interactions with the blackbody radiation field due to the temperature of the surrounding apparatus. For a room-temperature trap, the associated fractional blackbody Stark shift can be relatively large (few parts in 10^{16}) and the uncertainty in the absolute value of the correction is determined by the typical 30% uncertainties in the Stark shift coefficients. Amongst the room-temperature systems under investigation, the $^1D_0$-$^3P_0$ transition in $^{27}Al^+$ has the lowest fractional blackbody Stark shift (-8 ± 3 parts in 10^{18} for a temperature of 300 K). Since the blackbody Stark shift scales as T^4, for the $^{199}Hg^+$ standard, which is operated in a liquid helium cryostat, it is seven orders of magnitude lower than at room temperature, and hence negligible [765]. Recently, a new concept was developed to realize atomic clocks with suppressed (by 1-3 orders of magnitude) blackbody radiation shift [777]. Here, the suppression is based on the fact that in a system with two accessible clock transitions (with frequencies ν_1 and ν_2) which are exposed to the same thermal environment, there exists a *synthetic* frequency $\nu_{syn} \propto (\nu_1 - \varepsilon_{12}\nu_2)$ largely immune to the blackbody radiation shift.

- Exposure to the various light fields required to cool and probe the trapped ion. The main source is represented by the interaction of the probe-laser electric field with the light-field-induced electric dipole moment in the ion. However, for typical probe laser intensities, the associated clock transition shift is vanishingly small. An exception is represented by the extremely weak 467-nm electric-octupole transition in ^{171}Yb$^+$, for which a reasonable rate driving requires a significant light intensity. In this case, the unshifted value is determined by measuring the transition frequency *vs* the laser power and then extrapolating to zero power. Nevertheless, by narrowing the probe laser linewidth down to the hundreds-of-mHz level, it should be possible to use a lot less light intensity so as to make such ac Stark cease to dominate the uncertainty budget.

In conclusion, it is worth mentioning that the above list of systematic frequency shifts should not be considered as exhaustive, in the sense that previously unexamined effects may turn out to be more and more serious as the performance of optical frequency standards keeps on improving. Just as an example, when considering the gravitational redshift, in order to achieve 10^{-18} fractional accuracy, a to-1-cm knowledge of the height of the standard above the geoid is needed.

Finally, one should stress that, at such high levels of performance, an accurate characterization of all the systematic frequency shifts as well as of the system reproducibility can be carried out only through direct comparison between two or more independent, nominally similar systems. In this frame, for instance, a relative difference of $3.8(6.1) \cdot 10^{-16}$ was found when comparing the frequencies of the $^2S_{1/2}(F=0)$ - $^2D_{3/2}(F=2)$ reference transition in ^{171}Yb$^+$ for two single ions stored in independent traps [778].

8.1.2 Neutral atoms optical lattice clocks

A second valuable option for developing optical frequency standards hinges on ensembles of cold atoms. In this case, according to Equation 8.1, the promising benefit is the $\sqrt{N}$ enhancement in stability offered by ensembles of N atoms within a cloud, an optical lattice, or a continuous beam.

As discussed in Chapter 5, one of the most deeply investigated atomic beam standards is the two-photon 1S-2S hydrogen transition at 243 nm, which is characterized by a natural linewidth of 1.3 Hz. As already described thoroughly, such two-photon transition frequency has been measured with a fractional uncertainty of $4.2 \cdot 10^{-15}$. Since the predominant sources of uncertainty are presently associated with the relatively high mean velocity of the hydrogen atoms in the beam, further significant improvements in precision are closely dependent on the development of effective techniques for laser cooling and trapping atomic hydrogen [715].

In this context, although both first and second order Doppler shifts are heavily reduced in a magneto-optical trap (MOT), appreciable ac Stark and Zeeman shifts of the atomic energy levels arise due to the use of laser beams and magnetic field as required by the MOT operation. Thus, turning off of these fields during the clock transition frequency measurement is required, but this inevitably causes the atoms to expand ballistically under gravity.

As a consequence, detrimental velocity-related systematic frequency shifts appear and, even more fundamental, the maximum interrogation time is limited, thereby restricting clock operation to relatively broad transitions. Just as an example, in the MOT-based calcium atom optical frequency standard, the clock transition is the 657-nm $^1S_0$-$^3P_1$ intercombination line that has a natural linewidth of about 400 Hz.

The paradigm changed when a clever scheme was devised to increase the interrogation time without causing recognizable perturbations to the clock transition frequency [779]. In such a novel approach, called an "optical lattice clock," while utilizing a big number N of neutral atoms to substantially ameliorate the quantum projection noise limit, a single ion in a Paul trap is simulated. Actually, this unique combination is accomplished by trapping millions of neutral atoms in an optical lattice where well-controlled perturbations are achieved by appropriate design of the light-shift potentials (Figure 8.4). Also, the sub-wavelength localization of a single atom in each lattice site suppresses the first-order Doppler shift and collisional shift.

In more detail, cancelling out of the light-field perturbation is attained by means of the so-called "magic wavelength" protocol. In essence, although the electronic states of atoms trapped in an optical lattice are, in general, considerably energy-shifted, such light-shift perturbation can be eliminated if the lattice field provides exactly the same amount of light shift for the two states used in the clock transition. Now, the clock transition frequency of atoms exposed to a lattice laser with an electric field amplitude E is given by the sum of the unperturbed transition frequency ν_0 and the differential light shift ν_{ac}. In the electric dipole (E1) approximation, this can be expressed as [779]

$$\nu(\lambda_L, e_L) = \nu_0 - \frac{\Delta\alpha_{E1}(\lambda_L, e_L)}{2h}E^2 + O(E^4) \tag{8.5}$$

where $\Delta\alpha_{E1}(\lambda_L, e_L) = \alpha_b(\lambda_L, e_L) - \alpha_a(\lambda_L, e_L)$ is the difference between the E1 polarizabilities of the upper ($|b\rangle$) and lower ($|a\rangle$) states. By tuning the laser wavelength λ_L and polarization e_L to satisfy $\Delta\alpha_{E1}(\lambda_L, e_L) = 0$, the observed atomic transition frequency ν will be equal to ν_0 regardless of the lattice laser intensities ($\propto E^2$), as long as higher-order corrections $O(E^4)$ are negligible. This particular wavelength is referred to as the magic wavelength, λ_m. Such scheme can be applied quite generally to atoms in groups II and IIB such as He, Be, Mg, Ca, Yb, Zn, Cd, and Hg, which exhibit the $J = 0 \rightarrow J = 0$ transition between long-lived states. Ultimately, fourth and higher-order E1 light shift, multipolar atom-lattice interactions, blackbody radiation shifts as well as collisional interactions restrict the effectiveness of an optical lattice clock. In the following, we take isotopes of Sr as examples to gain a deeper insight. Figure 8.5 shows the light shift for $|a\rangle \equiv |^1S_0\rangle$ (blue line) and $|b\rangle \equiv |^3P_0\rangle$ (red line) states as a function of the laser frequency $\nu_L = c/\lambda_L$, which are calculated by summing up the dipole moments that couple $|a\rangle$ and $|b\rangle$ to the relevant states (assuming a laser intensity of $I = 10$ kW cm^{-2}). The crossed points of blue and red lines correspond to the magic wavelengths λ_m. While higher-order light shifts are ineludible at red-detuned magic wavelengths, a blue-detuned lattice may overcome this drawback. One such magic wavelength falls on the blue side of the 5s5p $^3P_0$ to 5s6d $^3D_1$ transition at 394 nm (see Figure 8.5). Indeed, since the atoms are trapped near the nodes, the effective light intensity they experience is about one-tenth of that at the antinodes. Then, assuming a trap depth of 10 μK, the fourth-order light shift is estimated to be 0.1 mHz, which corresponds to a fractional uncertainty of $2 \cdot 10^{-19}$.

However, even in a blue-detuned optical lattice clock, things get more complicated as soon as atomic multipolar interactions are considered. Let us examine, for instance, the case of a linearly polarized (e_z) standing-wave electric field $\boldsymbol{E} = \boldsymbol{e}_z \sin(ky)\cos(\omega t)$ with wave number k and frequency $\omega = ck$. From the Maxwell equation $\boldsymbol{\nabla} \times \boldsymbol{E} = -(\partial\boldsymbol{B}/\partial t)/c$, the

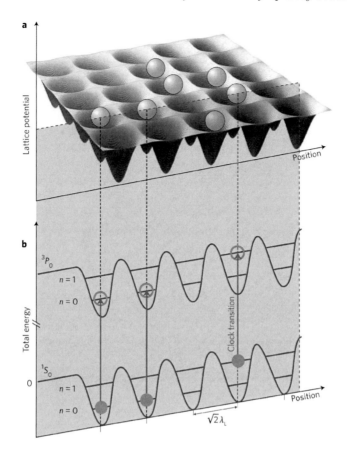

FIGURE 8.4

A one-dimensional spatially periodic intensity pattern, that is an optical lattice with period $\lambda/2$, is created by interference between two counter-propagating laser beams, both of wavelength λ. Laser-cooled atoms are trapped in the lattice by virtue of the interaction between the radiation-induced dipole moment in the atom and the intensity gradient of the laser radiation (recall from Chapter 7 that, if the frequency of the light is lower/higher than the resonance frequency of the atom, the atom will be attracted to the regions of highest/lowest intensity). The lattice potential can be tailored to confine atoms in a region much smaller than the relevant optical wavelength λ_L. Then, atoms are excited on the $^1S_0$-$^3P_0$ clock transition, where the $^1S_0$ and $^3P_0$ states are equally energy shifted by the lattice potential. Here, n represents the vibrational states of atoms in the lattice potential. (Courtesy of [779].)

corresponding magnetic field is given by $\boldsymbol{B} = -\boldsymbol{e}_x E_0 \cos(ky) \sin(\omega t)$, which indicates that the electric and magnetic field amplitudes are out of phase in space by a quarter of the wavelength. This means that the magnetic dipole (M1) interaction is largest at the nodes of the electric field. In addition, since the electric quadrupole (E2) interaction is proportional to the gradient of the electric field, it is also largest at the nodes of the electric field. As a result, the energy shift of atoms in an optical lattice is obtained by the second-order perturbation in the E1, M1, and E2 interactions, which vary as $V_{E1} \sin^2(ky)$, $V_{M1} \cos^2(ky)$, and $V_{V2} \cos^2(ky)$, respectively. Therefore, due to the spatial mismatch of the M1 and E2 interactions with the E1 one, which induces an atomic-motion-dependent light shift, it is no longer possible to perfectly cancel out the light shift in two clock states. To give an idea,

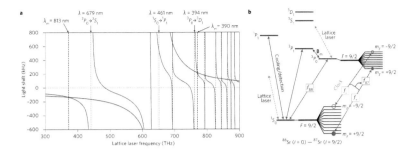

FIGURE 8.5 (SEE COLOR INSERT)
Illustration of the magic wavelength protocol in the case of fermionic ^{87}Sr and bosonic ^{88}Sr isotopes. f_{88} represents the clock transition frequency for ^{88}Sr, where an applied magnetic field B_m mixes the $^3P_0$ and $^3P_1$ states to permit the clock transition. $f_{87} = (f_+ + f_-)/2$ provides the clock transition frequency for ^{87}Sr. (Courtesy of [779].)

in the specific case of Sr, the contributions from the M1 and E2 interactions are 6-7 orders of magnitude smaller than that arising from the E1 one. Nevertheless, their effects cannot be neglected when addressing a 10^{-18}-level uncertainty [780]. So, denoting the differential polarizabilities with $\Delta\alpha_X(\lambda_L)$ and the corresponding spatial distributions with $q_X(\boldsymbol{r})$ for $X = E1, M1$, and $M2$ interactions, the transition frequency of atoms in the optical lattice can be generalized as

$$\nu(\lambda_L) = \nu_0 - \frac{[\Delta\alpha_{E1}(\lambda_L)q_{E1}(\boldsymbol{r}) + \Delta\alpha_{M1}(\lambda_L)q_{M1}(\boldsymbol{r}) + \Delta\alpha_{E2}(\lambda_L)q_{E2}(\boldsymbol{r})]E^2}{2h} \tag{8.6}$$

where the fourth- and higher-order terms as well as light polarization dependence are omitted. Now, in order to eliminate the atomic-motion-dependent light shift (caused by multipolar interactions), one has to choose a particular combination of optical lattice geometry and electric field polarization forcing the $q_{M1}(\boldsymbol{r})$ and/or $q_{E2}(\boldsymbol{r})$ terms to be either in or out of phase with the spatial dependence of $q_{E1}(\boldsymbol{r})$. For example, in the case of a 1-D lattice with the E1 spatial dependence $q_{E1}(\boldsymbol{r}) = \sin^2(ky)$, the corresponding M1 and E2 interactions are $q_{M1}(\boldsymbol{r}) = q_{E2}(\boldsymbol{r}) = \cos^2(ky) = 1 - q_{E1}(\boldsymbol{r})$. Thus, by taking $\Delta\alpha_{EM} = \Delta\alpha_{E1} - \Delta\alpha_{M1} - \Delta\alpha_{E2}$ and $\Delta\alpha_0 = \Delta\alpha_{M1} + \Delta\alpha_{E2}$, equation 8.6 can be rewritten as $\nu(\lambda_L) = \nu_0 - \Delta\alpha_{EM}(\lambda_L)q_{E1}(\boldsymbol{r})E^2/(2h) - \Delta\alpha_0 E^2/(2h)$. This suggests that the magic wavelength can be redefined as that satisfying $\Delta\alpha_{EM}(\lambda_m) = 0$. It is worth noting that the last term just represents a spatially constant offset (typically 10 mHz or below) that only depends on the total laser intensity ($\propto E^2$) giving life to the lattice. This offset frequency can be precisely determined by measuring the atomic vibrational frequencies Ω in the lattice.

The last two issues relevant to the design of optical lattice clocks concern the collisional frequency shift and the vector/tensor light shift:

- In ultracold atoms, the collisional shift is related to the mean field energy shift $4\pi\hbar^2 a\rho g^{(2)}(0)/m$ of the relevant electronic state, where a is the s-wave scattering length, ρ is the atomic density, m is the atomic mass, and $g^{(2)}(0)$ is the two-particle correlation function at zero distance (which is zero for identical fermions and 1-2 for distinguishable or bosonic atoms). So, collisional shifts are suppressed for ultracold spin-polarized fermions, whereas they are intrinsically unavoidable for bosons;

- The vector/tensor light shift arises from the coupling between the clock states and the light polarization of the lattice field. Such coupling is caused by a non-zero total angular momentum $F = I + J$ of the clock states, which can be zero for bosons, but not for fermions (indeed, since optical lattice clocks employ atoms with $J = 0$ electronic states, isotopes with a nuclear spin of $I = 0$ or any integer obey Bose statistics, and those with half-integer nuclear spin obey Fermi statistics). This drawback can be overcome, for instance, in a 1-D lattice composed of a single electric field vector. Here, the resulting spatially uniform light polarization allows the vector light shift to be cancelled out by alternately interrogating the transition frequencies $f_{\pm}$ corresponding to the $^1S_0(m_F = \pm F)$ - $^3P_0(m_F = \pm F)$ transition to obtain $f_{87} = (f_+ + f_-)/2$. This cancellation technique simultaneously removes the first-order Zeeman shift to realize virtual spin-zero atoms.

In the frame of magic lattice spectroscopy, absolute frequency measurements on the highly forbidden $^1S_0$-$^3P_0$ clock transition of ^{87}Sr have been carried out by groups at Tokyo-NMIJ, JILA [781] and SYRTE [782]. Spectroscopy is usually performed by using a ~ 100 ms Rabi pulse which, when on resonance, transfers a fraction of the atoms into the $^3P_0$ state. After applying the clock pulse, atoms remaining in the $^1S_0$ state are detected by measuring fluorescence on the strong $^1S_0$-$^1P_1$ transition. The population in the $^3P_0$ state is then measured by first pumping the atoms back to the $^1S_0$ state (through intermediate states not shown in figure) and then by again measuring the fluorescence on the $^1S_0$-$^1P_1$ transition (combining these two measurements gives a normalized excitation fraction insensitive to atomic number fluctuations from shot to shot). The agreement between the independent measurements by the above three groups led the Comité International des Poids et Mesures (CIPM) to adopt in 2006 the ^{87}Sr-based optical lattice clock transition frequency ($\nu_{Sr} = 429228004229873.7$) as a secondary representation of the second (with an uncertainty of $1 \cdot 10^{-15}$). Incidentally, the NIST group also measured the transition frequency of ^{171}Yb to be $\nu_{Yb} = 518295836590865.2(7)$ Hz [783].

Although experiments performed on fermionic ^{87}Sr and ^{171}Yb atoms trapped in 1-D lattices have revealed so far no prickly root of indeterminateness, investigating uncertainties beyond 10^{-17} may shed new light on a variety of intriguing phenomena. These include collisional shifts between spin-polarized fermions, tunnelling of atoms between lattice sites, hyperpolarizability effects, atomic multipolar interactions with lattices, and blackbody-radiation (BBR) shifts [779]. Addressing these issues is necessary to ascertain the effectiveness of the optical lattice clock scheme towards an uncertainty level of 10^{-18}. One of the most severe obstacles is represented by the BBR shift, which has a sensitivity of 29 (17) mHz K^{-1} for Sr (Yb) at a temperature of 293 K. Therefore, stabilization of the temperature to better than 0.1 K or the use of a cryogenic environment is required. Otherwise, one can resort to atomic elements like Hg whose BBR shift is an order of magnitude smaller than that of Sr or Yb.

Obviously, in order to evaluate the performance of optical clocks below uncertainties of 10^{-16}, references with equivalent or better performance are needed. Therefore, comparisons between two independent optical lattice clocks have been carried out [784]. However, contrary to expectations induced by the large number of the involved atoms, such optical lattice clocks exhibited stabilities of a few $10^{-15}\tau^{-1/2}$, essentially limited by the Dick effect (attributable to the local oscillator noise). Indeed, while quantum projection noise (QPN) limited stability (that is the ultimate measure of stability) has been demonstrated in cesium clocks and in single-ion optical clocks (where the quantum noise overwhelms the Dick effect), in well-designed optical lattice clocks ($N \simeq 1 \cdot 10^6$, $\Delta\nu/(\pi\nu_0) \simeq 10^{-15}$, $T \simeq 1$ Hz) the QPN limit can be calculated from Equation 8.1 as $\sigma_y(\tau) \simeq 10^{-18}\tau^{-1/2}$. Such level is masked by the Dick effect. Nevertheless, when synchronously evaluating the frequency

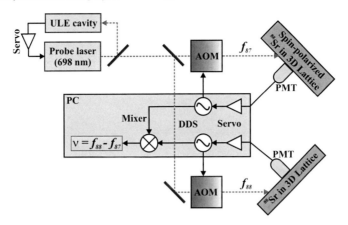

FIGURE 8.6

Experimental setup for synchronous comparison of two independent lattice clocks. Two probe beams derived from a single laser stabilized to an ULE cavity, with frequencies f_{87} and f_{88} (separated by a frequency difference of $\simeq 62$ MHz), simultaneously interrogate the clock transitions of the two isotopes ^{87}Sr and ^{88}Sr. Respective excitation probabilities are measured by PMTs to servo-lock laser frequencies f_{87} and f_{88} by steering AOMs driven by a direct digital synthesizer (DDS). The frequency difference $\nu = f_{88} - f_{87}$ is recorded in time series to evaluate the Allan deviation. (Adapted from [784].)

difference of two clocks by using a common clock laser, the stability degradation due to the Dick effect may be rejected as a common-mode noise, thereby allowing the two clocks to be compared at the QPN limit. Figure 8.6 shows the experimental setup for the frequency comparison of optical lattice clocks operated using ^{87}Sr and ^{88}Sr atoms [784]. With this approach, the Allan standard deviation reached $1 \cdot 10^{-17}$ in an averaging time of 1600 s by cancelling out the Dick effect to approach the QPN limit. More recently, the same group has also demonstrated fiber-based remote comparison of two ^{87}Sr clocks in 24-km distant labs with an instability of $5 \cdot 10^{-16}$ over an averaging time of 1000 s [785].

With the topic of optical atomic clocks we have inaugurated, in fact, the discussion on future trends in fundamental and applied physics. Just as an example, the gravitational redshift of the clock transition frequency could be exploited for refined measurements of the Earth's gravitational potential. Furthermore, when operated in space environment, optical atomic clocks could also find applications in Earth observation (geoscience) as well as in future generations of global satellite navigation systems [768]. However, it is likely that the most immediate applications will be restricted to fundamental science. In particular, as we will see in a short while, highly stable and reproducible atomic frequency standards play a key role in tests of fundamental physical theories such as general relativity and quantum electrodynamics. This scientific horizon, however, is but one aspect of the variegated speculative landscape opened by the field of laser-based frequency measurements. In this broader framework, we will try to outline the most exciting, current, and forthcoming research directions. In this spirit, some pertinent streams in the brand new realm of quantum metrology will be also sketched. It is likewise obvious that we are talking about research areas which are experiencing a veritable outburst. Therefore our survey will inevitably be incomplete, and only a few highlights will be touched on.

8.2   The hydrogen atom as an inexhaustible wellspring of advances in precision spectroscopy

Due to its simple structure (it is indeed the only stable neutral two-body system) and abundance in the universe, hydrogen is the most important atom. From a spectroscopic point of view, its relevance stems from the occurrence of an extremely rich spectrum of resonances extending from the radio frequency to the vacuum ultraviolet, several of them being particularly narrow and hence suitable for metrology. In addition, since its energy levels can be calculated with astonishing precision, sophisticated tests of fundamental theories can be inferred by comparison between theoretical and experimental data. Inspired by [786, 787, 788, 789], the aim of this section is precisely to show how studies on the hydrogen atom have always played a key role in the development of modern physics, and how increasingly precise spectroscopic measurements have gradually contributed to the improvement of existing physical theoretical models.

The experimental investigation of the atomic hydrogen spectrum can be traced back to 1885 when J. Balmer suggested an empirical formula to express the wavelengths of a particular series of lines. In 1889 J.R. Rydberg proposed independently the following more general relationship

$$\frac{1}{\lambda} = R_\infty \left(\frac{1}{p^2} - \frac{1}{n^2} \right) \tag{8.7}$$

where R_∞ is a constant and n and p any integers (the Balmer series is obtained for $p = 2$, while the case $p = 1$ corresponds to the Lyman one). From this wholly empirical formula, which was unexplainable in the frame of classical mechanics, the first value of the Rydberg constant R_∞ was determined. Afterwards, Rydberg's formula was put on solid theoretical foundations firstly by Bohr's theory and then by Schrödinger's equation. This led to the following theoretical expression for the Rydberg constant

$$\frac{1}{\lambda} = \frac{E_n - E_p}{hc} = \frac{me^4}{8\varepsilon_0^2 h^3 c} \left(\frac{1}{p^2} - \frac{1}{n^2} \right) \equiv R_\infty \left(\frac{1}{p^2} - \frac{1}{n^2} \right) \tag{8.8}$$

where m and e represent the mass and the charge of the electron. Equation 8.8 is derived by considering the motion of the electron around a fixed center of infinite mass and charge $-e$. If one takes into account the motion of the nucleus of mass M_p, the energy of the levels is easily generalized to

$$E_n = -hcR_H \frac{1}{n^2} \tag{8.9}$$

where

$$R_H = \frac{R_\infty}{1 + m/M_p} \tag{8.10}$$

Therefore, the knowledge of the electron-to-proton mass ratio is necessary to relate experimental measurements with the theoretical Rydberg constant R_∞.

In the early 1900's, the growing refinement in the spectroscopic sources and techniques soon revealed that, beyond the gross structure described by Equation 8.9, the hydrogen spectral lines possessed a fine structure, whose interpretation eventually motivated the introduction of the electron intrinsic spin concept. This was first accommodated in the phenomenological model by Uhlenbeck and Goudsmit and later shown by Dirac to gush out consistently in the frame of relativistic quantum mechanics. Indeed, the electron velocity v in the hydrogen atom is not so small compared to the speed of light c, the ratio being precisely on the order of the fine structure constant

$$\alpha = \frac{e^2}{4\pi\varepsilon_0 \hbar c} \simeq \frac{1}{137} \tag{8.11}$$

The main result of Dirac's theory is the relativistic wave equation which naturally introduces the electron spin s and yields energy levels characterized by the total angular momentum quantum number j ($\boldsymbol{j} = \boldsymbol{l} + \boldsymbol{s}$):

$$E_{nj} = -hcR_H \frac{1}{n^2} \left[1 + \frac{\alpha^2}{n^2} \left(\frac{n}{j+1/2} - \frac{3}{4} \right) + ... \right] \tag{8.12}$$

This result lifts the degeneracy in j of the levels calculated from the non-relativistic equation, and it gives the fine structure splitting which varies approximately as $1/n^3$; however, it preserves the degeneracy in l for two levels corresponding to the same j but with different values of l.

Later, using radio-frequency spectroscopy, Lamb and Retherford showed that, contrary to Dirac's prediction, the $2S_{1/2}$ and $2P_{1/2}$ energy levels were not degenerate in energy. The clarification of this tiny energy difference (less than 10^{-6} of the energy of the states) captured the research activity of several mid-twentieth-century famous physicists (Bethe, Dyson, Feynmann, Schwinger, Tomonaga, Weisskopf above all), culminating in the birth of quantum electrodynamics (QED). In particular, the detailed character of the interaction between the electron and the vacuum fluctuations was recognized as responsible for the deviation of experimental findings from Dirac's theoretical predictions. The same coupling accounts for the de-excitation of atomic levels by spontaneous decay, even in the absence of an externally applied field. A simple explanation (Welton model) can be provided as follows [788]. Due to the residual energy of the electromagnetic-field empty modes (the energy $\hbar\omega/2$ of the harmonic oscillators), the electron is subjected to the fluctuations of the vacuum field which induce fluctuations in its position. This modifies the Coulomb potential seen by the electron and is particularly significant for the S states (with respect to the P ones), for which the electron has a large probability to be inside the nucleus. As a result, the energy of the S states slightly increases. This is the reason for the splitting between the $2S_{1/2}$ and $2P_{1/2}$ levels. In an alternative picture, the same effect is related to the self-energy corresponding to the emission and re-absorption of virtual photons by the bound electron. Such self-energy term is but one of the many contributions which arise when calculating the hydrogen spectrum in the frame of quantum electrodynamics. Actually, except for the Dirac energy with reduced mass corrections presented above, $E_{n,j}^{Dirac}$, and the hyperfine splitting discussed below, $E_{n,l,j,F}^{HFS}$, all relativistic and QED corrections to the hydrogen energy are lumped into the Lamb shift, $E_{n,l,j}^{LS}$, such that the complete expression for the hydrogen energy levels can be written as

$$E_H = E_{n,j}^{Dirac} + E_{n,l,j}^{LS} + E_{n,l,j,F}^{HFS} \tag{8.13}$$

As detailed in [788, 790, 789] and references therein, each of these $E_{n,l,j}^{LS}$ terms is generally treated as a power series in $Z\alpha$ and $\ln(Z\alpha)$. The most important terms include vacuum polarization, recoil corrections, two-photon corrections, radiative-recoil corrections, three-photon corrections, finite nuclear size effect, nuclear-size correction to self energy and vacuum polarization, and nucleus self energy. Among these, particularly significant is the term related to non-zero size of the nucleus. Indeed, the attractive center was assumed as a dimensionless point so far. In fact, the electric charge of the nucleus (the proton in the case of hydrogen) has a small, but finite volume which can be characterized by the mean square value of the radius of this charge. Even if the RMS radius of the proton is only $\sqrt{\langle r_p^2 \rangle} \simeq 0.8$

fm, at extremely high levels of precision we can no longer ignore the slight reduction of the binding energy due to the electron penetrating inside the volume of the nucleus. Such nuclear size correction is important only for S states which have a significant probability density near the center of attraction. This small diminution of the negative binding energy increases slightly the energy of the S level, but not the P level, by an amount of relative order

$$\frac{\Delta E_n}{|E_n|} = \frac{\langle r_p^2 \rangle}{(\hbar/mc)^2} \frac{4\alpha^2}{n^2} \tag{8.14}$$

where $\hbar/mc$ is the Compton wavelength (divided by 2π). This gives a relative contribution whose maximum value for the ground level ($n = 1$) is of order of $3 \cdot 10^{-10}$ (the effect is reduced by a factor n for other levels).

Finally, one has to consider the interaction between the electron and the magnetic moment of the nucleus, which splits the fine-structure energy levels in two (or more) distinct hyperfine sub-levels. This effect is much more important than many of the corrections discussed above, as the corresponding energy differences are of the order of 1 GHz in the ground level $n = 1$ and 100 MHz in the second level $n = 2$. Nevertheless, in practice this effect is quite decoupled from the preceding calculations and can be easily eliminated in the measurement of energy levels. The additional energies of the hyperfine sub-levels of angular momenta $\boldsymbol{F} = \boldsymbol{I} + \boldsymbol{j}$ are given by the following formula [789]

$$\Delta E_{n,l,j,F}^{HFS} = \frac{\alpha^2 g_N R_\infty hc(1+\delta_N)}{n^3(m_e/M_p)} \frac{F(F+1) - I(I+1) - j(j+1)}{j(j+1)(2l+1)} \tag{8.15}$$

where g_N is the nuclear g-factor relating the magnetic moment μ to the nuclear spin via $\mu = g_N \mu_N I$ and δ_N is a small relativistic correction. The hyperfine structure is only well known experimentally for the low-lying states with $n = 1, 2$. In particular, the determination of the transition frequency at 1420 MHz between the $F = 1$ and $F = 0$ sub-levels has attained a very high level of precision (10^{-12}) thanks to the hydrogen maser. For higher lying states, the hyperfine structure is typically approximated from that of the low-lying states using the above equation. Accurate calculations of the hyperfine splittings are not available due to nuclear effects such as Zemach radius and polarizability. Nevertheless, in the particular combination $D_{21} = 8f_{HFS}(2S) - f_{HFS}(1S)$ nuclear size effects cancel to a large extent thus allowing to test theory against experiments [789].

To summarize, Figure 8.7 shows the energy levels of atomic hydrogen for successive steps of the theory. Clearly resolving the $2S$ Lamb shift (i.e., the splitting between the $2S_{1/2}$ and $2P_{1/2}$ states) in the optical spectrum, the observation in 1972 [791] of a Doppler-free saturation spectrum on the red hydrogen Balmer$-\alpha$ line marked, in fact, the beginning of a long adventure in precision spectroscopy of the simple hydrogen atom, which permits unique confrontations between experiment and theory. This quest continues today. It has inspired many advances in spectroscopic techniques, including the first proposal for laser cooling of atomic gases and, most recently, the femtosecond laser frequency comb [53].

8.2.1 Determination of the Rydberg constant and of the proton radius

As illustrated in Figure 8.8, thanks to these developments, the accuracy of the Rydberg constant R_∞ has been improved by several orders of magnitude in three decades. The hydrogen frequency measurements which are currently used for the determination of the Rydberg constant are [788]:

- Lamb shift of the $2S_{1/2}$ level - Since the historic measurement of Lamb and Retherford, a number of direct measurements of the $2S_{1/2} - 2P_{1/2}$ splitting have been reported. Also,

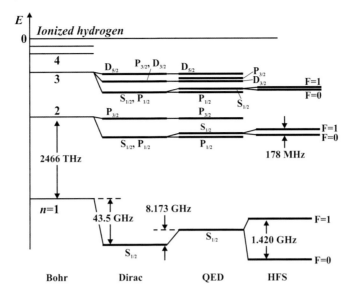

FIGURE 8.7
Energy levels of atomic hydrogen for successive steps of the theory. (Adapted from [789].)

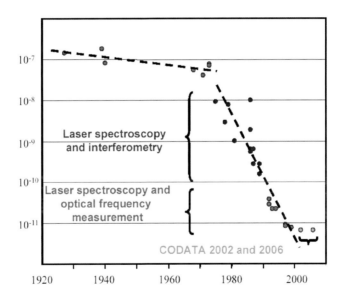

FIGURE 8.8 (SEE COLOR INSERT)
Relative precision of the Rydberg constant from 1920 to the present, clearly showing the improvements due to laser spectroscopy and optical frequency measurements. (Courtesy of [788].)

indirect determinations exist, based on measurements of the $2S_{1/2} - 2P_{3/2}$ splitting in conjunction with theoretical calculations of the $2P_{1/2} - 2P_{3/2}$ splitting. Now, by taking into account both direct and indirect determinations, a mean value of 1057.8439(72) MHz is obtained for the $2S_{1/2}$ Lamb shift;

- $1S - 2S$ transition - As already discussed in Chapter 5, from an experimental point of view, the highest resolution can be achieved on the ultraviolet $1S - 2S$ two-photon resonance with a natural linewidth of only 1 Hz. The first Doppler-free spectra were recorded in 1975 by W.T. Hänsch et al. Since then, in a long series of experiments, Hänsch has considerably improved the precision of the measurement of the $1S - 2S$ frequency. In particular, in the era of optical frequency measurements, Hänsch used a frequency chain which linked the $1S - 2S$ frequency (about 2466 THz) to a transportable CH_4-stabilized He-Ne frequency standard at 88 THz. Now, this complex frequency chain has been superseded by a femtosecond laser frequency comb, which links in one fell swoop the Cs clock at 9 GHz to the optical frequency. Thanks to this technique, Hänsch's group has recently succeeded in measuring the $1S - 2S$ interval with respect to a transportable Cs atomic fountain clock from the SYRTE. The latest value obtained is

$$f_{1S-2S} = 2466061413187035(10)\text{Hz} \tag{8.16}$$

- $2S - nS$ and $2S - nD$ transitions - In Paris, Biraben's group began to study the $2S - nS$ and $2S - nD$ transitions in 1983. These experiments are complementary to those of the $1S - 2S$ measurements, as the Lamb shift of the $2S$ level has been measured precisely, and, consequently, it is straightforward to extract the Rydberg constant from the $2S - nS/D$ interval. The two-photon $2S - nS/D$ transitions are induced in a metastable $2S$ atomic beam of hydrogen or deuterium collinear with the counter-propagating laser beams. The excitation wavelength is in the near-infrared, for example 778 nm for the $2S - 8S/D$ transitions. A relative uncertainty of $7.6 \cdot 10^{-12}$ was obtained for the $2S_{1/2} - 8D_{5/2}$, essentially limited by the natural width of the $8D$ level (572 kHz) and the inhomogeneous light shift experienced by the atoms passing through the Gaussian profile of the laser beams

$$f_{2S_{1/2}-8D_{5/2}} = 770649561581.1(5.9)\text{kHz} \tag{8.17}$$

Then, the procedure used to extract the Rydberg constant consists of comparing these experimental data for the transition frequencies with the calculated energy differences. It is possible to make such least-squares adjustment with only the hydrogen data, the values of the fine structure constant α, and the electron-to-proton mass ratio m_e/m_p being given a priori. Starting with α, it is known very accurately from the electron $g - 2$ experiment [792]. Indeed, QED predicts a relationship between the dimensionless magnetic moment of the electron (g) and the fine structure constant (α). Thus, the latest measurement of g using a one-electron quantum cyclotron, together with a QED calculation involving 891 eighth-order Feynman diagrams, have determined $\alpha^{-1} = 137.035999084(51)$. Concerning m_e/m_p, this has been inferred precisely in Penning traps from an accurate measurement of the cyclotron frequencies [790]. Then, the value reported in the 2006 CODATA adjustment is

$$R_\infty = 10973731.568527(73)\text{m}^{-1} \tag{8.18}$$

with a relative uncertainty of $6.6 \cdot 10^{-12}$. The present accuracy is such that, when comparing theory and experiment, one cannot dispense with considering the details of the proton charge distribution. Indeed, the different scaling of the Rydberg constant and the rms proton charge radius $\langle r_p^2 \rangle$ ($\propto n^{-2}$ and $\propto n^{-3}$, respectively) allows their simultaneous determination by measuring different transition frequencies in hydrogen. In this context, a major problem is represented by the fact that every other transition than $1S - 2S$ cannot be measured with similar accuracy. Thus, QED tests with hydrogen have been limited to the 10^{-12} level due to an insufficient knowledge of the proton charge radius. To gain a deeper knowledge of $\langle r_p^2 \rangle$, spectroscopists turned their attention to muonic hydrogen (μp). The latter is obtained by replacing the electron with a negative muon (muonic hydrogen atoms are much smaller than typical hydrogen atoms because the much larger mass of the muon gives it a much smaller ground-state wavefunction than is observed for the electron). Here, by virtue of to-the-third-power scaling with the orbiting particle's mass, the finite-size effect is much more pronounced, thus permitting a more precise determination of $\langle r_p^2 \rangle$ with less experimental accuracy. For this reason, for more than forty years, a measurement of the μp Lamb shift has been considered one of the fundamental experiments in atomic spectroscopy, but only recent progress in muon beams and laser technology made such an experiment feasible [793]. In that work, the $2S_{1/2}^{F=1} - 2P_{3/2}^{F=2}$ transition frequency in μp was measured to be 49881.88(76) GHz by means of pulsed laser spectroscopy. Then, by comparing this experimental value with its theoretical prediction based on bound-state QED, a proton radius value of $r_p = 0.84184(67)$ fm was determined. This new value is an order of magnitude preciser than previous results but disagrees by 5 standard deviations from the CODATA (0.8768(69) fm) and the electron-proton scattering values. Such discrepancy has been baptized as *proton size puzzle*. Indeed, it is not clear whether there is a problem in the electron scattering data, ordinary hydrogen data or theory for either ordinary hydrogen or muonic hydrogen which is needed to extract the charge radius from the measured transition frequencies. An overview of the present effort attempting to solve the observed discrepancy is given in [794].

Besides QED tests, the $1S - 2S$ resonance has been used to put constraints on possible variations of fundamental constants and violations of Lorentz boost invariance (see below for both issues). Furthermore, it may be used in severe tests of the charge conjugation/parity/time reversal (CPT) theorem by comparison with the same transition in anti-hydrogen (i.e., an H atom made entirely of antiparticles) [795].

8.3 Spectroscopy of cold, trapped metastable helium

Helium is the simplest atom after hydrogen, composed of only three particles (the nucleus and two electrons) and, due to its simplicity, *ab initio* calculations of its atomic structure are still possible with a very high accuracy. Atomic spectroscopy, performed on helium samples at room temperature, in beams or in Bose-Einstein condensates, has provided high precision measurements of key parameters, like lifetimes, Lamb shifts, or fine structure separations [796]. Accurate frequency measurements of He transitions can be used as a stringent test of the QED theory for a bound three-body system. Alternatively, comparison between theoretical calculations and experimental measurements for the 2^3P fine structure splittings has been used to give an accurate determination, based on spectroscopy, of the fine structure constant, α [797, 798]. Assuming the QED theory of fine structure energies of an atomic system to be correct, a determination of α is possible by frequency measurements of these fine structure splittings, which are proportional to α^2. Looking at the Periodic Table of Elements, He belongs to the family of noble gases. Noble gases have resonance lines from the ground state to other levels at wavelengths far in the ultraviolet. However, these atoms have metastable excited states which are connected to higher-lying levels by allowed transitions that are accessible to existing lasers. Such metastable states have lifetimes ranging from 15 seconds to about 8000 seconds (the longer lifetime being for the He 2^3S level) and represent "effective" ground states for optical manipulation and detection. As an example, Figure 8.9 shows the relevant level structure of helium. In these atoms, decay to the $1S_0$ ground state from metastable triplet states is doubly forbidden. Indeed, transitions to the ground state are not electric-dipole allowed and spin-flip (triplet-singlet) forbidden.

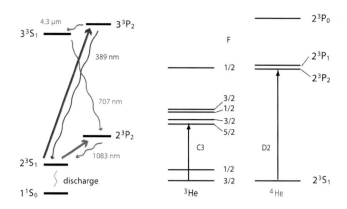

FIGURE 8.9
Energy levels for helium: (left) principal helium transitions for laser cooling and trapping. The long-lived metastable 2^3S_1 state (20 eV above the ground state) is used as an effective ground state and the $2^3S_1 - 2^3P_2$ transition at 1083 nm transition is used for laser cooling and trapping. The $2^3S_1 - 3^3P_2$ transition at 389 nm is used in some studies. (right) Excited-state manifold for the 1083 nm cooling transition in both ^{3}He and ^{4}He (not to scale). For ^{4}He, the $2^3S_1 - 2^3P_2$ D2 transition is used and for ^{3}He, which shows a hyperfine structure as a result of the $I = 1/2$ nuclear spin, the C3 transition ($F = 3/2 - F = 5/2$) is used. (Courtesy of [796].)

It is interesting to note that, in spite of their internal energy (in the order of 20 eV) that is large compared to kinetic energy values around quantum degeneracy (about 10^{-10} eV), Bose-Einstein condensation of metastable helium atoms was demonstrated in 2001 [799, 800] and the first degenerate Fermi gas of metastable helium was observed in 2006 [801].

As mentioned above, low-lying triplet states of He have been the subject of many spectroscopic measurements and theoretical calculations. Due to the easier access with available laser sources, He 2^3P and 3^3P manifolds have been widely investigated, and high precision determinations have been possible after the introduction of optical frequency combs (see [641] for a recent review and [802]). In particular, many experiments, over more than one decade, have targeted the transitions between the 2^3S and 2^3P states in ^{4}He, around 1083 nm wavelength, to get information on fine structure separations, with the perspective to get an accurate α value [803, 804, 805, 797, 806, 807, 808, 809]. Indeed, due to the wider fine structure separations, as compared to hydrogen, helium has long been considered an optimal candidate for a spectroscopy-based α determination (see upper frame of Figure 8.10 for a comparison of updated α determinations). Measurement accuracy for the frequency separations of the 2^3P energy manifold, exploiting different techniques ([803, 810] using optically pumped magnetic resonance microwave spectroscopy), ([797, 807, 809] taking frequency differences of the 1083 nm transitions) was pushed to the sub-kHz level (9 ppb for the largest interval connecting $J = 0$ to $J = 2$ [807]). However, though recent theoretical calculations have resolved previous discrepancies between theory and experiment [811, 812, 813], theoretical uncertainty is still about one order of magnitude worse than experimental values.

Similarly to what happened for hydrogen ([814, 815]), the history and the perspectives of spectroscopic measurements of He transitions were changed by pioneering work on pure frequency measurements, i.e., with a direct link to the atomic cesium primary frequency standard, replacing traditional interferometric techniques, prone to systematic effects [816]. For He, the first pure frequency measurement was performed to estimate another key parameter, the Lamb shift of the 2^3S_1 level, by measuring the absolute frequency of the 2^3S_1 - 3^3P_0 transition of ^{4}He around 389 nm wavelength [816].

The nuclear charge radius (r_c) is another important parameter that can be calculated from precise spectroscopic measurements of He triplet frequencies. Isotopic shift frequency measurements, e.g., comparing transition frequencies of ^{4}He and ^{3}He, provide an estimation of nuclear mass and volume differences between the atomic isotopes. When proper calculation singles out the nuclear mass contribution, the nuclear volume contribution can be determined, and thus the difference between the square charge radii for the two isotopes [817, 818]. The r_c of ^{4}He is equivalent to that of the α-particle, whose radius has been precisely measured [819, 820]. Therefore, a measurement of the isotopic shift of lines of other He isotopes with respect to ^{4}He directly gives a determination of the nuclear charge radius for that isotope. Since nuclear volume contribution accounts for less than 100 ppm for a light atom, like He, the required measurement accuracy must be better than this value. Uncertainties in the order of few tens of ppb could indeed be achieved for the isotope shift measurements, determining an uncertainty on r_c of $1 \cdot 10^{-3}$ fm, where the uncertainty limitation comes from the existing determination of the α-particle radius. Previous results using the isotopic shift to determine r_c were obtained using transitions at 1083 nm [821, 822] or transitions at 389 nm [823], and a satisfactory agreement among them is achieved, though the measurements come from different transitions. Comparison of the laser spectroscopy result for the ^{3}He r_c with estimations from calculation using nuclear theory [824] or electron-nucleus scattering measurements [825] shows that the spectroscopic result has a much lower uncertainty (Figure 8.10, lower graph). Further spectroscopic results for r_c were more recently obtained on short-lived radioactive isotopes of He, namely ^{6}He and ^{8}He [826, 827].

As witnessed by the *proton charge radius puzzle* discussed above, the combined progress of metrological-grade laser spectroscopic techniques and *ab initio* QED calculations of atomic properties is providing clues unveiling possible discrepancies between theory and

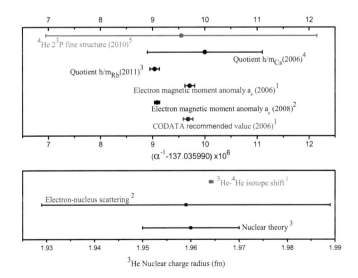

FIGURE 8.10
Top: Comparison of most updated α determinations. Bottom: Comparison of the ^{3}He nuclear charge radius measured from the ^{3}He - ^{4}He isotope shift, by electron-nucleus scattering and calculated with nuclear theory. (Courtesy of [796].)

experiment. These could change our view of elementary interactions or take to a change of the fundamental physical constants. In this frame, two very recent experiments on He have shown a significant discrepancy in the estimation of the difference of the squared nuclear charge radii of ^{3}He and ^{4}He (δr^2) by more than 4 standard deviations [828, 829]. The first experiment was done in a ultracold (sub-microkelvin temperature) mixture of ^{3}He and ^{4}He, where quantum degeneracy is reached by sympathetic cooling of metastable ^{3}He atoms with a Bose-Einstein condensate of metastable ^{4}He. In the second (and most recent) one, instead, the sample was an uncooled atomic beam of metastable (excited to the 2^3S level by electron collisions) ^{3}He atoms. In this work, the absolute frequency of seven out of the nine allowed transitions between the 2^3S and 2^3P hyperfine manifolds was measured by using an optical frequency comb synthesizer-assisted spectrometer [802] (incidentally, a relative uncertainty of $5 \cdot 10^{-12}$ was achieved in the frequency measurements, that is the most precise result for any optical ^{3}He transition to date). The resulting 2^3P - 2^3S centroid frequency was found to be 276702827204.8(2.4) kHz. Comparing this value with the known result for the ^{4}He centroid and performing *ab initio* QED calculations of the ^{4}He-^{3}He isotope shift, the difference of the squared nuclear charge radii δr^2 of ^{3}He and ^{4}He was eventually extracted. Finally, it is interesting to note that there is a third, and older, estimation of δr^2 [822]. The three values are respectively:

$$\delta r^2 = 1.028(11) \text{ fm}^2 \tag{8.19}$$

$$\delta r^2 = 1.074(3) \text{ fm}^2 \tag{8.20}$$

$$\delta r^2 = 1.059(3) \text{ fm}^2 \tag{8.21}$$

Such disagreements can be due to systematic effects depending on the specific sample phase or transitions investigated or by some new physics still to unveil. In this respect, it is interesting to note that such disagreements, as well as the proton radius, will probably soon be tested by a new experiment at the Paul Scherrer Institute, targeting a Lamb shift measurement in muonic helium [830].

8.4 Measurements of fundamental constants

As already explained in Chapter 1, today we are witnessing a strong tendency to relate the base units to fundamental constants [40]. As a shining example, this was done in 1983 by fixing the velocity of light c and thus defining the length unit from the time one. Other units, which are still linked to artifacts (that are far from being invariant in space and time), could follow the same line. In the following we focus, in particular, on the Boltzmann constant and the Newton gravitational constant.

8.4.1 Boltzmann constant k_B

Fixed by the temperature (273.16 K) of the triple point of water, the current definition of the Boltzmann constant implies a particular property of macroscopic matter. Indeed, the current value of k_B, $1.3806504(24) \cdot 10^{-23}$ JK^{-1}, recommended by the Committee on Data for Science and Technology (CODATA), comes from the ratio between the molar gas constant R and the Avogadro number N_A. Its relative uncertainty, $1.7 \cdot 10^{-6}$, is mostly due to the uncertainty on R, whose accepted value is that obtained in 1988 using acoustic gas thermometry in argon. In this case, the microscopic interpretation of temperature may provide the key to overcome such artifact. In fact, the mean energy E per particle and per degree is related to the sample temperature T through the well-known relationship $E = (1/2)k_B T$. In turn, this energy may be related to a frequency via the Planck constant. Thus, by directly measuring such a frequency in a gas at a well-defined temperature and fixing the value of k_B would connect temperature and time units.

In this frame, Doppler-broadened laser-absorption spectroscopy can serve as a primary thermometric method, with the advantage of being conceptually simple, applicable to any gas at any temperature, in whatever spectral region. It is well known that the Doppler width, $\Delta \nu_D$ (FWHM), of a line (with a center frequency ν) in an absorbing molecular gas at thermodynamic equilibrium depends on the temperature T through the equation

$$\Delta \nu_D^2 = 8 \ln 2 \frac{k_B T \nu^2}{mc^2} \equiv \frac{k_B}{\alpha} T \quad \text{with} \quad \alpha = \frac{mc^2}{8\nu^2 \ln 2} \tag{8.22}$$

where m is the mass of the molecule. The first spectroscopic determination of k_B was performed in the mid-infrared by Daussy et al. on the ν_2 Q(6,3) rovibrational line of $^{14}NH_3$ at a frequency of 28953694 MHz [831]. The absorption profile was observed in the pressure range between 1 and 10 Pa, at a temperature of 273.15 K, using a CO_2 laser, frequency stabilized on an OsO_4 line. Under these low-pressure conditions, close to the Doppler limit, by measuring the width of the absorption line as a function of the pressure and extrapolating to zero pressure, it was possible to deduce the Boltzmann constant with a relative uncertainty of $1.9 \cdot 10^{-4}$. This approach allows for a very simple spectral analysis, but requires an accurate determination of the gas pressures. Later on, a quite different implementation of laser-absorption spectroscopy for primary gas thermometry was reported [832]. In this work, Casa et al. demonstrated that it is possible to retrieve the gas temperature from a molecular absorption profile even when the gas pressure is sufficiently high that the line shape is far from the Doppler limit, but sufficiently small that one can neglect the averaging effect of velocity-changing collisions, so that the line shape is given by the exponential of a Voigt convolution. Absorption spectroscopy was performed in the near-infrared on a CO_2 gas sample at thermodynamic equilibrium using a distributed feedback (DFB) diode laser, probing the R(12) component of the $\nu_1 + 2\nu_2^0 + \nu_3$ combination band (in contrast to NH_3 this molecular target does not exhibit any hyperfine structure).

The absorption cell (10.5 cm long) was housed inside a stainless steel vacuum chamber (both the cell and the vacuum chamber were equipped with a pair of AR-coated BK7 windows). Consisting of a cylindrical cavity inside an aluminum block, with inner and external surfaces carefully polished, the cell was temperature stabilized by means of four Peltier elements. Three precision platinum resistance thermometers (Pt100) measured the temperature of the cell's body while a proportional integral derivative controller was used to keep the temperature uniform along the cell and constant within 40 mK, over a time interval of ~ 2 h. This active system also permitted to vary the gas temperature between 270 and 330 K. The Pt100 thermometers were calibrated at the triple point of water and at the gallium melting point with an overall accuracy better than 0.01 K. Both during the calibration and when placed in the absorption cell, the thermometers were fed by a 1 mA current and the correction due to the self-heating effect was applied together with the associated uncertainty. The sample cell was filled with CO_2 gas (with a nominal purity of 99.999%) at a pressure between 70 and 130 Pa, measured using a 1300 Pa full-scale capacitance gauge with a 0.25% accuracy. Doppler width determinations were repeated as a function of the gas temperature, in the range between the triple point of water and the gallium melting point. Then, the experimental values of the quantity $\alpha \Delta \nu_D^2$ were plotted as a function of the gas temperature. In this way, the slope of the weighted best-fit line directly provided the value of k_B with a relative accuracy of $\sim 1.6 \cdot 10^{-6}$. Obviously, in this approach special attention must be paid to the frequency calibration of the spectra. For this purpose, a broad laser frequency scan (0.5 cm^{-1}) was performed in order to observe a pair of CO_2 absorption lines, namely, the R(8) $v_1 + 2v_2^0 + v_3$ and R(27) $v_1 + 3v_2^1 - v_2^1 + v_3$ transitions, whose center frequencies, as well as that of the R(12) line under investigation, are accurately known in literature (i.e., with relative uncertainties well below that associated with the Doppler width determination).

More recently, quantitative atomic spectroscopy for primary thermometry was also carried out [833]. In this case, by using a conventional platinum resistance thermometer and the Doppler-thermometry technique, the Doppler broadening of atomic transitions in ^{85}Rb vapor was accurately measured, thus determining k_B with a relative uncertainty of $4.1 \cdot 10^{-4}$ and with a deviation of $2.7 \cdot 10^{-4}$ from the expected value. This experiment, using an effusive vapor, departs significantly from the two Doppler-broadened thermometry techniques discussed above (which rely on weakly absorbing molecules in a diffusive regime) and, in fact, very different systematic effects, like magnetic sensitivity and optical pumping, are dominant.

8.4.2   Newton gravitational constant G

Despite being one of the most measured fundamental physical constants, the Newtonian constant of gravity G is the least precisely known. Besides the obvious metrological interest, many physical theories and areas would benefit from an improved knowledge of G. Shining examples in this sense are represented by the theory of gravitation, astrophysics, and cosmology, as well as geophysical models. The major adversity in carrying out a very accurate measurement of G resides in the extreme weakness of the gravitational force in conjunction with the difficulty of shielding the effects of gravity [834]. The traditional torsion-pendulum method involves a well-characterized moving source mass that induces a torque on a test mass attached to a long fiber. Measurement of the test mass displacement, together with knowledge of the pendulum mechanics and of the source-test mass gravitational force, determines G [835]. In the last decade, several groups have devised new experiments based on various concepts (and hence with radically different systematics) including a beam-balance system [836], a laser interferometry measurement of the acceleration of a freely falling test mass [837] as well as experiments based on Fabry-Perot or microwave cavities [838, 839]. Despite all these efforts, the most precise measurements available today still show substantial discrepancies, limiting the accuracy of the 2006 CODATA recommended value for G to 1 part in 10^4. Therefore, the realization of additional conceptually different experiments is highly desirable in order to identify still hidden systematic effects. In this frame, more recently, two determinations of G were reported, based on cold-atom interferometry [835, 834]. Here, freely falling atoms act as probes of the gravitational field and an atom-interferometry scheme is used to measure the effect of nearby well-characterized source masses.

In the following, we focus on the experiment performed by the group of G.M. Tino [840, 834, 841]. Schematically, with reference to Figure 8.11, ^{87}Rb atoms, trapped and cooled in a magneto-optical trap (MOT), are launched upwards (and cooled down to 2.5 μ K) in a vertical vacuum tube with a moving optical molasses scheme, producing an atomic fountain. Near the apogee of the atomic trajectory, a measurement of their vertical acceleration is performed by a Raman interferometry scheme. External source masses are positioned in two different configurations and the induced phase shift is measured as a function of masses' positions. In order to suppress common mode noise and to reduce systematic effects, the vertical acceleration is simultaneously measured in two vertically separated positions with two atomic samples, that are launched in rapid sequence with a juggling method. From the differential acceleration measurements as a function of the position of source masses, and from the knowledge of the mass distribution, the value of G can be determined.

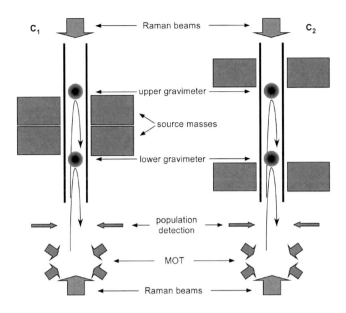

FIGURE 8.11
Schematic of the experiment showing the gravity gradiometer setup with the Raman beams propagating along the vertical direction. During the G measurement, the position of the source masses is alternated between configuration C1 (left) and C2 (right). (Courtesy of [842].)

In more detail, three main stages can be identified:

1. State preparation - After the launch, the atoms are selected both in velocity and by their m_F state. The selection procedure uses laser vertical beams (and a uniform vertical bias field of 250 mG which defines the quantization axis) so that the state preparation can take place simultaneously on both clouds. After this selection, the atoms end up in the $F = 1, m_F = 0$ state with a horizontal temperature of 4 μK and a vertical temperature of 40 nK, corresponding to velocity distribution widths (HWHM), respectively, of $3.3v_{rec}$ and $0.3v_{rec}$ ($v_{rec} = 6$ mm/s for Rb resonance transition);

2. Raman interferometer and detection - In a vertical Raman atom interferometer, free-falling atoms are illuminated by pulses of two laser beams with frequencies ω_1 and ω_2 (see Figure 8.12). The beams counter-propagate along the z-axis with wave vectors $\boldsymbol{k}_1 = k_1\hat{z}$ and $\boldsymbol{k}_2 = k_2\hat{z}$ ($k_i = \omega_i/c$, $i = 1, 2$). ω_1 and ω_2 are close to transitions of a three-level atom in which two ground or metastable states $|a\rangle$ and $|b\rangle$ share a common excited state $|e\rangle$ ($\omega_1 - \omega_2 = \Delta\omega_{ab}$). During the pulses, the lasers drive two-photon Raman transitions between $|a\rangle$ and $|b\rangle$ that lead to Rabi oscillations: the probability of finding atoms in one specific state oscillates temporally between zero and one. A π-pulse, which is a pulse of duration $\tau = \pi/\Omega$, switches the atomic state coherently from $|a\rangle$ to $|b\rangle$ or vice versa. On the contrary, a $\pi/2$-pulse with duration $\tau = \pi/(2\Omega)$ splits the atom wave packet into an equal superposition of $|a\rangle$ and $|b\rangle$ (note that for given transitions, the characteristic Rabi frequency Ω depends only on the light parameters). Thus, an interferometer is realized when the atoms are made to interact with the Raman lasers on the three-pulse sequence $\pi/2 - \pi - \pi/2$, which splits, redirects and recombines the atomic wavepackets. In the specific case of ^{87}Rb under discussion, $|a\rangle$ and $|b\rangle$

coincide respectively with the $F = 1$ and $F = 2$ hyperfine levels in the $5^2S_{1/2}$ state. The Raman laser beams are generated by two ECDLs (both detuned from the D_2 line by 3.4 GHz), and an optical PLL keeps their frequency difference in resonance with the transition between these two hyperfine levels ($\sim$ 6.8 GHz). At the end of this interferometer, the probability of detecting the atoms in the state $F = 2$ is given by $P_2 = (1/2)(1 - \cos\Phi)$, where Φ represents the phase difference accumulated by the wave packets along the two interferometer arms. As commonly accomplished in atomic fountains, probability P_2 is measured using normalized fluorescence detection. This returns the value of Φ. In the presence of a gravity field, atoms experience a phase shift $\Phi = (\boldsymbol{k}_1 - \boldsymbol{k}_2) \cdot \boldsymbol{g}T^2 \equiv \boldsymbol{k}_{eff} \cdot \boldsymbol{g}T^2$ depending on the local gravitational acceleration $\boldsymbol{g}$.

3. Gravity gradiometer - As already mentioned, two spatially separated atomic clouds in free fall along the same vertical axis are simultaneously interrogated by the same Raman beams to provide a measurement of the differential acceleration induced by gravity on the two samples. If $\boldsymbol{g}_{low}$ and $\boldsymbol{g}_{up}$ are the gravity acceleration values at the position of the lower and upper interferometers, then the differential phase shift is given by

$$\delta\Phi \equiv \Phi_{low} - \Phi_{up} = \boldsymbol{k}_{eff} \cdot (\boldsymbol{g}_{low} - \boldsymbol{g}_{up})T^2 \tag{8.23}$$

4. G measurement - When repeated for the two configurations of the source masses (C1 and C2), the above procedure yields two values for the phase difference $\delta\Phi$: $\delta\Phi^{C1}$ and $\delta\Phi^{C2}$. Finally, the difference $\delta\Phi^{C1} - \delta\Phi^{C2}$ is modelled by a complex numerical simulation (taking into account the evolution of the atomic wave-packets and the distribution of the source masses) having G as a unique free parameter [843] (recall that g and G are linked via the law of universal gravitation). In this respect, many efforts were devoted to the control of systematic effects related to atomic trajectories, positioning of source masses, and stray fields. In particular, the high density of tungsten (used for the source masses) was crucial to compensate for the Earth's gravity gradient. In both configurations of the source masses, atom interferometers could therefore be operated in spatial regions where the overall acceleration was slowly varying and the sensitivity of the measurement to the initial position and velocity of the atoms was strongly reduced.

After a careful analysis of the various error sources affecting the measurement, the value obtained for G was $G = 6.667 \cdot 10^{-11} \mathrm{m}^3 \mathrm{Kg}^{-1} \mathrm{s}^{-2}$ (consistent with the 2006 CODATA value with 1 standard deviation) with a statistical uncertainty of $\pm 0.011 \cdot 10^{-11} \mathrm{m}^3 \mathrm{Kg}^{-1} \mathrm{s}^{-2}$ and a systematic uncertainty of $\pm 0.003 \cdot 10^{-11} \mathrm{m}^3 \mathrm{Kg}^{-1} \mathrm{s}^{-2}$. Work is in progress to push the G measurement precision below the 100 ppm level [842].

At the end of this section, we should encourage the reader to refer to [844] for a comprehensive review on interferometry with cold atoms and molecules, which represents by now a rich branch of physics.

8.5 Constancy of fundamental constants

In the last few years there has been a strong interest in the possibility that fundamental physical constants might show variations over cosmological timescales [845, 846]. Such an

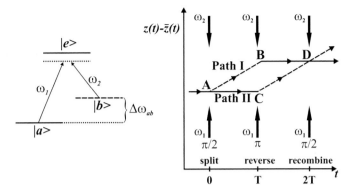

FIGURE 8.12
Graphical illustration of the Raman atom interferometer. (Left) Three-level atom with lower states $|a\rangle$ and $|b\rangle$ that share an excited state $|e\rangle \equiv |5^2P_{3/2}, F = 3\rangle$. (Right) Scheme of the Raman interferometer. On the vertical axis the atomic vertical position $z(t)$ after subtracting the free-fall trajectory $\bar{z}(t)$ is plotted (Adapted from [841].)

effect, as incompatible as it seems with the present foundations of physics, arises quite naturally in modern theories attempting to unify gravity and other interactions. The first conjecture dates back to Dirac's big-numbers hypothesis, which suggested in particular that Newton's gravitational constant G decreased with the inverse of time [847]. On the other hand, generalized Kaluza-Klein models [848] and more recent string theories [849], aiming at incorporating gravitational physics into the Standard Model, postulate the existence of additional compactified dimensions the size of which determines the strength of the fundamental forces. However, since the size of these extra dimensions is on the order of the Planck length, overwhelming energies are needed to experimentally reveal their presence. Nevertheless, theory predicts that if a change in the extent of these dimensions should happen over time, it would manifest as a variation in the fundamental constants of our 4-dimensional world. Actually, such an idea is contained in the widely credited Inflation model of the Universe according to which at a very early stage of evolution a phase transition caused by cooling of the Universe as a result of expansion drastically changed several properties of elementary particles among which the mass and the charge of the electron [850]. But, to our current knowledge, the Universe is still expanding and cooling and this should lead, as well, to some tiny variation of fundamental constants including the fine structure constant α and the proton-to-electron mass ratio $\beta \equiv m_p/m_e$ [851]. These parameters appear prominently both in atomic and molecular transition energies. One possible route to constrain the fractional temporal variation of α (β) is to compare wavelengths of atomic/molecular lines as measured at present epoch on the Earth with the corresponding ones from astronomical objects at redshifts around $z_Q \simeq 3$ (look-back time ~ 12 Gyr) [852, 853, 854]. Recall, incidentally, that such enormous cosmological redshifts imply that the astronomical object has a huge velocity component in the direction away from us; thus, according to Hubble's law, stating that the distance d of the astronomical object is proportional to its recessional velocity v, $d = H_0 v$, we see that the object is very distant from earth and hence we are observing light from very deep into the past. This approach relies on three main ingredients:

1. High-quality quasar absorption spectra of a selected molecular species M (H_2 for example); then, if λ_i is the wavelength of the ith spectral line as collected by the telescope (i.e., the redshifted line position) and λ_i^0 the corresponding rest-frame wavelength as calibrated in the lab (zero-redshift line position), we have

$\lambda_i/\lambda_i^0 = z_Q + 1$. Now, a possible mass dependence of a certain line can be added to this equation by the introduction of the so-called sensitivity coefficient K_i for the ith line, to obtain the corresponding wavelength λ_i'

$$\frac{\lambda_i'}{\lambda_i^0} = (1 + z_Q)\left(1 + \frac{\Delta\beta}{\beta}K_i\right) \tag{8.24}$$

where $K_i \equiv (\beta/\lambda_i)(d\lambda_i/d\beta)$ represents the induced shift of a spectral line as a function of the variation of β. Analogous relationships hold for α.

2. A highly accurate laboratory database of the M spectrum;

3. A precise semi-empirical or *ab initio* calculation of the sensitivity parameters K_i.

Absorption spectra from the high-redshift galaxies are collected by the largest optical telescopes on Earth, e.g., the Very Large Telescope (VLT) or the Keck telescope. These are equipped with high-resolution spectrometers, the Ultraviolet and Visible Echelle Spectrograph (UVES) and the High Resolution Echelle Spectrometer (HIRES), respectively. The accuracies of the astrophysical observations actually limit the present comparisons between laboratory and astrophysical data. In the case of VLT/UVES, for instance, the achieved transition wavelength accuracies are in the order of 10^{-6} - 10^{-7} in the 329-451 nm range. A recent preliminary application of frequency comb technologies in the calibration of astronomical spectra has been shown to improve the accuracy [554]. An additional drawback is that astrophysical data are retrieved from uncontrolled environments and, indeed, results obtained so far are inconclusive and sometimes contradictory [855]. An updated review on the activity searching for α variations in quasar absorption spectra is given in [853], which includes studies on optical atomic spectra, molecular rotational spectra, comparisons of hydrogen hyperfine and UV transitions as well as comparisons involving hyperfine and molecular rotational transitions. The recent paper [856] and references therein provide, instead, an overview of the current constraints for β.

A different, complementary approach relies on precision molecular spectroscopy experiments carried out in laboratory in a self-consistent way, which, while covering only short time spans, offers the inherent advantage to manipulate all the relevant parameters accurately and to investigate possible systematic effects in detail. In the following, we focus on this second road. We anticipate that, contrary to the astrophysical observations, all present laboratory experiments are compatible with constancy of constants to within 1σ. As we will see in a short while, the α (β) parameter is most conveniently addressed in atomic (molecular) systems.

8.5.1 Fine structure constant α

The electromagnetic interaction, whose strength is characterized by the constant α, is of primary importance for the macroscopic structure of matter and in a plethora of observed phenomena. The current limits on the evolution of α are established by laboratory measurements, studies of the abundances of radioactive isotopes and those of fluctuations in the cosmic microwave background, as well as other cosmological constraints [857].

In a typical all-laboratory experiment searching for temporal changes of fundamental constants, an atomic transition frequency f_{at} is measured with respect to a cesium clock (i.e., the frequency ratio f_{at}/f_{Cs} is determined, where $f_{Cs} \equiv 9192631770$ Hz). After a few years the measurement is repeated and changes in the frequency ratio are looked for [858]. Recently, this procedure has been accomplished for: the ground state hyperfine frequency of ^{87}Rb [859], the $1S \to 2S$ two-photon transition in atomic hydrogen [860], the transition

$^2S_{1/2} \to^2 D_{5/2}$ at 1065 THz in Hg$^+$ [861], and the $^2S_{1/2} \to^2 D_{3/2}$ transition at 688 THz in ^{171}Yb$^+$ [858]. A significant change in the α value would produce a clear indication, as it would influence differently the transition frequencies in these experiments. For a better understanding, we describe in certain detail the experiment performed at NIST [861]. With reference to Figure 8.13, a single Hg$^+$ ion was confined in an rf Paul trap and laser cooled to ~ 1.7 mK via the strongly allowed first resonance line at 194 nm. The reference for the optical clock was the $^2S_{1/2}(F = 0) \to^2 D_{5/2}(F = 2, m_F = 0)$ electric quadrupole allowed transition at 282 nm with a natural lifetime of 86 ms. Transitions to the $^2D_{5/2}$ level were detected via electron shelving using the 194-nm radiation. Spectroscopy of the narrow clock transition was performed by a frequency quadrupled fiber laser, where a portion of the light was used for stabilization to a low drift-rate (<1 Hz/s) high-finesse optical cavity at 563 nm. The transition frequency of the Hg$^+$-ion clock was compared to the 9192631770 Hz ground state hyperfine splitting of Cs as realized by a Cs-fountain clock. For this purpose, a Ti:Sa-based femtosecond laser frequency comb performed the optical-to-microwave synthesis that compared the rates of the Cs and Hg$^+$ clocks.

Then, the six-year time record of the absolute frequency measurements of the clock transition in Hg$^+$ was used to search for possible time-dependent variations of α, according to the following model [861, 858]. While the α dependence of the optical transition of atom j obeys $\nu_j \propto R_y F_j(\alpha)$, for the Cs hyperfine interval one has $\nu_{Cs} \propto \alpha^2 R_y(\mu_{Cs}/\mu_B)F_{Cs}(\alpha)$. Here R_y is the Rydberg constant, μ_B the Bohr magneton, and μ_{Cs} the magnetic dipole moment of the Cs nucleus. The factor $F_j(\alpha) \propto \alpha^{N_j}$ contains the relativistic correction as a power dependence on α, the value of N_j being different for diverse atomic transitions. Since, in principle, the ratio μ_{Cs}/μ_B could accompany any changes in α, the relationship between their coupled fractional variations is determined by analyzing the time dependence of the natural logarithm of the ratio between the optical and microwave frequencies:

$$\frac{d}{dt} \ln \frac{\nu_{Hg}}{\nu_{Cs}} = \frac{\frac{d}{dt}(\nu_{Hg}/\nu_{Cs})}{\nu_{Hg}/\nu_{Cs}} = \frac{\dot{\alpha}}{\alpha}N - \frac{d}{dt}\ln\frac{\mu_{Cs}}{\mu_B} \tag{8.25}$$

where $N = N_{Hg} - N_{Cs} - 2$. Next, the values $N_{Hg} = -3.2$ and $N_{Cs} = 0.8$ calculated for the Cs and Hg$^+$ clock transitions are used, whereupon

$$\frac{d}{dt}\ln\frac{\nu_{Hg}}{\nu_{Cs}} = -6\frac{\dot{\alpha}}{\alpha} - \frac{d}{dt}\ln\frac{\mu_{Cs}}{\mu_B} \tag{8.26}$$

Finally, any drift in the ratio of the clock frequencies is retrieved by a weighted least squares linear fit to the historical time record, which gives (0.39 ± 0.42) Hz·yr^{-1}, or fractionally, $d/dt[\ln(\nu_{Hg}/\nu_{Cs})] = (0.37 \pm 0.39) \cdot 10^{-15}$ yr^{-1}. Assuming $d/dt[\ln(\mu_{Cs}/\mu_B)] = 0$, this yields a $1 - \sigma$ limit of $\dot{\alpha}/\alpha = (-6.2 \pm 6.5) \cdot 10^{-17}$ yr^{-1}.

Moreover, by using measurements involving at least another optical transition, like the aforementioned $1S - 2S$ hydrogen transition or the $^2S_{1/2} \to^2 D_{3/2}$ ^{171}Yb$^+$ transition, one can also constrain the ratio μ_{Cs}/μ_B. To see this, let us re-write Equation 8.26 for atom j as follows

$$y = (N_j - 2.8)x - \xi_j \quad \text{with} \quad x \equiv \frac{\dot{\alpha}}{\alpha} \quad y \equiv \frac{d}{dt}\ln\frac{\mu_{Cs}}{\mu_B} \quad \xi_j \equiv \frac{d}{dt}\ln\frac{\nu_j}{\nu_{Cs}} \tag{8.27}$$

where the N_j (ξ_j) coefficients are known from calculations (measurements). A linear system of two (or more) equations in the two variables x and y is thus obtained, which simultaneously places limits on the fractional temporal variation of α and μ_{Cs}/μ_B. When this is accomplished, a coupled constraint of $-1.5 \cdot 10^{-15} < \dot{\alpha}/\alpha < 0.4 \cdot 10^{-15}$ and $-2.7 \cdot 10^{-15} < d/dt[\ln(\mu_{Cs}/\mu_B)] < 8.6 \cdot 10^{-15}$ is obtained [861]. More recently, based on a 6-year record of increasingly precise atomic clock frequency comparisons at NIST (Al^+ vs Hg^+, Hg^+ vs Cs, H vs Cs, Yb^+ vs Cs) an upgraded coupled constraint of $\dot{\alpha}/\alpha = (-1.6\pm2.3)\cdot10^{-17}$ yr^{-1} and $d/dt[\ln(\mu_{Cs}/\mu_B)] = (-1.9 \pm 4.0) \cdot 10^{-16}$ yr^{-1} was given [862].

In the near future, by implementing a synchronous interrogation scheme to remove the Dick effect, frequency comparisons of Sr/Yb and Sr/Hg clocks could, respectively, achieve stabilities of $2.4 \cdot 10^{-17}/\sqrt{\tau}$ and $1.5 \cdot 10^{-16}/\sqrt{\tau}$, assuming 10^6 atoms with an excitation linewidth of 8 Hz and use of a thermal-noise-limited laser operated at $5 \cdot 10^{-16}$. Therefore, an uncertainty of 10^{-18} would be reached with a few hours averaging, corresponding to an investigation $\Delta\alpha/\alpha \simeq 10^{-18}$ [779].

8.5.1.1 Proton-to-electron mass ratio β

While α is basically related to the quantum electrodynamics (QED) scale, β is primarily sensitive to the quantum chromodynamics (QCD) one. As a consequence, its secular change should be much larger than that of α. Indeed, within the framework of Grand Unification

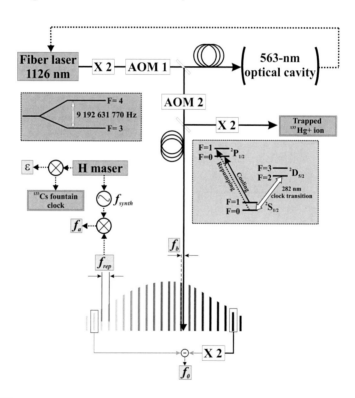

FIGURE 8.13
Experimental setup to constrain the fractional temporal variation of α through comparing the single-Hg^+ clock transition to the Cs microwave one. The two acousto-optic modulators stabilize the fiber laser to a high-finesse optical cavity and to the Hg^+ transition, respectively. The frequency comb repetition rate, f_{rep}, is measured with respect to the Cs-fountain clock via a synthesizer stabilized against a H maser. The absolute Hg^+-transition frequency is calculated from $\nu_{Hg} = 2 \cdot [n \cdot (f_a + f_{synht}) - f_0 - f_b](1 + \varepsilon)$, where f_0 is the comb offset frequency, f_b the beat between comb line n and the Hg^+-clock laser, $f_a = f_{rep} - f_{synht}$, and ε is the fractional frequency deviations between the H maser and the Cs-clock transition frequencies. (Adapted from [861].)

and also within string theories, a proportionality relation is derived between possible changes of α and β

$$\frac{\Delta\beta}{\beta} = S\frac{\Delta\alpha}{\alpha} \tag{8.28}$$

with the proportionality factor S varying between 20 and 40. The β parameter appears prominently both in atomic and molecular transition energies. In the case of atoms, however, constraints on $\dot{\beta}/\beta$ cannot be inferred in a strictly direct way, as the poorly accurate Schmidt model must be invoked for the nuclear magnetic moment [773]. Hence, the search for a possible time variation of β is currently focusing on molecular systems. In fact, it is well known that β defines the scales of electronic, vibrational, and rotational intervals in molecular spectra $E_{el} : E_{vib} : E_{rot} \sim 1 : \sqrt{\beta} : \beta$. Also, molecules may exhibit fine and hyperfine structures, Λ-doubling, hindered rotation, and other peculiar features which open up additional possibilities. All these effects have different dependences on β. Furthermore, it can be shown that sensitivity to temporal variation of both α and β may be strongly enhanced in coupling between molecular levels and continuum, as well as transitions between close-lying levels of different types. In this framework, different schemes have been proposed which involve either directly cooled stable molecules or dimers created by photoassociation of ultracold alkali atoms close to the quantum degeneracy. One design is based on a molecular fountain seeded by a Stark decelerated beam of $^{15}NH_3$ [863]. In this case, the β parameter is addressed by probing the inversion transition around 22.6 GHz in a microwave cavity according to a Ramsey-type measurement. In a different approach, diatomic molecular hydrogen ions (H_2^+, D_2^+, HD^+, HT^+) confined in a radio-frequency trap are sympathetically cooled by atomic ions ($^9Be^+$) [864]. Then, either two-photon spectroscopy with counter-propagating beams on a cold ensemble or quantum logic spectroscopy on a single ion could be performed to yield highly accurate values for β, as well as for the nuclear mass ratios. Other interesting but quite complex schemes consider molecules formed by ultracold atoms like Sr and Cs to constrain temporal variations of β at a $10^{-15}\cdot yr^{-1}$-level or better. In the former case, vibrationally excited Sr_2 in the electronic ground state is produced via photoassociation in an optical lattice [865]. Then, Raman spectroscopy aided by a femtosecond optical frequency comb will interrogate the energy spacings between deeply bound vibrational levels and those closer to the dissociation limit. In the latter, proposing the use of ultracold Cs_2, it is shown that a sensitive search for variation of β could be accomplished by measuring the change in the hyperfine splitting between a nearly degenerate pair of molecular vibrational levels, each associated with a different electronic potential [866]. In such a scheme, the Cs_2 molecules should be produced in the desired levels via Feshbach resonance followed by stimulated pumping processes.

So far, the most stringent test of the time variation of β has been performed by comparing a given vibrational transition in SF_6 with a cesium fountain over 2 years, which has resulted in a limit of $\dot{\beta}/\beta = (-3.8 \pm 5.6) \cdot 10^{-14} \text{ yr}^{-1}$ [455]. Since ν_{Cs} and ν_{SF_6} scale, respectively, as

$$\nu_{Cs} = K_2 \left(\frac{\mu_{Cs}}{\mu_B}\right) \alpha^2 F(\alpha) R_y = K_2 \left(\frac{\mu_{Cs}}{\mu_B}\right) \alpha^2 \alpha^{0.8} R_y \tag{8.29}$$

$$\nu_{SF_6} = K_1 \sqrt{\frac{1}{\beta}} R_y, \tag{8.30}$$

in this case one has

$$\frac{d}{dt} \ln \frac{\nu_{SF_6}}{\nu_{Cs}} = -\frac{1}{2}\frac{\dot{\beta}}{\beta} - 2.8\frac{\dot{\alpha}}{\alpha} - \frac{\dot{\mu}_r}{\mu_r} \simeq -\frac{1}{2}\frac{\dot{\beta}}{\beta} \tag{8.31}$$

where $\mu_r \equiv \mu_{Cs}/\mu_B$ has been defined, and $\dot{\beta}/\beta \gg \dot{\alpha}/\alpha$ (see Equation 8.28) and

$\dot{\alpha}/\alpha \sim \dot{\mu}_r/\mu_r$ (as obtained from atomic clock experiments) have been exploited. Hence, a spectroscopic frequency measurement carried out over a few years period directly translates into an assessment of the stability of the proton-to-electron mass ratio. In this experiment, already described in Chapter 5, the SF_6 transition (namely the $P(4)E^0$ transition in the $2v_3$ band) was accessed using a CO_2 laser to interrogate two-photon Ramsey fringes on a supersonic beam.

8.5.2 Speed of light c

Another intriguing question for both physics an metrology is the following: if α does vary, which of the quantities in its definition, namely, c, $\hbar$, e, or ε_0, is really varying? Indeed, a change of α over time could be interpreted as supporting some non-standard cosmological theories that invoke varying the speed of light or the electronic charge [50, 867, 868]. In this frame, Bekenstein has also shown that a varying-c cosmology can be rephrased as a varying-e theory through changes to standard units [869]. If attention is restricted to electromagnetic phenomena, then either c or e could account equally well for the variation in α. However, there may exist theoretical reasons of more fundamental character, concerned with gravitation, to incline to believe a varying c over varying e. To include gravitation into the current discussion, one can resort to the theory of black hole thermodynamics (see [867] and references therein). In this frame, entropy is associated with the area of a black hole's event horizon, leading to a generalized second law of thermodynamics according to which the event horizon's area may only diminish if there is a corresponding augmentation in the conventional entropy of the black hole's environment. Now, in the case of a non-rotating black hole with electric charge Q and mass M, the area A_H of its event horizon, as obtained in conventional general relativity theory from the Reissner-Nordström solution of Einstein's field equations, is given by

$$A_H = 4\pi r^2 \tag{8.32}$$

where

$$r = \frac{G}{c^2}[M + \sqrt{M^2 - Q^2/G}] \tag{8.33}$$

As a result, an increase in the magnitude Q, with c and G remaining constant, implies a reduction in A_H, whereas a decrease in c would heighten the event-horizon area. Therefore, these two contending alternatives (for an increase in α) lead to opposite end results. In conclusion, since a reduction in A_H would violate the generalized second law of thermodynamics, the fundamental electric charge cannot increase.

8.5.2.1 Frequency dependence of c and the mass of the photon

The great success of Maxwell's electromagnetism and quantum electrodynamics (QED) has naturally led to the consent on the concept of massless photon. Nevertheless, serious attempts have been made to establish experimentally, either directly or indirectly, whether the photon mass is zero or finite. From a theoretical point of view, apart from the loss of gauge invariance, classical electromagnetism and QED would not be overly troubled by a nonzero photon rest mass. In addition, this latter circumstance would be perfectly compatible also with the general principles of elementary particle physics. Thus, the final answer can come only through observations. As we will see in a short while, the most direct consequence of a finite-mass photon would be that electromagnetic waves propagate in free space with a frequency-dependent velocity. In turn, the question of whether the speed of light varies with frequency impacts on other key issues. For instance, Maxwell's electromagnetism implies that all electromagnetic radiation travels (in vacuum) at the speed of light c and, indeed, this has been experimentally confirmed so far over a wide range of frequencies to a high degree of accuracy. A second shining example is provided by Einstein's postulate of the invariance of light. Being a keystone in much of modern physics, such a postulate is continually tested in more and more stringent experiments.

One straightforward manner to consider some implications of a massive photon is to use the simplest relativistic generalization of Maxwell's equations, namely the Proca equations [43]. In this formalism, the electric and magnetic fields in free space are given by

$$A_\nu \sim \exp[i(\boldsymbol{k} \cdot \boldsymbol{r} - \omega t)] \tag{8.34}$$

where $\boldsymbol{k}$, ω, and the rest mass m_γ satisfy the Klein-Gordon equation

$$k^2 c^2 = \omega^2 - m_\gamma^2 c^2 \tag{8.35}$$

Here m_γ has the units of wave numbers, being related to the mass M_γ in grams by the relation $m_\gamma = c M_\gamma / \hbar$. Then, the phase and the group velocity of a free massive wave would take the form

$$u = \frac{\omega}{k} = \frac{c}{\sqrt{1 - \dfrac{m_\gamma^2 c^2}{\omega^2}}} \tag{8.36}$$

$$v_g = \frac{d\omega}{dk} = c\sqrt{1 - \frac{m_\gamma^2 c^2}{\omega^2}} \tag{8.37}$$

where $k = |\boldsymbol{k}| = 2\pi \lambda^{-1}$, λ being the wavelength. The above formulas clearly show that a nonzero photon mass would cause the group velocity to differ from the phase one as well as a frequency dependence in both quantities. In this case, c becomes the limiting velocity as the frequency approaches infinity. Next, let us consider two wave packets with different propagating frequencies, ω_1 and ω_2. Assuming $\omega_1 > \omega_2 \gg m_\gamma c$, then the velocity differential between them takes the form

$$\frac{v_{g1} - v_{g2}}{c} = \frac{m_\gamma^2}{8\pi^2}(\lambda_2^2 - \lambda_1^2) + \mathcal{O}[(m_\gamma \lambda_1)^4] \tag{8.38}$$

Thus, when moving through the same distance L, the time interval between the arrivals of the two waves is given by

$$\Delta t \equiv \frac{L}{v_{g1}} - \frac{L}{v_{g2}} \simeq \frac{L}{8\pi^2 c^2}(\lambda_2^2 - \lambda_1^2) m_\gamma^2 \tag{8.39}$$

in which the terms of order higher than $(m_\gamma \lambda_1)^4$ are neglected. Equations 8.38 and 8.39 represent the starting point for detecting a photon-rest-mass-related dispersion effect both in terrestrial and extraterrestrial approaches. In this frame, the most stringent limit ($\Delta c/c < 3.3 \cdot 10^{-7}$) with terrestrial means was set in 1958 by Froome using a radio-wave interferometer [43]. Later on, a much more severe constraint was placed by Schaefer, hinging on explosive astrophysical events at high redshift [870]. Based on the simultaneous arrival of a flare in GRB 930229 with a rise time of 220 ± 30 μs for photons of 30 and 200 keV, $\Delta c/c < 6.3 \cdot 10^{-21}$ was found. In this kind of experiment, in order to obtain tighter limits, one should use waves of lower and lower frequencies propagating over longer and longer distances. Unfortunately, however, lowering the wave energy makes measurements more problematic due to the dissipation occurring over the long pathway in the medium.

Of course, some other relevant implications are predicted on the basis of a non-zero photon mass, including deviations from exactness in Coulomb's law and Ampere's law, the existence of longitudinal electromagnetic waves, as well as the addition of a Yukawa component to the potential of magnetic dipole fields [43]. All these potential effects offer alternative, additional tools for setting an upper limit on the photon mass in both laboratory experiments and astrophysical/cosmological observations. Needless to say, fundamental properties of the photon, including a possible finite rest mass, are of great concern to the physics of elementary particles too. The upper limit currently accepted by the Particle Data Group on the photon rest mass is $M_\gamma \leq 1 \cdot 10^{-49}$ g (almost 22 orders of magnitude less than the electron mass), that should be compared with the bound of $M_\gamma \leq 4.2 \cdot 10^{-44}$ g set by the latest dispersion-of-light experiment [870]. In spite of these impressively low values, the search for a nonzero photon mass is relentless as several intriguing, fundamental questions (charge conservation and quantization, the possibility of charged black holes, the existence of magnetic monopoles and so on) are related to this issue.

A comprehensive review on the ongoing experiments in this field (both terrestrial and extraterrestrial) is given in [43].

8.5.3 Newton's gravitational constant

Possible temporal variations of Newton's gravitational constant G are more effectively pursued in space experiments rather than on Earth. For example, a decreasing gravitational constant, as suggested by P.A.M. Dirac in 1937, should dilate a planet's semimajor axis, a, as $\dot{a}/a = -\dot{G}/G$ (via coupling with angular momentum conservation). The corresponding orbital phase change expands quadratically with time, providing for strong sensitivity to the effect on $\dot{G}$ [871]. Recent analysis of Lunar Laser Ranging (LLR) data strongly limits such variations and constrains a local scale expansion of the solar system as $\dot{a}/a = -\dot{G}/G = -(5 \pm 6) \cdot 10^{-13}$ yr^{-1} (in the LLR technique, lasers on Earth are aimed at retro-reflectors planted on the Moon during the Apollo program, and the time for the reflected light to return is determined). Finally, high-accuracy timing measurements of binary and double pulsars could also provide a good test of the variability of the gravitational constant [872].

8.6 Tests of fundamental physics laws

8.6.1 Spectroscopic tests of spin-statistic connection and symmetrization postulate

The symmetrization postulate and the spin-statistic connection are at the basis of a quantum-mechanical description of systems composed of identical particles. Due to this pivotal role, several experiments have been performed to test them, mainly based on electrons and photons [873]. Here, instead, we only focus on spectroscopic tests, inspired by [874, 875, 876]. Preliminarily, we have to illustrate a few concepts.

The **symmetrization postulate** (SP) asserts that only wave functions that are completely symmetric or antisymmetric in the permutation of the particles' labels can describe physical states. In the former (latter) case, the particles are called bosons (fermions) and obey Bose-Einstein (Fermi-Dirac) statistics. Also, experiments indicate that particles with integral values of spin are bosons, while particles with half-integral spin are fermions. The **spin-statistic theorem**, proved by W. Pauli from the basic principles of quantum field theory and special relativity, states that given the choice between Bose and Fermi statistics, integral-spin particles must obey Bose statistics and half-integral spin particles must obey Fermi statistics. In principle, however, there would be no arguments against the existence of states with different symmetries (although they lack some of the properties which are peculiar to completely symmetric and antisymmetric states). The only strict requirement that can be derived in a formal way in quantum mechanics is the so-called *superselection rule*, which forbids transitions between different symmetry classes. In other words, it is not possible to consider states given by a coherent superposition of states with different permutation symmetries. As a consequence, symmetry violations in systems including identical particles can only be described in terms of an incoherent mixture which is represented by a density matrix. In the case of two integral-spin particles, for example, the density matrix taking into account small symmetry violations is [874]

$$\rho_2 = \left(1 - \frac{\beta^2}{2}\right)\rho_s + \frac{\beta^2}{2}\rho_a \qquad (8.40)$$

where ρ_s (ρ_a) is the symmetric (antisymmetric) two-particle density matrix. In the case of particles with half-integral values of spin, ρ_s and ρ_a are interchanged. In literature, the quantity $\beta^2/2$ is known as the *symmetry-violation* parameter. However, its real meaning needs to be specified from case to case, according to the specific physical system and theoretical model considered.

Although both the SP and the spin-statistics connection seem to hold in the physical world, a number of theories have been developed allowing for small deviations from conventional symmetry which might have been masked in the experiments performed so far. Indeed, the possibility of theories going beyond the Bose and Fermi statistics, arising from the SP, has long been recognized [877]. More recently, theories allowing small SP violations have been developed, basically following two different approaches. In the first approach, trilinear commutation relations are defined instead of the usual bilinear Bose and Fermi commutators [878], whereas in the second a slight deformation of the bilinear commutation relations is introduced, using a parameter which can continuously turn each statistic into the other [879]. Then, precise experiments are needed to discriminate between these theories. In this respect, it is worth noting that an experimental evidence of SP violation in a system of bosons is harder to be observed. This is due to the fact that, while there are several systems in which a violation of the Pauli exclusion principle (which is a particular case of the SP) would be detected as a signal on a zero background, the effect of a small violation for particles following Bose-Einstein statistics would usually manifest itself as a small change in the properties of a many-particle system. Just for this reason, spectroscopic tests represent the ideal tools to accurately and authentically (i.e., free from misunderstandings) seek for possible SP violations. In this frame, two independent spectroscopic tests on $^{16}O_2$, containing two identical ^{16}O nuclei (spin-0 bosons), set the bound to a SP violation, respectively, to $5 \cdot 10^{-7}$ [880] and $1.3 \cdot 10^{-6}$ [881]. The choice of a suitable test molecule/transition both depends on bare physics and on the availability of a proper source of coherent radiation at the corresponding wavelength. From the point of view of the general physical properties, diatomic homonuclear molecules, such as $^{16}O_2$, are not the best choice for the test, because of their lack of active electric dipole transitions in the infrared (IR). Higher sensitivity to look for the existence of exchange-antisymmetric states can be achieved by investigation of polyatomic molecules containing a pair of identical bosonic nuclei. Among them the $^{12}C^{16}O_2$ molecule is one of the best candidates because its rovibrational IR transitions may have line strengths up to $3.5 \cdot 10^{-18}$ cm/molecule. Indeed, the first test on $^{12}C^{16}O_2$ (performed at 2 μm wavelength with a DFB diode laser), due to the stronger allowed transitions involved in the measurement, improved upon the previous ones by more than 2 orders of magnitude, setting an upper limit of $2.1 \cdot 10^{-9}$ to SP violations [882]. Later on, a DFG source at 4.25 μm was employed to carry out a more sensitive search for the existence of exchange-antisymmetric states in spin-0 particles [876]. This was accomplished by investigating the spectrum of the $00^01 - 00^00$ fundamental band of $^{12}C^{16}O_2$, which is about 2000 stronger than the $12^01 - 00^00$

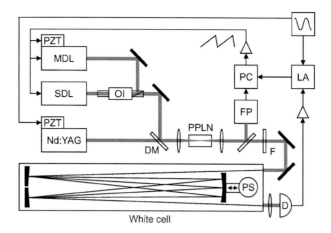

FIGURE 8.14
Experimental setup for a spectroscopic test of the symmetrization postulate. MDL: master diode laser; SDL: slave diode laser; PZT: piezoelectric translator; OI: optical isolator; DM: dichroic mirror; PPLN: periodically poled $LiNbO_3$ crystal; PC: computer; LA: lock-in amplifier; FP: Fabry-Perot cavity; F: Ge filter; PS: phase-scrambler; D: InSb detector.

band at 2 μm. According to the Born-Oppenheimer approximation, the total wave function Ψ of a single molecule can be factorized in the form $\Psi = \psi_e \psi_v \psi_r \psi_n$, where ψ_e, ψ_v, ψ_r, and ψ_n are the electronic, vibrational, rotational, and nuclear wave functions, respectively. The ground electronic and vibrational wave functions of the $^{12}C^{16}O_2$ molecule are symmetric in the exchange of the two ^{16}O nuclei. Also, the nuclear wave function is symmetric, since the nuclear spin $I(^{16}O) = 0$. The rotational wave function is symmetric for even values of J, antisymmetric for odd ones. The SP requires the total wave function to be symmetric in the exchange of the two ^{16}O nuclei. Therefore, rotational states with odd values of J are forbidden in the ground vibrational state. A similar argument could show that the situation is reversed in the 00^01 vibrational state, for which even values of J are forbidden, because the vibrational wave function ψ_v in the excited state is antisymmetric. As a consequence, detection of a weak transition of the rovibrational form $00^01 - 00^00$ $R(J)$ with an odd value of J could indicate a SP violation and its amount. The experimental setup is shown in Figure 8.14. The 4.25-μm radiation emerging from the non-linear crystal is coupled into a White-type multi-pass absorption cell, with an effective absorption path length of 130 m.

The investigated forbidden transition was chosen as the $00^01 - 00^00$ $R(25)$ line at 2367.265 cm^{-1} for the following reasons: (i) It is close to the strongest R-branch lines (at room temperature); (ii) it is far from any strong allowed line; (iii) a weak line to be used as a frequency and sensitivity marker (the $02^21 - 02^20$ $R(80)$ line of $^{12}C^{16}O_2$ at 2367.230 cm^{-1}) is well separated from the forbidden line and can be recorded within the frequency span. Pure carbon dioxide (99.99%) was used to fill the cell at pressures ranging from 270 to 330 Pa. These pressures were chosen in order to minimize the absorption due to the Lorentzian wings of the strong $R(24)$ and $R(26)$ lines and still maintain a good S/N ratio in the spectral region of interest. A first-derivative recording of the marker line is shown in Figure 8.15. In the same figure, an arrow indicates the calculated position of the forbidden line. In this case, the violation parameter $\beta^2/2$ can be expressed as

$$\frac{\beta^2}{2} < \frac{A_{R(25)}}{A_{R(80)}} \frac{S_{R(25)}}{S_{R(80)}} \tag{8.41}$$

where $A_{R(80)} = 14.7$ V $(S_{R(80)} = 1.81 \cdot 10^{-25}$ cm/molecule for $T = 296$ K) is the signal (the line strength) corresponding to the marker line. The analogous quantities for the forbidden line are calculated as follows. $A_{R(25)} = 3.7$ mV is estimated by measuring the root mean square (rms) noise in the spectral range where the forbidden line is expected. Thus we are left with the calculation of $S_{R(25)}$. To this aim, we start by observing that the main dependence of the line strength of a rovibrational R-branch line on J and T is

$$S(J,T) \propto (J+1)\exp\left[-\frac{h\nu_0 + E_r(J)}{k_B T}\right] \tag{8.42}$$

where $E_r(J)$ is the rotational energy, such that

$$S(25,T) \simeq S(24,T)\frac{26}{25}\exp\left(-\frac{50hB}{k_B T}\right) \tag{8.43}$$

with $B = 0.389$ cm^{-1} and $S_{R(24)} = 2.78 \cdot 10^{-18}$ cm/molecule for $T = 296$ K. Hence, $\beta^2/2 < 1.7 \cdot 10^{-11}$ was finally inferred for the transition connecting the forbidden states $(00^00, J = 25)$ and $(00^00, J = 26)$.

Substantial improvements to this sensitivity could be obtained by use of cavity-enhanced spectroscopic techniques, such as, for instance, the Saturated-Absorption Cavity Ring-Down Spectroscopy (SCAR) technique described in Chapter 5. Furthermore, it would be also interesting to look for possible small SP violations by investigating rovibrational transitions in symmetrical molecules containing three identical nuclei. In this frame, a high-sensitivity spectroscopic study of simple molecules such as SO_3, BH_3, and NH_3 could lead to the first test of the symmetrization postulate for spin-0 and spin-1/2 nuclei.

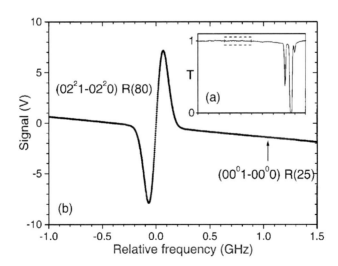

FIGURE 8.15
Transmission spectrum (10 GHz total frequency span) of the White cell filled with 18 Pa of CO_2. Absorption peaks of three CO_2 lines around 2367.4 cm^{-1} are shown. (b) Plot of the first-derivative demodulated signal. This scan, corresponding to the dashed rectangle in (a), contains the $R(80)$ marker line and the region where the forbidden $R(25)$ line is expected. A fit curve (solid line) is superimposed on the experimental points.

8.6.2   Search for the electron dipole moment

The electron is predicted to be slightly aspheric [883], with a distortion characterized by a permanent electric dipole moment (EDM), d_e. However, no experiment so far has ever detected this deviation. While the standard model (SM) of particle physics predicts that d_e is far too small to detect ($|d_e| < 10^{-38}$) [884], many extensions to the SM (including Higgs models, Supersymmetric models, and Lepton-Flavor-Changing models) naturally predict much larger values of d_e that should be detectable ($10^{-29} < |d_e| < 10^{-26}$) [885]. Moreover, most models of particle physics predict that, if some undiscovered particle interaction breaks the symmetry between matter and antimatter (that could be the reason for which the Universe has so little antimatter), this should manifest itself into a measurable EDM [886]. Therefore, the search for the electron EDM represents an important step towards the discovery of new physics. In a recent experiment, cold polar molecules were used to measure the electron EDM at the highest level of precision reported so far, providing $d_e = (-2.4 \pm 5.7_{stat} \pm 1.5_{syst}) \cdot 10^{-28} e$ cm, where e is the charge on the electron, which sets a new upper limit of $|d_e| < 10.5 \cdot 10^{-28}$ cm with 90 percent confidence. This result, consistent with zero, indicates that the electron is spherical at this improved level of precision [887]. The concept behind this experiment can be understood as follows. If any, a permanent electron EDM must lie along its spin, $\boldsymbol{\sigma}$, that is, $\boldsymbol{d} = d_e \boldsymbol{\sigma}$, whereupon the electron energy in an applied electric field, $\boldsymbol{E}$, is given by $-\boldsymbol{d} \cdot \boldsymbol{E} = -d_e \boldsymbol{\sigma} \cdot \boldsymbol{E}$. Then, for an atom or molecule with an unpaired valence electron, the interaction between the electron EDM and $\boldsymbol{E}$ manifests itself into an energy difference between two states that differ only in their spin orientation. A sensitive method of measuring such energy difference, which is proportional to d_e and changes sign when the direction of the field is reversed, consists of measuring the quantum interference between the two spin states. In this case, the EDM should appear as an interferometer phase shift that changes sign when the electric field is reversed. Compared to atoms, diatomic polar molecules naturally possess orbitals that are strongly polarized along the internuclear axis $\hat{\lambda}$, which originates a significant internal effective electric field $E_{int}\hat{\lambda}$. This can be particularly large when one of the atoms is heavy and has strong $s-p$ hybridization of its orbitals. However, due to the molecular rotation, this strong field averages to zero unless an external (even modest) field $E\hat{z}$ is applied to polarize $\hat{\lambda}$ along $\hat{z}$. In that case, the EDM interaction energy takes the form $V = -d_e E_{int} \langle \hat{\lambda} \cdot \hat{z} \rangle \langle \hat{\sigma} \cdot \hat{z} \rangle \equiv -d_e E_{eff} \langle \hat{\sigma} \cdot \hat{z} \rangle$ [888]. The relevant energy levels and the experimental apparatus are shown in Figures 8.16 and 8.17, respectively.

Pulses of YbF molecules are emitted by the source, and molecules in the $F = 0$ and $F = 1$ hyperfine levels of the ground state are used. First, the molecules pass through the pump fluorescence detector, which simultaneously measures and empties out the $F = 1$ population. Then they enter a pair of electric field plates, between which are static electric and magnetic fields $(E, B)\hat{z}$. A radio-frequency (rf) pulse is applied to transfer molecules from $|F, m_F\rangle = |0, 0\rangle$ to the state $(1/\sqrt{2})(|1, 1\rangle + |1, -1\rangle)$, where m_F is the component of the total angular momentum, F, along the z-axis. The molecules then evolve freely for a time T, during which the $m_F = \pm 1$ components develop a phase difference of $2\phi = 2(\mu_B B - d_e E_{eff})T/\hbar$, where μ_B is the Bohr magneton. This is due to the Zeeman shift $\mu_B B m_F$ and to the EDM shift expressed by the effective interaction $-d_e E_{eff} m_F$. A second rf pulse is then applied, which drives a transition back to the $F = 0$ state. The state amplitude $(1/\sqrt{2})(e^{i\phi} + e^{-i\phi})$ results in a final $F = 0$ population proportional to $\cos^2 \phi$, which the second fluorescence detector subsequently measures. Thus, when scanning the phase difference via the magnetic field, the fluorescence signal shows typical interference fringes $I = I_0 \cos^2 \phi$. So, if δB is a small, carefully calibrated change in the magnitude of

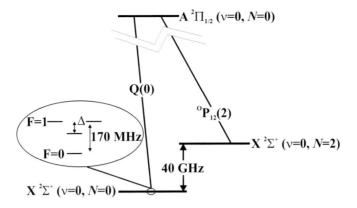

FIGURE 8.16
Optical transitions in ^{174}YbF at 553 nm, which are 40 GHz apart. Inset: ground state hyperfine levels $F = 0$, $F = 1$, 170 MHz apart. In static electric field, the $m_F = 0$ sublevel of $F = 1$ is lower than the $m_F = \pm 1$ sublevels by an amount Δ. The detection laser is tuned to the $F = 0$ component in the $Q(0)$ line, while a second dye laser is tuned to the $^{O}P_{12}(2)$ transition. Together, these two beams pump molecules into the $F = 0$ ground state, while emptying out $F = 1$. (Adapted from [888].)

the applied field, then the maximum detector count changes from I_0 to $I_0 + \delta I_1$ with

$$\delta I_1 = \frac{dI}{d\phi}\delta\phi_1 = \frac{dI}{d\phi}\frac{T}{\hbar}\mu_B \delta B \tag{8.44}$$

Moreover, since d_e is expected to be vanishingly small, a reversal in the applied electric field, which reverses E_{eff}, is also associated with a small change in the maximum detector count

$$\delta I_2 = \frac{dI}{d\phi}\delta\phi_2 = \frac{dI}{d\phi}\frac{T}{\hbar}2d_e E_{eff} \tag{8.45}$$

Finally, we have

$$\frac{\delta I_2}{\delta I_1} = \frac{2d_e E_{eff}}{\mu_B \delta B} \tag{8.46}$$

from which the EDM is eventually derived by using the measured $\delta I_2/\delta I_1$ ratio and theoretical value $E_{eff} = -14.5$ GV·cm^{-1}. Although there is some uncertainty in the theoretical calculation, even an uncertainty of 10% would have no impact on the error at the reported level. In practice, a more complex (with respect to Equation 8.46) correlation analysis is applied to extract the d_e value. This procedure, together with full particulars on the experimental apparatus, can be found in [887].

In conclusion, it should be mentioned that another EDM experiment is in progress, based on electron spin resonance spectroscopy on trapped molecular ions [889].

8.6.3 Parity violation in chiral molecules

Let us start by recalling that the inversion symmetry of a physical system, described by the wave function ψ, can be characterized through the parity operator P. The latter acts by inverting the spatial coordinates of ψ, such that $P\psi(\boldsymbol{r}) = \psi(-\boldsymbol{r})$. Then, if $P\psi(\boldsymbol{r}) = \pm\psi(\boldsymbol{r})$, we assert that ψ has well-defined parity (even parity if $+$, odd parity if $-$). Also, in a given physical interaction described by the Hamiltonian H, parity is conserved if H commutes with

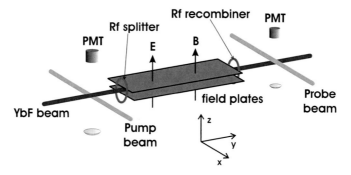

FIGURE 8.17
Schematic diagram of the pulsed molecular beam apparatus used for the EDM experiment. (Adapted from [887].)

P. Together with charge conjugation (C) and time reversal (T), the parity operation enters the well-known CPT theorem: a "mirror-image" of our universe, with all objects having their positions reflected by an imaginary plane (corresponding to a parity inversion), all momenta reversed (corresponding to a time inversion), and with all matter replaced by antimatter (corresponding to a charge inversion), would evolve under exactly our physical laws. In other words, the combined CPT symmetry is preserved by all physical phenomena (i.e., by the four fundamental interactions: Strong, Electromagnetic, Weak, and Gravitation), albeit individual symmetries may be broken.

Inspired by [890, 891, 892, 893, 894], here we focus on parity violation phenomena that may occur in processes involving the weak force, as observed in 1957 by Wu et al. in the β-decay of cobalt-60. In β-decay parity violation, a neutron is converted into a proton, while an electron and an electron anti-neutrino are created $(n \rightarrow p + e + \bar{\nu}_e)$. Indeed, until the 1970s, all processes discovered implying weak interactions were associated with an exchange of electric charge between the involved particles and thereupon by an alteration of their identity, such transformations being brought about by the gauge bosons W^+ and W^-. Analogous to the photons that mediate the electromagnetic interaction, these particles communicate the weak interaction. Unlike the photon, however, the above gauge bosons carry a unit of electric charge; this, together with the consideration that the stability of an atom is guaranteed only by interactions which preserve the identity of the particles, seemed to suggest that the weak interaction and its associated parity violation were not relevant to the physics of stable atoms. Later, due to some inconsistencies in the theoretical model of weak interactions, Glashow, Weinberg, and Salam independently proposed unification of the weak and electromagnetic interactions into a single fundamental interaction, namely the electroweak one. The most important "prophecy" of such revolutionary theory, usually referred to as the Standard Model of electroweak interactions, was the existence of a third gauge boson (Z^0) which, being associated with a neutral current, could communicate a new kind of weak interaction without changing the identity of the interacting particles. In the case of atoms, such neutral current is responsible for weak interactions between electrons and nucleons as well as among electrons: $e+p \rightarrow e+p$, $e+n \rightarrow e+n$, and $e+e \rightarrow e+e$ (in other words, the electron-electron current acts via the Z^0 boson with either the proton-proton, neutron-neutron, or electron-electron current).

Parity non-conservation (PNC) in stable atoms is precisely a revelation of the weak interaction between atomic electrons and the nucleus. Being complementary to high-energy experiments, it emerges in high-precision measurements which assay the symmetry properties of the optical-absorption process. Nonetheless, due to their extreme faintness, these effects can be detected only by resorting to very specific systems: first, highly forbidden transitions must be exploited in order to prevent the electromagnetic interaction from completely submerging the weak one; second, by virtue of the Z^3 enhancement factor (Z denotes here the number of protons), heavy atoms should be preferred. In principle, there is also a contribution to atomic parity violation arising due to Z^0 exchange between electrons, but this effect is negligibly small for heavy atoms. Instead, another appreciable contribution arises from the so-called anapole moment, which appears due to the presence of parity violating weak interactions between nucleons, and manifests itself in atoms through the usual electromagnetic interaction with atomic electrons [893]. Just to better grasp the concept of anapole moment, we mention here that, besides the usual electric and magnetic moments (monopole, dipole, quadrupole,...), other terms come out in the electromagnetic-multipole-moment expansion. These are usually not dealt with, as they give rise to contact, rather than long-range, potentials. The anapole is such a moment. Formally, it originates from an expansion of the vector potential as a series of R^{-1}, where R is the distance from the center of the charge distribution, and corresponds to a toroidal current distribution. The anapole moment obeys time reversal invariance but violates parity conservation and charge conjugation invariance, i.e., T-even and P- and C-odd.

In all atomic PNC experiments searching for the inversion asymmetry, the measured quantity is an electric dipole moment $\mathcal{E}_{PNC}$ activated by a force which disobeys parity conservation. When this force is also time-reversal violating, then $\mathcal{E}_{PNC}$ can be a permanent electric dipole moment, which produces an energy shift of the atom in an external electric field. By contrast, if the force is time-reversal preserving, $\mathcal{E}_{PNC}$ is only observable through its interference with some other atomic moment in radiative transitions. Of interest here is the PNC interaction which exhibits time-reversal symmetry. While both the two relevant PNC contributions (i.e., that originating from weak interaction between the nucleus and the electrons and that resulting from usual electromagnetic interaction between the nuclear anapole moment and the electrons) have been first observed in Cs atoms [895], in the following we describe a more recent experiment performed on ^{174}Yb, reporting the largest atomic PNC amplitude yet observed (2 orders of magnitude larger than cesium) [896]. With reference to Figure 8.18, the idea is to excite the forbidden 408-nm transition ($6s^2$ $^1S_0 \equiv |\Psi_1\rangle \rightarrow 5d6s\,^3D_1 \equiv |\Psi_2\rangle$) with resonant laser light in the presence of a quasi-static electric field ($\boldsymbol{E}_s$). In the absence of electric fields and weak neutral currents, an electric dipole (E1) transition between $|\Psi_1\rangle$ and $|\Psi_2\rangle$ is forbidden by the parity selection rule: $\Pi(|\Psi_1\rangle) \cdot \Pi_{phot} \neq \Pi(|\Psi_2\rangle)$, where the parity of photon is $\Pi_{phot} = -1$ and the atomic states, both with even angular momentum L, have the same parity $\Pi(|\Psi_1\rangle) = \Pi(|\Psi_2\rangle) = 1$. However, the weak neutral current interaction, associated with the Hamiltonian H_W, mixes a small amount of atomic wave functions with opposite parity into the $|\Psi_1\rangle$ and $|\Psi_2\rangle$ states. In this specific case the $6s6p\,^1P_1$ state is mixed into the $|\Psi_2\rangle$ one, thus leading to a small opposite-parity admixture in the original $|\Psi_2\rangle$ atomic state: $|\Psi_2\rangle \rightarrow |\tilde{\Psi}_2\rangle = |\Psi_2\rangle + |\delta\Psi_2\rangle$. This mixing gives rise to E1 transitions between states of the same nominal parity, resulting in a PV E1 transition amplitude given by

$$A_{PNC} = \langle\tilde{\Psi}_2|H_{E1}|\Psi_1\rangle = \langle\delta\Psi_2|H_{E1}|\Psi_1\rangle \tag{8.47}$$

where H_{E1} is the electric-dipole Hamiltonian (here we are not considering for simplicity the anapole contribution). The purpose of the applied electric field is to provide an observable (transition rate R) that is first order in this amplitude through the interference of the reference transition amplitude A_E (due to Stark mixing of the same states) with the PV amplitude:

$$R = |A_E + A_{PNC}|^2 = |A_E|^2 + |A_{PNC}|^2 + 2\Re[A_E A_{PNC}^*]$$
$$\simeq |A_E|^2 + 2\Re[A_E A_{PNC}^*] \equiv R_E + \Delta R \tag{8.48}$$

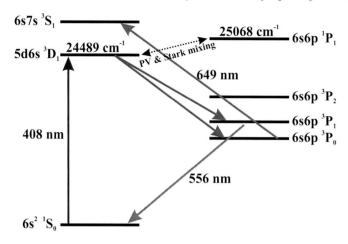

FIGURE 8.18
Energy levels and transitions which are relevant to the Yb parity-violation experiment.
(Adapted from [896].)

Then, the Stark-PV interference term (ΔR) is isolated from the dominant Stark-induced one (R_E) through a harmonic modulation of $\boldsymbol{E}_s$. In this way, R_E has a static component and a component oscillating at twice the modulation frequency, while ΔR oscillates at the first harmonic, which permits frequency discrimination using lock-in amplifiers. A schematic of the Yb-APV apparatus is shown in Figure 8.19. An effusive beam of Yb enters the main interaction region where the Stark- and PV-induced transitions take place. Frequency doubling the output of a Ti:Sa laser provides the 408.345-nm wavelength light (80 mW). The latter is coupled into a 9000-finesse power build-up cavity (PBC) inside the vacuum chamber. The laser is locked to the PBC via the PDH technique. A uniform magnetic field (up to 100 G) is generated by a pair of rectangular coils, while additional coils (placed outside of the vacuum chamber) compensate for the external magnetic fields down to 10 mG at the interaction region. The electric field is generated by two wire-frame electrodes: an ac voltage up to 10 kV (at a frequency of 76.2 Hz) is supplied to them to provide the field $\tilde{E} = 5$ kV/cm (an additional dc electric field of 40 V/cm is also used). The 556-nm wavelength light emitted from the interaction region is conveyed by a light guide to a photomultiplier tube, and the resulting signal used for initial selection of the atomic resonance as well as for monitoring purposes. The 649-nm light-induced fluorescence from the probe region is eventually collected by the photodetector PD (the 649-nm excitation light is derived from a single-frequency diode laser saturating the 6s6p $^3P_0 \to$ 6s7s $^3S_1$ transition). Finally, the signal from the PD is fed into a lock-in amplifier for frequency discrimination and averaging. Since $J = 1$ ($J = 0$) for the $^3D_1$ ($^1S_0$) state and hyperfine components are not resolved, for an arbitrary angle of polarization θ ($\theta = \pi/4$ is typically used), both the first- and second-harmonic signals contain three Zeeman components ($\Delta M_J = -1, 0, 1$) in the B-field split 408-nm spectral line. The emergence of a first-order harmonic spectrum provides the signature for atomic parity violation. One can show that the extent to which the PV phenomenon occurs, that is the ratio $\Delta R/R_E$, can be evaluated from the amplitudes of the Zeeman components, as measured in the first and second-harmonic spectrum, according to the following combination [896]

$$\frac{\mathcal{A}_{-1}^{1st}}{\mathcal{A}_{-1}^{2nd}} + \frac{\mathcal{A}_{+1}^{1st}}{\mathcal{A}_{+1}^{2nd}} - 2\frac{\mathcal{A}_{0}^{1st}}{\mathcal{A}_{0}^{2nd}} = \frac{16\mathcal{I}}{\tilde{E}} \tag{8.49}$$

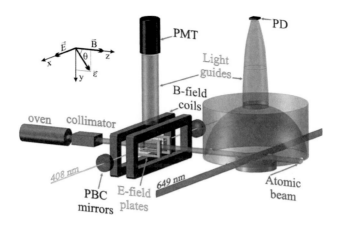

FIGURE 8.19
Experimental setup for the PV-Stark interference experiment. Light is applied collinearly with x, and θ is the angle between the light polarization and the magnetic field. (Courtesy of [896].)

where the PV-interference parameter $\mathcal{I}$ has been introduced. In the case of ^{174}Yb, $\mathcal{I} = 39(4)_{stat}(5)_{syst}$ mV/cm is found, which is in agreement with theoretical predictions and about 2 order of magnitude larger than in Cs.

The weak interaction is also active in molecular systems, especially in the interplay between electrons and nuclei. As a consequence, the standard model predicts a petite energy difference between enantiomers of chiral molecules [890]. More in detail, as shown in Figure 8.20, the potential energy surface of a chiral molecule exhibits two minima, corresponding to the left L and right R enantiomers. For a very high interconversion barrier, these can be regarded as energy eigenstates which, in the absence of the weak force, are degenerate. By contrast, in the presence of the weak force, a small parity-violation energy difference (PVED), ΔE_{PV}, is predicted to arise between the ground states (as well as excited states) of the two enantiomers. In this case, right- and left-handed molecules are no longer the exact mirror images of each other (in other words, enantiomers become diastereomers). However, such parity-violation frequency differences are predicted to be vanishingly small. Just to give an idea, it is approximately 30 mHz for the chiral molecule CHFClBr. For a comprehensive theoretical treatment of PV violation in chiral molecules, the reader is referred to [897, 898].

The idea proposed by Letokhov in 1975 to detect such effect is illustrated in Figure 8.21. This approach consists of measuring the difference $\Delta\nu_{PV} = \nu_L - \nu_R = (\Delta E_{PV}^* - \Delta E_{PV})/h$ between the frequencies of a given ro-vibrational transition in two enantiomers of a suitable chiral molecule. Later on, theoretical studies have indicated that $\Delta\nu_{PV}/\nu$ is in the range 10^{-16} - 10^{-19}, depending on the considered molecular transition (it is worth noting that $\Delta\nu_{PV}$ is much smaller than the molecular intrinsic line width). The first high-resolution experiment was carried out in 1999 on CHFClBr by the group of Chardonnet using laser-saturated absorption spectroscopy in two separate Fabry-Perot cavities, containing the S-(+)-**1** and R-(-)-**1** configuration, respectively (Figure 8.21). For this purpose, a high-performance CO_2 laser-based spectrometer was developed to probe a hyperfine component of the C-F stretching fundamental band of $CHF^{37}Cl^{81}Br$ at 30 THz. In that experiment, a mean difference $\nu_{R(-)} - \nu_{S(+)}$ of 9.4 Hz was obtained with statistical and

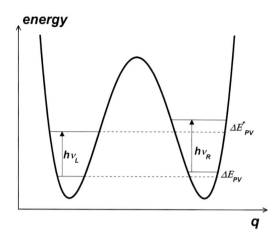

FIGURE 8.20
Potential energy surface of a chiral molecule as a function of the inversion coordinate q. The weak interaction is responsible for a small difference between the ro-vibrational frequencies ν_L and ν_R of the left and right enantiomers. (Adapted from [890].)

systematic uncertainties of 5.1 and 12.7 Hz, respectively. This corresponded to an upper limit of $\Delta\nu/\nu = 4 \cdot 10^{-13}$ for the PV effect [899]. Repeated in 2002, this experiment set the limit $5 \cdot 10^{-14}$. Work is in progress in the same group to tackle the challenge of observing PV in chiral molecules. To this aim, a consortium of physicists, chemists, theoreticians, and spectroscopists has been organized. On one hand, one has to enhance resolution implementing the two-photon Ramsey-fringes technique on an alternate supersonic beam of right- and left-handed molecules. Also, an ultrastable laser source (ultimately referenced to an atomic frequency standard) should be employed. On the other hand, one has to synthesize the *ideal* chiral molecule which should satisfy all the following requirements: (i) exhibit a large $\Delta\nu_{PV}$; (ii) be attainable in huge enantiomeric excess; (iii) evading nuclei with a quadrupole moment to avoid large hyperfine structure; (iv) possess an appropriate two-photon transition connecting a state in the fundamental vibrational level, $v = 0$ to one in the $v = 2$ level; (v) permit the production of molecular beams.

8.7 Perspectives for precision spectroscopy of cold molecules

Cold molecules are currently subject to intense research efforts as they promise to lead major advances in precision measurement, quantum control, and cold chemistry. As already discussed in Chapters 5 and 7, a number of techniques has already been demonstrated to reach temperatures in the mK regime, while different schemes are being examined to access the μK threshold. Essentially, by increasing considerably the interaction time between the interrogating spectroscopic source and the molecular sample, such cooling/decelerating techniques may lead to unprecedented spectral resolution and accuracy levels (recall that the cooling process also strongly simplifies the molecule spectrum). Furthermore, the field of ultrahigh-resolution molecular spectroscopy has known, in recent years, an ulterior revolu-

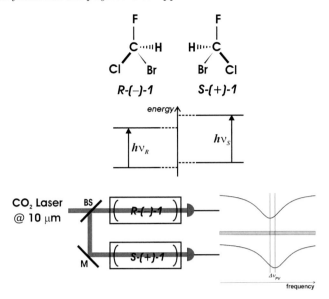

FIGURE 8.21
Principle of the PV test on CHFClBr: spectra of the two enantiomers are recorded simultaneously using identical cavities. (Adapted from [890].)

tion marked by the appearance of optical frequency comb synthesizers based on femtosecond mode-locked lasers. Thus, a new generation of more and more severe tests of fundamental physics laws is today possible by bringing together the two main thrusts of modern atomic, molecular, and optical (AMO) physics: the cold and the ultraprecise. In this frame, the check of the constancy of fundamental constants, the search for parity and time reversal-violating permanent electric dipole moments, the assay of the spin-statistic relation, and the pursuit of parity-violating electroweak interactions are just examples of intriguing challenges where molecular spectroscopy may play a crucial role [687].

As anticipated, another extremely alluring trend of experiments hinges on the realization of colder molecules in a relentless race towards ensembles with the highest possible phase space density. One can envisage that this stream of experiments, gradually approaching the quantum degeneracy regime, will take on and further expand the role of cold atomic physics and chemistry in the discovery of new phenomena. With concern to low-temperature chemistry, the quantum nature of ultracold collisions may provide a detailed probe of fundamental chemical reactions. At higher temperatures, measurements of the rates of chemical processes are subject to a large amount of averaging which obscures the intricate details of molecular collisions. By contrast, at ultracold temperatures, averaging over quantum states is minimized; the populations of the internal quantum states collapse into the lowest one or two states and the collisional angular momentum becomes highly restricted, ultimately reaching the limit of a single active collisional angular momentum state (pure s-wave or p-wave scattering). Chemical reactions at low collision energies can also be profoundly influenced by the long-range intermolecular forces which control the orientation of the reactants during collisions. Experiments with cold molecules offer a way to sensitively probe this part of the intermolecular potential surface. Moreover, the weakness of these long-range interactions suggests that external fields are likely to exert a strong influence on the dynamics of chemical processes, potentially allowing fine control over the rates and outcomes of molecular collisions. These considerations suggest that the sub-Kelvin world of chemical reactions will be one in which the refined control and detailed interrogation of chemical processes should be possible. Experimental measurements should provide considerable challenges to quantum theories of chemical reaction rates, both in the calculation of *ab initio* potential energy surfaces and the solution of the equations for quantum reactive scattering. And although these temperatures lie well below those occurring naturally, studies of cold and ultracold chemical processes will ultimately provide a better understanding of thermally averaged dynamics at higher temperatures, for example in the 10 to 20 K range found in interstellar gas clouds. In the area that is more purely attributable to Physics, we would like to mention here a few breakthrough perspectives which, following the tradition inaugurated by ultracold atoms, are at the boundary with the condensed-matter domain, in the continued advance of our capabilities in the precise study, control, and measurement of increasingly complex quantum systems. As an example, the study of electric dipole-dipole interaction in quantum degenerate fermionic molecular gases would give a new insight into the Bardeen-Cooper-Schrieffer (BCS) pairing. Even more cardinal, ultracold polar molecules may shed light on the character of fundamental interactions between particles. So far, only forces between pairs of particles have been deeply investigated and, as a consequence, our understanding of the plethora of phenomena in condensed-matter physics rests on models involving effective two-body interactions. On the other hand, exotic quantum phases, such as topological phases or spin liquids, are often identified as ground states of Hamiltonians with three or more-body terms. Although the study of these phases and the properties of their excitations is currently one of the most exciting developments in theoretical condensed-matter physics, it is difficult to identify real physical systems exhibiting such properties. It has been recently recognized, though, that polar molecules in optical lattices driven by microwave fields naturally give rise to Hubbard models with strong nearest-neighbor three-body interactions, whereas the two-body terms can be tuned with external fields [686]. This may open a new route for an experimental study of exotic quantum phases with quantum degenerate molecular gases. The growing interest in states of matter with topological order also lies in the fact that these are characterized by highly stable ground states robust to arbitrary perturbations and which support excitations with so-called anyonic statistics. Topologically ordered states can arise in two-dimensional lattice-spin models, which have been proposed as the basis for a new class of quantum computation. Again, the relevant Hamiltonians for such spin lattice models can be systematically engineered with polar molecules stored in optical lattices, where the spin is represented by a single-valence electron of a heteronuclear molecule. The combination of microwave excitation with dipole-dipole interactions and spin-rotation couplings enables building a complete toolbox for effective two-spin interactions with designable range, spatial anisotropy, and coupling strengths significantly larger than relevant decoherence rates. In this framework, two paradigmatic models have been theoretically demonstrated: one with an energy gap providing for error-resilient qubit encoding, and another leading to topologically protected quantum memory. Therefore, ultracold polar molecules trapped in optical lattices are believed to provide a promising platform for quantum information processing [687]. About this, it is worth mentioning that a Stark decelerator, made of gold electrodes deposited on a glass substrate, has already been realized on a chip. This can be seen as the very first step towards the development of more complex hybrid quantum devices where decelerated molecules are trapped just above the surface of mesoscopic features patterned on a chip. The proximity of molecules to the surface can lead to strong couplings between the internal and/or motional states of the molecules and the quantum states of objects on the chip itself. For example, in the case of molecules trapped by static electric fields formed by on-chip electrodes, the molecular rotational states heavily couple to microwave photons confined in the strip-line geometry.

In conclusion, although the field of ultracold molecules can be considered still in its infancy at the time of writing, the exceptional growing rate of the emerging techniques for cooling/trapping of molecules in conjunction with the flourishing of convincing proposals for both fundamental and applied research portends that *Molecular Science* will be the supreme protagonist in the immediate future.

8.8 Tests of general relativity: from ground-based experiments to space missions

Currently, the fundamental physical laws of Nature are contained in two major theories, namely the Standard Model and Einstein's General Relativity (GR). The former designates the families of fermions (leptons and quarks) and describe their interactions by vector fields which communicate the strong, electromagnetic, and weak forces. The latter is a tensor field theory of gravity. In spite of its beauty and ability to describe cosmological phenomena, GR is a classical theory and hence fundamentally incomplete. Notwithstanding recent progress in string theory, the road to merge gravity and quantum mechanics, so as to have a unified description of all particle interactions, seems long yet (Figure 8.22). There are even scientists who speculate that a new theory of quantum gravity must be radically different from standard GR and quantum theory, and must thus violate one or more of the principles underlying these theories [768]. In this frame, a number of theoretical groups is at work to perfect the formulation of alternative theories [900]. In particular, fresh developments in scalar-tensor extensions of gravity suggest a few tests to look for deviations (from Einstein's theory) which are three to five orders of magnitude below the level currently tested by experiments. In conjunction with the cutting-edge technologies developed in ground-based labs, offering variable gravity potentials, large distances, as well as high velocity and low acceleration regimes, space environment may provide a unique playground for many of these searches. Just to give an idea, for a practical height of a laboratory clock, the interaction time in an atomic fountain is limited to about 1 s. This can instead be increased up to 10 s in a microgravity environment, like that provided by a satellite orbiting Earth. In this respect, space-borne atomic clocks, both microwave and optical, will provide major benefits to address a variety of fundamental physical issues. In the following we shall examine the most popular ones.

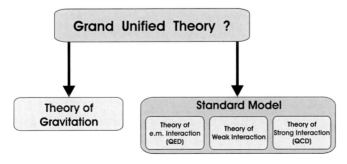

FIGURE 8.22
Classical theories of gravitation like GR are fundamentally incomplete. The Grand Unified Theory, aiming at merging the theory of gravity with the Standard Model, is the dream of all physicists. (Adapted from [768].)

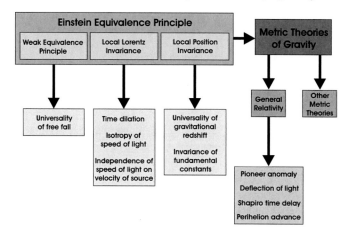

FIGURE 8.23
The Einstein EP is at the basis of the concept that space-time is curved. Only metric theories of gravity satisfy it: the most celebrated one is General Relativity, but there are other examples [901, 900, 902]. Some experiments are also indicated, which can be performed to test certain aspects of the theory. (Adapted from [768].)

8.8.1 Testing the Einstein Equivalence Principle

Being at the foundation of the general theory of relativity, the Equivalence Principle (EP) contains three hypotheses (see also Figure 8.23) [768, 900]:

1. The trajectory of a freely falling *test* body (one not acted upon by forces such as electromagnetism and too small to be affected by tidal gravitational forces) is independent of its internal structure and composition. This is known as the weak equivalence principle (WEP). When dropping two different test bodies in a gravitational field, the WEP implies that the bodies fall with the same acceleration. This is known as the Universality of Free Fall (UFF);

2. The outcome of any local non-gravitational experiment is independent of the velocity of the freely falling reference frame in which it is performed. This is termed Local Lorentz Invariance (LLI) and suggests that clocks' rates are independent of the clocks' velocities;

3. The outcome of any local non-gravitational experiment is independent of where and when in the universe it is performed. This is referred to as Local Position Invariance (LPI) and postulates that clocks' rates are also independent of their space-time positions.

8.8.1.1 Tests of UFF

Precise laboratory tests of UFF are typically performed by comparing the free-fall accelerations, a_1 and a_2, of two different test bodies. When these are at the same distance from the gravity source, the expression for the EP can be cast in the form

$$\frac{\Delta a}{a} = \frac{2(a_1 - a_2)}{a_1 + a_2} = \left[\frac{m_G}{m_I}\right]_1 - \left[\frac{m_G}{m_I}\right]_2 \tag{8.50}$$

where m_G and m_I denote the gravitational and inertial masses of each body, respectively

[900]. Then, the sensitivity of the experiment is fixed by the precision of the differential-acceleration measurement divided by the extent to which the test bodies differ (e.g., composition). Within this approach, the fractional differential acceleration between beryllium and titanium test bodies was recently given as $\Delta a/a = (1.0 \pm 1.4) \cdot 10^{-13}$, based on an Eöt-Wash rotating-torsion-balance experiment [903]. With respect to Earth-based laboratories, experiments in space can take advantage of free fall as well as dramatically suppressed contributions related to seismic, thermal, and other nongravitational noise. Among the many experiments that have been proposed to test the EP in space, we only mention the cold atom-based Quantum Interferometer Test of the Equivalence Principle (QuITE). This aims at measuring with an accuracy of 1 part in 10^{16} the absolute single-axis differential acceleration by employing two colocated-matter wave interferometers with different atomic species [900]. QuITE will improve the current EP limits placed by ground-based experiments working on the same principle, like that by Fray and co-workers [904]. Here, a matter wave interferometer was developed, based on the diffraction of atoms from effective absorption gratings of light: in a setup with cold rubidium atoms in an atomic fountain, the gravitational acceleration of the two isotopes ^{85}Rb and ^{87}Rb was compared, yielding a difference $\Delta g/g = (1.2 \pm 1.7) \cdot 10^{-7}$.

8.8.1.2 Tests of LLI

Before being incorporated into the theory of General Relativity, Local Lorentz Invariance had been at the basis of the formulation of Special Relativity (SR). As such, LLI has been intensively scrutinized in an experimental perspective, and various test theories have been proposed to account for possible deviations from SR. For the purpose of quantifying experimental results, the one by Robertson, Mansouri, and Sexl (RMS) is the most widely used. In the following, we present it following the brief formulation given in [905].

Essentially, this model abandons Einstein's postulates and assumes an isotropic speed of light c only for a hypothetical preferred reference frame $\Sigma(T, \mathbf{X})$. Then, general linear transformations between Σ and a frame $S(t, \mathbf{x})$ moving at a velocity V along $\mathbf{X}$ (with respect to Σ) are derived. By virtue of Einstein synchronization, such generalized Lorentz transformations take the form

$$T = \Gamma \left(\frac{t}{\hat{a}} + \frac{Vx}{\hat{b}c^2} \right) \tag{8.51}$$

$$X = \Gamma \left(\frac{x}{\hat{b}} + \frac{Vt}{\hat{a}} \right) \quad Y = \frac{y}{\hat{d}} \quad Z = \frac{z}{\hat{d}} \tag{8.52}$$

with $\Gamma = 1/\sqrt{1 - V^2/c^2}$. The three velocity-dependent test functions $\hat{a}(V^2)$, $\hat{b}(V^2)$, $\hat{d}(V^2)$ parametrize time dilation as well as Lorentz contraction in the longitudinal and transverse directions. For special relativity, they are all unity. By expanding these functions in powers of V^2/c^2, that is, $\hat{a}(V^2) = [1 + \hat{\alpha}V^2/c^2 + \mathcal{O}(c^{-4})]$ and so on, three test parameters $\hat{\alpha}(V^2)$, $\hat{\beta}(V^2)$, $\hat{\delta}(V^2)$ are obtained:

- The Ives-Stilwell experiment tests $\hat{\alpha}$, that is the time dilation parameter;

- The Kennedy-Thorndike experiment tests $|\hat{\alpha} - \hat{\beta}|$, that is the boost dependence of the speed of light;

- The Michelson-Morley experiment tests $|\hat{\beta} - \hat{\delta}|$, that is the isotropy of the speed of light.

Ives-Stilwell experiments

In this type of experiments, the idea is to measure the Doppler shift of light emitted from moving particles [905]. The first time-dilation measurement, carried out by Ives and Stilwell in 1938, used a collinear setup in which the Doppler-shifted frequencies of light emitted from hydrogen canal rays were measured both in forward and backward directions relative to the atomic motion. If $\beta = v/c$ and $\gamma = 1/\sqrt{1 - \beta^2}$, for the simultaneous measurement of the Doppler-shifted frequencies $\nu_p = \nu_0/[\gamma(1 - \beta)]$ parallel and $\nu_a = \nu_0/[\gamma(1 + \beta)]$ antiparallel to the particles' motion in the laboratory frame, one obtains

$$\nu_0^2 = \nu_a \nu_p \tag{8.53}$$

assuming plane waves. As discussed above, the β-independence of this relation is a consequence of time dilation in special relativity, regardless of the chosen clock synchronization scheme. On the other hand, an analysis of Doppler shift experiments within the RMS test theory shows that a non-vanishing $\hat{\alpha}$ would modify the outcome of the Ives-Stilwell experiment as

$$\frac{\nu_a \nu_p}{\nu_0^2} = 1 + 2\hat{\alpha}(\beta^2 + 2\boldsymbol{\beta}_{lab} \cdot \boldsymbol{\beta}) + \mathcal{O}(c^{-4}) \simeq 1 + 2\hat{\alpha}\beta^2 \tag{8.54}$$

where $\boldsymbol{\beta}_{lab} = \boldsymbol{V}_{lab}/c$ with $V_{lab} = 350$ km/s (Σ is, in fact, generally assumed to be the cosmic-microwave-background rest frame). Recently, an Ives-Stilwell experiment was carried with fast optical atomic clocks [905]. More in detail, $^7\text{Li}^+$ ions are accelerated by a tandem Van de Graaff accelerator and injected into the test storage ring (TSR) shown in Figure 8.24. In the TSR, $^7\text{Li}^+$ can be stored at velocities ranging from $\beta_1 \simeq 0.030$ to $\beta_2 \simeq 0.064$. Neglecting the sidereal term, the Doppler-shifted frequencies $\nu_a^{1,2}$ and $\nu_p^{1,2}$ measured at β_1 and β_2 can be combined using Equation 8.54 to

$$\frac{\nu_a^{(2)} \nu_p^{(2)}}{\nu_a^{(1)} \nu_p^{(1)}} = \frac{1 + 2\hat{\alpha}\beta_2^2}{1 + 2\hat{\alpha}\beta_1^2} \simeq 1 + 2\hat{\alpha}(\beta_2^2 - \beta_1^2) \tag{8.55}$$

The moving clocks are read using laser saturation spectroscopy on the strong 2s $^3S_1$ ($F = 5/2$) $\rightarrow$ 2p $^3P_2$ ($F = 7/2$) transition at 548 nm. The laboratory frequencies ν_p and ν_a of the parallel and antiparallel laser beams (with respect to the ion beam) must obey Equation 8.54 for resonance, which is indicated by a dip in the fluorescence spectrum. Through permanent cooling of the ions by a cold electron beam, a Doppler width of 2.5 GHz (FWHM) is obtained. Then, to overcome this broadening in a saturation-spectroscopy scheme, two lasers are overlapped parallel and antiparallel with the ion beam, respectively, and excite the clock transition. The co-propagating laser (a Nd:YAG laser at 532 nm for β_1 and an argon-ion laser at 514 nm for β_2) is frequency-locked to a well-known iodine (I_2) line, whereas the counter-propagating light is generated by a tunable dye laser (at 565 nm and 585 nm for β_1 and β_2, respectively). The dye laser frequency is referenced to a second, I_2-stabilized dye laser by a tunable frequency-offset lock. The iodine lines are calibrated using an optical frequency comb. All laser frequencies are known absolutely to 70 kHz during the whole experiment. Taking all systematic errors into account, the transition frequencies ν_a and ν_p measured for $\beta_1 = 0.030$ and $\beta_2 = 0.064$ yield

$$\hat{\alpha} = (-4.8 \pm 8.4) \cdot 10^{-8} \tag{8.56}$$

which is consistent with special relativity. Later, the same group also placed a limit on the second-order time dilation parameter ($\sqrt{\nu_a \nu_p}/\nu_0 = 1 + \hat{\alpha}\beta^2 + \hat{\alpha}_2\beta^4$): $|\hat{\alpha}_2| < 1.2 \cdot 10^{-5}$ [906].

Space missions with optical clocks in their payloads also represent an enormous potential to severely assess LLI. Indeed, besides the gravitational redshift term, the frequency difference between a space-borne clock and a terrestrial one will generally contain a contribution that depends on the coordinate velocities of the two clocks. In fact, a precise measurement

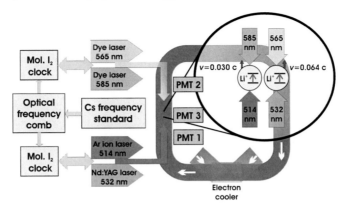

FIGURE 8.24
Schematic setup of the Ives-Stilwell experiment with fast optical atomic clocks. (Adapted from [905].)

of such term is equivalent to address time dilation [768]. Just as an example, let us consider the SAGAS (Search for Anomalous Gravitation using Atomic Sensors) mission proposal, which is intended, in essence, to fly highly sensitive atomic sensors (an optical clock and a cold atom accelerometer) on a Solar System escape trajectory [907]. Now, at the end of the SAGAS mission, the above time-dilation term is about $4 \cdot 10^{-9}$. Thus, with the proposed clock accuracy of 10^{-17}, it could be measured with a relative uncertainty of approximately $3 \cdot 10^{-9}$. This would represent an improvement by approximately a factor of 30 compared to the best (ground-based) current limit given above ($|\hat{\alpha}| \leq 8.4 \cdot 10^{-8}$).

Kennedy-Thorndike experiments

The best Kennedy-Thorndike experiment to date has been performed by continuously comparing, over a period of greater than six years (from September 2002 to December 2008), the resonant frequency (11.932 GHz) of a cryogenic sapphire oscillator (CSO) with the 100-MHz output of a hydrogen maser [908]. In such a terrestrial experiment, the laboratory velocity $V(t)$ is modulated daily by the rotation of the Earth about its axis (amplitude ~ 300 m/s, depending on latitude) and annually by Earth's orbital motion around the Sun (amplitude ~ 30 km/s). Since the resonance frequency of the cavity (the CSO in this case) is proportional to $c(V)$, the frequency difference between the frequency standard (the H maser) and the cavity will be modulated in a corresponding way if Lorentz invariance is violated. Due to the long-term operation, both sidereal and annual modulations could be investigated. By properly modelling the boost with respect to the cosmic microwave background (CMB) frame, these results gave $P_{KT} \equiv \hat{\beta} - \hat{\alpha} - 1 = -1.7(4.0) \cdot 10^{-8}$ for the sidereal and $-23(10) \cdot 10^{-8}$ for the annual term, with a weighted mean of $-4.8(3.7) \cdot 10^{-8}$.

Michelson-Morley experiments

Concerning the isotropy of c, an improved laboratory test was recently performed by comparing the resonance frequencies of two orthogonal optical resonators implemented in a single block of fused silica and rotated continuously on a precision air bearing turntable [909]. More in detail, the resonance frequencies of the two crossed optical Fabry-Perot (FP) resonators are compared by stabilizing (via the PDH technique) two Nd:YAG lasers to these cavities and taking a beat note measurement (see Figure 8.25). Recalling that the resonance frequency ν of a linear FP cavity (length L) depends on the speed of light c along its optical

 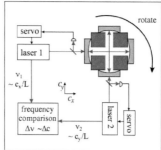

FIGURE 8.25

Improved Michelson-Morley test. Picture of the crossed high-finesse fused-silica resonators and basic principle of the experiment. The frequencies of two lasers, each stabilized to one of two orthogonal cavities, are compared during active rotation of the setup. (Courtesy of [909].)

axes as given by $\nu = mc/(2L)$ (with m an integer number), in order to detect an anisotropy $\Delta c = c_x - c_y$, a modulation of the beat frequency $\Delta\nu$ is sought, while continuously rotating the setup. Since the light in the cavities travels in both directions and c refers to the two-way speed of light, such an isotropy violation indicating modulation would occur at twice the rotation rate. An analysis of data recorded over the course of one year set a limit on an anisotropy of the speed of light of $\Delta c/c \sim 1 \cdot 10^{-17}$ or, equivalently, a limit $P_{MM} \equiv \hat{\delta} - \hat{\beta} + 1/2 = (4 \pm 8) \cdot 10^{-12}$ on the RMS isotropy parameter. This was a factor of 10 more stringent as compared to the value given in [910], based on the comparison of two orthogonal CSOs rotating in the lab.

Possessing the inherent advantages of high orbital velocity and strongly reduced cavity deformation in the microgravity environment, space-borne optical clocks are also expected to substantially improve the detection sensitivity in both Kennedy-Thorndike and Michelson-Morley experiments.

8.8.1.3 Tests of LPI

Such principle is effectively assayed in gravitational redshift experiments. Indeed, as a consequence of LPI, a clock proceeds more slowly the closer it is to a massive body. Therefore, when measuring the fractional frequency shift between two identical clocks located at different heights in a static gravitational field, one should find

$$Z \equiv \frac{\Delta\nu}{\nu} = \frac{\Delta U}{c^2} \tag{8.57}$$

where ΔU is the difference in the Newtonian gravitational potential between the clocks. By contrast, if LPI were not valid, the above equation would modify to

$$Z \equiv \frac{\Delta\nu}{\nu} = (1 + \alpha')\frac{\Delta U}{c^2} \tag{8.58}$$

with the parameter α' being related to the internal structure of the particular clocks employed [715].

Before focusing on tests based on atomic clocks, we mention that a precise, all-ground-based measurement of such gravitational redshift was recently carried out, hinging on a re-interpretation of cold-atom interferometry experiments previously used to measure the acceleration of free fall. This determination provided the value $\alpha' = (7\pm7)\cdot10^{-9}$, compatible with General Relativity within the standard error [911].

Now, let us come back to atomic clocks. We start by observing that, although being routinely accounted for in satellite-based navigation systems, the gravitational redshift effect is exceptionally small in the range of length scales encountered in our daily life. For example, if two identical clocks are separated vertically by 1 km near the surface of Earth, the higher clock emits about three more second-ticks than the lower one in a million years. Nevertheless, due to the astonishing performance of optical atomic clocks, the detection of relativistic time dilation due to a change in height of 33 cm was recently reported [774]. In this experiment, two optical clocks based on individual trapped $^{27}Al^+$ ions, with reported systematic frequency uncertainties of $8.6 \cdot 10^{-18}$ and $2.3 \cdot 10^{-17}$, were compared to each other. Now, for small height changes on the surface of Earth, Equation 8.57 specifies to

$$\frac{\Delta\nu}{\nu} = \frac{g\Delta h}{c^2} \tag{8.59}$$

where $g = 9.80$ m/s^2 is the local acceleration due to gravity and Δh denotes the height distance between the two clocks. Hence, the gravitational shift corresponds to a clock shift of about $1.1\cdot10^{-16}$ per meter of change in height. To observe this shift, the frequencies of the two Al$^+$ clocks were first compared at the original height difference of $\Delta h = h(Mg - Al) - h(Be - Al) = -17$ cm, which was measured with a laser level. Then, the optical table of the Mg-Al clock was elevated, supporting it on platforms that increased the height by 33 cm. By comparing the two frequencies again, a fractional frequency change of $(4.1 \pm 1.6) \cdot 10^{-16}$ was found. When interpreted as a measurement of the change in height of the Mg-Al clock, the result 37 ± 15 cm well agreed with the known value of 33 cm.

Concerning space-based tests, the most accurate measurement so far dates back to the Gravity Probe A experiment performed in 1976 [913]. By comparing the frequency of two hydrogen masers, one on the ground and other on board a spacecraft launched to an altitude of 10000 km, this set a limit of $|\alpha'| < 7 \cdot 10^{-5}$. An improvement of more than one order of magnitude should come from the ACES (Atomic Clock Ensemble in Space) mission which, planned for launch in 2014, will install a cold-atom cesium clock (PHARAO) and a space hydrogen maser (SHM) on board the International Space Station (ISS) [914]. By exploiting the high accuracy of PHARAO (10^{-16}) in conjunction with the gravitational redshift of the ISS orbit ($Z \simeq 4.5\cdot10^{-11}$), a precision of few ppm could be reached in the measurement of the frequency difference with a high-performance ground-based clock (having an accuracy better than 10^{-16}). Obviously, optical clocks could provide further appreciable gain in precision. In this frame, one proposed experiment is contained in the Einstein Gravity Explorer (EGE) mission, which uses a highly elliptical Earth orbit to make an absolute measurement of the frequency difference between the ground and the satellite clocks when the satellite is at apogee (hence, in this approach, the dominant contribution to the redshift is from the earth's gravitational potential) [915]. This provides a terrestrial gravitational potential difference of $\Delta U/c^2 \sim 6.5 \cdot 10^{-10}$. For an accuracy of the optical clock of $2 \cdot 10^{-17}$, a measurement of the gravitational redshift with $3 \cdot 10^{-8}$ relative uncertainty is expected. In a complementary manner, an extremely accurate measurement of the solar gravitational redshift may be accomplished in the SAGAS mission whose escape trajectory exhibits a gravitational-potential change of $\Delta U/c^2 \sim 10^{-8}$. In particular, a fractional uncertainty of 10^{-9} should be achieved with an optical clock of 10^{-17} accuracy [907].

Null redshift tests

LPI also entails the so-called universality of the gravitational redshift (UGR) according to which the frequencies ν_A and ν_B of two atomic clocks with different structure must undergo identical redshifts as they move together through a changing gravitational potential $U(t)$ [768, 908]. In such a null-redshift experiment, if LPI were violated, from Equation 8.58 we may write

$$\frac{\nu_A(t) - \nu_B(t)}{\nu} = (\alpha'_A - \alpha'_B)\frac{U(t)}{c^2} \qquad (8.60)$$

For ground-based clocks, the dominant contribution to changes in $U(t)$ comes from the annual elliptical orbit of the Earth about the Sun: $U(t)/c^2 = -[Gm_s/(ac^2)]e\cos(\Omega_\oplus t_\oplus)$, where G is the gravitational constant, m_s the mass of the sun, a the semimajor axis, e the eccentricity of the orbit, $\Omega_\oplus$ the angular frequency of a sidereal year, and $t_\oplus$ the time elapsed with respect to a recorded Aphelion. This implies variations in U/c^2 of $3.3 \cdot 10^{-10}$. Recently, on the basis of a seven-year comparison between cesium-fountain primary frequency standards and hydrogen masers, the bound $|\alpha'_H - \alpha'_{Cs}| = (0.1 \pm 1.6) \cdot 10^{-6}$ was reported [862]. Of course, much more sensitive tests can be performed in space, where larger values of $\Delta U/c^2$ can be exploited (considering again the SAGAS mission, for instance, the change in U/c^2 is approximately 10^{-8}). In conclusion, it is worth mentioning that null-redshift experiments are inherently more accurate than absolute gravitational-redshift measurements for a twofold reason: first, there is no demand for a high-precision knowledge of the gravitational potential along the orbit; second, dependence of the measurement on a link to ground clocks is avoided.

8.8.2 Test of post-Newtonian gravity

Although a metric theory is generally identified by ten parameters, only two of them, known as the Eddington-Robertson-Schiff ones, β and γ, are needed to describe an isotropic Universe in which conservation laws for total momentum and angular momentum hold. Physically, β is a measure of "nonlinearity" in the superposition law for gravity, whereas γ defines the extent of space-time curvature produced by unit rest mass. Since $\beta = \gamma = 1$ in general relativity, precise measurements of such parameters can play a key role in discriminating between GR and other metric theories of gravity [768, 916].

In the framework of optical clocks, perspectives for improved measurements essentially concern the γ parameter. Its first determination dates back to the celebrated experiment by Dyson and co-workers, who measured the deflection of light by the Sun's gravitational field from observations made at the total eclipse of May 29, 1919 [917]. Later, more refined investigations involved evaluating the so-called Shapiro delay, i.e., measuring the time delay (to and from an interplanetary spacecraft) suffered by radio signals as a consequence of the Sun's gravitational field. Such increase Δt in the time taken for the electromagnetic radiation to travel the round-trip between the ground antenna and the spacecraft can be expressed as [768]

$$\Delta t = 2(1 + \gamma)\frac{GM_S}{c^3}\ln\left(\frac{4r_1 r_2}{b^2}\right) \qquad (8.61)$$

where M_S is the mass of Sun, r_1 and r_2 are the respective distances of the ground antenna and the spacecraft from the Sun, and b is the impact parameter (i.e., the perpendicular distance between the radiation path and the center of the field created by an object that the radiation is approaching). In practice, the associated fractional frequency shift $\Delta\nu/\nu$ is rather measured

$$\frac{\Delta\nu}{\nu} = \frac{d(\Delta t)}{dt} = -4(1 + \gamma)\frac{GM_S}{c^3 b}\frac{db}{dt} \qquad (8.62)$$

Now, if $r_2 \gg r_1$, $db/dt \simeq 30$ km/s (namely the Earth's orbital velocity) and $b \simeq 7 \cdot 10^8$ m (at grazing incidence), whereupon the maximum size corresponding to this effect is $1.7 \cdot 10^{-9}$. So far, the most severe constraint, $\gamma = 1 + (2.1 \pm 2.3) \cdot 10^{-5}$, was placed by Doppler ranging to the Cassini mission (during solar occultation). Much lower fractional uncertainties (down to 10^{-7} in the case of SAGAS) could be reached by space missions making use of optical atomic clocks.

8.8.3 Tests of the gravitational inverse square law

A complementary exploration scope is represented by the Physics of the short-range (sub-millimeter) world, where new interactions may arise due to the existence of extra dimensions. This suggestive hypothesis is contained in several modern theories of gravity, ranging from string to brane world ones [871]. In particular, these predictions have triggered a significant trend of experiments to search for possible departures from Newton's gravitational inverse-square on ranges from 1 mm to 1 μm [918]. In a recent ground-based torsion-balance experiment, the gravitational inverse-square law was found to hold down to 55 μm, which corresponded to constrain the size of a possible extra-dimension to less than 44 μm [919]. The most striking feature of this experiment is that it was able to probe distances less than the dark-energy length scale $\lambda_d = \sqrt[4]{\hbar c / u_d} \simeq 85$ μm, with energy density $u_d \simeq 3.8$ keV/cm^3. This means that laboratory-setting experiments are now competitive with particle physics research. More recently, a cryogenic microcantilever was used as the force sensor, its displacement being measured with a fiber interferometer. Such apparatus was capable of measuring atto-Newton forces between masses separated by distances on the order of few microns, thus allowing to constrain Yukawa-type deviations from Newtonian gravity, $V_Y(r) = -G_N(m_1 m_2/r)(1 + \alpha e^{-r/\lambda})$, with a 95% confidence exclusion of forces with $|\alpha| > 14000$ at $\lambda = 10\mu$m [920].

While in torsion-pendulum-type experiments the role of lasers is possibly relegated to traditional interferometers and/or autocollimators used to read rotations and translations, far more important *missions* will be entrusted to lasers in radically different experiments. For example, Dimopoulos and Geraci in consultation with Mark Kasevich have proposed a probe of submicron-range forces using interferometry of Bose-Einstein condensed atoms [921]. In this approach, two Bose-Einstein condensates (having a well-defined initial relative phase) would be loaded into two laser-trap regions at different distances from a source mass. Then, the difference in phase evolution rate of the two BECs would provide a measure of the different sitting potentials. In other similar proposals, sensitive measurements of forces at micron scale would be performed by exploiting Bloch oscillations of ultracold atoms in optical lattices [922, 923].

In conclusion, we mention that theoreticians have also considered the possibility that non-compact extra dimensions may produce deviations from the inverse-square law (ISL) at astronomical distances [871]. Currently, however, the gravitational ISL is only assayed through precise measurements of the Moon's orbit about the Earth (analysis of Lunar Laser Ranging data tests the gravitational inverse-square law to $3 \cdot 10^{-11}$ of the gravitational field strength), whereas interplanetary laser ranging could substantially extend the distance-scale of the test.

8.8.3.1 Detection of gravitational waves

Gravitational waves (GWs), a key prediction of Einstein's general theory of relativity, are perturbations (ripples) in the curvature of space-time caused by accelerated masses. The first (indirect) evidence for their existence, coming from binary pulsar investigation, can be traced back to 1974. Such discovery earned the 1993 Nobel Prize in physics to Russell Hulse and Joseph Taylor. However, although gravitational wave detectors have been built and constantly improved since the 1960s, no direct observation has been reported so far. A discussion on GWs is beyond the scope of this book, and only a brief account on detectors based on laser interferometers will be given in the following inspired by [375] (this same review represents a good starting point for the reader who wants to know more about GWs). Just to have an idea of the experimental challenge, the gravitational wave amplitude (or strain) is usually given as the dimensionless quantity

$$h = \frac{2\delta l_{GW}}{l} \tag{8.63}$$

where δl_{GW} is the change in the proper distance l between two space-time events, caused by the gravitational wave. It is related to the reduced quadrupole moment I of the source as well as to the distance r from the source

$$h = \frac{2G}{c^4} \frac{1}{r} \frac{\partial^2 I}{\partial t^2} \tag{8.64}$$

where the factor $2G/c^4$ is precisely responsible for the exceptionally small values of the gravitational wave amplitude, which can be compensated for solely by compact cosmic objects with large accelerations and quadrupole moments. But even for violent events, like a supernova explosion in the Milky Way, the strain h is as low as 10^{-20}. Therefore, if the amplitude spectral density $\tilde{h} = \sqrt{S_h(f)}[1/\sqrt{\text{Hz}}]$ (that is the square root of the power spectral density) is used to characterize the performance of a GW detector, then the effective detection sensitivity can be expressed as $h = \tilde{h}\Delta f$ for a detector with bandwidth Δf (where $\tilde{h} \simeq$ const has been assumed).

In the case of a laser-interferometer detector, let us say a Michelson-type one, a GW changes two perpendicular proper distances (corresponding to the two interferometer arms) by the same amount δl_{GW}, but with different sign, provided that the orientation of the test masses (that are the masses of the interferometer mirrors) is optimum. As a result, if two light beams (of wavenumber $k = 2\pi/\lambda$) travel these distances, a phase shift

$$\delta\phi_{GW} = \frac{4\pi}{\lambda}\delta l_{GW} \tag{8.65}$$

arises between them, which can be detected as a change in the output interference signal. To give some numbers, for an arm length of $l = 1$ km and a gravitational wave amplitude of $h = 10^{-21}$, the detector response is $2 \cdot \delta l_{GW} \sim h \cdot l \sim 10^{-18}$ m. The corresponding phase shift ($\delta\phi_{GW} \sim 6 \cdot 10^{-12}$, with a Nd:YAG laser as the light source) is, in practice, monitored by a nulling method: one keeps the laser light returning from the two arms always 180° out of phase so that the output is dark; then, the error signals (from the automatic control) applied to the end mirrors in order to maintain the dark fringe are directly proportional to the exertion by the gravitational wave. In this respect, the advantage of laser interferometers (compared to resonant-mass detectors) is the broad detection band, from about 10 Hz to 5 kHz. The ultimate sensitivity obviously depends on the arm length and the amount of light energy stored in the arms. In order to increase the storage time, Fabry-Perot cavities are usually employed in the interferometer arms. Also, to suppress acoustic disturbances as well as fluctuations in the local refractive index, the interferometer must be operated in ultra-high vacuum (at a pressure of $\sim 10^{-7}$ Pa). Moreover, a cw laser with great output power and exceptional frequency and intensity stability is required. Most often, a non-planar ring oscillator Nd:YAG laser is used, which typically delivers an output power of about 1 W at a wavelength of 1064 nm. This is amplified in a second laser resonator by injection-locking (master-slave scheme) or in a combination of master-oscillator/power-amplifier (MOPA) to about 10 or 20 W. This light is then coupled into one or two ring resonators (mode cleaners) to prepare the TEM_{00} mode.

The present-day generation of large laser interferometric GW detectors (GEO600, LIGO, TAMA300, Virgo) is close to the targeted sensitivity for detecting GWs from sources in the Milky Way. In case of failure, however, the second generation of advanced detectors (based on 100-W power cw lasers, massive mirrors of high-Q materials and cryogenic coolers) will open new possibilities. Moreover, allowing for larger arm lengths (50000 or 5 million km), space-based gravitational wave observatories, such as the planned Laser Interferometer Space Antenna (LISA), will offer access to a range of the gravitational wave frequency spectrum (the mHz frequency band) that is not accessible on Earth. Consisting of an array of three spacecraft orbiting the sun, each separated from its neighbor by about 5 million kilometers, laser beams will be used to measure the minute changes in distance between the spacecraft induced by passing gravitational waves (for this purpose, the spacecraft have to be drag-free, a requirement common for many fundamental physics missions). LISA expects to detect gravitational waves from the merger of massive black holes in the centers of galaxies or stellar clusters at cosmological distances, and from stellar mass compact objects as they orbit and fall into massive black holes. Finally, since both emission and propagation properties of gravitational waves are altered in modified-gravity models, a powerful insight for these theories may be gained through LISA [871].

8.9   Quantum-enhanced time and frequency measurements

8.9.1   Standard quantum limit in physical measurements

Measurement is a physical process, and the accuracy to which measurements can be performed is governed by the laws of physics. In particular, the behavior of systems at small scales is governed by the laws of quantum mechanics, which place limits on the accuracy to which measurements can be performed.

In general, any measurement process is plagued by statistical or systematic errors. The source of statistical errors can be accidental (for example, resulting from insufficient control of the probes or the measured system) or fundamental (for example, resulting from Heisenberg uncertainty relations). Whatever their origin, one can reduce the effect of statistical errors by repeating the measurement and averaging the outcomes. This is a consequence of the central limit theorem, which states that the average of a large number n of independent measurements (each having a standard deviation $\Delta\sigma$) will converge to a Gaussian distribution with standard deviation $\Delta\sigma/\sqrt{n}$, so that the error on average scales as $n^{-1/2}$, where one can generally consider n as the number of times that the system is sampled. The statistical scaling of errors with $n^{-1/2}$ is referred to as the standard quantum limit (SQL) or shot noise in quantum optics, and only assumes at most classical statistical correlations between different probes, which, on the other hand, are usually uncorrelated in a typical experiment.

Usual measurement strategies therefore fail to fully exploit the quantum nature of the system and probes. A more favorable statistical scaling of errors can be achieved if quantum effects such as entanglement are used to correlate the probes before letting them interact with the system to be measured, allowing the SQL boundary to be surpassed through non-classical strategies [924, 925]. Nevertheless, quantum mechanics still sets ultimate limits in

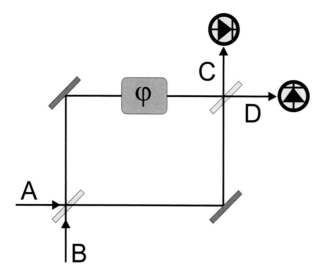

FIGURE 8.26
Scheme of a Mach-Zehnder-type interferometer for the measurement of an unknown phase φ.

precision through Heisenberg-like uncertainty relations, which are typically referred to as Heisenberg bounds and scale as n^{-1}. Quantum metrology is a recently born new discipline that aims to study these bounds and the quantum strategies that allow us to attain them.

We can gain some insight on the limits on the precision of classical measurement strategies by examining the prototype case of a Mach-Zehnder interferometer for the measurement of an unknown phase. Note that Mach-Zehnder interferometry is formally equivalent to Ramsey spectroscopy; so the main conclusions derived here are also valid for the prototype of frequency measurements that we are interested in.

The interferometer acts in the following way: a light beam impinges through the input port A on a semitransparent mirror (a beam splitter), which divides it into a reflected and a transmitted part. These two components travel along different paths and then are recombined by a second beam splitter. Information on the phase difference φ between the two optical paths of the interferometer can be extracted by monitoring the two output beams at the outputs C and D, typically by measuring their intensity (the photon number). To see how this works, suppose that a classical coherent beam with N average photons enters the interferometer through the input A. If the phase difference φ is zero, all the photons will exit the apparatus at output D. On the other hand, if $\varphi = \pi$ radians, all the photons will exit at output C. In the intermediate situations, a fraction $\cos^2(\varphi/2)$ of the photons will exit at the output D and a fraction $\sin^2(\varphi/2)$ at the output C. By measuring the intensity at the two output ports, one can estimate the value of φ with a statistical error proportional to $\sqrt{n}$.

Let us see how this is obtained. In a typical setup the intensity difference M between the two outputs C and D is usually computed

$$M(\varphi) = I_A \cos(\varphi) \tag{8.66}$$

where I_A is the input intensity, proportional to the mean number of photons in the field. If we set the phase close to the point $\varphi = \pi/2$, the two output intensities are balanced and $M \approx 0$. This is the point where the curve 8.66 crosses the horizontal axis and its slope is the steepest, so the sensitivity of the interferometer to phase changes is maximum here. For small changes of the phase around this position the change in the output intensity difference is thus

$$\Delta M = \frac{\partial M}{\partial \varphi} \Delta \varphi = I_A \sin(\varphi) \Delta \varphi \tag{8.67}$$

and

$$\Delta \varphi \approx \frac{\Delta M}{I_A} \tag{8.68}$$

therefore, if one could measure the intensity unbalance M with infinite precision ($\Delta M = 0$), also the phase would be perfectly determined, that is $\Delta \varphi = 0$. As far as electromagnetic waves are concerned, nothing fundamentally prevents ΔM being zero and it would appear that, if all technical imperfections are eliminated, any minimum phase shift can be detected, no matter how small it is.

The problem is that the simple classical arguments we used above do not take into account the effects of quantum mechanics [926]. Specifically they do not take into account the fact that the intensity of the light field is not a constant, which can be measured with infinite precision, but that it fluctuates about some average value, and those fluctuations have their origin in the vacuum fluctuations of the quantized electromagnetic field. Hence the uncertainty always has some finite value and the consequent phase φ will always have its finite related uncertainty $\Delta \varphi \neq 0$.

Due to the quantized nature of the electromagnetic field and to the Poissonian statistics of classical coherent light emitted by a laser, the number of photons in a coherent state having a mean number of photons $\bar{n}$ has a probability distribution

$$p(n) = \frac{\bar{n}^n e^{-\bar{n}}}{n!} \tag{8.69}$$

with a $\sqrt{\bar{n}}$ width. So, given the field intensity I_A proportional to $\bar{n}$, and the minimum measurable intensity difference ΔM, corresponding to a signal-to-noise ratio of 1, proportional to $\sqrt{\bar{n}}$, from Equation 8.68, one obtains that the minimum uncertainty on the phase determination is

$$\Delta \varphi \approx \frac{1}{\sqrt{\bar{n}}} \tag{8.70}$$

Hence quantum mechanics puts a quantitative limit on the uncertainty of the optical intensity, and that reflects itself in a consequent quantitative uncertainty of the phase measurement. This is a consequence of the fact that photons in a classical coherent beam are uncorrelated and do not present any cooperative behavior. In fact, the same $1/\sqrt{n}$ dependence can be obtained if, instead of using a classical beam with n average photons, one uses n separate single-photon beams. In this case, $\cos^2(\varphi/2)$ is the probability of the photon exiting at output C, while $\sin^2(\varphi/2)$ is the probability of the photon exiting at output D.

Anyway, by looking at Equation 8.70, it would seem that any desired sensitivity $\Delta \varphi$ could be attained by simply increasing the laser power and hence $\bar{n}$. However, since $\Delta \varphi$ scales only slowly as $1/\sqrt{\bar{n}}$, the laser power rapidly becomes so large that the power fluctuations at the interferometer's mirrors introduce additional noise terms that eventually limit the device's overall sensitivity. This is the common situation in the case of gravitational wave detectors based on interferometric setups [927], where this kind of quantum noise is currently the main limit to the achievable sensitivity.

8.9.2 Using nonclassical light states for quantum-enhanced measurements

The $1/\sqrt{n}$ bound on the precision (n being the number of photons used, either one by one or as the average photon number in a coherent beam) is referred to as the shot noise or standard quantum limit. The term shot noise comes from the notion that the photon-number fluctuations arise from the scatter in arrival times of the photons at the beam splitter, much like buckshot from a shotgun ricocheting off a metal plate. However, the shot noise limit is not fundamental and is only a consequence of the employed classical detection strategy, where neither the state preparation nor the readout takes advantage of quantum correlations.

In quantum mechanics, the outcomes x_j of the measurements of a physical quantity x are statistical variables; that is, they are randomly distributed according to a probability determined by the state of the system. The Heisenberg uncertainty relation states that when simultaneously measuring incompatible observables such as position x and momentum p, the product of the spreads is lower-bounded: $\Delta x \Delta p \geq \hbar/2$, where $\hbar$ is Planck's constant. The same is true when measuring one of the observables (say x) on a set of particles prepared with a spread Δp on the other observable. In the general case, when we are measuring two observables A and B the lower bound is given by the expectation value of the commutator between the quantum operators associated to A and B.

One can gain an intuitive picture of the situation of measurement in quantum mechanics by looking at the representation of quantum and classical states of the field in the phase space. In Figure 8.27 the Wigner function (a so-called quasi-probability distribution that completely describes the state and allows one to extract the probability distribution of any observable) of a coherent state is shown. Differently from what one might guess for a classical state of an harmonic oscillator, where the position and momentum are always perfectly measurable (at least in principle) and the corresponding Wigner function should look like a Dirac δ-function in phase space, a coherent state has finite and equal spreads in position and momentum: $\Delta x = \Delta p$. A coherent state is also a minimum uncertainty state, in the sense that the combined spread of position and momentum is the minimum achievable according to the Heisenberg relation.

In Figure 8.27, two so-called squeezed states are also shown; they have reduced fluctuations in one of the two incompatible observables at the expense of increased fluctuations in the other. The Heisenberg relation only states that the product $\Delta x \Delta p$ must be larger than $\hbar/2$; therefore these states are also minimum uncertainty states that are perfectly allowed by quantum mechanics.

When dealing with the quantum description of electromagnetic fields, that is in quantum optics, the position and momentum observables x and p are replaced by the in-phase and out-of-phase amplitudes of the field, usually called the "field quadratures." It is then easy to see that a coherent state, the kind of light state normally generated by a well-above-threshold laser, has characteristic uncertainties both in its intensity and phase, approximately indicated by the spreads of the Wigner distribution in the radial and tangential directions of Figure 8.28. The Heisenberg uncertainty principle then tells us that both phase and intensity cannot be measured simultaneously with infinite precision. In particular, the Heisenberg relation can be cast in this case in the form of a number-phase uncertainty as

$$\Delta\varphi\Delta n \geq 1 \qquad (8.71)$$

For a coherent light state, where the photon number distribution is Poissonian with a spread $\Delta n = \sqrt{\bar{n}}$, one easily obtains Equation 8.70 from here.

However, as shown above, squeezed states are also allowed, and in particular the so-called phase-squeezed states, that can decrease the uncertainty in the phase determination at the expense of a larger spread in the intensity. Rather intuitively, the maximum spread that one can obtain in the photon number distribution (i.e., in the laser field intensity) must be of the order of the mean photon number $\bar{n}$ itself, which can be obtained by squeezing the uncertainty ellipse of Figure 8.28 along the radial direction all the way to the origin

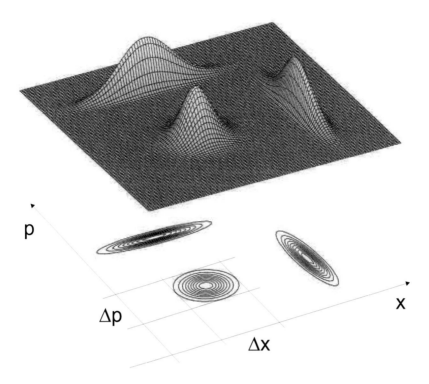

FIGURE 8.27
Wigner function quasi-probability phase-space distributions and contour plots for a coherent
state $|\alpha\rangle$, having equal uncertainties in x and p, and two squeezed states, with reduced
fluctuations in either of the two conjugate observables.

of the phase space. In these conditions one can reach $\Delta n = \bar{n}$ and, using 8.71, recover the
Heisenberg limit for phase measurements

$$\Delta\varphi = \frac{1}{\bar{n}} \tag{8.72}$$

It is interesting to note that also in the case of a vacuum state, where the average field
amplitude is zero, quantum fluctuations are nonetheless present, as indicated by its Wigner
function, which has the same Gaussian shape and width of a coherent state but is centered
at the origin of the quadrature phase space. In an accurate analysis of the limits of phase
sensitivity for an interferometer fed with coherent light, Caves [926] found that it is actually
the quantum fluctuations associated to the vacuum field entering the unused interferometer
port that are responsible for the standard quantum limit proportional to $1/\sqrt{n}$. No matter
what state of the photon field injected in port A, so long as nothing (quantum vacuum) is
put in port B, one will always recover the standard quantum limit. Note that this effect can
be understood also by invoking the partition noise in the distribution of photons between
the output modes operated by the first beam splitter.

In his 1981 paper [926], Caves also showed a natural solution to this problem, that was
to plug the unused port B with squeezed light (squeezed vacuum to be exact). In that
case, with coherent laser light in port A as before and in the limit of infinite squeezing,
the phase sensitivity can asymptotically approach the Heisenberg limit $1/n$ for large mean
photon numbers, proportional to the optical input power. This is a great achievement in
that the total laser power required for a given amount of phase sensitivity is greatly reduced.

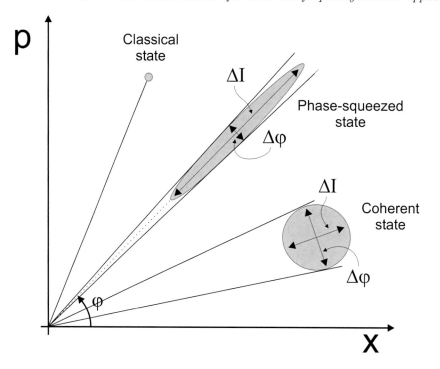

FIGURE 8.28
Phase-space diagram showing quantum fluctuations of different states of the electromagnetic field. A "classical" field is represented as a point in phase-space, with perfectly defined intensity and phase, and no fluctuations. A coherent state, the closest quantum approximation to a classical field, is a disk, with equal fluctuations in intensity and phase. Also shown is a particular squeezed state, a "phase-squeezed" one, showing decreased fluctuations along the angular direction at the expense of increased fluctuations in the radial (corresponding to the intensity) coordinate. Such a phase-squeezed state can be used to beat the shot-noise limit in a phase measurement.

For a typical milliwatt laser power in an optical interferometer gravity wave detector, this could amount to about an eight order-of-magnitude increase in phase-shift sensitivity of the interferometer from the quadratic increase in the power law alone.

Using squeezed states is not the only possible way to beat the standard quantum limit in measurements. Instead of such an "analog" approach, one can follow other routes that exploit discrete photon number and path-entangled optical states, where the number of photons is strictly fixed. The first use of states with a well-defined number of photons, or Fock states, for improving interferometric phase sensitivity was proposed in 1986 by Yurke et al. [928] and Yuen [929]. They showed that phase sensitivity can reach the Heisenberg $1/n$ limit for large n if suitably correlated states are used at the interferometer inputs. A typical state of this kind has the form:

$$|\Psi\rangle = \frac{1}{\sqrt{2}}(|n_+\rangle_A |n_-\rangle_B + |n_-\rangle_A |n_+\rangle_B) \qquad (8.73)$$

where $n_\pm = (n \pm 1)/2$ and the subscripts A and B label the two input modes. This is a highly nonclassical state, where the correlations between the inputs at A and B cannot be described by a local statistical model. It is therefore called a path-entangled state, in the sense that the state at each single input is random, i.e., it can contain $(n + 1)/2$ or

$(n-1)/2$ photons with equal probabilities, but the global state at the two inputs is strictly correlated. If input mode A contains $(n+1)/2$ photons, then input mode B necessarily contains $(n-1)/2$, and vice versa.

In 1993, Holland and Burnett [930] proposed the use of so-called dual-Fock states, where the number of photons at each input mode of the interferometer is exactly the same, such that the input quantum state can be written as

$$|\Psi\rangle = |n/2\rangle_A |n/2\rangle_B \tag{8.74}$$

In this case, although the final measurement strategy at the interferometer outputs turned out to be more complicated than simply taking the photocount difference, the same improvement over the standard quantum limit was proved possible.

Although it was demonstrated that these and other similar states [931] can reach the Heisenberg $1/n$ scaling for sensitivity in the limit of Fock states with large number of photons [932], it was soon found that another kind of heavily path-entangled state could offer the same advantage for all values of n. The so-called N00N state was first discussed by Sanders in 1989 [933], but it was later rediscovered in 2000 in the context of quantum imaging as a tool to circumvent the Rayleigh diffraction limit [934, 935]. The idea is that one still uses a fixed number n of photons in the two arms of the interferometer (after the first beam splitter) but arranged in such a way that they are either all in the upper or all in the lower arm, but one cannot tell - not even in principle - which is which. Naming the mode of the upper (lower) interferometer arm as A (B), the state of all up and none down is written $|all\ up\rangle = |n\rangle_A |0\rangle_B$, and the state of all down and none up is similarly $|all\ down\rangle = |0\rangle_A |n\rangle_B$. A quantum superposition of these two extreme situations can thus be described as

$$|N00N\rangle = \frac{1}{\sqrt{2}}(|all\ up\rangle + |all\ down\rangle) = \frac{1}{\sqrt{2}}(|n\rangle_A |0\rangle_B + |0\rangle_A |n\rangle_B) \tag{8.75}$$

hence the name, N00N state. The reason why the N00N states are performing so well compared to coherent or even Fock states resides in the particular way they respond to phase shifts. When a coherent state passes through a phase shifter φ, such as depicted in Figure 8.26, it picks up a phase of φ. This is a property of a classical monochromatic light beam that coherent states inherit quantum mechanically. However, the behavior of Fock states in a phase shifter is radically different. When a monochromatic beam of number states passes through a phase shifter, the phase shift is directly proportional to n, the number of photons. The evolution for a state passing through a phase shifter φ can thus be shown to have the following two different effects on coherent versus number states:

$$|\alpha\rangle \to |e^{i\varphi}\alpha\rangle \tag{8.76}$$

$$|n\rangle \to e^{in\varphi}|n\rangle \tag{8.77}$$

Notice that the phase shift for the coherent state is independent of n (if n is the average number of photons), but that there is an n dependence in the exponential for the number state. The number state then evolves in phase n times more rapidly than the coherent state. After the phase shifter, the N00N state can be thus shown to evolve into

$$|n\rangle |0\rangle + |0\rangle |n\rangle \to e^{in\varphi}(|n\rangle |0\rangle + |0\rangle |n\rangle) \tag{8.78}$$

which is the origin of the quantum improvement in phase sensitivity. If one now uses an n-photon detector, the obtained signal at the interferometer output is of the form

$$M_{N00N}(\varphi) = I_A \cos(n\varphi) \tag{8.79}$$

and it is simple to see that the N00N signal oscillates n times faster than that for a coherent state of the same wavelength and same mean number of photons. The distance between peaks in the interference fringes becomes n times smaller, which allows one to beat the Rayleigh diffraction limit by a factor of n in quantum lithographic schemes; moreover, the slope of zero crossing also gets larger by the same amount, which implies an increase of phase sensitivity. It turns out that, combined with the different detection scheme necessary for N00N states, this amounts to a net minimum uncertainty for the phase

$$\Delta\varphi_{N00N} = \frac{1}{n} \tag{8.80}$$

which is precisely the Heisenberg limit. Beating of the standard quantum limit in interferometric schemes by the use of N00N states has been proposed and demonstrated experimentally by several groups but, while generating N00N states with $n = 1$ or 2 is straightforward, moving to larger n, or high-N00N states, is very difficult [936, 937, 938, 939, 940, 941, 942]. Besides the technical issues involved in their production, this kind of quantum states is also extremely fragile. The loss of a single photon, or its escape to the environment, is sufficient to reveal which of the two arms of the interferometer is populated and thus destroy the superposition state.

8.9.3 Applications to time and frequency measurements

8.9.3.1 Quantum logic spectroscopy

As already extensively explained, in order to realize atomic clocks with better and better accuracy and stability performance, four key requirements are as follows: (i) selection of an atom with a good reference or spectroscopy transition, which is suitably narrow and relatively immune to environmental perturbations; (ii) cooling to minimize velocity-induced frequency shifts; (iii) reliable initial state preparation; and (iv) efficient state detection. Before the demonstration of quantum logic spectroscopy, to satisfy these requirements, an atom was chosen that simultaneously had a good spectroscopy transition and other, more strongly allowed transitions to accomplish requirements (ii) to (iv). There are, however, excellent candidate ions not yet used for optical frequency standards since their transitions for cooling the ion and detecting the excitation by the electron shelving technique are in the deep ultraviolet. A method has been devised that allows one to overcome these limitations by using two ions in the same trap where besides the clock or spectroscopy ion there is a second one, called the logic ion, which is used for cooling the clock ion and detecting transitions in the clock ion [770].

As an example, let us consider two ions in a linear Paul trap. The Coulomb interaction between the ions couples their motional modes. We are interested in one of the resulting axial normal modes with excitation quantum number m. This mode can be cooled to the ground state ($m \simeq 0$) by use of Raman sideband cooling on the logic ion. We consider only two internal levels of the spectroscopy (S) and logic (L) ion, which we denote $|\downarrow\rangle_{S,L}$ and $|\uparrow\rangle_{S,L}$. Figure 8.29 illustrates the coherent transfer process:

1. Initially, we assume both ions to be in the internal ground state (frame a)

$$\Psi_0 = |\downarrow\rangle_S |\downarrow\rangle_L |0\rangle_m \tag{8.81}$$

2. Then, a laser pulse is applied to interrogate the transition of interest in the spectroscopy ion, thus creating a superposition of ground and excited state with

state amplitudes α and β (frame b)

$$\Psi_0 \longmapsto \Psi_1 = (\alpha \left|\downarrow\right\rangle_S + \beta \left|\uparrow\right\rangle_S) \left|\downarrow\right\rangle_L \left|0\right\rangle_m$$
$$= (\alpha \left|\downarrow\right\rangle_S \left|0\right\rangle_m + \beta \left|\uparrow\right\rangle_S \left|0\right\rangle_m) \left|\downarrow\right\rangle_L \qquad (8.82)$$

with $|\alpha|^2 + |\beta|^2 = 1$.

3. The internal state superposition is coherently mapped to a motional superposition by applying a laser pulse of appropriate length and frequency on the spectroscopy ion: this so-called red sideband (RSB) pulse transfers the excited state to the ground state while adding an excitation to the transfer mode, which is shared by both ions (frame c)

$$\Psi_1 \longmapsto \Psi_2 = (\alpha \left|\downarrow\right\rangle_S \left|0\right\rangle_m + \beta \left|\downarrow\right\rangle_S \left|1\right\rangle_m) \left|\downarrow\right\rangle_L$$
$$= \left|\downarrow\right\rangle_S \left|\downarrow\right\rangle_L (\alpha \left|0\right\rangle_m + \beta \left|1\right\rangle_m) \qquad (8.83)$$

Note that $\left|\downarrow\right\rangle_S \left|0\right\rangle_m$ component of the wave function is unaffected by this operation, because the state $\left|\uparrow\right\rangle \left|-1\right\rangle_m$ does not exist.

4. This mapping process can be reversed on the logic ion by applying laser light resonant to a red sideband transition in the logic ion (frame d).

Overall, the electronic superposition created in the spectroscopy ion has then been transferred to an electronic superposition in the logic ion, where it can be efficiently detected via the usual electron shelving technique.

This quantum-logic spectroscopy was first demonstrated by using $^9\text{Be}^+$ as the logic ion and $^{27}\text{Al}^+$ as the spectroscopy or clock ion (Figure 8.30).

8.9.3.2 Time and frequency quantum metrology

The general idea that one can extract from the above discussion is that one can surpass classical measurement limits and approach the more fundamental quantum boundaries by using its resources in a clever way. Typically, using a highly correlated input state and a collective measurement strategy does the trick (see Figure 8.31).

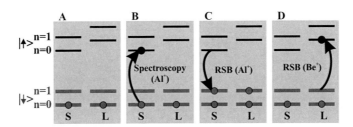

FIGURE 8.29

Spectroscopy and transfer scheme for spectroscopy (S) and logic (L) ions sharing a common normal mode of motion, the transfer mode m, with excitation n ($\left|n\right\rangle_m$ denotes the corresponding harmonic oscillator state). (A) Initialization to the ground internal and transfer-mode states. (B) Interrogation of the spectroscopy transition. (C) Coherent transfer of the internal superposition state of the spectroscopy ion into a motional superposition state by use of an RSB π pulse on the spectroscopy ion. (D) Coherent transfer of the motional superposition state into an internal superposition state of the logic ion by use of an RSB π pulse on the logic ion. (Adapted from [770].)

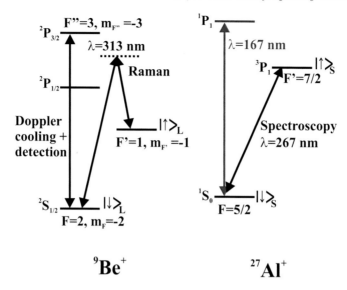

FIGURE 8.30
Partial ^{9}Be$^+$ and ^{27}Al$^+$ energy level diagrams (not to scale). Shown are the relevant transitions for Doppler and Raman cooling on the ^{9}Be$^+$ ion, the spectroscopy transition, and the hard-to-reach Doppler cooling transition at 167 nm on the ^{27}Al$^+$ ion. (Adapted from [770].)

Consider for example Ramsey interferometry, which is a strict counterpart of what is described above for a Mach-Zehnder interferometer. The aim is to measure an unknown relative phase φ picked up by two orthogonal states (like the ground and an excited state, $|g\rangle$ and $|e\rangle$) of an atomic probing system. In a conventional setup, probe preparation consists of producing each atom in the superposition $|\psi_{in}\rangle = 1/\sqrt{2}(|g\rangle + |e\rangle)$, which yields the output state $|\psi_\varphi\rangle = 1/\sqrt{2}(|g\rangle + e^{i\varphi}|e\rangle)$ after the probing stage. Readout consists of checking whether $|\psi_\varphi\rangle$ is still in the initial state $|\psi_{in}\rangle$, which occurs with probability $p = |\langle\psi_{in}||\psi_\varphi\rangle|^2 = (1 - cos\varphi)/2$. Thus, by taking the ratio between the number of successes and the total number of readouts, we can recover the phase φ. If we repeat this measurement n times, the associated error on our phase estimation can then be shown to scale as $1/\sqrt{n}$. The quantum-enhanced version of the Ramsey techniques implies preparing the n atoms in a highly entangled state where they are either all excited or all in the ground state, similarly to the situation of Equation 8.78. If the proper detection scheme is adopted, the phase sensitivity can thus scale much more favorably with the number of used atoms [943, 944].

The Ramsey scheme is very relevant in the accurate time or frequency determination based on atomic transitions, that is, in the context of atomic clocks. To measure time or frequency accurately, one can start with n cold ions or atoms in the ground state and apply an electromagnetic pulse that creates independently in each particle an equally weighted superposition $1/\sqrt{2}(|g\rangle + |e\rangle)$ of the ground and of an excited state. A subsequent free evolution of the particles for a time t introduces a phase factor between the two states that can be measured at the end of the interval by applying a second, identical electromagnetic pulse and measuring the probability that the final state is $|g\rangle$ or $|e\rangle$. This is just what is explained above, but here the phase factor is time-dependent and is equal to $\varphi = \omega t$, where ω is the frequency of the transition $|g\rangle \leftrightarrow |e\rangle$. Hence, the same analysis applies: From the n independent atoms or ions we can recover the pursued frequency ω with an error $\Delta\omega = 1/(\sqrt{n}t)$, if the interaction time t is well known. Conversely, one can use a reproducible atomic frequency standard to accurately measure time intervals.

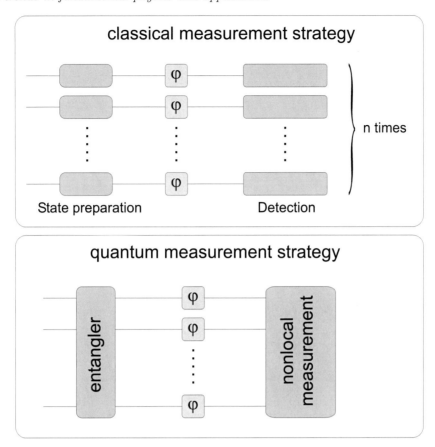

FIGURE 8.31
Classical and quantum measurement strategy.

Instead of acting independently on each particle, one can also start from the entangled states introduced above. In this case, the error in the determination of the frequency is $\Delta\omega = 1/(nt)$, with an enhancement of the square root of the number n of entangled ions over the previous strategy. It boils down to the fact that the phase or frequency estimation is based on a state that, after the first interaction with the field, is a coherent superposition of two energy eigenstates whose energies differ by more than $\hbar\omega$. For such an entangled state of n particles, the eigenstate $|all\,excited\rangle$ and $|all\,ground\rangle$ provide the largest energy difference, with an accumulated phase difference over the free precession period which is n times greater than that for a single atom or ion.

Using entangled ensembles of atoms or ions the precision of atomic clocks can be increased significantly (see [945, 946, 947, 948, 949, 950, 951, 952, 953, 954] for some recent results). Further applications include highly sensitive atom interferometers for the detection of extremely weak forces and the realization of a quantum gate, a key element in future quantum computers.

8.9.3.3 Quantum positioning and clock synchronization

To find out the position of an object, one can measure the time it takes for some light signals to travel from that object to some known reference points. The best classical strategy is to measure the travel times of the single photons in the beam and to calculate their average.

This allows one to determine the travel time with an error proportional to $1/(\Delta\omega\sqrt{n})$, where $\Delta\omega$ is the signal bandwidth. The minimum time duration $1/\Delta\omega$ for each photon thus determines an equal spread in their time of arrival. The accuracy of the travel time measurements thus depends on the spectral distribution of the employed signal.

Similarly to the discussion presented above, one can imagine that the use of some form of quantum correlation may improve the precision also in this kind of measurement [955]. One of the proposed tricks is therefore that of exchanging the series of n single-photon pulses (or, correspondingly, a coherent pulse with a n mean number of photons) of bandwidth $\Delta\omega$, with a single frequency-entangled state whose frequency spread scales as $n\Delta\omega$. This result can also be seen by noting that the entanglement in frequency translates into the bunching of the times of arrival of the different photons; although the individual times of arrival are random, their average is highly peaked. Similar frequency-entangled quantum states can be readily generated in the simple case $n = 2$ by means of spontaneous parametric down-conversion pumped by a continuous wave laser source. In this case, the emitted frequencies of the idler and signal photons are individually random but are strictly anti-correlated, because, for energy conservation, they must sum up to the frequency of the pump photons. Therefore, even if the arrival times of the two individual photons to distant locations have a large spread, their difference is perfectly defined and can be used for precise timing [956].

The problem of localization is intimately connected with the problem of synchronizing distant clocks [957, 958, 959]. In fact, by measuring the time it takes for a signal to travel to known locations, it is possible to synchronize clocks at these locations. This immediately tells us that the above quantum protocols can give a quantum improvement in the precision of synchronization of distant clocks.

Recent works have proposed new ways of using quantum resources to improve accurate space-time positioning. Lamine et al. [960] first derived the standard quantum limit for the precise measurement of the delay between two light pulses. This is the fundamental step for both ranging and clock synchronization and is based on comparing the arrival time of pulses emitted from a local and a remote clock. They found that, by combining both time-of-flight information derived from the pulse envelopes and interferometric information derived from the relative pulse phases, a minimum classical uncertainty in timing can be expressed as:

$$\Delta t_{SQL} = \frac{1}{2\sqrt{n}\sqrt{\omega_0^2 + \Delta\omega^2}} \tag{8.84}$$

where n is the total number of photons of central frequency ω_0 and frequency spread $\Delta\omega$ used in the experiment during the detection time. Although reaching this standard quantum limit (of the order of tens of yoctoseconds, or 10^{-21} - 10^{-24} s) is already far from current technological capabilities, the authors also proposed the use of appropriately squeezed light states for going even further.

Finally, quantum effects can also be useful in avoiding the detrimental effects of dispersion [961]. The speed of light in dispersive media has a frequency dependence, so that narrow signals (which are constituted by many frequencies) tend to spread out during their travel. This effect ruins the sharp timing signals transmitted. Using the nonlocal correlations of entangled signals, we can engineer frequency-entangled pulses that are not affected by dispersion and that allow clock synchronization [962].

8.10   Environmental metrology

We would like to close this chapter with a topic which, although apparently different, is, in fact, intimately linked to the matter of frequency metrology, precisely in the sense discussed so far. This subject goes under the name of environmental metrology and includes, in the widest meaning, climate changes modelling, pollution control, biomedical applications, environmental-friendly engines, geophysical survey, and homeland security. Indeed, the outstanding precision and reliability levels characterizing today's laser-based systems, mainly achieved in fundamental physical research, are engendering very sophisticated measurements in such *less traditional* disciplines too, meeting the rising public concern towards quality-of-life issues. In the following, we select a few but representative applications of precise laser-based techniques to environmental monitoring.

8.10.1   Geophysical survey of volcanic areas

Needless to emphasize, accurate and time-resolved information on partial pressures of gases is a key for studies in many different environments [963, 336, 964, 965, 966]. Over the last few decades, growing attention has been devoted to infrared laser methodologies for molecular gas analysis. Whereas the NIR spectral region took advantage from the long run of telecom-oriented semiconductor manufacturing technologies, with huge fall outs in terms of combined low cost/high-tech lasers and components for spectroscopic detection, the big gap in detection sensitivity, as compared to mid-/far-IR spectral regions, suggested to move to longer wavelengths. Indeed, NIR spectroscopy can, in general, only interrogate molecules exciting forbidden vibrational overtones. To fix the ideas, a key molecule like carbon dioxide (e.g., the most abundant $^{12}C^{16}O_2$) has a transition linestrength about five orders of magnitude stronger in the fundamental ro-vibrational band around 4.3 micron wavelength, as compared to the overtone band in the telecom window, around 1.57 microns. In principle, this means that a comparably lower number of molecules can be detected if moving from the telecom window. Moreover, Doppler-limited molecular linewidths scale linearly with frequency. This means that, when detecting gases at low pressure (i.e., causing no appreciable collisional broadening), the spectral resolution is lower when exciting overtone rather than fundamental bands. Although interrogation of fundamental bands is very advantageous, extension to this spectral range of well-established spectroscopic techniques used in the visible/NIR has been, until recently, a formidable challenge. A critical point has been for a long time the lack of widely tunable mid-IR sources and optical components of sufficiently high quality. Experimental demonstration of quasi-phase-matched (QPM) periodically poled non-linear crystals [967, 968] paved the way to highly efficient coherent sources based on frequency mixing, like optical parametric oscillators (OPOs) [969] and difference-frequency generators (DFGs) [970]. Another key event was quantum-well engineering of semiconductor structures and the subsequent successful construction of quantum cascade lasers (QCLs) [264]. By virtue of these achievements, high-performance sensors incorporating laser sources are under continuous development in laboratories worldwide. For environmental studies, most efforts have been focused on detection of trace gases, such as NO, NO_2, HCl, CO, NH_3, C_2H_2, CH_4 as well as on quantitative chemical analysis of samples containing H_2O, CO_2, O_2, and others. A number of high sensitivity optical techniques, relying on long-path schemes or aiming at achieving shot-noise limited sensitivity, have been developed (see Chapter 5). Semiconductor diode lasers still offer unique features to develop reliable and compact spectrometers, at a reasonable cost. Also, they retain several advantages in terms of tunability, spectral selectivity, and power consumption. In addition, they can be easily coupled into low-loss optical fibers providing, e.g., a unique tool for developing fiber-based nets with the potential to monitor large areas.

A survey of active volcanic areas is a very demanding task for any optical instrumentation, due to the extreme conditions (high humidity, large temperature variations, fumigation with chemically aggressive gases, presence of dust and small rocks) that are generally found in these environments. However, the analysis of volcanic gases is widely recognized as a powerful tool to validate geological models relevant to reliable predictions of volcanic events [971, 972]. Indeed, the chemical composition of certain gases in volcanic effluxes may provide direct information on deep magmatic processes and hydrothermal circulation [973]. In particular, the absolute concentration and the isotopic content of some molecular species gives an indication of sources and sinks of volcanic gases, whereas its time variation may be a consequence of changes in the status of a volcano. These changes can be due to chemical reactions of magmatic gases with rocks or fluids occurring along their path to the surface. Such strong motivations together with the difficulty of insitu monitoring led to the development of many different techniques for the remote sensing of gases emitted from volcanoes. They include ground-based and shipborne spectrometers using incoherent sources [974, 975, 972, 976] or powerful lasers [977] to record molecular spectra, from which retrieval of gas concentration and/or gas fluxes is possible. These instruments have mostly been used in campaigns, due to the difficulty of continuous gas monitoring of volcanic areas. The first prototypes of laser-based instruments for long-standing, ground-based gas measurements were tested around the end of the '90s in the easier telecom window [978]. In that case, optical fibers could be used to take all the relevant instrumentation far enough from the measurement points. One of the major problems, anyway, was represented by the geometry of the gas-laser interaction region.

Figure 8.32 shows the setup used in the first demonstration of monitoring of volcanic gases with a distributed feedback (DFB) diode laser-based spectrometer [979]. The latter was tested at Solfatara, a volcano near Naples, Italy. There, H_2O (about 80%-90%) and CO_2 (about 10%-20%) account for most of the gas emission. Indeed, two lasers were used, each tuned, respectively, on the 1.393 μm (0,0,0)-(1,0,1) vibro-rotational combination band of H_2O and the 1.578 micron (0,0,0)-(3,0,1) band of CO_2. After coupling to a diplexer, both lasers were launched into a 30-m long single-mode fiber, ending on the fumarolic flow. There, the collimated radiation was passed through a specially designed glass cell, relaunched into a fiber, and then coupled out for detection, using again a diplexer.

A major difficulty associated with such an optical scheme was the high content of sulfur and water vapor in the emitted flows. As soon as the temperature of the fumarolic flow decreased, sulfur and water vapor condensation took place, which coated/flooded the cell, eventually blocking any transmitted laser radiation. This problem was solved by a clever design of the interaction region (see Figure 8.32), aimed to avoid contact of the ejected volcanic fluid with the atmosphere and to minimize its temperature variations, in order to rule out any condensation process before interaction with laser light. Concerning data processing, an *ad hoc*-developed method for simultaneous and accurate determination of partial pressures of several gases in direct absorption spectroscopic measurements (without using any reference cell) was adopted [980].

From this quick description, it emerges that a setup like this could evolve towards fiber-based nets, to be deployed for monitoring large areas. However, despite their really great potential, such devices still need a deep engineering work to become really competitive with simpler systems (e.g., based on incoherent sources) and be suitable for long-term outdoor operation by non-expert end users. Although for the toughest environment-related difficulties proper solutions were found with this dual-wavelength spectrometer, the measurement accuracy for carbon dioxide concentration was quite unsatisfactory, especially if compared

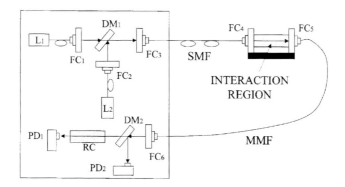

FIGURE 8.32
Sketch of the portable spectrometer. L_1 and L_2 represent the 1.393 μm and 1.578 μm DFB lasers. FC stands for input/output fiber port, DM for dichroic mirror, SMF for single mode fiber, MMF for multimode fiber, RC for reference cell, and PD for photodetector.

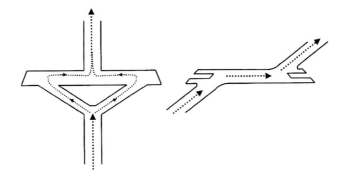

FIGURE 8.33
Sketch of the two glass cells in which the fumarolic gas flowed.

to the few-percent overall uncertainty achieved for water vapor. The two reasons for that were the much lower concentration of CO_2, as compared to water vapor, and the low absorption intensity of this overtone band in the telecom window. Therefore, a setup still using DFB diode lasers and optical fibers, but emitting at longer wavelengths (around 2 micron) to excite a lower order CO_2 overtone, was first built and tested in the lab [981]. Then it was field-deployed, for several measurement campaigns in Vulcano island (Aeolian islands, Sicily), one of the most active volcanoes of the Mediterranean area. Such spectrometer was devoted to the simultaneous monitoring of CO_2 and H_2O concentrations as well as to flux measurements of carbon dioxide diffusely released from soil, using an accumulation chamber configuration (indeed, significant soil-released CO_2 fluxes may represent a serious threat for the health of people and animals). These pioneering results may pave the way to future deployment of other 2-μm-wavelength sources, solid-state or fiber based, emitting a single (tunable) frequency or even octave-spanning combs.

8.10.2 Detection of very rare isotopes

Some of the aforementioned challenges in the field of environmental monitoring require quantitative measurements of extremely small amounts of molecular gases. Presently, molecular concentration measurements are limited to the parts-per-trillion (ppt) range, even when the strongest rovibrational molecular absorptions in the fingerprint region are targeted. This limitation has prevented optical detection of rare-isotope-containing molecules, due to the very low abundance of such species. Among them, concentration measurements down to parts per billion of ubiquitous long-lived radioisotopes are relevant in a broad range of scientific and technological fields. In particular, nature marks every living being with a specific isotope, ^{14}C, which has acquired great importance after the discovery of the radiocarbon dating method in the early 1950s. Radiocarbon, produced in the upper atmosphere by secondary cosmic rays, has a natural abundance $^{14}C/^{12}C \simeq 1.2 \cdot 10^{-12}$. Because of its historical-scale radioactive half-life (about 5730 years), it is an ideal marker for age assessment of samples of biological origin. At present, high energy accelerator mass spectrometry (AMS), with $^{14}C/^{12}C$ sensitivity hitting 10^{-15}, represents the only method capable of dating very old samples (about 50000 years), at the price of big size, cost, and complexity of the instrumentation. Apart from dating, radiocarbon detection has become an essential tool in modern science, such as biomedicine [982] or environmental and earth sciences [983]. AMS technology is also used for these applications but with more relaxed sensitivity requirements. In this frame, radiocarbon optical detection well below natural abundance in CO_2 was recently reported, by exploiting saturated-absorption cavity ring-down (SCAR) spectroscopy (discussed in Chapter 5) in combination with an OFCS-referenced, high-power DFG source emitting at 4.5 μm [420] (see Figure 8.34 for a sketch of the experimental apparatus). In this work, the elusive $^{14}C^{16}O_2$ molecule was detected at concentrations below 43 parts per quadrillion (ppq). This represents the lowest pressure ever detected for a gas of simple molecules and corresponds to a minimum detectable pressure (for radiocarbon dioxide) of $5 \cdot 10^{-16}$ bar.

For applications not requiring the ultimate AMS sensitivity, SCAR spectroscopy can be already superior to AMS, due to the wider dynamic range for the $^{14}C/^{12}C$ ratio (important for biomedical applications) or for the potential future portability of an infrared spectroscopy setup, essential for environmental applications. Another important difference of a SCAR setup, as compared to AMS, is that, in mass spectrometry, an ion from a sample is counted only once, because its measurement neutralizes it, while a spectroscopic measurement does not destroy the sample, allowing it to be repeatedly analyzed.

8.10.3 Stratospheric survey with tunable diode laser spectrometers

The understanding and modelling of stratospheric dynamics and chemistry are vital aspects for the knowledge and prediction of climate changes, so that there has been an increased interest in the possibility to realize analyzers for in situ monitoring of the atmosphere. Since the last decades of the 20th century an increasing number of instruments have been employed for trace gas measurements in the stratosphere and in the troposphere, using both aircraft and balloons as platform. This kind of instrumentation must fulfill several and very demanding requirements. First of all, a high selectivity is required to identify a single molecule in a probe of air where a large number of different species is present: the concentration measurement of the molecule of interest must be neither positively nor negatively influenced by the other species. Also, the instrument must allow concentration measurements from the ground to the stratosphere (namely from zero to tens of km) where a large range of values is possible (from the ppm to the ppb or ppt level, depending on the species). Then, a high precision is required to identify small variations in the concentration

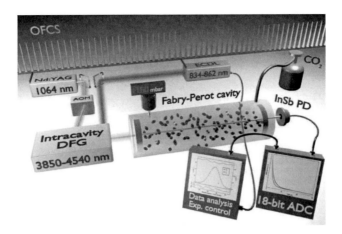

FIGURE 8.34
Schematics of the experimental setup. The IR radiation, tunable between 3850 and 4540 nm, is delivered by a DFG process inside a Ti:Sapphire laser cavity. The Ti:Sapphire laser, optically injected by an external-cavity diode laser (ECDL), and a Nd:YAG laser are mixed in a nonlinear crystal. Both lasers, and hence the IR radiation, are frequency-locked to the OFCS. After switching off-resonance the IR radiation by an acousto-optic modulator (AOM), the light transmitted by a Fabry-Perot cavity is detected by an InSb photodiode (PD). The signal is acquired by an analog-to-digital converter (ADC).

of trace gases measured at the same altitude, connected with air transport mechanisms in the atmosphere. Moreover, a fast time response is needed to detect fine structures due to air transport and mixing from the stratosphere to the troposphere and vice versa. For instance, if the measurement is performed from an aircraft having a cruise speed of hundreds of m/s, a time response of a few seconds is necessary to observe structures of few kilometers. Even faster time responses (less than one second) are necessary to perform flux measurements. In addition, the need to work aboard an aircraft or a balloon under harsh ambient conditions (presence of vibrations, electromagnetic interference, large excursion of pressure and temperature between ground and stratosphere, humidity) and without an operator leads to the requirement of robustness and totally unattended operation. Finally, the instrument must meet the criteria of compactness, lightness, and portability, related to the platform where it will be mounted.

Having the necessary characteristics to fulfill the challenging requirements listed above, tunable diode lasers (TDLs) represent ideal sources to realize instruments for in situ trace gas measurements. The first stratospheric balloon-borne TDL spectrometer was realized in 1983, based on a cryogenic diode laser and a multi-pass White cell [984]. Since then, many other TDL spectrometers have been realized and employed during several measurements campaigns, and significant advances have been achieved in the last years. In particular, the increasing development and optimization of quantum cascade lasers and difference frequency generators is making possible a gradual replacement of cryogenic lead-salt diode lasers (emitting in the mid-infrared) for airborne/balloon measurements too. The first in situ measurement of trace gas concentrations using a QCL was performed in 1999. The spectrometer, which measured CH_4 and N_2O concentrations on board the NASA ER-2 aircraft, was based on a quantum cascade DFB laser, cooled with liquid nitrogen, emitting at 7.9 μm [985]. In the following years, other QCL spectrometers have been deployed aboard stratospheric aircrafts. Their main advantages, with respect to lead-salt diode lasers, are the higher power, with a resulting increase of the instrument sensitivity, and the possibility of room-temperature operation, with a reduction of spectrometer weight and size. Recently, also DFG technology has been used to increase the sensitivity of airborne spectrometers. In 2006, an instrument based on DFG at 3.5 μm by mixing a DFB diode laser at 1562 nm and a DFB fiber laser at 1083 nm in a periodically poled LiNbO$_3$ crystal was used in 3 different campaigns for highly sensitive measurements of formaldehyde [986]. An overview of the last studies and developments of TDL spectrometers in airborne measurements can be found in [987].

Despite the large development of new technologies, TDL analyzers based on lead-salt diode lasers are still employed and provide reliable, high-sensitivity and precise concentration measurements. As an example, the COLD (Cryogenically Operated Laser Diode) spectrometer was developed for in situ Carbon Monoxide (CO) concentration measurements on board the M55 Geophysica, a Russian stratospheric aircraft able to operate at an altitude up to 21-22 km [988]. The COLD spectrometer is based on a Pb lead-salt diode laser, cooled with liquid nitrogen, emitting around 4.6 μm, used in combination with an astigmatic Herriott multi-pass cell providing an optical path of 36 m. A direct absorption detection technique, which does not need in-flight calibration, is used in conjunction with fast sweep integration, to allow absolute concentration measurements of CO. Sensitivities achieved during in-flight operation are of few ppb, with a time resolution of 4 s (corresponding to a local resolution of about 800 m for an average cruising speed of 750 km/h) and a precision of 1%. COLD was deployed during 3 tropical campaigns between 2005 and 2006 (TROCCINOX-2, SCOUT-O3, AMMA) and during a polar campaign (RECONCILE) in 2010. The goal of the campaigns was the analysis of the chemistry and of the transport processes in the upper troposphere (UT) and in the lower stratosphere (LS). In particular, trace gas measurements can be very useful for the study of the tropical convection mechanism, to get a deeper understanding on the region of tropical tropopause layer (during tropical campaigns) and for the analysis of structure and composition of the polar vortex as well as for the study of the ozone depletion (during polar campaigns). In both cases, measurements of tracers recorded during campaigns provide valuable input for the validation of atmospheric models, with possible important implications for the global climate [989, 990, 991, 992, 993, 994]. Typical flight data, registered by the instrument COLD for CO during the TROCCINOX-2 campaign, are reported in Figure 8.35. The CO profile follows inversely the M55 aircraft altitude, reaching the lower concentration levels, about 20 ppb, in the stratosphere, over 18-19 km and the higher level, about 120 ppb, close to ground . The spectrometer time response of 4 s makes possible to resolve fine structures as the one shown in the circle of frame (a). For an aircraft speed of about 210 m/s, the COLD instrument provides 10-20 values of the CO concentrations in a space interval of 10-20 km, so that structures contained in this interval can be clearly identified. Details of the structure, which was observed at a constant aircraft altitude of about 16 km, are displayed in frame (b). The fast CO decrease at a constant altitude is a clear indicator of direct injection of air from the LS, which is poor of CO, into the UT normally richer of CO. For an accurate analysis of these fine structures, transport models at very short time scales are needed.

8.10.4 Fiber sensing of physical and chemical parameters

Optical fiber systems have had great impact in the field of sensing thanks to the growth of the optoelectronics and fiber-optic communication industries. Most components used in these markets were developed with benefits of outstanding technologies in material design, optical fibers, and light sources. The inherent advantages of optical fibers include light weight, low cost, small size, and ruggedness, making it possible to directly install and integrate them in hard-to-access environments or when minimally invasive techniques are required. Their

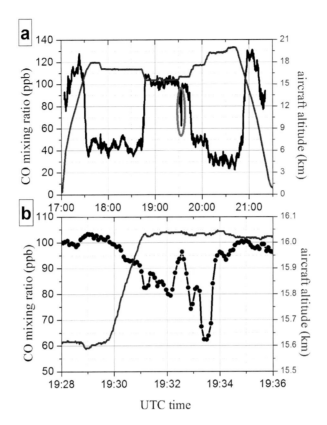

FIGURE 8.35
CO data recorded by the COLD spectrometer during the flight of the 05/02/2005 (TROCCINOX-2 campaign) from Aracatuba (Brazil). (a) CO mixing ratio (black line) and aircraft altitude (blue line) vs. universal time (UTC). (b) Zoom of the fine CO structure, included in the circle of figure (a).

immunity to electromagnetic interference and high sensitivity are crucial for accurate and precise sensing in various applications, while their low optical loss and wide bandwidth ensure the data transmission over long distance to realize sensor networks covering large areas. The past 20 years have witnessed an intensive research effort on optical fiber sensors to measure different physical and chemical parameters [995, 996].

Among fiber-optic sensors, fiber Bragg gratings (FBGs) have shown promising features as mechanical probes for a number of applications. Several interrogation systems have been developed so far, often based on broad-emission radiation sources in conjunction with either optical spectrum analyzers or filters [997, 998]. More recently, sophisticated schemes based on narrow-band laser sources and laser-frequency stabilization methods were devised, achieving ultrahigh strain sensitivity for quasi-static and dynamic monitoring [999, 1000]. Among them, a significant contribution came from the use of FBG-based resonant structures whose highly-dispersive power near resonance is exploited to measure sub-pm length perturbations over a wide range of frequencies. On the other hand, optical resonators based on high-reflectivity FBGs, fiber loops, and silica microspheres have been employed for refractive index measurements and chemical sensing. Cavity-enhanced and ring-down techniques

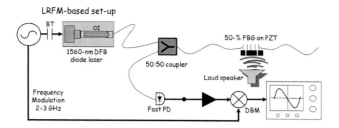

FIGURE 8.36

RF-modulation-based FBG interrogation setup. PD: photodiode; DBM: double-balanced mixer; BT: bias-tee.

enable the extraction of information on light-matter interaction in direct or evanescent-wave spectroscopy schemes [1001, 1002, 1003].

8.10.4.1   Strain sensing

A frequency-modulated telecom diode laser can be used for static and dynamic interrogation of single FBGs with improved sensitivity, as shown in Figure 8.36 [1004]. The system relies on radio-frequency (RF) sideband generation on the laser beam, via current modulation, and heterodyne detection of the FBG reflected light. If the sideband frequency is high enough compared to the FBG width, its reflection spectrum can be treated as a molecular absorption line. Demodulation at that frequency is performed by a double-balanced mixer which yields a highly-dispersive signal with a zero-crossing around the Bragg's resonance in quiescent conditions. This can be employed as a discriminator (error) for Pound-Drever-Hall (PDH) frequency locking of the laser onto the Bragg grating's peak for continuous tracking of the sensor.

A different kind of apparatus can be developed with high-finesse in-fiber Bragg-grating Fabry-Perot (FFP) resonators as strain sensors. The resonator is formed by two high-reflectivity single-mode FBGs at a relative distance of 100 mm. Small optical pathlength variations in the intracavity fiber are turned into frequency shifts of the narrow resonance. A first demonstration of its sensitivity to strain signals is given below (Figure 8.37). Similarly to the previous one, a diode laser is actively locked to the resonator by an optical-electronic loop based on PDH technique.

Although the interrogating laser can be frequency controlled again by the PDH method [1000], different schemes, such as those based on polarization-spectroscopy (PS), can be devised. The PS technique rests on the birefringence induced by FBG fabrication in the resonator [1005, 1006]. The error signal is obtained by adjusting the state of polarization (SOP) of the laser beam at 45° to the fiber birefringence axis and analyzing the cavity-reflected field with a polarization analyzer [1008]. In this way, an excellent performance was obtained without using any RF laser modulation or sophisticated electronics. The minimum detectable strain level may be as low as 1 pε/Hz, in the acoustic range (recall that the strain ε is defined as the fractional length variation $\Delta L/L$).

In the past few years, intensive work has been done to reduce the effect of free-running laser jitter via pre-stabilization on an optical-frequency reference [1007, 1008]. In a recent work, a (fiber-based) frequency-comb (OFC) stabilized diode-laser system for strain interrogation of a high-finesse fiber Bragg-grating cavity was devised (Figure 8.38). Thanks to the exceptional frequency stability of the OFC, a strain resolution of 10^{-13} was achieved from 20 mHz up to 1.5 kHz, approaching a noise floor that is possibly due to thermal effects

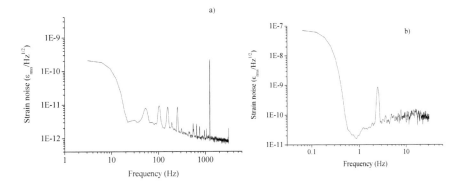

FIGURE 8.37
Noise spectral density of the FBG-resonator locking signal for different excitation frequencies in the SM-fiber cavity: (a) a sharp peak is evident at 1.2 kHz with a noise increase towards low frequencies and spurious oscillations due to harmonics of the AC line frequency; (b) the system is capable of detecting deformations down to 2.4 Hz.

in the fiber [1009]. A Pound-Drever-Hall frequency locking technique was implemented for low-noise and wide dynamic range readout of the sensor.

8.10.4.2 Acceleration measurements

The monitoring of seismic signals is essential to the study of volcanoes, for surveillance of seismic areas, or even in anti-intrusion systems for homeland-security. Seismic waves, both longitudinal and transverse, cause vibrations that can occur over a large frequency span ranging from quasi-static (below 1 mHz) to acoustic frequencies (above 100 Hz). At present, commercial accelerometers generally operate below 100 Hz. To fully understand the link between the seismic occurrences and seismic signals, detectors with high bandwidth at high sensitivity are required. Velocities and accelerations can be efficiently measured with fiber-optic sensors, provided that the mechanical response of the sensor element is known. The possibility of strain-to-acceleration transduction was previously demonstrated using a massive flexural beam sensor monitored by a FBG with basic demodulation schemes [1010, 1011]. A possible approach consists in a long flexural-beam horizontal accelerometer, containing three different FBGs, which were able to monitor deformations of a rigid cylinder in all directions within the horizontal plane. A large mass is placed on top of the cylinder while its base is anchored to the ground using a special screw. Three FBG elements are glued into the cylinder internal surface, parallel to the vertical axis, and placed at angles of 120 degrees. Mechanical waves can be detected in the horizontal plane by at least two sensors for determination of their intensity and direction [1012]. A customized laser-spectroscopic interrogation technique improved the sensitivity and dynamic response of the system. A laboratory test demonstrated successful operation along two directions in the plane for subsequent acceleration pulses. As a further development, three separate π-shifted FBGs (PSFBGs) can be attached to three cantilever beams that flex in orthogonal directions (Figure 8.39) [1013]. The PSFBGs present a characteristic response which is quite similar to common optical resonators [1014]. The phase defect in the periodic structure indeed modifies its photonic bandgap and creates a sharp resonance exactly at the Bragg wavelength. That strongly improves the capability of detecting small shifts caused by mechanical action on the fiber. The sensors were interrogated by three distributed feedback lasers actively locked with PDH technique to their central resonance using a radio-frequency modulation technique

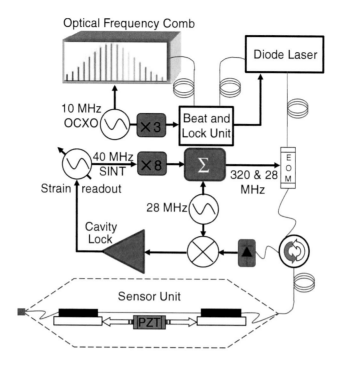

FIGURE 8.38

Experimental setup for ultrasensitive strain measurements. The system is composed of a sensor unit, a diode laser, an OFC with a reference oscillator (OCXO), and a laser-comb phase-lock unit. All optical units are connected by optical fibers. The sensing element is a Fabry-Perot fiber resonator formed by two identical single-mode 99% FBG reflectors, placed at a relative distance $L = 130$ mm along an acrylic-coated silica fiber. A PZT is used to modulate the cavity length. The resonator is thermally and acoustically shielded from the environment. A seismic insulation in the horizontal and vertical planes, at frequencies above 0.7 Hz, is provided with a latex-cord pendulum suspension. SINT, synthesizer; EOM, electro-optic modulator.

to obtain high sensitivity over a wide dynamic range while preserving a large frequency bandwidth. With stainless steel cantilevers, a bandwidth exceeding 1 kHz and a dynamic range of approximately 50 g can be achieved. Such system reaches a sensitivity noise floor ranging between 10 and 900 g/Hz, in the 10-1000 Hz interval, with similar performance along different orthogonal directions. The detection limit can be further improved either employing a different laser or decreasing its free-running frequency noise by pre-stabilization onto an external cavity (e.g., a fiber ring resonator). It is remarkable that such fiber-based sensors find application in many very different fields. For example, microphones for acoustic-waves detection can now be replaced by fiber sensors [1015].

8.10.4.3 Chemical sensing by optical-fiber ring resonators

Chemical sensors using fiber-optic methodology are the focus of extensive research and development activity with potential applications in industrial, environmental, and biomedical monitoring [1016, 1017, 1018]. In this context, miniaturized chemical sensors combining laser spectroscopy and state-of-the-art optical fiber technology may be suitable for in situ, non-invasive gas and liquid analysis with high selectivity and sensitivity. This can

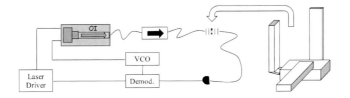

FIGURE 8.39
Schematic diagram of the accelerometer's head. Stainless steel cantilevers are clamped together using aluminum plates. All cantilevers have the same dimensions and nominal resonant frequencies of about 1.5 kHz.

be based on either direct or indirect (indicator-based) detection techniques [1019, 1020]. In recent years, interrogation techniques have further advanced with the use of spatially resolved spectroscopy [1021], evanescent-wave spectroscopy [1022, 1023, 1024] as well as surface-plasmon resonance [1025, 1026, 1027, 1028]. Sensors have also been incorporated into passive optical cavities consisting of fiber loops or linear fiber cavities defined, e.g., by two identical FBGs [1029, 1030, 1031]. Optical microresonators, of different geometries, have been also used as label-free and ultrasensitive chemical sensors over the past several years [1032, 1033, 1034, 1035]. In all cases above, a change in ambient refractive index may lead to a wavelength shift of the cavity modes, if part of the evanescent wave of the mode is exposed to the environment. On the other hand, if the molecules exhibit absorption lines or bands in the vicinity of the resonance wavelength, the cavity lifetime, namely the ring-down time (RDT), will be reduced, leading also to a reduction in power transmitted through the resonator and in the quality (Q-) factor.

An interesting example is a passive optical-fiber ring (OFR) resonator. As is well known, a light leakage from the fiber changes the resonator finesse. If the fiber core is exposed along a short region of the fiber within the loop, the presence of a liquid analyte can be measured, for example, by monitoring the light loss due to optical absorption or the refractive index changes through evanescent-wave interaction. A possible experimental setup to interrogate the fiber resonator and extract the absorption information was described in [132] and a

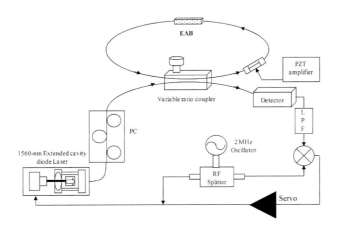

FIGURE 8.40
Schematic of the fiber-ring resonator sensor. PC: polarization controller; EAB: evanescent-access block; LPF: low-pass filter.

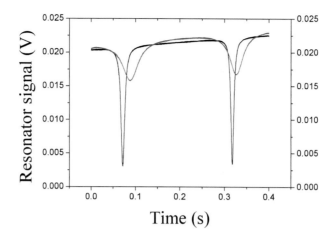

FIGURE 8.41
Cavity transmission signal for a free-running laser with no sample on the EAB (black line)
and with a sample (glycerol diluted at 99.5% with D_2O) causing a small index overlay (gray
line).

sketch is shown in Figure 8.40. The OFR is made of SMF-28 fiber and a variable-ratio fiber
coupler (1-99 percent) to inject near-infrared radiation into the cavity. A fiber evanescent-
field access block (EAB) allows the (evanescent) cavity field to interact with the external
environment. The transmission peaks observed on cavity resonances over a laser-frequency
sweep are shown in Figure 8.41. A PDH signal is generated to frequency lock the laser to the
cavity resonance. In this way, the laser could be frequency-stabilized to a cavity mode for
several hours, without suffering from thermal fluctuations of the environment. A first test
of the sensors response consists of covering the EAB with a solution with high refractive
index. When the sample was applied to the EAB, the cavity resonances were observed
to remain symmetric while the width of the signal increased, as expected due to finesse
degradation. Hence a fraction of the guided core mode leaked out from the cavity due to
the high-index glycerol cladding. Other experimental schemes are possible if the absorption
process is detected through the measurement of the so-called cavity ring-down time, which is
directly related to the internal optical loss (and thus to the absorption coefficient of external
analytes). Furthermore, broadband radiation sources, such as supercontinuum lasers and
optical synthesizers, can be employed to reconstruct multiple spectra of gas samples or to
investigate wide absorption bands by liquid species.

Bibliography

[1] J. Jespersen and J. Fitz-Randolph *From sundials to atomic clocks. Understanding time and frequency (Second Revised Edition)* Dover, Washington D.C. (1999)

[2] http://www.nasa.gov/topics/earth/features/japanquake/earth20110314.html

[3] D.D. McCarthy and K.P. Seidelmann *Time. From Earth rotation to atomic physics* Wiley-VCH, Darmstadt (2009)

[4] J.C. Bergquist, S.R. Jefferts and D. J. Wineland *Time measurement at the millennium* Physics Today **March 2001**, 37–42 (2001)

[5] S.A. Diddams, J.C. Bergquist, S.R. Jefferts and C.W. Oates *Standards of time and frequency at the outset of the 21st century* Science **306**, 1318–1324 (2004)

[6] M.A. Lombardi, T.P. Heavner and S.R. Jefferts *NIST primary frequency standards and the realization of the SI second* NCSL International Measure **2**, 74–89 (2007)

[7] C. Audoin and B. Guinot *The measurement of time. Time, Frequency and the Atomic Clock* Cambridge University Press, Cambridge (2001)

[8] W.M. Itano, L.L. Lewis and D.J. Wineland *Shift of $^2S_{1/2}$ hyperfine splittings due to blackbody radiation* Phys. Rev. A **25**, 1233-1235 (1982)

[9] R.J. MacKay and R.W. Oldford *Scientific method, statistical method and the speed of light* Statistical Science **15**, 254–278 (2000)

[10] I.B. Cohen *Roemer and the first determination of the velocity of light (1676)* Isis **31**, 327–379 (1940)

[11] A. Wroblewski *de Mora Luminis: A spectacle in two acts with a prologue and an epilogue* Am. J. Phys. **53**, 620–630 (1985)

[12] O. Morizot, A. Sellé, S. Ferri, D. Guyomarc'h, J.M. Laugier and M Knoop *A modern Fizeau experiment for education and outreach purposes* Eur. J. Phys. **32**, 161–168 (2011)

[13] M.E.J.G. de Bray *The velocity of light* Isis **25**, 437–448 (1936)

[14] K.S. Mendelson *The story of c* Am. J. Phys. **74**, 995-997 (2006)

[15] J.C. Maxwell *On the electromagnetic field* Philos. Trans. R. Soc. London **155**, 499 (1865)

[16] J.G. Ferguson and B.W. Bartlett *The measurement of capacitance in terms of resistance and frequency* Bell System Technical Journal **7**, 420-437 (1928)

[17] E.B. Rosa and N.E. Dorsey *A comparison of the various methods of determining the ratio of the electromagnetic to the electrostatic unit of electricity* Bulletin of the Bureau of Standards **3**, 605-622 (1907)

[18] L. Essen and A.C. Gordon-Smith *The velocity of propagation of electromagnetic waves derived from the resonant frequencies of a cylindrical cavity resonator* Proceedings of the Royal Society of London. Series A, Mathematical and Physical Sciences **194**, 348-361 (1948)

[19] A.A. Michelson *Measurement of the velocity of light between mount Wilson and mount San Antonio* The Astrophysical Journal **65**, 1-22 (1927)

[20] L. Essen *Velocity of electromagnetic waves* Nature **159**, 611–612 (1947)

[21] K. Bol *A determination of the speed of light by the resonant cavity method* Phys. Rev. **80**, 298 (1950)

[22] C.I. Aslakson *Velocity of electromagnetic waves* Nature **164**, 711-712 (1949)

[23] E. Bergstrand *Velocity of light and measurement of distances by high-frequency light signalling* Nature **163**, 338 (1949)

[24] J. Rogers, R. McMillan, R. Pickett, and R. Anderson *A determination of the speed of light by the phase-shift method* Am. J. Phys. **37**, 816-822 (1969)

[25] E. Bergstrand *Distance measuring by means of modulated light* Bulletin Geodesique **24**, 243-249 (1952)

[26] K.M. Evenson and F.R. Petersen *Laser frequency measurements, the speed of light, and the meter* **p.352** in "Laser Spectroscopy of Atoms and Molecules. Topics in Applied Physics (Volume 2)." Edited by H. Walther. Published by Springer-Verlag, Berlin (1976)

[27] K.D. Froome *A new determination of the free-space velocity of electromagnetic waves* Proceedings of the Royal Society of London. Series A, Mathematical and Physical Sciences **247**, 109-122 (1958)

[28] D.H. Rank *The band spectrum method of measuring the velocity of light* J. Mol. Spectrosc. **17**, 50-57 (1965)

[29] A. Javan, W.R. Bennett, and D.R. Herriott *Population inversion and continuous optical maser oscillation in a gas discharge containing a He-Ne mixture* Phys. Rev. Lett. **6**, 106-110 (1961)

[30] D.J.E. Knight and P.T. Woods *Application of nonlinear devices to optical frequency measurement* Journal of Physics E: Scientific Instruments **9**, 898-916 (1976)

[31] D.A. Jennings, K.M. Evenson and D.J.E. Knight *Optical frequency measurements* Proceedings of the IEEE **74**, 168-179 (1986)

[32] K.M. Evenson, G.W. Day, J.S. Wells and L.O. Mullen *Extension of absolute frequency measurements to the cw He-Ne laser at 88 THz (3.39 μm)* Appl. Phys. Lett. **20**, 133-134 (1972)

[33] R.L. Barger and J.L. Hall *Wavelength of the 3.39-μm laser-saturated absorption line of methane* Appl. Phys. Lett. **22**, 196-199 (1973)

[34] K.M. Evenson, J.S. Wells, F.R. Petersen, B.L. Danielson, G.W. Day, R.L. Barger and J.L. Hall *Speed of light from direct frequency and wavelength measurements of the methane-stabilized laser* Phys. Rev. Lett. **29**, 1346-1349 (1972)

[35] T.G. Blaney, C.C. Bradley, G.J. Edwards, B.W. Jolliffe, D.J.E. Knight, W.R.C. Rowley, K.C. Shotton and P.T. Woods *Measurement of the speed of light* Nature **251**, 46 (1974)

[36] B.W. Jolliffe, W.R.C. Rowley, K.C. Shotton, A.J. Wallard and P.T. Woods *Accurate wavelength measurement on up-converted CO_2 laser radiation* Nature **251**, 46-47 (1974)

[37] J.-P. Monchalin, M.J. Kelly, J.E. Thomas, N.A. Kurnit, A. Szöke, A. Javan, F. Zernike and P.H. Lee *Determination of the speed of light by absolute wavelength measurement of the R(14) line of the CO_2 9.4-μm band and the known frequency of this line* Opt. Lett. **1**, 5-7 (1977)

[38] D.A. Jennings, C.R. Pollock, F.R. Peterson, R.E. Drullinger, K.M. Evenson, J.S. Wells, J.L Hall and H.P. Layer *Direct frequency measurement of the I_2-stabilized He-Ne 473-THz (633-nm) laser* Opt. Lett. **8**, 136-138 (1983)

[39] C.R. Pollock, D.A. Jennings, F.R. Peterson, J.S. Wells, R.E. Drullinger, E.C. Beaty, and K.M. Evenson *Direct frequency measurements of transitions at 520 THz (576 nm) in iodine and 260 THz (1.15 μm) in neon* Opt. Lett. **8**, 133-135 (1983)

[40] J. Flowers *The route to atomic and quantum standards* Science **306**, 1324-1330 (2004)

[41] R.J.A. Lambourne *Relativity, gravitation and cosmology* Cambridge University Press, New York (2010)

[42] G.F.R. Ellis and J-P. Uzan *c is the speed of light, isn't it?* Am. J. Phys. **73**, 240-247 (2005)

[43] L.-C. Tu, J. Luo and G.T. Gillies *The mass of the photon* Rep. Prog. Phys. **68**, 77 (2005)

[44] G. Amelino-Camelia *Burst of support for relativity* Nature **462**, 291-292 (2009)

[45] The OPERA collaboration *Measurement of the neutrino velocity with the OPERA detector in the CNGS beam* arXiv:1109.4897v2 (2011)

[46] J.-L. Basdevant and J. Dalibard *Quantum mechanics* Springer-Verlag, Berlin (2002)

[47] H.G. Winful *Tunneling time, the Hartman effect, and superluminality: a proposed resolution of an old paradox* Phys. Rep. **436**, 1-69 (2006)

[48] R. Onofrio *Casimir forces and non-Newtonian gravitation* New Journal of Physics **8**, 237 (2006)

[49] K. Wynne *Causality and the nature of information* Opt. Commun. **209**, 85-100 (2002)

[50] J.D. Barrow *Cosmologies with varying light speed* Phys. Rev. D **59**, 043515-8 (1999)

[51] P. Gill *Optical frequency standards* Metrologia **42**, S125-S137 (2005)

[52] J.L. Hall *Nobel lecture: Defining and measuring optical frequencies* Rev. Mod. Phys. **78**, 1279-1295 (2006)

[53] T.W. Hänsch *Nobel lecture: Passion for precision* Rev. Mod. Phys. **78**, 1297-1309 (2006)

[54] R.J. Glauber *Nobel lecture: One hundred years of light quanta* Rev. Mod. Phys. **78**, 1267-1278 (2006)

[55] D. D'Humieres, M.R. Beasley, B.A. Huberman and A. Libchaber *Chaotic states and routes to chaos in the forced pendulum* Phys. Rev. A **26**, 3483-3496 (1982)

[56] M. Bennett, M.F. Schatz, H. Rockwood and K. Wiesenfeld *Huygens's clocks* Proc. R. Soc. Lond. A **458**, 563-579 (2002)

[57] E. Rubiola *The Leeson effect. Phase noise in quasilinear oscillators* arXiv:physics/0502143v1 (2005)

[58] J. Millman and A. Grabel *Microelectronics (Second Edition)* McGraw-Hill, New York (1987)

[59] J.R. Vig *Quartz Crystal Resonators and Oscillators for Frequency Control and Timing Applications - A Tutorial* IEEE International Frequency Control Symposium Tutorials (2004)

[60] D.A. Howe *Frequency domain stability measurements: a tutorial introduction* Nat. Bur. Stand. (U.S.) Tech Note **679**, 27 pages (1976)

[61] S.R. Stein *Frequency and time. Their measurement and characterization* from **Precision Frequency Control Vol. 2**, edited by E.A. Gerber and A. Ballato (Academic Press, New York), pp. 191-416 (1985)

[62] R.F.C. Vessot *Frequency and time standards* Methods in Experimental Physics **12** Part C, pp. 198-227 (1976)

[63] L.S. Cutler and C.L. Searle *Some aspects of the theory and measurements of frequency fluctuations in frequency standards* Proc. IEEE **54**, 136-154 (1966)

[64] *Characterization of clocks and oscillators* D.B. Sullivan, D.W. Allan, D.A. Howe and F.L. Walls, Eds., **NIST technical note 1337** (1990)

[65] F. Riehle *Frequency standards* Wiley-VCH, Weinheim (2004)

[66] B.P. Lathi and Z. Ding *Modern digital and analog communication systems* Oxford University Press (2009)

[67] W. Schottky *Über spontane Stromschwankungen in verschiedenen Elektrizitätsleitern* Ann. der Phys. **362**, 541-567 (1918)

[68] A. Yariv and P. Yeh *Photonics. Optical electronics in modern communications (Sixth Edition)* Oxford University Press, New York (2007)

[69] J.L. Staff *Johnson noise and shot noise: The determination of the Boltzmann constant, absolute zero temperature and the charge of the electron* jLab E-Library, URL http://web.mit.edu/8.13/www/JLExperiments/JLExp43.pdf (2011)

[70] J.B. Johnson *Thermal agitation of electricity in conductors* Phys. Rev. **32**, 97 (1928)

[71] G. Clark and J. Kirsch *Junior physics laboratory: Johnson noise*, MIT, Cambridge, MA (1997)

[72] E. Milotti $1/f$ *noise: a pedagogical review* arXiv:physics/0204033, 26 pages (2002)

[73] A. Van der Ziel *Noise in measurements* Wiley, New York (1976)

[74] Yu.B. Rumer, M.S. Ryvkin and S. Semyonov *Thermodynamics, Statistical Physics and Kinetics* Central Books Ltd (1981)

[75] M. Hansen *Achieving accurate on-wafer flicker noise measurements through 30 MHz* Cascade Microtech Inc., White paper (2009)

[76] A. Crisanti and F. Ritort *Violation of the fluctuation-dissipation theorem in glassy systems: basic notions and the numerical evidence* J. Phys. A: Math. Gen. **36**, R181-R290 (2003)

[77] A. Hajimiri and T.H. Lee *A general theory of phase noise in electrical oscillators* IEEE Journal of Solid-state circuit **33**, 179–194 (1998)

[78] D.W. Allan *Statistics of atomic frequency standards* Proc. IEEE **54**, 221-230 (1966)

[79] D.W. Allan and J.A. Barnes *A modified Allan variance with increased oscillator characterization ability* Proc. 35th Freq. Cont. Symp. pp. 470-474 (1981)

[80] P. Lesage and T. Ayi *Characterization of frequency stability: analysis of the modified Allan variance and properties of its estimate* IEEE Trans. Instrum. Meas. **33**, 332-336 (1984)

[81] D.S. Elliott, R. Roy and S.J. Smith *Extra-cavity laser band-shape and bandwidth modification* Phys. Rev. A **26**, 12-18 (1982)

[82] A. Godone and F. Levi *About the radio-frequency spectrum of a phase noise modulated carrier* in Proceedings of the 12th European Frequency and Time Forum, pp. 392-396 (1998)

[83] G.D. Rovera and O. Acef *Optical frequency measurements relying on a mid-infrared frequency standard* in Frequency Measurement and Control. Advanced Techniques and Future Trends, A.N. Luiten, Ed., pp. 249-272 Springer (2001)

[84] F.L. Walls and A. De Marchi *RF spectrum of a signal after frequency multiplication; measurement and comparison with a simple calculation* IEEE Trans. Instrum. Meas. **IM-24**, 210-217 (1975)

[85] F.L. Walls *Phase noise issues in femtosecond lasers* Proc. SPIE **4269**, 170-177 (2001)

[86] J. Ye and S.T. Cundiff *Femtosecond optical frequency comb technology. Principle, operation and application*, Springer (2005)

[87] W.F. Egan *Phase-lock basics (Second Edition)* Wiley, Hoboken (2008)

[88] A.N. Luiten *Accurate optical frequency synthesis* in Frequency Measurement and Control. Advanced Techniques and Future Trends, A.N. Luiten Ed., pp. 249-272 Springer (2001)

[89] M. Kourogi, C.H. Shin and M. Ohtsu *A 134 MHz bandwidth homodyne optical phase locked loop of semiconductor lasers* IEEE Phot. Tech. Lett. **3**, 270-207 (1991)

[90] M. Prevedelli, T. Freegarde and T.W. Hänsch *Phase locking of grating-tuned diode lasers* Appl. Phys. B **60**, S241-S248 (1995)

[91] L. Cacciapuoti, M. de Angelis, M. Fattori, G. Lamporesi, T. Petelski, M. Prevedelli, J. Sthuler and G.M. Tino *Analog+digital phase and frequency detector for phase locking of diode lasers* Rev. Sci. Instrum. **76**, 053111 (2005)

[92] D. Höckel, M. Scholz and O. Benson *A robust phase-locked diode laser system for EIT experiments in cesium* Appl. Phys. B **94**, 429-435 (2009)

[93] B. Razavi *A study of injection locking and pulling in oscillators* IEEE Journal of Solid-State Circuits **39**, 1415-1424 (2004)

[94] B. Razavi *A study of locking phenomena in oscillators* Proc. IEEE **61**, 1380-1385 (1973)

[95] *Fundamentals of the electronic counters* Agilent Technologies, Application note 200 Electronic Counter Series (1997)

[96] E. Rubiola *On the measurement of frequency and of its sample variance with high-resolution counters* Rev. Sci. Instrum. **76**, 054703 (2005)

[97] S.T. Dawkins, J.J. McFerran and A.N. Luiten *Considerations on the measurement of the stability of oscillators with frequency counters* IEEE Trans. Ultrason. Ferroelectr. Freq. Control. **54**, 918-925 (2007)

[98] E. Rubiola, E. Salik, S. Huang, N. Yu and L. Maleki *Photonic-delay technique for phase-noise measurement of microwave oscillators* J. Opt. Soc. Am. B **22**, 987-997 (2005)

[99] E. Rubiola *The measurement of AM noise in oscillators* arXiv:physics/0512082v1 (2005)

[100] J. Ye and T.W. Lynn *Applications of optical cavities in modern atomic, molecular, and optical physics* in Advances in Atomic, Molecular and Optical Physics, Vol. 49, B. Bederson and H. Walther, Eds., Academic Press, pp. 1-83 (2003)

[101] J.D. Jackson *Classical Electrodynamics (Third Edition)* Wiley (1999)

[102] E. Hecht *Optics (Fourth Edition)* Addison Wesley (2002)

[103] H. Kogelnik and T. Li *Laser beams and resonators* Appl. Opt. **5**, 1550-1567 (1966)

[104] B.E.A. Saleh and M.C. Teich *Fundamentals of photonics* Wiley, New York (1991)

[105] G.R. Fowles *Introduction to modern optics (Second Edition)* Dover (1975)

[106] M.J. Thorpe *Cavity-enhanced direct frequency comb spectroscopy* PhD thesis (2009)

[107] K.K. Lehmann, P.S. Johnston and P. Rabinowitz *Brewster angle prism retro-reflectors for cavity enhanced spectroscopy* Appl. Opt. **48**, 2966-2978 (2009)

[108] P.S. Johnston and K.K. Lehmann *Cavity enhanced absorption spectroscopy using a broadband prism cavity and a super-continuum source* Optics Express **16**, 15013-15023 (2008)

[109] J.-P Taché *Ray matrices for tilted interfaces in laser resonators* Appl. Opt. **26**, 427-429 (1987)

[110] K. Numata, A. Kemery and J. Camp *Thermal-noise limit in the frequency stabilization of lasers with rigid cavities* Phys. Rev. Lett. **93**, 250602 (2004)

[111] M. Notcutt, L.S. Ma, A.D. Ludlow, S.M. Foreman, J. Ye and J.L. Hall *Contribution of thermal noise to frequency stability of rigid optical cavity via Hertz-linewidth lasers* Phys. Rev. A **73**, 031804(R) (2006)

[112] K. Numata *Direct measurement of mirror thermal noise* Thesis (2002)

[113] K. Yamamoto *Study of the thermal noise caused by inhomogeneously distributed loss* PhD Thesis (2000)

[114] Y. Levin *Internal thermal noise in the LIGO test masses: a direct approach* Phys. Rev. D **57**, 659-663 (1998)

[115] L. D. Landau, L. P. Pitaevskii, E.M. Lifshitz and A. M. Kosevich *Theory of elasticity (Third Edition)* Butterworth-Heinemann (1986)

[116] Y.K. Liu and K.S. Thorne *Thermoelastic noise and homogeneous thermal noise in finite sized gravitational-wave test masses* Phys. Rev. D **62**, 122002 (2000)

[117] N. Nakagawa, A.M. Gretarsson, E.K. Gustafson and M.M. Fejer *Thermal noise in half-finite mirrors with non-uniform loss: a slab of excess loss in a half-finite mirror* Phys. Rev. D **65**, 102001 (2002)

[118] S.A. Webster, M. Oxborrow and P. Gill *Vibration insensitive optical cavity* Phys. Rev. A **75**, 011801(R) (2007)

[119] T. Nazarova, F. Riehle and U. Sterr *Vibration-insensitive reference cavity for an ultra-narrow-linewidth laser* Appl. Phys. B **83**, 531-536 (2006)

[120] A.D. Ludlow, X. Huang, M. Notcutt, T. Zanon-Willette, S.M. Foreman, M.M. Boyd, S. Blatt, and J. Ye *Compact, thermal-noise-limited optical cavity for diode laser stabilization at* $1 \cdot 10^{-15}$ Opt. Lett. **32**, 641-643 (2007)

[121] J. Millo, D.V. Magalhaes, C. Mandache, Y. Le Coq, E.M.L. English, P.G. Westergaard, J. Lodewyck, S. Bize, P. Lemonde and G. Santarelli *Ultrastable lasers based on vibration insensitive cavities* Phys. Rev. A **79**, 053829 (2009)

[122] L. Chen, J.L. Hall, J. Ye, T. Yang, E. Zang, and T. Li *Vibration-induced elastic deformation of Fabry-Perot cavities* Phys. Rev. A **74**, 053801 (2006)

[123] D.R. Leibrandt, M.J. Thorpe, M. Notcutt, R.E. Drullinger, T. Rosenband and J.C. Bergquist *Spherical reference cavities for frequency stabilization of lasers in non-laboratory environments* Optics Express **19**, 3471-3482 (2011)

[124] L.F. Stokes, M. Chodorow and H.J. Shaw *All-single-mode fiber resonator* Opt. Lett. **7**, 288-290 (1982)

[125] H.-P. Loock, J.A. Barnes, G. Gagliardi, R. Li, R.D. Oleschuk, and H. Wächter *Absorption detection using optical waveguide cavities* Can. J. Chem. **88**, 401-410 (2010)

[126] D. Hunger, T. Steinmetz, Y. Colombe, C. Deutsch, T.W. Hänsch and J. Reichel *A fiber Fabry-Perot cavity with high finesse* New Journal of Physics **12**, 065038 (2010)

[127] C.J. Hood, H.J. Kimble and J. Ye *Characterization of high-finesse mirrors: Loss, phase shifts, and mode structure in an optical cavity* Phys. Rev. A **64**, 033804 (2001)

[128] T. Erdogan *Fiber grating spectra* Journal of Lightwave Technology **15**, 1277-1294 (1997)

[129] V. Finazzi and M.N. Zervas *Effect of periodic background loss on grating spectra* Appl. Opt. **41**, 2240-2250 (2002)

[130] Y.O. Barmenkov, D. Zalvidea, S. Torres-Peiro, J.L. Cruz and M.V. Andres *Effective length of short Fabry-Perot cavity formed by uniform fiber Bragg gratings* Optics Express **14**, 6394-6399 (2006)

[131] J. Canning *Fibre gratings and devices for sensors and lasers* Laser & Photon. Rev. **2**, 275-289 (2008)

[132] G. Gagliardi, M. Salza, P. Ferraro, E. Chehura, R.P. Tatam, T.K. Gangopadhyay, N. Ballard, D. Paz-Soldan, J.A. Barnes, H.-P. Loock, T.T.-Y. Lam, J.H. Chow and P. De Natale *Optical fiber sensing based on reflection laser spectroscopy* Sensors **10**, 1823-1845 (2010)

[133] A.B. Matsko and V.S. Ilchenko *Optical resonators with whispering-gallery modes-Part I: Basics* IEEE Journal of Selected Topics in Quantum Electronics **12**, 3-14 (2006)

[134] T.J.A. Kippenberg *Nonlinear optics in ultra-high-Q whispering-gallery optical microcavities* PhD thesis (2004)

[135] A.N. Oraevsky *Whispering-gallery waves* Quantum Electronics **32**, 377-400 (2002)

[136] J.A. Stratton *Electromagnetic theory* McGraw-Hill (1941)

[137] W.K.H. Panofsky and M. Phillips *Classical electricity and magnetism (Second Edition)* Addison-Wesley (1962).

[138] S. Schiller and R.L. Byer *High-resolution spectroscopy of whispering gallery modes in large dielectric spheres* Opt. Lett. **16**, 1138-1140 (1991)

[139] E. Feenberg and K.C. Hammack *A note on Rainwater's spheroidal nuclear model* Phys. Rev. **81**, 285 (1950)

[140] G. Lin *Fabrication and characterization of optical microcavities functionalized by rare-earth oxide nanocrystals: realization of a single-mode ultra-low threshold laser* PhD thesis (2010)

[141] A.A. Savchenkov, V.S. Ilchenko, A.B. Matsko and L. Maleki *Kilo-Hertz optical resonances in dielectric crystal cavities* Phys. Rev. A **70**, 051804 (2004)

[142] D.W. Vernooy, V.S. Ilchenko, H. Mabuchi, E.W. Streed and H.J. Kimble *High-Q measurements of fused-silica microspheres in the near infrared* Opt. Lett. **23**, 247-249 (1998)

[143] L. Collot, V. Lefevre-Seguin, M. Brune, J.-M. Raimond, and S. Haroshe *Very high-Q whispering-gallery mode resonances observed in fused silica microspheres* Europhys. Lett. **23**, 327-334 (1993)

[144] M.L. Gorodetsky, A.D. Pryamikov, and V.S. Ilchenko *Rayleigh scattering in high-Q microspheres* J. Opt. Soc. Am. B **17**, 1051-1057 (2000)

[145] S.C. Hill and R.E. Benner, *Morphology-dependent resonances associated with stimulated processes in microspheres* J. Opt. Soc. Am. B **3**, 1509-1514 (1986)

[146] S. Arnold and L.M. Folan *Energy transfer and the photon lifetime within an aerosol particle* Opt. Lett. **14**, 387-389 (1989)

[147] S. Uetake, R.S.D. Sihombing and K. Hakuta *Stimulated Raman scattering of a high-Q liquid-hydrogen droplet in the ultraviolet region* Opt. Lett. **27**, 421-423 (2002)

[148] D.K. Serkland, R.C. Eckardt and R.L. Byer *Continuous-wave total internal- reflection optical parametric oscillator pumped at 1064 nm* Opt. Lett. **19**, 1046-1048 (1994)

[149] S.J. Choi, K. Djordjev, S.J. Choi, P.D. Dapkus, W. Lin, G. Griffel, R. Menna and J. Connolly *Microring resonators vertically coupled to buried heterostructure bus waveguides* IEEE Photon. Technol. Lett. **16**, 828-830 (2004)

[150] B.E. Little, S.T. Chu, P.P. Absil, J.V. Hryniewicz, F.G. Johnson, F. Seiferth, D. Gill, V. Van, O. King and M. Trakalo *Very high-order microring resonator filters for WDM applications* IEEE Photon. Technol. Lett. **16**, 2263-2265 (2004)

[151] B. Gayral and J. M. Gerard *Strong Purcell effect for InAs quantum boxes in high-Q wet-etched microdisks* Physica E **7**, 641-645 (2000)

[152] M.L. Gorodetsky and V. S. Ilchenko *Optical microsphere resonators: Optimal coupling to high-Q whispering-gallery modes* J. Opt. Soc. Am. B **16**, 147-154 (1999)

[153] D.S. Weiss, V. Sandoghdar, J. Hare, V. Lefevre-Seguin, J.-M. Raimond, and S. Haroche *Splitting of high-Q Mie modes induced by light backscattering in silica microspheres* Opt. Lett. **20**, 1835-1837 (1995)

[154] T. Kippenberg, S. Spillane and K. Vahala *Modal coupling in traveling-wave resonators* Opt. Lett. **27**, 1669-1671 (2002).

[155] K.J. Vahala *Optical microcavities* Nature **424**, 839-846 (2003)

[156] G.C. Righini, Y. Dumeige, P. Feron, M. Ferrari, G. Nunzi Conti, D. Ristic and S. Soria *Whispering gallery mode microresonators: Fundamentals and applications* Rivista Del Nuovo Cimento **34**, 435-488 (2011)

[157] V.S. Ilchenko and A.B. Matsko *Optical Resonators with Whispering-Gallery Modes - Part II: Applications* IEEE Journal of Selected Topics in Quantum Electronics **12**, 15-32 (2006)

[158] A.L. Schawlow and C.H. Townes *Infrared and optical masers* Phys. Rev. **112**, 1940-1949 (1958)

[159] *The invention of the laser at Bell Laboratories: 1958–1998* http://www.belllabs.com/hitory/laser/

[160] J.P. Gordon, H.J. Zeiger and C.H. Townes *Molecular microwave oscillator and new hyperfine structure in the microwave spectrum of NH_3* Phys. Rev. **95**, 282-285 (1954)

[161] T.H. Maiman *Stimulated Optical Radiation in Ruby* Nature **187**, 493 (1960)

[162] J.P. Gordon, H.J. Zeiger and C.H. Townes *The maser-new type of microwave amplifier, frequency standard, and spectrometer* Phys. Rev. **99**, 1264-1274 (1955)

[163] F.G. Major *The quantum beat. Principles and applications of atomic clocks (Second Edition)* Springer (2007)

[164] H.M. Goldenberg, D. Kleppner and N.F. Ramsey *Atomic hydrogen maser* Phys. Rev. Lett. **5**, 361-362 (1960)

[165] D. Kleppner, H.C. Berg, S.B. Crampton, N.F. Ramsey, R.F.C. Vessot, H.E. Peters and J. Vanier *Hydrogen-maser principles and techniques* Phys. Rev. **138**, A972-A983 (1965)

[166] G. Grynberg, A. Aspect and C. Fabre *Introduction to quantum optics. From the semi-classical approach to quantized light* Cambridge University Press, New York (2010)

[167] H.C. Berg *Spin exchange and surface relaxation in the atomic hydrogen maser* Phys. Rev. **137**, A1621-A1635 (1965)

[168] O. Svelto *Principles of lasers (Fourth Edition)* Springer, New York (1998)

[169] P.W. Milonni and J.H. Eberly *Laser physics* Wiley, Hoboken (2010)

[170] W. Koechner and M. Bass *Solid-state lasers: A graduate text* Springer (2003)

[171] A.E. Siegman *Lasers* University Science Books, Mill Valley (1986)

[172] P.K. Das *Lasers and optical engineering* Springer (1990)

[173] http://www.rp-photonics.com/nonplanarringoscillators.html

[174] R. Loudon *The Quantum Theory of Light (Second Edition)* Oxford (1995)

[175] G. Di Domenico, S. Schilt, and P. Thomann *Simple approach to the relation between laser frequency noise and laser line shape* Appl. Opt. **49**, 4801-4807 (2010)

[176] T. Day, E.K. Gustafson, and R.L. Byer *Sub-Hertz relative frequency stabilization of two-diode laser-pumped Nd:YAG lasers locked to a Fabry-Perot interferometer* IEEE J. Quantum Electronics **28**, 1106-1117 (1992)

[177] G. Kirchmair *Frequency stabilization of a Titanium-Sapphire laser for precision spectroscopy on calcium ions* Diploma Thesis (2006)

[178] J.L. Hall, M.S. Taubman and J. Ye *Laser stabilization Handbook of Optics IV*, (Second Edition), McGraw Hill (2001)

[179] K.J. Astrom and T. Hagglund *Revisiting the Ziegler-Nichols step response method for PID control* Journal of Process Control **14**, 635-650 (2004)

[180] M. Maric and A. Luiten *Power-insensitive side locking for laser frequency stabilization* Opt. Lett. **30**, 1153-1155 (2005)

[181] E.D. Black *An introduction to Pound-Drever-Hall laser frequency stabilization* Am. J. Phys. **69**, 79-87 (2000)

[182] R.W.P. Drever, J.L. Hall, F.V. Kowalski, J. Hough, G.M. Ford, A.J. Munley, and H. Ward *Laser phase and frequency stabilization using an optical resonator* Appl. Phys. B **31**, 97-105 (1982)

[183] B. Willke, N. Uehara, E.K. Gustafson, and R.L. Byer, P.J. King, S.U. Seel, and R. L. Savage Jr. *Spatial and temporal filtering of a 10-W Nd:YAG laser with a Fabry-Perot ring-cavity premode cleaner* Opt. Lett. **23**, 1704-1706 (1998)

[184] U. Sterr, T. Legero, T. Kessler, H. Schnatz, G. Grosche, O. Terra and F. Riehle *Ultrastable lasers - new developments and applications* Proc. of SPIE **7431**, 74310A (2009)

[185] Y.Y. Jiang, A.D. Ludlow, N.D. Lemke, R.W. Fox, J.A. Sherman, L.-S. Ma, and C.W. Oates *Making optical atomic clocks more stable with $10^{-16}-$level laser stabilization* Nature Photonics **5**, 158-161 (2011)

[186] H.J. Kimble, B.L. Lev and J. Ye *Optical interferometers with reduced sensitivity to thermal noise* Phys. Rev. Lett. **101**, 260602 (2008)

[187] M.L. Gorodetsky *Thermal noises and noise compensation in high-reflection multilayer coating* Phys. Lett. A **372**, 6813 (2008)

[188] D. Meiser, J. Ye, D.R. Carlson and M.J. Holland *Prospects for a Millihertz-linewidth laser* Phys. Rev. Lett. **102**, 163601 (2009)

[189] S. Seel, R. Storz, G. Ruoso, J. Mlynek and S. Schiller *Cryogenic optical resonators: a new tool for laser frequency stabilization at the 1 Hz level* Phys. Rev. Lett. **78**, 4741 (1997)

[190] J. Alnis, A. Schliesser, C.Y. Wang, J. Hofer, T.J. Kippenberg and T.W. Hänsch *Thermal-noise-limited crystalline whispering-gallery-mode resonator for laser stabilization* Phys. Rev. A **84**, 011804(R) (2011)

[191] I. Fescenko, J. Alnis, A. Schliesser, C.Y. Wang, T.J. Kippenberg and T.W. Hänsch *Dual-mode temperature compensation technique for laser stabilization to a crystalline whispering gallery mode resonator* Optics Express **20**, 19185-19193 (2012)

[192] T.W. Hänsch and B. Couillaud *Laser frequency stabilization by polarization spectroscopy of a reflecting reference cavity* optics communications **35**, 441 (1980)

[193] Y.T. Chen *Use of single-mode optical fiber in the stabilization of laser frequency* Appl. Opt. **28**, 2017-2021 (1989)

[194] G.A. Cranch *Frequency-noise reduction in erbium-doped fiber distributed-feedback lasers by electronic feedback* Opt. Lett. **27**, 1114-1116 (2002)

[195] F. Kefelian, H. Jiang, P. Lemonde and G. Santarelli *Ultralow-frequency-noise stabilization of a laser by locking to an optical fiber-delay line* Opt. Lett. **34**, 914-916 (2009)

[196] C.J. Buczek, R.J. Freiberg and M.L. Skolnick *Laser injection locking* Proceedings of the IEEE **61**, 1411-1431 (1973)

[197] H.L. Stover *Locking of laser oscillators by light injection* Appl. Phys. Lett. **8**, 91-93 (1966)

[198] G. Hadley *Injection locking of diode lasers* IEEE Journal of Quantum Electronics **22**, 419-426 (1986)

[199] S. Borri, I. Galli, F. Cappelli, A. Bismuto, S. Bartalini, P. Cancio, G. Giusfredi, D. Mazzotti, J. Faist and P. De Natale *Direct link of a mid-infrared QCL to a frequency comb by optical injection* Optics Letters **37**, 1011-1013 (2012)

[200] P. Kwee, B. Willke and K. Danzmann *New concepts and results in laser power stabilization* Appl. Phys. B **102**, 515-522 (2011)

[201] P. Kwee, B. Willke and K. Danzmann *Optical ac coupling to overcome limitations in the detection of optical power fluctuations* Opt. Lett. **33**, 1509-1511 (2008)

[202] H. Tsuchida *Simple technique for improving the resolution of the delayed self-heterodyne method* Opt. Lett. **15**, 640-642 (1990)

[203] P. Horak and W.L. Loh *On the delayed self-heterodyne interferometric technique for determining the linewidth of fiber lasers* Optics Express **14**, 3923-3928 (2006)

[204] X. Chen *Ultra-narrow laser linewidth measurement* PhD thesis (2006)

[205] A. Yariv *Quantum electronics (Third Edition)* Wiley (1989)

[206] *Encyclopedia of Modern Optics* B.D. Guenther, Ed. Elsevier, (2005)

[207] M. Zhu and J.L Hall *Stabilization of optical/frequency of a laser system: application to a commercial dye laser with an external stabilizer* J. Opt. Soc. Am. B **10**, 802-816 (1993)

[208] Ch. Salomon, D. Hills, and J.L Hall *Laser stabilization at the millihertz level* J. Opt. Soc. Am. B **5**, 1576-1587 (1988)

[209] B.C. Young, F.C. Cruz, W.M. Itano, and J.C. Bergquist *Visible lasers with sub-hertz linewidths* Phys. Rev. Lett. **82**, 3799-3802 (1999)

[210] G. Huber, C. Kränkel and K. Petermann *Solid-state lasers: status and future* J. Opt. Soc. Am. B **27**, B93-B105 (2010)

[211] S.A. Webster, M. Oxborrow and P.Gill *Subhertz-linewidth Nd:YAG laser* Opt. Lett. **29**, 1497 (2004)

[212] R.G. Hunsperger *Integrated Optics. Theory and Technology (Fifth Edition)* Springer (2002)

[213] E. Desurvire *Erbium-doped fiber amplifiers: basic physics and characteristics* in *Rare-earth-doped fiber lasers and amplifiers (Second Edition)* M.J.F. Digonnet, Ed., Marcel Dekker (2001)

[214] P.F. Wysocki *Erbium-doped fiber amplifiers: advanced topics* in *Rare-earth-doped fiber lasers and amplifiers (Second Edition)* M.J.F. Digonnet, Ed., Marcel Dekker (2001)

[215] J. Ota, A. Shirakawa and K. Ueda *High-power Yb-doped double-clad fiber laser directly operating at 1178 nm* Jpn. J. Appl. Phys. **45**, L117-L119 (2006)

[216] A.S. Kurkov, V.M. Paramonov, and O.I. Medvedkov *Ytterbium fiber emitting at 1160 nm* Laser Phys. Lett. **3**, 503-506 (2006)

[217] A. Shirakawa, H. Maruyama, K. Ueda, C.B. Olausson, J.K. Lyngsøand, and J. Broeng *High-power Yb doped photonic bandgap fiber amplifier at 1150-1200 nm* Opt. Express **17**, 447-454 (2009)

[218] R. Goto, E.C. Magi, and S.D. Jackson *Narrow-linewidth, Yb^{3+}-doped, hybrid microstructured fibre laser operating at 1178 nm* Electron. Lett. **45**, 877-878 (2009)

[219] M.P. Kalita, S. Alam, C. Codemard, S. Yoo, A.J. Boyland, M. Ibsen, and J.K. Sahu *Multi-watts narrow-linewidth all fiber Yb-doped laser operating at 1179 nm* Optics Express **18**, 5920-5925 (2010)

[220] E. Snitzer, H. Po, F. Hakimi, R. Tumminelli and B. C. McCollum *Double-clad offset core Nd fiber laser* in Optical Fiber Sensors, 1988 OSA Tech. Dig. Ser. Washington, DC: Opt. Soc. Amer., 1988, vol. 2, paper PD5.

[221] H.M. Pask, J.L. Archambault, D.C. Hanna, L. Reekie, P. St.J. Russell, J.E. Townsend, and A.C. Tropper *Operation of cladding-pumped Yb^{3+}-doped silica fiber lasers in 1 micron region* Electron. Lett. **30**, 863-865 (1994)

[222] R. Paschotta, D.C. Hanna, P. De Natale, G. Modugno, M. Inguscio, and P. Laporta *Power amplifier for 1083 nm using ytterbium doped fibre* Optics Commun. **136**, 243-246 (1997)

[223] S.V. Chernikov, J.R. Taylor, N.S. Platonov, V.P. Gapontsev, P.J. Nacher, G. Tastevin, M. Leduc, and M.J. Barlow *1083 nm ytterbium doped fibre amplifier for optical pumping of helium* Electron. Lett. **33**, 787-789 (1997)

[224] D.J.E. Knight, F. Minardi, P. De Natale, and P. Laporta *Frequency doubling of a fibre-amplified 1083 nm DBR laser* European Physical Journal D **3**, 211-216 (1998)

[225] P. Cancio, P. Zeppini, A. Arie, P. De Natale, G. Giusfredi, G. Rosenman, and M. Inguscio *Sub-Doppler spectroscopy of molecular iodine around 541 nm with a novel solid state laser source* Opt. Comm. **176**, 453-458 (2000)

[226] N. Picqué, P. Cancio, G. Giusfredi, and P. De Natale *High-stability diode-laser-based frequency reference at 1083 nm with iodine lines at 541.5 nm* Journal of the Optical Society of America B **18**, 692-697 (2001)

[227] P. Cancio, P. Zeppini, P. De Natale, S. Taccheo, and P. Laporta *Noise characteristics of a high power Ytterbium-doped fibre amplifier at 1083 nm* Appl. Phys. B **70**, 763-768 (2000)

[228] R. Paschotta, J. Nilsson, A.C. Tropper, and D.C. Hanna *Ytterbium-doped fiber amplifiers* IEEE J. Quantum Electron. **33**, 1049 (1997)

[229] N. Langford *Narrow-linewidth fiber lasers* in *Rare-Earth-Doped Fiber Lasers and Amplifiers (Second Edition)* M.J.F. Digonnet, Ed., Marcel Dekker (2001)

[230] N. Park, J.W. Dawson, K.J. Vahala, and C. Miller *All-fiber, low-threshold, widely tunable single-frequency, erbium-doped fiber ring laser with a tandem fiber Fabry-Perot filter* Appl. Phys. Lett. **59**, 2369 (1991)

[231] Ch. Spiegelberg, J. Geng, Y. Kaneda, Y. Hu, Shibin Jiang and N. Peyghambarian *Narrow-linewidth fiber lasers* in *high power narrow linewidth fiber lasers* OMD1 OSA/OAA (2004)

[232] W.W. Chow and S.W. Koch *Semiconductor-laser fundamentals. Physics of the gain Materials* Springer (1999)

[233] G. Lutz *Semiconductor radiation detectors* Springer (1999)

[234] D. Sands *Diode lasers* IOP (2005)

[235] Z. Alferov *Double heterostructure lasers: early days and future perspectives* IEEE Journal on Selected Topics in Quantum Electronics **6**, 832-840 (2000)

[236] H.R. Telle *Stabilization and modulation schemes of laser diodes for applied spectroscopy* Spectrochimica Acta Rev. **15**, 301-327 (1993)

[237] C.H. Henry *Theory of the linewidth of semiconductor lasers* IEEE Journal of Quantum Electronics **QE18**, 259-264 (1982)

[238] J.C. Camparo *The diode laser in atomic physics* Contemp. Phys. **26**, 443-477 (1985)

[239] C.E. Wieman and L. Hollberg *Using diode lasers for atomic physics* Rev. Sci. Instrum. **62**, 1-20 (1991)

[240] G.M. Tino *Atomic spectroscopy with diode lasers* Physica Scripta **T51**, 58-66 (1994)

[241] B. Dahmani, L. Hollberg, and R. Drullinger *Frequency stabilization of semiconductor lasers by resonant optical feedback* Opt. Lett. **12**, 876-878 (1987)

[242] Ph. Laurent, A. Clairon, and Ch. Breant *Frequency noise analysis of optically self-locked diode lasers* IEEE J. of Quantum Electronics **25**, 1131-1142 (1989)

[243] L. Ricci, M. Weidemüller, T. Esslinger, A. Hemmerich, C. Zimmermann, V. Vuletic, W. König, and T.W. Hänsch *A compact grating-stabilized diode laser system for atomic physics* Opt. Comm. **117**, 541-549 (1995)

[244] C.J. Hawthorn, K.P. Weber, and R.E. Scholten *Littrow configuration tunable external cavity diode laser with fixed direction output beam* Rev. Sci. Instrum. **72**, 4477-4479 (2001)

[245] X. Baillard, A. Gauguet, S. Bize, P. Lemonde, Ph. Laurent, A. Clairon, and P. Rosenbusch *Interference-filter-stabilized external-cavity diode lasers* Opt. Comm. **266**, 609-613 (2006)

[246] M. de Angelis, G.M. Tino, P. De Natale, C. Fort, G. Modugno, M. Prevedelli and C. Zimmermann *Tunable frequency controlled laser source in the near UV based on doubling of a semiconductor diode laser* Applied Physics B **62**, 333-338 (1996)

[247] A. Wicht, M. Rudolf, P. Huke, R.H. Rinkleff, and K. Danzmann *Grating enhanced external cavity diode laser* Appl. Phys. B **78**, 137-144 (2004)

[248] K. Döringshoff, I. Ernsting, R.-H. Rinkleff, S. Schiller and A. Wicht *Low-noise, tunable diode laser for ultra-high-resolution spectroscopy* Opt. Lett. **32**, 2876-2878 (2007)

[249] H. Yoshida, Y. Yamashita, M. Kuwabara and H. Kan *A 342-nm ultraviolet AlGaN multiple-quantum-well laser diode* Nature Photonics **2**, 551-554 (2008)

[250] T. Lehnhardt, M. Hümmer, K. Rössner, M. Müller, S. Höfling and A. Forchel *Continuous wave single mode operation of GaInAsSb/GaSb quantum well lasers emitting beyond 3 μm* Appl. Phys. Lett. **92**, 183508 (2008)

[251] J.A. Gupta, P.J. Barrios, J. Lapointe, G.C. Aers, C. Storey and P. Waldron *Modal gain of 2.4-μm InGaAsSb-AlGaAsSb complex-coupled distributed-feedback lasers* IEEE Photonics Technology Letters **21**, 1532-1534 (2009)

[252] A. Joullié, P. Christol, A.N. Baranov and A. Vicet *Mid-infrared 2-5 μm heterojunction laser diodes* in *Solid-State Mid-Infrared laser sources* I.I. Sorokina and K.L. Vodopyanov, Eds., Springer (2003)

[253] A. Bachmann, S. Arafin and K. Kashani-Shirazi *Single-mode electrically pumped GaSb-based VCSELs emitting continuous-wave at 2.4 and 2.6 μm* New Journal of Physics **11**, 125014 (2009)

[254] J.N. Walpole *Semiconductor amplifiers and lasers with tapered gain regions* Opt. Quantum Electron. **28**, 623-645 (1996)

[255] A.C. Wilson, J.C. Sharpe, C.R. McKenzie, P.J. Manson and D.M. Warrington *Narrow-linewidth master-oscillator power amplifier based on a semiconductor tapered amplifier* Appl. Opt. **37**, 4871-4875 (1998)

[256] V. Bolpasi and W. von Klitzing *Double-pass tapered amplifier diode laser with an output power of 1 W for an injection power of only 200 μW* Rev. Sci. Instrum. **81**, 113108 (2010)

[257] W. Tsang, M. Wu, Y. Chen, F. Chou, R. Logan, S. Chu, A. Sergant, P. Magill, K. Reichmann and C. Burrus *Long-wavelength InGaAsP/InP multiquantum well distributed feedback and distributed Bragg reflector lasers grown by chemical beam epitaxy* IEEE J. **QE-30**, 1370 (1994)

[258] L. Naehle, S. Belahsene, M.von. Edlinger, M. Fischer, G. Boissier, P. Grech, G. Narcy, A. Vicet, Y. Rouillard, J. Koeth and L. Worschech *Continuous-wave operation of type-I quantum well DFB laser diodes emitting in 3.4 μm wavelength range around room temperature* Electronics Letters **47**, 46-47 (2011)

[259] D.I. Babic, K. Streubel, R.P. Mirin, N.M. Margalit, J.E. Bowers, E.L. Hu, D.E. Mars, L. Yang and K. Carey *Room-temperature continuous-wave operation of 1.54-μm vertical-cavity lasers* IEEE Photonics Technology Letters **7**, 1225 (1995)

[260] J.P. Tourrenc, P. Signoret, M. Myara, R. Alabedra, F. Marin and K. D. Choquette *Frequency noise in 850-nm Selectively Oxidized VCSELs* Fluctuation and Noise Letters **3**, L407-L412 (2003)

[261] J-P. Hermier, I. Maurin, E. Giacobino, P. Schnitzer, R. Michalzik, K.J. Ebeling, A. Bramati and A.Z. Khoury *Quantum noise in VCSELs* New Journal of Physics **2**, 26.1-26.13 (2000)

[262] J. Kitching, S. Knappe, N. Vukicevic, L. Hollberg, R. Wynands and W. Weidmann *A microwave frequency reference based on VCSEL-driven dark line resonances in Cs vapor* IEEE Transactions on Instrumentation and Measurement **49**, 1313-1317 (2000)

[263] Federico Capasso, Claire Gmachl, Deborah L. Sivco and Alfred Y. Cho *Quantum Cascade Lasers* Physics Today **55**, 34 (2002)

[264] J. Faist, F. Capasso, D. L. Sivco, A. L. Hutchinson, and A. Y. Cho *Quantum Cascade Laser* Science **264**, 553 (1994)

[265] R.F. Kazarinov and R.A. Suris *Possibility of amplification of electromagnetic waves in a semiconductor with a superlattice* Sov. Phys. Semicond. **5**, 207 (1971)

[266] M. Razeghi *High-power high-wall plug efficiency mid-infrared quantum cascade lasers based on InP/GaInAs/InAlAs material system* Proc. SPIE 7230, 723011 (2009)

[267] A. Lyakh et al. *3 W continuous-wave room temperature single-facet emission from quantum cascade lasers based on nonresonant extraction design approach* Appl. Phys. Lett. **95**, 141113 (2009)

[268] M. Razeghi *High-Performance InP-Based Mid-IR Quantum Cascade Lasers* IEEE J. Sel. Top. Quantum Electron. **15**, 941 (2009)

[269] J. Faist et al. *Distributed feedback quantum cascade lasers* Appl. Phys. Lett. **70**, 2670 (1997)

[270] Robert F. Curl, Federico Capasso, Claire Gmachl, Anatoliy A. Kosterev, Barry Mc-Manus, Rafal Lewicki, Michael Pusharsky, Gerard Wysocki and Frank K. Tittel *Quantum cascade lasers in chemical physics* Chemical Physics Letters **487**, 1 (2010)

[271] G. Wysocki, R.F. Curl, F.K. Tittel, R. Maulini, J.M. Bulliard and J. Faist *Widely tunable mode-hop free external cavity quantum cascade laser for high resolution spectroscopic applications* Appl. Phys. B **81**, 769 (2005)

[272] R. Maulini, A. Mohan, M. Giovannini, J. Faist and E. Gini *External cavity quantum-cascade laser tunable from 8.2 to 10.4 μm using a gain element with a heterogeneous cascade* Appl. Phys. Lett. **88**, 201113 (2006)

[273] G. Wysocki et al. *Widely tunable mode-hop free external cavity quantum cascade lasers for high resolution spectroscopy and chemical sensing* Appl. Phys. B **92**, 305 (2008)

[274] A. Wittmann, A. Hugi, E. Gini, N. Hoyler and J. Faist *Heterogeneous High-Performance Quantum-Cascade Laser Sources for Broad-Band Tuning* IEEE J. Quantum Electron. **44**, 1083 (2008)

[275] C. Sirtori et al. *$GaAs/Al_x Ga_{1-x}$ As quantum cascade lasers* Appl. Phys. Lett. **73**, 3486 (1998)

[276] J. Devenson, O. Cathabard, R. Teissier and A. N. Baranov *High temperature operation of 3.3 μm quantum cascade lasers* Appl. Phys. Lett. **91**, 141106 (2007)

[277] D.G. Revin, J.W. Cockburn, M.J. Steer, R.J. Airey, M. Hopkinson, A.B. Krysa, L.R. Wilson and S. Menzel *InGaAs/AlAsSb/InP quantum cascade lasers operating at wavelengths close to 3 μm* Appl. Phys. Lett. **90**, 021108 (2007)

[278] J. Devenson, O. Cathabard, R. Teissier and A. N. Baranov *InAs/AlSb quantum cascade lasers emitting at 2.75-2.97 μm* Appl. Phys. Lett. **91**, 251102 (2007)

[279] O. Cathabard, R. Teissier, J. Devenson, J. C. Moreno and A. N. Baranov *Quantum cascade lasers emitting near 2.6 μm* Appl. Phys. Lett. **96**, 141110 (2010)

[280] D. Barate, R. Teissier, Y. Wang and A.N. Baranov *Short wavelength intersubband emission from InAs/AlSb quantum cascade structures* Appl. Phys. Lett. **87**, 051103 (2005)

[281] D.J. Paul *Si/SiGe heterostructures: from material and physics to devices and circuits* Semicond. Sci. Technol. **19**, R75 (2004)

[282] K. Driscoll and R. Paiella *Silicon-based injection lasers using electronic intersubband transitions in the L valleys* Appl. Phys. Lett. 89, 191110 (2006)

[283] M. De Seta, G. Capellini, Y. Busby, F. Evangelisti, M. Ortolani, M. Virgilio, G. Grosso, G. Pizzi, A. Nucara and S. Lupi *Conduction band intersubband transitions in Ge/SiGe quantum wells* Appl. Phys. Lett. **95**, 051918 (2009)

[284] R. Köhler, A. Tredicucci, F. Beltram, H. E. Beere, E. H. Linfield, A. Giles Davies, D. A. Ritchie, R. C. Iotti and F. Rossi *Terahertz semiconductor-heterostructure laser* Nature **417**, 156 (2002)

[285] A. Barkan, F.K. Tittel, D.M. Mittleman, R. Dengler, P. H. Siegel, G. Scalari, L. Ajili, J. Faist, H.E. Beere, E. H. Linfield, A.G. Davies and D.A. Ritchie *Linewidth and tuning characteristics of terahertz quantum cascade lasers* Opt. Lett. **29**, 575 (2004)

[286] J. Xu, J.M. Hensley, D.B. Fenner, R. P. Green, L. Mahler, A. Tredicucci, M. G. Allen, F. Beltram, H. E. Beere and D. A. Ritchie *Tunable terahertz quantum cascade lasers with an external cavity* Appl. Phys. Lett. **91**, 121104 (2007)

[287] A. W. M. Lee, B. S. Williams, S. Kumar, Q. Hu and J. L. Reno *Tunable terahertz quantum cascade lasers with external gratings* Opt. Lett. **35**, 910 (2010)

[288] S. Fathololoumi, E. Dupont, C.W.I. Chan, Z.R. Wasilewski, S.R. Laframboise, D. Ban, A. Matyas, C. Jirauschek, Q. Hu and H. C. Liu *Terahertz quantum cascade lasers operating up to ∼ 200 K with optimized oscillator strength and improved injection tunneling* Opt. Express **20**, 3866–3876 (2012)

[289] A. Wade, G. Fedorov, D. Smirnov, S. Kumar, B.S.Williams, Q. Hu and J. Reno *Magnetic-field-assisted terahertz quantum cascade laser operating up to 225 K* Nature Photon. **3**, 41 (2009)

[290] M. S. Vitiello, G. Scamarcio, V. Spagnolo, S. S. Dhillon and C. Sirtori *Terahertz quantum cascade lasers with large wall-plug efficiency* Appl. Phys. Lett. **90**, 191115 (2007)

[291] S. Bartalini et al. *Observing the intrinsic linewidth of a quantum-cascade laser: beyond the Schawlow-Townes limit* Phys. Rev. Lett. **104**, 083904 (2010)

[292] S. Borri, S. Bartalini, P. Cancio, I. Galli, G. Giusfredi, D. Mazzotti, M. Yamanishi and P. De Natale *Frequency-noise dynamics of mid-infrared quantum cascade lasers* IEEE J. Quantum Electron. **47**, 984–988 (2011)

[293] L. Tombez et al. *Frequency noise of free-running 4.6 µm distributed feedback quantum cascade lasers near room temperature* Opt. Lett. **36**, 3109-3111 (2011)

[294] S. Bartalini et al. *Measuring frequency noise and intrinsic linewidth of a room-temperature DFB quantum cascade laser* Opt. Express **19**, 17996 (2011)

[295] M. Vitiello, L. Consolino, S. Bartalini, A. Taschin, A. Tredicucci, M. Inguscio and P. De Natale *Quantum-limited frequency fluctuations in a terahertz laser* Nature Photonics **6**, 525-528 (2012)

[296] L. Consolino, A. Taschin, P. Bartolini, S. Bartalini, P. Cancio, A. Tredicucci, H.E. Beere, D.A. Ritchie, R. Torre, M.S. Vitiello and P. De Natale *Phase-locking to a free-space terahertz comb for metrological-grade terahertz lasers* Nature Communications **3**, 1040 (2012)

[297] M. A. Belkin, F. Capasso, A. Belyanin, D. L. Sivco, A. Y. Cho, D. C. Oakley, C. J. Vineis and G. W. Turner *Terahertz quantum-cascade-laser source based on intracavity difference-frequency generation* Nature Photonics **1**, 288 (2007)

[298] M. A. Belkin, F. Capasso, F. Xie, A. Belyanin, M. Fischer, A. Wittmann and J. Faist *Room temperature terahertz quantum cascade laser source based on intracavity difference-frequency generation* Appl. Phys. Lett. **92**, 201101 (2008)

[299] C. Sirtori, F. Capasso, J. Faist, L. N. Pfeiffer and K. W. West *Far-infrared generation by doubly resonant difference frequency mixing in a coupled quantum well two-dimensional electron gas system* Appl. Phys. Lett. **65**, 445 (1994)

[300] G. Gagliardi, S. Viciani, M. Inguscio, P. De Natale, C. Gmachl, F. Capasso, D.L. Sivco, J.N. Baillargeon, A.L. Hutchinson and A.Y. Cho *Generation of tunable far-infrared radiation using a quantum cascade laser* Opt. Lett. **27**, 521 (2002)

[301] B.S. Williams *Terahertz qunatum-cascade lasers* Nature **1**, 517-525 (2007)

[302] M.S. Vitiello and A. Tredicucci *Tunable emission in THz quantum cascade lasers* IEEE Transactions on Terahertz Science and Technology **1**, 76-84 (2011)

[303] T. Liu and Q. J. Wang *Fundamental frequency noise and linewidth broadening caused by intrinsic temperature fluctuations in quantum cascade lasers* Phys. Rev. B **84**, 125322 (2011)

[304] T. Gensty, W. Elsässer and C. Mann *Intensity noise properties of quantum cascade lasers* Optics Express **13**, 2032 (2005)

[305] F. Rana, P. Mayer and R.J. Ram *Scaling of the photon noise in semiconductor cascade lasers* J. Opt. B: Quantum Semiclass. Opt. **6**, S771 (2004)

[306] M. Yamanishi, T. Edamura, K. Fujita, N. Akikusa and H. Kan *Theory of the Intrinsic Linewidth of Quantum-Cascade Lasers: Hidden Reason for the Narrow Linewidth and Line-Broadening by Thermal Photons* IEEE J. Quantum Electron. **44**, 12 (2008)

[307] D. Weidmann and G. Wysocki *High-resolution broadband (>100 cm^{-1}) infrared heterodyne spectro-radiometry using an external cavity quantum cascade laser* Opt. Express **17**, 248 (2009)

[308] Y. Takagi, N. Kumazaki, M. Ishihara, K. Kasahara, A. Sugiyama, N. Akikusa, and T. Edamura *Relative intensity noise measurements of 5 µm quantum cascade laser and 1.55 µm semiconductor laser* Electron. Lett. **44**, 860 (2008)

[309] D. Weidmann, K. Smith and B. Ellison *Experimental investigation of high-frequency noise and optical feedback effects using a 9.7 µm continuous-wave distributed-feedback quantum-cascade laser* Appl. Opt. **46**, 947 (2007)

[310] R. N. Hall, G. E. Fenner, J.D. Kingsley, T. J. Soltys and R. O. Carlson *Coherent Light Emission from GaAs Junctions* Phys. Rev. Lett. **9**, 366 (1962)

[311] M. Lax *Classical Noise. V. Noise in Self-Sustained Oscillators* Phys. Rev. **160**, 290 (1967)

[312] R. Lang, M. O. Scully and W. E. Lamb *Why Is the Laser Line So Narrow? A Theory of Single-Quasimode Laser Operation* Phys. Rev. A **7**, 1788 (1973)

[313] J. von Staden, T. Gensty, W. Elsässer, G. Giuliani and C. Mann *Measurements of the α factor of a distributed-feedback quantum cascade laser by an optical feedback self-mixing technique* Opt. Lett. **31**, 2574 (2006)

[314] T. Aellen, R. Maulini, R. Terazzi, N. Hoyler, M. Giovannini, J. Faist, S. Blaser and L. Hvozdara *Direct measurement of the linewidth enhancement factor by optical heterodyning of an amplitude-modulated quantum cascade laser* Appl. Phys. Lett. **89**, 091121 (2006)

[315] N. Kumazaki, Y. Takagi, M. Ishihara, K. Kasahara, A. Sugiyama, N. Akikusa, and T. Edamura *Detuning characteristics of the linewidth enhancement factor of a mid-infrared quantum cascade laser* Appl. Phys. Lett. **92**, 121104 (2008)

[316] I. Vurgaftman, W.W. Bewley, C.L. Canedy, C.S. Kim, M. Kim, C.D. Merritt, J. Abell, J.R. Lindle and J.R. Meyer *Rebalancing of internally generated carriers for mid-infrared interband cascade lasers with very low power consumption* Nature Communications 2:585 (2011)

[317] A.N. Chryssis *Design and fabrication of high-performance interband cascade tunable external cavity lasers*, PhD thesis (2010)

[318] D. Caffey, T. Day, C.S. Kim, M. Kim, I. Vurgaftman, W.W. Bewley, J.R. Lindle, C.L. Canedy, J. Abell and J.R. Meyer *Performance characteristics of a continuous-wave compact widely tunable external cavity interband cascade lasers* Optics Express **18**, 15691 (2010)

[319] G. Wysocki, Y. Bakhirkin, S. So, F.K. Tittel, C.J. Hill, R.Q. Yang and M.P. Fraser *Dual interband cascade laser based trace-gas sensor for environmental monitoring* Appl. Opt. **46**, 8202 (2007)

[320] K.R. Parameswaran, D.I. Rosen, M.G. Allen, A.M. Ganz and T.H. Risby *Off-axis integrated cavity output spectroscopy with a mid-infrared interband cascade laser for real-time breath ethane measurements* Appl. Opt. **48**, B73 (2009)

[321] P.A. Folkes *Interband cascade laser photon noise* J. Phys. D: Appl. Phys. **41**, 245109 (2008)

[322] R.W. Boyd *Nonlinear optics (Third Edition)* Academic Press, Burlington (2008)

[323] W. Chen, J. Cousin, E. Poullet, J. Burie, D. Boucher, X. Gao, M.W. Sigrist and F.K. Tittel *Continuous-wave mid-infrared laser sources based on difference frequency generation* C. R. Physique, doi:10.1016/j.crhy.2007.09.011 (2007)

[324] D. Mazzotti *A tunable and narrow-linewidth difference-frequency spectrometer around 4.25 µm for CO_2 high-resolution spectroscopy* PhD Thesis (1999)

[325] I. Ricciardi, M. De Rosa, A. Rocco, P. Ferraro, A. Vannucci, P. Spano and P. De Natale *Sum-frequency generation of cw ultraviolet radiation in periodically poled $LiTaO_3$* Opt. Lett. **34**, 1348-1350 (2009)

[326] U. Strössner *Development of Optical Synthesizers*, Dissertation of Univerität Konstanz (2001)

[327] A.K.Y. Ngai *Spectroscopic applications of continuous wave optical parametric oscillators* PhD Thesis, Radboud University Nijmegen (2008).

[328] R.G. Batchko, D.R. Weise, T. Plettner, G.D. Miller, M.M. Fejer and R.L. Beyer *Continuous-wave 532-nm-pumped singly resonant optical parametric oscillator based on periodically poled lithium niobate* Opt. Lett. **23**, 168-170 (2008).

[329] S.J. Brosnan and R.L. Byer *Optical Parametric Oscillator Threshold and Linewidth Studies* IEEE J. of Quant. Electr. **15**, 415-431 (1979)

[330] L.E. Myers, R.C. Eckardt, M.M. Fejer and R.L. Byer *Quasi-phase-matched optical parametric oscillators in bulk periodically poled $LiNbO_3$* J. Opt. Soc. Am. B **12**, 2102-2116 (1995)

[331] H.P. Li, D.Y. Tang, S.P. Ng and J. Kong *Temperature-tunable nanosecond optical parametric oscillator based on periodically poled MgO:LiNbO₃*, Optics and Laser Technology **28**, 192-195 (2006)

[332] R. Graham and H. Haken *The quantum-fluctuations of the optical parametric oscillator. I* Zeit. F. Physik **210**, 276-302 (1968)

[333] N.C. Wong *Optical frequency division using an optical parametric oscillator* Opt. Lett. **15**, 1129-1131 (1990)

[334] I. Ricciardi, E. De Tommasi, P. Maddaloni, S. Mosca, A. Rocco, J.-J. Zondy, M. De Rosa and P. De Natale *A narrow-linewidth optical parametric oscillator for mid-infrared high-resolution spectroscopy* Molecular Physics, DOI:10.1080/00268976.2012.699640

[335] I. Ricciardi, E. De Tommasi, P. Maddaloni, S. Mosca, A. Rocco, J.-J. Zondy, M. De Rosa and P. De Natale *Frequency-comb-referenced singly-resonant OPO for sub-Doppler spectroscopy* Opt. Express **20**, 9178-9186 (2012)

[336] P. Cancio, S. Bartalini, S. Borri, I. Galli, G. Gagliardi, G. Giusfredi, P. Maddaloni, P. Malara, D. Mazzotti and P. De Natale *Frequency-comb-referenced mid-IR sources for next-generation environmental sensors* Appl. Phys. B **102**, 255-269 (2011)

[337] D. Richter, A. Fried and P. Weibring *Difference frequency generation laser based spectrometers* Laser and Photon. Rev. **3**, 343-354 (2009)

[338] M. Asobe, O. Tadanaga, T. Yanagawa, T. Umeki, Y. Nishida and H. Suzuki *High-power mid-infrared wavelength generation using difference frequency generation in damage-resistant Zn:LiNbO₃ waveguide* Electron. Lett. **44**, 288-290 (2008)

[339] I. Galli, S. Bartalini, S. Borri, P. Cancio, G. Giusfredi, D. Mazzotti and P. De Natale *Ti:sapphire laser intracavity difference-frequency generation of 30 mW cw radiation around 4.5 μm* Opt. Lett. **35**, 3616 (2010)

[340] I. Galli, S. Bartalini, P. Cancio, G. Giusfredi, D. Mazzotti and P. De Natale *Ultra-stable, widely tunable and absolutely linked mid-IR coherent source* Optics Express **17**, 9582 (2009)

[341] D.D. Bicanic, B.F.J. Zuidberg and A. Dymanus *Generation of continuously tunable laser sidebands in the submillimeter region* Appl. Phys. Lett. **32**, 367-369 (1978)

[342] W.A.M. Blumberg, H.R. Fettermand, D. Peck and P.F. Goldsmith *Tunable submillimeter sources applied to the excited state rotational spectroscopy and kinetics of CH₃F* Appl. Phys. Lett. **35**, 582-585 (1979)

[343] J. Farhoomand, G.A. Blake, M.A. Frerking and H.M. Pickett *Generation of tunable laser sidebands in the far-infrared region* J. Appl. Phys. **57**, 1763-1766 (1985)

[344] G. Piau, F.X. Brown, D. Dangoisse and P. Glorieux *Heterodyne detection of tunable FIR sidebands* IEEE J. Quantum Electron. **QE-23**, 1388-1391 (1987)

[345] K.M. Evenson, D.A. Jennings and F.R. Petersen *Tunable far-infrared spectroscopy* Appl. Phys. Lett. **44**, 576-578 (1984)

[346] L.R. Zink, P. De Natale, F.S. Pavone, M. Prevedelli, K.M. Evenson and M. Inguscio *Rotational far infrared spectrum of ¹³CO¹* Journal of Molecular Spectroscopy **143**, 304-310 (1990)

[347] I.G. Nolt, J.V. Radostitz, G. Di Lonardo, K.M. Evenson, D.A. Jennings, K.R. Leopold, M.D. Vanek, L.R. Zink, A. Hinz and K.V. Chance *Accurate rotational constants of CO, HCl, and HF: Spectral standards for the 0.3- to 6-THz (10- to 200-cm^{-1}) region* J. Mol. Spectrosc. **125**, 274-287 (1987)

[348] C. Freed and A. Javan *Standing-wave saturation resonances in the CO_2 10.6-μ transitions observed in a low-pressure room-temperature absorber gas* Appl. Phys. Lett. **17**, 53-56 (1970)

[349] F.R. Petersen, E.C. Beatty and C.R. Pollock *Improved rovibrational constants and frequency tables for the normal laser bands of $^{12}C^{16}O_2$* J. Mol. Spectrosc. **102**, 112-122 (1983)

[350] L. C. Bradley, K.L. Soohoo and C. Freed *Absolute frequencies of lasing transitions in nine CO_2 isotopic species* IEEE J. Quantum Electron. **QE-22**, 234-267 (1986)

[351] F. Matsushima, H. Odashima, D. Wang, S. Tsunekawa and K. Takagi, *Far-Infrared Spectroscopy of LiH using a Tunable Far-Infrared Spectrometer* Jpn. J. Appl. Phys. **33**, 315-318 (1994)

[352] L. Fusina, P. De Natale, M. Prevedelli and L.R. Zink *The submillimeter rotation spectrum of DCl* J. Mol. Spectrosc. **152**, 55 (1992)

[353] G. Di Lonardo, L. Fusina, P. De Natale et al. *The pure rotation spectrum of HBr in the submillimeter-wave region* J. Mol. Spectrosc. **148**, 86 (1991)

[354] M. Bellini, P. De Natale, G. Di Lonardo et al. *Tunable far infrared spectroscopy of $^{16}O_3$ ozone* J. Mol. Spectrosc. **152**, 256 (1992)

[355] G. Di Lonardo, L. Fusina, M. Bellini, P. De Natale, G. Buffa and O. Tarrini *Air-Broadening of Rotational Lines of Ozone in the 1.5-THz Region* Journal of Molecular Spectroscopy **161**, 581-584 (1993)

[356] K. Chance, P. De Natale, M. Bellini, M. Inguscio, G. Di Lonardo and L. Fusina *Pressure Broadening of the 2.4978-THz Rotational Lines of HO_2 by N_2 and O_2* Journal of Molecular Spectroscopy **163**, 67-70 (1994)

[357] G. Cazzoli, L. Cludi, G. Cotti, L. Dore, C. Degli Esposti, M. Bellini and P. De Natale *The Rotational Spectrum of CHF_3 in the Submillimeter-Wave and Far-Infrared Region: Observation of the K = 3 Line Splitting* Journal of Molecular Spectroscopy **163**, 521-528 (1994).

[358] P. De Natale, M. Bellini, M. Inguscio, G. Buffa and O. Tarrini *Far-Infrared Collisional Lineshapes of Lithium Hydride and Deuteride Perturbed by H_2 and D_2* Journal of Molecular Spectroscopy **163**, 510-514 (1994)

[359] M. Bellini, P. De Natale, M. Inguscio, E. Fink, D. Galli and F. Palla *Laboratory measurements of rotational transitions of lithium hydride in the Far-Infrared* Astrophysical Journal **424**, 507-509 (1994)

[360] M. Bellini, P. De Natale, L. Fusina and G. Modugno *The Pure Rotation Spectrum of HOCl in the Submillimeter-Wave Region* Journal of Molecular Spectroscopy **172**, 559-562 (1995)

[361] P. De Natale, L. Lorini, M. Inguscio, G. Di Lonardo, L. Fusina, P.A.R. Ade and A.G. Murray *Improved sensitivity of tunable far-infrared spectroscopy: application to the detection of HBr in the v = 1 state* Applied Optics **36**, 5822-5826 (1997)

[362] P. De Natale, L. Lorini, M. Inguscio, I.G. Nolt, J. Park, G. Di Lonardo, L. Fusina, P.A.R. Ade and A.G. Murray *Accurate frequency measurements for H_2O and $^{16}O_3$ in the 119-cm^{-1} OH atmospheric window* Applied Optics, **36**, 8526-8532 (1997)

[363] P. De Natale, L. Lorini, M. Inguscio, G. Di Lonardo and L. Fusina *High-sensitivity detection of the rotation spectrum of HCl in the v = 1 state by tunable FIR spectroscopy* Chemical Physics Letters, **273**, 253-258 (1997)

[364] P. De Natale et al. *Hyperfine structure and isotope shift in the far-infrared ground-state transitions of atomic oxygen* Phys. Rev. A **48**, 3757 (1993)

[365] M. Inguscio, P. De Natale and L. Veseth *Isotopic shift in atomic fine structure theory and experiment for oxygen transitions in the far infrared* Comments At. Mol. Phys. **30**, 3-13 (1994)

[366] J.M. Brown, K.M. Evenson and L.R. Zink *Laser magnetic-resonance measurement of the $^3P_1$-$^3P_2$ fine-structure splittings in ^{17}O and ^{18}O* Phys. Rev. A **48**, 3761-3763 (1993)

[367] G. Modugno et al. *Precise measurement of molecular dipole moments with a tunable far-infrared Stark spectrometer: application to HOCl* J. Opt. Soc. Am. B **13**, 1645 (1996)

[368] M. Bellini et al. *Stark and Frequency Measurements in the FIR Spectrum of H_2O_2* Journal of Molecular Spectroscopy **177**, 115-123 (1996)

[369] S. Viciani et al. *Noise characterization of a coherent tunable far infrared spectrometer* Review Of Scientific Instruments **69**, 372-376 (1998)

[370] P. De Natale, L. Gianfrani, S. Viciani and M. Inguscio *Spectroscopic observation of the Faraday effect in the far infrared* Opt. Lett. **22**, 1896 (1997)

[371] S. Viciani et al. *Magnetic-field effects on molecular transitions in the far-infrared region: prospects for more-sensitive spectrometers* J. Opt. Soc. Am. B **16**, 301 (1999)

[372] H. Odashima, L.R. Zink and K.M. Evenson *Tunable far-infrared spectroscopy extended to 9.1 THz* Opt. Lett. **24**, 406-407 (1999)

[373] L. Novotny *From near-field optics to optical antennas* Physics Today, July (2011).

[374] M.C. Wanke et al. *Monolithically integrated solid-state terahertz transceivers* Nature Photonics **4**, 565-569 (2010)

[375] P. Aufmuth and K. Danzmann *Gravitational wave detectors* New Journal of Physics **7**, 202 (2005)

[376] P.J. Fox, R.E. Scholten, M.R. Walkiewicz and R. E. Drullinger *A reliable, compact, and low-cost Michelson wavemeter for laser wavelength measurement* Am. J. Phys. **67**, 624–630 (1999)

[377] T.M. Niebauer, J.E. Faller, H.M. Godwin, J.L. Hall and R. L. Barger *Frequency stability measurements on polarization-stabilized He-Ne lasers* Appl. Opt. **27**, 1285–1289 (1988)

[378] C. Reiser and R.B. Lopert *Laser wavemeter with solid Fizeau wedge interferometer* Appl. Opt. **27**, 3656 (1988)

[379] W. Demtröder *Laser spectroscopy (Third Edition)* Springer-Verlag, Berlin (2003)

[380] B. Barbieri, N. Beverini and A. Sasso *Optogalvanic Spectroscopy* Rev. Mod. Phys. **62**, 603-644 (1990)

[381] E. Arimondo, M. Inguscio and P. Violino *Experimental determinations of the hyperfine structure in the alkali atoms* Rev. Mod. Phys. **49**, 31-75 (1977)

[382] N.C. Shand, C.-L. Ning and J. Pfab *Laser photoionisation (REMPI) spectroscopy of the 3s ← n Rydberg transition of jet-cooled acetaldehyde (CH$_3$CHO)* Chemical Physics Letters **247**, 32-37 (1995)

[383] A.A. Mills, B.M. Siller and B.J. McCall *Precision cavity enhanced velocity modulation spectroscopy* Chem. Phys. Lett. **501**, 1-5 (2010)

[384] A.R.W. McKellar *High-resolution infrared spectroscopy with synchrotron sources* Journal of Molecular Spectroscopy **262**, 1-10 (2010)

[385] B.A. Thrush *Laser magnetic resonance spectroscopy and its application to atmospheric chemistry* Acc. Chem. Res. **14**, 116-122 (1981)

[386] A.A. Kosterev, Y.A. Bakhirkin and F.K. Tittel *Ultrasensitive gas detection by quartz-enhanced photoacoustic spectroscopy in the fundamental molecular absorption bands region* Appl. Phys. B **80**, 133-138 (2005)

[387] J.M. Supplee, E.A. Whittaker and W. Lenth *Theoretical description of frequency modulation and wavelength modulation spectroscopy* Appl. Opt. **33**, 6294 (1994)

[388] C.R. Webster, R.T. Menzies and E.D. Hinkley *Infrared laser absorption: theory and measurements* R.M. Measures, Ed., Laser Remote Chemical Analysis, Wiley, New York, NY (1987)

[389] G.C. Bjorklund *Frequency-modulation spectroscopy: a new method for measuring weak absorptions and dispersions* Opt. Lett. **5**, 15 (1980)

[390] G.C. Bjorklund, M.D. Levenson, W. Lenth and C. Ortiz *Frequency modulation (FM) spectroscopy: theory of lineshapes and signal-to-noise analysis* Opt. Lett. **5**, 15 (1983)

[391] M. Gehrtz, G.C. Bjorklund and E.A. Whittaker *Quantum-limited laser frequency-modulation spectroscopy* J. Opt. Soc. Am. B **2**, 1510 (1985)

[392] G.R. Janik, C.B. Carlisle and T.F. Gallagher *Two-tone frequency-modulation spectroscopy* J. Opt. Soc. Am. B **3**, 1070 (1986)

[393] G. Litfin, C.R. Pollock, R.F. Curl Jr. and F.K. Tittel *Sensitivity enhancement of laser absorption spectroscopy by magnetic rotation effect* J. Chem. Phys. **72**, 6602-6605 (1980)

[394] A. Hinz, J. Pfeiffer, W. Bohle and W. Urban *Mid-infrared laser magnetic resonance using the Faraday and Voigt effects for sensitive detection* Mol. Phys **45**, 1131-1139 (1982)

[395] R.J. Brecha, L.M. Pedrotti and D. Krause *Magnetic rotation spectroscopy of molecular oxygen with a diode laser* J. Opt. Soc. Am. B **14**, 1921-1930 (1997)

[396] T. Fritsch, M. Horstjann, D. Halmer, Sabana, P. Hering and M. Mürtz *Magnetic Faraday modulation spectroscopy of the 1-0 band of ^{14}NO and ^{15}NO* Appl. Phys. B **93**, 713-723 (2008)

[397] J.U. White *Long Optical Paths of Large Aperture* J. Opt. Soc. Am. **32**, 285 (1942)

[398] D.R. Herriott, H. Kogelnik and R. Kompfner *Off-Axis Paths in Spherical Mirror Interferometers* Appl. Opt. **3**, 523 (1964)

[399] A. Foltynowicz, F.M. Schmidt, W. Ma and O. Axner *Noise-immune cavity-enhanced optical heterodyne molecular spectroscopy: current status and future potential* Appl. Phys. B **92**, 313-326 (2008)

[400] V. Motto-Ros, M. Durand and J. Morville *Extensive characterization of the optical feedback cavity enhanced absorption spectroscopy (OF-CEAS) technique: ringdown-time calibration of the absorption scale* Appl. Phys. B **91**, 203-211 (2008)

[401] J. Morville, S. Kassi, M. Chenevier and D. Romanini *Fast, low-noise, mode-by-mode, cavity-enhanced absorption spectroscopy by diode-laser self-locking* Appl. Phys. B **80**, 1027-1038 (2005)

[402] A. O'Keefe, J.J. Scherer and J.B. Paul *cw Integrated cavity output spectroscopy* Chem. Phys. Lett. **307**, 343 (1999)

[403] P. Maddaloni, G. Gagliardi, P. Malara and Paolo De Natale *Off-axis integrated-cavity-output spectroscopy for trace-gas concentration measurements: modeling and performance* J. Opt. Soc. Am. B **23**, 1938-1945 (2006)

[404] P. Malara, P. Maddaloni, G. Gagliardi and P. De Natale *Combining a difference-frequency source with an off-axis high-finesse cavity for trace-gas monitoring around 3 μm* Optics Express **14**, 1304-1313 (2006)

[405] G.S. Engel, W.S. Drisdell, F.N. Keutsch, E.J. Moyer and J.G. Anderson *Ultrasensitive near-infrared integrated cavity output spectroscopy technique for detection of CO at 1.57 μm: new sensitivity limits for absorption measurements in passive optical cavities* Appl. Opt. **45**, 9221 (2006)

[406] D.S. Baer, J.B. Paul, J.B. Gupta and A. O'Keefe *Sensitive absorption measurements in the near-infrared region using off-axis integrated-cavity-output spectroscopy* Appl. Phys. B **75**, 261 (2002)

[407] J. Ye, L.-S. Ma and J.L. Hall *Ultrasensitive detections in atomic and molecular physics: demonstration in molecular overtone spectroscopy* J. Opt. Soc. Am. B **15**, 6-15 (1998)

[408] V.L. Kasyutich, C.E. Canosa-Mas, C. Pfrang, S. Vaughan and R.P. Wayne *Off-axis continuous-wave cavity-enhanced absorption spectroscopy of narrow-band and broadband absorbers using red diode lasers* Appl. Phys. B **75**, 755 (2002)

[409] L.-S. Ma, J. Ye, P. Dube and J.L. Hall *Ultrasensitive frequency-modulation spectroscopy enhanced by a high-finesse optical cavity: theory and application to overtone transitions of C_2H_2 and C_2HD* J. Opt. Soc. Am. B **16**, 2255-2268 (1999)

[410] R.G. DeVoe and R.G. Brewer *Laser-frequency division and stabilization* Physical Review A **30**, 2827-2829 (1984)

[411] A. O'Keefe and D.A.G. Deacon *Cavity ring-down optical spectrometer for absorption measurements using pulsed laser sources* Rev. Sci. Instrum. **59**, 2544 (1988)

[412] D. Romanini, A.A. Kachanov, N. Sadeghi and F. Stoeckel *CW cavity ring down spectroscopy* Chem. Phys. Lett. **264**, 316 (1997)

[413] M. Mazurenka, A.J. Orr-Ewing, R. Peverall and G.A.D. Ritchie *Cavity ring-down and cavity enhanced spectroscopy using diode lasers* Annu. Rep. Prog. Chem., Sect. C **101**, 100-142 (2005)

[414] K.K. Lehmann and D. Romanini *The superposition principle and cavity ring-down spectroscopy* J. Chem. Phys. **105**, 15 (1996)

[415] B.A. Paldus, C.C. Harb, T.G. Spence, B. Wilke, J. Xie, J.S. Harris, and R.N. Zare *Cavity-locked ring-down spectroscopy* Journal of Applied Physics **83**, 3991 (1998)

[416] T.G. Spence, C.C. Harb, B.A. Paldus, R.N. Zare, B. Wilke and R.L. Byer *A laser-locked cavity ring-down spectrometer employing an analog detection scheme* Rev. Sci. Instrum. **71**, 347 (2000)

[417] J. Ye and J.L. Hall *Cavity ringdown heterodyne spectroscopy: high sensitivity with microwatt light power* Phys. Rev. A **61**, 061802(R) (2000)

[418] R. Engeln, G. von Helden, G. Berden and G. Meijer *Phase shift cavity ring down absorption spectroscopy* Chemical Physics Letters **262**, 105-109 (1996)

[419] G. Giusfredi, S. Bartalini, S. Borri, P. Cancio, I. Galli, D. Mazzotti and P. De Natale *Saturated-absorption cavity ring-down spectroscopy* Phys. Rev. Lett. **104**, 110801 (2010)

[420] I. Galli, S. Bartalini, S. Borri, P. Cancio, D. Mazzotti, P. De Natale and G. Giusfredi *Molecular gas sensing below parts per trillion: Radiocarbon-dioxide optical detection* Phys. Rev. Lett. **107**, 270802 (2011)

[421] N. Zare *Ultrasensitive radiocarbon detection* Nature **482**, 312 (2012)

[422] D. Mazzotti, P. Cancio, G. Giusfredi, P. De Natale and M. Prevedelli *Frequency-comb-based absolute frequency measurements in the mid-infrared with a difference-frequency spectrometer* Opt. Lett. **30**, 997 (2005)

[423] K. Shimoda *Line-broadening and Narrowing Effects* in *High-resolution Laser Spectroscopy*, Springer (1976)

[424] R.J. Butcher *Sub-Doppler laser spectroscopy* Optical and Quantum Electronics **25**, 79-95 (1993)

[425] S. Borri, S. Bartalini, I. Galli, P. Cancio, G. Giusfredi, D. Mazzotti, A. Castrillo, L. Gianfrani and P. De Natale *Lamb-dip-locked quantum cascade laser for comb-referenced IR absolute frequency measurements* Opt. Express **16**, 11637 (2008)

[426] G. Gagliardi, G. Rusciano and L. Gianfrani *Sub-Doppler spectroscopy of $H_2^{19}O$ at 1.4 μm* Appl. Phys. B **70**, 883-888 (2000)

[427] D. Mazzotti, S. Borri, P. Cancio, G. Giusfredi and P. De Natale *Low-power Lamb-dip spectroscopy of very weak CO_2 transitions near 4.25 μm* Opt. Lett. **27**, 1256-1258 (2002)

[428] A. Foltynowicz, W. Ma and O. Axner *Characterization of fiber-laser-based sub-Doppler NICE-OHMS for quantitative trace gas detection* Opt. Express **16**, 14689-14702 (2008)

[429] O. Axner, W. Ma and A. Foltynowicz *Sub-Doppler dispersion and noise-immune cavity-enhanced optical heterodyne molecular spectroscopy revised* J. Opt. Soc. Am. B **25**, 1166 (2008)

[430] J. Ye, L.-S. Ma and J.L. Hall *Sub-Doppler optical frequency reference at 1.064 μm by means of ultrasensitive cavity-enhanced frequency modulation spectroscopy of a C_2HD overtone transition* Opt. Lett. **21**, 1000 (1996)

[431] C. Wieman, and T.W. Hänsch *Doppler-free laser polarization spectroscopy* Phys. Rev. Lett. **36**, 1170 (1976)

[432] C.P. Pearman, C.S. Adams, S.G. Cox, P.F. Griffin, D.A. Smith and I.G. Hughes *Polarization spectroscopy of a closed atomic transition: applications to laser frequency locking* J. Phys. B: At. Mol. Opt. Phys. **35**, 5141-5151 (2002)

[433] S. Bartalini, S. Borri and P. De Natale *Doppler-free polarization spectroscopy with a quantum cascade laser at 4.3 μm* Opt. Express **17**, 7440 (2009)

[434] Y. Yoshikawa, T. Umeki, T. Mukae, Y. Torii and T. Kuga *Frequency stabilization of a laser diode with use of light-induced birefringence in an atomic vapour* Appl. Opt. **42**, 6645 (2003)

[435] L.S. Vasilenko, V.P. Chebotaev and A.V. Schishaev *Line Shape of Two-photon Absorption in a Standing-Wave Field in a gas* Pis'ma Zh. Eksp. Teor. Fiz **12**, 161 (1970)

[436] V.S. Letokhov *Nonlinear High Resolution Laser Spectroscopy* Science **190**, 344-351 (1975)

[437] S.N. Bagayev, A.E. Baklanov, V.P. Chebotayev and A.S. Dychkov *Superhigh resolution spectroscopy in methane with cold molecules* Appl. Phys. B **48**, 31-35 (1989)

[438] S.N. Bagayev, V.P. Chebotayev, A.K. Dmitriyev, A.E. Om and Yu.V. Nekrasov *Second-order Doppler-free spectroscopy* Appl. Phys. B **52**, 63-66 (1991)

[439] S.N. Bagayev, V.P. Chebotayev and E. A. Titov *Saturated absorption lineshape under the transit-time conditions* Laser Physics **4**, 224-292 (1994)

[440] Ch. Chardonnet, F. Guernet, G. Charton and Ch. Bordé *Ultrahigh-resolution saturation spectroscopy using slow molecules in an external cell* Appl. Phys. B **59**, 333 (1994)

[441] P.E. Durand, G. Nogues, V. Bernard, A. Amy-Klein and Ch. Chardonnet *Slow-molecule detection in Doppler-free two-photon spectroscopy* Europhys. Lett. **37**, 103-108 (1997)

[442] C.G. Parthey, A. Matveev, J. Alnis, B. Bernhardt, A. Beyer, R. Holzwarth, A. Maistrou, R. Pohl, K. Predehl, T. Udem, T. Wilken, N. Kolachevsky, M. Abgrall, D. Rovera, Ch. Salomon, Ph. Laurent and T.W. Hänsch *Improved measurement of the hydrogen 1S-2S transition frequency* Phys. Rev. Lett. **107**, 203001 (2011)

[443] S.Y.T. Van de Meerakker, H.L. Bethlem and G. Meijer *Taming molecular beams* Nature Physics **4**, 595-602 (2008)

[444] M.H. Havenith *Infrared Spectroscopy of Molecular Clusters: An Introduction to Intermolecular Forces* Springer (2002)

[445] D.H. Levy *The spectroscopy of very cold gases* Science **214**, 263-269 (1981)

[446] S.E. Maxwell, N. Brahms, R. deCarvalho, D.R. Glenn, J.S. Helton, S.V. Nguyen, D. Patterson, J. Petricka, D. DeMille and J. M. Doyle *High-Flux Beam Source for Cold, Slow Atoms or Molecules* Phys. Rev. Lett. **95**, 173201 (2005)

[447] Sarah Margaretha Skoff *Buffer gas cooling of YbF molecules* Imperial College London, PhD thesis (2011)

[448] N.F. Ramsey *Experiments with separated oscillatory fields and hydrogen masers* Science **248**, 1612-1619 (1990)

[449] N.F. Ramsey *A molecular beam resonance method with separated oscillating fields* Phys. Rev. **78**, 695-699 (1950)

[450] Ch.J. Bordé, Ch. Salomon, S. Avrillier, A. Van Lerberghe, Ch. Breant, D. Bassi and G. Scoles *Optical Ramsey fringes with traveling waves* Phys. Rev. A **30**, 1836-1848 (1984)

[451] L.F. Constantin, R.J. Butcher, P.E. Durand, A. Amy-Klein and Ch. Chardonnet *2.3-kHz two-photon Ramsey fringes at 30 THz* Phys. Rev. A **60**, R753-R756 (1999)

[452] A. Shelkovnikov, C. Grain, R.J. Butcher, A. Amy-Klein, A. Goncharov and Ch. Chardonnet *Two-photon Ramsey fringes at 30 THz referenced to an H maser/Cs fountain via an optical-frequency comb at the 1-Hz level* IEEE Journal of Quantum Electronics **40**, 1023-1029 (2004)

[453] Ye.V. Baklanov, V.P. Chebotayev and B.Ya Dubetsky *The resonance of two-photon absorption in separated optical fields* Phys. Rev. A **60**, R753-R756 (1976)

[454] S.R. Lundeen, P.E. Jessop and F.M. Pipkin *Measurement of the hyperfine structure of $2^2 P_{1/2}$ state in hydrogen* Phys. Rev. Lett. **34**, 377 (1975)

[455] A. Shelkovnikov et al. *Stability of the Proton-to-Electron Mass Ratio* Phys. Rev. Lett. **100**, 150801 (2008)

[456] F. Riehle and J. Helmcke *Optical frequency standards based on neutral atoms and molecules* in *Frequency Measurement and Control* A.N. Luiten, Ed., Springer 95-129 (2001)

[457] A. Onae, K. Okomura, Y. Miki, T. Kurosawa, E. Sakuma, J. Yoda and K. Nakagawa *Saturation spectroscopy of an acetylene molecule in the 1550 nm region using an erbium doped fiber amplifier* Opt. Commun. **142**, 41-44 (1997)

[458] M. de Labachelerie, K. Nakagawa and M. Ohtsu *Ultranarrow $^{13}C_2H_2$ saturated-absorption lines at 1.5 μm* Opt. Lett. **19**, 840-842 (1994)

[459] J.L. Hall, C.J. Borde and K. Uehara *Direct optical resolution of the recoil effect using saturated absorption spectroscopy* Phys. Rev. Lett. **37**, 1339-1342 (1976)

[460] V. Bernard, C. Daussy, G. Nogues, L. Constantin, P. E. Durand, A. Amy-Klein, A. Van Lerberghe and C. Chardonnet *CO_2 laser stabilization to 0.1-Hz level using external electrooptic modulation* IEEE Journal of Quantum Electronics **33**, 1282 (1997)

[461] C. Chardonnet and C. J. Borde *Hyperfine interactions in the v_3 band of osmium tetroxide: accurate determination of the spin-rotation constant by crossover resonance spectroscopy* J. Mol. Spectrosc. **167**, 71-98 (1994)

[462] O. Acef *CO_2/OsO_4 lasers as frequency standards in the 29 THz range* IEEE Transactions on Instrumentation and Measurement **46**, 162 (1997)

[463] A. Godone, M. P. Sassi and E. Bava *High-accuracy capabilities of an OsO_4 molecular-beam frequency standard* Metrologia **26**, 1-8 (1989)

[464] Th. Udem et al. *Accuracy of optical frequency comb generators and optical frequency interval divider chains* Opt. Lett. **23**, 1387 (1998)

[465] C. Cohen-Tannoudji, J. Dupont-Roc and G. Grynberg *Atom-photon Interactions. Basic Processes and Applications* Wiley (1992)

[466] R. Grimm, M. Weidemüller and Y.B. Ovchinnikov *Optical dipole traps for neutral atoms* Advances in Atomic, Molecular, and Optical Physics **42**, 95-170 (2000)

[467] R. L. Barger *Influence of second-order Doppler effect on optical Ramsey fringe profiles* Opt. Lett. **6**, 145-147 (1981)

[468] P. Kersten, F. Mensing, U. Sterr and F. Riehle *A transportable optical calcium frequency standard* Appl. Phys. B **68**, 27-38 (1999)

[469] B. Mallick, A. Lakshmanna, V. Radhalakshmi and S. Umapathy *Design and development of stimulated Raman spectroscopy apparatus using a femtosecond laser system* Current Science **95**, 1551-1559 (2008)

[470] S. Nath, D.C. Urbanek, S.J. Kern and M.A. Berg *High-resolution Raman spectra with femtosecond pulses: an example of combined time- and frequency-domain spectroscopy* Phys. Rev. Lett. **97**, 267401 (2006)

[471] J. Mandon, G. Guelachvili and N. Picqué *Fourier transform spectroscopy with a laser frequency comb* Nature Photonics, **3(2)**, 99–102 (2009)

[472] J.B. Bates *Transform infrared spectroscopy* Science **191**, 31-37 (1976)

[473] http://infrared.phy.bnl.gov/pdf/homes/fir.pdf

[474] J.W. Cooley and J.W. Tukey *An algorithm for the machine computation of the complex Fourier series* Math. Computation **19**, 297-301 (1965)

[475] B.C. Smith *Fundamentals of Fourier Transform Infrared Spectroscopy (Second Edition)* CRC Press (2011)

[476] D.A. Long *The Raman Effect: A Unified Treatment of the Theory of Raman Scattering by Molecules* Wiley (2002)

[477] J.L. McHale *Molecular spectroscopy* Prentice Hall, Upper Saddle River (1998)

[478] W.M. Tolles, J.W. Nibler, J.R. McDonald and A.B. Harvey *A review of the theory and application of coherent Anti-Stokes Raman Spectroscopy (CARS)* Applied Spectroscopy **31**, 253-271 (1977)

[479] Fouad El-Diasty *Coherent anti-Stokes Raman scattering: spectroscopy and microscopy* Vibrational Spectroscopy **55**, 1-37 (2011)

[480] M.A. Henesian, M.D. Duncan, R.L. Byer and A.D. May *Absolute Raman frequency measurement of the $Q(2)$ line in D_2 using cw CARS* Opt. Lett. **1**, 149 (1977)

[481] A. Owyoung *Coherent Raman gain spectroscopy using cw laser sources* IEEE J. Quantum Electronics **QE14**, 192-203 (1978)

[482] A. Weber *High-resolution Raman Spectroscopy of Gases* in *Handbook of High-resolution Spectroscopy* Wiley (2011)

[483] C.A. Codemard et al. *High-power continuous-wave cladding-pumped Raman fiber laser* Opt. Lett. **31**, 2290 (2006)

[484] P. Dekker et al. *All-solid-state 704 mW continuous-wave yellow source based on intracavity, frequency-doubled crystalline Raman laser* Opt. Lett. **32**, 1114 (2007)

[485] O. Kitzler et al. *Continuous-wave wavelength conversion for high-power applications using an external cavity diamond Raman laser* Opt. Lett. **37**, 2790 (2012)

[486] H. Rong et al. *A cascaded silicon Raman laser* Nat. Photonics **2**, 170 (2008)

[487] T.J. Kippenberg et al. *Ultralow-threshold microcavity Raman laser on a microelectronic chip* Opt. Lett. **29**, 1224 (2004)

[488] F. Couny et al. *Subwatt threshold cw Raman fiber-gas laser based on H_2-filled hollow-core photonic crystal fiber* Phys. Rev. Lett. **99**, 143903 (2007)

[489] M. Bellini, A. Bartoli and T. W. Hänsch *Two-photon Fourier spectroscopy with femtosecond light pulses* Optics Letters, **22**, 540 (1997)

[490] J. N. Eckstein, A. I. Ferguson and T. W. Hänsch *High-resolution two-photon spectroscopy with picosecond light pulses* Phys. Rev. Lett. **40(13)**, 847–850 (1978)

[491] Ivan P. Christov, Margaret M. Murnane, Henry C. Kapteyn, Jianping Zhou and Chung P. Huang *Fourth-order dispersion-limited solitary pulses* Optics Letters, **19(18)**, 1465–1467 (1994)

[492] R. Szipocs, K. Ferencz, C. Spielmann and F. Krausz *Chirped multilayer coatings for broadband dispersion control in femtosecond lasers* Opt. Lett. **19**, 201 (1994)

[493] D. H. Sutter, G. Steinmeyer, L. Gallmann, N. Matuschek, F. Morier-Genoud, U. Keller, V. Scheuer, G. Angelow and T. Tschudi *Semiconductor saturable-absorber mirror-assisted kerr-lens mode-locked ti:sapphire laser producing pulses in the two-cycle regime* Optics Letters **24(9)**, 631–633 (1999)

[494] A. Bartels, D. Heinecke and S. A. Diddams *Passively mode-locked 10 GHz femtosecond ti:sapphire laser* Optics Letters **33(16)**, 1905–1907 (2008)

[495] J. Rauschenberger, T. M. Fortier, D. J. Jones, J. Ye and S. T. Cundiff *Control of the frequency comb from a modelocked erbium-doped fiber laser* Optics Express **10(24)**, 1404–1410 (2002)

[496] F. Tauser, A. Leitenstorfer and W. Zinth *Amplified femtosecond pulses from an Er:fiber system: nonlinear pulse shortening and self-referencing detection of the carrier-envelope phase evolution* Optics Express **11(6)**, 594–600 (2003)

[497] B. R. Washburn, S. A. Diddams, N. R. Newbury, J. W. Nicholson, M. F. Yan and C. G. Jörgensen *Phase-locked, erbium-fiber-laser-based frequency comb in the near infrared* Optics Letters **29(3)**, 250–252 (2004)

[498] T. R. Schibli, K. Minoshima, F. Hong, H. Inaba, A. Onae, H. Matsumoto, I. Hartl and M. E. Fermann *Frequency metrology with a turnkey all-fiber system* Optics Letters **29(21)**, 2467–2469 (2004)

[499] F. Adler, K. Moutzouris, A. Leitenstorfer, H. Schnatz, B. Lipphardt, G. Grosche and F. Tauser *Phase-locked two-branch erbium-doped fiber laser system for long-term precision measurements of optical frequencies* Optics Express **12**(24), 5872–5880 (2004)

[500] E. Desurvire, D. Bayart, B. Desthieux and S. Bigo *Erbium-Doped Fiber Amplifiers: Device and System Developments* Wiley, New York (2002)

[501] T. Okuno, M. Onishi, T. Kashiwada, S. Ishikawa and M. Nishimura *Silica-based functional fibers with enhanced nonlinearity and their applications* IEEE Journal on Selected Topics in Quantum Electronics, **5(5)**, 1385–1391 (1999)

[502] T. R. Schibli, I. Hartl, D. C. Yost, M. J. Martin, A. Marcinkeviius, M. E. Fermann and J. Ye *Optical frequency comb with submillihertz linewidth and more than 10 W average power* Nature Photonics **2(6)**, 355–359 (2008)

[503] I. Hartl, T. R. Schibli, A. Marcinkevicius, D. C. Yost, D. D. Hudson, M. E. Fermann and Jun Ye *Cavity-enhanced similariton Yb-fiber laser frequency comb:* $3 \cdot 10^{14}$ W/cm^2 *peak intensity at 136 MHz* Optics Letters **32**, 2870–2872 (2007)

[504] D. C. Yost, T. R. Schibli, J. Ye, J. L. Tate, J. Hostetter, M. B. Gaarde and K. J. Schafer *Vacuum-ultraviolet frequency combs from below-threshold harmonics* Nature Physics **5(11)**, 815–820 (2009)

[505] K. D. Moll, R. J. Jones and J. Ye *Output coupling methods for cavity-based high-harmonic generation* Optics Express **14(18)**, 8189–8197 (2006)

[506] Arman Cingoz, Dylan C. Yost, Thomas K. Allison, Axel Ruehl, Martin E. Fermann, Ingmar Hartl and Jun Ye *Direct frequency comb spectroscopy in the extreme ultraviolet* Nature **482**, 68–71 (2012)

[507] F. Adler, K. C. Cossel, M. J. Thorpe, I. Hartl, M. E. Fermann and J. Ye *Phase-stabilized, 1.5 W frequency comb at 2.8-4.8 μm* Optics Letters **34**(9), 1330–1332 (2009)

[508] E. P. Ippen, H. A. Haus and L. Y. Liu *Additive pulse mode locking* J. Opt. Soc. Am. B, **6(9)**, 1736–1745 (1989)

[509] K. Tamura, E. P. Ippen, H. A. Haus and L. E. Nelson *77-fs pulse generation from a stretched-pulse mode-locked all-fiber ring laser* Optics Letters **18(13)**, 1080–1082 (1993)

[510] J. Peng, H. Ahn, R. Shu, H. Chui and J. W. Nicholson *Highly stable, frequency-controlled mode-locked erbium fiber laser comb* Applied Physics B: Lasers and Optics, **86(1)**, 49–53 (2007)

[511] M. Bellini and T. W. Hänsch *Phase-locked white-light continuum pulses: toward a universal optical frequency-comb synthesizer* Optics Letters **25(14)**, 1049–1051 (2000)

[512] J. K. Ranka, R. S. Windeler and A. J. Stentz *Visible continuum generation in air-silica microstructure optical fibers with anomalous dispersion at 800 nm* Optics Letters **25(1)**, 25–27 (2000)

[513] T. A. Birks, W. J. Wadsworth and P. S. J. Russell *Supercontinuum generation in tapered fibers* Optics Letters **25(19)**, 1415–1417 (2000)

[514] D. J. Jones, S. A. Diddams, J. K. Ranka, A. Stentz, R. S. Windeler, J. L. Hall and S. T. Cundiff *Carrier-envelope phase control of femtosecond mode-locked lasers and direct optical frequency synthesis* Science **288(5466)**, 635–639 (2000)

[515] S. A. Diddams, D. J. Jones, J. Ye, S. T. Cundiff, J. L. Hall, J. K. Ranka, R. S. Windeler, R. Holzwarth, T. Udem and T. W. Hänsch *Direct link between microwave and optical frequencies with a 300 THz femtosecond laser comb* Physical Review Letters **84(22)**, 5102–5105 (2000)

[516] R. Holzwarth, Th Udem, T. W. Hänsch, J. C. Knight, W. J. Wadsworth and P. S. J. Russell *Optical frequency synthesizer for precision spectroscopy* Physical Review Letters **85(11)**, 2264–2267 (2000)

[517] Motonobu Kourogi, Ken'ichi Nakagawa and Motoichi Ohtsu *Wide-span optical frequency comb generator for accurate optical frequency difference measurement* IEEE Journal of Quantum Electronics **29(10)**, 2693–2701 (1993)

[518] Th. Udem, J. Reichert, R. Holzwarth and T. W. Hänsch *Accurate measurement of large optical frequency differences with a mode-locked laser* Optics Letters **24(13)**, 881–883 (1999)

[519] Th. Udem, J. Reichert, R. Holzwarth and T. W. Hänsch *Absolute optical frequency measurement of the cesium d1 line with a mode-locked laser* Physical Review Letters **82(18)**, 3568–3571 (1999)

[520] S. A. Diddams, D. J. Jones, L. Ma, S. T. Cundiff and J. L. Hall *Optical frequency measurement across a 104-THz gap with a femtosecond laser frequency comb* Optics Letters **25(3)**, 186–188 (2000)

[521] J. Reichert, M. Niering, R. Holzwarth, M. Weitz, T. Udem and T. W. Hänsch *Phase coherent vacuum-ultraviolet to radio frequency comparison with a mode-locked laser* Physical Review Letters **84(15)**, 3232–3235 (2000)

[522] J. Reichert, R. Holzwarth, T. Udem and T. W. Hänsch *Measuring the frequency of light with mode-locked lasers* Optics Communications **172(1)**, 59–68 (1999)

[523] Ye V. Baklanov and V. P. Chebotayev *Narrow resonances of two-photon absorption of super-narrow pulses in a gas* Applied Physics **12(1)**, 97–99 (1977)

[524] M. J. Snadden, A. S. Bell, E. Riis and A. I. Ferguson *Two-photon spectroscopy of laser-cooled Rb using a mode-locked laser* Optics Communications **125(1-3)**, 70–76 (1996)

[525] A. Marian, M. C. Stowe, J. R. Lawall, D. Felinto and J. Ye *United time-frequency spectroscopy for dynamics and global structure* Science **306(5704)**, 2063–2068 (2004)

[526] V. Gerginov, C. E. Tanner, S. A. Diddams, A. Bartels, and L. Hollberg *High-resolution spectroscopy with a femtosecond laser frequency comb* Optics Letters **30(13)**, 1734–1736 (2005)

[527] J. E. Stalnaker, V. Mbele, V. Gerginov, T. M. Fortier, S. A. Diddams, L. Hollberg and C. E. Tanner *Femtosecond frequency comb measurement of absolute frequencies and hyperfine coupling constants in cesium vapor* Physical Review A **81(4)**, 043840 (2010)

[528] A. Marian, M. C. Stowe, D. Felinto and J. Ye *Direct frequency comb measurements of absolute optical frequencies and population transfer dynamics* Physical Review Letters **95(2)**, 023001 (2005)

[529] V. A. Sautenkov, Y. V. Rostovtsev, C. Y. Ye, G. R. Welch, O. Kocharovskaya and M. O. Scully *Electromagnetically induced transparency in rubidium vapor prepared by a comb of short optical pulses* Physical Review A **71(6)**, 063804 (2005)

[530] L. Arissian and J. Diels *Repetition rate spectroscopy of the dark line resonance in rubidium* Optics Communications **264(1)**, 169–173 (2006)

[531] V. Gerginov, C. E. Tanner, S. Diddams, A. Bartels and L. Hollberg *Optical frequency measurements of 6s2s$_{1/2}$-6p 2p$_{3/2}$ transition in a ^{133}Cs atomic beam using a femtosecond laser frequency comb* Physical Review A **70(4)**, 042505 (2004)

[532] V. Gerginov, K. Calkins, C. E. Tanner, J. J. McFerran, S. Diddams, A. Bartels and L. Hollberg *Optical frequency measurements of 6s $^2S_{1/2}$-6p $^2P_{1/2}$ (D1) transitions in ^{133}Cs and their impact on the fine-structure constant* Physical Review A **73(3)**, 032504 (2006)

[533] T. M. Fortier, Y. Le Coq, J. E. Stalnaker, D. Ortega, S. A. Diddams, C. W. Oates and L. Hollberg *Kilohertz-resolution spectroscopy of cold atoms with an optical frequency comb* Physical Review Letters **97(16)**, 163905 (2006)

[534] F. C. Cruz, M. C. Stowe and J. Ye *Tapered semiconductor amplifiers for optical frequency combs in the near infrared* Optics Letters **31(9)**, 1337–1339 (2006)

[535] D. Aumiler, T. Ban, H. Skenderovic and G. Pichler *Velocity selective optical pumping of Rb hyperfine lines induced by a train of femtosecond pulses* Physical Review Letters **95(23)**, 23300 (2005)

[536] T. Ban, D. Aumiler, H. Skenderovic and G. Pichler *Mapping of the optical frequency comb to the atom-velocity comb* Physical Review A **73(4)**, 043407 (2006)

[537] P. Fendel, S. D. Bergeson, Th. Udem and T. W. Hänsch *Two-photon frequency comb spectroscopy of the 6s-8s transition in cesium* Optics Letters **32(6)**, 701–703 (2007)

[538] Vela Mbele, Jason E. Stalnaker, Vladislav Gerginov, Tara M. Fortier, Scott A. Diddams, Leo Hollberg and Carol E. Tanner *Direct two-photon resonant excitation and absolute frequency measurement of cesium transitions using a femtosecond comb* 2007 Digest of the IEEELEOS Summer Topical Meetings, pages 147–148 (2007)

[539] M. J. Thorpe, K. D. Moll, J. R. Jones, B. Safdi and J. Ye *Broadband cavity ringdown spectroscopy for sensitive and rapid molecular defection* Science **311(5767)**, 1595–1599 (2006)

[540] S. A. Diddams, L. Hollberg and V. Mbele *Molecular fingerprinting with the resolved modes of a femtosecond laser frequency comb* Nature **445(7128)**, 627–630 (2007)

[541] M. J. Thorpe and J. Ye *Cavity-enhanced direct frequency comb spectroscopy* Applied Physics B: Lasers and Optics **91(3-4)**, 397–414 (2008)

[542] M. J. Thorpe, D. Balslev-Clausen, M. S. Kirchner and J. Ye *Cavity-enhanced optical frequency comb spectroscopy: Application to human breath analysis* Optics Express **16(4)**, 2387–2397 (2008)

[543] C. Gohle, B. Stein, A. Schliesser, T. Udem and T. W. Hänsch *Frequency comb Vernier spectroscopy for broadband, high-resolution, high-sensitivity absorption and dispersion spectra* Physical Review Letters **99(26)**, 263902 (2007)

[544] P.R. Griffiths and J.A. De Haseth *Fourier Transform Infrared Spectrometry* Wiley, Hoboken (2007)

[545] K. A. Tillman, R. R. J. Maier, D. T. Reid and E. D. McNaghten *Mid-infrared absorption spectroscopy of methane using a broadband femtosecond optical parametric oscillator based on aperiodically poled lithium niobate* Journal of Optics A: Pure and Applied Optics **7(6)**, S408–S414, 2005.

[546] J. Mandon, G. Guelachvili, N. Picqué, F. Druon and P. Georges *Femtosecond laser Fourier transform absorption spectroscopy* Optics Letters **32(12)**, 1677–1679 (2007)

[547] S. Schiller *Spectrometry with frequency combs* Optics Letters **27(9)**, 766–768 (2002)

[548] F. Keilmann, C. Gohle, and R. Holzwarth *Time-domain mid-infrared frequency-comb spectrometer* Optics Letters **29(13)**, 1542–1544 (2004)

[549] I. Coddington, W. C. Swann and N. R. Newbury *Coherent multiheterodyne spectroscopy using stabilized optical frequency combs* Physical Review Letters **100**, 013902 (2008)

[550] B. Bernhardt, A. Ozawa, P. Jacquet, M. Jacquey, Y. Kobayashi, T. Udem, R. Holzwarth, G. Guelachvili, T. W. Hänsch and N. Picqué *Cavity-enhanced dual-comb spectroscopy* Nature Photonics **4(1)**, 55–57 (2010)

[551] A. Bartels, D. Heinecke and S. A. Diddams *10-GHz self-referenced optical frequency comb* Science **326(5953)**, 681 (2009)

[552] M. S. Kirchner, D. A. Braje, T. M. Fortier, A. M. Weiner, L. Hollberg and S. A. Diddams *Generation of 20 GHz, sub-40 fs pulses at 960 nm via repetition-rate multiplication* Optics Letters **34(7)**, 872–874 (2009)

[553] C. Li, A. J. Benedick, P. Fendel, A. G. Glenday, F. X. Kärtner, D. F. Phillips, D. Sasselov, A. Szentgyorgyi and R. L. Walsworth *A laser frequency comb that enables radial velocity measurements with a precision of 1 cm s^{-1}* Nature **452(7187)**, 610–612 (2008)

[554] T. Steinmetz, T. Wilken, C. Araujo-Hauck, R. Holzwarth, T. W. Hänsch, L. Pasquini, A. Manescau, S. D'Odorico, M. T. Murphy, T. Kentischer, W. Schmidt and T. Udem *Laser frequency combs for astronomical observations* Science **321(5894)**, 1335–1337 (2008)

[555] D. A. Braje, M. S. Kirchner, S. Osterman, T. Fortier and S. A. Diddams *Astronomical spectrograph calibration with broad-spectrum frequency combs* European Physical Journal D **48(1)**, 57–66 (2008)

[556] P. Del'Haye, A. Schliesser, O. Arcizet, T. Wilken, R. Holzwarth and T. J. Kippenberg *Optical frequency comb generation from a monolithic microresonator* Nature **450(7173)**, 1214–1217 (2007)

[557] A. A. Savchenkov, A. B. Matsko, V. S. Ilchenko, I. Solomatine, D. Seidel and L. Maleki *Tunable optical frequency comb with a crystalline whispering gallery mode resonator* Physical Review Letters **101(9)**, 093902 (2008)

[558] L. Razzari, D. Duchesne, M. Ferrera, R. Morandotti, S. Chu, B. E. Little and D. J. Moss *Cmos-compatible integrated optical hyper-parametric oscillator* Nature Photonics **4(1)**, 41–45 (2010)

[559] D. Braje, L. Hollberg and S. Diddams *Brillouin-enhanced hyperparametric generation of an optical frequency comb in a monolithic highly nonlinear fiber cavity pumped by a cw laser* Physical Review Letters **102(19)**, 193902 (2009)

[560] T. Wilken, C. Lovis, A. Manescau, T. Steinmetz, L. Pasquini, G. L. Curto, T. W. Hansch, R. Holzwarth and T. Udem *High-precision calibration of spectrographs* Monthly Notices of the Royal Astronomical Society: Letters **405(1)**, L16–L20 (2010)

[561] A. J. Benedick, G. Chang, J. R. Birge, L. Chen, A. G. Glenday, C.Li, D. F. Phillips, A. Szentgyorgyi, S. Korzennik, G. Furesz, R. L. Walsworth and F. X. Kärtner *Visible wavelength astro-comb* Optics Express **18(18)**, 19175–19184 (2010)

[562] F. Quinlan, G. Ycas, S. Osterman and S. A. Diddams *A 12.5 GHz-spaced optical frequency comb spanning >400 nm for near-infrared astronomical spectrograph calibration* Review of Scientific Instruments **81(6)**, 063105 (2010)

[563] I. Coddington, W. C. Swann, L. Lorini, J. C. Bergquist, Y. Le Coq, C. W. Oates, Q. Quraishi, K. S. Feder, J. W. Nicholson, P. S. Westbrook, S. A. Diddams and N. R. Newbury *Coherent optical link over hundreds of metres and hundreds of terahertz with subfemtosecond timing jitter* Nature Photonics **1(5)**, 283–287 (2007)

[564] N. R. Newbury, P. A. Williams and W. C. Swann *Coherent transfer of an optical carrier over 251 km* Optics Letters **32(21)**, 3056–3058 (2007)

[565] J. Kim, J. A. Cox, J. Chen and F. X. Kärtner *Drift-free femtosecond timing synchronization of remote optical and microwave sources* Nature Photonics **2(12)**, 733–736 (2008)

[566] J. J. McFerran, E. N. Ivanov, A. Bartels, G. Wilpers, C. W. Oates, S. A. Diddams and L. Hollberg *Low-noise synthesis of microwave signals from an optical source* Electronics Letters **41(11)**, 650–651 (2005)

[567] A. Bartels, S. A. Diddams, C. W. Oates, G. Wilpers, J. C. Bergquist, W. H. Oskay and L. Hollberg *Femtosecond-laser-based synthesis of ultrastable microwave signals from optical frequency references* Optics Letters **30(6)**, 667–669 (2005)

[568] Z. Jiang, C. Huang, D. E. Leaird and A. M. Weiner *Optical arbitrary waveform processing of more than 100 spectral comb lines* Nature Photonics **1(8)**, 463–467 (2007)

[569] N. K. Fontaine, R. P. Scott, J. Cao, A. Karalar, W. Jiang, K. Okamoto, J. P. Heritage, B. H. Kolner and S. J. B. Yoo *32 phase x 32 amplitude optical arbitrary waveform generation* Optics Letters **32(7)**, 865–867 (2007)

[570] S. T. Cundiff and A. M. Weiner *Optical arbitrary waveform generation* Nature Photonics **4(11)**, 760–766 (2010)

[571] J. Jin, Y. Kim, Y. Kim, S. Kim and C. Kang *Absolute length calibration of gauge blocks using optical comb of a femtosecond pulse laser* Optics Express **14(13)**, 139–145 (2006)

[572] N. Schuhler, Y. Salvadé, S. Lévêque, R. Dändliker and R. Holzwarth *Frequency-comb-referenced two-wavelength source for absolute distance measurement* Optics Letters **31(21)**, 3101–3103 (2006)

[573] J. Ye *Absolute measurement of a long, arbitrary distance to less than an optical fringe* Optics Letters **29(10)**, 1153–1155 (2004)

[574] K. Joo, Y. Kim, and S. Kim *Distance measurements by combined method based on a femtosecond pulse laser* Optics Express **16(24)**, 19799–19806 (2008)

[575] M. Cui, M. G. Zeitouny, N. Bhattacharya, S. A. Van Den Berg, H. P. Urbach and J. J. M. Braat *High-accuracy long-distance measurements in air with a frequency comb laser* Optics Letters **34(13)**, 1982-1984 (2009)

[576] I. Coddington, W. C. Swann, L. Nenadovic and N. R. Newbury *Rapid and precise absolute distance measurements at long range* Nature Photonics **3(6)**, 351–356 (2009)

[577] J. Lee, Y. Kim, K. Lee, S. Lee, and S. Kim *Time-of-flight measurement with femtosecond light pulses* Nature Photonics **4(10)**, 716–720 (2010)

[578] R. Ell, U. Morgner, F. X. Kärtner, J. G. Fujimoto, E. P. Ippen, V. Scheuer, G. Angelow, T. Tschudi, M. J. Lederer, A. Boiko and B. Luther-Davies *Generation of 5-fs pulses and octave-spanning spectra directly from a ti:sapphire laser* Optics Letters **26(6)**, 373–375 (2001)

[579] T. M. Fortier, D. J. Jones and S. T. Cundiff *Phase stabilization of an octave-spanning ti:sapphire laser* Optics Letters **28(22)**, 2198–2200 (2003)

[580] T. M. Fortier, A. Bartels and S. A. Diddams *Octave-spanning ti:sapphire laser with a repetition rate >1 GHz for optical frequency measurements and comparisons* Optics Letters **31(7)**, 1011–1013 (2006)

[581] H. A. Haus and Antonio Mecozzi *Noise of mode-locked lasers* IEEE Journal of Quantum Electronics **29(3)**, 983–996 (1993)

[582] N. R. Newbury and W. C. Swann *Low-noise fiber-laser frequency combs (invited)* Journal of the Optical Society of America B: Optical Physics **24(8)**, 1756–1770 (2007)

[583] R. P. Scott, T. D. Mulder, K. A. Baker and B. H. Kolner *Amplitude and phase noise sensitivity of modelocked ti:sapphire lasers in terms of a complex noise transfer function* Optics Express **15(14)**, 9090–9095 (2007)

[584] N. R. Newbury, B. R. Washburn, K. L. Corwin and R. S. Windeler *Noise amplification during supercontinuum generation in microstructure fiber* Optics Letters **28(11)**, 944–946 (2003)

[585] K. L. Corwin, N. R. Newbury, J. M. Dudley, S. Coen, S. A. Diddams, K. Weber and R. S. Windeler *Fundamental noise limitations to supercontinuum generation in microstructure fiber* Physical Review Letters **90(11)**, 113904 (2003)

[586] J. N. Ames, S. Ghosh, R. S. Windeler, A. L. Gaeta, and S. T. Cundiff *Excess noise generation during spectral broadening in a microstructured fiber* Applied Physics B: Lasers and Optics **77**(2-3), 279–284 (2003)

[587] D. C. Heinecke, A. Bartels, T. M. Fortier, D. A. Braje, L. Hollberg and S. A. Diddams *Optical frequency stabilization of a 10 GHz ti:sapphire frequency comb by saturated absorption spectroscopy in 87 rubidium* Physical Review A **80**(5), 053806 (2009)

[588] T. J. Kippenberg, S. M. Spillane, and K. J. Vahala *Kerr-nonlinearity optical parametric oscillation in an ultrahigh-Q toroid microcavity* Physical Review Letters **93**(8), 083904-1–083904-4 (2004)

[589] D. K. Armani, T. J. Kippenberg, S. M. Spillane and K. J. Vahala *Ultra-high-Q toroid microcavity on a chip* Nature **421**(6926), 925–928 (2003)

[590] P. Del'Haye, T. Herr, E. Gavartin, M. L. Gorodetsky, R. Holzwarth and T. J. Kippenberg *Octave spanning tunable frequency comb from a microresonator* Physical Review Letters **107**(6), 063901 (2011)

[591] P. Del'Haye, O. Arcizet, A. Schliesser, R. Holzwarth and T. J. Kippenberg *Full stabilization of a microresonator-based optical frequency comb* Physical Review Letters **101**(5), 053903 (2008)

[592] A. A. Savchenkov, A. B. Matsko, D. Strekalov, M. Mohageg, V. S. Ilchenko and L. Maleki *Low threshold optical oscillations in a whispering gallery mode CaF_2 resonator* Physical Review Letters **93**(24), 243905 (2004)

[593] J. S. Levy, A. Gondarenko, M. A. Foster, A. C. Turner-Foster, A. L. Gaeta and M. Lipson *Cmos-compatible multiple-wavelength oscillator for on-chip optical interconnects* Nature Photonics **4**(1), 37–40 (2010)

[594] M. A. Foster, J. S. Levy, O. Kuzucu, K. Saha, M. Lipson and A. L. Gaeta *Silicon-based monolithic optical frequency comb source* Optics Express **19**(15), 14233–14239 (2011)

[595] A. McPherson, G. Gibson, H. Jara, U. Johann, T. S. Luk, I. A. McIntyre, K. Boyer and C. K. Rhodes *Studies of multiphoton production of vacuum-ultraviolet radiation in the rare gases* J. Opt. Soc. Am. B **4**(4), 595–601, Apr 1987.

[596] M. Ferray, A. L'Huillier, X. F. Li, L. A. Lompre, G. Mainfray and C. Manus *Multiple-harmonic conversion of 1064 nm radiation in rare gases* Journal of Physics B: Atomic, Molecular and Optical Physics **21**(3), L31–L35 (1988)

[597] P. Salieres, A. L'Huillier, P. Antoine and M. Lewenstein *Study of the spatial and temporal coherence of high-order harmonics* Adv. At., Mol., Opt. Phys. **41**(C), 83–142 (1999)

[598] P. B. Corkum *Plasma perspective on strong field multiphoton ionization* Physical Review Letters **71**(13), 1994–1997 (1993)

[599] M. Bellini, C. Lynga, A. Tozzi, M. B. Gaarde, T. W. Hänsch, A. L'Huillier and C. Wahlström *Temporal coherence of ultrashort high-order harmonic pulses* Physical Review Letters **81**(2), 297–300 (1998)

[600] C. Lynga, M. B. Gaarde, C. Delfin, M. Bellini, T. W. Hänsch, A. L'Huillier and C. Wahlström *Temporal coherence of high-order harmonics* Physical Review A - Atomic, Molecular, and Optical Physics **60(6)**, 4823–4830 (1999)

[601] M. Herrmann, M. Haas, U. D. Jentschura, F. Kottmann, D. Leibfried, G. Saathoff, C. Gohle, A. Ozawa, V. Batteiger, S. Knünz, N. Kolachevsky, H. A. Schüssler, T. W. Hänsch and Th Udem *Feasibility of coherent XUV spectroscopy on the 1s-2s transition in singly ionized helium* Physical Review A - Atomic, Molecular, and Optical Physics **79(5)**, 052505 (2009)

[602] P. A. Heimann, M. Koike, C. W. Hsu, D. Blank, X. M. Yang, A. G. Suits, Y. T. Lee, M. Evans, C. Y. Ng, C. Flaim and H. A. Padmore *Performance of the vacuum ultraviolet high-resolution and high-flux beamline for chemical dynamics studies at the advanced light source* Review of Scientific Instruments **68(5)**, 1945–1951 (1997)

[603] K. S. E. Eikema, J. Walz, and T. W. Hänsch *Continuous coherent Lyman-α excitation of atomic hydrogen* Physical Review Letters **86(25)**, 5679–5682 (2001)

[604] C. Lynga, F. Ossler, T. Metz and J. Larsson *A laser system providing tunable, narrow-band radiation from 35 nm to 2 μm* Applied Physics B: Lasers and Optics **72(8)**, 913–920 (2001)

[605] W. Ubachs, K. S. E. Eikema, W. Hogervorst and P. C. Cacciani *Narrow-band tunable extreme-ultraviolet laser source for lifetime measurements and precision spectroscopy* Journal of the Optical Society of America B: Optical Physics **14(10)**, 2469–2476 (1997)

[606] F. Brandi, D. Neshev and W. Ubachs *High-order harmonic generation yielding tunable extreme-ultraviolet radiation of high spectral purity* Physical Review Letters **91(16)**, 1639011–1639014 (2003)

[607] M. Bellini and T. W. Hänsch *Measurement of the temporal coherence of ultrashort harmonic pulses: towards coherent spectroscopy in the extreme ultraviolet* Applied Physics B: Lasers and Optics **65(4-5)**, 677–680 (1997)

[608] R. Zerne, C. Altucci, M. Bellini, M. B. Gaarde, T. W. Hansen, A. L'Huillier, C. Lynga and C. Wahlström *Phase-locked high-order harmonic sources* Physical Review Letters **79(6)**, 1006-1009 (1997)

[609] D. Descamps, C. Lynga, J. Norin, A. L'Huillier, C. Wahlström, J. Hergott, H. Merdji, P. Salieres, M. Bellini and T. W. Hänsch *Extreme ultraviolet interferometry measurements with high-order harmonics* Optics Letters **25(2)**, 135–137 (2000)

[610] M. Bellini, S. Cavalieri, C. Corsi and M. Materazzi *Phase-locked, time-delayed harmonic pulses for high spectral resolution in the extreme ultraviolet* Optics Letters **26(13)**, 1010–1012 (2001)

[611] P. Salieres, L. Le Déroff, T. Auguste, P. Monot, P. D'Oliveira, D. Campo, J. Hergott, H. Merdji and B. Carré *Frequency-domain interferometry in the XUV with high-order harmonics* Physical Review Letters **83(26 I)**, 5483–5486 (1999)

[612] M. M. Salour *Quantum interference effects in two-photon spectroscopy* Reviews of Modern Physics **50(3)**, 667–681 (1978)

[613] John T. Fourkas, William L. Wilson, G. Wäckerle, Amy E. Frost, and M. D. Fayer *Picosecond time-scale phase-related optical pulses: measurement of sodium optical coherence decay by observation of incoherent fluorescence* J. Opt. Soc. Am. B **6(10)**, 1905–1910 (1989).

[614] R. Van Leeuwen, M. L. Bajema and R. R. Jones *Coherent control of the energy and angular distribution of autoionized electrons* Physical Review Letters **82(14)**, 2852–2855 (1999)

[615] M. Kovacev, S. V. Fomichev, E. Priori, Y. Mairesse, H. Merdji, P. Monchicourt, P. Breger, J. Norin, A. Persson, A. L'Huillier, C. Wahlström, B. Carré and P. Saliéres *Extreme ultraviolet Fourier-transform spectroscopy with high order harmonics* Physical Review Letters **95(22)**, 1–4 (2005)

[616] S. Cavalieri, R. Eramo, M. Materazzi, C. Corsi and M. Bellini *Ramsey-type spectroscopy with high-order harmonics* Physical Review Letters **89(13)**, 1330021–1330024 (2002)

[617] A. Pirri, E. Sali, C. Corsi, M. Bellini, S. Cavalieri and R. Eramo *Extreme-ultraviolet Ramsey-type spectroscopy* Physical Review A - Atomic, Molecular, and Optical Physics **78**, 043410 (2008)

[618] S. Cavalieri and R. Eramo *Time-delay spectroscopy of autoionizing resonances* Physical Review A - Atomic, Molecular, and Optical Physics **58(6)**, R4263–R4266 (1998)

[619] I. Liontos, S. Cavalieri, C. Corsi, R. Eramo, S. Kaziannis, A. Pirri, E. Sali and M. Bellini *Ramsey spectroscopy of bound atomic states with extreme-ultraviolet laser harmonics* Optics Letters **35(6)**, 832–834 (2010)

[620] I. Liontos, C. Corsi, S. Cavalieri, M. Bellini and R. Eramo *Split-pulse spectrometer for absolute XUV frequency measurements* Optics Letters **36(11)**, 2047–2049 (2011)

[621] R. Eramo, M. Bellini, C. Corsi, I. Liontos and S. Cavalieri *Improving Ramsey spectroscopy in the extreme-ultraviolet region with a random-sampling approach* Physical Review A - Atomic, Molecular, and Optical Physics **83(4)**, 041402 (2011)

[622] R. Eramo, S. Cavalieri, C. Corsi, I. Liontos and M. Bellini *Method for high-resolution frequency measurements in the extreme ultraviolet regime: Random-sampling Ramsey spectroscopy* Physical Review Letters **106(21)**, 213003 (2011)

[623] S. Witte, R. T. Zinkstok, W. Ubachs, W. Hogervorst and K. S. E. Eikema *Deep-ultraviolet quantum interference metrology with ultrashort laser pulses* Science **307(5708)**, 400–403 (2005)

[624] R.T. Zinkstok, S. Witte, W. Ubachs, W. Hogervorst and K.S.E. Eikema *Frequency comb laser spectroscopy in the vacuum-ultraviolet region* Physical Review A - Atomic, Molecular, and Optical Physics **73(6)**, 061801 (2006)

[625] D. Z. Kandula, C. Gohle, T. J. Pinkert, W. Ubachs and K. S. E. Eikema *Extreme ultraviolet frequency comb metrology* Physical Review Letters **105(6)**, 063001 (2010)

[626] C. Gohle, T. Udem, M. Herrmann, J. Rauschenberger, R. Holzwarth, H. A. Schuessler, F. Krausz and T. W. Hänsen *A frequency comb in the extreme ultraviolet* Nature **436(7048)**, 234–237 (2005)

[627] R. J. Jones, K. D. Moll, M. J. Thorpe and J. Ye *Phase-coherent frequency combs in the vacuum ultraviolet via high-harmonic generation inside a femtosecond enhancement cavity* Physical Review Letters **94(19)**, 193201 (2005)

[628] A. Ozawa, J. Rauschenberger, Ch Gohle, M. Herrmann, D. R. Walker, V. Pervak, A. Fernandez, R. Graf, A. Apolonski, R. Holzwarth, F. Krausz, T. W. Hänsch and Th Udem *High harmonic frequency combs for high resolution spectroscopy* Physical Review Letters **100(25)**, 253901 (2008)

[629] T. Brabec and F. Krausz *Intense few-cycle laser fields: Frontiers of nonlinear optics* Reviews of Modern Physics **72(2)**, 545–591 (2000)

[630] G. G. Paulus, F. Grasbon, H. Walther, P. Villoresi, M. Nisoli, S. Stagira, E. Priori and S. De Silvestri *Absolute-phase phenomena in photoionization with few-cycle laser pulses* Nature **414(6860)**, 182–184 (2001)

[631] A. Apolonski, A. Poppe, G. Tempea, Ch Spielmann, Th Udem, R. Holzwarth, T. W. Hänsch, and F. Krausz *Controlling the phase evolution of few-cycle light pulses* Physical Review Letters **85**(4), 740–743 (2000)

[632] R. Kienberger, E. Goulielmakis, M. Uiberacker, A. Baltuska, V. Yakovlev, F. Bammer, A. Scrinzi, Th. Westerwalbesioh, U. Kleineberg, U. Heinzmann, M. Drescher and F. Krausz *Atomic transient recorder* Nature **427(6977)**, 817–821 (2004)

[633] A. Baltuska, Th Udem, M. Uiberacker, M. Hentschel, E. Goulielmakis, Ch Gohle, R. Holzwarth, V. S. Yakovlev, A. Scrinzi, T. W. Hänsch and F. Krausz *Attosecond control of electronic processes by intense light fields* Nature **421(6923)**, 611–615 (2003)

[634] E. Goulielmakis, M. Uiberacker, R. Kienberger, A. Baltuska, V. Yakovlev, A. Scrinzi, Th Westerwalbesloh, U. Kleineberg, U. Heinzmann, M. Drescher and F. Krausz *Direct measurement of light waves* Science **305(5688)**, 1267–1269 (2004)

[635] I. Thomann, A. Bartels, K. L. Corwin, N. R. Newbury, L. Hollberg, S. A. Diddams, J. W. Nicholson and M. F. Yan *420-MHz Cr:forsterite femtosecond ring laser and continuum generation in the 1-2-μm range* Optics Letters **28(15)**, 1368–1370 (2003)

[636] P. De Natale, S. Borri, P. Cancio, G. Giusfredi, D. Mazzotti, M. Prevedelli, C. De Mauro and M. Inguscio *Extending the optical comb synthesizer to the infrared: From He at 1.083 μm to CO_2 at 4.2 μm* Laser Spectroscopy-Proc. 16th Int. Conf. (2004)

[637] D. Mazzotti, P. De Natale, G. Giusfredi, C. Fort, J. A. Mitchell and L. Hollberg *Saturated-absorption spectroscopy with low-power difference-frequency radiation* Optics Letters **25(5)**, 350–352 (2000)

[638] D. Mazzotti, S. Borri, P. Cancio, G. Giusfredi and P. De Natale *Low-power lamb-dip spectroscopy of very weak CO_2 transitions near 4.25 μm* Optics Letters **27(14)**, 1256–1258 (2002)

[639] S. M. Foreman, A. Marian, J. Ye, E. A. Petrukhin, M. A. Gubin, O. D. Mücke, F. N. C. Wong, E. P. Ippen and F. X. Kärtner *Demonstration of a $HeNe/CH_4$-- based optical molecular clock* Optics Letters **30(5)**, 570–572 (2005)

[640] J. H. Sun, B. J. S. Gale and D. T. Reid *Composite frequency comb spanning 0.4-2.4 μm from a phase-controlled femtosecond ti:sapphire laser and synchronously pumped optical parametric oscillator* Optics Letters **32(11)**, 1414–1416 (2007)

[641] P. Maddaloni, P. Malara, G. Gagliardi and P. De Natale *Mid-infrared fibre-based optical comb* New Journal of Physics **8**, 262 (2006)

[642] P. Malara, P. Maddaloni, G. Gagliardi and P. De Natale *Absolute frequency measurement of molecular transitions by a direct link to a comb generated around 3-μm* Optics Express **16(11)**, 8242–8249 (2008)

[643] S. Bartalini, P. Cancio, G. Giusfredi, D. Mazzotti, P. De Natale, S. Borri, I. Galli, T. Leveque and L. Gianfrani *Frequency-comb-referenced quantum-cascade laser at 4.4 μm* Optics Letters **32(8)**, 988–990 (2007)

[644] O. D. Mücke, O. Kuzucu, F. N. C. Wong, E. P. Ippen, F. X. Kärtner, S. M. Foreman, D. J. Jones, L. Ma, J. L. Hall, and J. Ye *Experimental implementation of optical clockwork without carrier-envelope phase control* Optics Letters **29(23)**, 2806–2808 (2004)

[645] A. Amy-Klein, A. Goncharov, C. Daussy, C. Grain, O. Lopez, G. Santarelli, and C. Chardonnet *Absolute frequency measurement in the 28-THz spectral region with a femtosecond laser comb and a long-distance optical link to a primary standard* Applied Physics B: Lasers and Optics **78**(1), 25–30 (2004)

[646] A. Amy-Klein, A. Goncharov, M. Guinet, C. Daussy, O. Lopez, A. Shelkovnikov and C. Chardonnet *Absolute frequency measurement of a SF_6 two-photon line by use of a femtosecond optical comb and sum-frequency generation* Optics Letters **30**(24), 3320–3322 (2005)

[647] T. Yasui, H. Takahashi, K. Kawamoto, Y. Iwamoto, K. Arai, T. Araki, H. Inaba and K. Minoshima *Widely and continuously tunable terahertz synthesizer traceable to a microwave frequency standard* Optics Express **19**, 4428–4437 (2011)

[648] M. Ravaro, C. Manquest, C. Sirtori, S. Barbieri, G. Santarelli, K. Blary, J.-F. Lampin, S.P. Khanna and E.H. Linfield *Phase-locking of a 2.5-THz quantum cascade laser to a frequency comb using a GaAs photomixer* Opt. Lett. **36**, 3969–3971 (2011)

[649] S. Barbieri, P. Gellie, G. Santarelli, L. Ding, W. Maineult, C. Sirtori, R. Colombelli, H. Beere and D. Ritchie *Phase-locking of a 2.7-THz quantum cascade laser to a mode-locked erbium-doped fibre laser* Nat. Photon. **4**, 636–640 (2010)

[650] J. Levine *Introduction to time and frequency metrology* Rev. Sci. Instrum. **70**, 2567–2596 (1999)

[651] W. Zhou *Time, Frequency Measurement and Control Technology* Xidian University Press (2006)

[652] E. Rubiola and V. Giordano *On the 1/f frequency noise in ultrastable quartz oscillators* IEEE Transactions on Ultrasonics, Ferroelectrics, and Frequency Control **54**, 15-22 (2007)

[653] A.G. Mann *Ultrastable Cryogenic Microwave Oscillators* in *Frequency Measurement and Control* A.N. Luiten, Ed., Springer-Verlag (2001)

[654] E.N. Ivanov and M.E. Tobar *Low phase-noise sapphire crystal microwave oscillators: current status* IEEE Transactions on Ultrasonics, Ferroelectrics, and Frequency Control **56**, 263 (2009)

[655] C. McNeilage, J.H. Searls, E.N. Ivanov, P.R. Stockwell, D.M. Green and M. Mossamaparast *A review of sapphire whispering gallery-mode oscillators including technical progress and future potential of the technology* in *Proceeding of: Frequency Control Symposium and Exposition* (2004)

[656] C.R. Locke, E.N. Ivanov, J.G. Hartnett, P.L. Stanwix and M.E. Tobar *Design techniques and noise properties of ultrastable cryogenically cooled sapphire-dielectric resonator oscillators* Rev. Sci. Instrum. **79**, 051301 (2008)

[657] R.C. Taber and C.A. Flory *Microwave Oscillators Incorporating Cryogenic Sapphire Dielectric Resonators* IEEE transactions on ultrasonics, ferroelectrics, and frequency control **42**, 111-119 (1995)

[658] J.G. Hartnett and M.E. Tobar *Frequency-temperature compensation techniques for high-Q microwave resonators* in *Frequency Measurement and Control* A.N. Luiten, Ed., Springer-Verlag (2001)

[659] V. Giordano, P.Y. Bourgeois, Y. Gruson, N. Boubekeur, R. Boudot, E. Rubiola, N. Bazin and Y. Kersale *New advances in ultrastable microwave oscillators* Eur. Phys. J. Appl. Phys **32**, 133-141 (2005)

[660] S. Grop, P.-Y. Bourgeois, R. Boudot, Y. Kersale, E. Rubiola and V. Giordano *10 GHz cryocooled sapphire oscillator with extremely low phase noise* Electronics Letters **46**, 420 (2010)

[661] A.B. Matsko, A.A. Savchenkov, N. Yu and L. Maleki *Whispering-gallery-mode resonators as frequency references. I. Fundamental limitations* J. Opt. Soc. Am. B **24**, 1324-1335 (2007)

[662] A.A. Savchenkov, A.B. Matsko, V.S. Ilchenko, N. Yu and L. Maleki *Whispering-gallery-mode resonators as frequency references. II. Stabilization* J. Opt. Soc. Am. B **24**, 2988-2997 (2007)

[663] A.A. Savchenkov, E. Rubiola, A.B. Matsko, V.S. Ilchenko and L. Maleki *Phase noise of whispering gallery photonic hyper-parametric microwave oscillators* Optics Express **16**, 4130 (2008)

[664] T.M. Fortier, M.S. Kirchner, F. Quinlan, J. Taylor, J.C. Bergquist, T. Rosenband, N. Lemke, A. Ludlow, Y. Jiang, C.W. Oates and S.A. Diddams *Generation of ultrastable microwaves via optical frequency division* Nature Photonics **5**, 425-429 (2011)

[665] G. M. Tino and M. Inguscio *Experiments on Bose-Einstein condensation* Rivista del Nuovo Cimento **22**, N.4 (1999)

[666] W.D. Phillips *Laser cooling and trapping of neutral atoms* Reviews of Modern Physics **70**, 721-741 (1998)

[667] C.N. Cohen-Tannoudji *Manipulating atoms with photons* Reviews of Modern Physics **70**, 707-719 (1998)

[668] W. Ketterle, D.S. Durfee, and D.M. Stamper-Kurn *Making, probing and understanding Bose-Einstein condensates* in Bose-Einstein condensation in atomic gases, Proceedings of the International School of Physics "Enrico Fermi," Course CXL, M. Inguscio, S. Stringari and C.E. Wieman, Eds., IOS Press, Amsterdam (1999)

[669] C.S. Adams and E. Riis *Laser cooling and trapping of neutral atoms* Prog. Quant. Electr. **21**, 1-79 (1997)

[670] T.W. Hänsch and A.L. Schawlow *Cooling of gases by laser radiation* Opt. Commun. **13**, 68 (1975)

[671] S. Chu, L. Hollberg, J.E. Bjorkholm, A. Cable and A. Ashkin *Three-dimensional viscous confinement and cooling of atoms by resonance radiation pressure* Phys. Rev. Lett. **55**, 48 (1985)

[672] P.D. Lett, R.N. Watts, C.I. Westbrook, W.D. Phillips, P.L. Gould and H.J. Metcalf *Observation of atoms laser cooled below the Doppler limit* Phys. Rev. Lett. **61**, 169 (1988)

[673] J. Dalibard and C. Cohen-Tannoudji *Laser cooling below the Doppler limit by polarization gradients: simple theoretical models* J. Opt. Soc. Am. **6**, 2023 (1989)

[674] K.J. Arnold and M.D. Barrett *All-optical Bose-Einstein condensation in a 1.06 μm dipole trap* Optics Communications **284**, 3288-3291 (2011)

[675] M. Inguscio, G. Modugno and G. Roati *Fermi-Bose and Bose-Bose K-Rb quantum degenerate mixtures* Laser Physics **13**, 401-406 (2003)

[676] W. Ketterle and N.J. Van Druten *Evaporative cooling of trapped atoms* Advances in Atomic, Molecular and Optical Physics **37**, 181-236 (1996)

[677] E. Nugent *Novel traps for Bose-Einstein condensates* PhD thesis, University of Oxford (2009)

[678] Y.N. Martinez de Escobar *Bose-Einstein condensation of ^{84}Sr* PhD thesis, Rice University (2010)

[679] L. You and M. Holland *Ballistic expansion of trapped thermal atoms* Phys. Rev. A **53**, R1-R4 (1996)

[680] F. Dalfovo, S. Giorgini, L.P. Pitaevskii and S. Stringari *Theory of Bose-Einstein condensation in trapped gases* Rev. Mod. Phys. **71**, 463-512 (1999)

[681] C. Chin, R. Grimm, P. Julienne and E. Tiesinga *Feshbach resonances in ultracold gases* Rev. Mod. Phys. **82**, 1225-1286 (2010)

[682] Y. Takahashi, Y. Takasu, K. Maki, K. Komori, T. Takano, K. Honda, A. Yamaguchi, Y. Kato, M. Mizoguchi, M. Kumakura and T. Yabuzaki *Bose-Einstein condensation of ytterbium atoms* Laser Physics **14**, 621-623 (2004)

[683] W. Hänsel, P. Hommelhoff, T.W. Hänsch and J. Reichel *Bose-Einstein condensation on a microelectronic chip* Nature **413**, 498-501 (2011)

[684] www.bec2011.ethz.ch

[685] J. Klaers, J. Schmitt, F. Vewinger and M. Weitz *Bose-Einstein condensation of photons in an optical microcavity* Nature **468**, 545-548 (2010)

[686] A. Micheli et al. *A toolbox for lattice-spin models with polar molecules* Nature Phys. **2**, 341 (2006)

[687] L.D. Carr, D. DeMille, R.V. Krems and J. Ye *Cold and ultracold molecules: science, technology and applications* New Journal of Physics **11**, 055049 (2009)

[688] M.T. Bell and T.P. Softley *Ultracold molecules and ultracold chemistry* Mol. Phys. **107**, 99-132 (2009)

[689] R. Fulton et al. *Controlling the motion of cold molecules with deep periodic optical potential* Nature Phys. **2**, 465 (2006)

[690] A. Marian, H. Haak, P. Geng and G. Meijer *Slowing polar molecules using a wire Stark decelerator* Eur. Phys. J. D **59**, 179-181 (2010)

[691] S.A. Meek, H. Conrad and G. Meijer *A Stark decelerator on a chip* New Journal of Physics **11**, 055024 (2009)

[692] A. Osterwalder et al. *Deceleration of neutral molecules in macroscopic travelling traps* Phys. Rev. A **81**, 051401 (2010)

[693] B.C. Sawyer et al. *Magnetoelectrostatic trapping of ground state OH molecules* Phys. Rev. Lett. **98**, 253002 (2007)

[694] B.L. Lev et al. *Prospects for the cavity-assisted laser cooling of molecules* Phys. Rev. A **77**, 023402 (2008)

[695] B.K. Stuhl et al. *Magneto-optical trap for polar molecules* Phys. Rev. Lett. **101**, 243002 (2008)

[696] M. Zeppenfeld et al. *Opto-electrical cooling of polar molecules* Phys. Rev. A **80**, 041401(R) (2009)

[697] G. Modugno et al. *Bose-Einstein Condensation of Potassium Atoms by Sympathetic Cooling* Science **294**, 1320 (2001)

[698] W. Paul *Electromagnetic traps for charged and neutral particles* Rev. Mod. Phys. **62**, 531 (1990)

[699] J.L.K. Koo *Laser cooling and trapping of Ca^+ ions in a Penning trap* PhD thesis, Imperial College London (2003)

[700] A.A. Madej and J.E. Bernard *Single-Ion Optical Frequency Standards and Measurement of their Absolute Optical Frequency* in Frequency Measurement and Control. Advanced Techniques and Future Trends, A.N. Luiten, Ed., Springer, pp. 249-272 (2001)

[701] D. Leibfried, R. Blatt, C. Monroe and D. Wineland *Quantum dynamics of single trapped ions* Reviews of Modern Physics **75**, 281 (2003)

[702] D.J. Berkeland, J.D. Miller, J.C. Bergquist, W.M. Itano and D.J. Wineland *Minimization of ion micromotion in a Paul trap* Journal of Applied Physics **83**, 5025 (1998)

[703] L.S. Brown and G. Gabrielse *Geonium theory: physics of a single electron or ion in a Penning trap* Rev. Mod. Phys. **58**, 233 (1986)

[704] http://www.mpi-hd.mpg.de/blaum/

[705] G. Werth, V.N. Gheorghe and F.G. Major *Charged Particle Traps. II Applications* Springer (2009)

[706] F. Diedrich, J.C. Bergquist, W.M. Itano and D.J. Wineland *Laser cooling to the zero-point energy of motion* Phys. Rev. Lett. **62**, 403 (1989)

[707] Ch. Roos, Th. Zeiger, H. Rohde, H.C. Nägerl, J. Eschner, D. Leibfried, F. Schmidt-Kaler and R. Blatt *Quantum State Engineering on an optical transition and decoherence in a Paul trap* Phys. Rev. Lett. **83**, 4713 (1999)

[708] C. Monroe, D.M. Meekhof, B.E. King, S.R. Jefferts, W.M. Itano, D.J. Wineland and P. Gould *Resolved-sideband Raman cooling of a bound atom to the 3D zero-point energy* Phys. Rev. Lett. **75**, 4011 (1995)

[709] R.E. Drullinger, D.J. Wineland and J.C. Bergquist *High-resolution optical spectra of laser cooled ions* Appl. Phys. **22**, 365 (1980)

[710] D.J. Larson, J.C. Bergquist, J.J. Bollinger, W.M. Itano and D.J. Wineland *Sympathetic cooling of trapped ions: a laser-cooled two-species nonneutral ion plasma* Phys. Rev. Lett. **57**, 70 (1986)

[711] C. Zipkes, S. Palzer, C. Sias and M. Köhl *A trapped single ion inside a Bose-Einstein condensate* Nature **464**, 388 (2010)

[712] R.H. Dicke *The effect of collisions upon the Doppler width of spectral lines* Phys. Rev. **89**, 472-473 (1953)

[713] J.C. Bergquist *Doppler-free Saturation Spectroscopy* in *Experimental Methods in the Physical Sciences* Academic Press 255-272 (1996)

[714] W. Nagourney, J. Sandberg and H. Dehmelt *Shelved optical electron amplifier: Observation of quantum jumps* Phys. Rev. Lett. **56**, 2797-2799 (1986)

[715] H.S. Margolis *Optical frequency standards and clocks* Contemporary Physics **51**, 37-58 (2010)

[716] J.C. Bergquist, W.M. Itano and D.J. Wineland *Recoilless optical absorption and Doppler sidebands of a single trapped ion* Phys. Rev. A **36**, R428 (1987)

[717] http://www.symmetricom.com

[718] J. Vanier *The active hydrogen maser: state of the art and forecast* Metrologia **18**, 173-186 (1982)

[719] H.E. Peters *Hydrogen masers using cavity frequency-switching servos: present system design (2006) and possible improvements* Metrologia **43**, 353-360 (2006)

[720] R.L. Walsworth Jr., I.F. Silvera, H. P. Godfried, C.C. Agosta, R.F.C. Vessot and E.M. Mattison *Hydrogen maser at temperatures below 1 K*, Phys. Rev. A **2550R** (1986)

[721] M.D. Hürlimann, W.N. Hardy, A.J. Berlinsky and R.W. Cline *Recirculating cryogenic hydrogen maser* Phys. Rev. A **1605R** (1986)

[722] D.B. Sullivan, J.C. Bergquist, J.J. Bollinger, R.E. Drullinger, W.M. Itano, S.R. Jefferts, W.D. Lee, D. Meekhof, T.E. Parker, F.L. Walls and D.J. Wineland *Primary Atomic Frequency Standards at NIST* J. Res. Natl. Inst. Stand. Technol. **106**, 47-63 (2001)

[723] P. Lemonde, Ph. Laurent, G. Santarelli, M. Abgrall, Y. Sortais, S. Bize, Ch. Nicolas, S. Zhang, A. Clairon, N. Dimarcq, P. Petit, A.G. Mann, A.N. Luiten, S. Chang and Ch. Salomon *Cold-Atom Clocks on Earth and in Space* in Frequency Measurement and Control. Advanced Techniques and Future Trends, A.N. Luiten, Ed., Springer, pp. 249-272 (2001)

[724] R. Wynands and S. Weyers *Atomic fountain clocks* Metrologia **42** S64-S79 (2005)

[725] D. Chambon, S. Bize, M. Lours, F. Narbonneau, H. Marion, A. Clairon, G. Santarelli, A. Luiten and M. Tobar *Design and realization of a flywheel oscillator for advanced time and frequency metrology* Rev. Sci. Instrum. **76**, 094704 (2005)

[726] Jocelyne Guena, Michel Abgrall, Daniele Rovera, Philippe Laurent, Baptiste Chupin, Michel Lours, Giorgio Santarelli, Peter Rosenbusch, Michael E. Tobar, Ruoxin Li, Kurt Gibble, Andre Clairon and Sebastien Bize *Progress in Atomic Fountains at LNE-SYRTE* IEEE Transactions on Ultrasonics, Ferroelectrics, and Frequency Control **59**, 391-410 (2012)

[727] W.M. Itano et al. *Quantum projection noise: population fluctuations in two-level systems* Phys. Rev. A **47**, 3554 (1993)

[728] P. Westergaard, J. Lodewyck and P. Lemonde *Minimizing the Dick effect in an optical lattice clock* IEEE Transactions on Ultrasonics, Ferroelectrics and Frequency Control **57**, 623 (2010)

[729] A. Joyet *Aspects metrologiques d'une fontaine continue a atoms froids*, PhD Thesis, University of Neuchatel, Switzerland (2003)

[730] G. Santarelli, Ph. Laurent, P. Lemonde, A. Clairon, A.G. Mann, S. Chang, A.N. Luiten and C. Salomon *Quantum projection noise in an atomic fountain: a high stability cesium frequency standard* Phys. Rev. Lett. **82**, 4619-4622 (1999)

[731] S. Bize, Y. Sortais, C. Mandache, A. Clairon and C. Salomon *Cavity Frequency Pulling in Cold Atom Fountains* IEEE Transactions on Instrumentation and Measurement **50**, 503-506 (2001)

[732] Y. Ovchinnikov and G. Marra *Accurate rubidium atomic fountain frequency standard* Metrologia **48**, 87-100 (2011)

[733] J.H. Shirley, W.D. Lee, G.D. Rovera and R.E. Drullinger *Rabi pedestal shifts as a diagnostic tool in primary frequency standards* IEEE Transactions on Instrumentation and Measurement **44**, 136-139 (1995)

[734] K. Gibble *Difference between a Photon's Momentum and an Atom's Recoil* Phys. Rev. Lett. **97**, 073002 (2006)

[735] Jon H. Shirley, Filippo Levi, Thomas P. Heavner, Davide Calonico, Dai-Hyuk Yu and Steve R. Jefferts *Microwave Leakage-Induced Frequency Shifts in the Primary Frequency Standards NIST-F1 and IEN-CSF1* IEEE transactions on ultrasonics, ferroelectrics, and frequency control **53**, 2376-2385 (2006)

[736] W.D. Lee, J.H. Shirley, F.L. Walls and R.E. Drullinger *Systematic errors in cesium beam frequency standards introduced by digital control of the microwave excitation* in Proc. 1995 IEEE Int. Symp. Freq. Control, IEEE Catalogue No. 95CH35752, 113-117 (1995)

[737] J. Vanier and C. Audoin, Metrologia **42**, S31-S42 (2005)

[738] T.P. Heavner, T.E. Parker, J.H. Shirley, P. Kunz and S.R. Jefferts *NIST F1 and F2* 42nd Annual Precise Time and Time Interval (PTTI) Meeting pp. 457-464 (2010)

[739] Y. Sortais, S. Bize, C. Nicolas, A. Clairon, C. Salomon and C. Williams *Cold collision frequency shifts in a ^{87}Rb atomic fountain* Physical Review Letters **85**, 3117 (2000)

[740] S. Bize, Y. Sortais, M.S. Santos, C. Mandache, A. Clairon and C. Salomon *High-accuracy measurement of the ^{87}Rb ground-state hyperfine splitting in an atomic fountain* Europhys. Lett. **45**, 558 (1999)

[741] H. Haken and H.C. Wolf *The physics of atoms and quanta. Introduction to experiments and theory (Third Edition)* Springer-Verlag, Berlin (1993)

[742] J. Brossel and F. Bitter *A New "Double Resonance" Method for Investigating Atomic Energy Levels. Application to Hg $^{3}P_1$* Phys. Rev. **86**, 311 (1952)

[743] A. Kastler *Optical Methods of Atomic Orientation and of Magnetic Resonance* Journal of the Optical Society of America **47**, 460 (1957)

[744] T. Bandi, C. Affolderbach, C.E. Calosso and G. Mileti *High-performance laser-pumped rubidium frequency standard for satellite navigation* Electronics Letters **47**, No. 12 (2011)

[745] J. Vanier, M. Levine, S. Kendig, D. Janssen, C. Everson and M. Delaney *Practical Realization of a Passive Coherent Population Trapping Frequency Standard* Int. Freq. Control Symposium, Montreal, Canada (2004)

[746] V. Shah *Microfabricated atomic clocks based on coherent population trapping* PhD Thesis, University of Colorado (2002)

[747] J. Vanier, A. Godone and F. Levi *Coherent population trapping in cesium: dark lines and coherent microwave emission* Phys. Rev. A **58**, 2345 (1998)

[748] S. Knappe, P.D.D. Schwindt, V. Shah, L. Hollberg, J. Kitching, L. Liew and J. Moreland *A chip-scale atomic clock based on ^{87}Rb with improved frequency stability* Optics Express **13**, 1249 (2005)

[749] J.D. Prestage, M. Tu, S.K. Chung and P. MacNeal *Compact microwave mercury ion clock for space applications* 39th Annual Precise Time and Time interval (PTTI) Meeting (2007)

[750] J.D. Prestage, R.L. Tjoelker and L. Maleki *Recent Developments in Microwave Ion Clocks* in Frequency Measurement and Control. Advanced Techniques and Future Trends, A.N. Luiten, Ed., Springer, pp. 249-272 (2001)

[751] D.J. Berkeland, J.D. Miller, J.C. Bergquist, W.M. Itano and D.J. Wineland *Laser-cooled mercury ion frequency standard* Phys. Rev. Lett. **80**, 2089 (1998)

[752] P.T.H. Fisk, M.A. Lawn and C. Coles *Laser cooling of $^{171}Yb^+$ ions in a linear Paul trap* Appl. Phys. B **57**, 287-291 (1993)

[753] S.J. Park, P.J. Manson, M.J. Wouters, R.B. Warrington, M.A. Lawn and P.T.H. Fisk $^{171}Yb^+$ *Microwave Frequency Standard* Frequency Control Symposium, 2007 Joint with the 21st European Frequency and Time Forum. IEEE International, pp. 613-616 (2007)

[754] B. Guinot *Application of general relativity to metrology* Metrologia **34**, 261-290 (1997)

[755] V.A. Brumberg, *Essential Relativistic Celestial Mechanics*, Bristol/Philadelphia/New York, Adam Hilger, (1991)

[756] P. Wolf and G. Petit, *Relativistic theory for clock syntonization and the realization of geocentric coordinate times* Astron. Astrophys. **304**, 653-661 (1995)

[757] T. Feldmann *Advances in GPS based Time and Frequency Comparisons for Metrological Use* Dipl.-Phys. Thesis, Von der Fakultät für Mathematik und Physik der Gottfried Wilhelm Leibniz Universität Hannover (2011)

[758] D. Piester, A. Bauch, L. Breakiron, D. Matsakis, B. Blanzano and O. Koudelka *Time transfer with nanosecond accuracy for the realization of international atomic Time* Metrologia **45**, 185-198 (2008)

[759] J. Ye, Jin-Long Peng, R. J. Jones, K.W. Holman, J.L. Hall, D.J. Jones, S.A. Diddams, J. Kitching, S. Bize, J.C. Bergquist, L.W. Hollberg, L. Robertsson and Long-Sheng Ma *Delivery of high-stability optical and microwave frequency standards over an optical fiber network* J. Opt. Soc. Am. B **20**, 1459 (2003)

[760] C. Daussy, O. Lopez, A. Amy-Klein, A. Goncharov, M. Guinet, C. Chardonnet, F. Narbonneau, M. Lours, D. Chambon, S. Bize, A. Clairon, G. Santarelli, M. E. Tobar and A. N. Luiten *Long-distance frequency dissemination with a resolution of* 10^{-17} Phys. Rev. Lett. **94**, 203904 (2005)

[761] G. Grosche, O. Terra, K. Predehl, R. Holzwarth, B. Lipphardt, F. Vogt, U. Sterr and H. Schnatz *Optical frequency transfer via 146 km fiber link with* 10^{-19} *relative accuracy* Opt. Lett. **34**, 2270 (2009)

[762] S.M. Foreman, A.D. Ludlow, M.H.G. de Miranda, J.E. Stalnaker, S.A. Diddams and Jun Ye *Coherent optical phase transfer over a 32-km fiber with 1 s instability at* 10^{-17} Phys. Rev. Lett. **99**, 153601 (2007)

[763] G. Marra, H.S. Margolis and D.J. Richardson *Dissemination of an optical frequency comb over fiber with* $3 \cdot 10^{-18}$ *fractional accuracy* Optics Express **20**, 1775 (2012)

[764] H.S. Margolis *Frequency metrology and clocks* J. Phys. B: At. Mol. Opt. Phys. **42**, 154017 (2009)

[765] S.A. Diddams, Th. Udem, J.C. Bergquist, E.A. Curtis, R.E. Drullinger, L. Hollberg, W.M. Itano, W.D. Lee, C.W. Oates, K.R. Vogel and D.J. Wineland *An optical clock based on a single trapped* $^{199}Hg^+$ *Ion* Science **293**, 825 (2001)

[766] Long-Sheng Ma, Zhiyi Bi, Albrecht Bartels, Kyoungsik Kim, Lennart Robertsson, Massimo Zucco, Robert S. Windeler, Guido Wilpers, Chris Oates, Leo Hollberg and Scott A. Diddams *Frequency uncertainty for optically referenced femtosecond laser frequency combs* IEEE Journal of Quantum Electronics **43**, 139 (2007)

[767] Leo Hollberg, Chris W. Oates, E. Anne Curtis, Eugene N. Ivanov, Scott A. Diddams, Thomas Udem, Hugh G. Robinson, James C. Bergquist, Robert J. Rafac, Wayne M. Itano, Robert E. Drullinger and David J. Wineland *Optical frequency standards and measurements* IEEE Journal of Quantum Electronics **37**, 1502-1513 (2001)

[768] P. Gill, H.Margolis, A. Curtis, H. Klein, S. Lea, S. Webster and P. Whibberley *Optical Atomic Clocks for Space* Technical Supporting Document, Version 1.7 (2008)

[769] N. Huntemann, M. Okhapkin, B. Lipphardt, S. Weyers, Chr. Tamm and E. Peik *High-accuracy optical clock based on the octupole transition in* $^{171}Yb^+$ Physical Review Letters **108**, 090801 (2012)

[770] P.O. Schmidt et al. *Spectroscopy using quantum logic* Science **309**, 749 (2005)

[771] W.H. Oskay, S.A. Diddams, E.A. Donley, T.M. Fortier, T.P. Heavner, L. Hollberg, W.M. Itano, S.R. Jefferts, M.J. Delaney, K. Kim, F. Levi, T.E. Parker and J.C. Bergquist *Single-atom optical clock with high accuracy* Phys. Rev. Lett. **97**, 020801 (2006)

[772] T. Rosenband, D.B. Hume, P.O. Schmidt, C.W. Chou, A. Brusch, L. Lorini, W.H. Oskay, R.E. Drullinger, T.M. Fortier, J.E. Stalnaker, S.A. Diddams, W.C. Swann, N.R. Newbury, W.M. Itano, D.J. Wineland and J. C. Bergquist *Frequency Ratio of* Al^+ *and* Hg^+ *Single-Ion Optical Clocks; Metrology at the 17th Decimal Place* Science **319**, 1808 (2008)

[773] E. Peik, T. Schneider and C. Tamm *Laser frequency stabilization to a single ion* J. Phys. B: At. Mol. Opt. Phys. **39**, 145-158 (2006)

[774] C.W. Chou, D.B. Hume, J.C.J. Koelemeij, D.J. Wineland and T. Rosenband *Frequency Comparison of Two High-Accuracy* Al^+ *Optical Clocks* Physical Review Letters **104**, 070802 (2010)

[775] P. Gill and F. Riehle *On secondary representations of the second* in Proceedings of the 20th European Frequency and Time Forum, 282-288 (2006)

[776] P. Dubé, A.A. Madej, J.E. Bernard, L. Marmet, J.S. Boulanger and S. Cundy *Electric quadrupole shift cancellation in single-ion optical frequency standards* Phys. Rev. Lett. **95**, 033001 (2005)

[777] V.I. Yudin, A.V. Taichenachev, M.V. Okhapkin, S.N. Bagayev, Chr. Tamm, E. Peik, N. Huntemann, T.E. Mehlstäubler and F. Riehle *Atomic clocks with suppressed blackbody radiation shift* Physical Review Letters **107**, 030801 (2011)

[778] T. Schneider, E. Peik and Chr. Tamm *Sub-hertz optical frequency comparisons between two trapped* $^{171}Yb^+$ *Ions* Physical Review Letters **94**, 230801 (2005)

[779] H. Katori *Optical lattice clocks and quantum metrology* Nature Photonics **5**, 203 (2011)

[780] Andrei Derevianko and Hidetoshi Katori *Colloquium: Physics of optical lattice clocks* Rev. Mod. Phys. **83**, 331-347 (2011)

[781] A.D. Ludlow et al. *Systematic study of the* ^{87}Sr *clock transition in an optical lattice* Phys. Rev. Lett. **96**, 033003 (2006)

[782] R. Le Targat et al. *Accurate optical lattice clock with* ^{87}Sr *atoms* Phys. Rev. Lett. **97**, 130801 (2006)

[783] N.D. Lemke et al. *Spin-1/2 optical lattice clock* Phys. Rev. Lett. **103**, 063001 (2009)

[784] M. Takamoto, T. Takano and H. Katori *Frequency comparison of optical lattice clocks beyond the Dick limit* Nature Photonics **5**, 288 (2011)

[785] Atsushi Yamaguchi, Miho Fujieda, Motohiro Kumagai, Hidekazu Hachisu, Shigeo Nagano, Ying Li, Tetsuya Ido, Tetsushi Takano, Masao Takamoto and Hidetoshi Katori *Direct comparison of distant optical lattice clocks at the 10^{-16} uncertainty* Applied Physics Express **4**, 082203 (2011)

[786] B. Cagnac, M.D. Plimmer, L. Julien and F. Biraben *The hydrogen atom, a tool for metrology* Rep. Prog. Phys. **57**, 853-893 (1994)

[787] F. Biraben et al. *Precision spectroscopy of atomic hydrogen* in S.G. Karshenboim et al., Eds., LNP 570, Springer, pp. 17-41, (2001)

[788] F. Biraben, *Spectroscopy of atomic hydrogen* The European Physical Journal - Special Topics **172**, 109-119 (2009)

[789] C.G. Parthey *Precision spectroscopy on atomic hydrogen* PhD Thesis, Ludwig-Maximilians-Universitat Munchen (2011)

[790] P.J. Mohr, B.N. Taylor, and D.B. Newell *CODATA recommended values of the fundamental physical constants: 2006* Rev. Mod. Phys. **80**, 633 (2008)

[791] T.W. Hänsch, I.S. Shahin and A.L. Schawlow *Optical resolution of the Lamb shift in atomic Hydrogen by laser saturation spectroscopy* Nature **235**, 63 (1972)

[792] D. Hanneke, S. Fogwell, and G. Gabrielse, *New measurement of the electron magnetic moment and the fine structure constant* Phys. Rev. Lett. **100**, 120801 (2008)

[793] R. Pohl et al. *The size of the proton* Nature **466**, 213 (2010)

[794] A. Antognini et al. *The Lamb shift in muonic hydrogen and the proton radius* Physics Procedia **17**, 10-19 (2011)

[795] The ALPHA Collaboration *Confinement of antihydrogen for 1000 seconds* Nature Physics **7**, 558 (2011)

[796] W. Vassen, C. Cohen-Tannoudji, M. Leduc, D. Boiron, C.I. Westbrook, A. Truscott, K. Baldwin, G. Birkl, P. Cancio and M. Trippenbach *Cold and trapped metastable noble gases* Rev. Mod. Phys. **84**, 175-210 (2012)

[797] G. Giusfredi, P. De Natale, D. Mazzotti, P. Cancio Pastor, C. de Mauro, L. Fallani, G. Hagel, V. Krachmalnicoff and M. Inguscio *Present status of the fine-structure frequencies of the 2^3P helium level* Canadian J. of Physics **83**, 301 (2005)

[798] V.W. Hughes, *The fine-structure constant* Comm. At. Mol. Phys. **1**, 5 (1969)

[799] F. Pereira Dos Santos et al., *Bose-Einstein Condensation of Metastable Helium* Phys. Rev. Lett. **86**, 3459 (2001)

[800] A. Robert *Realisation d'un condensat de Bose-Einstein d'Helium metastable* Ph.D. thesis, Universite Paris-Sud (2001)

[801] J.M. McNamara, T. Jeltes, A.S. Tychkov, W. Hogervorst and W. Vassen, *Degenerate Bose-Fermi Mixture of Metastable Atoms* Phys. Rev. Lett. **97**, 080404 (2006)

[802] L. Consolino, G. Giusfredi, P. De Natale, M. Inguscio and P. Cancio, *Optical frequency comb assisted laser system for multiplex precision spectroscopy* Opt. Express **19**, 3155 (2011)

[803] J.S. Borbely, M.C. George, L.D. Lombardi, M. Weel, D.W. Fitzakerley and E.A. Hessels *Separated oscillatory-field microwave measurement of the 2^3P_1-2^3P_2 fine-structure interval of atomic helium* Phys. Rev. A **79**, 060503 (2009)

[804] J. Castillega, D. Livingston, A. Sanders and D. Shiner *Precise Measurement of the J =1 to J=2 Fine Structure Interval in the 2^3P State of Helium* Phys. Rev. Lett. **84**, 4321 (2000)

[805] W. Frieze, E. Hinds, V. Hughes and F. Pichanick *Experiments on the 2^3P state of helium. IV. Measurement of the 2^3P_0-2^3P_2 fine-structure interval* Phys. Rev. A **24**, 279 (1981)

[806] F. Minardi, G. Bianchini, P. Cancio Pastor, G. Giusfredi, F. Pavone and M. Inguscio *Measurement of the Helium 2^3P_0-2^3P_1 Fine Structure Interval* Phys. Rev. Lett. **82**, 1112 (1999)

[807] M. Smiciklas and D. Shiner *Determination of the Fine Structure Constant Using Helium Fine Structure* Phys. Rev. Lett. **105**, 123001 (2010)

[808] C. Storry, M.C. George and E. Hessels *Precision Microwave Measurement of the 2^3P_1-2^3P_2 Interval in Atomic Helium* Phys. Rev. Lett. **84**, 3274 (2000)

[809] T. Zelevinsky, D. Farkas and G. Gabrielse *Precision Measurement of the Three 2^3P_J Helium Fine Structure Intervals* Phys. Rev. Lett. **95**, 203001 (2005)

[810] M.C. George, L.D. Lombardi and E.A. Hessels *Precision Microwave Measurement of the 2^3P_1-2^3P_0 Interval in Atomic Helium: A Determination of the Fine-Structure Constant* Phys. Rev. Lett. **87**, 173002 (2009)

[811] G.W.F. Drake *Progress in helium fine-structure calculations and the fine-structure constant* Can. J. Phys. **80**, 1195 (2002)

[812] K. Pachucki and V. A. Yerokhin *Fine Structure of Heliumlike Ions and Determination of the Fine Structure Constant* Phys. Rev. Lett. **104**, 070403 (2010)

[813] V.A. Yerokhin and K. Pachucki *Theoretical energies of low-lying states of light helium-like ions* Phys. Rev. A **81**, 022507 (2010)

[814] F. Nez et al. *First pure frequency-measurement of an optical-transition in atomic-hydrogen - better determination of the rydberg constant* Europhys. Lett. **24**, 635 (1993)

[815] Th. Udem et al. *Phase-Coherent Measurement of the Hydrogen 1S-2S Transition Frequency with an Optical Frequency Interval Divider Chain* Phys. Rev. Lett. **79**, 2646 (1997)

[816] F. Pavone, F. Marin, P. De Natale, M. Inguscio and F. Biraben *First Pure Frequency Measurement of an Optical Transition in Helium: Lamb Shift of the 2^3S_1 Metastable Level* Phys. Rev. Lett. **73**, 42 (1994)

[817] D.C. Morton, Q.X. Wu and G.W.F. Drake *Energy levels for the stable isotopes of atomic helium (^{4}He I and ^{3}He I)* Can. J. Phys. **84**, 83 (2006)

[818] D.C. Morton, Q.X. Wu and G.W.F. Drake *Nuclear charge radius for* ^3He Phys. Rev. A **73**, 034502 (2006)

[819] E. Borie, and G.A. Rinker *Improved calculation of the muonic-helium Lamb shift* Phys. Rev. A **18**, 324 (1978)

[820] I. Sick *Precise root-mean-square radius of* ^4He Phys. Rev. C **77**, 041302 (2008)

[821] P. Zhao, J.R. Lawall and F. M. Pipkin *High-precision isotope-shift measurement of* 2^3S-2^3P *transition in helium* Phys. Rev. Lett. **66**, 592 (1991)

[822] D. Shiner, R. Dixson and V. Vedantham *Three-Nucleon Charge Radius: A Precise Laser Determination Using* ^3He Phys. Rev. Lett. **74**, 3553 (1995)

[823] F. Marin, F. Minardi, F.S. Pavone, M. Inguscio and G.W.F. Drake *Hyperfine structure of the* 3^3P *state of* ^3He *and isotope shift for the* $2^3S-3^3P_0$ *transition* Z. Phys. D **32**, 285 (1995)

[824] S.C. Pieper and R.B. Wiringa *Quantum monte carlo calculations of light nuclei* Annu. Rev. Nucl. Part. Sci. **51**, 53 (2001)

[825] A. Amroun et al. ^3H *and* ^3He *electromagnetic form factors* Nucl. Phys. **A579**, 596 (1994)

[826] P. Mueller et al. *Nuclear Charge Radius of* ^8He Phys. Rev. Lett. **99**, 252501 (2007)

[827] L.-B. Wang et al. *Laser Spectroscopic Determination of the* ^6He *Nuclear Charge Radius* Phys. Rev. Lett. **93**, 142501 (2004)

[828] R. van Rooij et al. *Frequency Metrology in Quantum Degenerate Helium: Direct Measurement of the* $2^3S_1 \rightarrow 2^1S_0$ *Transition* Science **333**, 196 (2011)

[829] P. Cancio Pastor, L. Consolino, G. Giusfredi, P. De Natale, M. Inguscio, V.A. Yerokhin and K. Pachucki *Frequency metrology of helium around 1083 nm and determination of the nuclear charge radius* Phys. Rev. Lett. **108**, 143001 (2012)

[830] A. Antognini et al. *Illuminating the proton radius conundrum: the* μHe^+ *Lamb shift* Can. J. Phys. **89**, 47 (2011)

[831] C. Daussy, M. Guinet, A. Amy-Klein, K. Djerroud, Y. Hermier, S. Briaudeau, Ch.J. Borde and C. Chardonnet *Direct determination of the Boltzmann constant by an optical method* Phys. Rev. Lett. **98**, 250801 (2007)

[832] G. Casa, A. Castrillo, G. Galzerano, R. Wehr, A. Merlone, D. Di Serafino, P. Laporta and L. Gianfrani *Primary gas thermometry by means of laser-absorption spectroscopy: determination of the Boltzmann constant* Phys. Rev. Lett. **100**, 200801 (2008)

[833] G.-W. Truong, E.F. May, T.M. Stace and A.N. Luiten *Quantitative atomic spectroscopy for primary thermometry* Physical Review A **83**, 033805 (2011)

[834] G. Lamporesi, A. Bertoldi, L. Cacciapuoti, M. Prevedelli and G.M. Tino *Determination of the Newtonian gravitational constant using atom interferometry* Phys. Rev. Lett. **100**, 050801 (2008)

[835] J.B. Fixler et al. *Atom Interferometer Measurement of the Newtonian Constant of Gravity* Science **315**, 74 (2007)

[836] S. Schlamminger et al. *Measurement of Newton's gravitational constant* Phys. Rev. D **74**, 082001 (2006)

[837] J.P. Schwarz et al., *A Free-Fall Determination of the Newtonian Constant of Gravity* Science **282**, 2230 (1998)

[838] W.T. Ni et al. *The application of laser metrology and resonant optical cavity techniques to the measurement of G* Meas. Sci. Technol. **10**, 495 (1999)

[839] U. Kleinevoss et al. *Absolute measurement of the Newtonian force and a determination of G* Meas. Sci. Technol. **10**, 492 (1999)

[840] A. Bertoldi et al. *Atom interferometry gravity-gradiometer for the determination of the Newtonian gravitational constant G* Eur. Phys. J. D **40**, 271 (2006)

[841] M. Fattori, G. Lamporesi, T. Petelski, J. Stuhler and G.M. Tino *Towards an atom interferometric determination of the Newtonian gravitational constant* Phys. Lett. A **318**, 184 (2003)

[842] F. Sorrentino et al. *Sensitive gravity-gradiometry with atom interferometry: progress towards an improved determination of the gravitational constant* New Journal of Physics **12**, 095009 (2010)

[843] G. Lamporesi *Determination of the gravitational constant by atom interferometry* PhD thesis, University of Florence (2006)

[844] A.D Cronin, J. Schmiedmayer and D.E. Pritchard *Optics and interferometry with atoms and molecules* Rev. Mod. Phys. **81**, 1051 (2009)

[845] J.-P. Uzan *The fundamental constants and their variation: observational and theoretical status* Rev. Mod. Phys. **75**, 403 (2003)

[846] S.G. Karshenboim *Fundamental physical constants: looking from different angles* Can. J. Phys. **83**, 767-811 (2005)

[847] P.A.M. Dirac *The cosmological constants* Nature **139**, 323 (1937)

[848] J.M. Overduin and P.S. Wesson *Kaluza-Klein gravity* Physics Reports **283**, 303-378 (1997)

[849] B. Zwiebach *A First Course in String Theory (Second Edition)* Cambridge (2009)

[850] G. Borner *The Early Universe* Springer-Verlag (1993)

[851] S.G. Karshenboim *Time and space variation of fundamental constants: motivation and laboratory search* arXiv:physics/0306180v1, An invited talk at *Time and Matter symposium*, Venice (2002)

[852] P. Molaro and E. Vangioni *Are the fundamental constants varying with time?* Highlights of Astronomy **15**, 326 (2010)

[853] J.C. Berengut and V.V. Flambaum *Astronomical and laboratory searches for space-time variation of fundamental constants* Journal of Physics: Conference Series **264**, 012010 (2011)

[854] E.J. Salumbides *Laser precision metrology for probing variation of fundamental constants* Graduate thesis, Vrije Universiteit (2009)

[855] F. van Weerdenburget et al. *First Constraint on Cosmological Variation of the Proton-to-Electron Mass Ratio from Two Independent Telescopes* Phys. Rev. Lett. **106**, 180802 (2011)

[856] J. Bagdonaite, M.T. Murphy, L. Kaper and Wim Ubachs *Constraint on a variation of the proton-to-electron mass ratio from H_2 absorption towards quasar Q2348-011* Mon. Not. R. Astron. Soc. **421**, 419-425 (2012)

[857] K.A. Bronnikov and S.A. Kononogov *Possible variations of the fine structure constant α and their metrological significance* Metrologia **43**, R1-R9 (2006)

[858] E. Peik, B. Lipphardt, H. Schnatz, Chr. Tamm, S. Weyers and R. Wynands *Laboratory Limits on Temporal Variations of Fundamental Constants: An Update* arXiv:physics/0611088v1, Proceedings of the 11th Marcel Grossmann Meeting, Berlin (2006)

[859] S. Bize et al. *Cold atom clocks and applications* J. Phys. B: At. Mol. Opt. Phys. **38**, S449 (2005)

[860] M. Fischer et al. *New limits on the drift of fundamental constants from laboratory measurements* Phys. Rev. Lett. **92**, 230802 (2004)

[861] T.M. Fortier et al. *Precision atomic spectroscopy for improved limits on variation of the fine structure constant and local position invariance* Phys. Rev. Lett. **98**, 070801 (2007)

[862] L. Lorini, N. Ashby, A. Brusch, S. Diddams, R. Drullinger, E. Eason, T. Fortier, P. Hastings, T. Heavner, D. Hume, W. Itano, S. Jefferts, N. Newbury, T. Parker, T. Rosenband, J. Stalnaker, W. Swann, D. Wineland and J. Bergquist *Recent atomic clock comparisons at NIST* Eur. Phys. J. Special Topics **163**, 19-35 (2008)

[863] H.L. Bethlem et al. *Prospects for precision measurements on ammonia molecules in a fountain* Eur. Phys. J. Special Topics **163**, 55 (2008)

[864] D. Bakalov et al. *Precision spectroscopy of the molecular ion HD^+: control of Zeeman shifts* Phys. Rev. A **82**, 055401 (2010)

[865] T. Zelevinsky et al. *Precision test of mass-ratio variations with lattice-confined ultracold molecules* Phys. Rev. Lett. **100**, 043201 (2008)

[866] D. DeMille et al. *Enhanced Sensitivity to Variation of m_e/m_p in Molecular Spectra* Phys. Rev. Lett. **100**, 043202 (2008)

[867] P.C.W Davies, T.M. Davis and C.H. Lineweaver *Black holes constrain varying constants* Nature **418**, 602 (2002)

[868] Jacob D. Bekenstein and Marcelo Schiffer *Varying fine structure "constant" and charged black holes* Physical Review D **80**, 123508 (2009)

[869] J.D. Bekenstein *Fine-structure constant: Is it really a constant?* Phys. Rev. D **25**, 1527-1539 (1982)

[870] B.E. Schaefer *Severe limits on variations of the speed of light with frequency* Phys. Rev. Lett. **82**, 4964 (1999)

[871] S.G. Turyshev et al. *Space-based research in fundamental physics and quantum technologies* Int. J. Mod. Phys. D **16**, 1879-1925 (2007)

[872] M. Kramer et al. *Tests of general relativity from timing the double pulsar* Science **314**, 97 (2006)

[873] *Spin-Statistics Connection and Commutation Relations*, R.C. Hilborn and G.M. Tino, Eds., AIP Conf. Proc. No. 545, AIP (2000)

[874] G.M. Tino *Spectroscopic tests of spin-statistics connection and symmetrization postulate of quantum mechanics* Physica Scripta **T95**, 62-67 (2001)

[875] G. Modugno and M. Modugno *Testing the symmetrization postulate on molecules with three identical nuclei* Phys. Rev. A **62**, 022115 (2000)

[876] D. Mazzotti, P. Cancio, G. Giusfredi, M. Inguscio and P. De Natale *Search for exchange-antisymmetric states for spin-0 particles at the 10^{-11} Level* Phys. Rev. Lett. **86**, 1919 (2001)

[877] G. Gentile *Osservazioni sopra le statistiche intermedie* Nuovo Cimento **17**, 493 (1940)

[878] O.W. Greenberg and R. N. Mohapatra *Local Quantum Field Theory of Possible Violation of the Pauli Principle* Phys. Rev. Lett. **59**, 2507 (1987)

[879] O.W. Greenberg *Example of infinite statistics* Phys. Rev. Lett. **64**, 705 (1990)

[880] M. de Angelis et al. *Test of the Symmetrization Postulate for Spin-0 Particles* Phys. Rev. Lett. **76**, 2840 (1996)

[881] R.C. Hilborn and C.L. Yuca *Spectroscopic Test of the Symmetrization Postulate for Spin-0 Nuclei* Phys. Rev. Lett. **76**, 2844 (1996)

[882] G. Modugno, M. Inguscio and G. M. Tino *Search for Small Violations of the Symmetrization Postulate for Spin-0 Particles* Phys. Rev. Lett. **81**, 4790 (1998)

[883] I.B. Khriplovich, and S.K. Lamoreaux *CP Violation without Strangeness* Springer (1997)

[884] M. Pospelov and A. Ritz *Electric dipole moments as probes of new physics* Ann. Phys. **318**, 119-169 (2005)

[885] E.D. Commins *Electric dipole moments of leptons*. In Advances in Atomic, Molecular, and Optical Physics, Vol. 40, 1-56 (B. Bederson and H. Walther), Eds., Academic Press (1999)

[886] A.D. Sakharov *Violation of CP invariance, C asymmetry, and baryon asymmetry of the Universe* Sov. Phys. JETP Lett. **5**, 24-27 (1967)

[887] J.J. Hudson, D.M. Kara, I.J. Smallman, B.E. Sauer, M.R. Tarbutt and E.A. Hinds *Improved measurement of the shape of the electron* Nature **473**, 493 (2011)

[888] J.J. Hudson, B.E. Sauer, M.R. Tarbutt and E.A. Hinds *Measurement of the electron electric dipole moment using ybf molecules* Physical Review Letters **89**, 023003 (2002)

[889] A.E. Leanhardt, J.L. Bohn, H. Loh, P. Maletinsky, E.R. Meyer, L.C. Sinclair, R.P. Stutz and E.A. Cornell *High-resolution spectroscopy on trapped molecular ions in rotating electric fields: a new approach for measuring the electron electric dipole moment* Journal of Molecular Spectroscopy **270**, 1-25 (2011)

[890] B. Darquié et al. *Progress toward the first observation of parity violation in chiral molecules by high-resolution laser spectroscopy* Chirality **22**, 870-884 (2010)

[891] M.-A. Bouchiat and C. Bouchiat *Parity violation in atoms* Rep. Prog. Phys. **60**, 1351-1396 (1997)

[892] E.N. Fortson and L.L. Lewis *Atomic parity nonconservation experiments* Phys. Reports **113**, 289-344 (1984)

[893] J.S.M. Ginges and V.V. Flambaum *Atomic parity nonconservation experiments* Phys. Reports **397**, 63-154 (2004)

[894] J. Guena, M. Lintz and M.A. Bouchiat *Measurement of the parity violating 6S-7S transition amplitude in cesium achieved within $2 \cdot 10^{-13}$ atomic-unit accuracy by stimulated-emission detection* Phys. Rev. A **71**, 042108 (2005)

[895] C.S. Wood et al. *Measurement of parity nonconservation and an anapole moment in cesium* Science **275**, 1759 (1997)

[896] K. Tsigutkin, D. Dounas-Frazer, A. Family, J.E. Stalnaker, V.V. Yashchuk and D. Budker *Observation of a large atomic parity violation effect in ytterbium* Phys. Rev. Lett. **103**, 071601 (2009)

[897] M. Quack, J. Stohner and M. Willeke *High-resolution spectroscopic studies and theory of parity violation in chiral molecules* Annu. Rev. Phys. Chem. **59**, 741 (2008)

[898] F. Faglioni and P. Lazzeretti *Understanding parity violation in molecular systems* Physical Review E **65**, 011904 (2002)

[899] Ch. Daussy, T. Marrel, A. Amy-Klein, C.T. Nguyen, Ch.J. Bordé and Ch. Chardonnet *Limit on the parity nonconserving energy difference between the enantiomers of a chiral molecule by laser spectroscopy* Phys. Rev. Lett. **83**, 1554 (1999)

[900] S.G. Turyshev *Experimental tests of general relativity* Annu. Rev. Nucl. Part. Sci. **58**, 207 (2008)

[901] C.M. Will *Resource letter PTG-1: precision tests of gravity* Am. J. Phys. **82**, 1240 (2010)

[902] S.G. Karshenboim and E. Peik (Eds.) *Astrophysics, Clocks and Fundamental Constants*, Lect. Notes Phys. 648, Springer, Berlin Heidelberg (2004)

[903] S. Schlamminger, K.-Y. Choi, T.A. Wagner, J.H. Gundlach and E.G. Adelberger *Test of the equivalence principle using a rotating torsion balance* Phys. Rev. Lett. **100**, 041101 (2008)

[904] S. Fray, C. Alvarez Diez, T.W. Hänsch and Martin Weitz *Atomic interferometer with amplitude gratings of light and its applications to atom based tests of the equivalence principle* Phys. Rev. Lett. **93**, 240404 (2004)

[905] S. Reinhardt et al. *Test of relativistic time dilation with fast optical atomic clocks at different velocities* Nature **3**, 861 (2007)

[906] C. Novotny, G. Huber, S. Karpuk, S. Reinhardt, D. Bing, D. Schwalm, A. Wolf, B. Bernhardt, T.W. Hänsch, R. Holzwarth, G. Saathoff, Th. Udem, W. Nörtershäuser, G. Ewald, C. Geppert, T. Kühl, T. Stöhlker and G. Gwinner *Sub-Doppler laser spectroscopy on relativistic beams and tests of Lorentz invariance* Phys. Rev. A **80**, 022107 (2009)

[907] P. Wolf et al. *Quantum physics exploring gravity in the outer solar system: the SAGAS project* Exp Astron **23**, 651-687 (2009)

[908] M.E. Tobar, P. Wolf, Sebastien Bize, G. Santarelli and V. Flambaum *Testing local Lorentz and position invariance and variation of fundamental constants by searching the derivative of the comparison frequency between a cryogenic sapphire oscillator and hydrogen maser* Phys. Rev. D **81**, 022003 (2010)

[909] S. Herrmann, A. Senger, K. Möhle, M. Nagel1, E. V. Kovalchuk and A. Peters *Rotating optical cavity experiment testing Lorentz invariance at the 10^{-17} level* Phys. Rev. D **80**, 105011 (2009)

[910] P.L. Stanwix, M.E. Tobar, P. Wolf, C.R. Locke and E.N. Ivanov *Improved test of Lorentz invariance in electrodynamics using rotating cryogenic sapphire oscillators* Phys. Rev. D **74**, 081101(R) (2006)

[911] H. Müller, A. Peters and S. Chu *A precision measurement of the gravitational redshift by the interference of matter waves* Nature **463**, 926 (2010)

[912] C.W. Chou et al. *Optical clocks and relativity* Science **329**, 1630 (2010)

[913] R.F.C. Vessot, M.W. Levine, E.M. Mattison, E.L. Blomberg, T.E. Hoffman, G.U. Nystrom, B.F. Farrel, R. Decher, P.B. Eby, C.R. Baugher, J.W. Watts, D.L. Teuber and F.D. Wills *Test of relativistic gravitation with a space-borne hydrogen maser* Phys. Rev. Lett. **45**, 2081-2084 (1980)

[914] L. Cacciapuoti and Ch. Salomon *Space clocks and fundamental tests: the ACES experiment* Eur. Phys. J. Special Topics **172**, 57-68 (2009)

[915] S. Schiller, G. Tino, P. Gill, C. Salomon et al. *Einstein gravity explorer. A medium-class fundamental physics mission proposal for cosmic vision 2015-2025*, Experimental Astronomy **23**, 573-610 (2009)

[916] C.M. Will *The confrontation between general relativity and experiment* Living Rev. Relativity **9**, 3 (2006)

[917] F.W. Dyson, A.S. Eddington and C. Davidson *A determination of the deflection of light by the Sun's gravitational field, from observations made at the total eclipse of May 29, 1919* Phil. Trans. Roy. Soc. Lond. **A220**, 291-333 (1920)

[918] R.D. Newman, E.C. Berg and P.E. Boynton *Tests of the gravitational inverse square law at short ranges* Space Sci Rev **148**, 175-190 (2009)

[919] D.J. Kapner, T.S. Cook, E.G. Adelberger, J.H. Gundlach, B.R. Heckel, C.D. Hoyle and H.E. Swanson *Tests of the gravitational inverse-square law below the dark-energy length scale* Phys. Rev. Lett. **98**, 021101 (2007)

[920] A.A. Geraci, Sylvia J. Smullin, David M. Weld, John Chiaverini and Aharon Kapitulnik *Improved constraints on non-Newtonian forces at 10 microns* Physical Review D **78**, 022002 (2008)

[921] S. Dimopoulos and A.A. Geraci *Probing submicron forces by interferometry of Bose-Einstein condensed atoms* Phys. Rev. D **68**, 124021 (2003)

[922] I. Carusotto, L. Pitaevskii, S. Stringari, G. Modugno, and M. Inguscio *Sensitive Measurement of forces at the micron scale using Bloch oscillations of ultracold atoms* Phys. Rev. Lett. **95**, 093202 (2005)

[923] G. Ferrari, N. Poli, F. Sorrentino, and G.M. Tino *Long-lived Bloch oscillations with bosonic Sr atoms and application to gravity measurement at the micrometer scale* Phys. Rev. Lett. **97**, 060402 (2006)

[924] V. Giovannetti, S. Lloyd and L. Maccone *Quantum-enhanced measurements: beating the standard quantum limit* Science **306(5700)**, 1330–1336 (2004)

[925] B. C. Sanders and G. J. Milburn *Optimal quantum measurements for phase estimation* Physical Review Letters **75(16)**, 2944–2947 (1995)

[926] C. M. Caves *Quantum-mechanical noise in an interferometer* Physical Review D **23(8)**, 1693–1708 (1981)

[927] A. F. Pace, M. J. Collett and D. F. Walls *Quantum limits in interferometric detection of gravitational radiation* Physical Review A **47(4)**, 3173–3189 (1993)

[928] B. Yurke *Input states for enhancement of fermion interferometer sensitivity* Physical Review Letters **56(15)**, 1515–1517 (1986)

[929] H. P. Yuen *Generation, detection, and application of high-intensity photon-number-eigenstate fields* Physical Review Letters **56(20)**, 2176–2179 (1986)

[930] M. J. Holland and K. Burnett *Interferometric detection of optical phase shifts at the Heisenberg limit* Physical Review Letters **71(9)**, 1355–1358 (1993)

[931] M. Hillery and L. Mlodinow *Interferometers and minimum-uncertainty states* Physical Review A **48(2)**, 1548–1558 (1993)

[932] J. P. Dowling *Correlated input-port, matter-wave interferometer: quantum-noise limits to the atom-laser gyroscope* Physical Review A - Atomic, Molecular, and Optical Physics **57(6)**, 4736–4746 (1998)

[933] B. C. Sanders *Quantum dynamics of the nonlinear rotator and the effects of continual spin measurement* Physical Review A **40(5)**, 2417–2427 (1989)

[934] A. N. Boto, P. Kok, D. S. Abrams, S. L. Braunstein, C. P. Williams and J. P. Dowling *Quantum interferometric optical lithography: exploiting entanglement to beat the diffraction limit* Physical Review Letters **85(13)**, 2733–2736 (2000)

[935] M. D'Angelo, M. V. Chekhova and Y. Shih *Two-photon diffraction and quantum lithography* Physical Review Letters **87(1)**, 013602/1–013602/4 (2001)

[936] P. Kok, H. Lee and J. P. Dowling *Creation of large-photon-number path entanglement conditioned on photodetection* Physical Review A - Atomic, Molecular, and Optical Physics **65(5 A)**, 521041–521045 (2002)

[937] G. J. Pryde and A. G. White *Creation of maximally entangled photon-number states using optical fiber multiports* Physical Review A - Atomic, Molecular, and Optical Physics **68(5)**, 523151–523154 (2003)

[938] H. Cable and J. P. Dowling *Efficient generation of large number-path entanglement using only linear optics and feed-forward* Physical Review Letters **99(16)** (2007)

[939] M. W. Mitchell, J. S. Lundeen and A. M. Steinberg *Super-resolving phase measurements with a multiphoton entangled state* Nature **429(6988)**, 161–164 (2004)

[940] T. Nagata, R. Okamoto, J. L. O'Brien, K. Sasaki and S. Takeuchi *Beating the standard quantum limit with four-entangled photons* Science **316(5825)**, 726–729 (2007)

[941] R. Okamoto, H. F. Hofmann, T. Nagata, J. L. O'Brien, K. Sasaki and S. Takeuchi *Beating the standard quantum limit: phase super-sensitivity of n-photon interferometers* New Journal of Physics **10** (2008)

[942] I. Afek, O. Ambar and Y. Silberberg *High-noon states by mixing quantum and classical light* Science **328(5980)**, 879–881 (2010)

[943] J. J. Bollinger, W. M. Itano, D. J. Wineland and D. J. Heinzen *Optimal frequency measurements with maximally correlated states* Physical Review A - Atomic, Molecular, and Optical Physics **54(6)**, R4649–R4652 (1996)

[944] S. F. Huelga, C. Macchiavello, T. Pellizzari, A. K. Ekert, M. B. Plenio and J. I. Cirac *Improvement of frequency standards with quantum entanglement* Physical Review Letters **79(20)**, 3865–3868 (1997)

[945] D. J. Wineland, J. J. Bollinger, W. M. Itano, F. L. Moore and D. J. Heinzen *Spin squeezing and reduced quantum noise in spectroscopy* Physical Review A **46(11)**, R6797–R6800 (1992)

[946] M. Kitagawa and M. Ueda *Squeezed spin states* Physical Review A **47(6)**, 5138–5143 (1993)

[947] A. Sörensen, L. Duan, J. I. Cirac and P. Zoller *Many-particle entanglement with bose-einstein condensates* Nature **409(6816)**, 63–66 (2001)

[948] C. Orzel, A. K. Tuchman, M. L. Fenselau, M. Yasuda and M. A. Kasevich *Squeezed states in a Bose-Einstein condensate* Science **291(5512)**, 2386–2389 (2001)

[949] A. André, A. S. Sörensen and M. D. Lukin *Stability of atomic clocks based on entangled atoms* Physical Review Letters **92(23)**, 230801–1 (2004)

[950] D. Leibfried, M. D. Barrett, T. Schaetz, J. Britton, J. Chiaverini, W. M. Itano, J. D. Jost, C. Langer and D. J. Wineland *Toward Heisenberg-limited spectroscopy with multiparticle entangled states* Science **304(5676)**, 1476–1478 (2004)

[951] D. Oblak, P. G. Petrov, C. L. Garrido Alzar, W. Tittel, A. K. Vershovski, J. K. Mikkelsen, J. L. Sörensen and E. S. Polzik *Quantum-noise-limited interferometric measurement of atomic noise: towards spin squeezing on the cs clock transition* Physical Review A - Atomic, Molecular, and Optical Physics **71(4)** (2005)

[952] C. F. Roos, M. Chwalla, K. Kim, M. Riebe and R. Blatt *'Designer atoms' for quantum metrology* Nature **443(7109)**, 316–319 (2006)

[953] P. J. Windpassinger, D. Oblak, P. G. Petrov, M. Kubasik, M. Saffman, C. L. G. Alzar, J. Appel, J. H. Müller, N. Kjaergaard and E. S. Polzik *Nondestructive probing of Rabi oscillations on the cesium clock transition near the standard quantum limit* Physical Review Letters **100(10)**, 103601 (2008)

[954] J. Appel, P. J. Windpassinger, D. Oblak, U. B. Hoff, N. Kjaergaard and E. S. Polzik *Mesoscopic atomic entanglement for precision measurements beyond the standard quantum limit* Proceedings of the National Academy of Sciences of the United States of America **106(27)**, 10960–10965 (2009)

[955] V. Giovannetti, S. Lloyd and L. Maccone *Quantum-enhanced positioning and clock synchronization* Nature **412(6845)**, 417–419 (2001)

[956] A. Valencia, G. Scarcelli and Y. Shih *Distant clock synchronization using entangled photon pairs* Applied Physics Letters **85(13)**, 2655–2657 (2004)

[957] M. Jaekel and S. Reynaud *Time-frequency transfer with quantum fields* Physical Review Letters **76(14)**, 2407–2411 (1996)

[958] R. Jozsa, D. S. Abrams, J. P. Dowling and C. P. Williams *Quantum clock synchronization based on shared prior entanglement* Physical Review Letters **85(9)**, 2010–2013 (2000)

[959] V. Giovannetti, S. Lloyd and L. Maccone *Positioning and clock synchronization through entanglement* Physical Review A.Atomic, Molecular, and Optical Physics **65(2)**, 022309/1–022309/9 (2002)

[960] B. Lamine, C. Fabre and N. Treps *Quantum improvement of time transfer between remote clocks* Physical Review Letters **101(12)** (2008)

[961] J. D. Franson *Nonlocal cancellation of dispersion* Physical Review A **45(5)**, 3126–3132 (1992)

[962] V. Giovannetti, S. Lloyd, L. Maccone and F. N. C. Wong *Clock synchronization with dispersion cancellation* Physical Review Letters **87(11)**, 117902/1–117902/4 (2001)

[963] A.N. Patrinos and R.A. Bradley *Energy and Technology Policies for Managing Carbon Risk* Science **325** 949 (2009)

[964] M. De Rosa, G. Gagliardi, A. Rocco et al. *Continuous in situ measurements of volcanic gases with a diode-laser-based spectrometer: CO_2 and H_2O concentration and soil degassing at Vulcano (Aeolian islands: Italy)* Geochemical Transactions **8**, Article Number 5 (2007)

[965] A. Rocco, G. De Natale, P. De Natale et al. *A diode-laser-based spectrometer for in-situ measurements of volcanic gases* Applied Physics B **78** 235-240 (2004)

[966] S. Stry, S. Thelen, J. Sacher, D. Halmer, P. Hering and M. Mürtz *Widely tunable diffraction limited 1000 mW external cavity diode laser in Littman/Metcalf configuration for cavity ring-down spectroscopy* Appl. Phys. B **85**, 365 (2006)

[967] Y.H. Xue, N.B. Ming, J.S. Zhu and D. Feng *The second harmonic generation in $LiNbO_3$ crystals with period laminar ferroelectric domains* Chinese Physics **4**, 554-564 (1984)

[968] W.K. Burns, W. McElhanon and L. Goldberg *Second harmonic generation in field poled, quasi-phase-matched, bulk $LiNbO_3$* IEEE Photonics Technol. Lett. **6**, 252 (1994)

[969] M. Ebrahimzadeh In *Solid-State Mid-Infrared Laser Sources*, I.T. Sorokina and K.L. Vodopyanov. Eds.,Topics in Appl. Phys., vol. 89 (Springer, Berlin, 2003), pp. 179-218

[970] F.K. Tittel, D. Richter and A. Fried In *Solid-State Mid-Infrared Laser Sources*, I.T. Sorokina and K.L. Vodopyanov. Topics in Appl. Phys., vol. 89 (Springer, Berlin, 2003), pp. 445-510

[971] J.C. Baubron, P. Allard, J.C. Sabroux, D. Tedesco and J.P. Toutain *Soil gas emanations as precursory indicators of volcanic eruptions* J. Geol. Soc. London **148**, 571-576 (1991)

[972] K. Notsu, T. Mori, G. Igarashi, Y. Tohjima and H. Wakita *A new tool for remote measurement of SO of volcanic gas* Geochem. J. **27**, 361-366 (1993)

[973] G. Chiodini, M. Todesco, S. Caliro, C. Del Gaudio, G. Macedonio and M. Russo *Magma degassing as a trigger of bradyseismic events: the case of Phlegraean Fields (Italy)*, Geophys. Res. Lett. **30**, 1434 (2003)

[974] P. Weibring, H. Edner, S. Svanberg, G. Cecchi, L. Pantani, R. Ferrara and T. Caltabiano *Monitoring of volcanic sulphur dioxide emissions using differential absorption lidar (DIAL), differential optical absorption spectroscopy (DOAS), and correlation spectroscopy (COSPEC)* Appl. Phys. B **67**, 419 (1998)

[975] C. Oppenheimer, P. Francis, M. Burton, A.J.H. Maciejewski and L. Boardman *Remote measurement of volcanic gases by Fourier transform infrared spectroscopy* Appl. Phys. B **67**, 505 (1998)

[976] P. Francis, M. Burton and C. Oppenheimer *Remote measurements of volcanic gas compositions by solar occultation spectroscopy* Nature **396**, 567-570 (1998)

[977] H. Edner, P. Ragnarson, S. Svanberg, E.Wallinder, R. Ferrara, R. Cioni, B. Raco and G. Taddeucci, *Total fluxes of sulfur dioxide from the Italian volcanoes Etna, Stromboli, and Vulcano measured by differential absorption lidar and passive differential optical absorption spectroscopy* J. Geophys. Res. **99**, 18827 (1994)

[978] P. De Natale, L. Gianfrani, G. De Natale et al. *Gas concentration measurements with DFB lasers to monitor volcanic activity* Proc. SPIE **3491**, 783-787 (1998)

[979] L. Gianfrani, P. De Natale and G. De Natale *Remote sensing of volcanic gases with a DFB-laser-based fiber spectrometer* Appl. Phys. B **70**, 467-470 (2000)

[980] L. Gianfrani, M. Gabrysch, C. Corsi and P. De Natale *Detection of H_2O and CO_2 with distributed feedback diode lasers: measurement of broadening coefficients and assessment of the accuracy levels for volcanic monitoring* Appl. Opt. **36**, 9481 (1997)

[981] G. Gagliardi, R. Restieri, G. De Biasio, P. De Natale, F. Cotrufo and L. Gianfrani *Quantitative diode laser absorption spectroscopy near 2 µm with high precision measurements of CO_2 concentration* Rev. Sci. Instr. **72**, 4228-4233 (2001)

[982] H.-A. Synal and L. Wacker *Nucl. AMS measurement technique after 30 years: Possibilities and limitations of low energy systems* Instrum. Methods Phys. Res., Sect. **B 268**, 701 (2010)

[983] I. Levin et al. *Observations and modelling of the global distribution and long-term trend of atmospheric $^{14}CO_2$* Tellus, Ser. B, Chem. Phys. Meteorol. **62**, 26 (2010)

[984] R.T. Menzies, C.R. Webster and E.D. Hinkley *Balloon-borne diode-laser absorption spectrometer for measurements of stratospheric trace species* Applied Optics **22**, 2655-2664 (1983)

[985] C.R. Webster, G.J. Flesch, D.C. Scott, J.E. Swanson, R.D. May, W. Stephen Woodward, C. Gmachl, F. Capasso, D.L. Sivco, J.N. Baillargeon, A.L. Hutchinson and A.Y. Cho *Quantum-cascade laser measurements of stratospheric methane and nitrous oxide* Applied Optics **40**, 321-326 (2001)

[986] P. Weibring, D. Richter, J. G. Walega and A. Fried *First demonstration of a high performance difference frequency spectrometer on airborne platforms* Optics Express **15**, 13476-13495 (2007)

[987] A. Fried, G. Diskin, P. Weibring, D. Richter, J.G. Walega, G. Sachse, T. Slate, M. Rana and J. Podolske *Tunable infrared laser instruments for airborne atmospheric studies* Applied Physics B **92**, 409-417 (2008)

[988] S. Viciani, F. D'Amato, P. Mazzinghi, F. Castagnoli, G. Toci and P. Werle *A cryogenically operated laser diode spectrometer for airborne measurement of stratospheric trace gases* Applied Physics B **90**, 581-592 (2008)

[989] P. Konopka, G. Günther, R. Müller, F H.S. dos Santos, C. Schiller, F. Ravegnani, A. Ulanovsky, H. Schlager, C.M. Volk, S. Viciani, L.L. Pan, D.-S. McKenna and M. Riese *Contribution of mixing to the upward transport across the tropical tropopause layer (TTL)* Atmospheric Chemistry and Physics **7**, 3285-3308 (2007).

[990] F. Cairo, J.P. Pommereau, K.S. Law, H. Schlager, A. Garnier, F. Fierli, M. Ern, M. Streibel, S. Arabas, S. Borrmann, J.J. Berthelier, C. Blom, T. Christensen, F. D'Amato, G. Di Donfrancesco, T. Deshler, A. Diedhiou, G. Durry, O. Engelsen, F. Goutail, N.R.P. Harris, E.R.T. Kerstel, S. Khaykin, P. Konopka, A. Kylling, N. Larsen, T. Lebel, X. Liu, A.R. MacKenzie, J. Nielsen, A. Oulanowski, D.J. Parker, J. Pelon, J. Polcher, J. A. Pyle, F. Ravegnani, E.D. Riviere, A.D. Robinson, T. Rockmann, C. Schiller, F. Simoes, L. Stefanutti, F. Stroh, L. Some, P. Siegmund, N. Sitnikov, J. P. Vernier, C.M. Volk, C. Voigt, M. von Hobe, S. Viciani and V. Yushkov *An introduction to the SCOUT-AMMA stratospheric aircraft, balloons and sondes campaign in West Africa, August 2006: rationale and roadmap* Atmospheric Chemistry and Physics **10**, 2237-2256 (2010)

[991] C.D. Homan, C.M. Volk, A.C. Kuhn, A. Werner, J. Baehr, S. Viciani, A. Ulanovski and F. Ravegnani *Tracer measurements in the tropical tropopause layer during the AMMA/SCOUT-O3 aircraft campaign* Atmospheric Chemistry and Physics **10**, 3615-3627 (2010).

[992] S. Borrmann, D. Kunkel, R. Weigel, A. Minikin, T. Deshler, J. C. Wilson, J. Curtius, C.M. Volk, C.D. Homan, A. Ulanovsky, F. Ravegnani, S. Viciani, G.N. Shur, G. V. Belyaev, K. S. Law and F. Cairo *Aerosols in the tropical and subtropical UT/LS: in-situ measurements of submicron particle abundance and volatility* Atmospheric Chemistry and Physics **10**, 5573-5592 (2010)

[993] K.S. Law, F. Fierli, F. Cairo, H. Schlager, S. Borrmann, M. Streibel, E. Real, D. Kunkel, C. Schiller, F. Ravegnani, A. Ulanovsky, F. D'Amato, S. Viciani and C.M. Volk *Air mass origins influencing TTL chemical composition over West Africa during 2006 summer monsoon* Atmospheric Chemistry and Physics **10**, 10753-10770 (2010)

[994] C.R. Hoyle, V. Marcal, M.R. Russo, G. Allen, J. Arteta, C. Chemel, M.P. Chipperfield, F. D'Amato, O. Dessens, W. Feng, J.F. Hamilton, N.R.P. Harris, J.S. Hosking, A.C. Lewis, O. Morgenstern, T. Peter, J.A. Pyle, T. Reddmann, N. A. D. Richards, P. J. Telford, W. Tian, S. Viciani, A. Volz-Thomas, O. Wild, X. Yang and G. Zeng *Representation of tropical deep convection in atmospheric models Part 2: Tracer transport* Atmospheric Chemistry and Physics **11**, 8103-8131 (2011)

[995] A.D. Kersey *A review of recent developments in fiber optic sensor technology* Opt. Fiber Technol. **2**, 291-317 (1996)

[996] Y. J. Rao and S. Huang *Applications of Fiber Optic Sensors in Fiber Optic Sensors,* F.T.S. Yu and S. Yin, Eds., Marcel Dekker, Inc., New York, NY, pp. 449 (2002)

[997] Y.J. Rao *In-fibre Bragg grating sensors* Meas. Sci. Technol. **8**, 355-375 (1997)

[998] M.G. Xu, H. Geiger and J.P. Dakin *Interrogation of fiberoptic in interferometric sensors using acoustooptic tunable filter* Electron. Lett **31**, 1487-1488 (1995)

[999] B. Lissak, A. Arie and M. Tur *Highly sensitive dynamic strain measurements by locking lasers to fiber bragg gratings* Opt. Lett. **23**, 1930-1932 (1998)

[1000] G. Gagliardi, M. Salza, P. Ferraro and P. De Natale *Interrogation of FBG- based strain sensors by means of laser radio-frequency modulation techniques.* J. Opt. A-Pure Appl. Opt. **8**, S507-S513 (2006)

[1001] A.C.R. Pipino *Ultrasensitive surface spectroscopy with a miniature optical resonator* Phys. Rev. Lett. **83**, 3093-3096 (1999)

[1002] Z.G. Tong, A. Wright, T. McCormick, R.K. Li, R.D. Oleschuk and H.P. Loock *Phase-shift fiber-loop ring-down spectroscopy* Anal. Chem. **76**, 6594-6599 (2004)

[1003] Z.G. Tong, M. Jakubinek, A. Wright, A. Gillies and H.P. Loock *Fiber-loop ring-down spectroscopy: a sensitive absorption technique for small liquid samples* Rev. Sci. Instrum. **74**, 4818-4826 (2003)

[1004] G. Gagliardi, M. Salza, P. Ferraro and P. De Natale *Fiber Bragg-grating strain sensor interrogation using laser radio-frequency modulation* Opt. Express **13**, 2377-2384 (2005)

[1005] T. Erdogan and V. Mizrahi *Characterization of UV-induced birefringence in photosensitive Ge-doped silica optical fibers* J. Opt. Soc. Am. B **11**, 2100-2105 (1994)

[1006] K. Dossou, S. LaRochelle and M. Fontaine *Numerical analysis of the contribution of the transverse asymmetry in the photo-induced index change profile to the birefringence of optical fiber.* J. Lightwave Technol. **20**, 1463-1470 (2002)

[1007] J.H. Chow, D.E. McClelland, M.B. Gray and I.C.M. Littler *Demonstration of a Passive Subpicostrain Fiber Strain Sensor* Opt. Lett. **30**, 1923-1925 (2005)

[1008] G. Gagliardi, S. De Nicola, P. Ferraro and P. De Natale *Interrogation of fiber Bragg-grating resonators by polarization-spectroscopy laser-frequency locking* Opt. Express **15**, 3715-3728 (2007)

[1009] G. Gagliardi, M. Salza, S. Avino, P. Ferraro and P. De Natale *Probing the ultimate limit of fiber-optic strain sensing* Science **330**, 1081-1084 (2010)

[1010] T.A. Berkoff and A.D. Kersey *Experimental demonstration of a fiber Bragg grating accelerometer* IEEE Photon. Technol. Lett. **8**, 1677-1679 (1996)

[1011] M.D. Todd, G.A. Johnson, B.A. Althouse and S.T. Vohra *Flexural beam-based fiber Bragg grating accelerometers* IEEE Photon. Technol. Lett. **10**, 1605-1607 (1998)

[1012] G. Gagliardi, M. Salza, P. Ferraro, P. De Natale, A. Di Maio, A.; S. Carlino, G. De Natale and E. Boschi *Design and test of a laser-based optical-fiber Bragg-grating accelerometer for seismic applications* Meas. Sci. Technol. **19**, 085306 (2008)

[1013] G. Gagliardi, M. Salza, T.T.-Y. Lam, J.H. Chow and P. De Natale *3-Axis accelerometer based on lasers locked to π-shifted fibre bragg gratings* Proc. SPIE **7503**, pp. 75033X75031-75033X75034 (2009)

[1014] J. Canning and M.G. Sceats *π-phase-shifted periodic distributed structures in optical fibers by UV post-processing* Electron. Lett. **30**, 1344-1345 (1994)

[1015] S. Avino, J.A. Barnes, G. Gagliardi, X. Gu, D. Gutstein, J.R. Mester, C. Nicholaou and H.-P. Loock *Musical instrument pickup based on a laser locked to an optical fiber resonator* Optics Express **19**, 2505725065 (2011).

[1016] W.R. Seitz *Chemical sensors based on fiber optics* Anal. Chem. **56**, A16 (1984)

[1017] W.R. Seitz *New directions in fiber optic chemical sensors-sensors based on polymer swelling* J. Mol. Struct. **292**, 105-113 (1993)

[1018] O.S. Wolfbeis *Fiber-optic chemical sensors and biosensors* Anal. Chem. **78**, 3859-3873 (2006)

[1019] W.R. Seitz *Chemical sensors based on immobilized indicators and fiber optics* Crit. Rev. Anal. Chem. **19**, 135-173 (1988)

[1020] O.S. Wolfbeis *Fiber optic probes for determining enzyme-activities* Select. Electr. Rev. **10**, 41 (1988)

[1021] S. Caron, C. Pare, P. Paradis, J.M. Trudeau and A. Fougeres *Distributed fibre optics polarimetric chemical sensor* Meas. Sci. Technol. **17**, 1075-1081 (2006)

[1022] T. von Lerber and M.W. Sigrist *Cavity-ring-down principle for fiber-optic resonators: experimental realization of bending loss and evanescent-field sensing* Appl. Optics **41**, 3567-3575 (2002)

[1023] K.M. Zhou, X.F. Chen, L. Zhang and I. Bennion *Implementation of optical chemsensors based on HF-etched fibre Bragg grating structures* Meas. Sci. Technol. **17**, 1140-1145 (2006)

[1024] H. Golnabi, M. Bahar, M. Razani, M. Abrishami and A. Asadpour *Design and operation of an evanescent optical fiber sensor* Opt. Laser Eng. **45**, 12-18 (2007)

[1025] A.C.R. Pipino, J.T. Woodward, C.W. Meuse and V. Silin *Surface-plasmon-resonance-enhanced cavity ring-down detection* J. Chem. Phys. **120**, 1585-1593 (2004)

[1026] L. Ma, T. Katagiri and Y. Matsuura *Surface-plasmon resonance sensor using silica-core Bragg fiber* Opt. Lett. **34**, 1069-1071 (2009)

[1027] V.V.R. Sai, T. Kundu and S. Mukherji *Novel U-bent fiber optic probe for localized surface plasmon resonance based biosensor* Biosens. Bioelectron. **24**, 2804-2809 (2009)

[1028] B. Spackova, M. Piliarik, P. Kvasnicka, C. Themistos, M. Rajarajan and J. Homola *Novel concept of multi-channel fiber optic surface plasmon resonance sensor* Sensors and Actuators B - Chemical **139**, 199-203 (2009)

[1029] M. Gupta, H. Jiao and A- O'Keefe *Cavity-enhanced spectroscopy in optical fibers* Opt. Lett. **27**, 1878-1880 (2002)

[1030] H.P. Loock *Ring-down absorption spectroscopy for analytical microdevices* Trac-Trend Anal. Chem. **25**, 655-664 (2006)

[1031] S.A. Pu and X.J. Gu *Fiber loop ring-down spectroscopy with a long-period grating cavity* Opt. Lett. **34**, 1774-1776 (2009)

[1032] S. Arnold, M. Khoshsima, I. Teraoka, S. Holler and F. Vollmer *Shift of whispering-gallery modes in microspheres by protein adsorption* Opt. Lett. **28**, 272-274 (2003)

[1033] A.M. Armani and K.J. Vahala *Heavy water detection using ultra-high-Q microcavities* Opt. Lett. **31**, 1896-1898 (2006)

[1034] F. Vollmer and S. Arnold *Whispering-gallery-mode biosensing: label-free detection down to single molecules* Nat. Methods **5**, 591-596 (2008)

[1035] J. Barnes, B. Carver, J.M. Fraser, G. Gagliardi, H.P. Loock, Z. Tian, M.W.B. Wilson, S. Yam and O. Yastrubshak *Loss determination in microsphere resonators by phase-shift cavity ring-down measurements* Opt. Express **16**, 13158-13167 (2008)

Index

Printed and bound by CPI Group (UK) Ltd, Croydon, CR0 4YY

24/10/2024

01778309-0010